FOR OFFICIAL USE ONLY.

NOTE.

The information given in this Treatise is not to be communicated, either directly or indirectly, to the Press, or to any person not holding an official position in His Majesty's Service.

TREATISE

ON

AMMUNITION

10th EDITION.

WAR OFFICE, 1915.

The Naval & Military Press Ltd

Published by
The Naval & Military Press Ltd
5 Riverside, Brambleside, Bellbrook
Industrial Estate, Uckfield, East Sussex,
TN22 1QQ England
Tel: +44 (0) 1825 749494
Fax: +44 (0) 1825 765701
www.naval-military-press.com

CONTENTS.

(B 11123) Wt. w 6515—2475 5000 8/15 H & S P 15/36

LIST OF PLATES.

The Colour plates in the reprint are placed after this page.

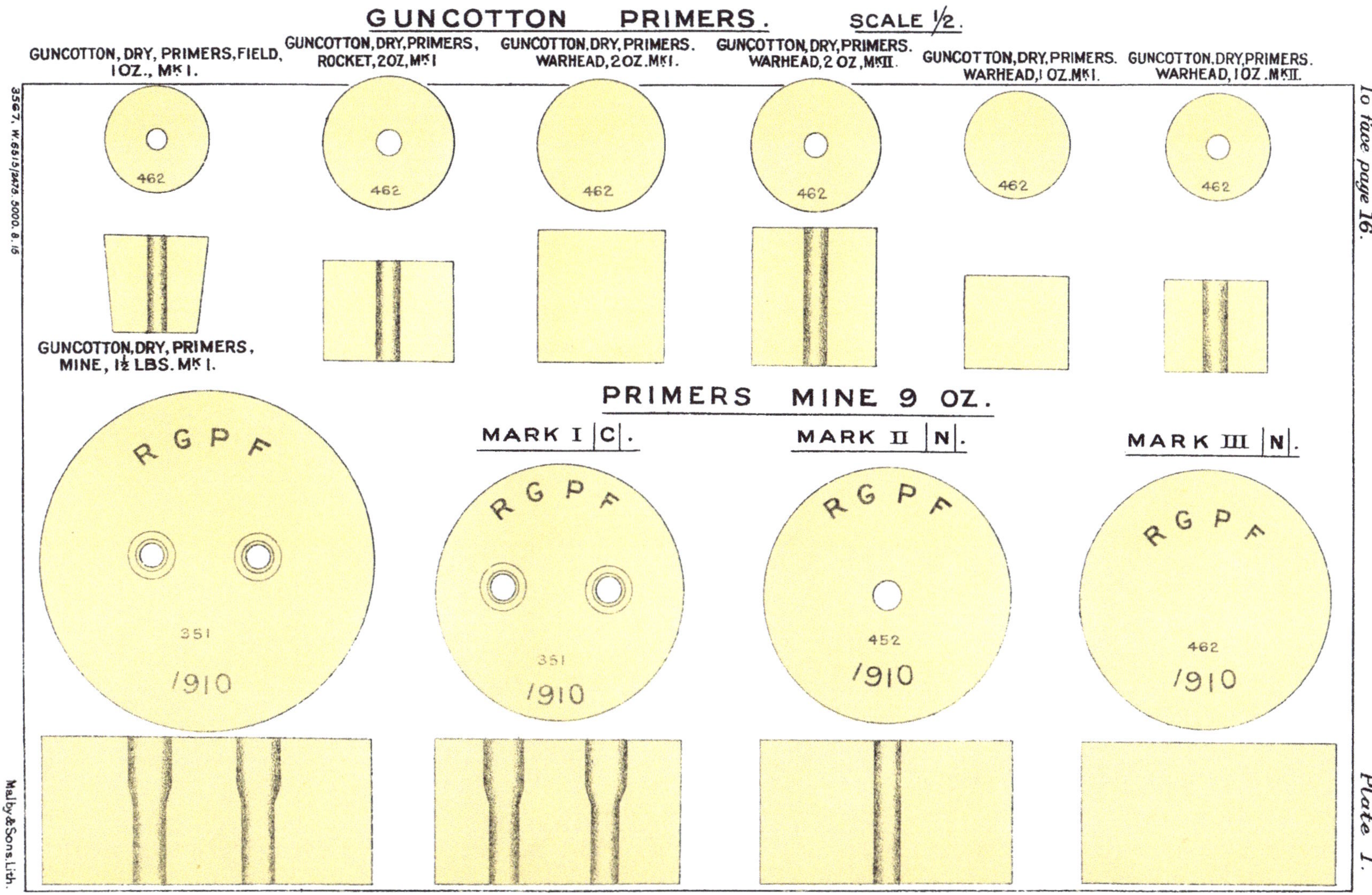

3567. W.6515/2475. 5000. 8. 15

Malby & Sons. Lith.

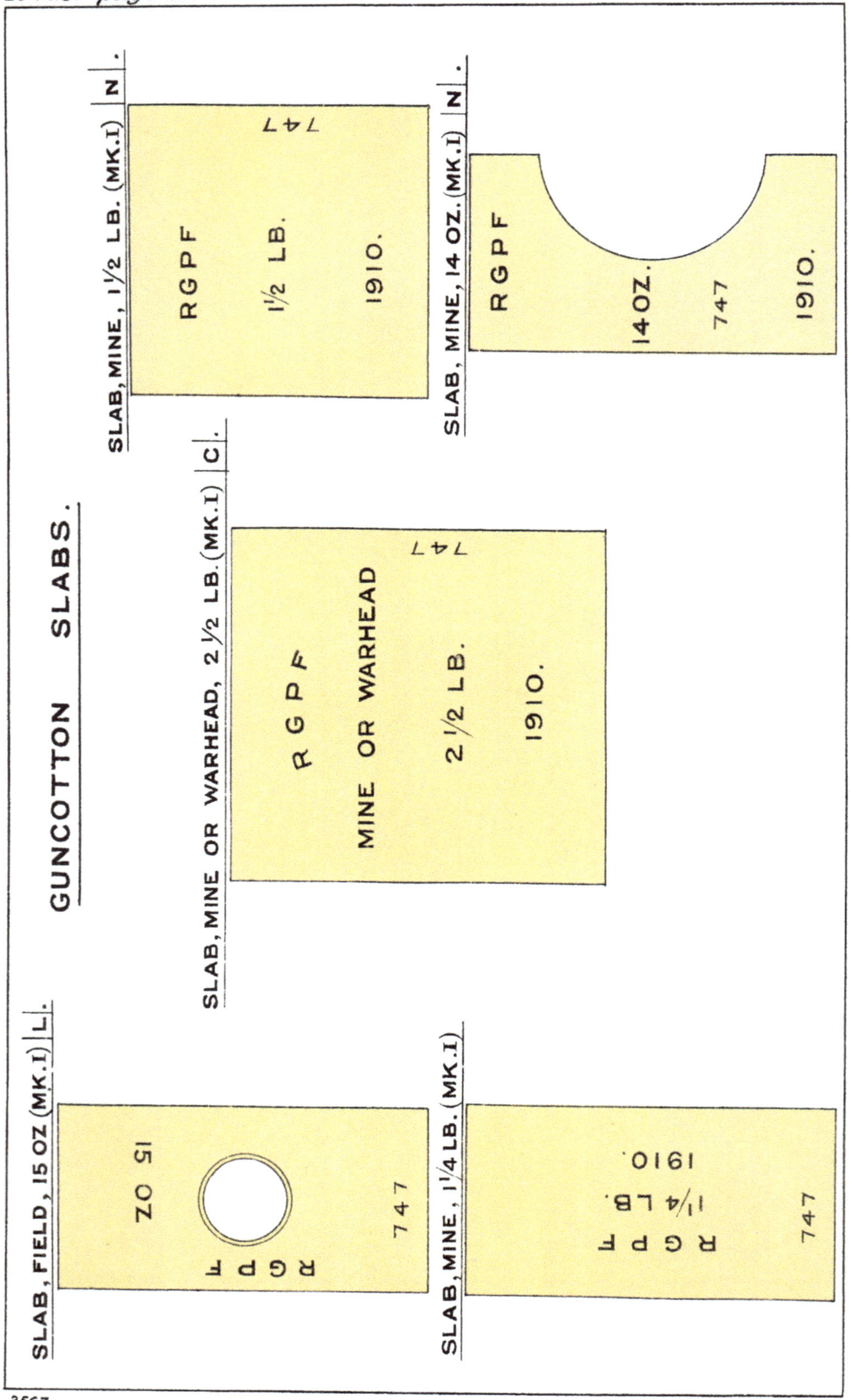

GUNCOTTON SLABS.

SLAB, FIELD, 15 OZ (MK.I) | L |.

RGPF
15 OZ
747

SLAB, MINE, 1¼ LB. (MK.I)

RGPF
1¼ LB.
1910.
747

SLAB, MINE OR WARHEAD, 2½ LB. (MK.I) | C |.

RGPF
MINE OR WARHEAD
2½ LB.
1910.
747

SLAB, MINE, 1½ LB. (MK.I) | N |.

RGPF
1½ LB.
1910.
747

SLAB, MINE, 14 OZ. (MK.I) | N |.

RGPF
14 OZ.
747
1910.

3567 Malby & Sons, Lith.

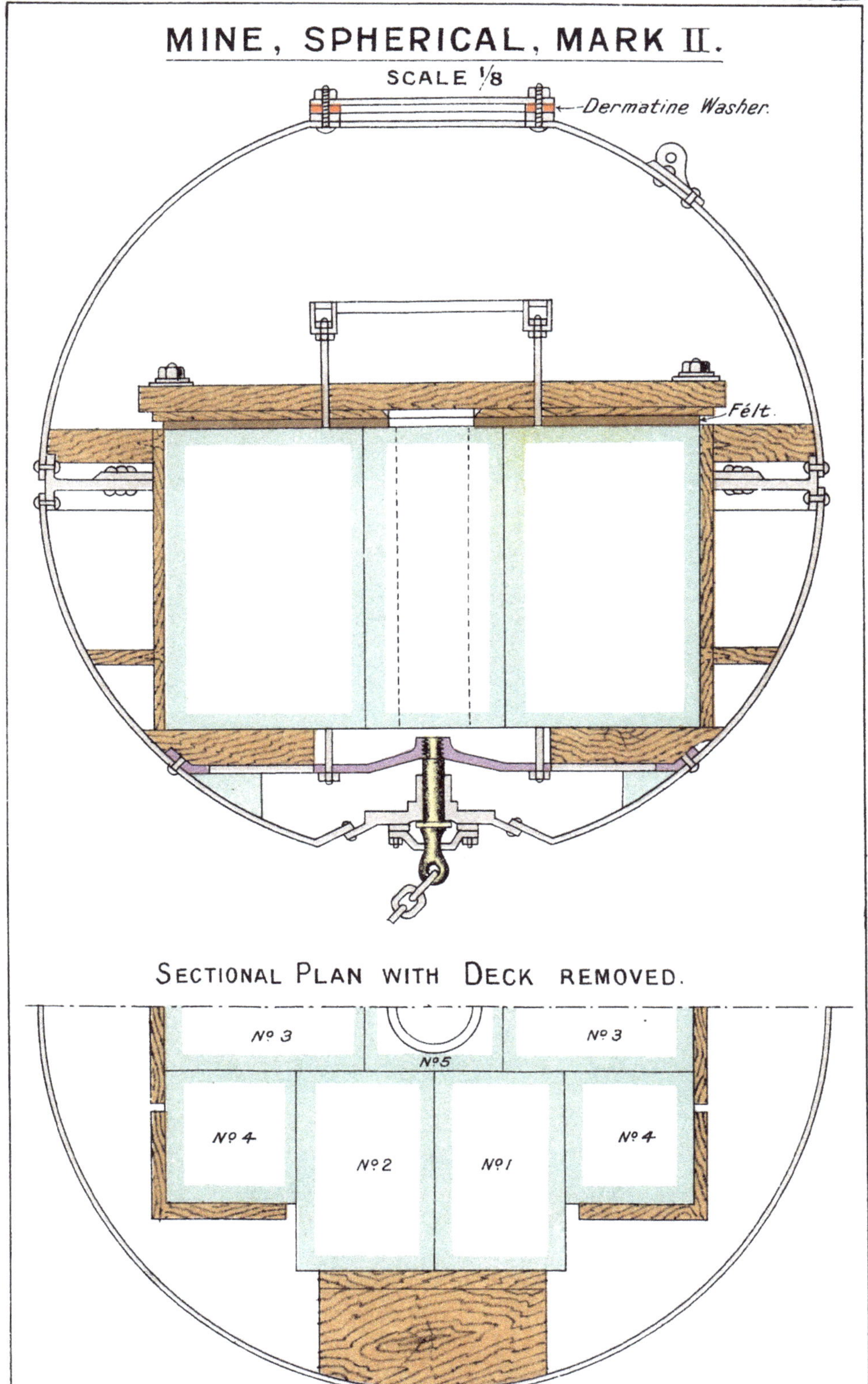
MINE, SPHERICAL, MARK II.
SCALE 1/8
Dermatine Washer.
Felt.
SECTIONAL PLAN WITH DECK REMOVED.
Nº 3
Nº 3
Nº 5
Nº 4
Nº 2
Nº 1
Nº 4

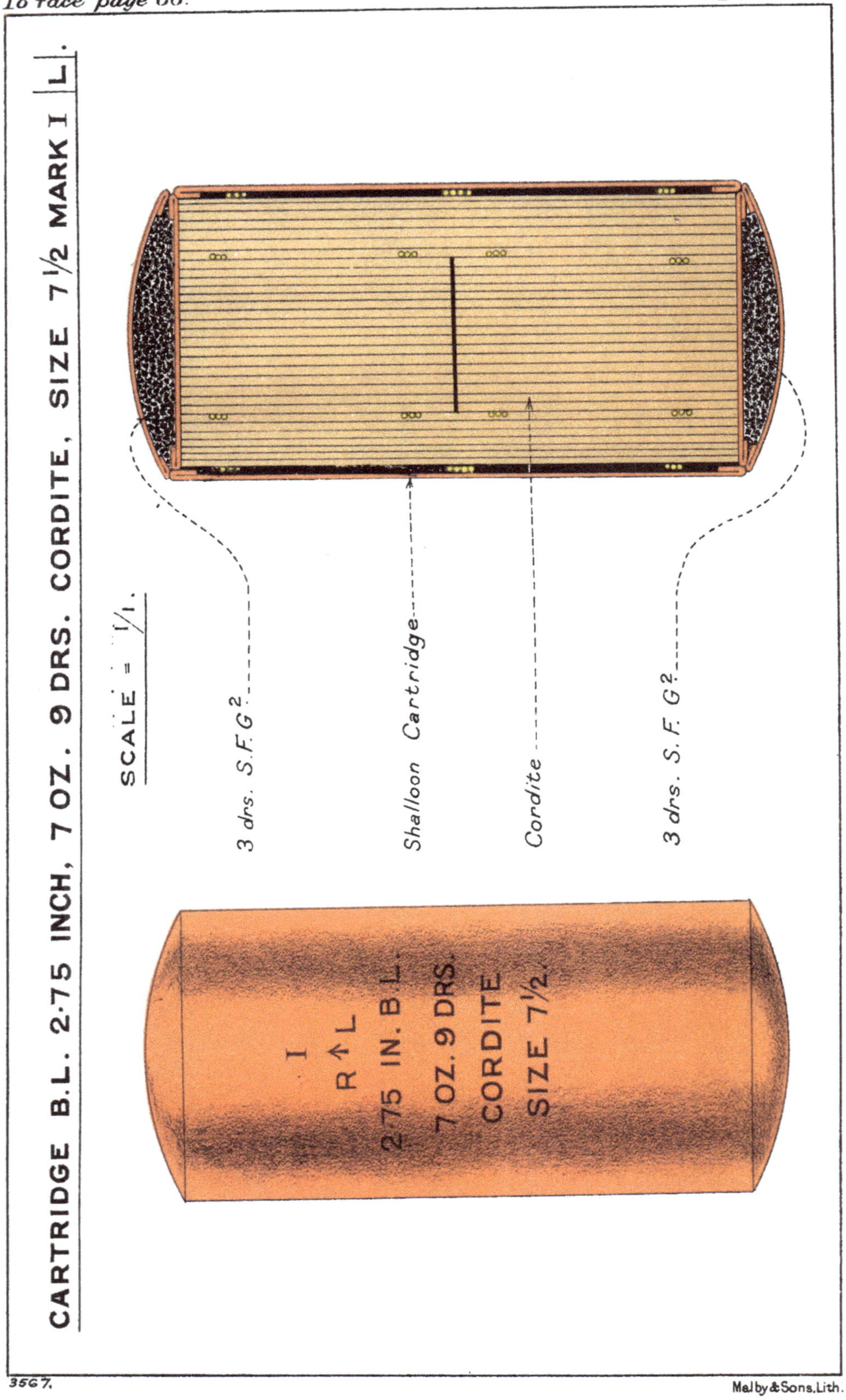

To face page 66.

CARTRIDGE B.L. 2·75 INCH, 7 OZ. 9 DRS. CORDITE, SIZE 7½ MARK I |L|.

SCALE = 1/1.

I
R ↑ L
2·75 IN. B.L.
7 OZ. 9 DRS.
CORDITE
SIZE 7½.

3 drs. S.F.G².

Shalloon Cartridge

Cordite

3 drs. S.F. G².

3567.

Malby & Sons, Lith.

CARTRIDGE B.L. 60 PR. 9 LBS. 12 OZ. CORDITE M.D. SIZE 16. MARK II.

SCALE = 1/3

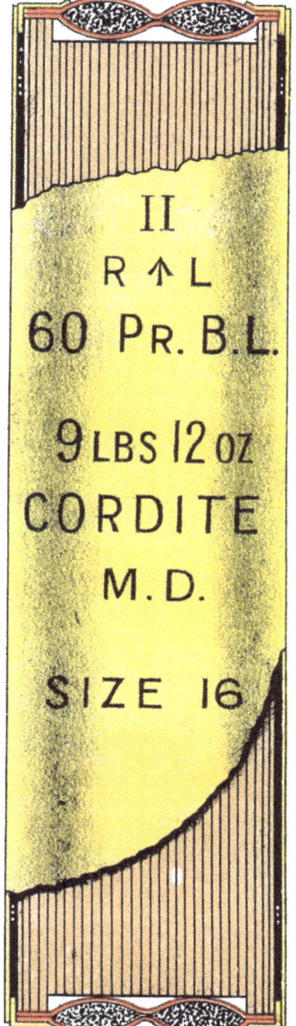

LOT
N. 246.A.C

W
2-14.

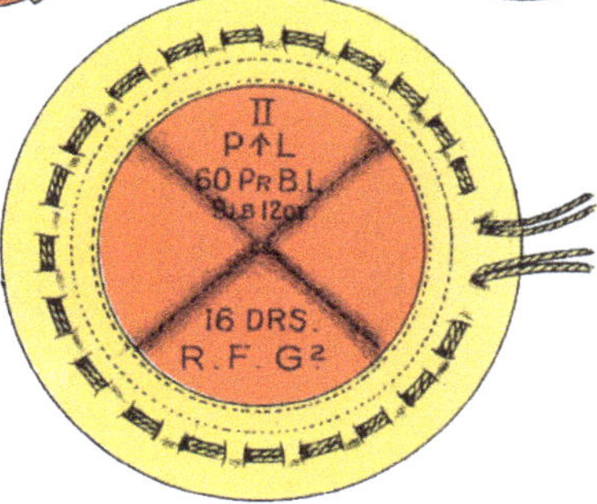

 Malby & Sons, Lith.

Plate VI.

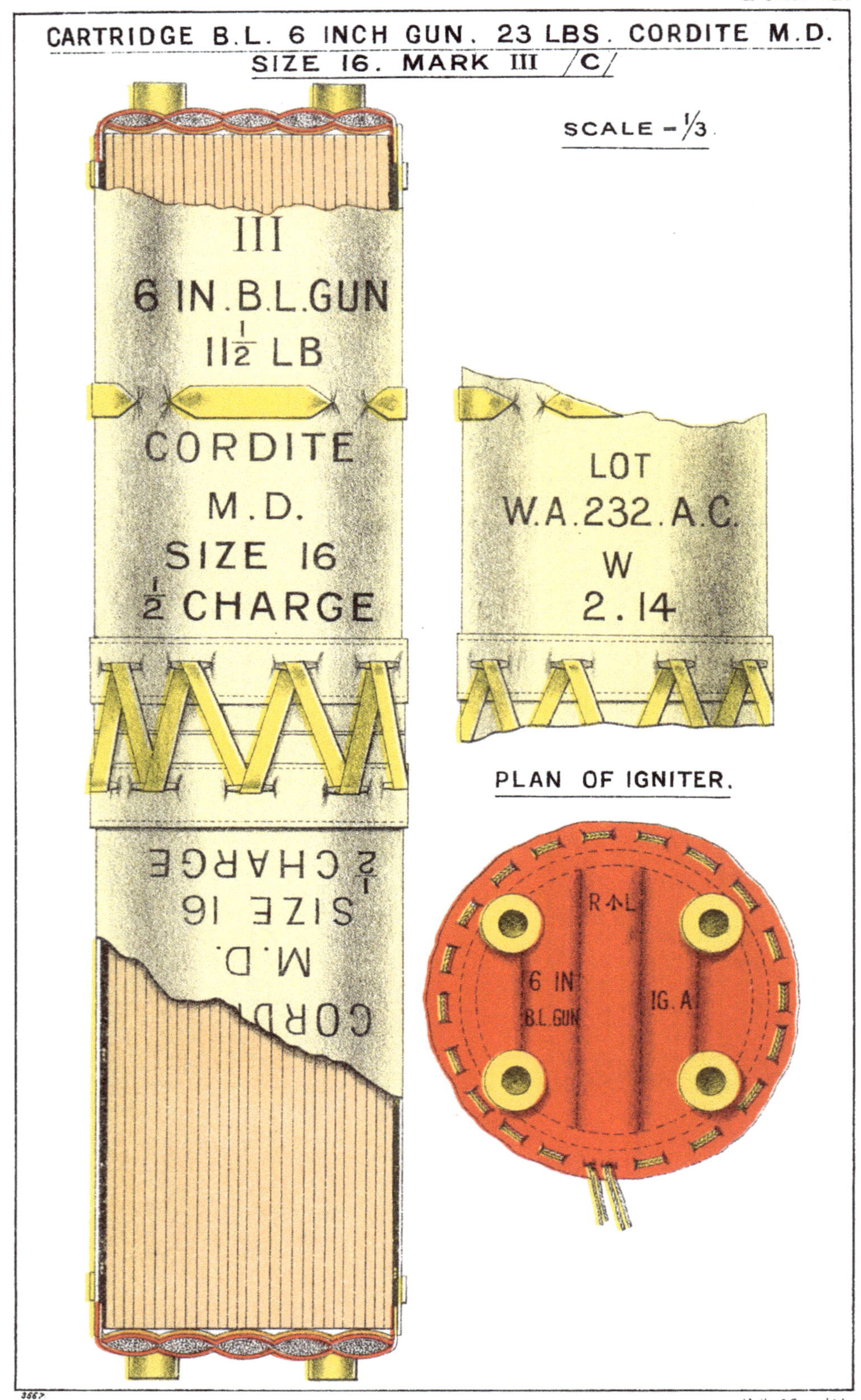

3557

Malby & Sons. Lith.

Plate VII.

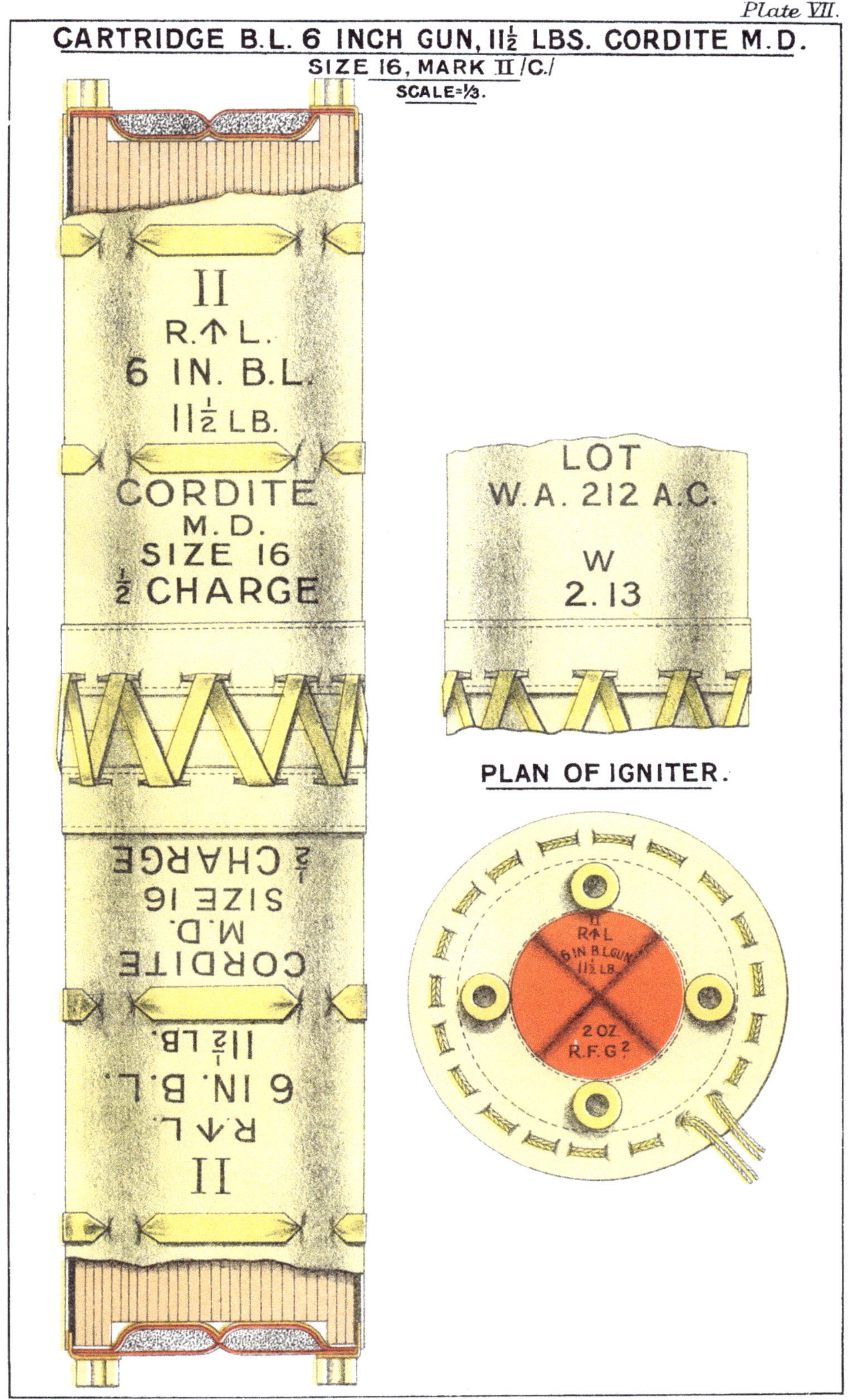

3567.

Malby & Sons, Lith.

CARTRIDGE B.L. 9·2 INCH, 60 LBS. CORDITE M.D.

SIZE 37, ½ CHARGE, MARK III /C/

SCALE = 1/5

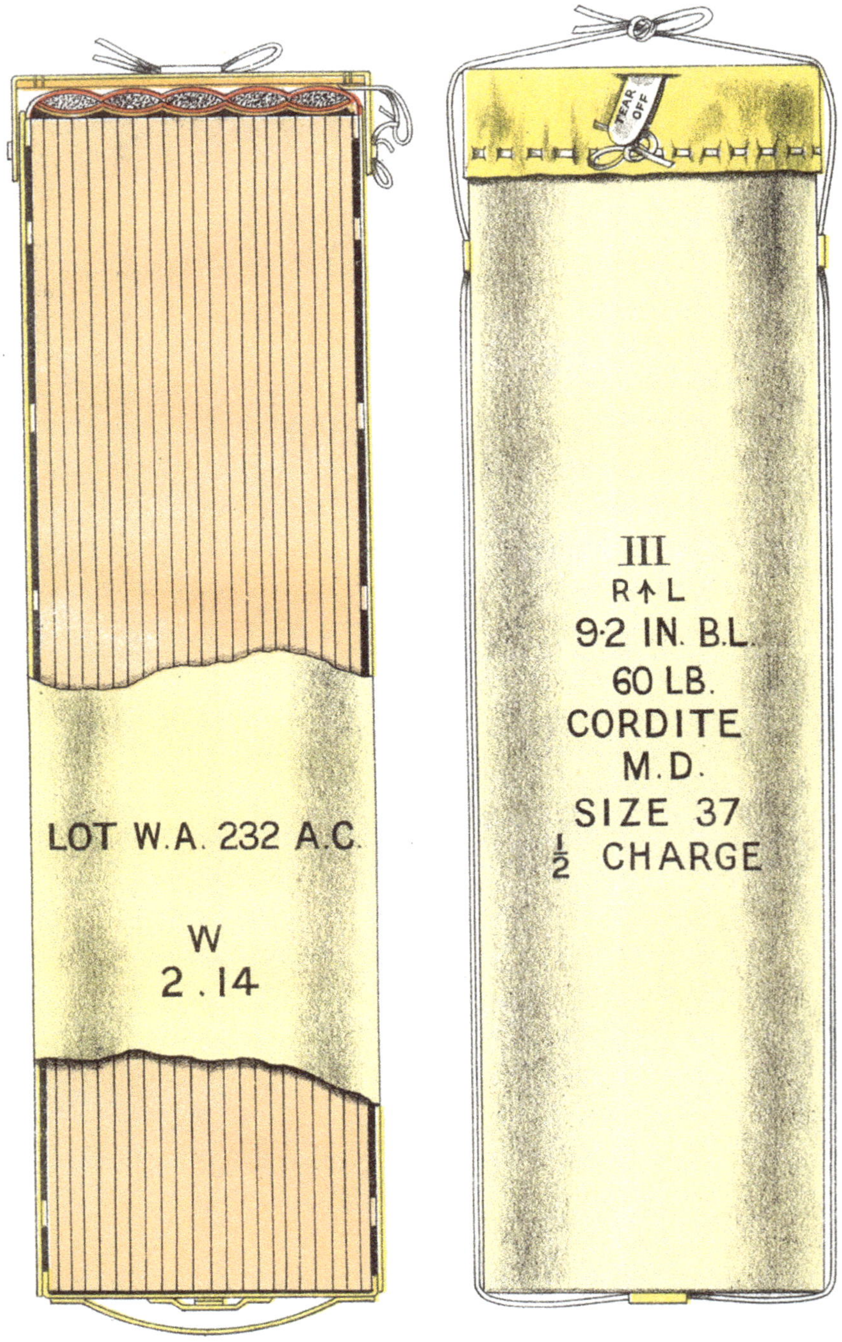

TEAR OFF

LOT W.A. 232 A.C.

W
2.14

III
R ↑ L
9·2 IN. B.L.
60 LB.
CORDITE
M.D.
SIZE 37
½ CHARGE

3567.

Malby & Sons, Lith.

Cartridge B.L. 9·2 in. 60 lbs. M.D. Cordite Size 37. Mark II (C). ½ Charge for Marks IX & X Guns.

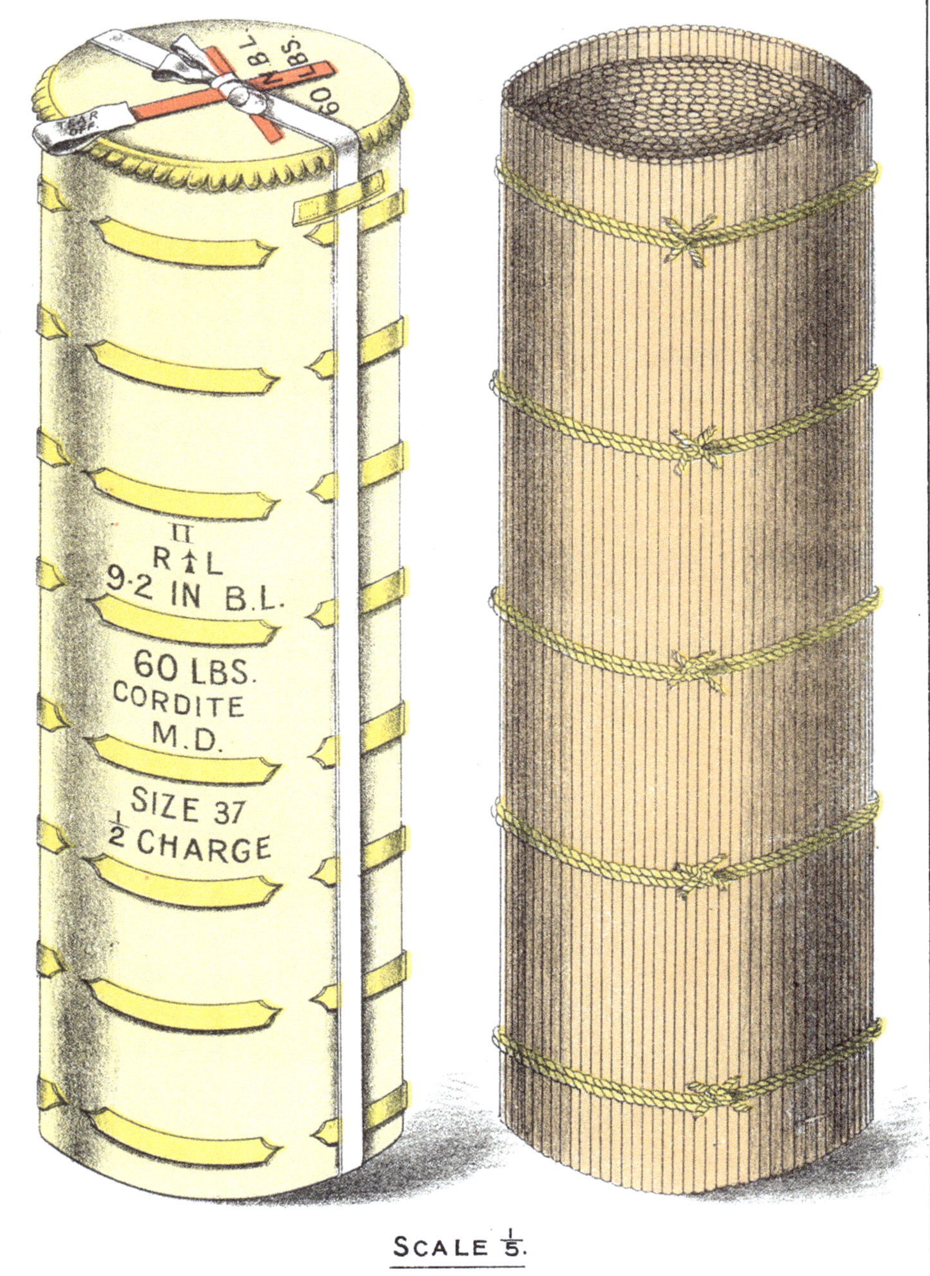

Malby & Sons, Lith.

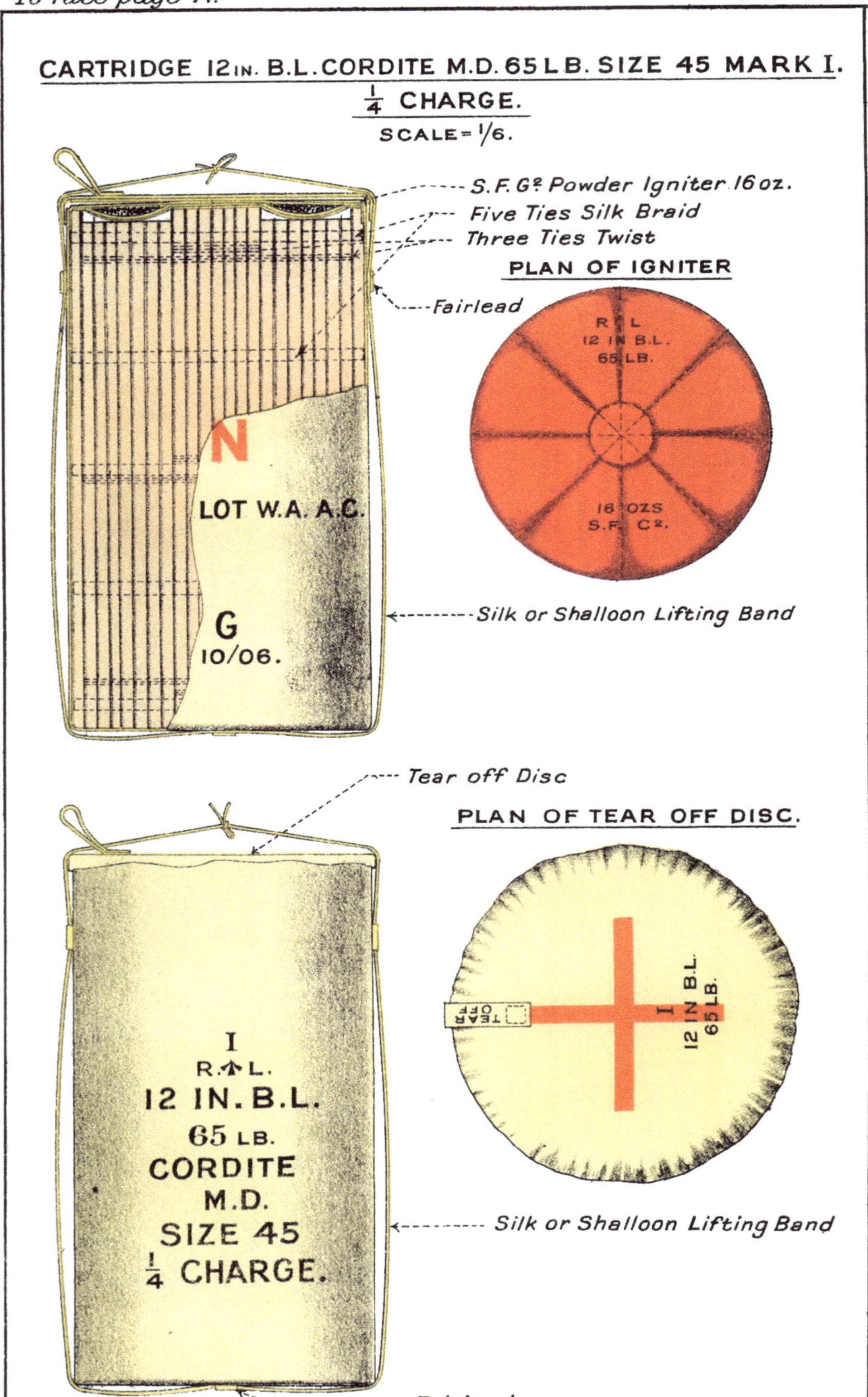
CARTRIDGE 12 IN. B.L. CORDITE M.D. 65 LB. SIZE 45 MARK I.
¼ CHARGE.
SCALE = 1/6.
S. F. G². Powder Igniter 16 oz.
Five Ties Silk Braid
Three Ties Twist
PLAN OF IGNITER
Fairlead
12 IN B.L.
65 LB.
16 OZS
N
LOT W.A. A.C.
G
10/06.
Silk or Shalloon Lifting Band
Tear off Disc
PLAN OF TEAR OFF DISC.
TEAR OFF
I
12 IN B.L.
65 LB.
I
R. L.
12 IN. B.L.
65 LB.
CORDITE
M.D.
SIZE 45
¼ CHARGE.
Silk or Shalloon Lifting Band
Fairlead

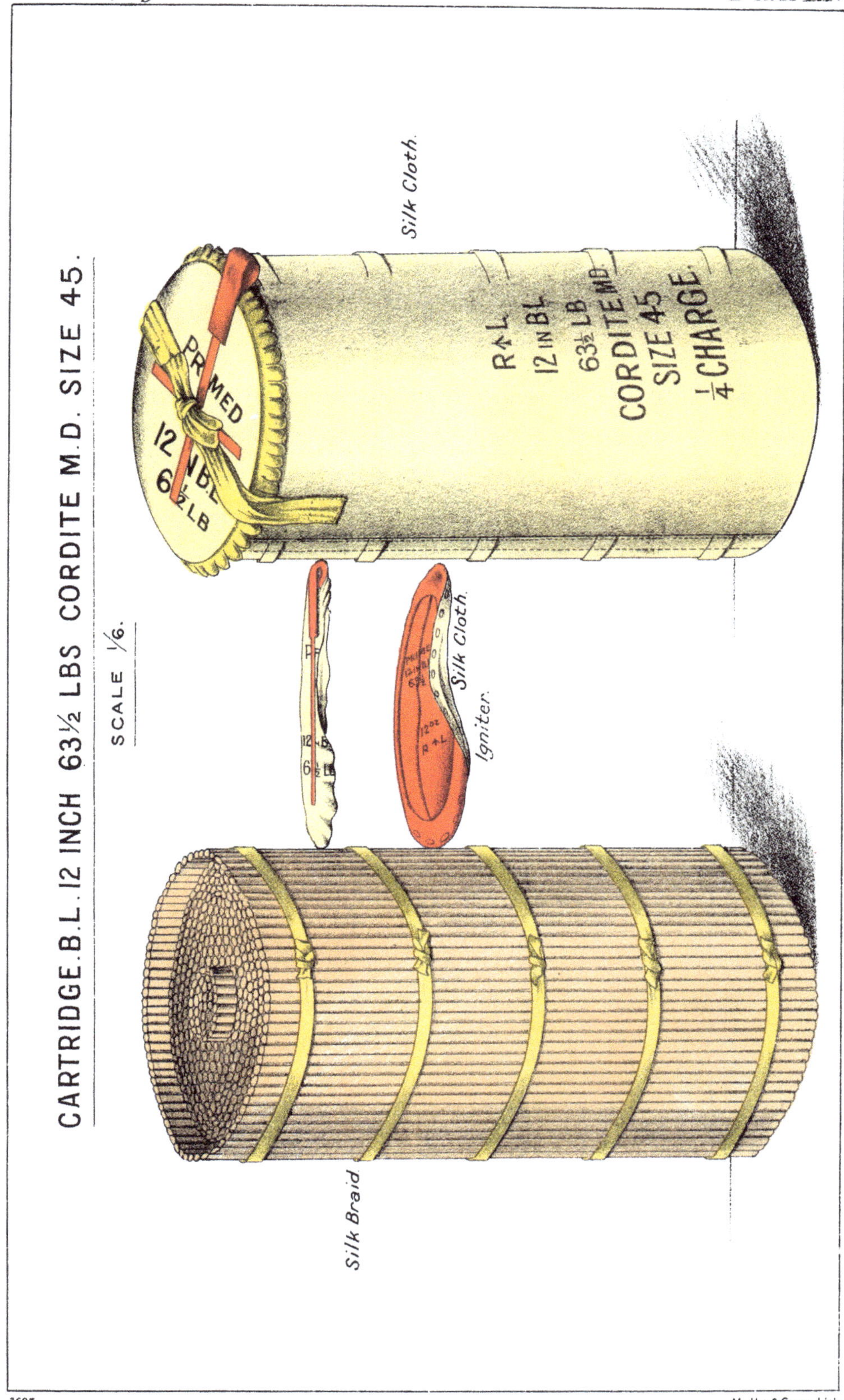

3567. Malby & Sons, Lith.

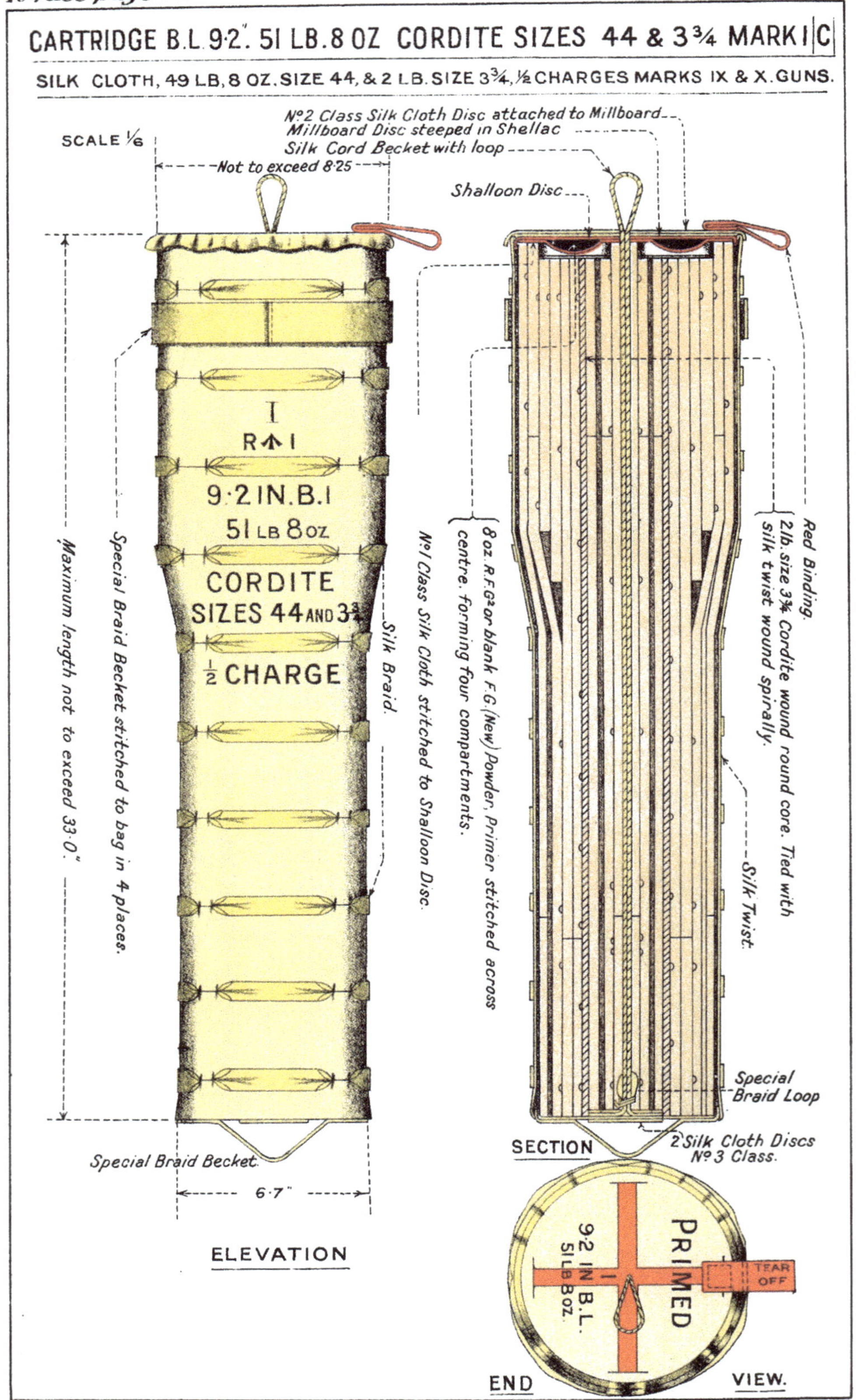
CARTRIDGE B.L. 9·2". 51 LB. 8 OZ CORDITE SIZES 44 & 3¾ MARK I|C
SILK CLOTH, 49 LB, 8 OZ. SIZE 44, & 2 LB. SIZE 3¾, ½ CHARGES MARKS IX & X. GUNS.
SCALE 1/6
Not to exceed 8·25
No 2 Class Silk Cloth Disc attached to Millboard
Millboard Disc steeped in Shellac
Silk Cord Becket with loop
Shalloon Disc
I
R ↑ I
9·2 IN. B.I
51 LB 8 OZ
CORDITE
SIZES 44 AND 3¾
½ CHARGE
Maximum length not to exceed 33·0"
Special Braid Becket stitched to bag in 4 places.
Silk Braid.
No 1 Class Silk Cloth stitched to Shalloon Disc.
8 oz. R.F.G2 or blank F.G. (New) Powder, Primer stitched across centre, forming four compartments.
2 lb. size 3¾ Cordite wound round core. Tied with silk twist wound spirally.
Red Binding.
Silk Twist.
Special Braid Loop
2 Silk Cloth Discs No 3 Class.
Special Braid Becket.
6·7"
ELEVATION
SECTION
PRIMED
I
9·2 IN B.L.
51 LB 8 OZ.
TEAR OFF
END VIEW.

CARTRIDGE. B. L. 6 INCH 30 CWT HOWITZER 2 LB 8½ OZ CORDITE, M.D. SIZE 4¼, MK I.

SCALE = ½

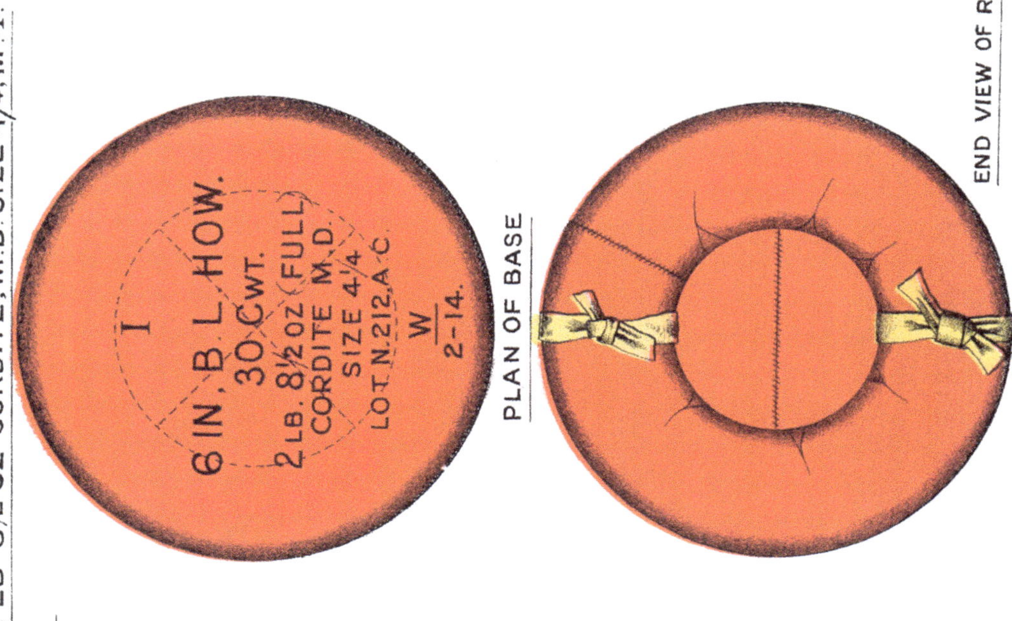

PLAN OF BASE

END VIEW OF RING.

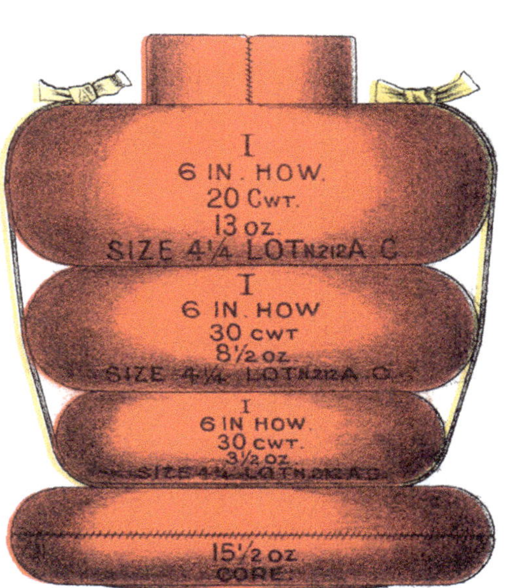

3567

Malby & Sons, Lith.

CARTRIDGE B.L. 5 INCH HOWITZER 11 7/16 OZ. CORDITE SIZE 3 3/4 MARK V.

CORE AND THREE RINGS.

SCALE = $\frac{1}{1}$

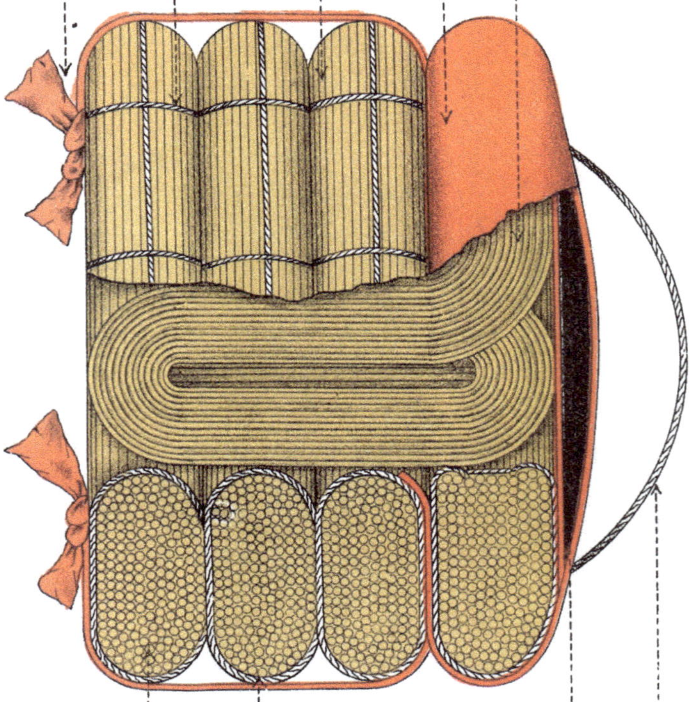
Shallon braid securing rings.
Silk Sewing No. 1.
Cordite Rings (each 2 9/16 Oz.)
Shalloon. Base.
Core (3 Oz. 12 Drams.)
Cordite Rings shown in section.
Silk Sewing No. 1.
8 Drams of S.F.G.² Gunpowder.
Silk Sewing Becket.

3867. Malby & Sons, Lith.

TRACER, SHELL, NIGHT, INTERNAL, (MARK I)/C/OR SHOT.

SCALE = 2/1

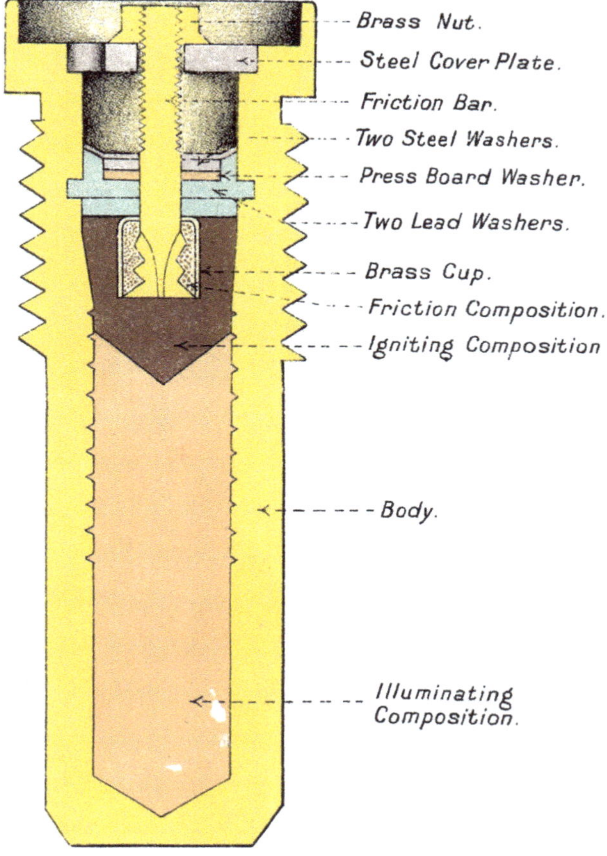

To face page 153

Plate XVI.

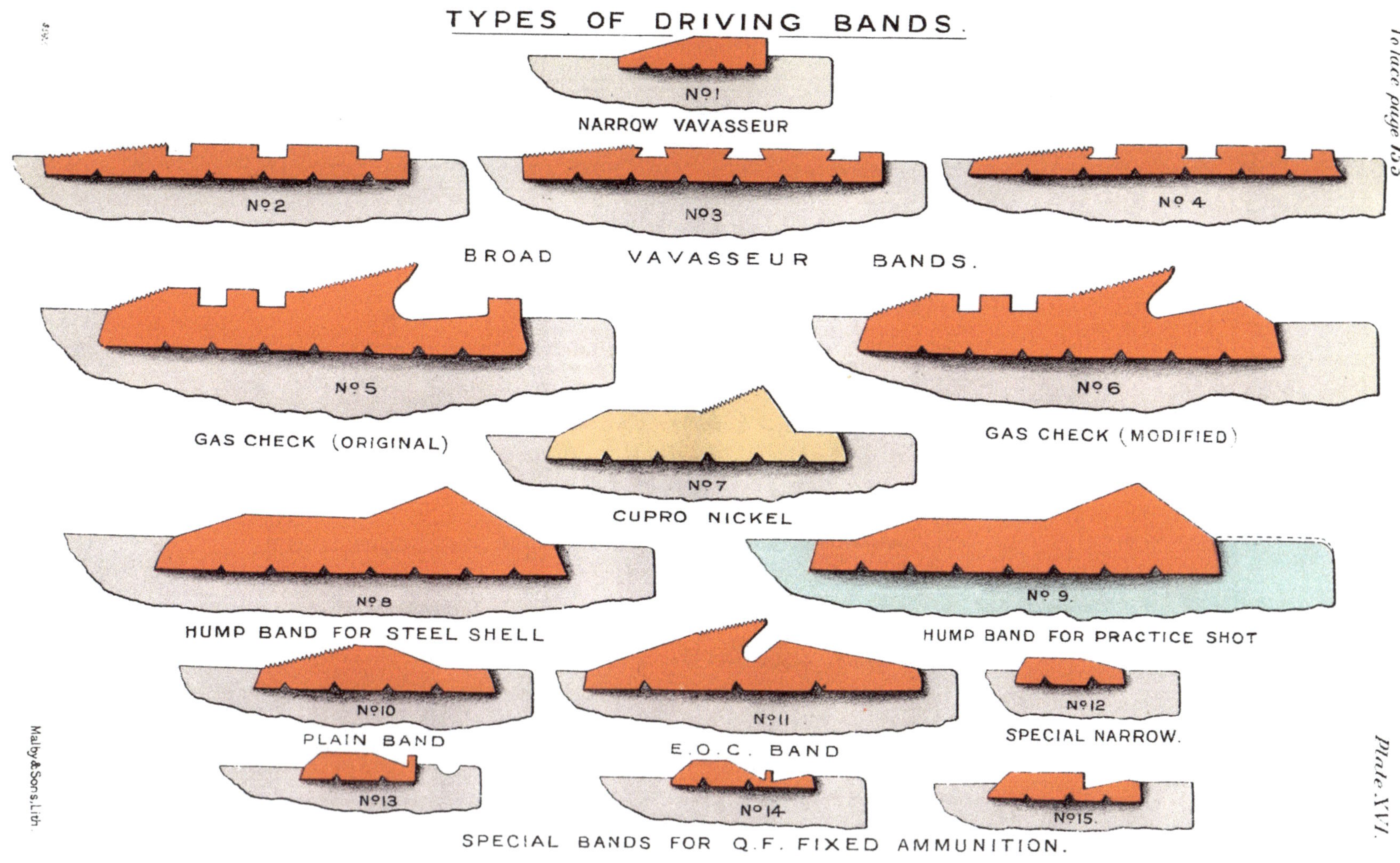

Malby & Sons, Lith.

To face page 159.

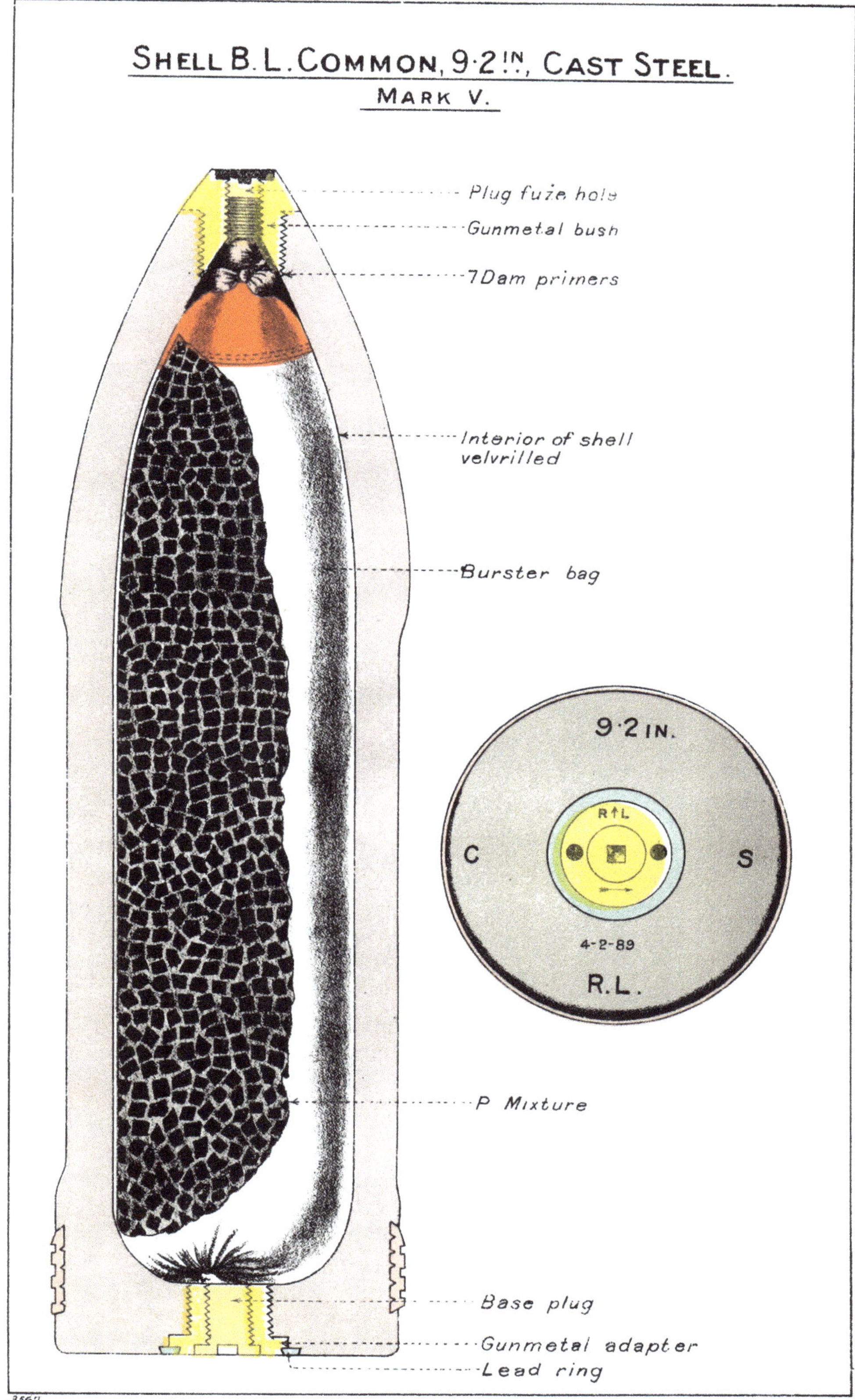

3567

Malby & Sons, Lith.

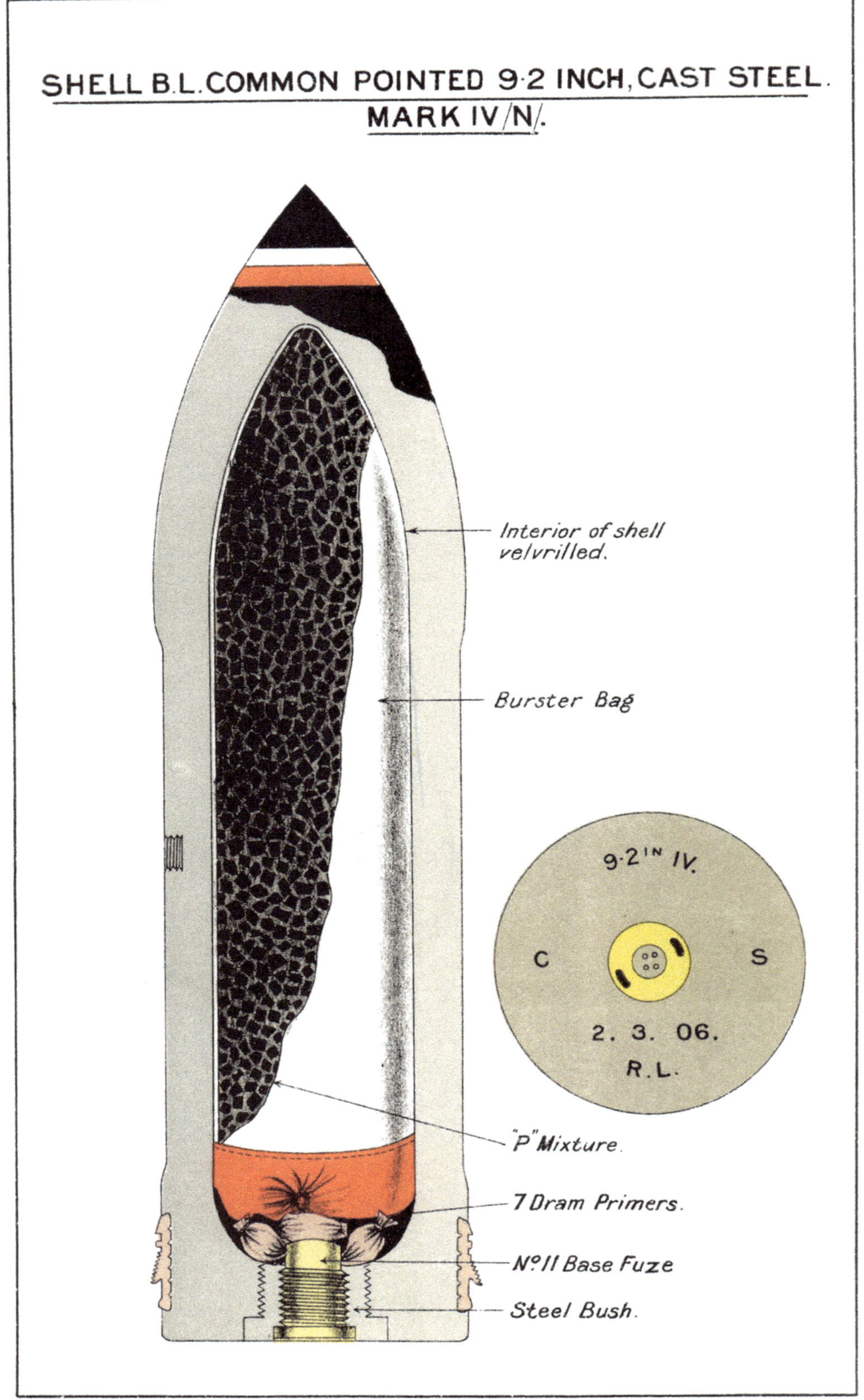

Malby & Sons, Lith

To face page 162.

SHELL, Q.F. COMMON POINTED 12 & 14 PRS. MARK II.

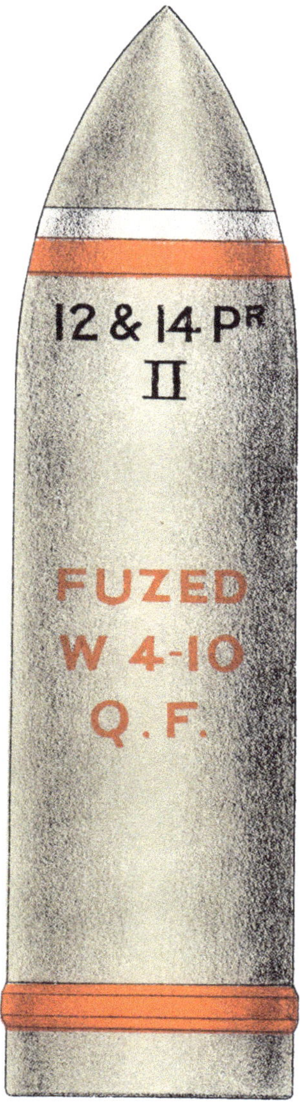

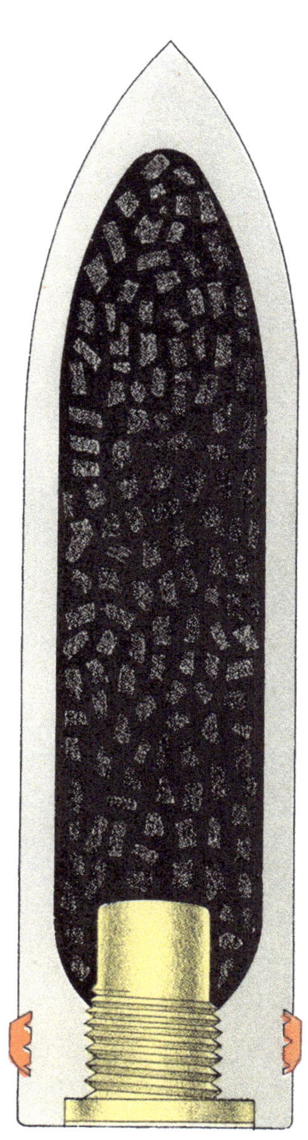

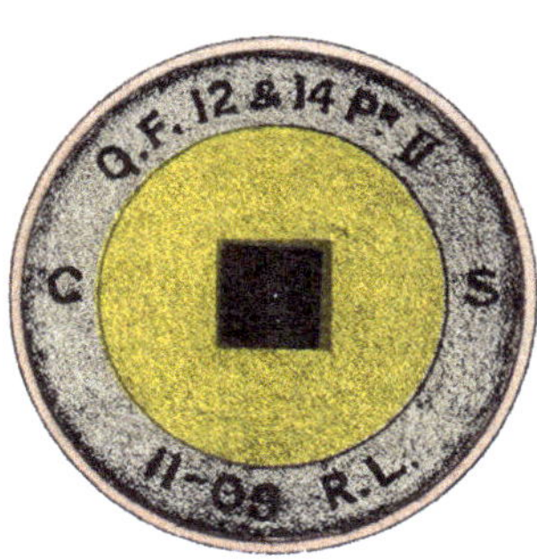

Malby & Sons, Lith.

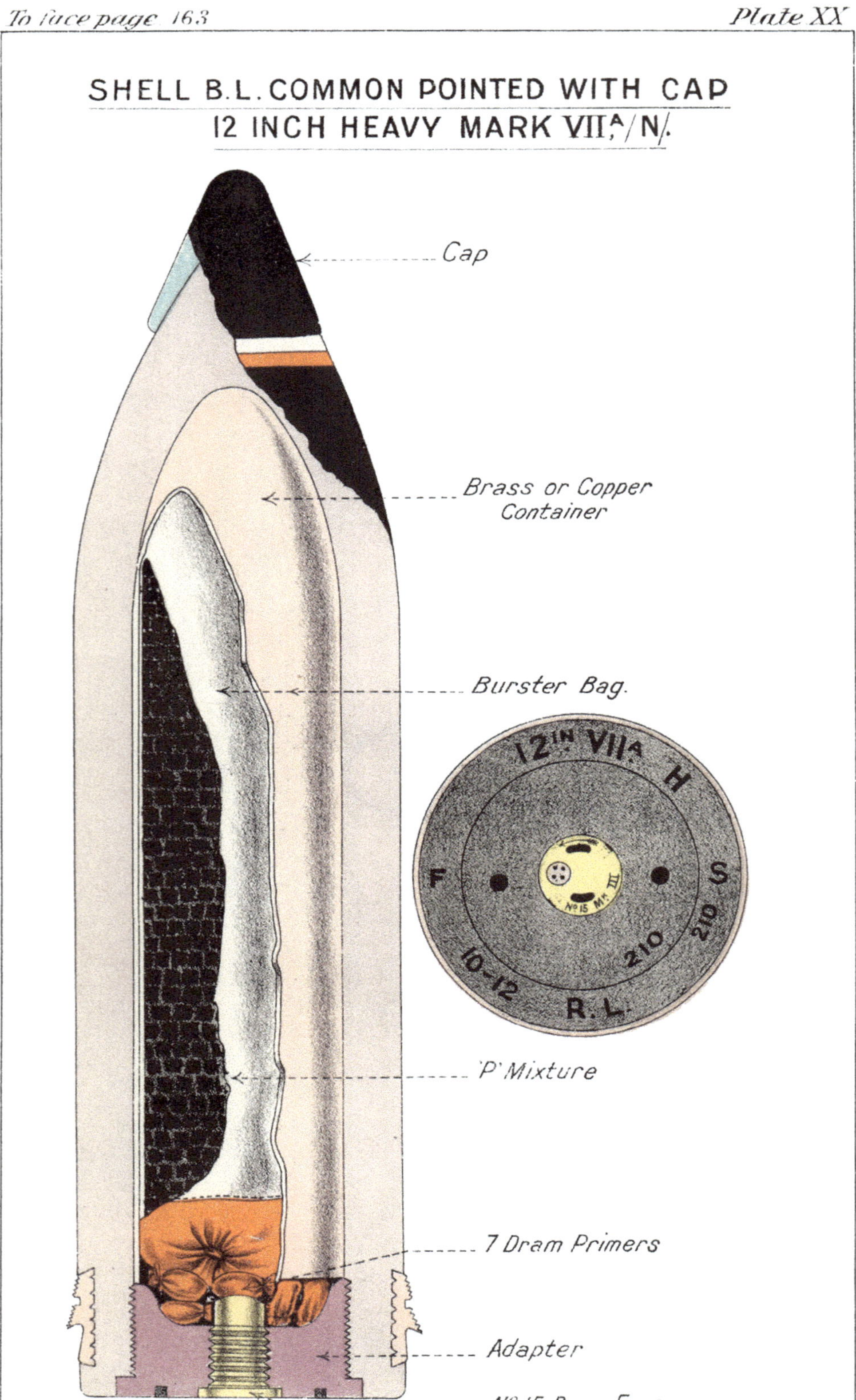

3567

Malby & Sons, Lith.

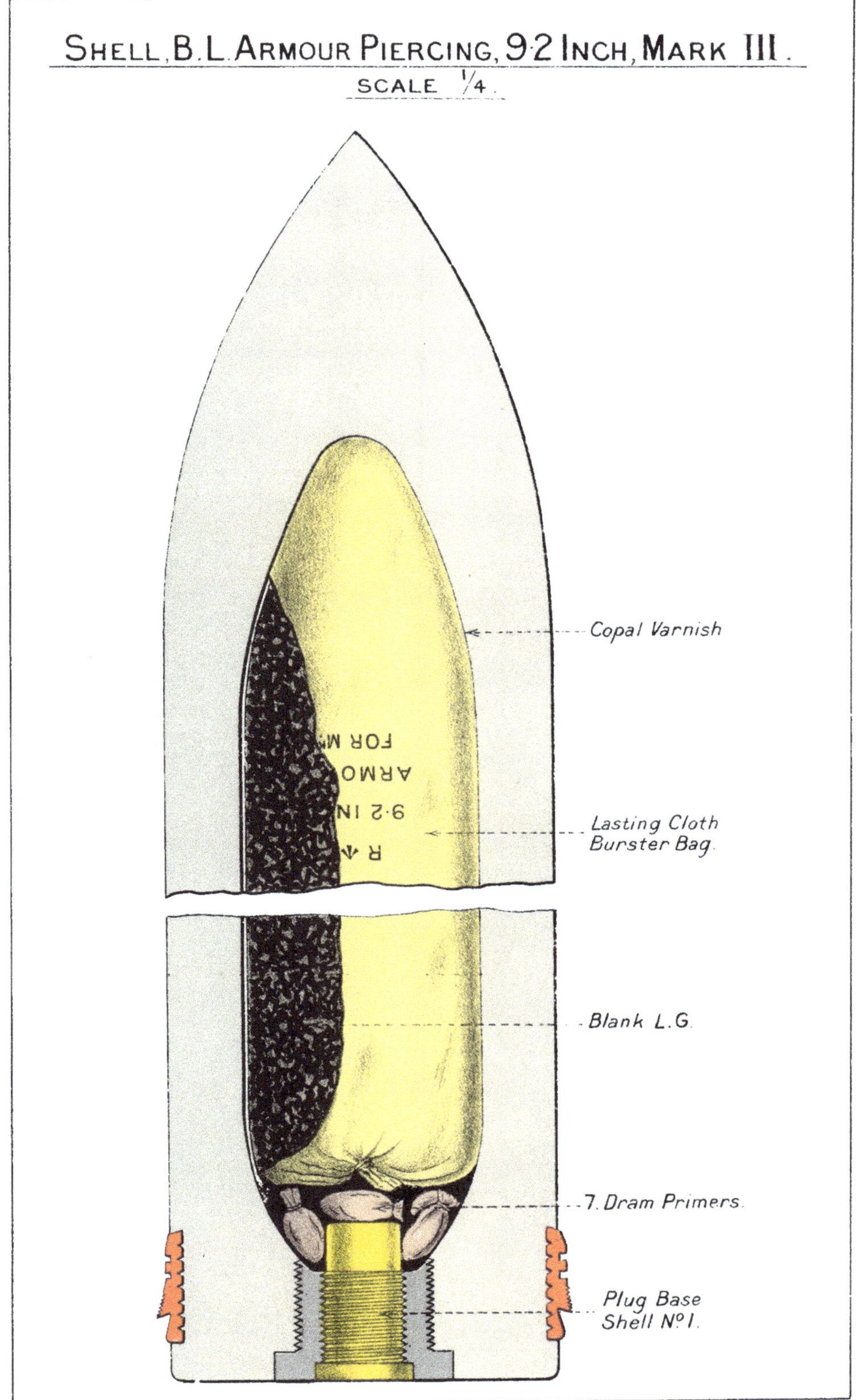

SHELL, B.L. ARMOUR PIERCING, 9·2 INCH, MARK III.

SCALE 1/4.

3567 Malby & Sons, Lith

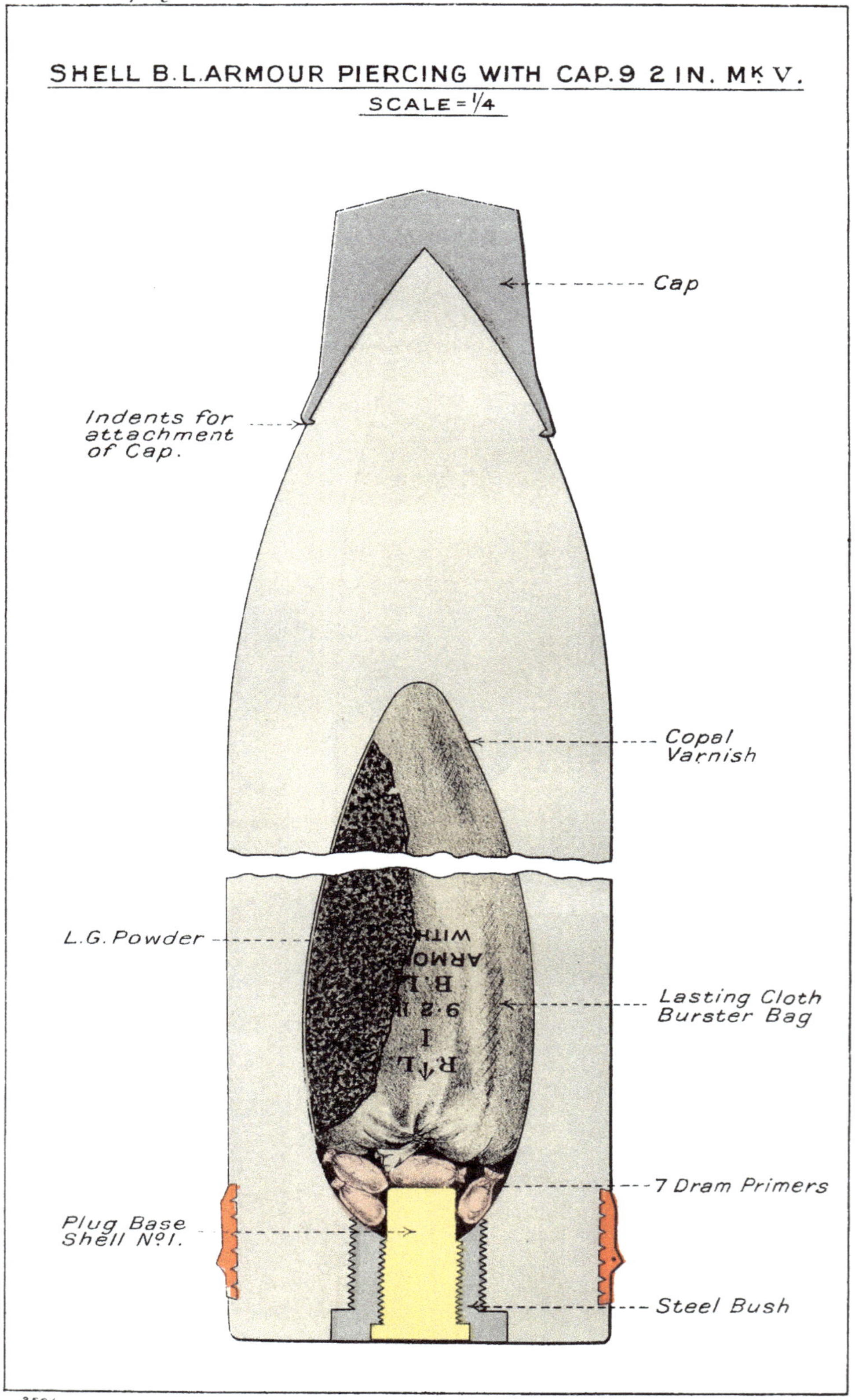

SHELL B. L. ARMOUR PIERCING WITH CAP. 9 2 IN. Mk. V.

SCALE = 1/4

Cap

Indents for attachment of Cap.

Copal Varnish

L. G. Powder

Lasting Cloth Burster Bag

7 Dram Primers

Plug Base Shell No. 1.

Steel Bush

Malby & Sons, Lith.

SHELL, B.L. B.L.C. OR Q.F. SHRAPNEL 15 PR (MARK VI) L

FORGED STEEL

(§ 11.235.)

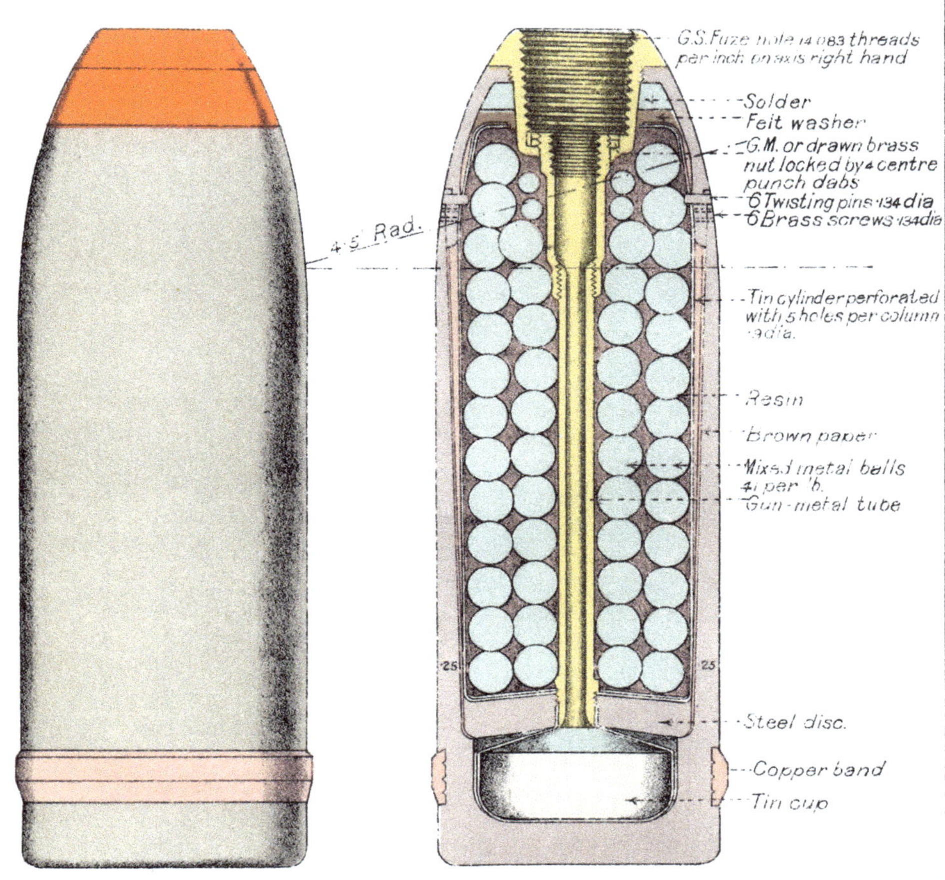

AVERAGE WEIGHT

	Lbs	Oz
Shell empty with band	13	1½
Bursting charge R.F.G.²		1½
Fuze T. & P.		13
Mean total weight	14	0

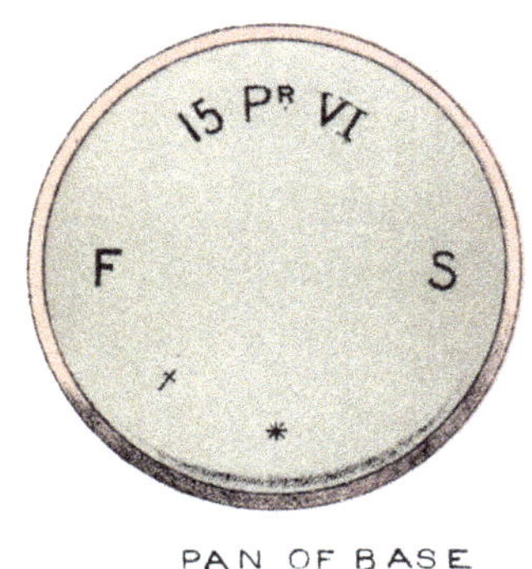

PAN OF BASE

* Contractors initials or recognised trade mark
+ Date of manufacture

Malby & Sons, Lith.

To face page 176. Plate XXIV.

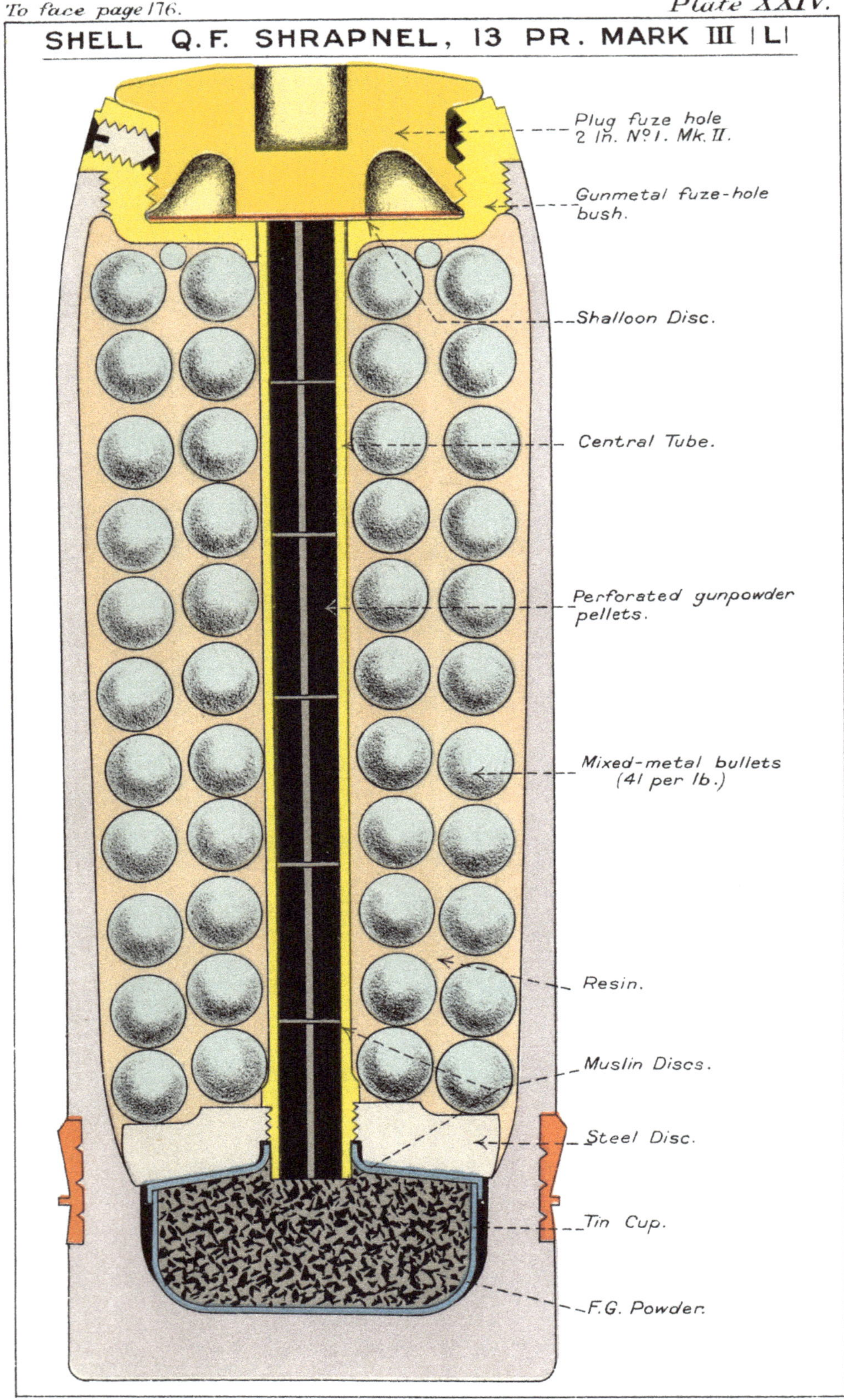

3567. Malby & Sons, Lith.

SHELL, B.L. SHRAPNEL, 2·75 INCH, MARK I.

Scale = 1/2.

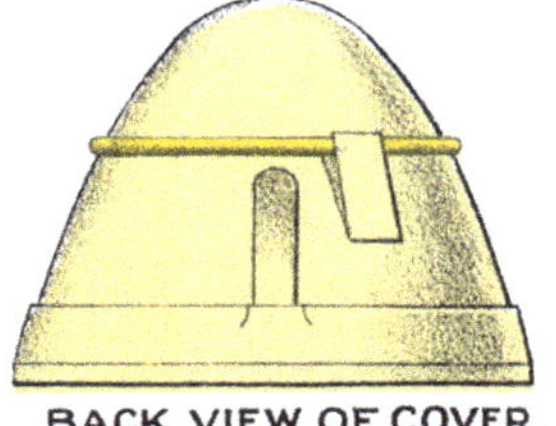

BACK VIEW OF COVER.

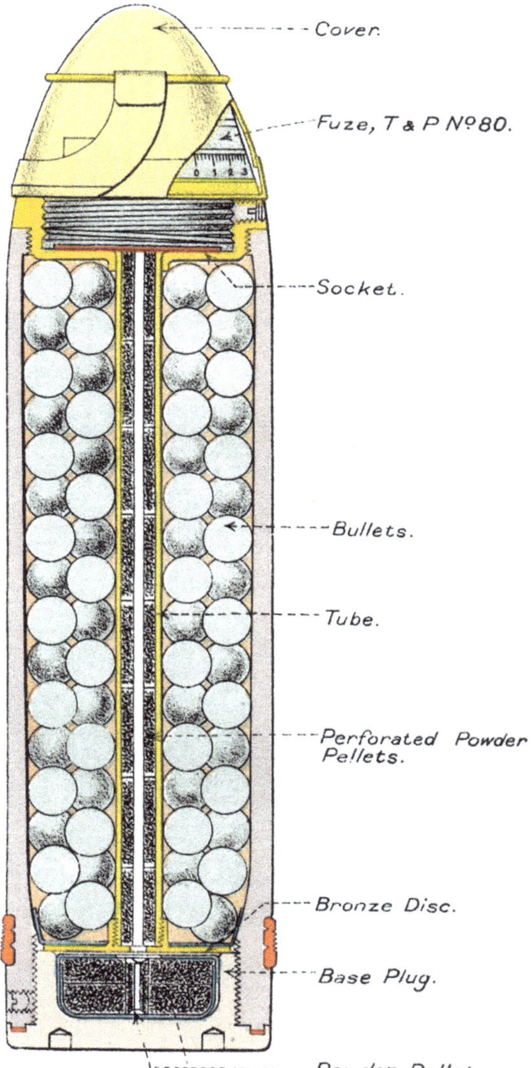

3567.

Malby & Sons, Lith.

SHELL, Q.F. SHRAPNEL, AND TRACER 3-INCH MARK I.

Scale = 2/3.

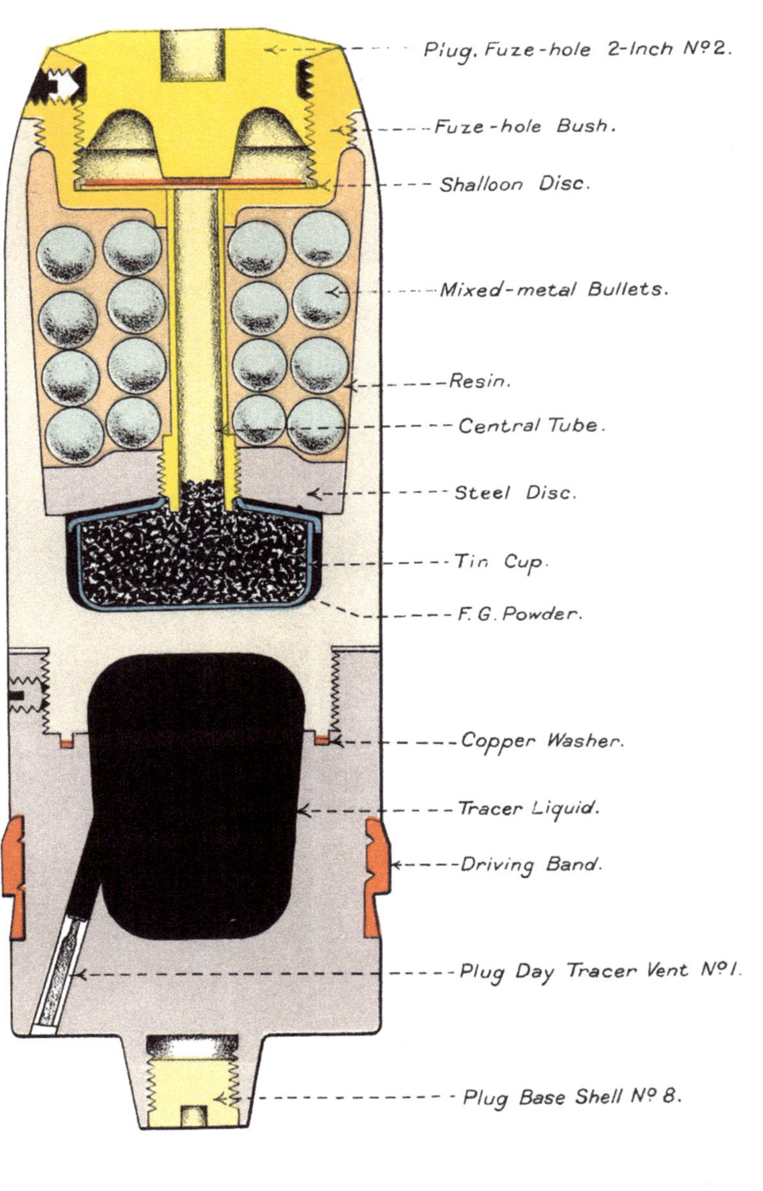

3567.

Malby & Sons, Lith.

Plate XXVII.

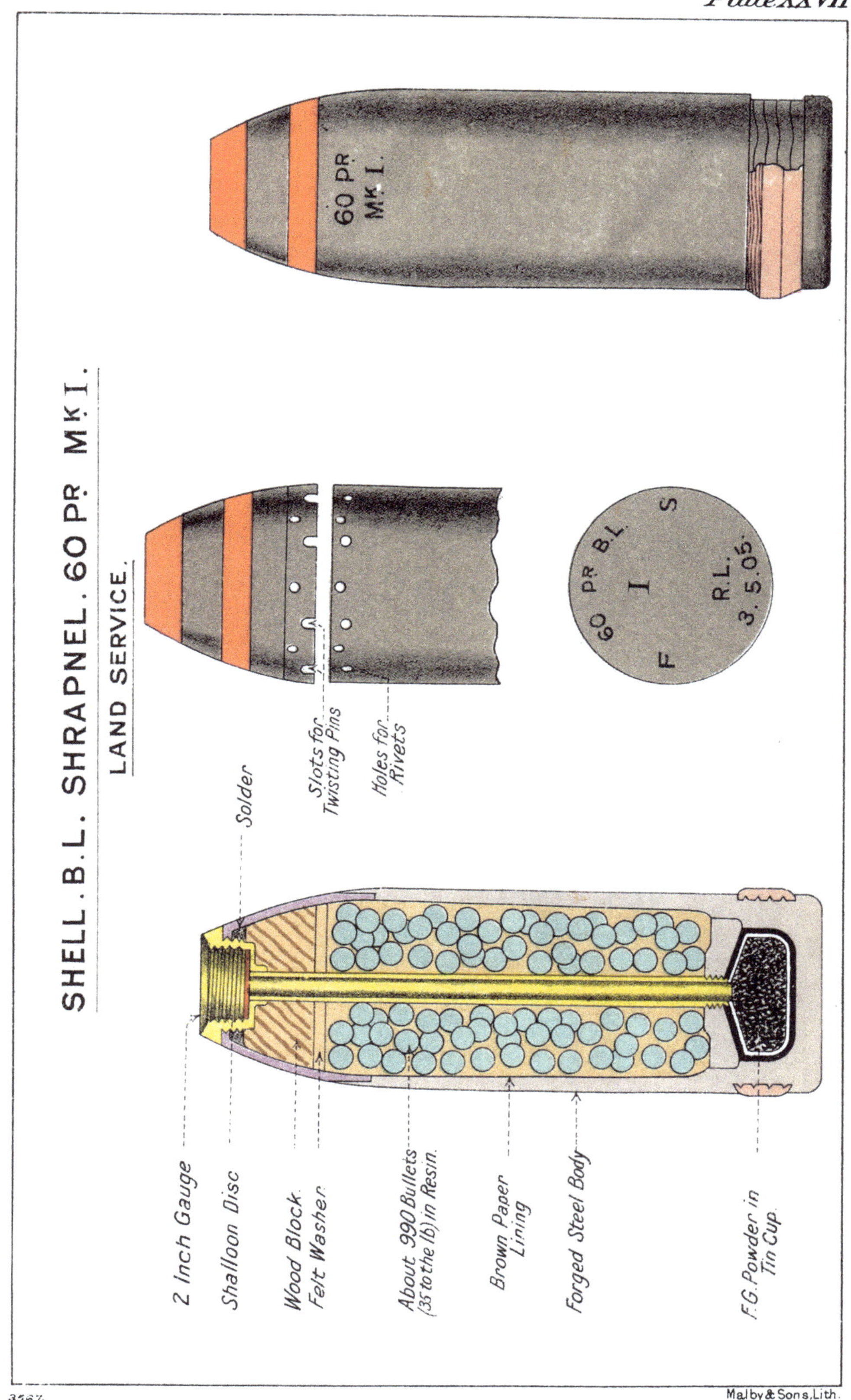

3567.

Malby & Sons, Lith.

SHELL, B.L. OR Q.F. SHRAPNEL 6 INCH GUN OR HOWITZER.

CAST STEEL, MARK IX |C|.

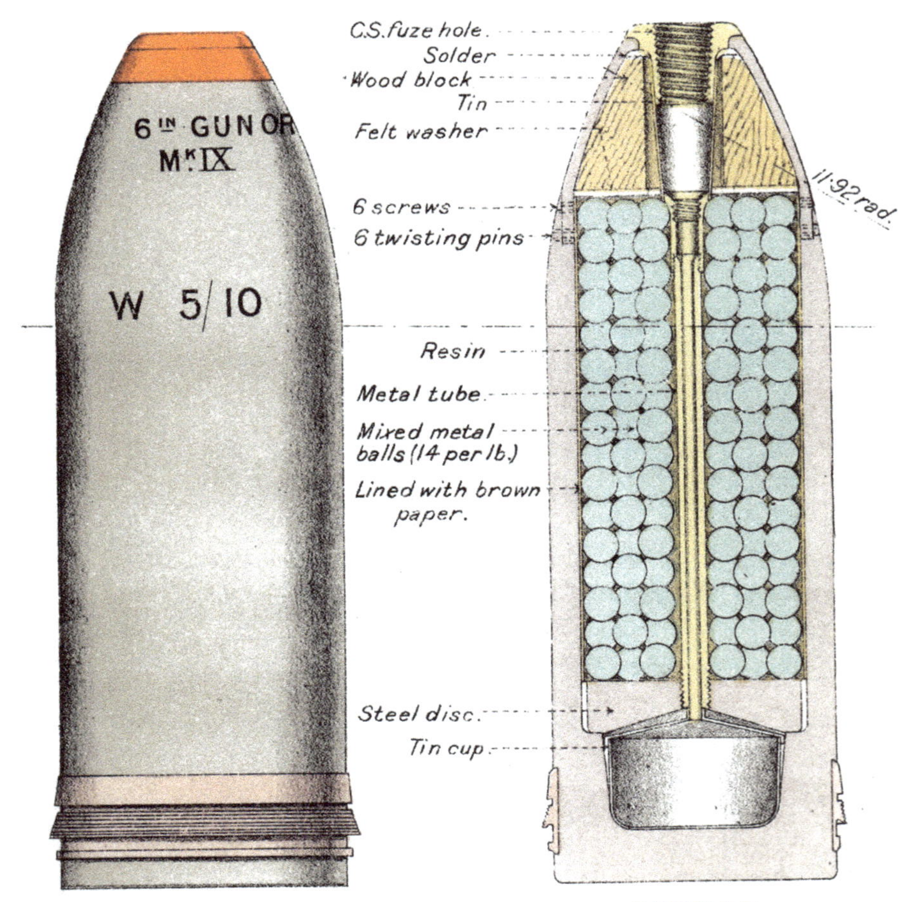

ELEVATION.

SECTION

AVERAGE WEIGHT OF SHELL	LB.	OZ.
Empty with band.	98.	9
Bursting charge, R.F.G.²		10½
Fuze T.& P. middle.	1.	5
Mean total weight.	100.	8½

Manufacturers initials or recognised trade mark.

PLAN OF BASE

3567

Malby & Sons, Lith.

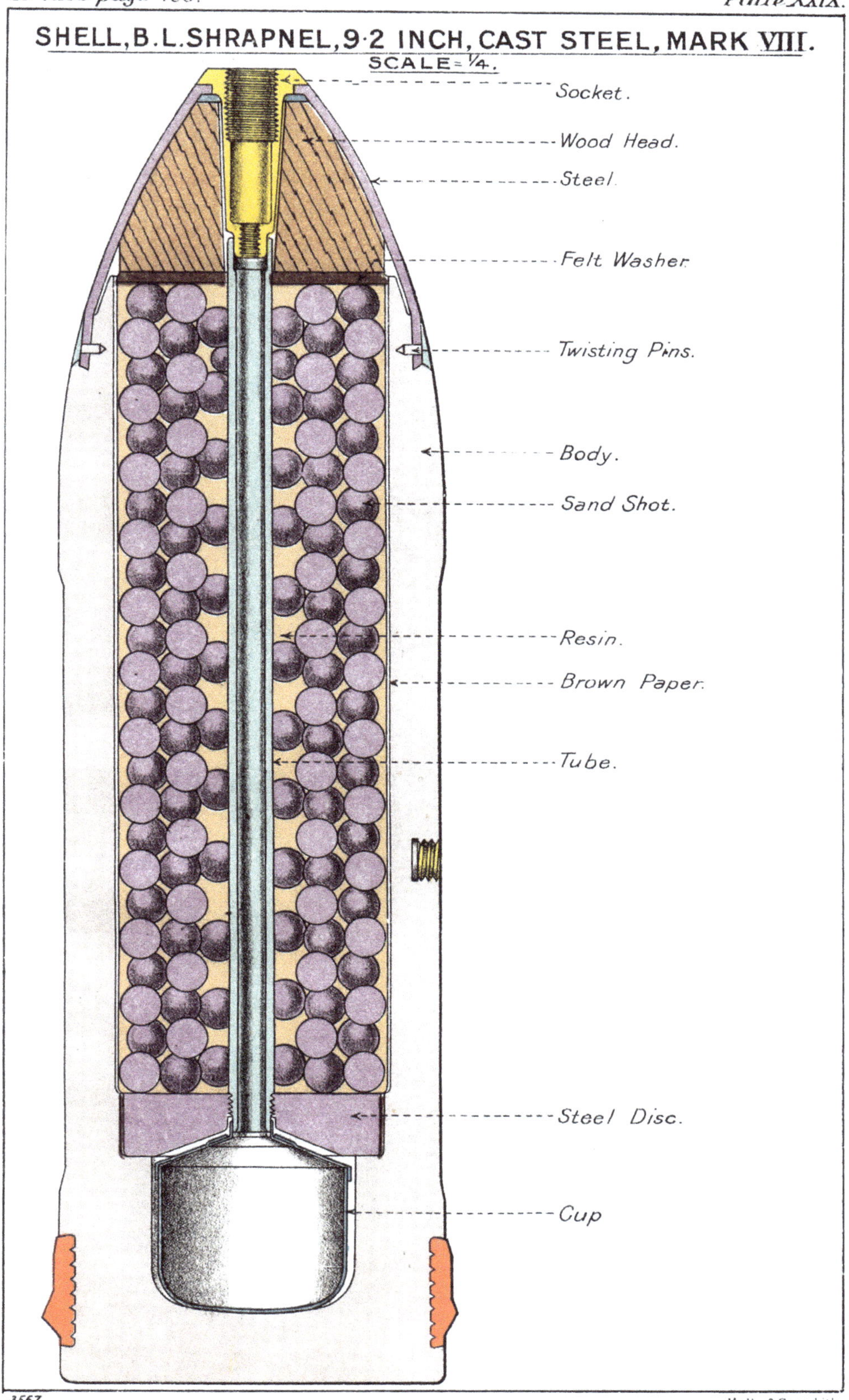
SHELL, B.L. SHRAPNEL, 9·2 INCH, CAST STEEL, MARK VIII.
SCALE = 1/4.
Socket.
Wood Head.
Steel.
Felt Washer.
Twisting Pins.
Body.
Sand Shot.
Resin.
Brown Paper.
Tube.
Steel Disc.
Cup
3567
Malby & Sons, Lith.

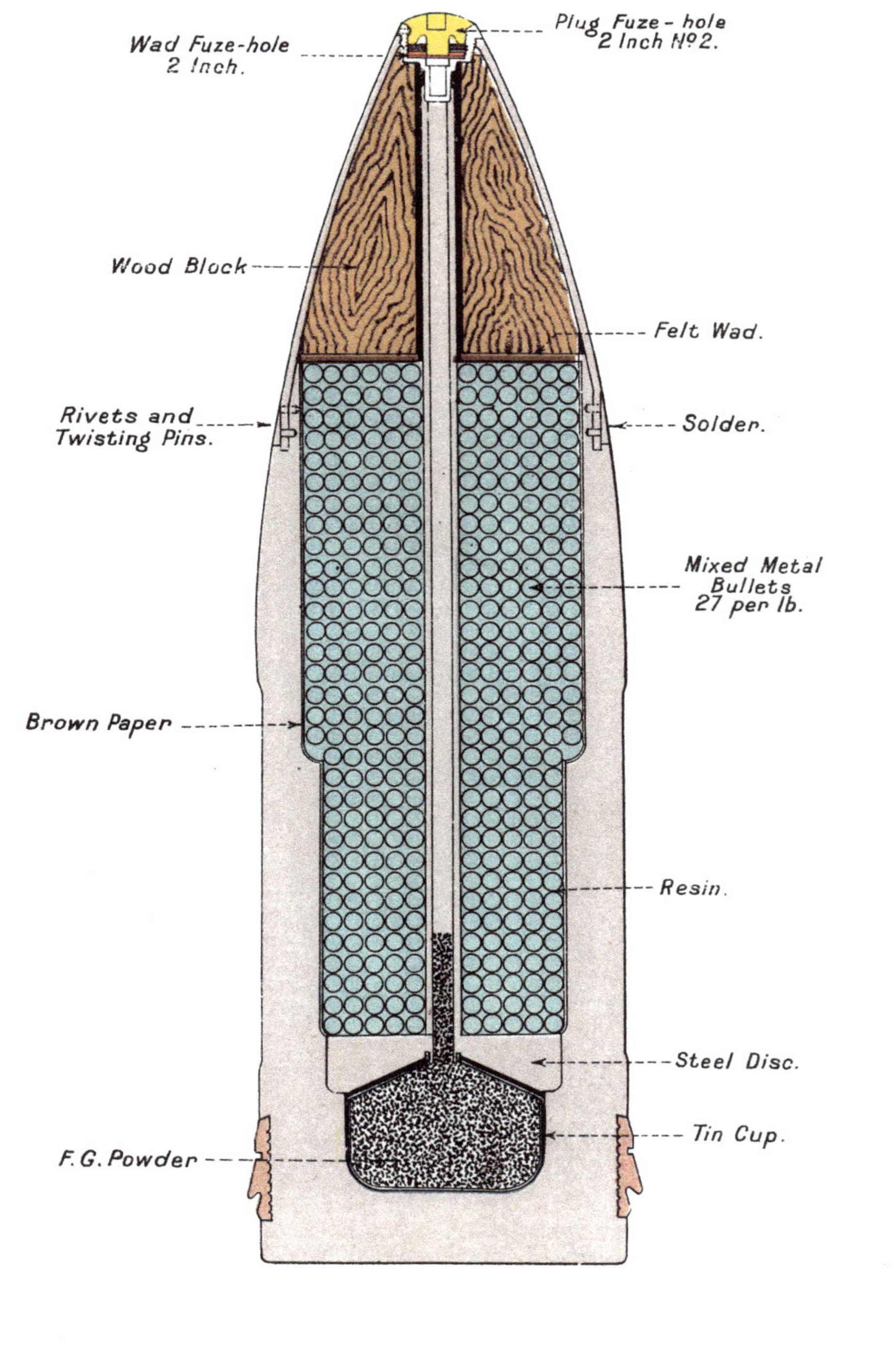

3567. Malby & Sons, Lith

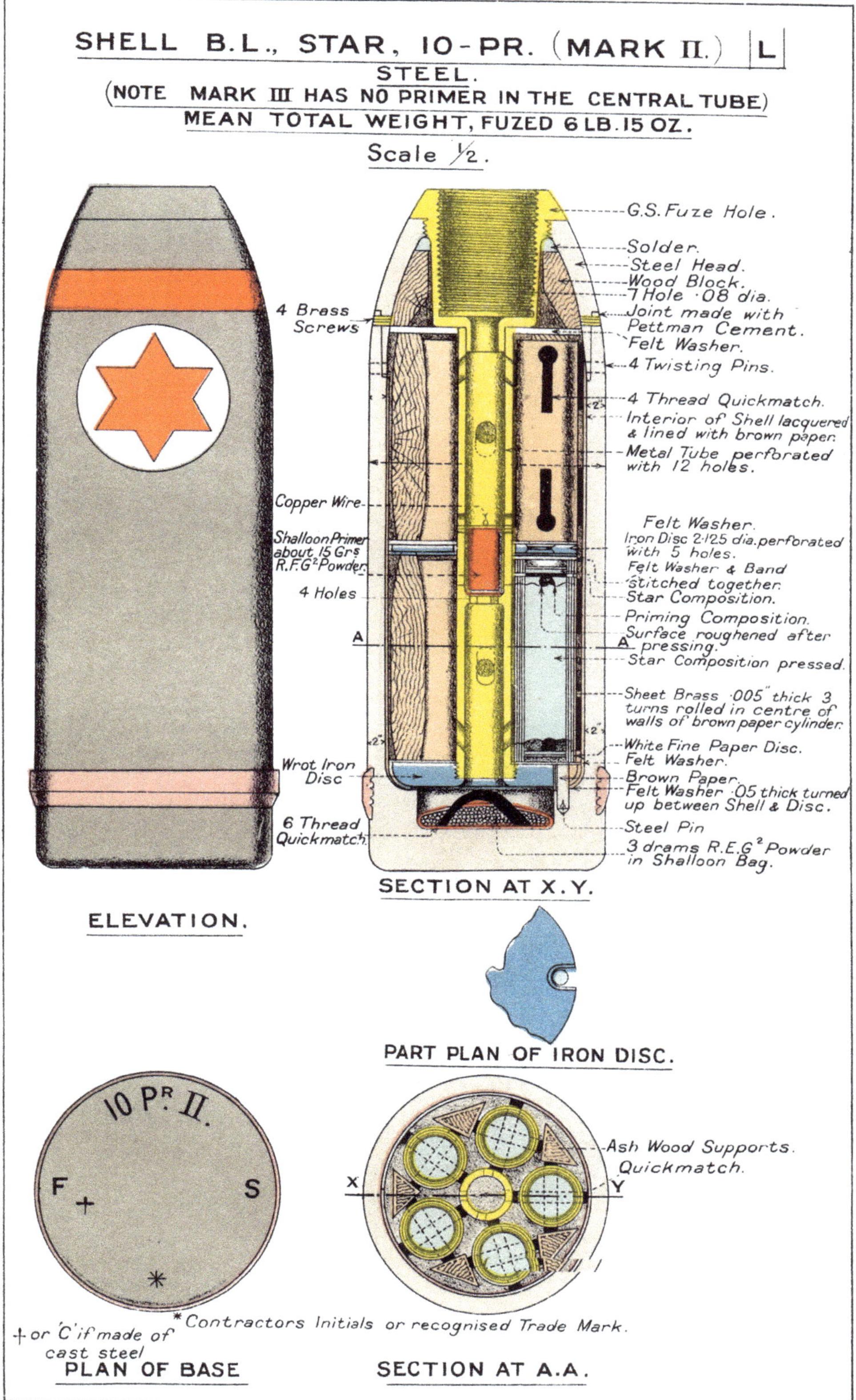

3567. Malby & Sons. Lith.

SHELL, B.L. STAR 6 INCH, HOWITZER, MARK I. |L|

SCALE = 1/3

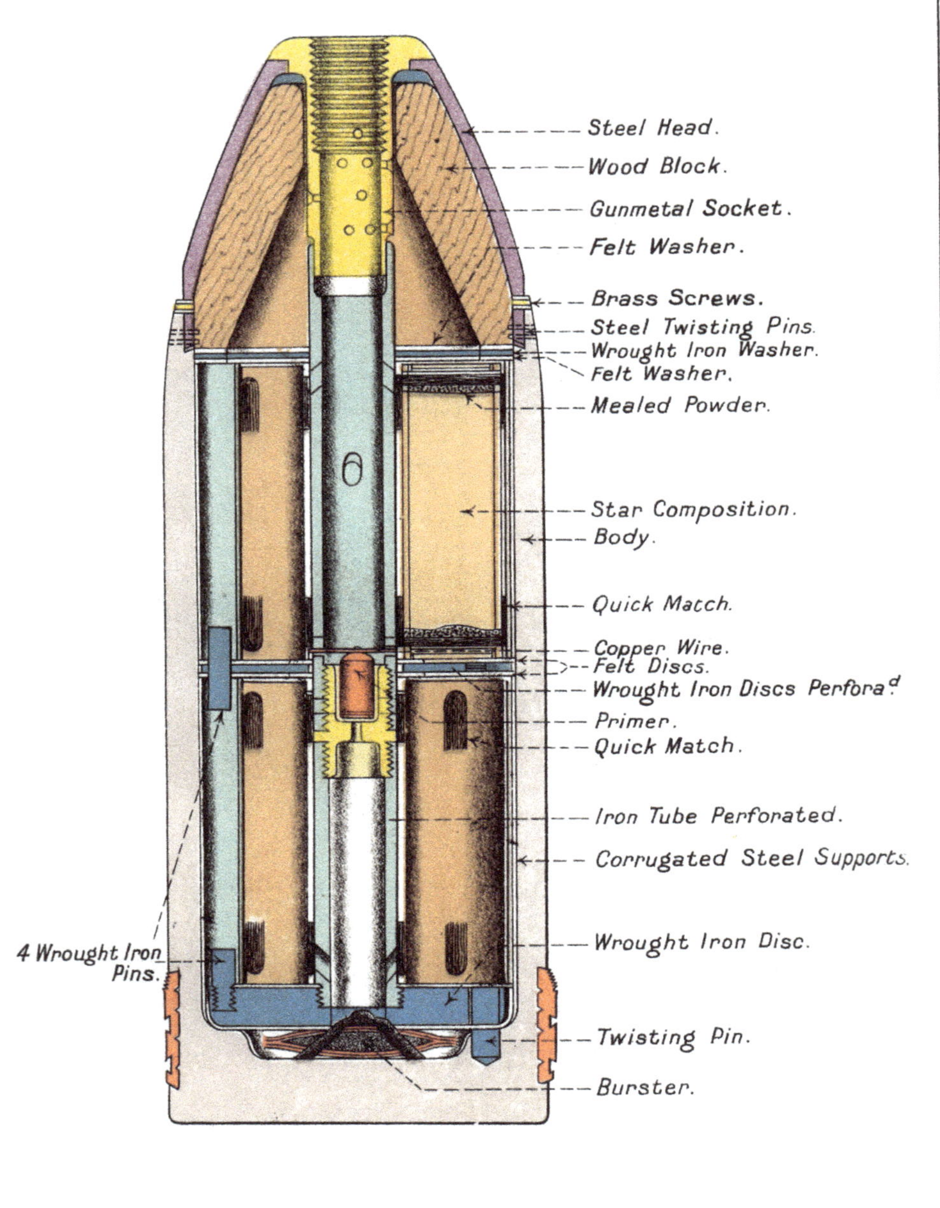

Malby & Sons, Lith.

SHELL B.L. 60 P^R COMMON LYDDITE MARK III.
FORGED STEEL.

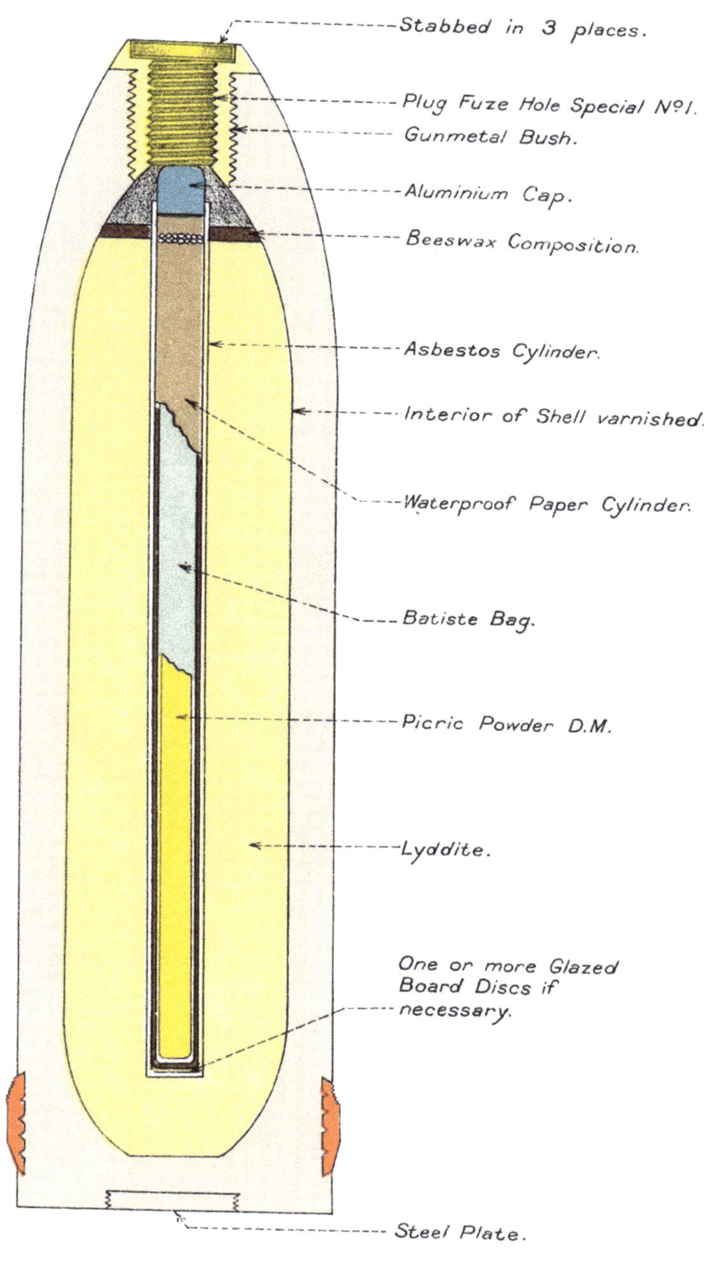

Malby & Sons, Lith.

SHELL B.L.Q.F. OR Q.F.C. COMMON LYDDITE, 6 INCH GUN, MARK IV. FORGED STEEL.

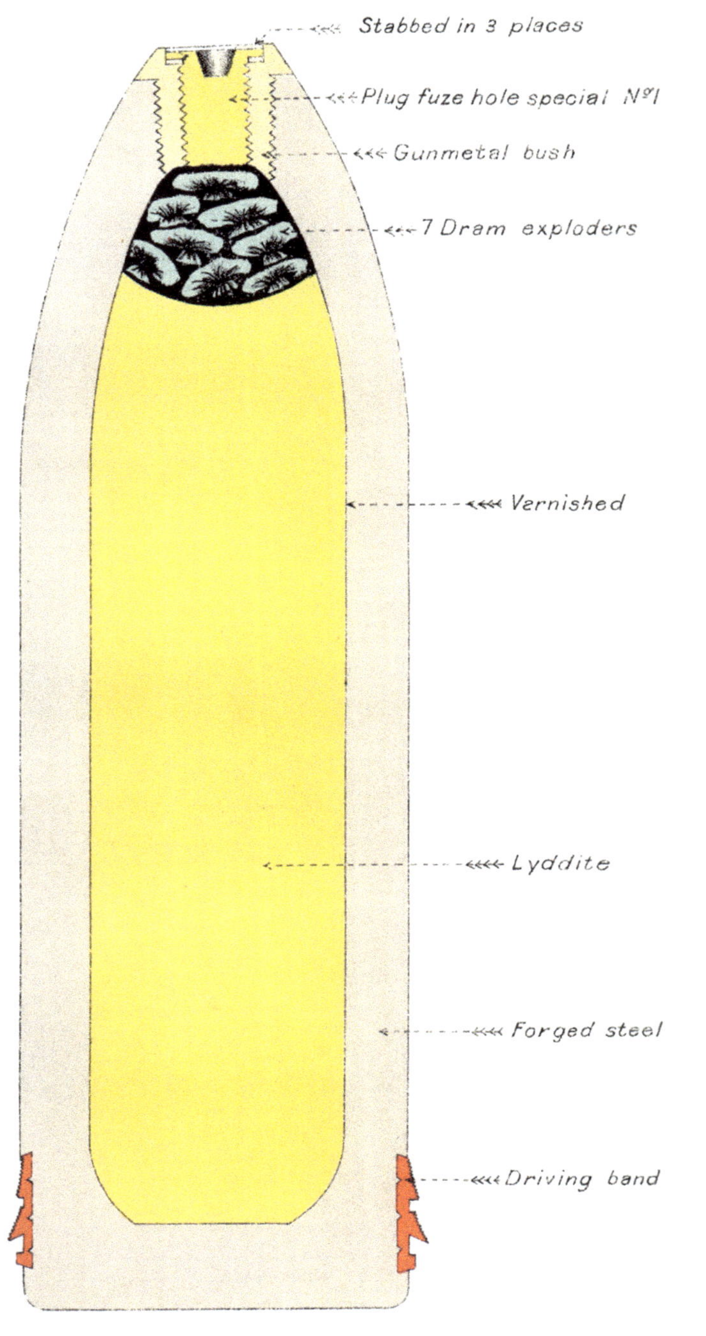

3567. Malby & Sons, Lith.

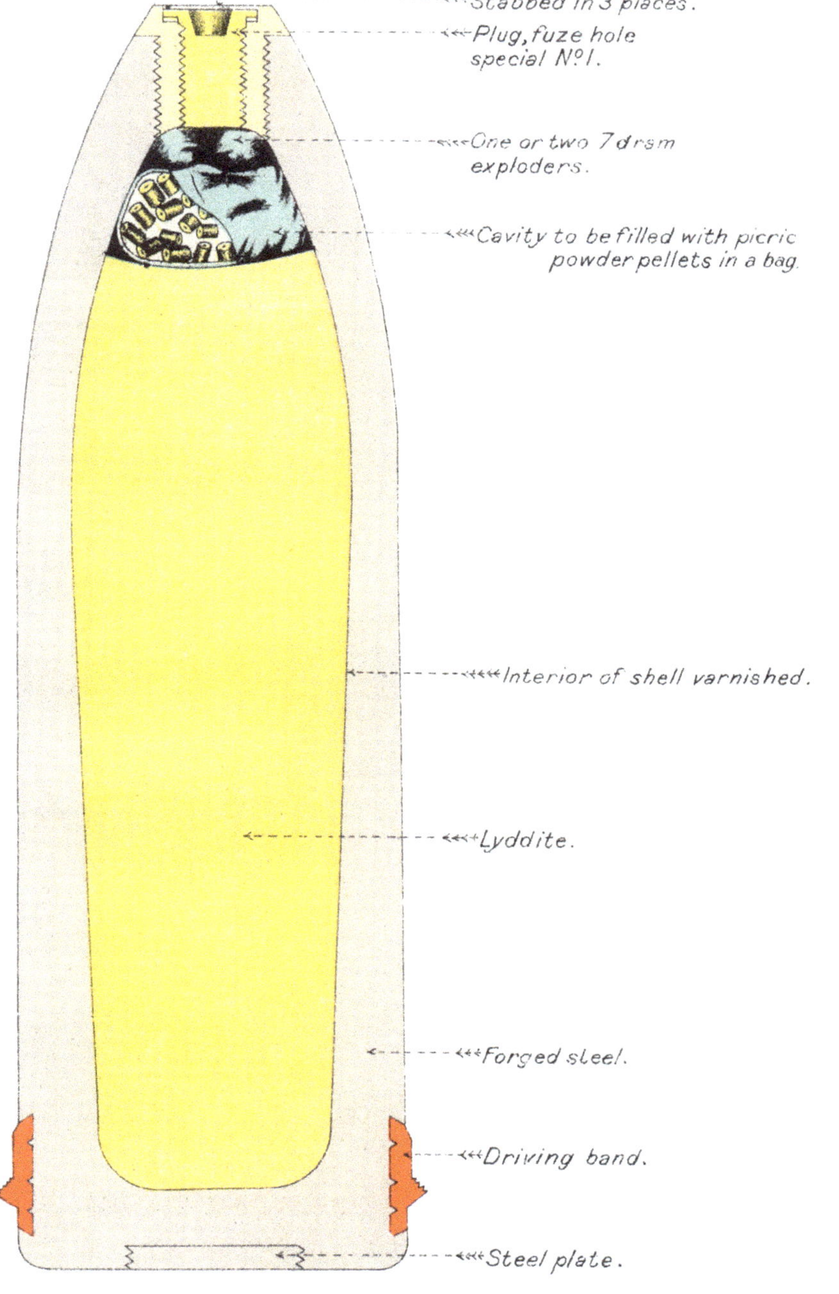

3567. Malby & Sons, Lith.

SHELL B.L. COMMON LYDDITE
6 INCH GUN MARK XII A.Q.N.T.

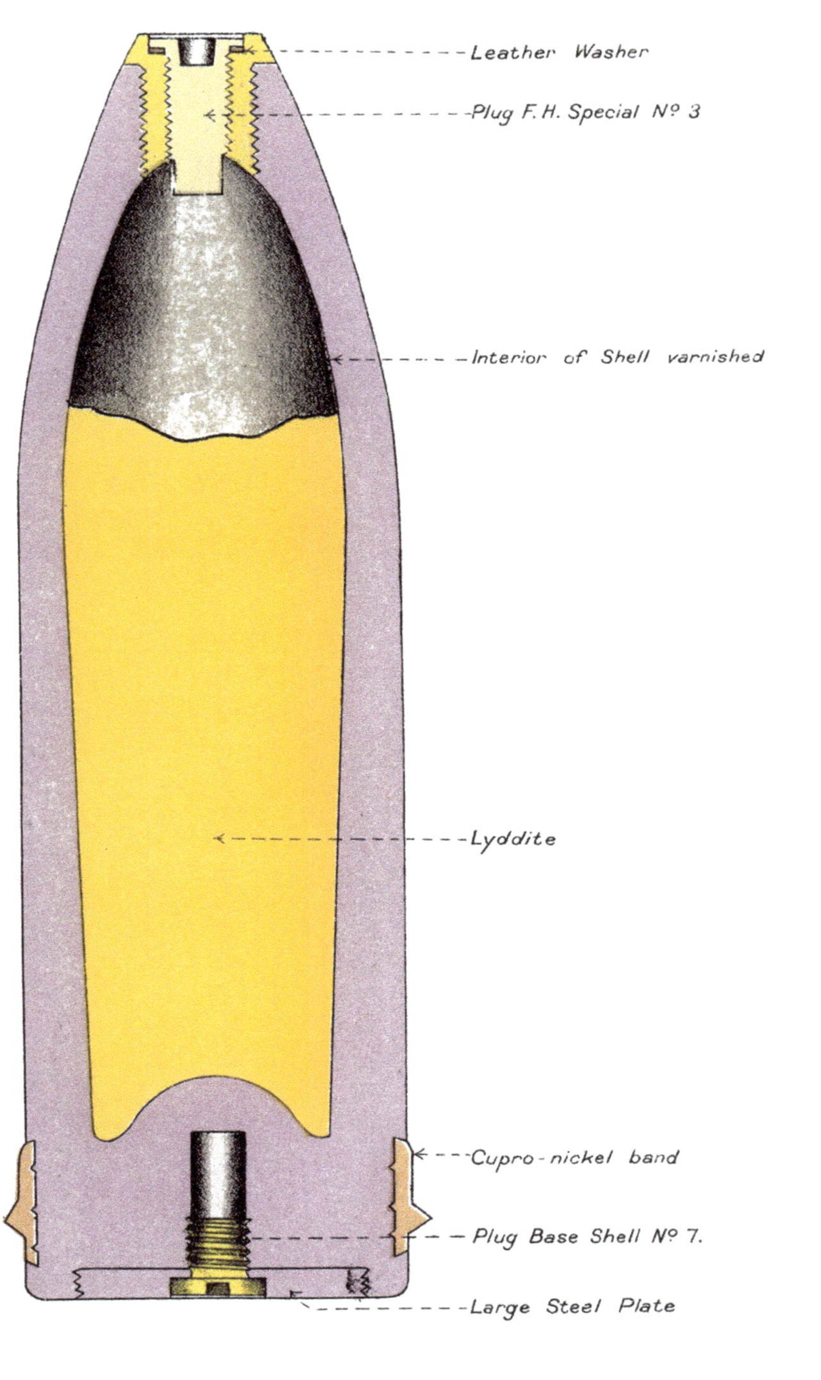

3567

Malby & Sons, Lith.

SHELL Q.F. COMMON LYDDITE, 12 AND 14 PR. MK IV.

Scale = 2/3.

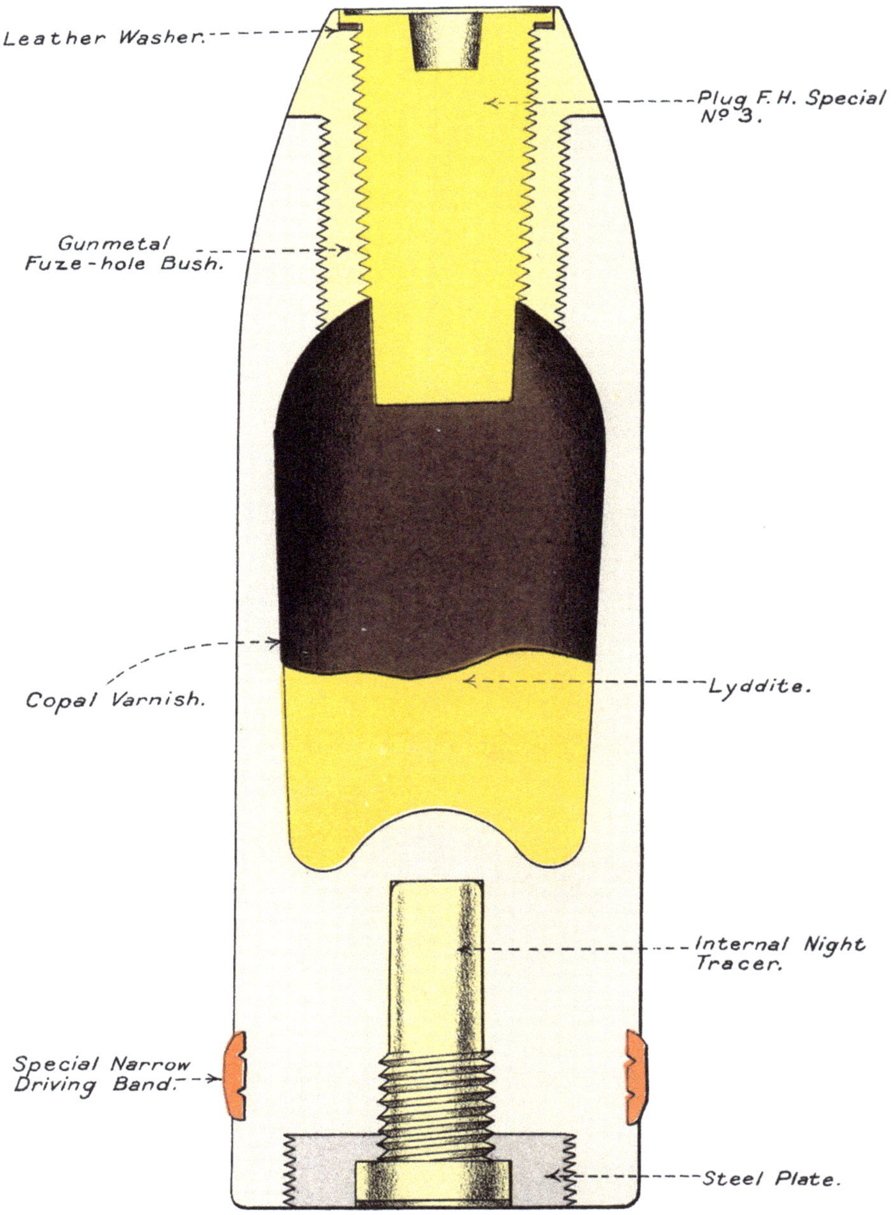

Leather Washer.

Plug F. H. Special No. 3.

Gunmetal Fuze-hole Bush.

Copal Varnish.

Lyddite.

Internal Night Tracer.

Special Narrow Driving Band.

Steel Plate.

3567. Malby & Sons, Lith.

SHELL. Q.F. COMMON LYDDITE 3 P^{R} MARK V. N.T.

FULL SIZE.

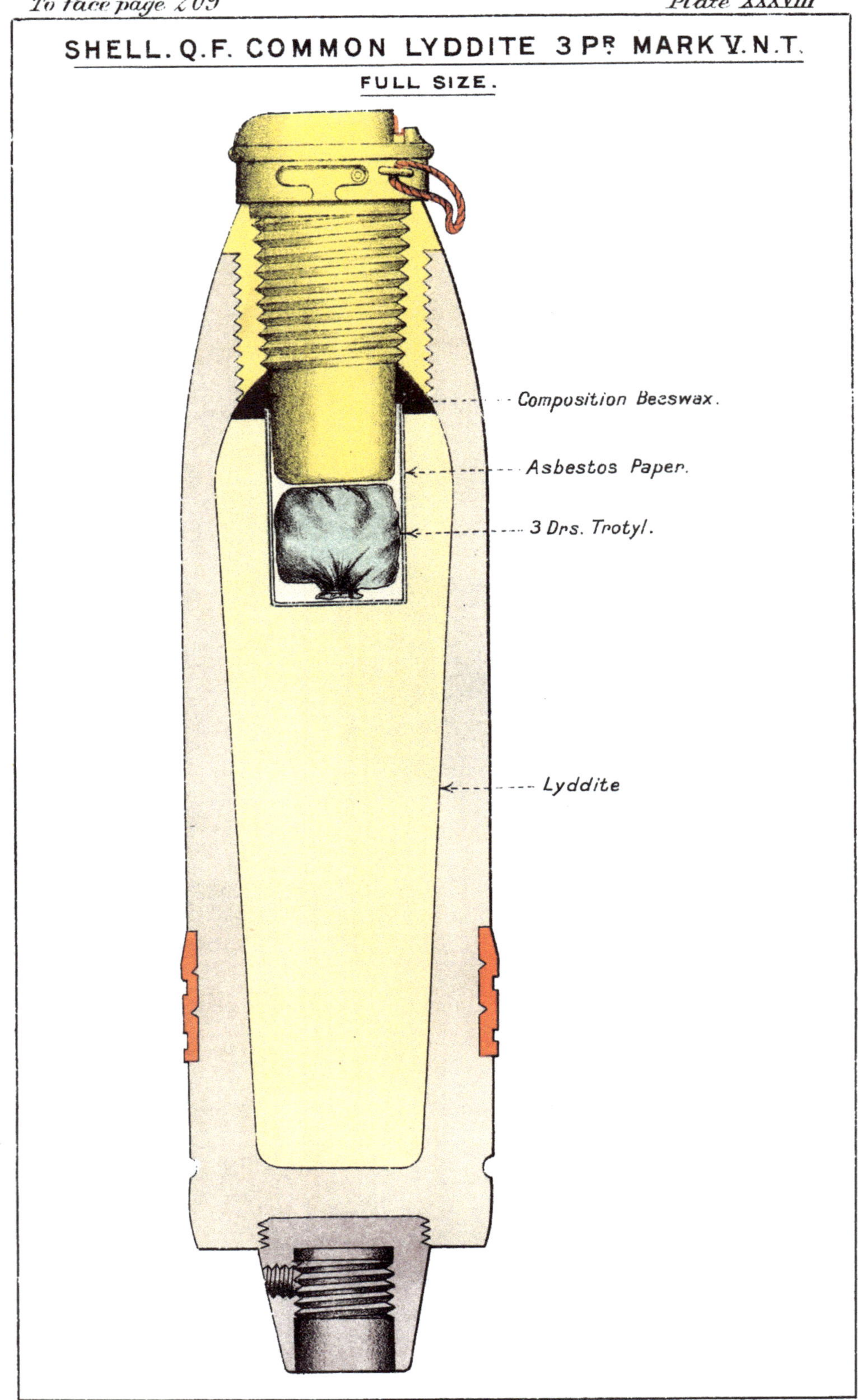

3587. Malby & Sons, Lith.

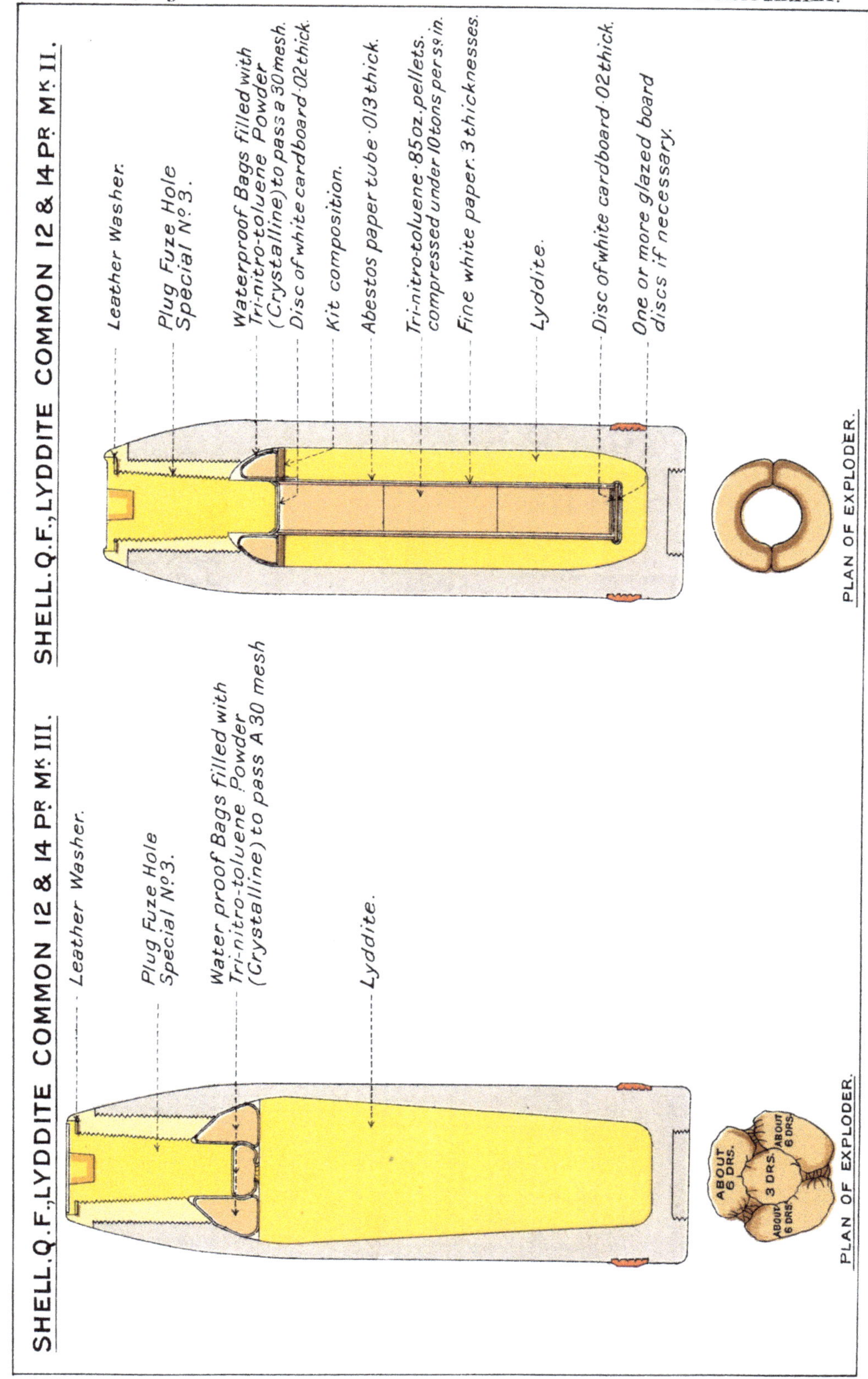
SHELL. Q.F., LYDDITE COMMON 12 & 14 PR. MK. III.
Leather Washer.
Plug Fuze Hole Special No. 3.
Water proof Bags filled with Tri-nitro-toluene Powder (Crystalline) to pass A 30 mesh
Lyddite.
ABOUT 6 DRS.
3 DRS.
ABOUT 6 DRS.
ABOUT 6 DRS.
PLAN OF EXPLODER.
SHELL. Q.F., LYDDITE COMMON 12 & 14 PR. MK. II.
Leather Washer.
Plug Fuze Hole Special No. 3.
Waterproof Bags filled with Tri-nitro-toluene Powder (Crystalline) to pass a 30 mesh.
Disc of white cardboard ·02 thick.
Kit composition.
Abestos paper tube ·013 thick.
Tri-nitro-toluene ·85 oz. pellets. compressed under 10 tons per sq. in.
Fine white paper. 3 thicknesses.
Lyddite.
Disc of white cardboard ·02 thick.
One or more glazed board discs if necessary.
PLAN OF EXPLODER.

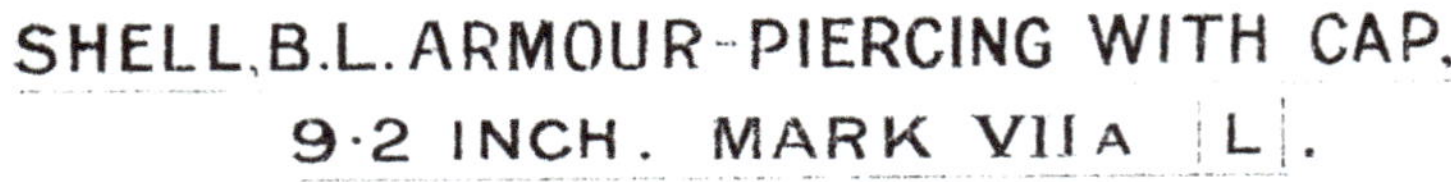

SHELL, B.L. ARMOUR-PIERCING WITH CAP.
9·2 INCH. MARK VII A [L].

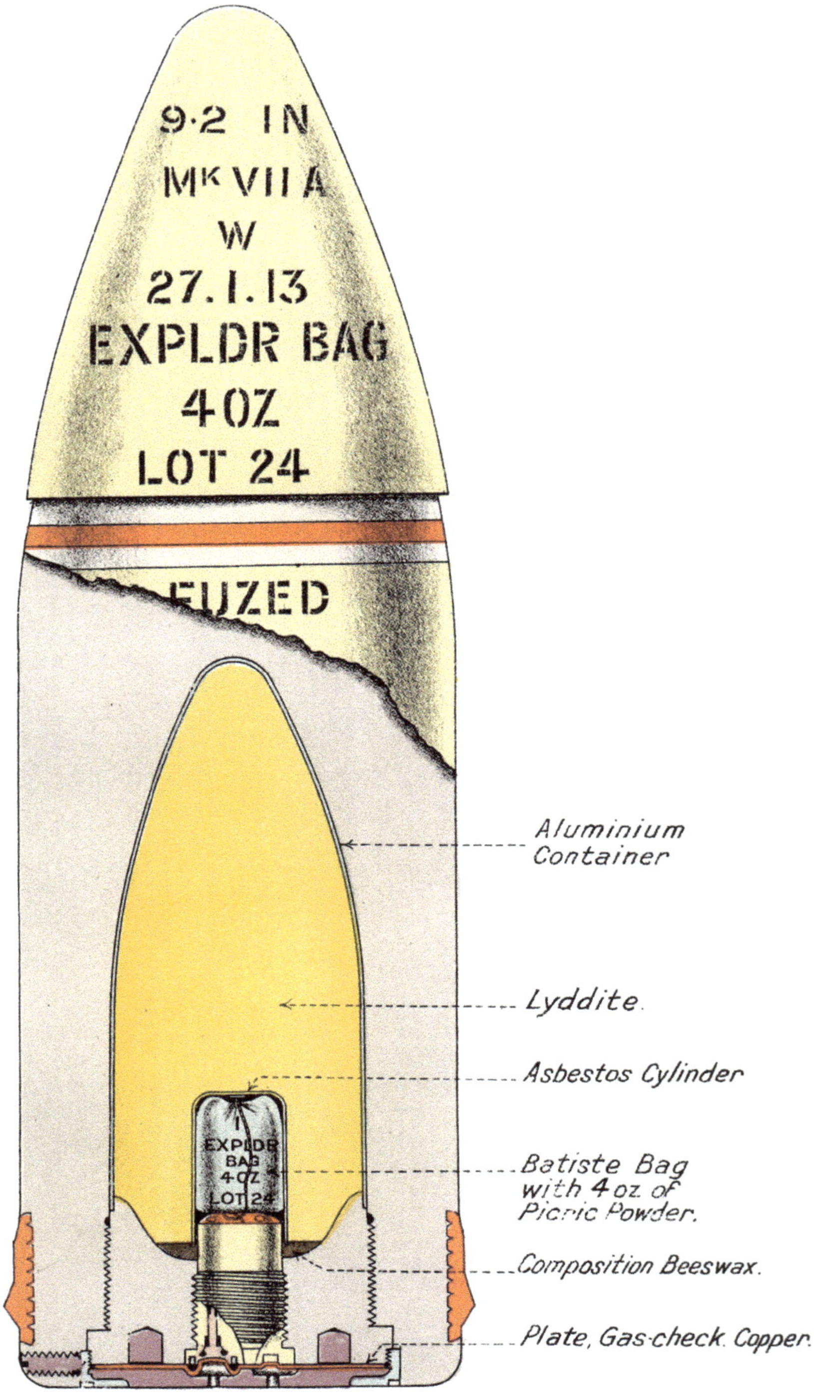

9·2 IN
M^K VII A
W
27.1.13
EXPLDR BAG
4 OZ
LOT 24

FUZED

EXPLDR
BAG
4 OZ
LOT 24

Aluminium Container

Lyddite.

Asbestos Cylinder

Batiste Bag with 4 oz. of Picric Powder.

Composition Beeswax.

Plate, Gas-check. Copper.

3567 Malby & Sons, Lith.

SHOT B.L. CASE 15 Pr. AND 12 Pr. MARK V | C.

§ 9456

SCALE ½

Manufacturers Initials or recognised Trade Mark.

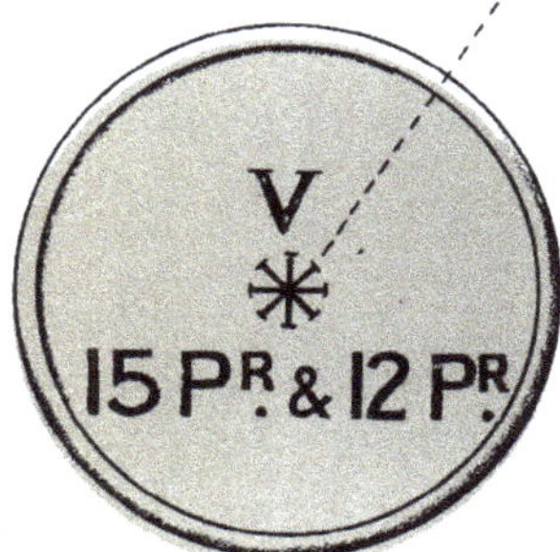

PLAN OF TOP

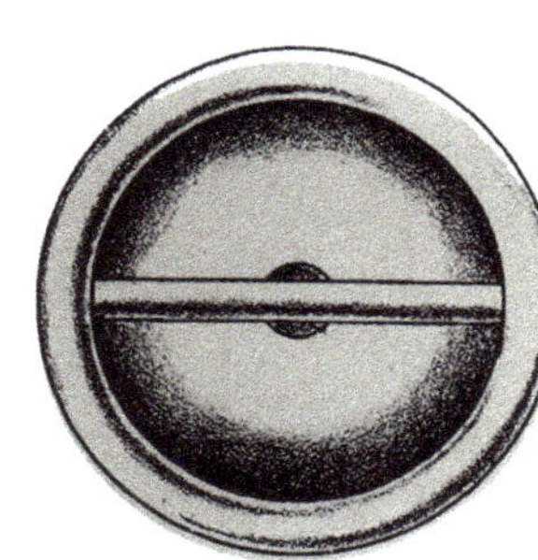

PLAN OF BASE

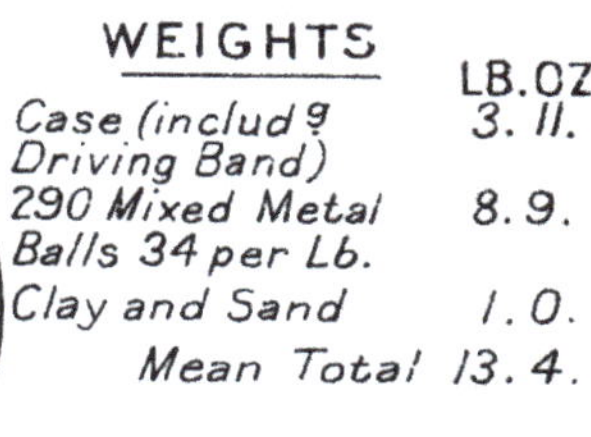

WEIGHTS	LB. OZ.
Case (includg Driving Band)	3. 11.
290 Mixed Metal Balls 34 per Lb.	8. 9.
Clay and Sand	1. 0.
Mean Total	13. 4.

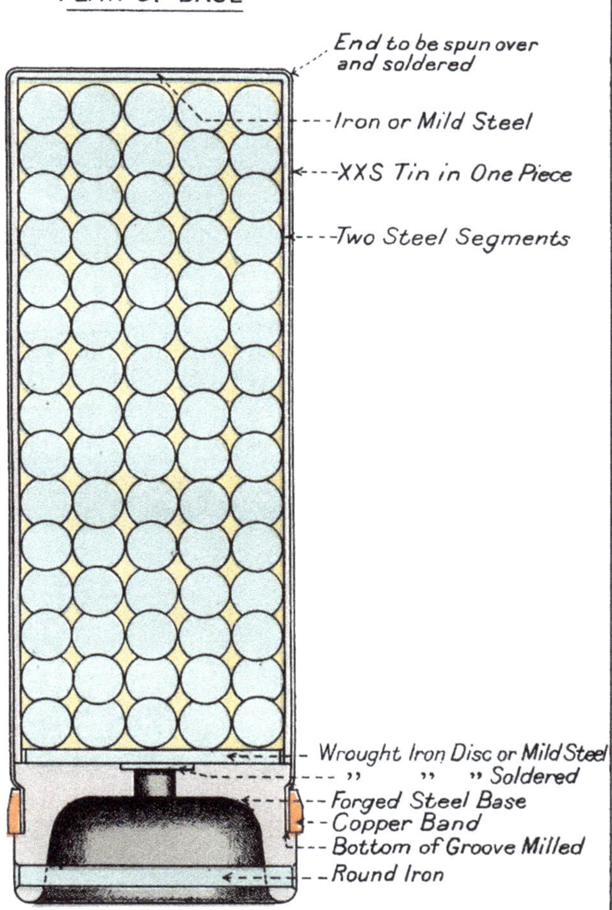

To face page 237.

DISTINGUISHING MARKS FOR SHOT AND SHELL.

Plate XLII.

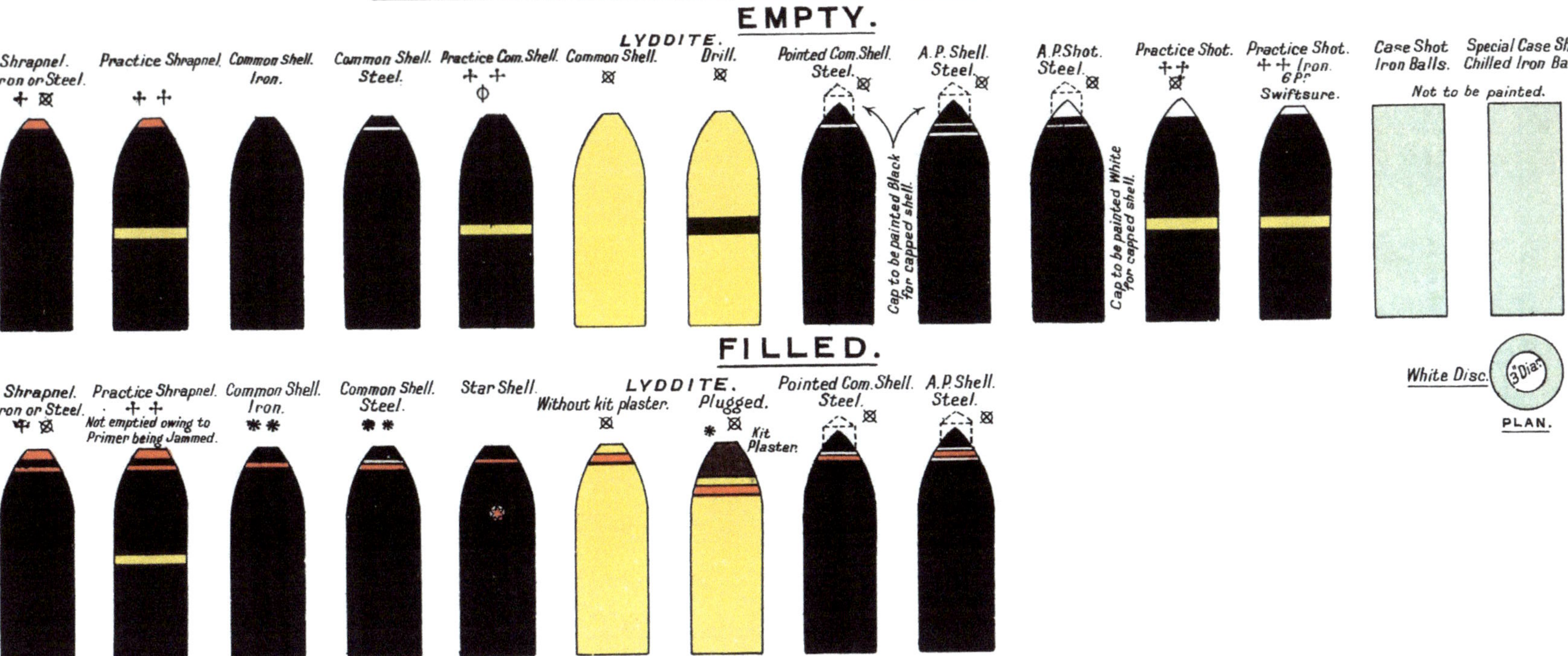

R.B.L. Projectiles coated with lead are not to be painted over the lead coating, except as provided for base and shoulder.
Naval Projectiles suitable only for use with Q.F. 12 Pr. 12 and 8 cwt. Guns to be painted lead colour.
Projectiles fitted with a Cupro Nickel driving band have a white band ½" in width immediately above the driving band.

✢ {15 Pr. Shrapnel shell body to be painted lead, and plug, Black. 13 Pr. and 18 Pr. Mark II Shrapnel shell, bodies to be painted lead colour.

ɸ Pointed common shell for Practice will be similarly marked.

✱ Kit Plaster required with No. 1 Mark I plug only.

✢ ✢ {Practice projectiles (except those referred to in footnote ✱✱), width of Yellow band to be 2 inches for 6 inch and over, and 1 inch for under 6 inch to 3 Pr.

✱✱ {Common Practice shell filled with Pitch and Powder, or with smoke producing mixture to represent shrapnel will have in addition two yellow bands. The bands will be 1 inch wide and 1 inch apart for 6 inch shell, and ½ inch wide and ½ inch apart for shell below 6 inch.

⊠ {Projectiles B.L. 12 inch and 13·5 inch will be stencilled ⊖ in 3 places equally spaced round the projectiles to indicate centre of gravity when filled.

3567.

Malby & Sons, Lith.

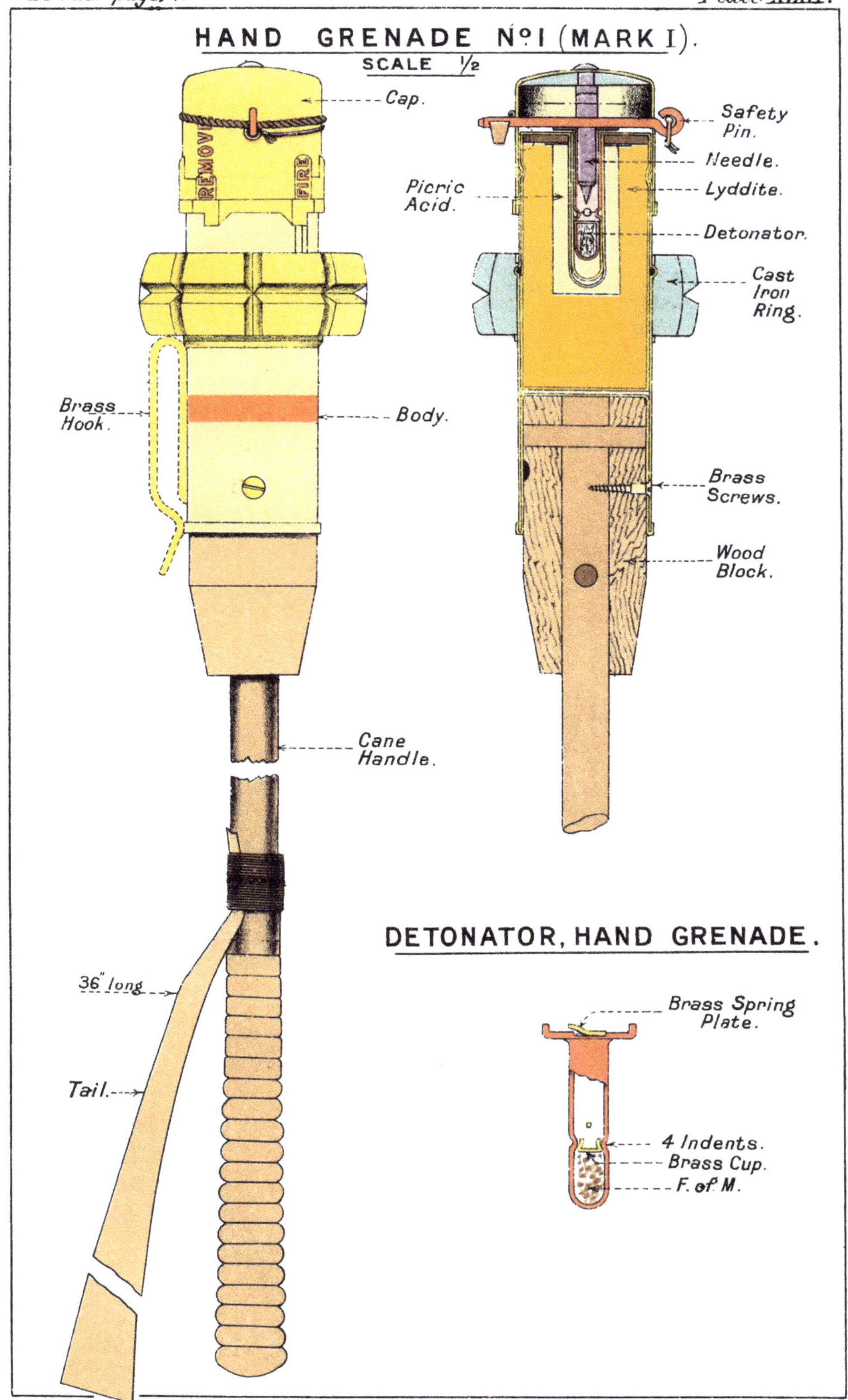
HAND GRENADE No. I (MARK I).
SCALE 1/2
Cap.
REMOVE
FIRE
Safety Pin.
Needle.
Picric Acid.
Lyddite.
Detonator.
Cast Iron Ring.
Brass Hook.
Body.
Brass Screws.
Wood Block.
Cane Handle.
DETONATOR, HAND GRENADE.
36" long
Tail.
Brass Spring Plate.
4 Indents.
Brass Cup.
F. of M.

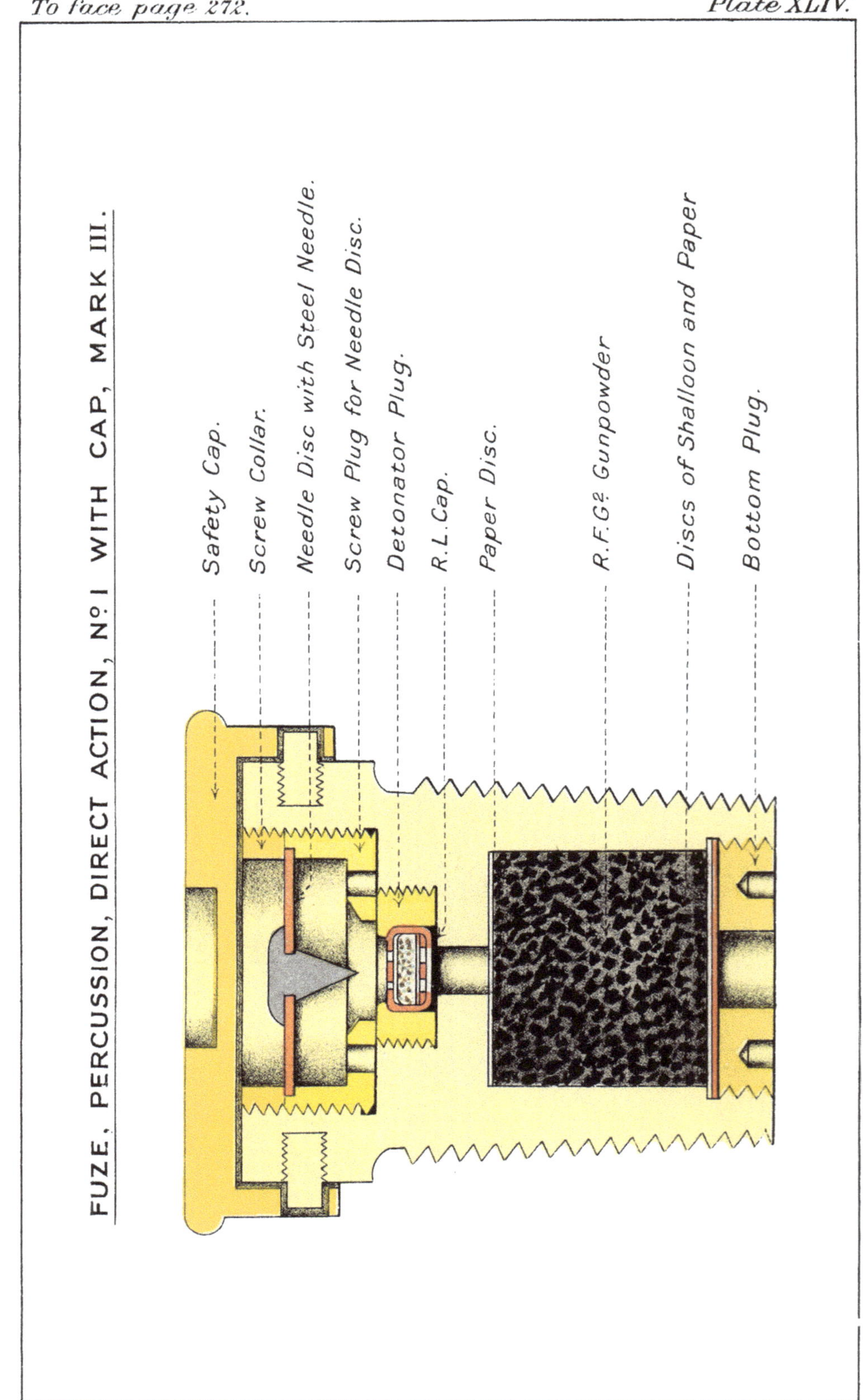

3567 Malby & Sons, Lith.

FUZE PERCUSSION. D.A. WITH CAP. Nº 17 MARK III

SCALE 2/1

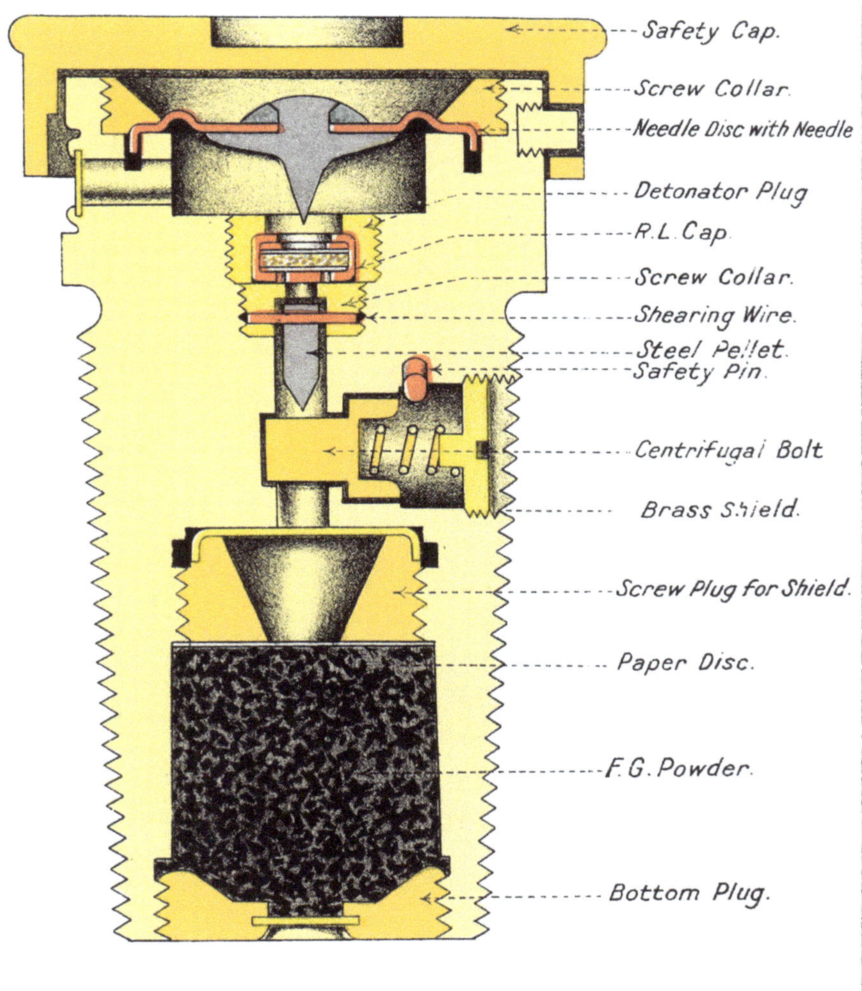

3567.

Malby&Sons, Lith.

To face page 278

Plate XLVI.

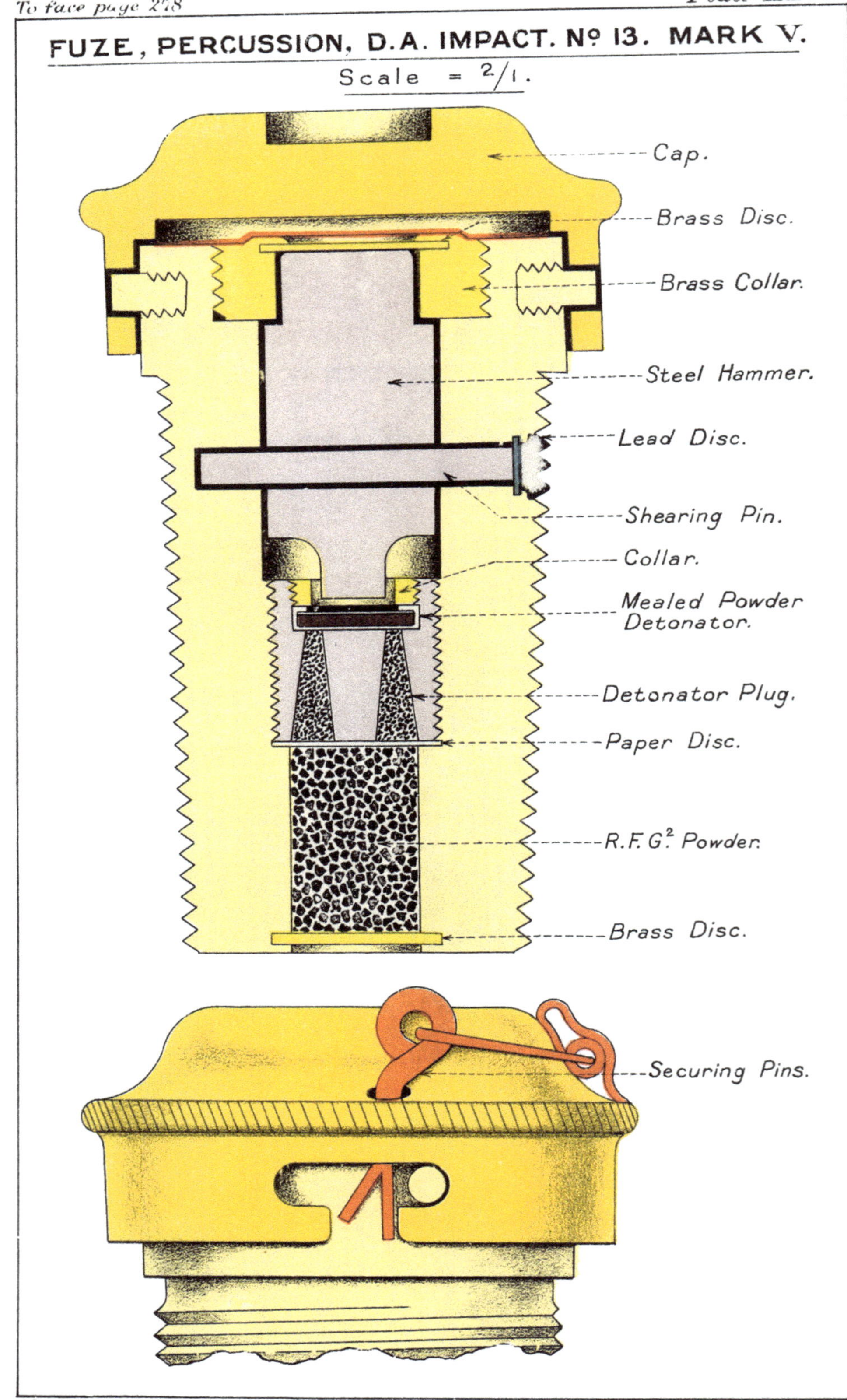

3567.

Malby & Sons, Lith.

FUZE, PERCUSSION, D. A. IMPACT No. 18 MARK II.

Scale = 2/1.

Safety Cap.
Brass Disc.
Securing Pin.
Steel Plug.
Shearing Pin.
Holder.
Detonator Plug.
Detonator.
Tinfoil Disc.
Loose C.E. stemmed in.
Paper Disc.
Pellet of C.E.
Shalloon Disc.
Bottom Plug.

Plate XLVIII.

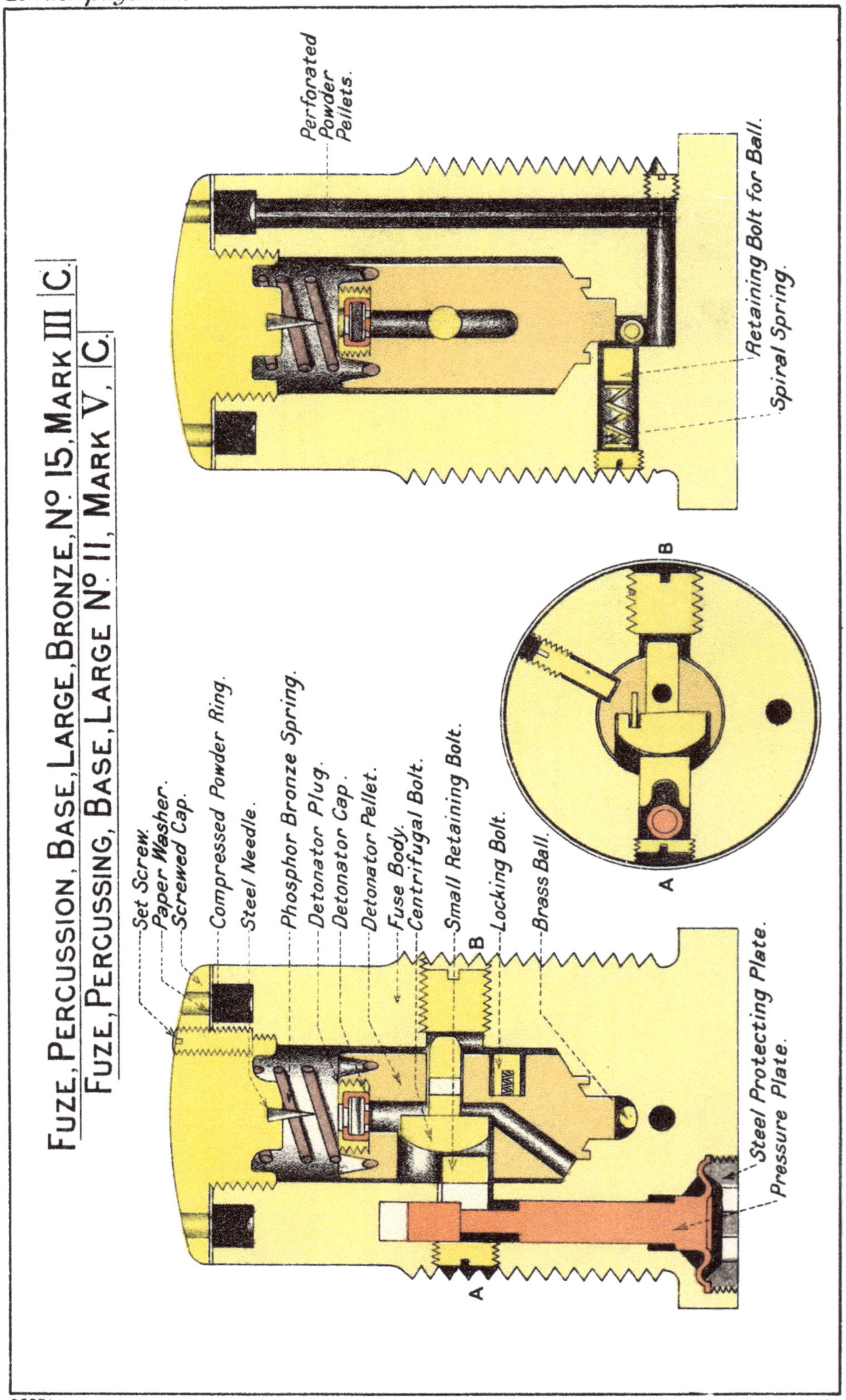

Fuze, Percussion, Base, Large, Bronze, No. 15, Mark III |C.|
Fuze, Percussing, Base, Large No. 11, Mark V, |C.|

Set Screw.
Paper Washer.
Screwed Cap.
Compressed Powder Ring.
Steel Needle.
Phosphor Bronze Spring.
Detonator Plug.
Detonator Cap.
Detonator Pellet.
Fuse Body.
Centrifugal Bolt.
Small Retaining Bolt.
Locking Bolt.
Brass Ball.
A
B
Steel Protecting Plate.
Pressure Plate.
A
B
Perforated Powder Pellets.
Retaining Bolt for Ball.
Spiral Spring.

3567.

Malby & Sons, Lith.

To face page 285.

FUZE. PERCUSSION. BASE. LARGE, No II MARK IV. | C. |

METAL.

FULL SIZE.

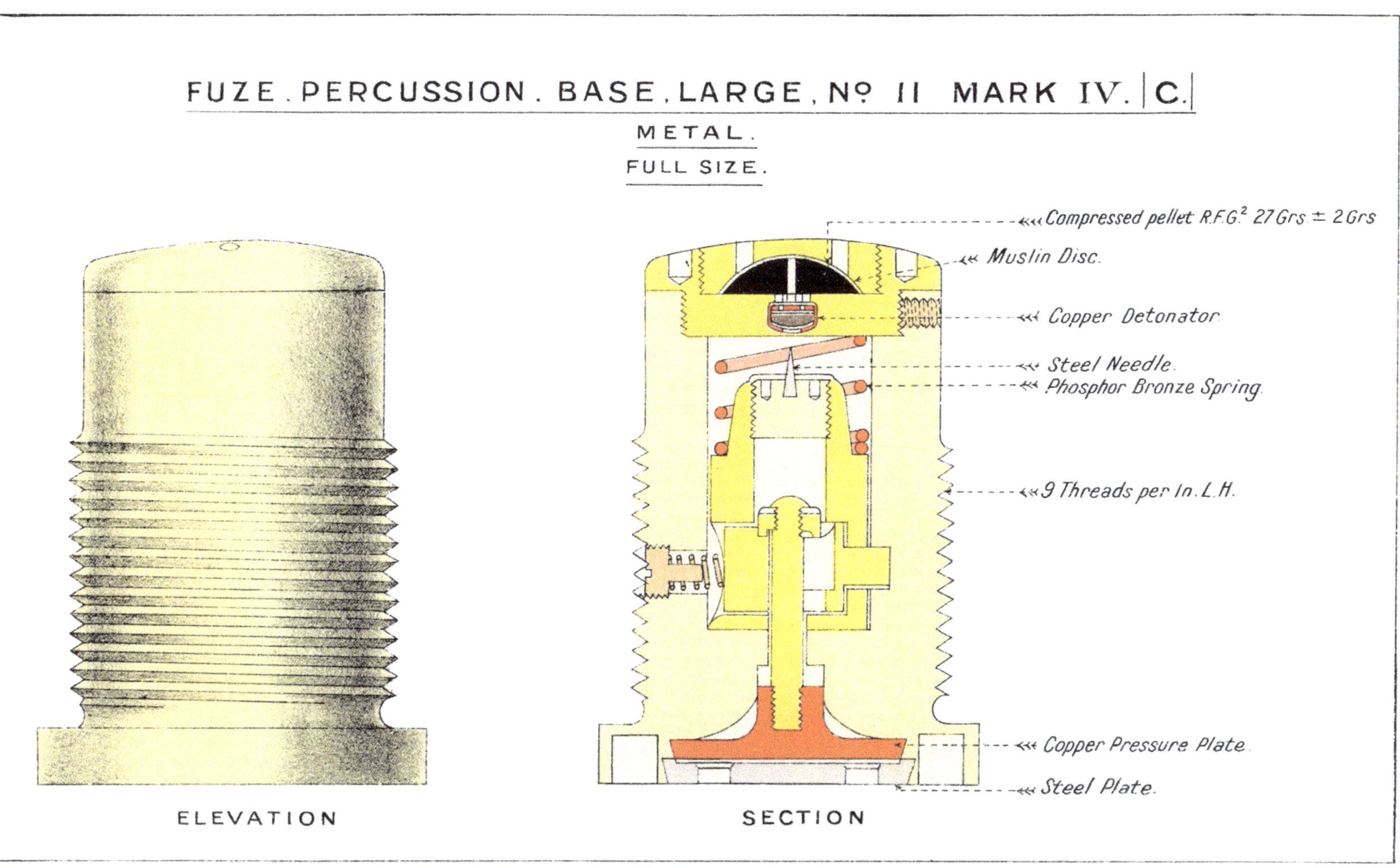

3567

Malby & Sons, Lith.

To face page 289.

Plate L.

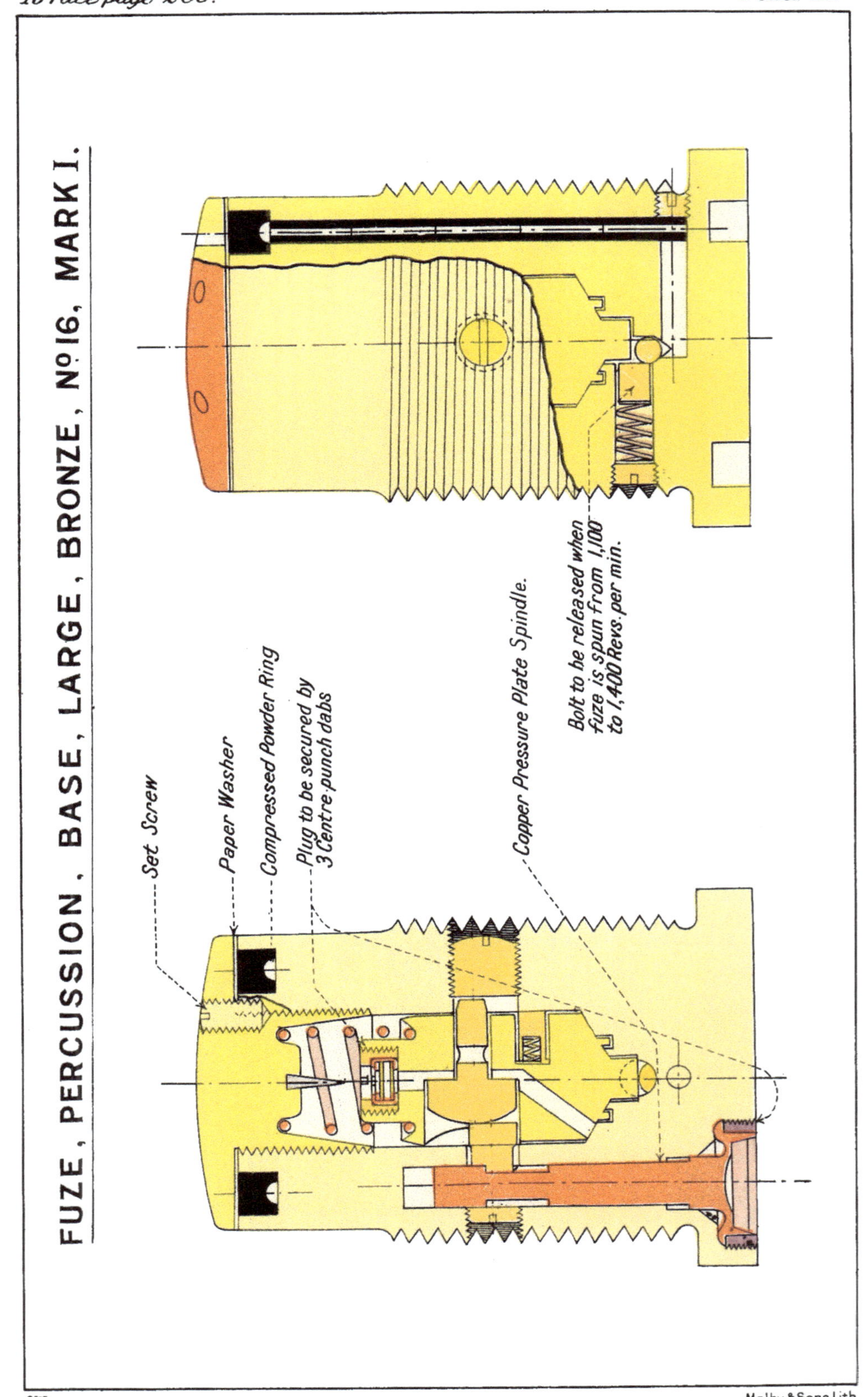

3667.

Malby & Sons, Lith.

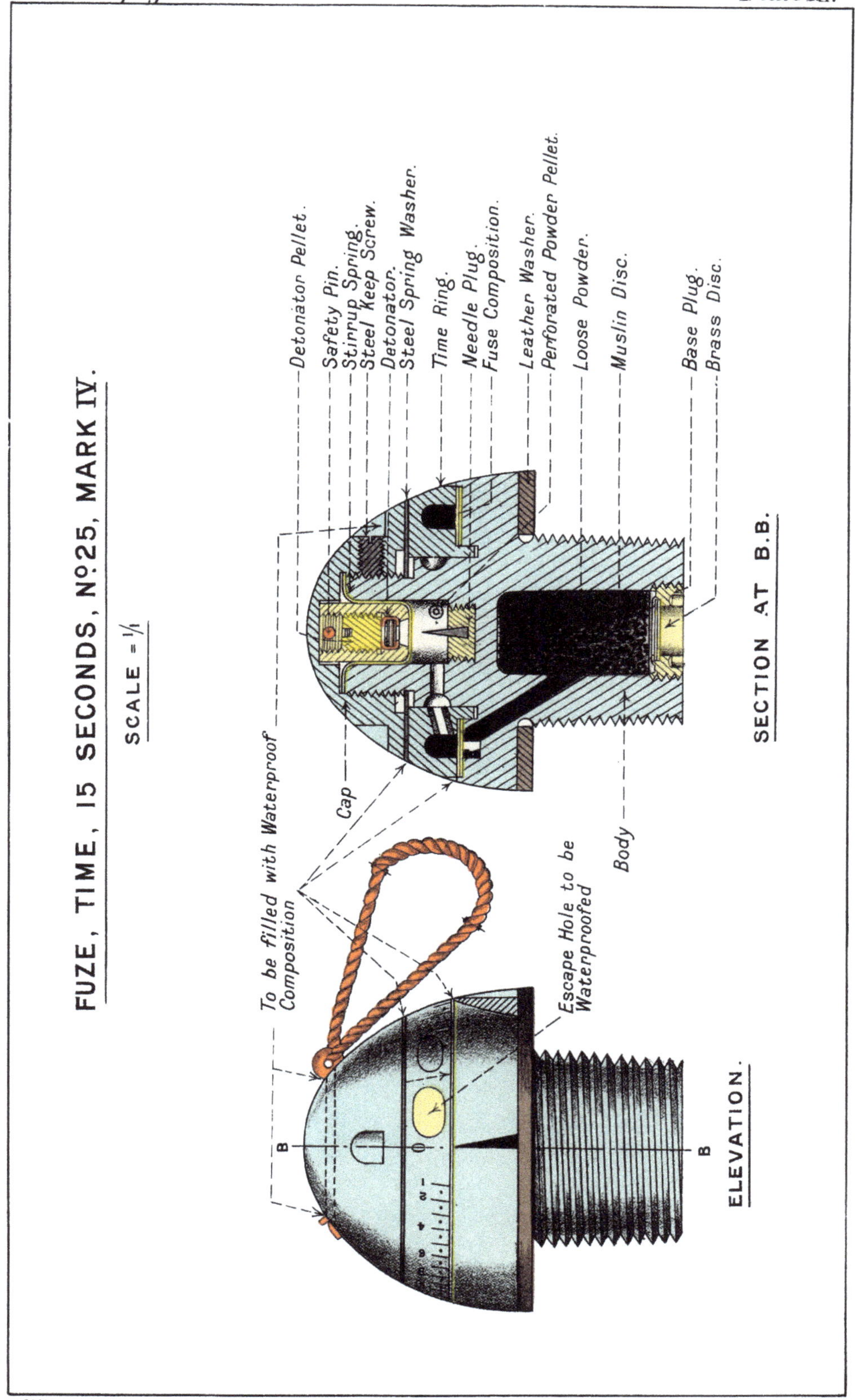
FUZE, TIME, 15 SECONDS, No. 25, MARK IV.
SCALE = 1/1
Detonator Pellet.
Safety Pin.
Stirrup Spring.
Steel Keep Screw.
Detonator.
Steel Spring Washer.
Time Ring.
Needle Plug.
Fuse Composition.
Leather Washer.
Perforated Powder Pellet.
Loose Powder.
Muslin Disc.
Base Plug.
Brass Disc.
To be filled with Waterproof Composition
Cap
Escape Hole to be Waterproofed
Body
B
B
ELEVATION.
SECTION AT B.B.

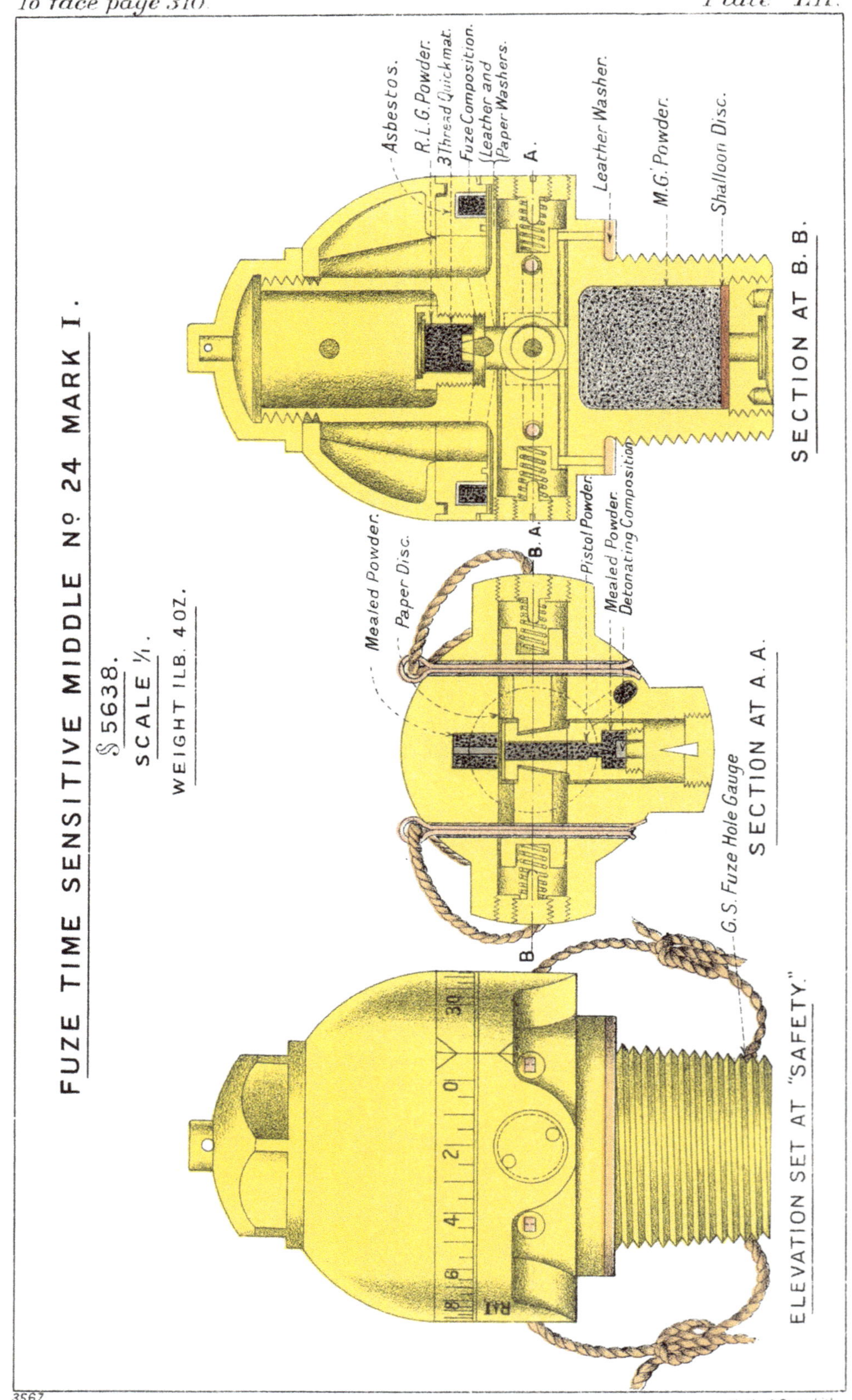

3567

Malby & Sons, Lith.

TIME, FUZE No. 30. MARK II /N/.

FULL SIZE.

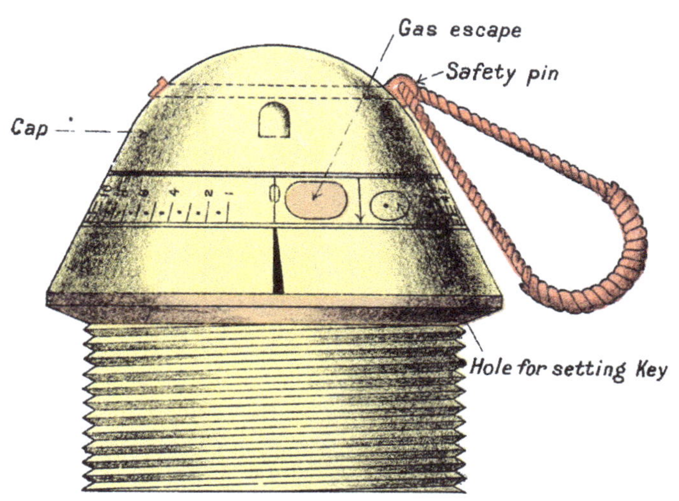

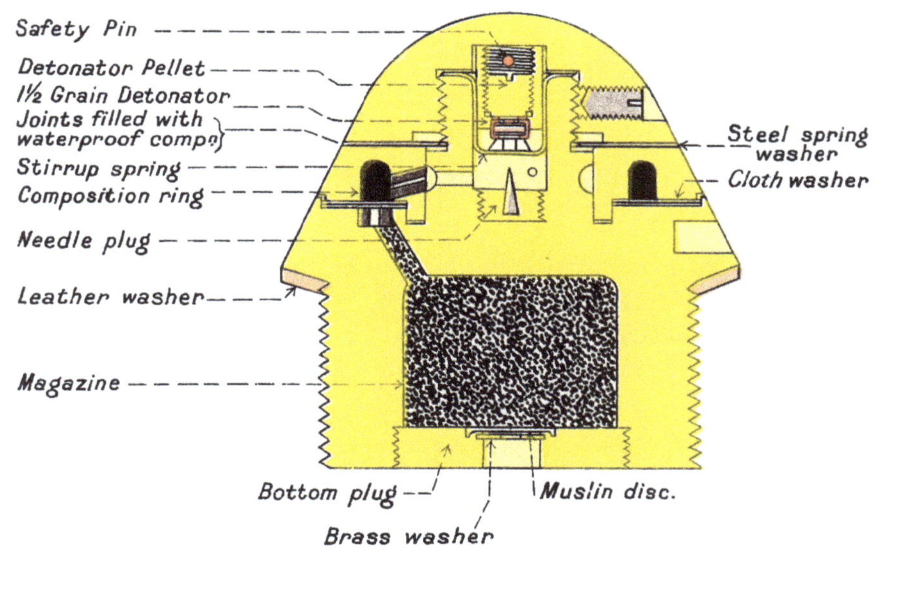

 Malby & Sons, Lith.

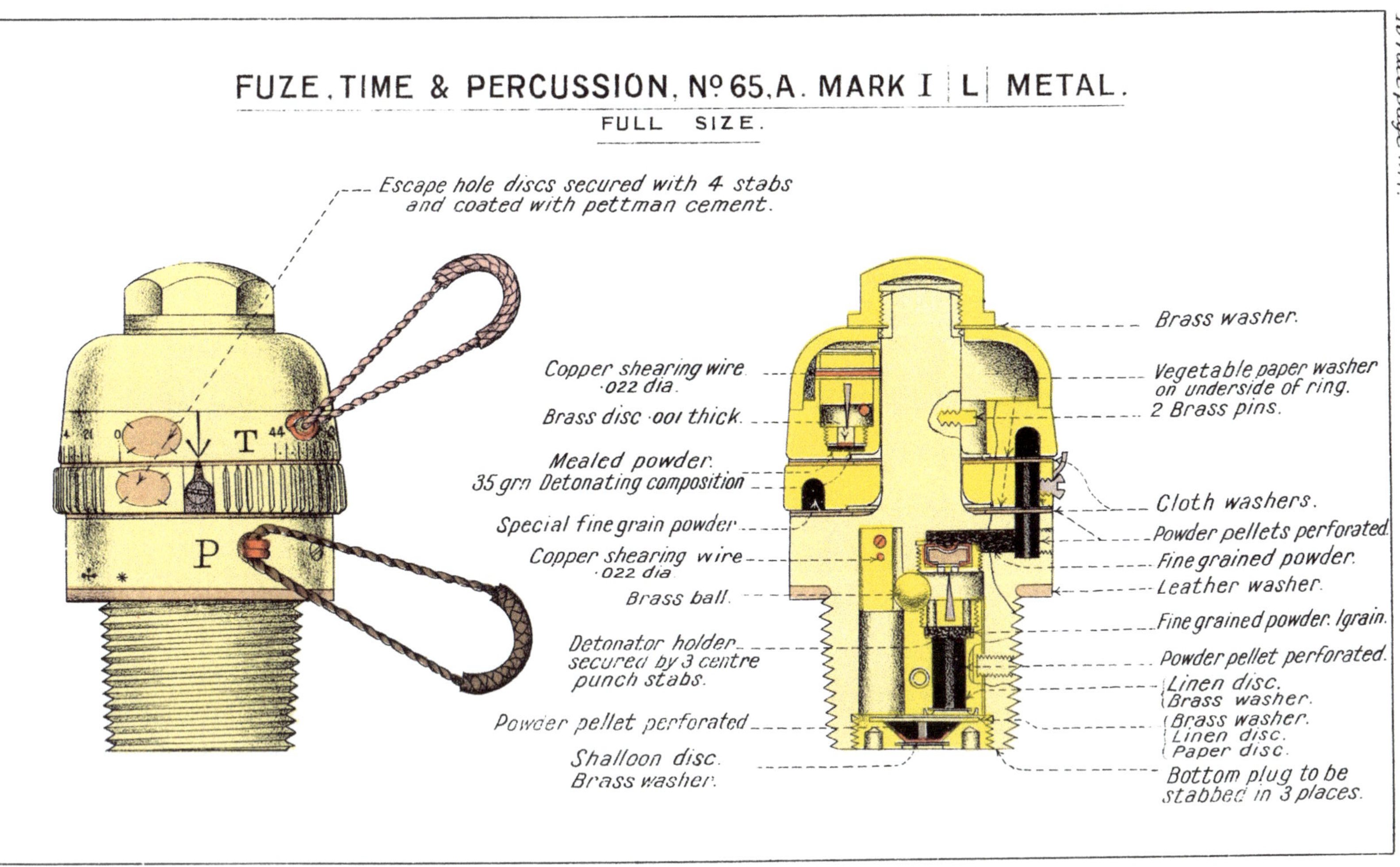

3567.

Malby & Sons, Lith.

FUZE, TIME & PERCUSSION, № 63, MARK I | C | METAL.

FULL SIZE.

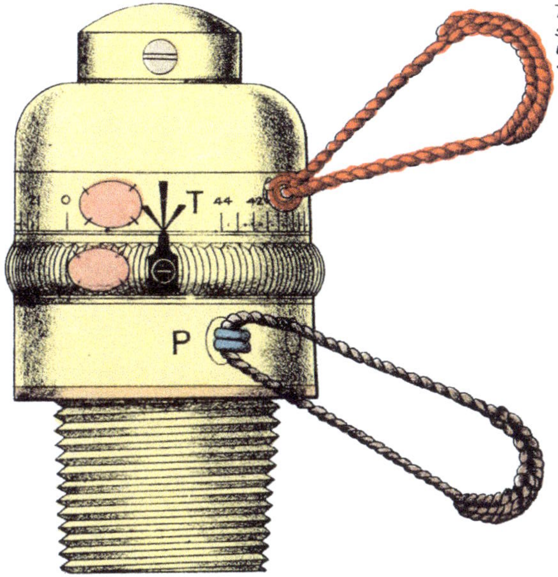

ELEVATION.

SET AT SAFETY.

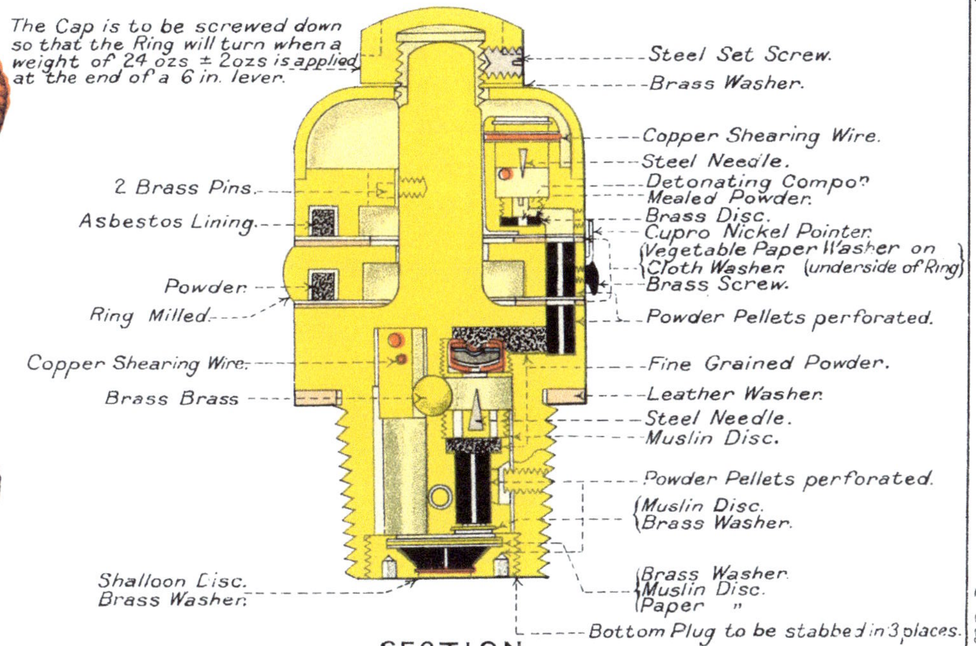

T 44 42

P

The Cap is to be screwed down so that the Ring will turn when a weight of 24 ozs ± 2ozs is applied at the end of a 6 in. lever.

2 Brass Pins.

Asbestos Lining.

Powder.

Ring Milled.

Copper Shearing Wire.

Brass Brass

Shalloon Disc.
Brass Washer.

Steel Set Screw.

Brass Washer.

Copper Shearing Wire.

Steel Needle.

Detonating Compo.n

Mealed Powder.

Brass Disc.

Cupro Nickel Pointer.

{Vegetable Paper Washer on
{Cloth Washer. (underside of Ring)

Brass Screw.

Powder Pellets perforated.

Fine Grained Powder.

Leather Washer.

Steel Needle.

Muslin Disc.

Powder Pellets perforated.

{Muslin Disc.
{Brass Washer.

{Brass Washer.
{Muslin Disc.
{Paper "

Bottom Plug to be stabbed in 3 places.

SECTION.

Malby & Sons Lith.

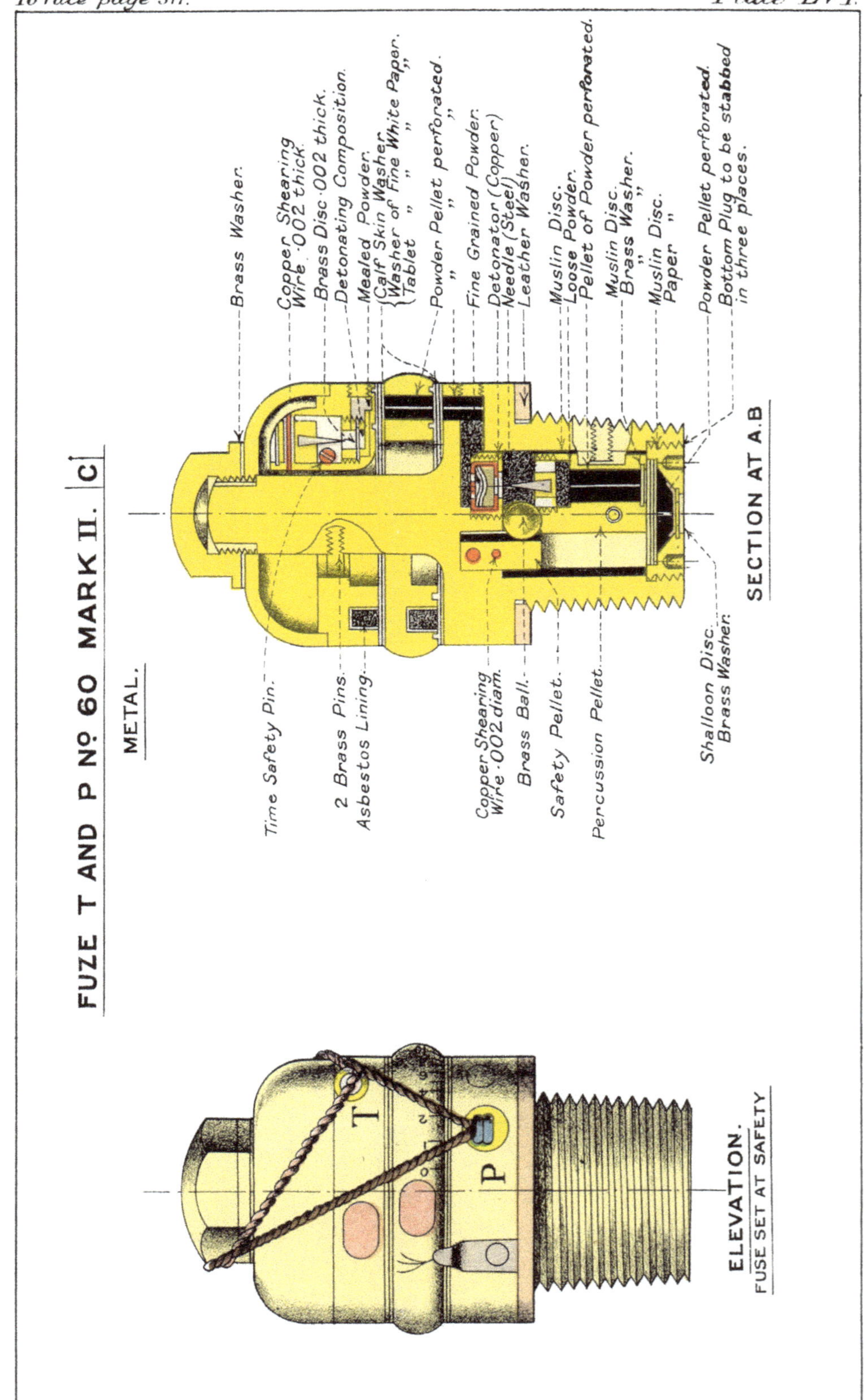
FUZE T AND P No 60 MARK II. C
METAL.
Brass Washer.
Copper Shearing Wire ·002 thick.
Brass Disc ·002 thick.
Detonating Composition.
Mealed Powder.
Calf Skin Washer.
Washer of Fine White Paper.
Tablet " " "
Powder Pellet perforated.
" "
Fine Grained Powder.
Detonator (Copper)
Needle (Steel)
Leather Washer.
Muslin Disc.
Loose Powder.
Pellet of Powder perforated.
Muslin Disc.
Brass Washer.
"
Muslin Disc.
Paper "
Powder Pellet perforated.
Bottom Plug to be stabbed in three places.
Time Safety Pin.
2 Brass Pins.
Asbestos Lining.
Copper Shearing Wire ·002 diam.
Brass Ball.
Safety Pellet.
Percussion Pellet.
Shalloon Disc.
Brass Washer.
SECTION AT A.B
T
P
ELEVATION.
FUSE SET AT SAFETY

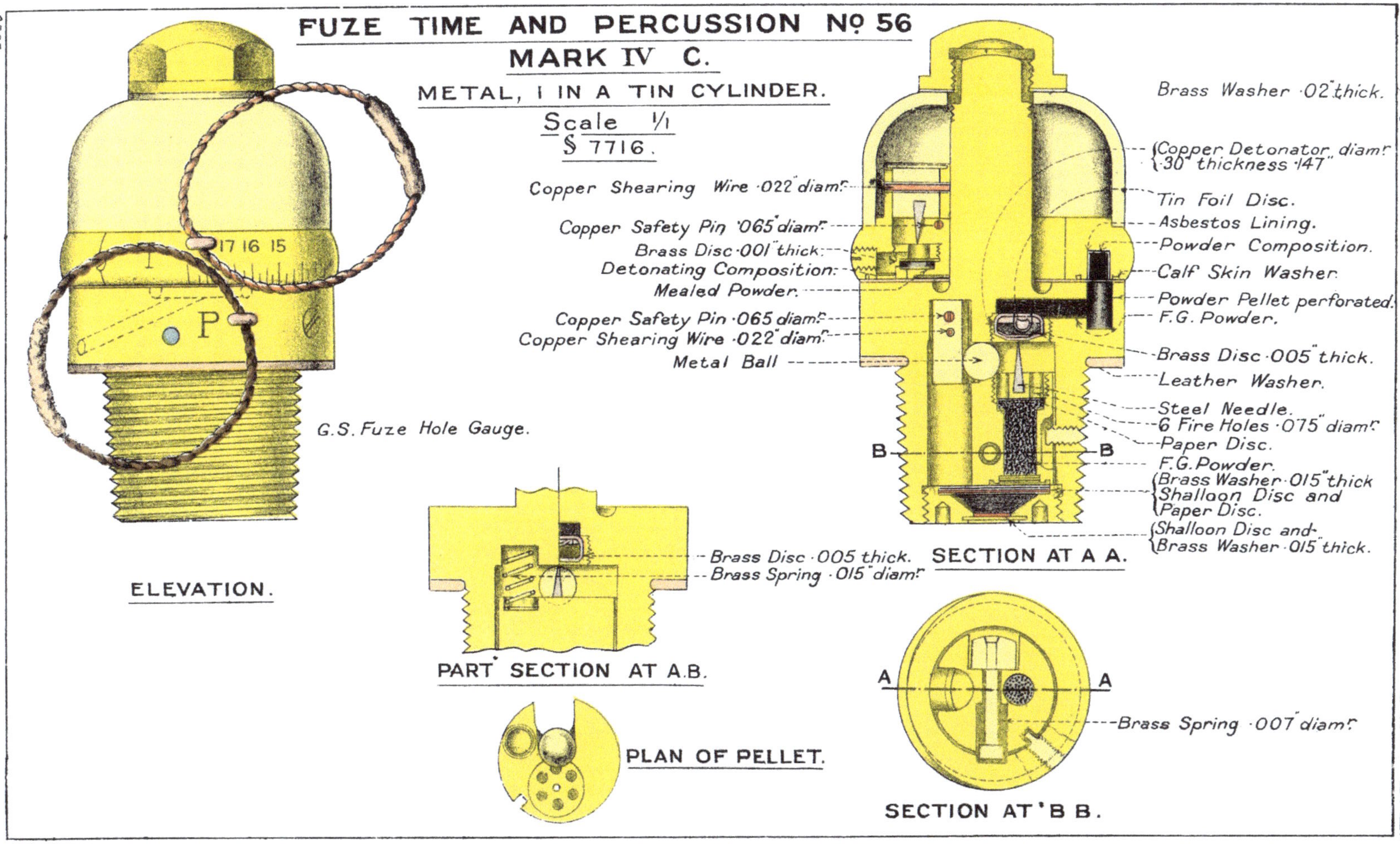

To face page 317.

FUZE TIME AND PERCUSSION No 56
MARK IV C.
METAL, 1 IN A TIN CYLINDER.

Scale 1/1
§ 7716.

Copper Shearing Wire ·022" diam.r
Copper Safety Pin ·065" diam.r
Brass Disc ·001" thick.
Detonating Composition.
Mealed Powder.
Copper Safety Pin ·065 diam.r
Copper Shearing Wire ·022" diam.r
Metal Ball

Brass Washer ·02" thick.
Copper Detonator, diam.r ·30" thickness ·147"
Tin Foil Disc.
Asbestos Lining.
Powder Composition.
Calf Skin Washer.
Powder Pellet perforated.
F.G. Powder.
Brass Disc ·005" thick.
Leather Washer.
Steel Needle.
6 Fire Holes ·075" diam.r
Paper Disc.
F.G. Powder.
Brass Washer ·015" thick Shalloon Disc and Paper Disc.
Shalloon Disc and Brass Washer ·015" thick.

B B

SECTION AT A A.

17 16 15

P

G.S. Fuze Hole Gauge.

ELEVATION.

Brass Disc ·005 thick.
Brass Spring ·015" diam.r

PART SECTION AT A.B.

PLAN OF PELLET.

A A

Brass Spring ·007" diam.r

SECTION AT B B.

3567.

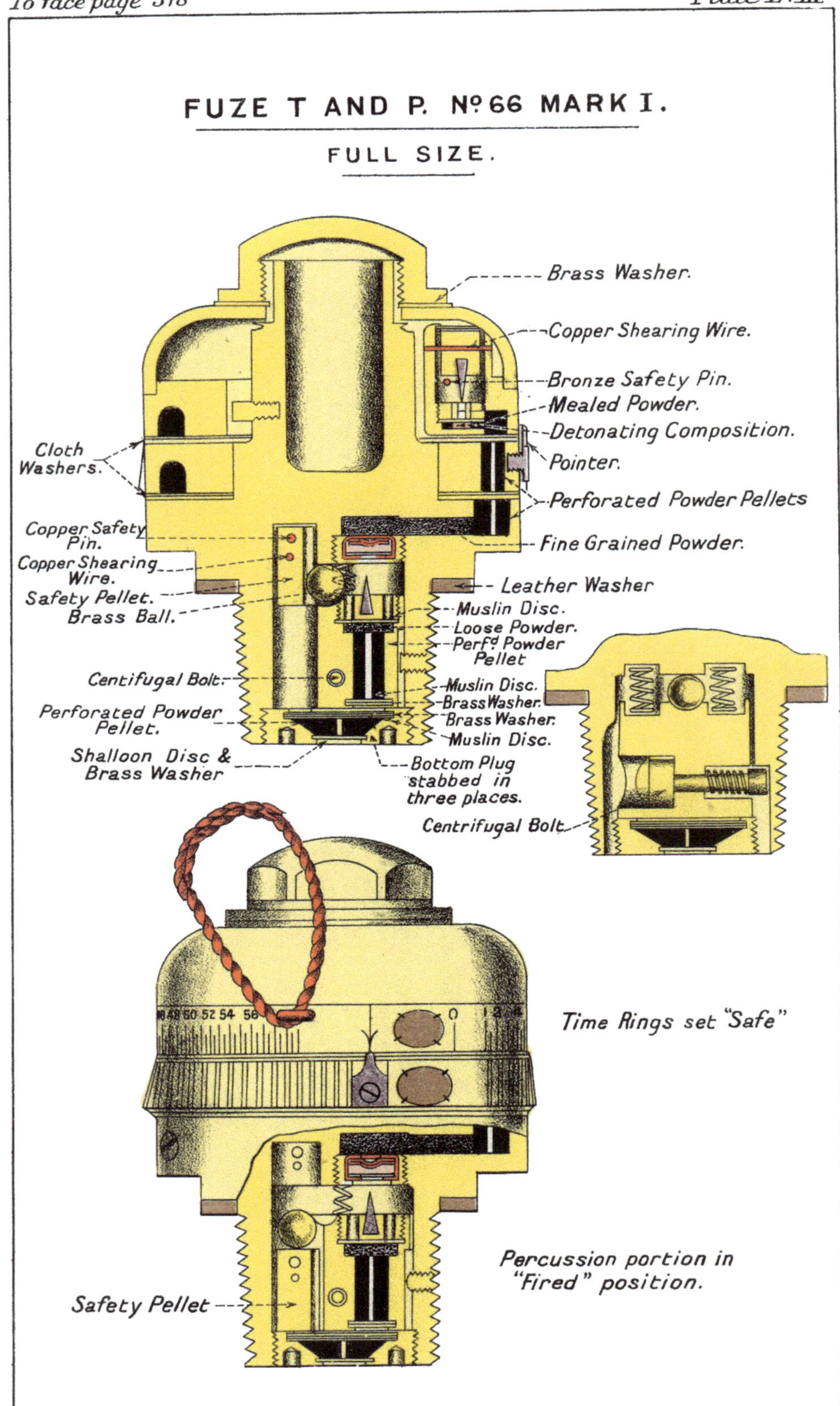
FUZE T AND P. No. 66 MARK I.
FULL SIZE.
Brass Washer.
Copper Shearing Wire.
Bronze Safety Pin.
Mealed Powder.
Detonating Composition.
Pointer.
Perforated Powder Pellets
Fine Grained Powder.
Cloth Washers.
Copper Safety Pin.
Copper Shearing Wire.
Safety Pellet.
Brass Ball.
Leather Washer
Muslin Disc.
Loose Powder.
Perfd. Powder Pellet
Centifugal Bolt.
Muslin Disc.
Brass Washer.
Brass Washer.
Muslin Disc.
Perforated Powder Pellet.
Shalloon Disc & Brass Washer
Bottom Plug stabbed in three places.
Centrifugal Bolt.
48 50 52 54 56
0 1 2 3 4
Time Rings set "Safe"
Safety Pellet
Percussion portion in "Fired" position.

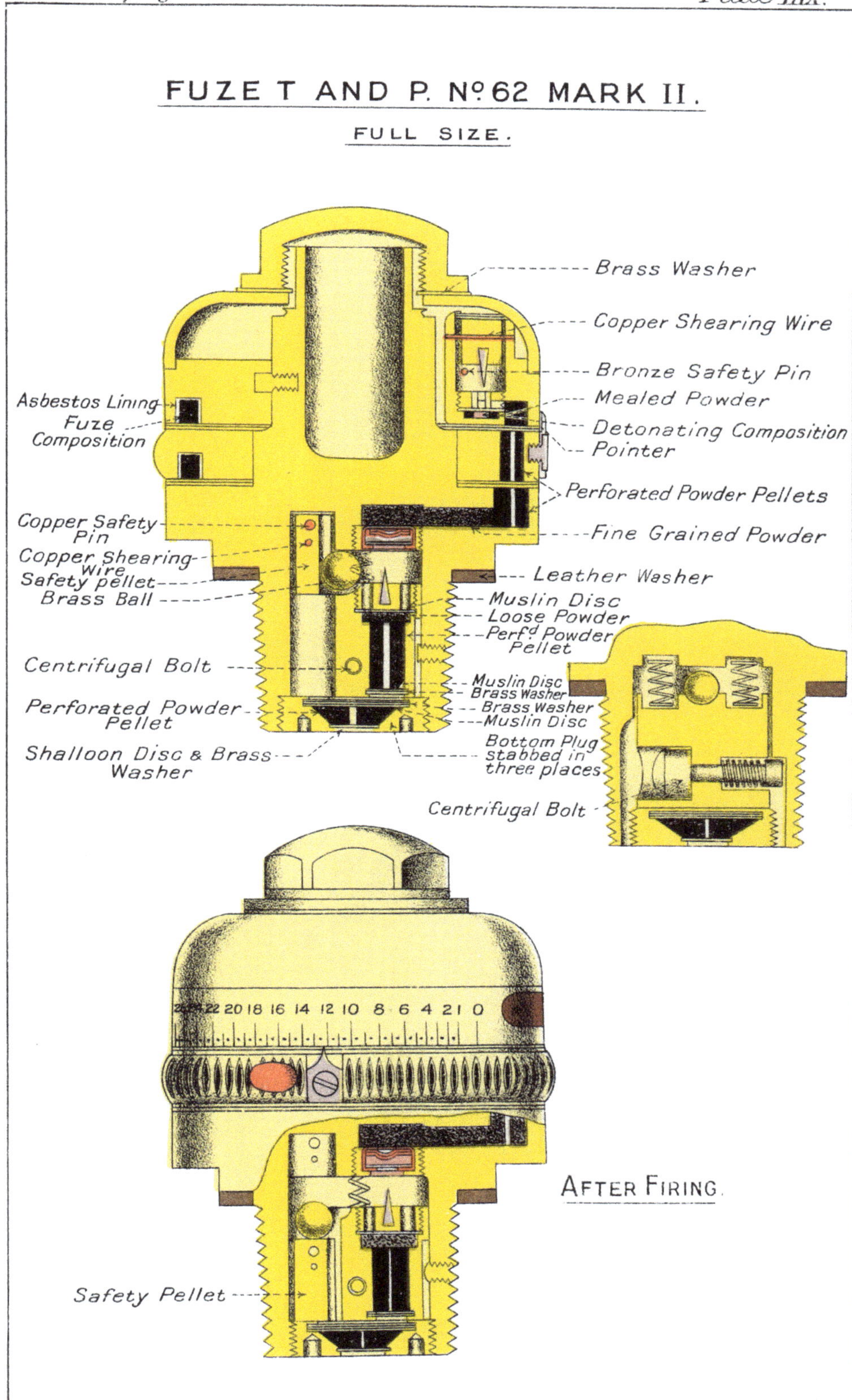

3567. Malby & Sons, Lith.

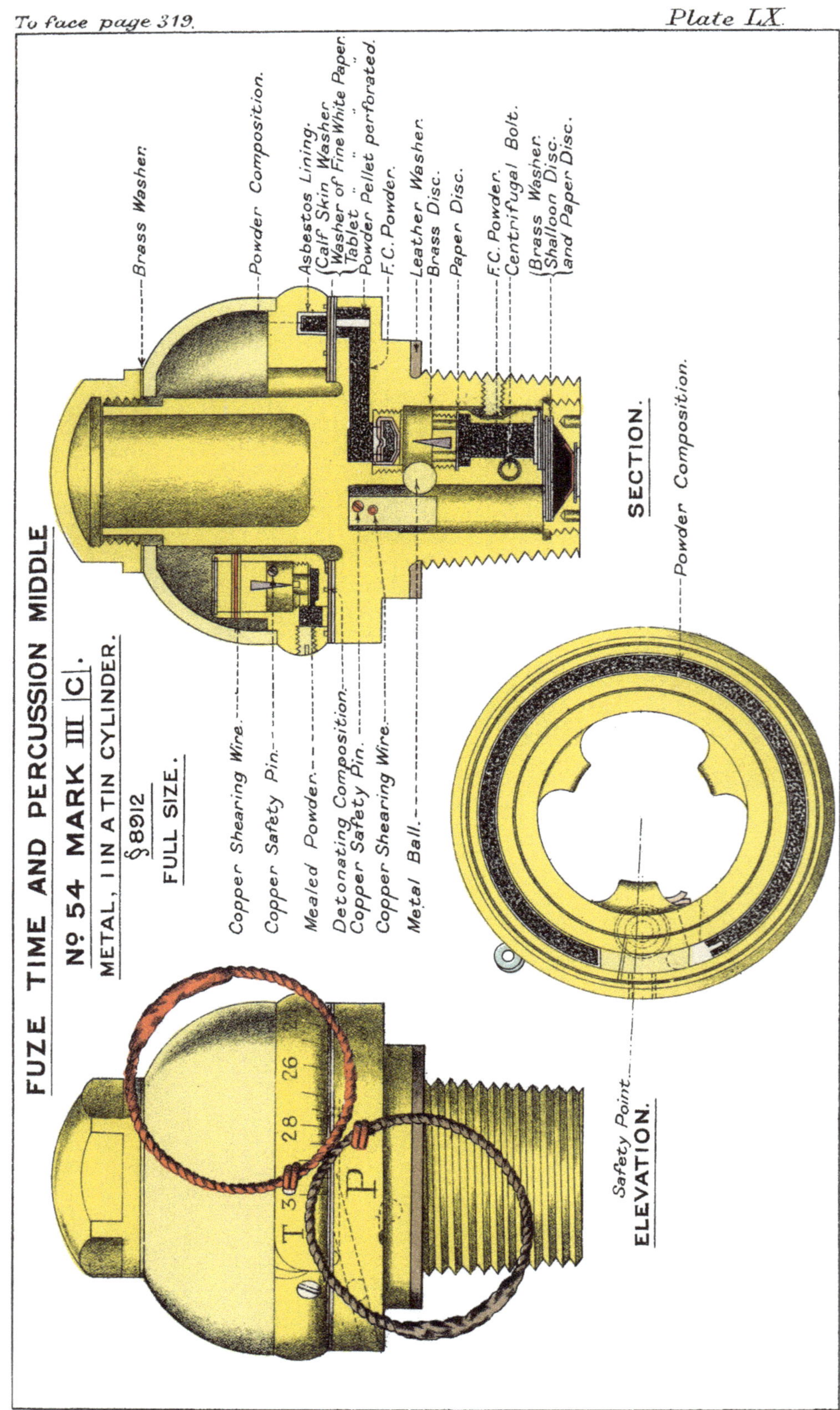

3567. Malby & Sons, Lith.

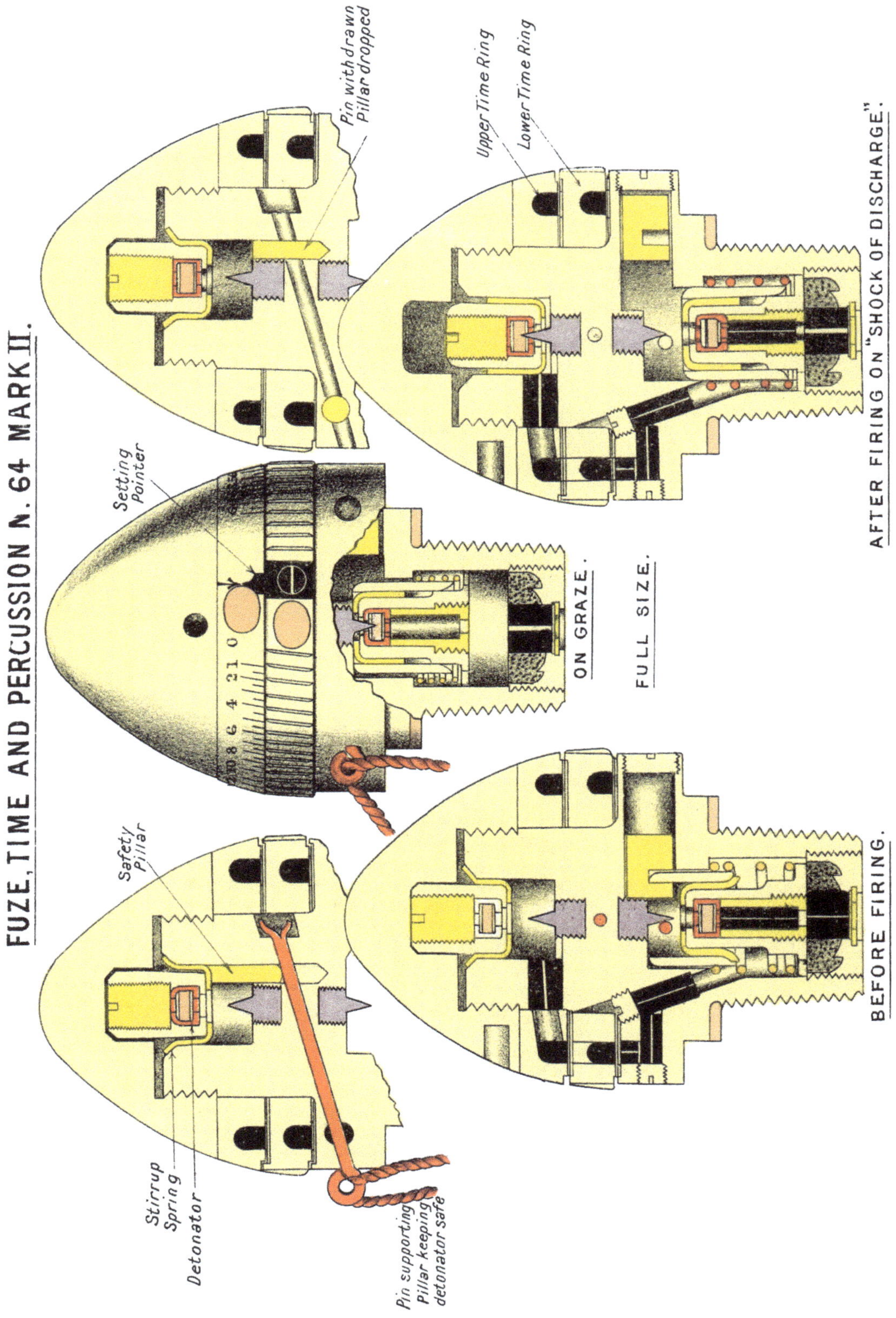

3567 Malby & Sons, Lith.

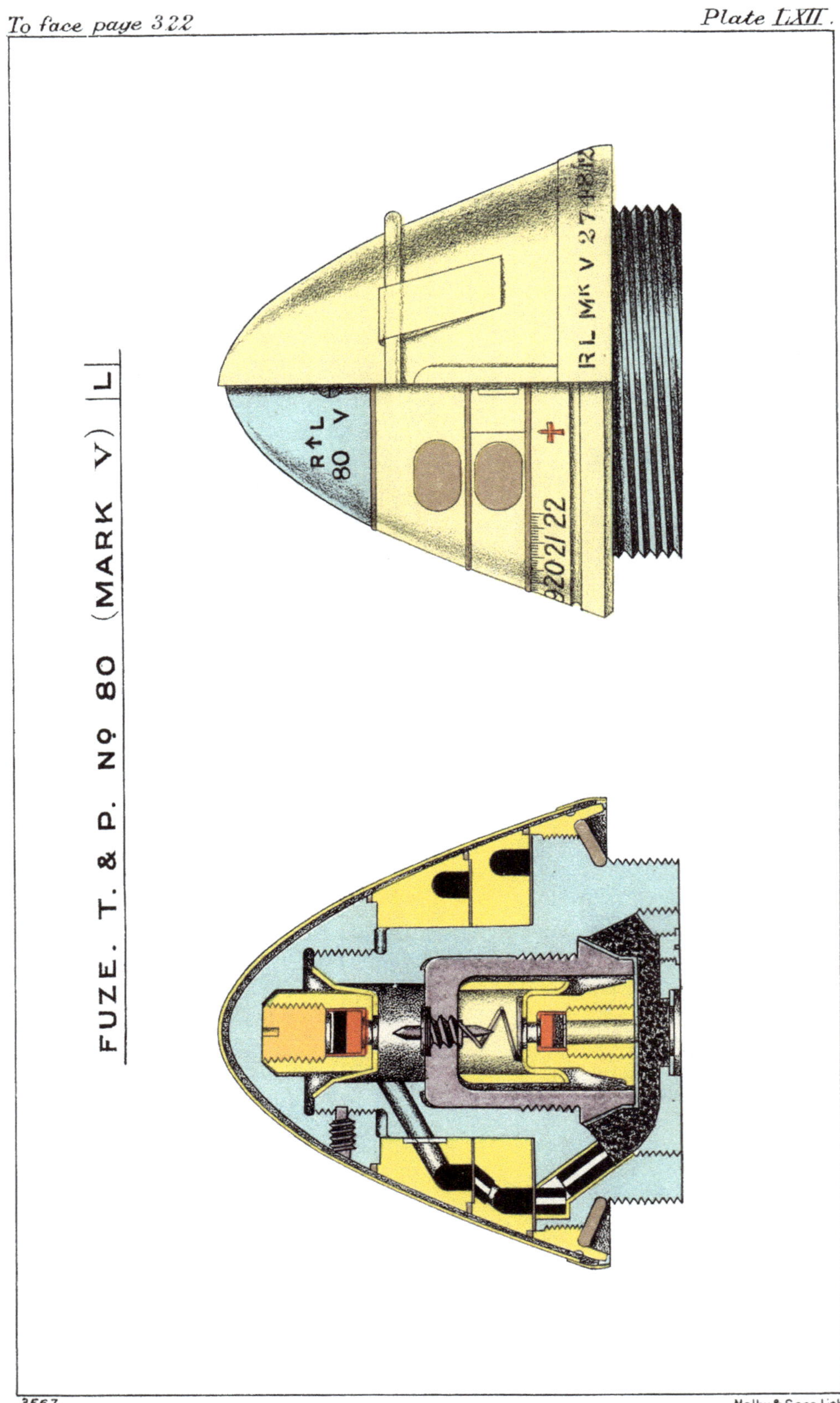

FUZE. T. & P. No 80 (MARK V) L

3567. Malby & Sons, Lith.

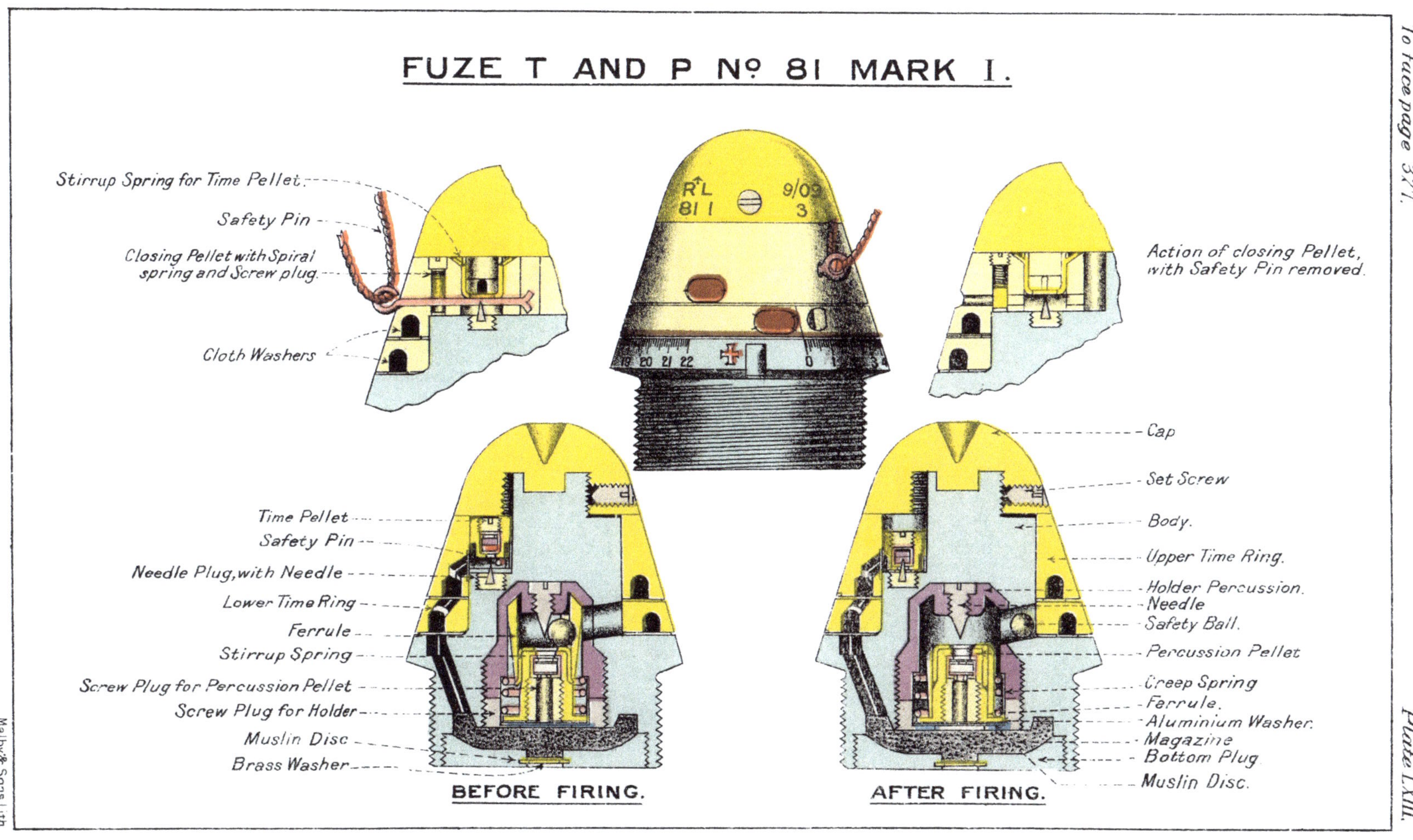
FUZE T AND P No. 81 MARK I.
Stirrup Spring for Time Pellet.
Safety Pin
Closing Pellet with Spiral spring and Screw plug.
Cloth Washers
R L
81 I
9/02
3
Action of closing Pellet, with Safety Pin removed.
Time Pellet
Safety Pin
Needle Plug, with Needle
Lower Time Ring
Ferrule
Stirrup Spring
Screw Plug for Percussion Pellet
Screw Plug for Holder
Muslin Disc
Brass Washer
BEFORE FIRING.
AFTER FIRING.
Cap
Set Screw
Body.
Upper Time Ring.
Holder Percussion.
Needle
Safety Ball.
Percussion Pellet
Creep Spring
Ferrule.
Aluminium Washer.
Magazine
Bottom Plug
Muslin Disc.
Malby & Sons. Lith

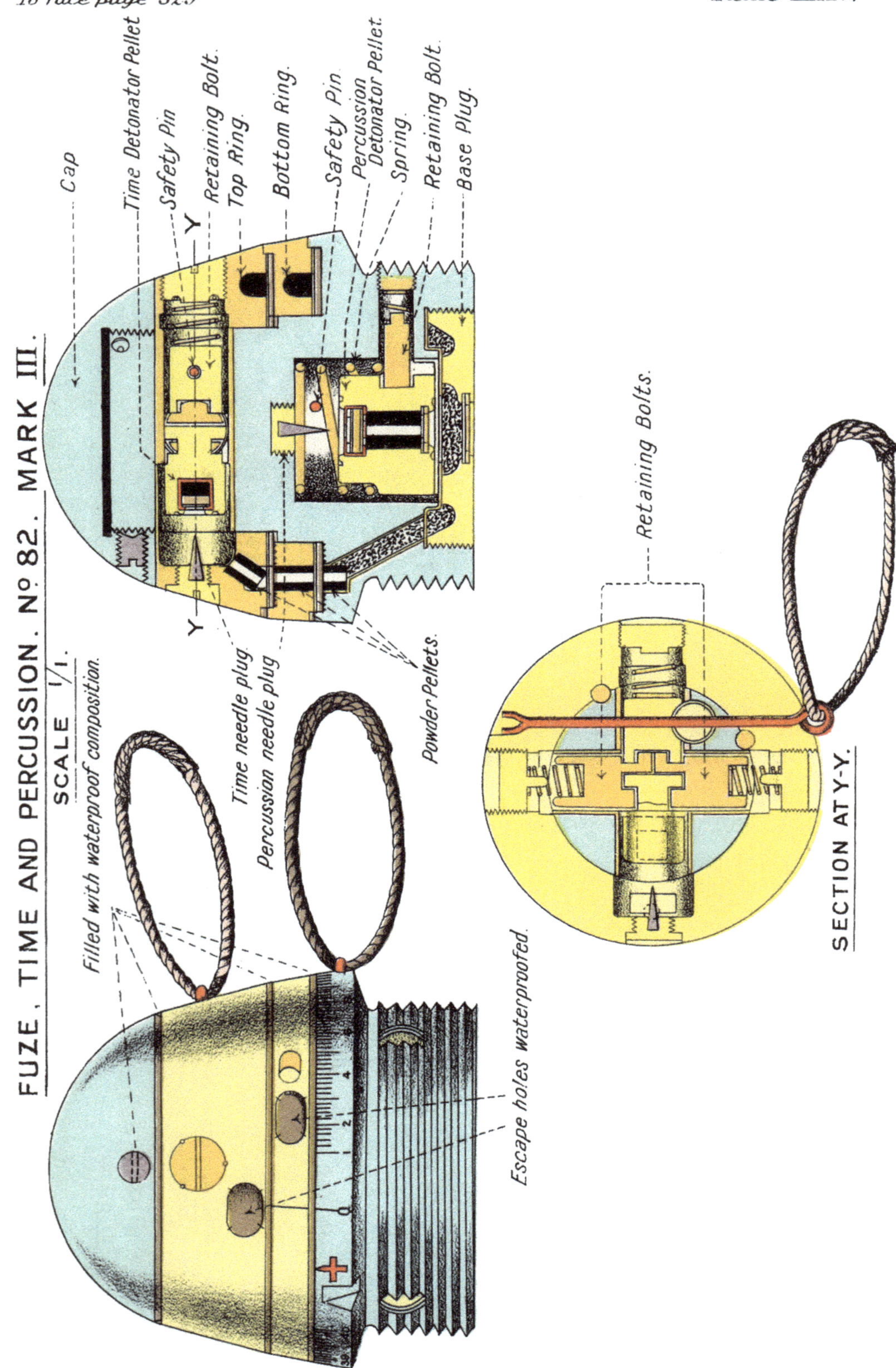

3567.

Malby & Sons, Lith.

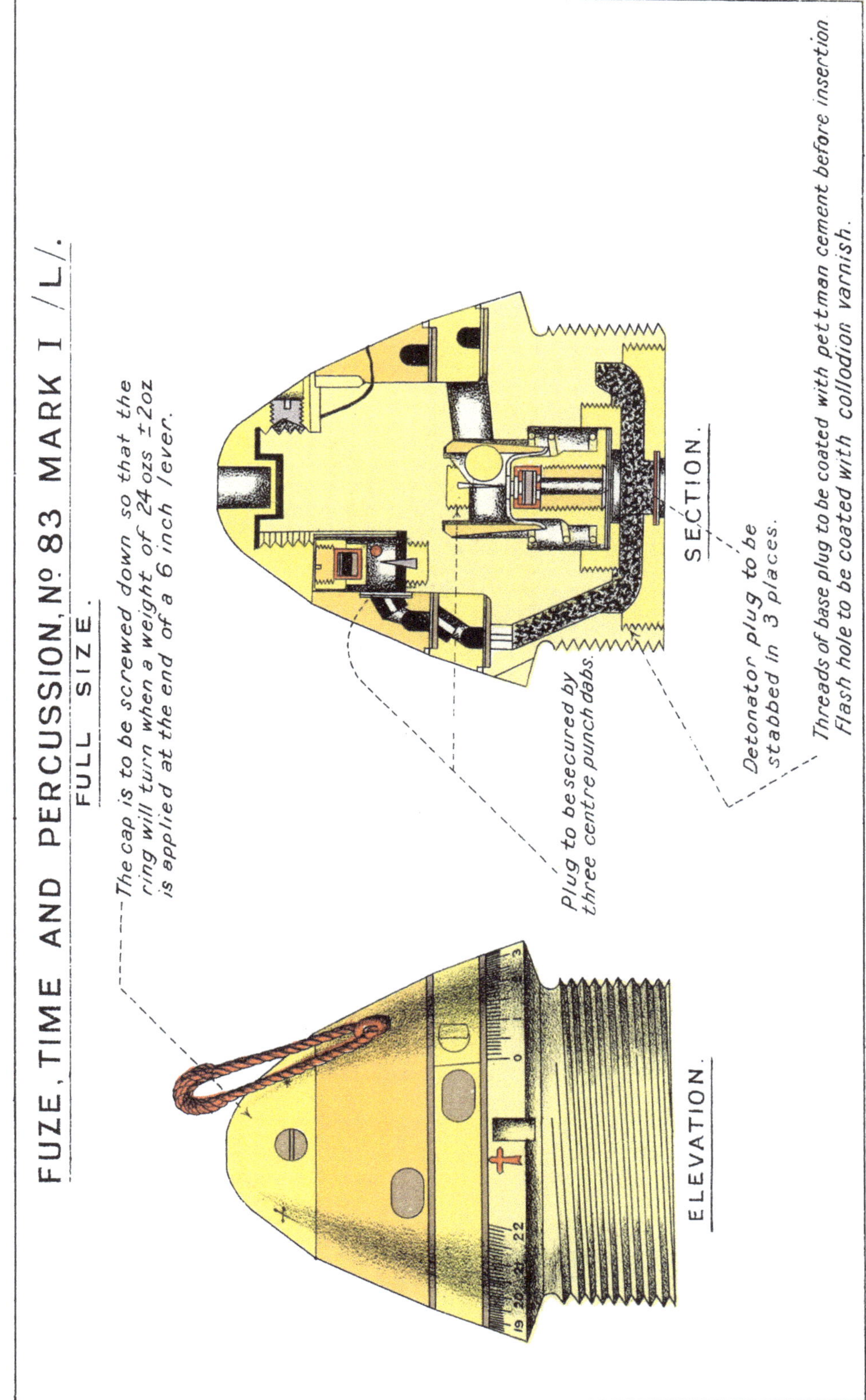

3657.
Malby & Sons, Lith.

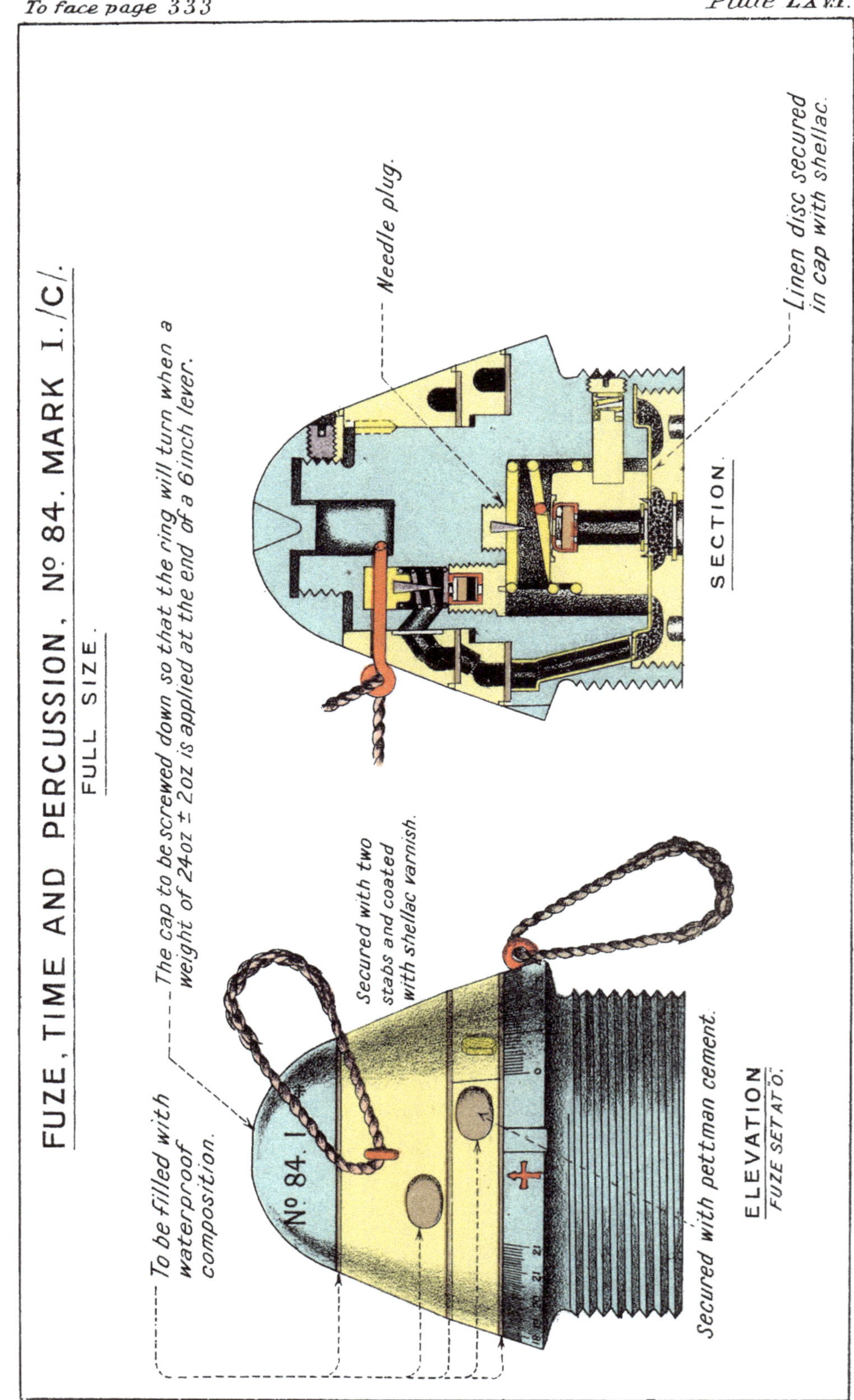

3567.

Malby & Sons, Lith.

TUBE, V. S. PERCUSSION MARK VII /C/.

SCALE 3/1.

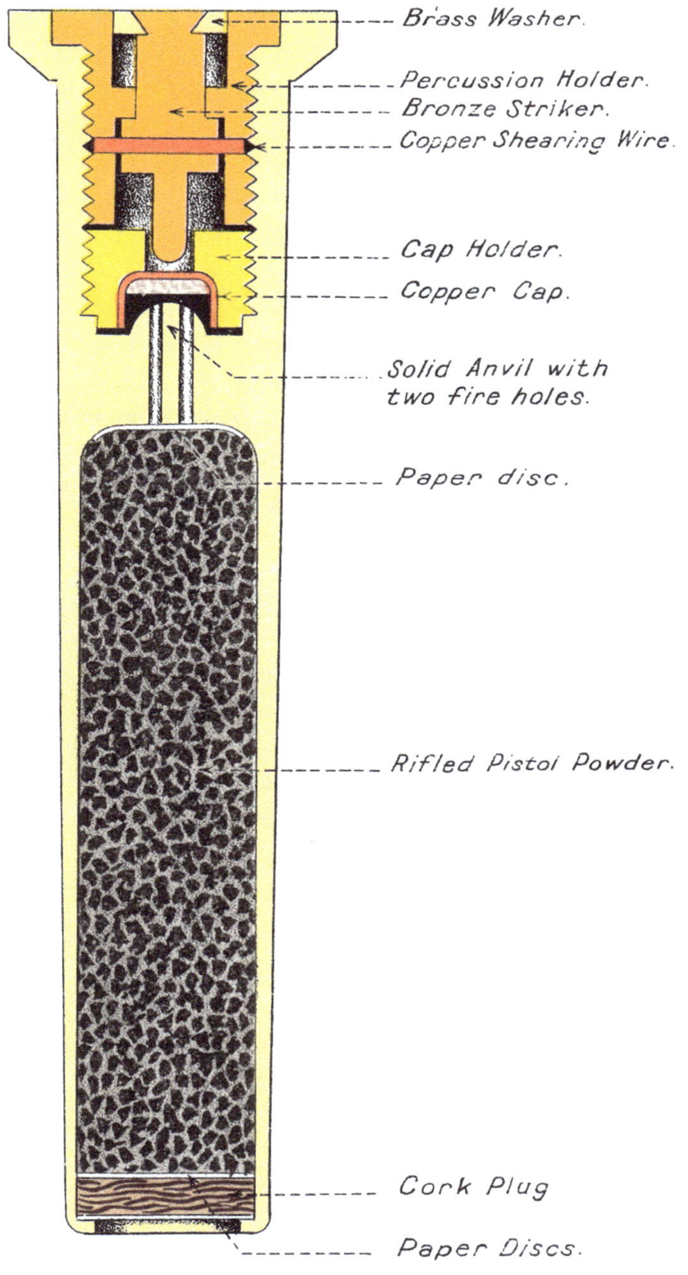

3567.

Malby & Sons, Lith.

TUBE. V.S. PERCUSSION MARK IV /C/.

SCALE 3/1.

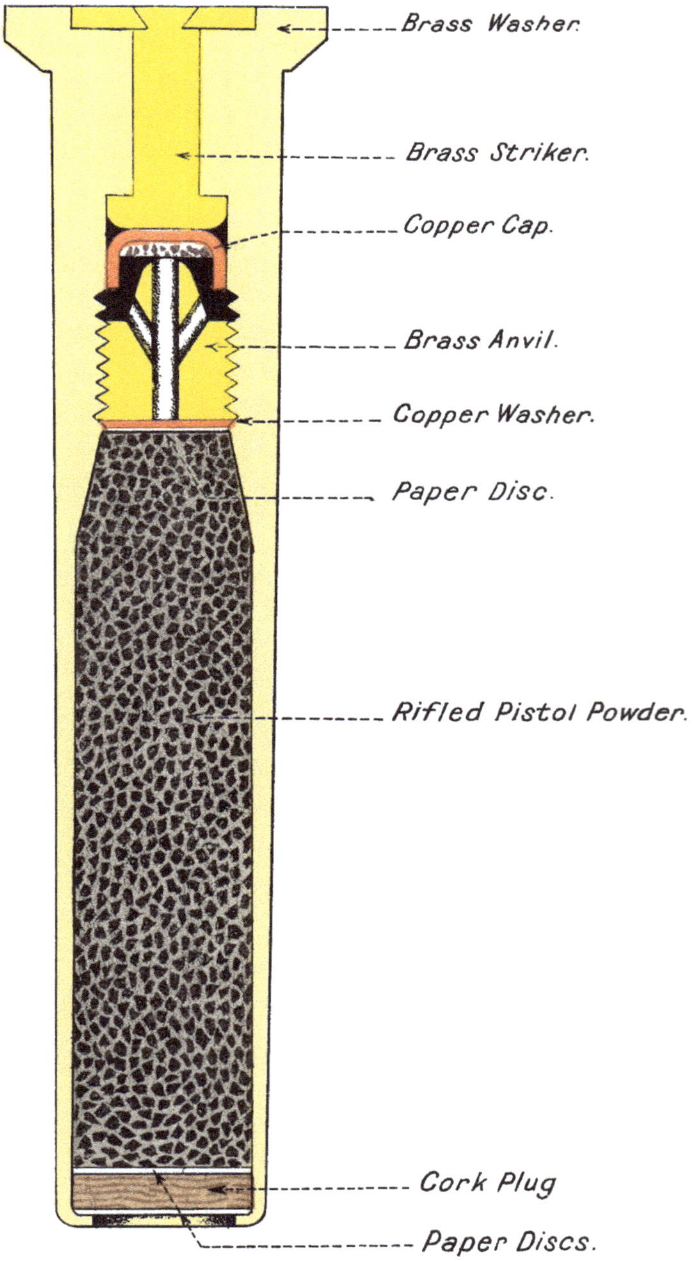

Brass Washer.
Brass Striker.
Copper Cap.
Brass Anvil.
Copper Washer.
Paper Disc.
Rifled Pistol Powder.
Cork Plug
Paper Discs.

3567.

Malby & Sons, Lith.

TUBE, V.S. ELECTRIC WIRELESS "P" MARK VI C

Scale = 3/1.

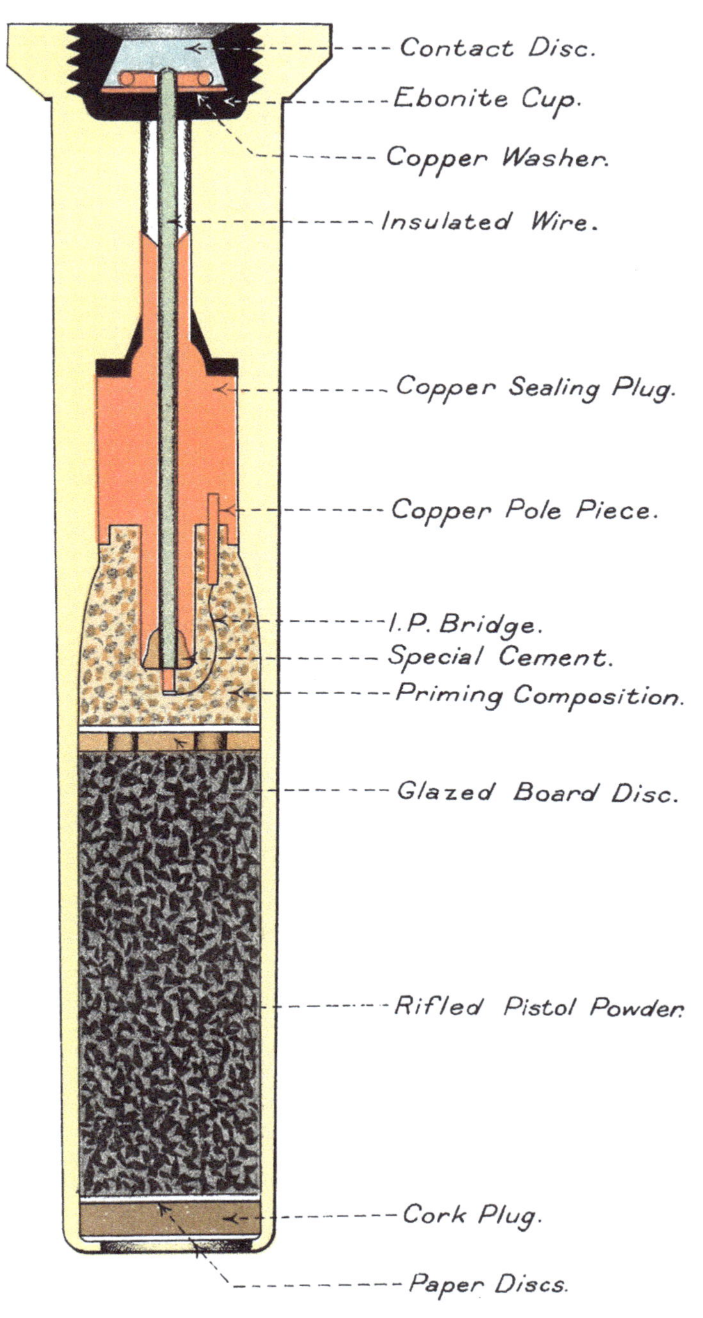

3567.

Malby & Sons, Lith.

TUBE, V.S. ELECTRIC, WIRELESS "P" MARK V.

Scale = 3/1.

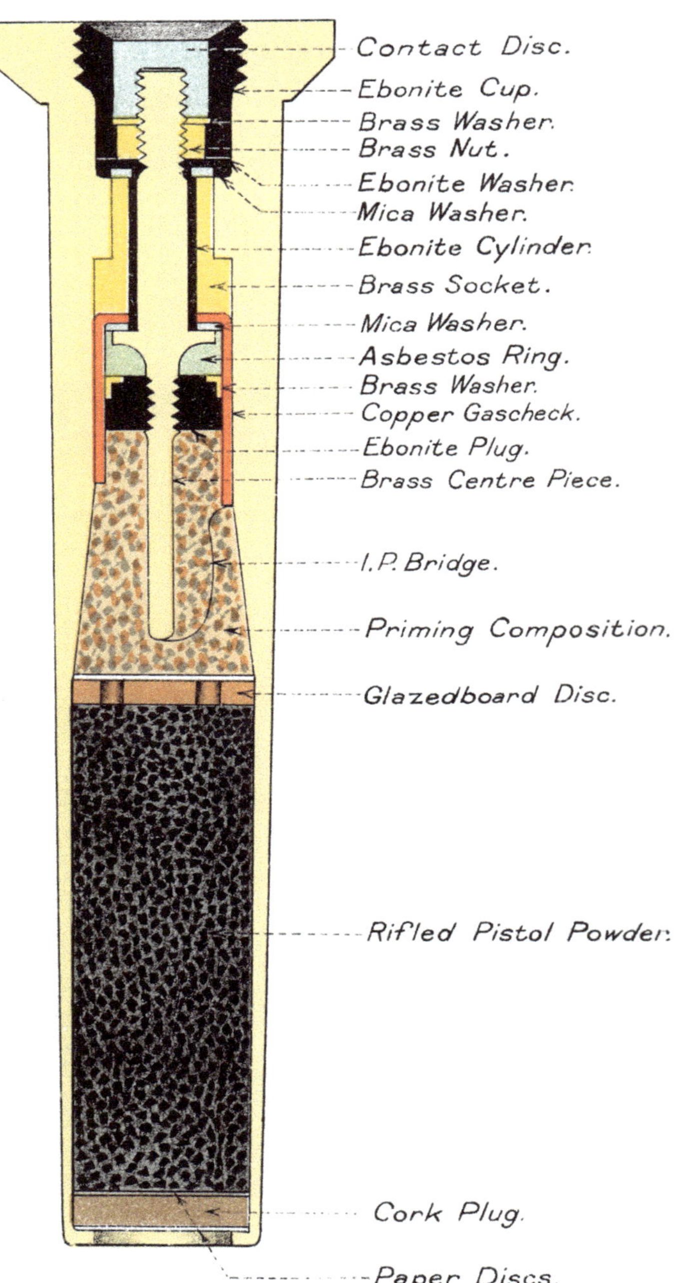

3567. Malby & Sons, Lith.

TUBE, V.S. ELECTRIC, WIRELESS "P" MARK IV.

Scale = 3/1.

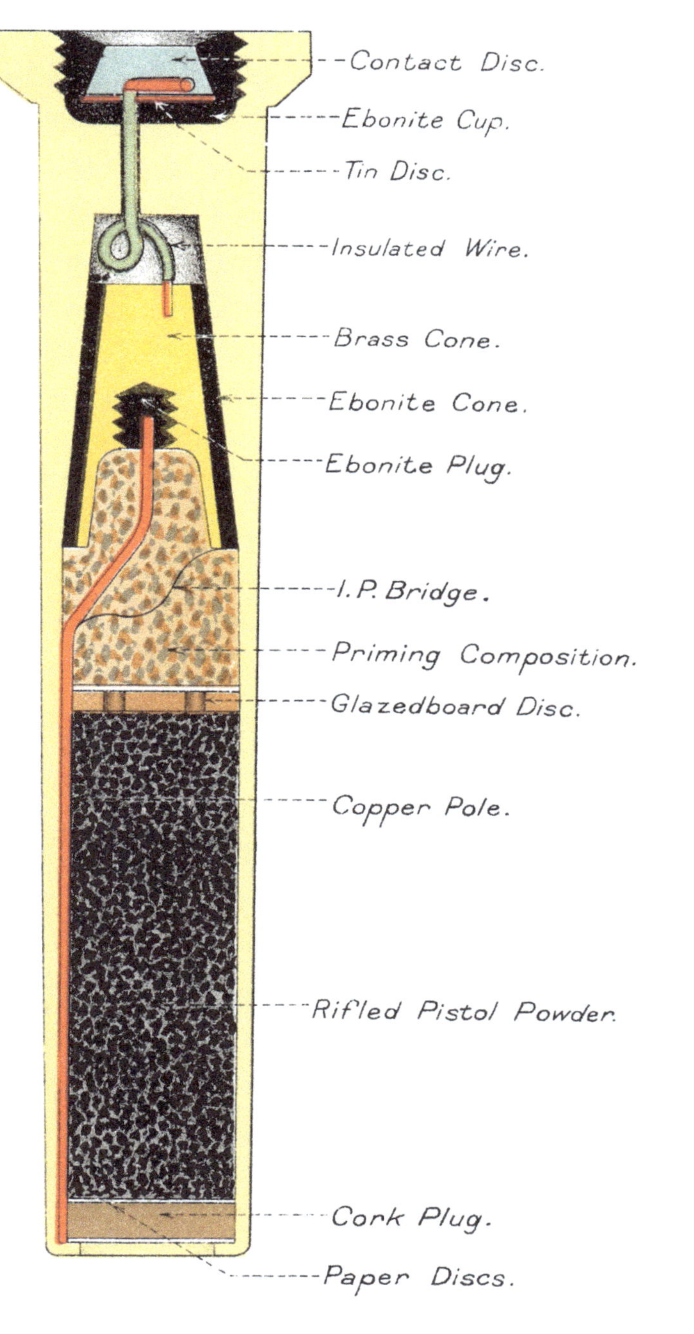

3567.

Malby & Sons, Lith.

TUBE, V.S. ELECTRIC "S" LARGE MARK I | N

Scale = 3/1.

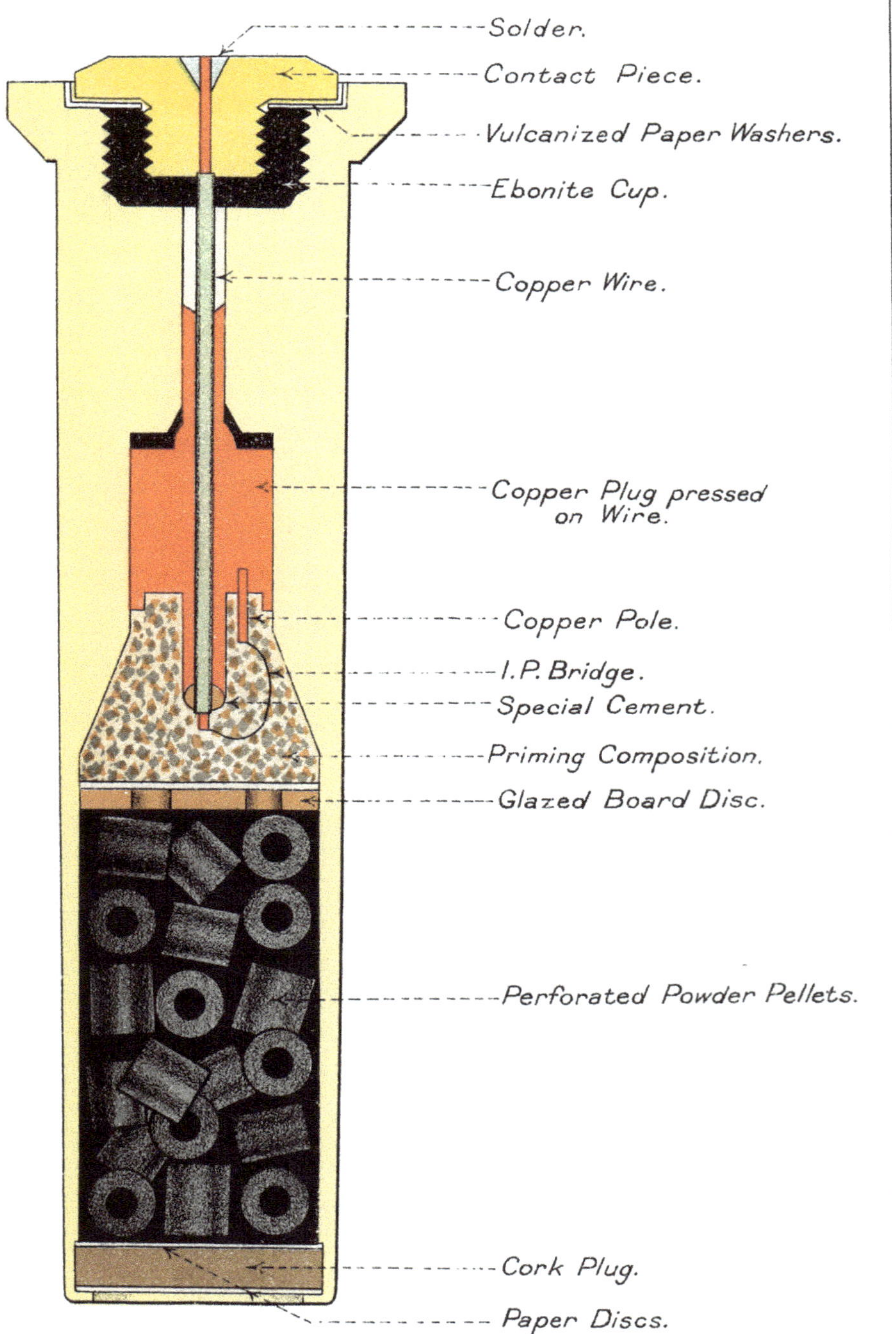

3567 Malby & Sons, Lith.

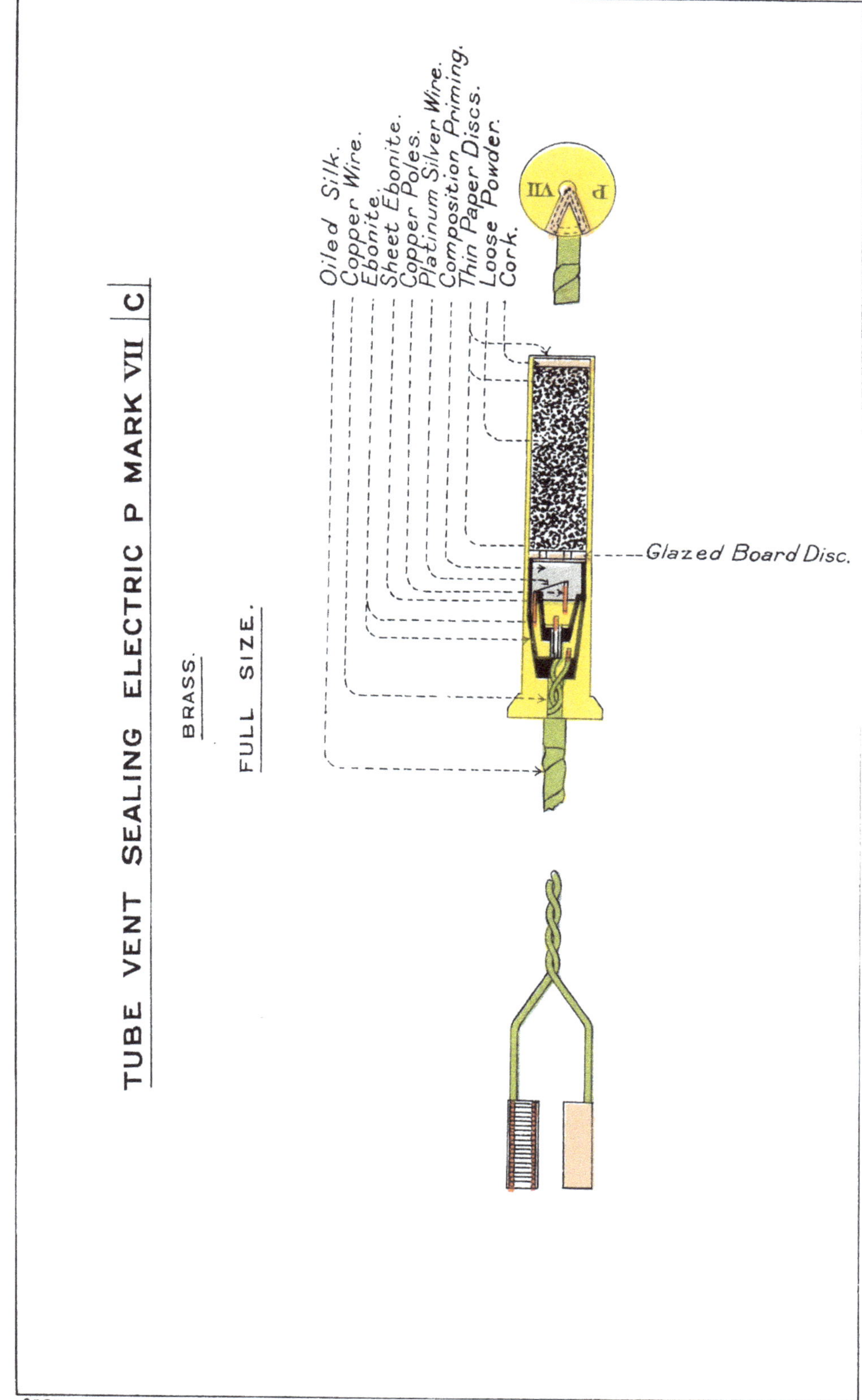

Malby & Sons, Lith.

TUBE, FRICTION, T. (MARK IV.) |L|.

FULL SIZE.

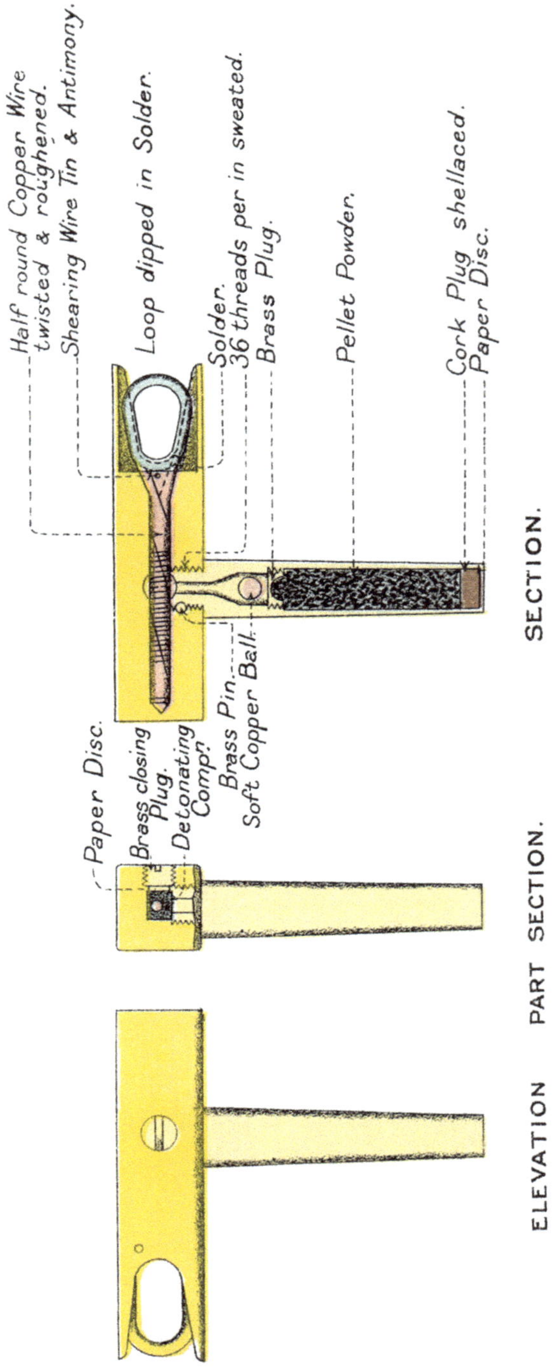

Malby & Sons, Lith.

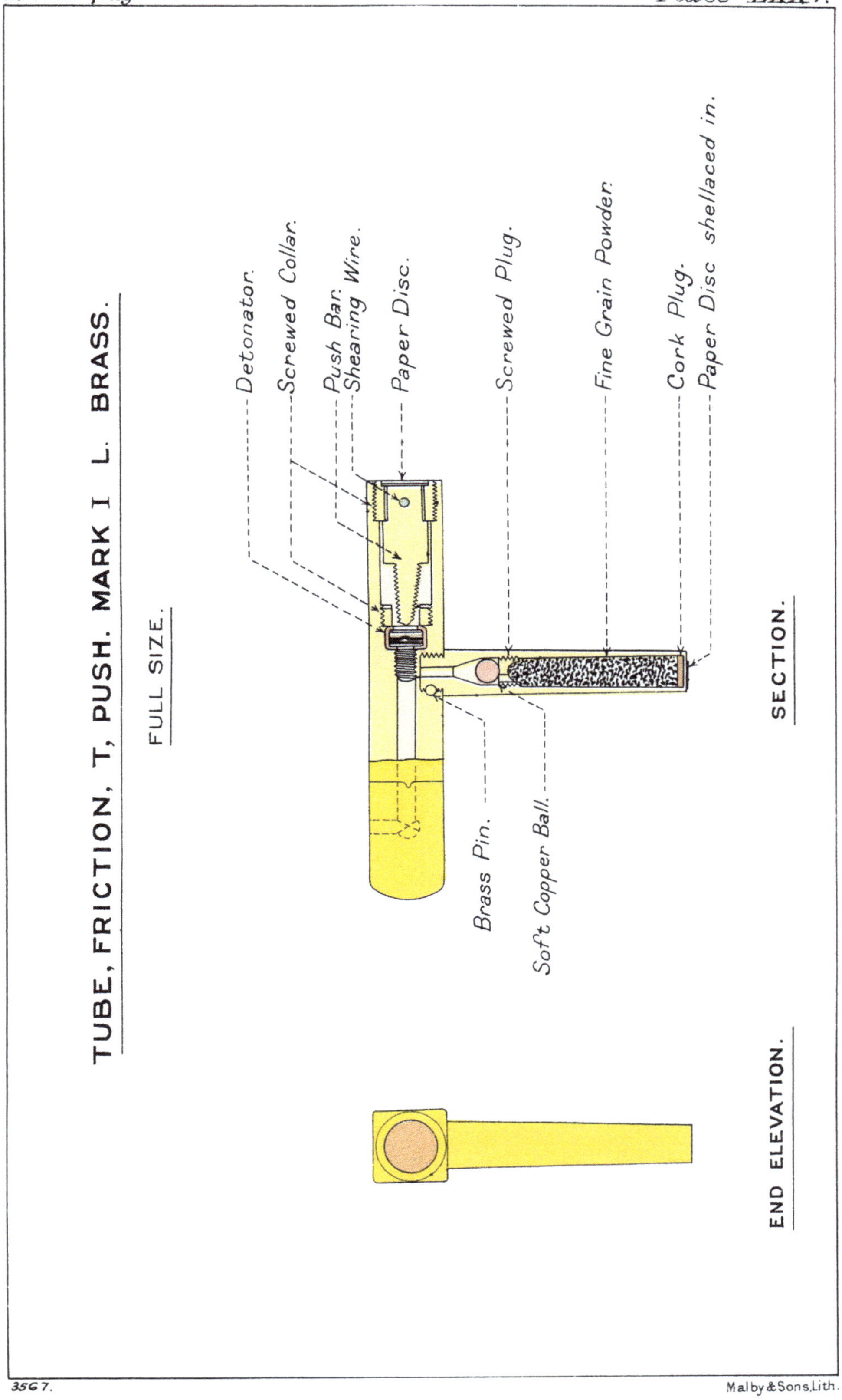

TUBE, FRICTION, T, PUSH. MARK I L. BRASS.

FULL SIZE.

END ELEVATION.

SECTION.

Malby & Sons, Lith.

CARTRIDGE Q.F. OR Q.F.C. PRIMER ELECTRIC LARGE MARK V|C|

MANGANESE OF ALUMINUM BRONZE.

FULL SIZE.

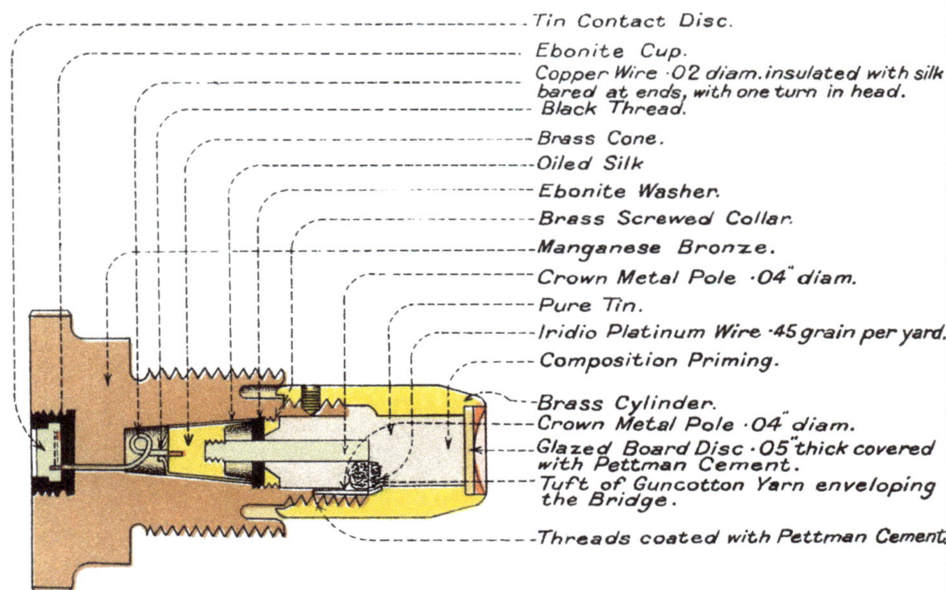

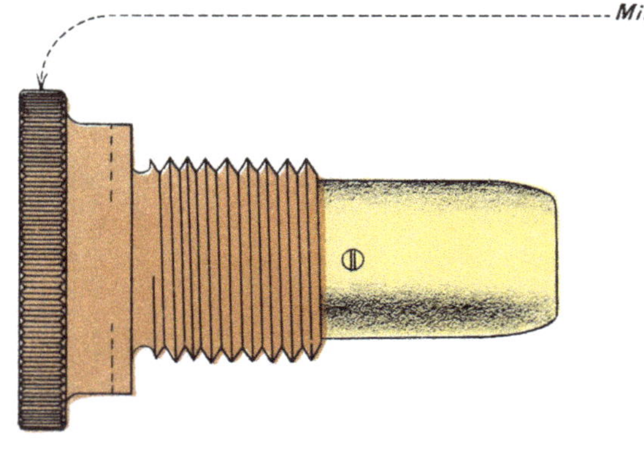

3567.

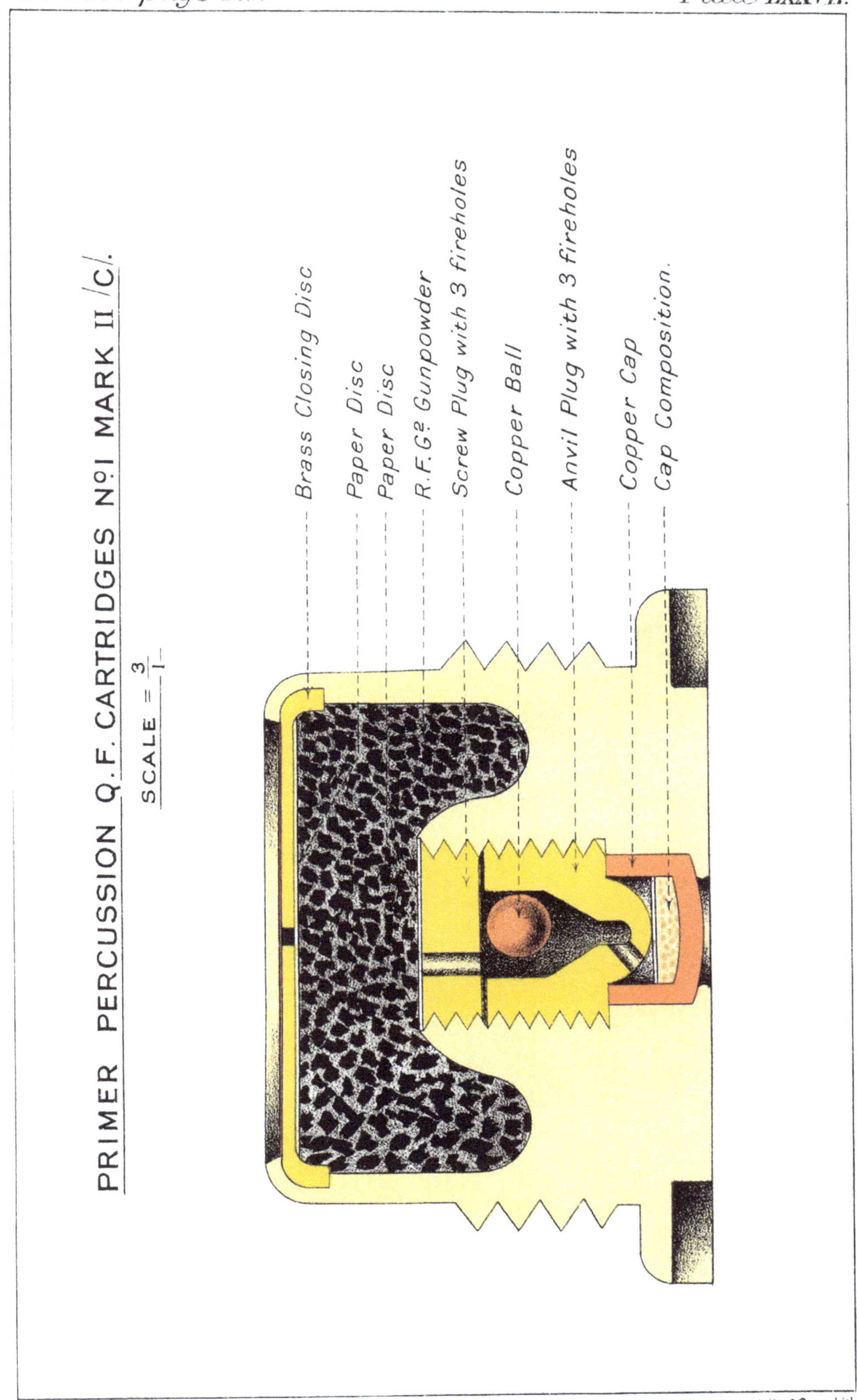

3567. Malby & Sons, Lith.

PRIMER, PERCUSSION, Q.F. CARTRIDGES Nº 2 MARK III | C |

FULL SIZE.

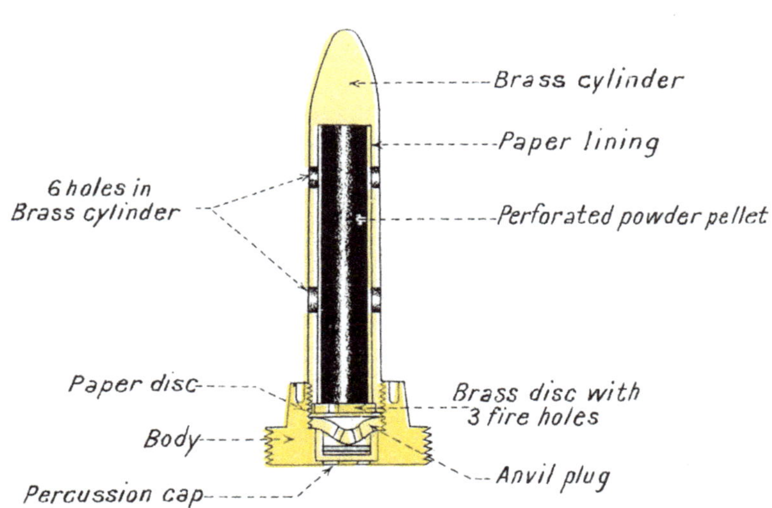

PRIMER, PERCUSSION Q.F. CARTRIDGES Nº 2 MARK IV | C |

FULL SIZE.

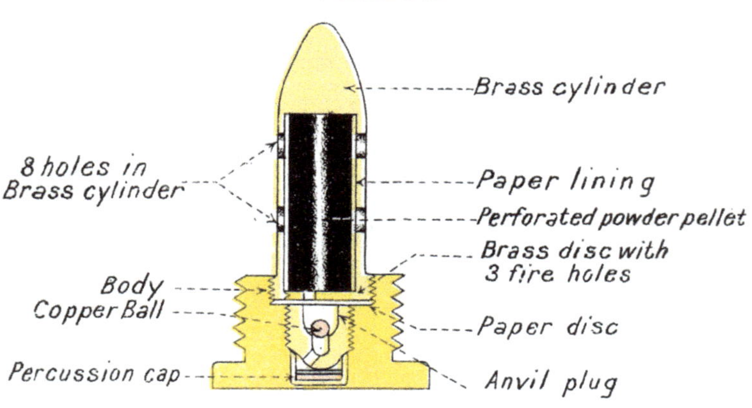

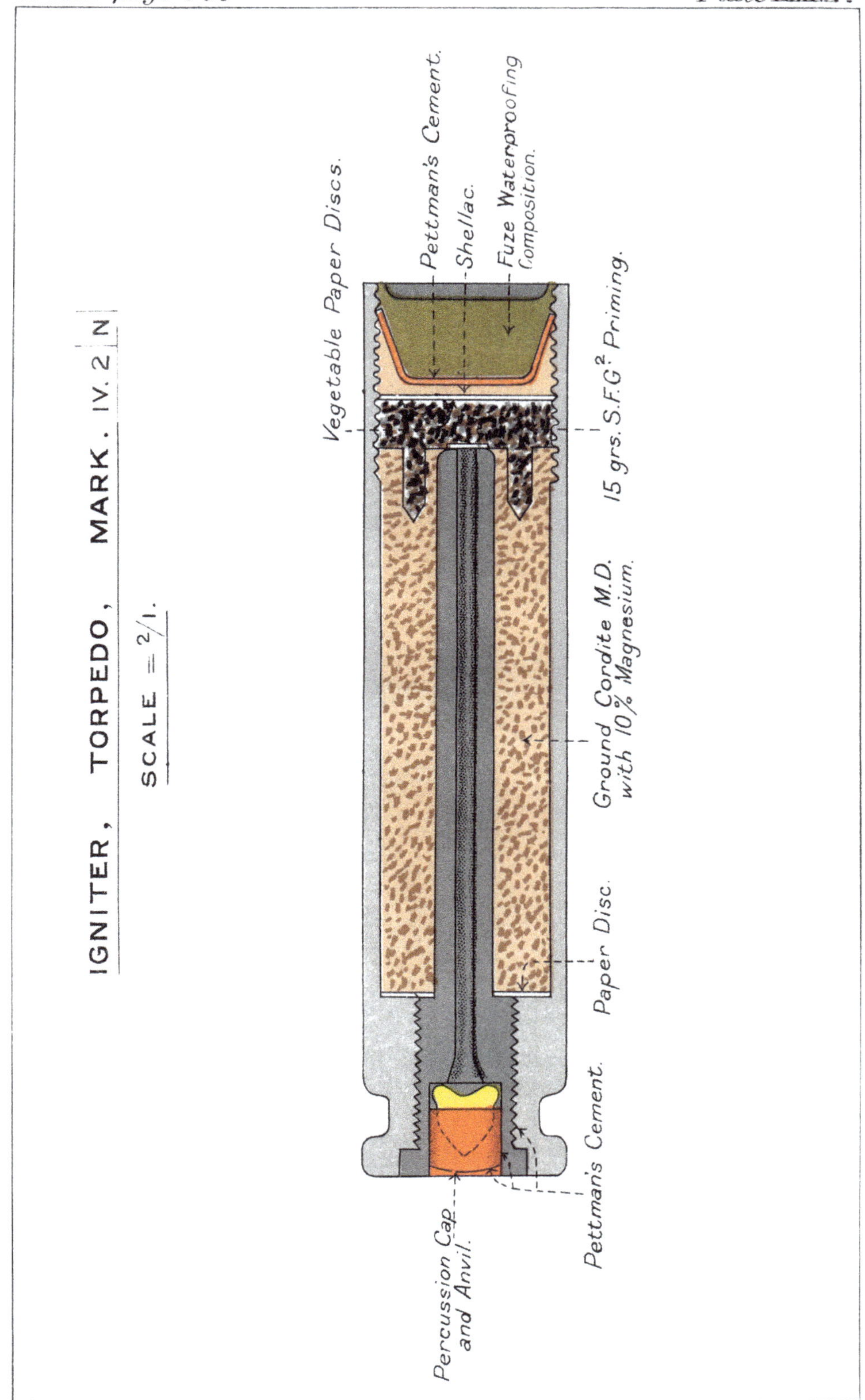
IGNITER, TORPEDO, MARK. IV. 2 N
SCALE = 2/1.
Vegetable Paper Discs.
Pettman's Cement.
Shellac.
Fuze Waterproofing Composition.
15 grs. S.F.G². Priming.
Ground Cordite M.D. with 10% Magnesium.
Paper Disc.
Pettman's Cement.
Pettman's Cement.
Percussion Cap and Anvil.

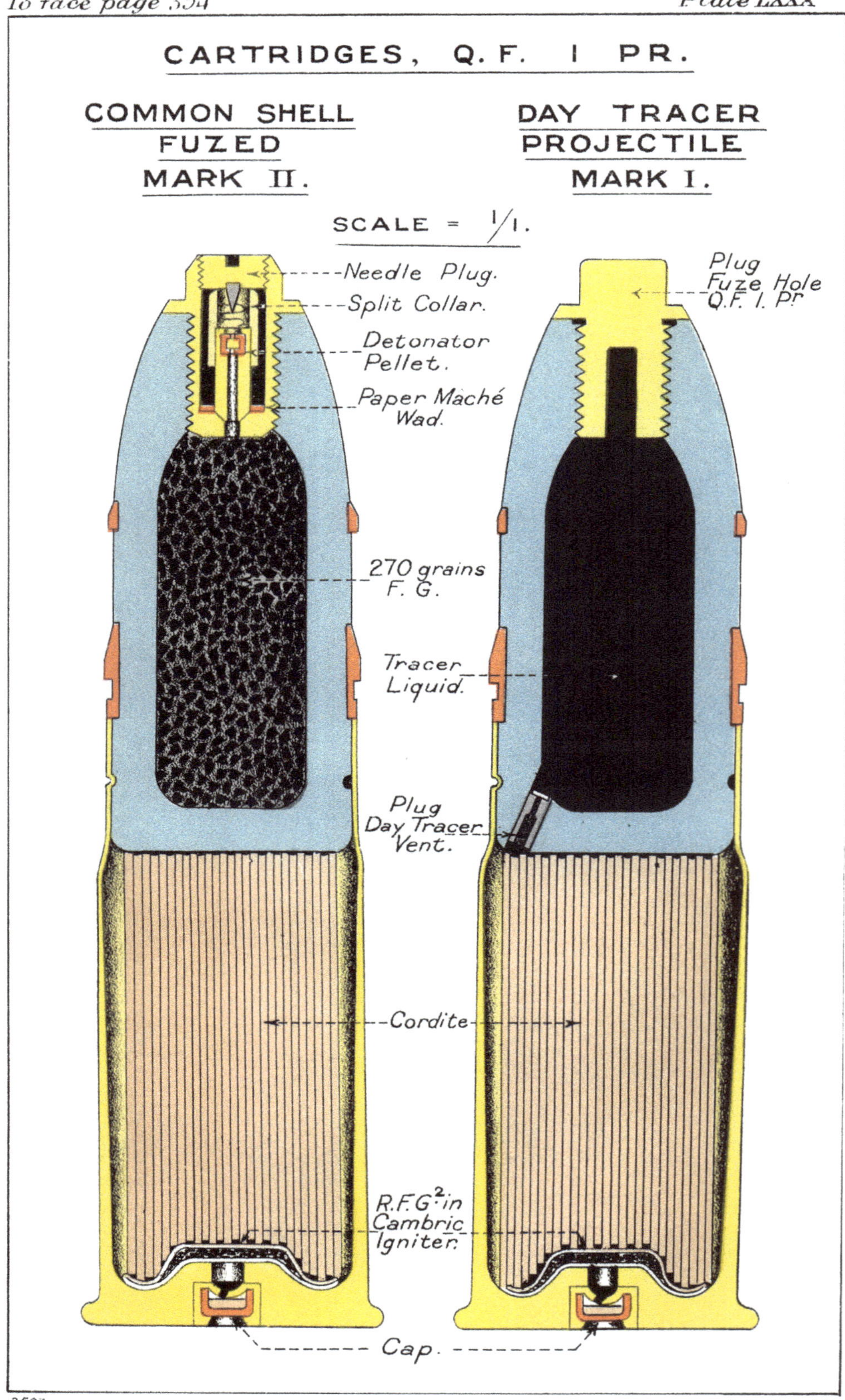
CARTRIDGES, Q.F. 1 PR.
COMMON SHELL FUZED MARK II.
DAY TRACER PROJECTILE MARK I.
SCALE = 1/1.
Needle Plug.
Split Collar.
Detonator Pellet.
Paper Maché Wad.
270 grains F.G.
Tracer Liquid.
Plug Day Tracer Vent.
Plug Fuze Hole Q.F. 1. Pr
Cordite
R.F.G.2 in Cambric Igniter.
Cap.

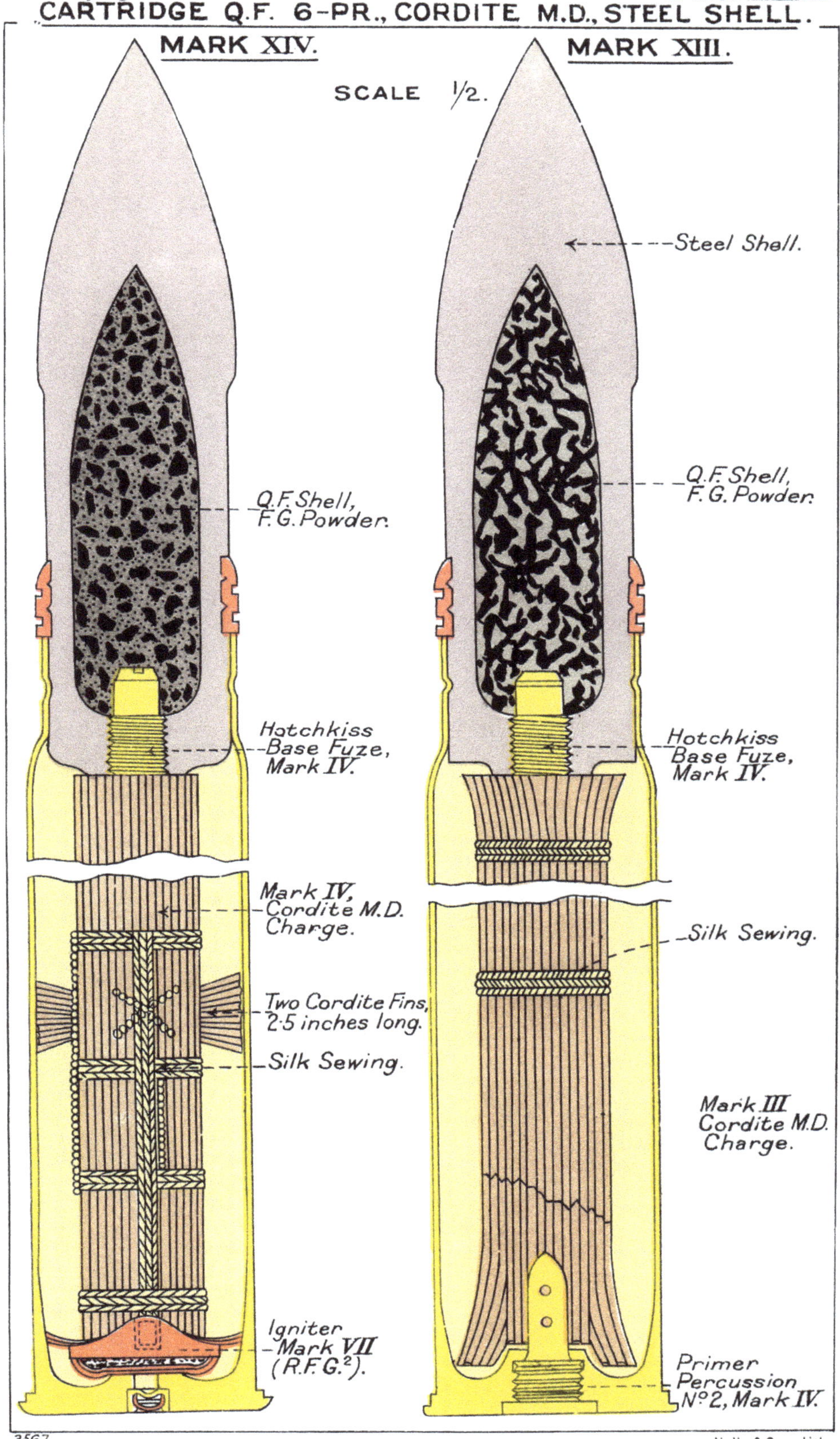
CARTRIDGE Q.F. 6-PR., CORDITE M.D., STEEL SHELL.
MARK XIV.
MARK XIII.
SCALE 1/2.
Steel Shell.
Q.F. Shell, F.G. Powder.
Q.F. Shell, F.G. Powder.
Hotchkiss Base Fuze, Mark IV.
Hotchkiss Base Fuze, Mark IV.
Mark IV, Cordite M.D. Charge.
Silk Sewing.
Two Cordite Fins, 2·5 inches long.
Silk Sewing.
Mark III Cordite M.D. Charge.
Igniter Mark VII (R.F.G.²).
Primer Percussion No. 2, Mark IV.
3567.
Malby & Sons, Lith.

Plate LXXXII

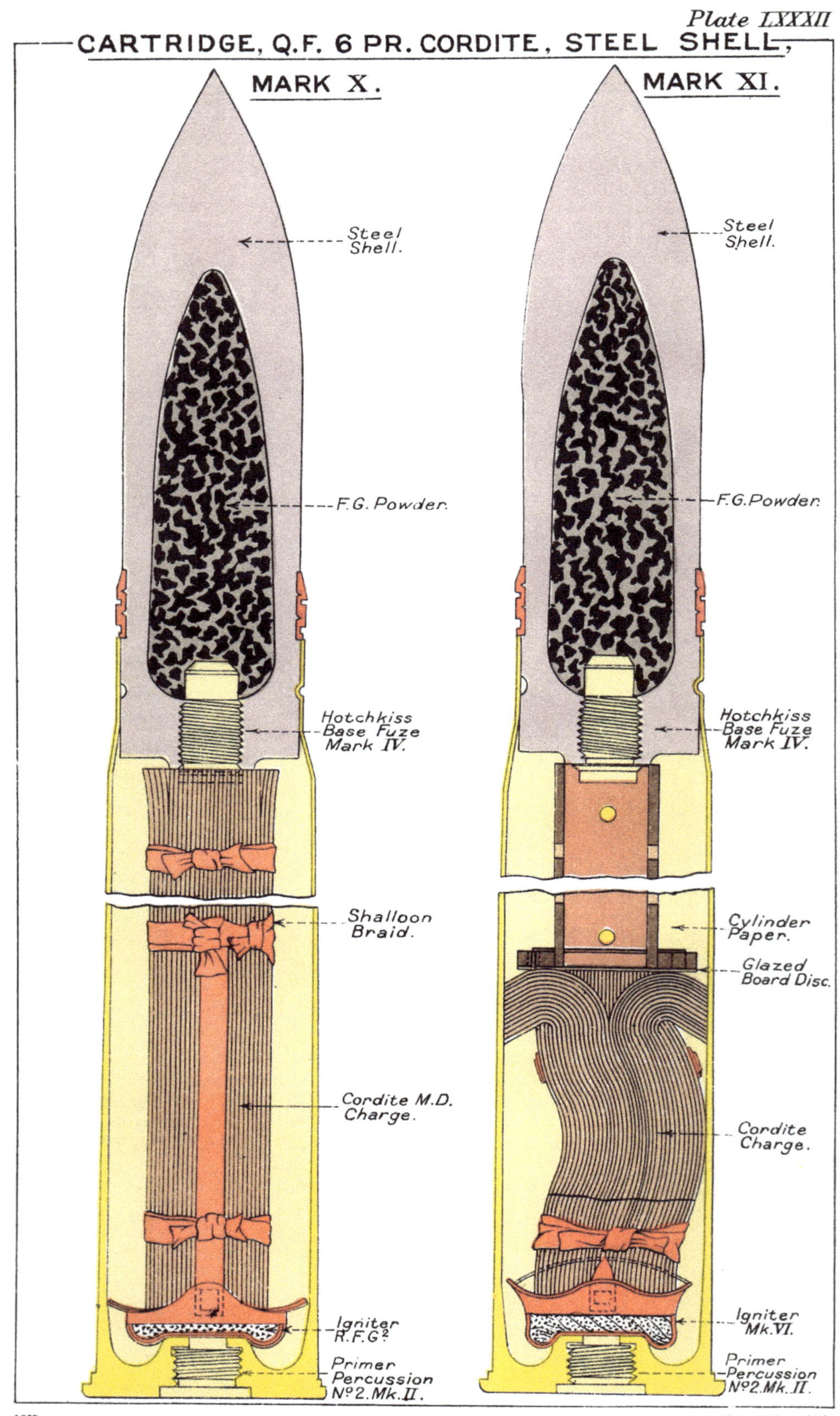

3567.

Malby & Sons, Lith.

CARTRIDGE, Q.F. 13 P^{R} SHRAPNEL, MARK II.

Scale = 1/3.

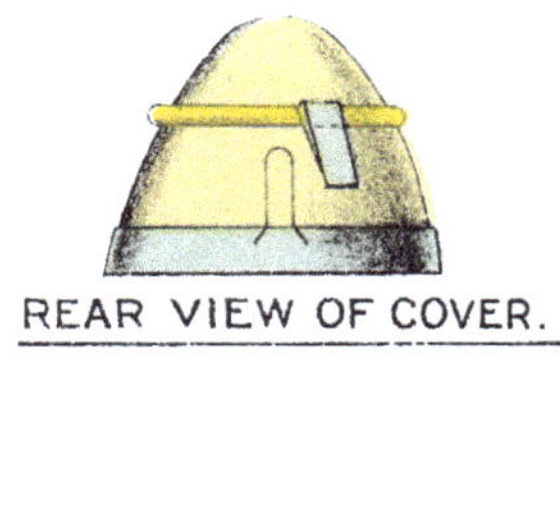

REAR VIEW OF COVER.

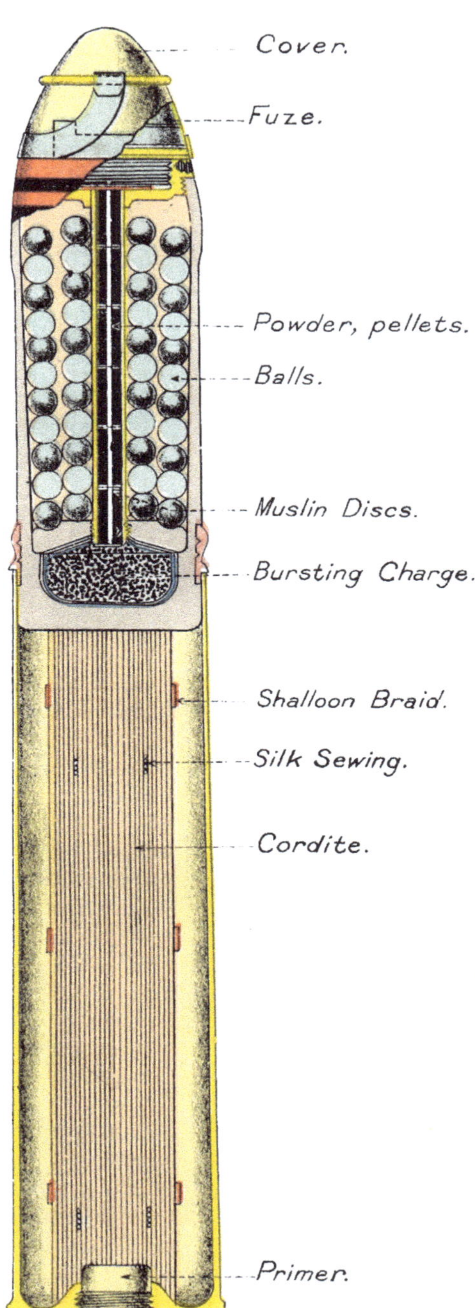

Malby & Sons, Lith.

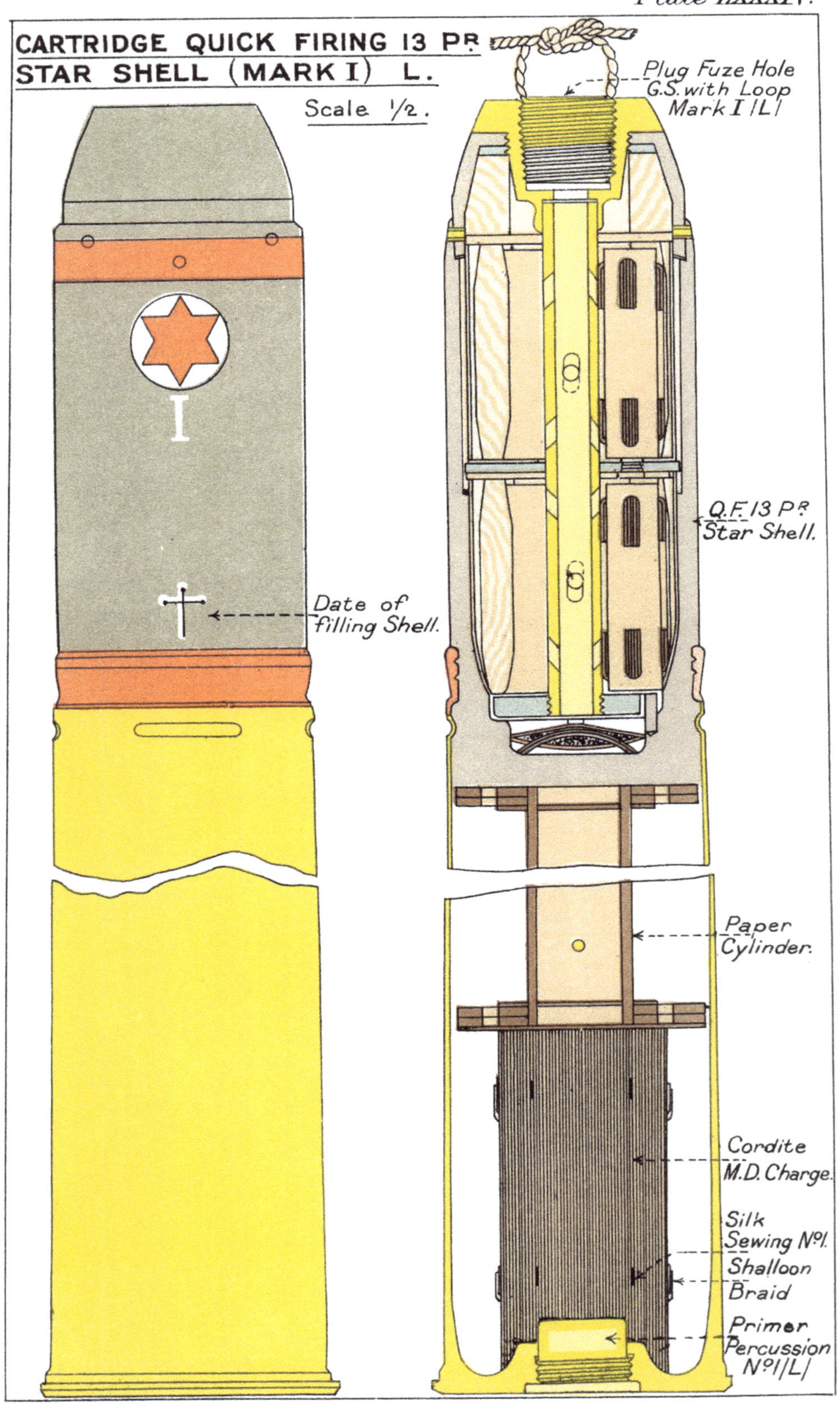

3567.

Malby & Sons, Lith.

Plate LXXXV.

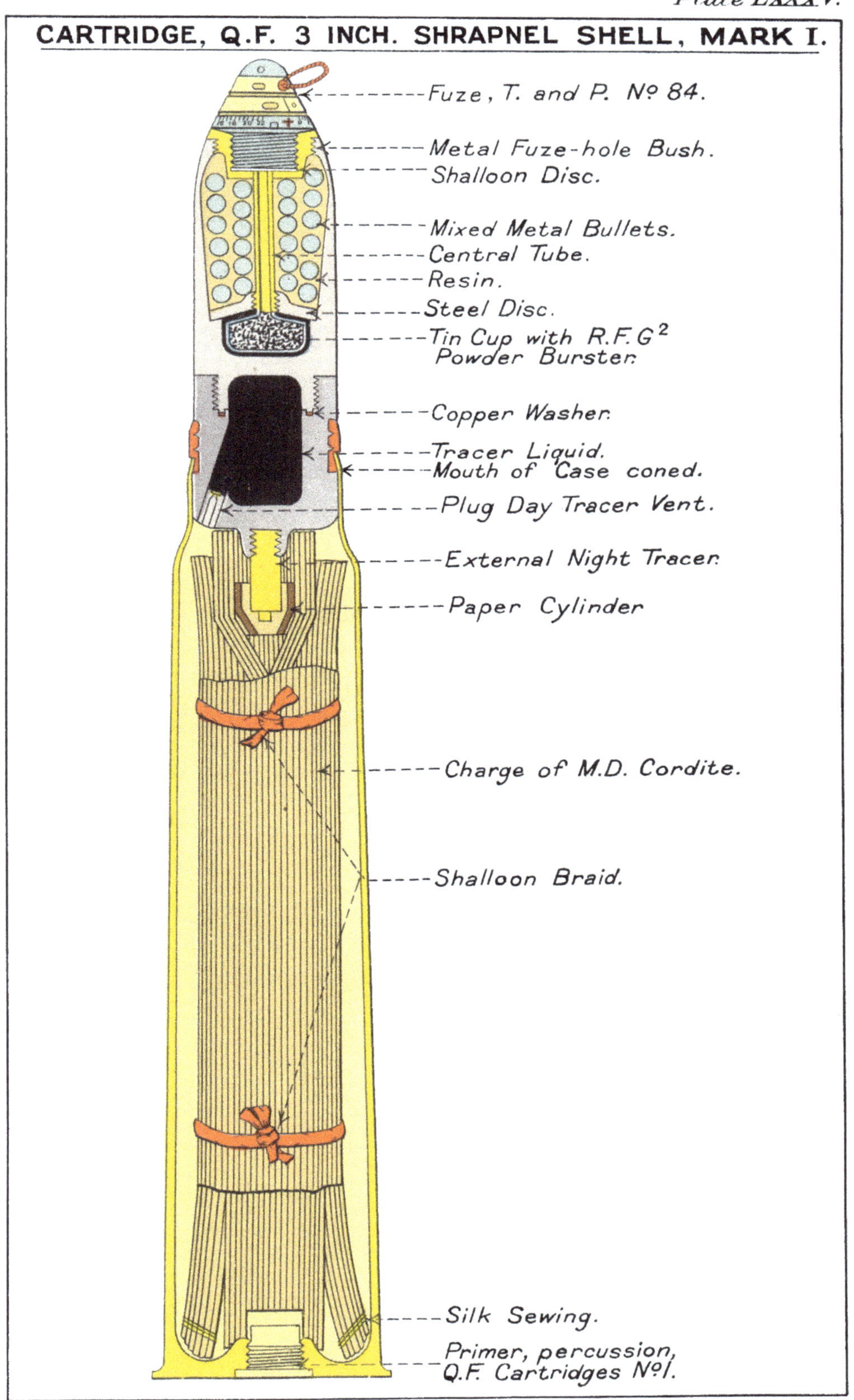

3567.

Malby & Sons, Lith.

Plate LXXXVI

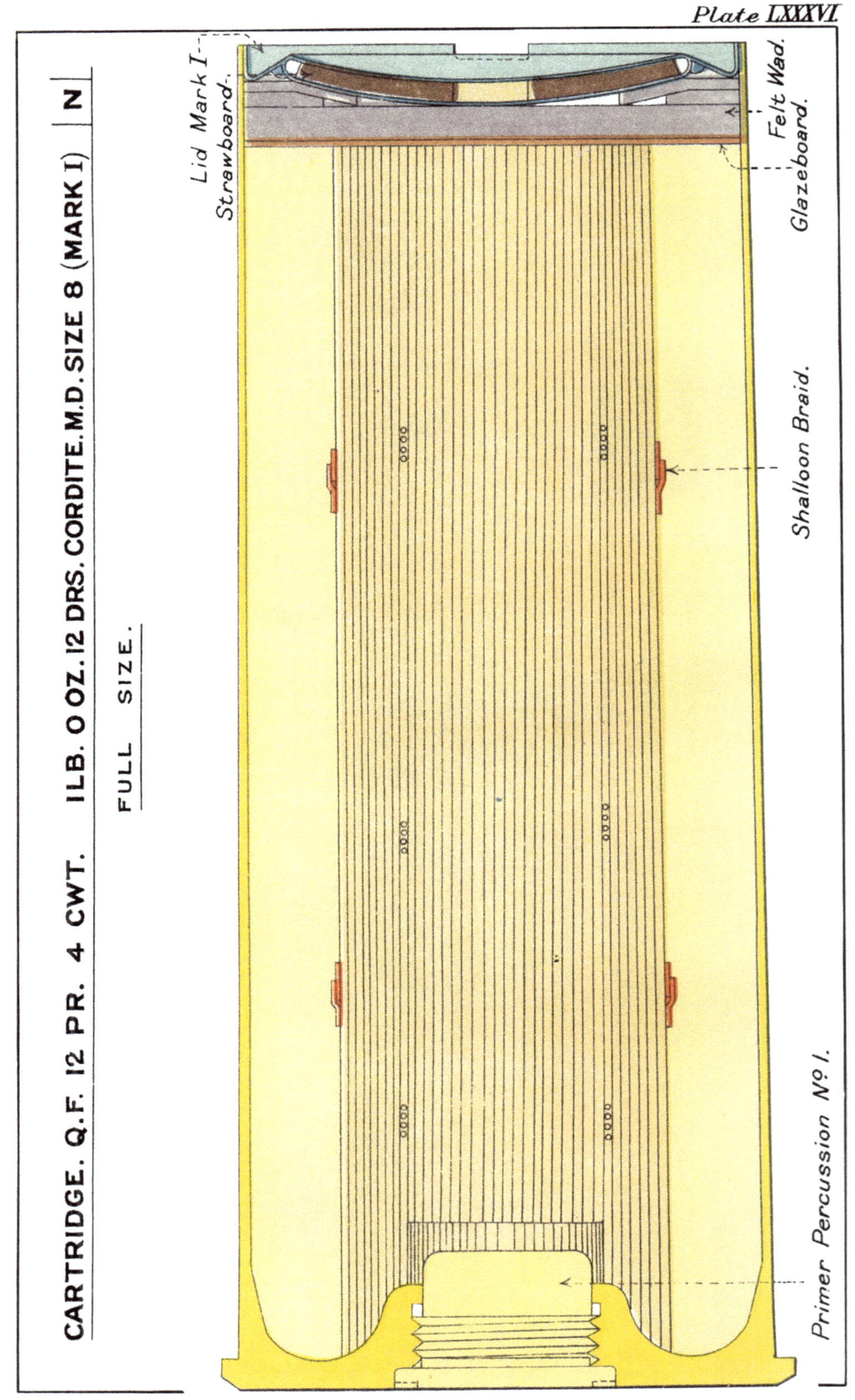

3567.

Malby & Sons, Lith.

Plate LXXXVII.

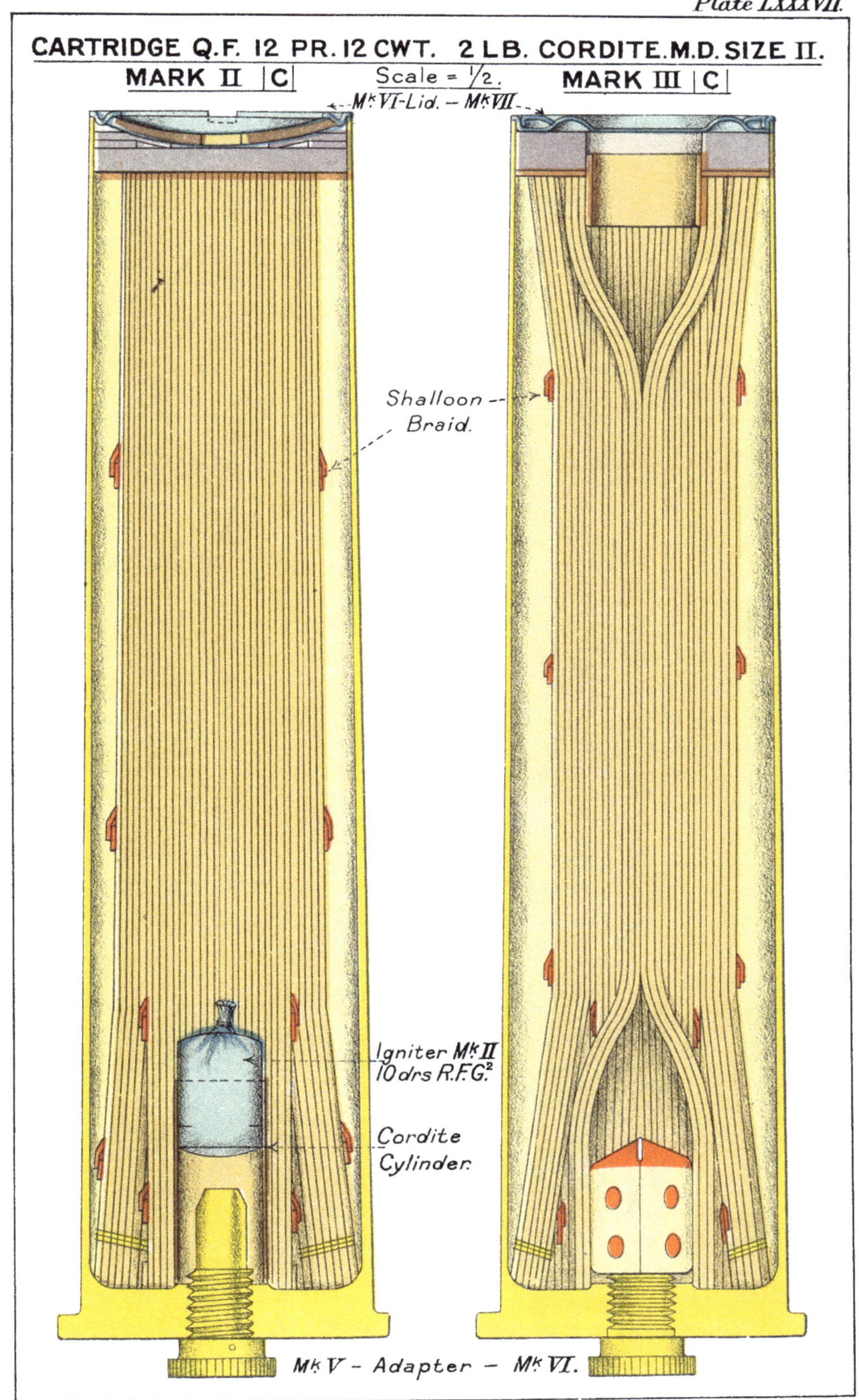

3567.

Malby & Sons, Lith.

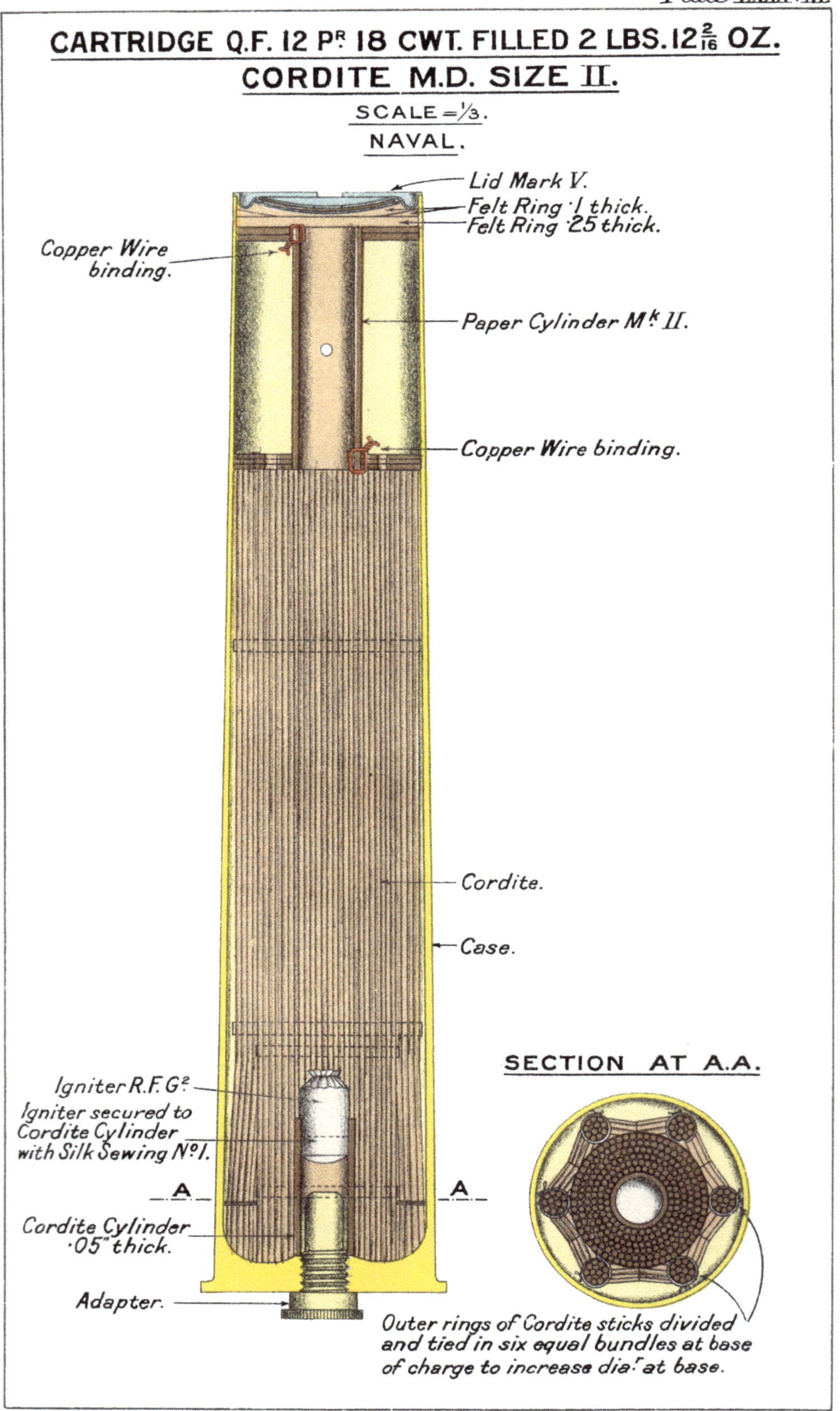

Malby & Sons, Lith.

Plate LXXXIX.

CARTRIDGE Q.F. 15 PR. MARK III | L |
1LB. 2OZ. 10DRS. CORDITE M.D.T. SIZE 20-10.

FULL SIZE.

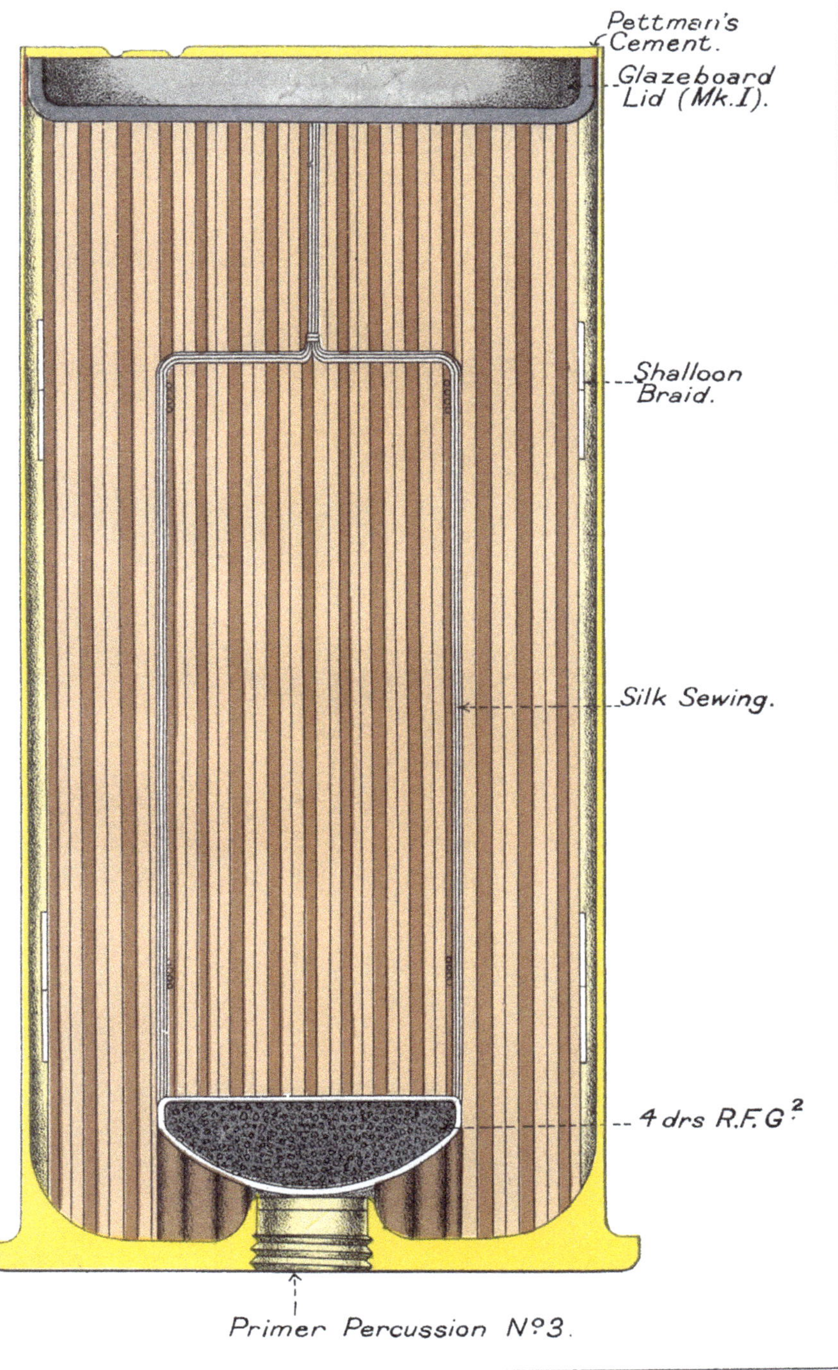

3567.

Malby & Sons, Lith.

Plate XC.

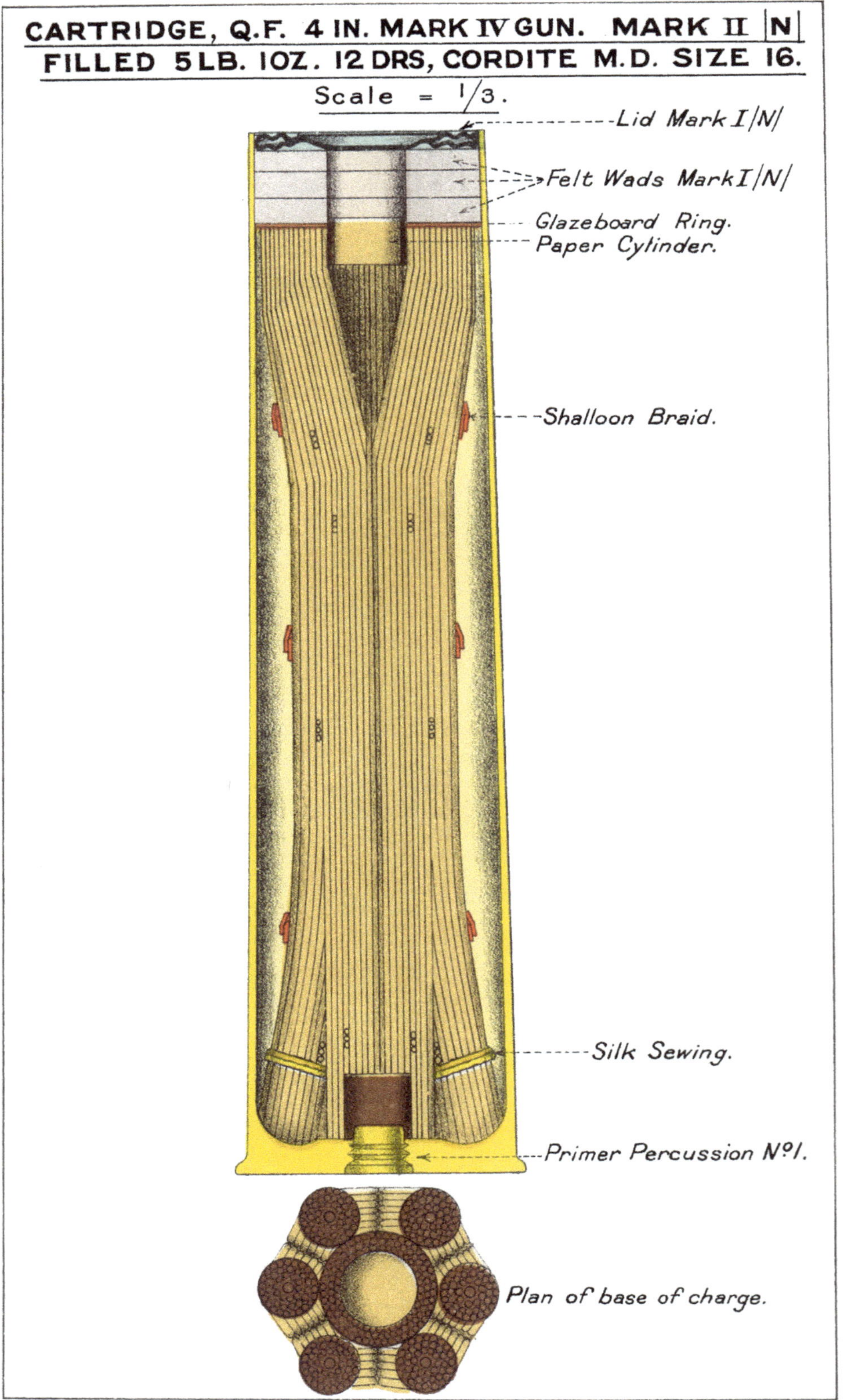

3567.

Malby & Sons, Lith.

Plate XCI.

CARTRIDGE, Q.F. 4 IN. MARK V. GUN, FILLED 7 LB. 11 OZ. CORDITE M.D. SIZE 16 MARK I |N|

Full Size.

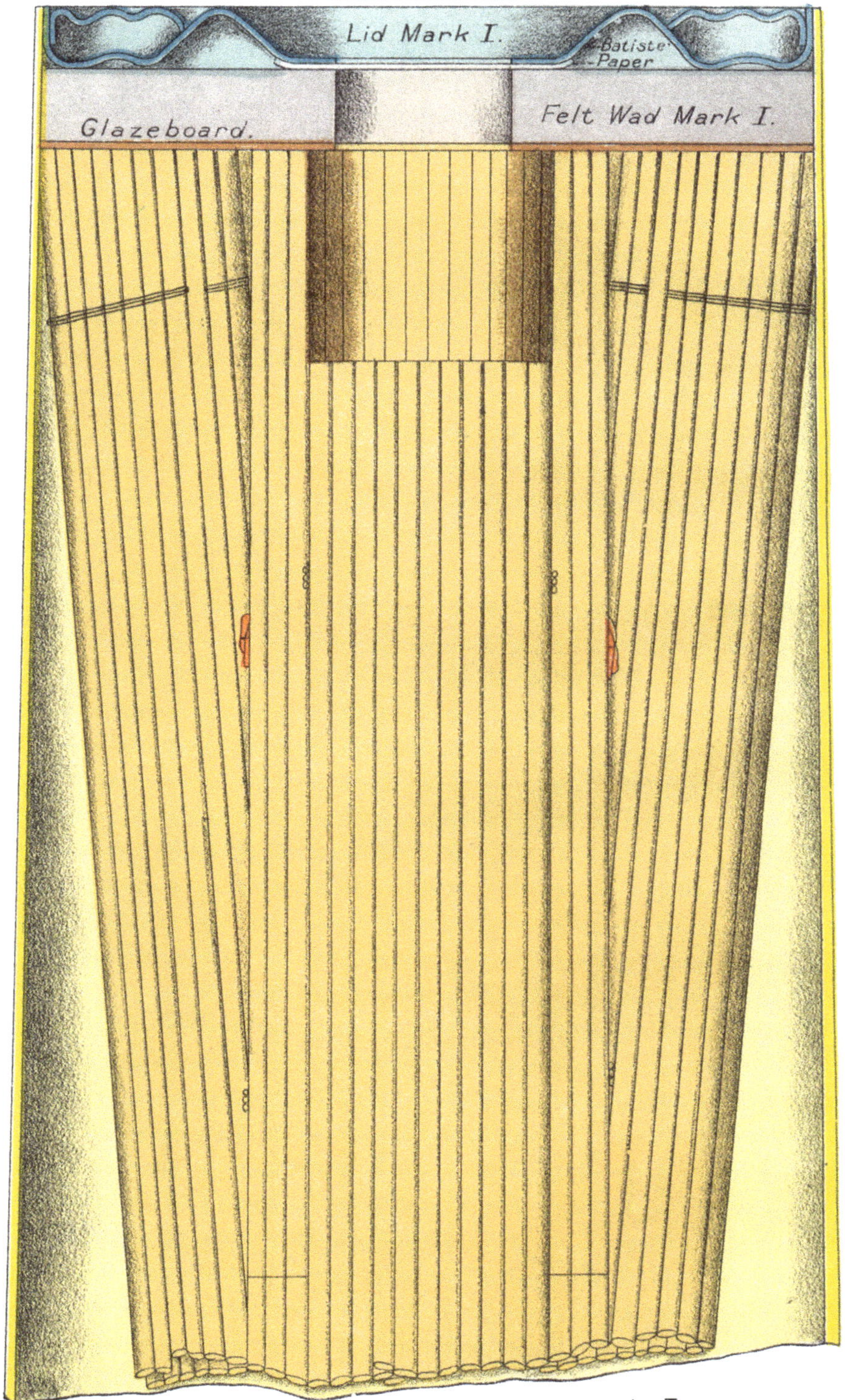

Lid Mark I.

Batiste Paper

Glazeboard.

Felt Wad Mark I.

NOTE:- Base of charge is made up in same way as the Top.

3567.

Malby & Sons, Lith.

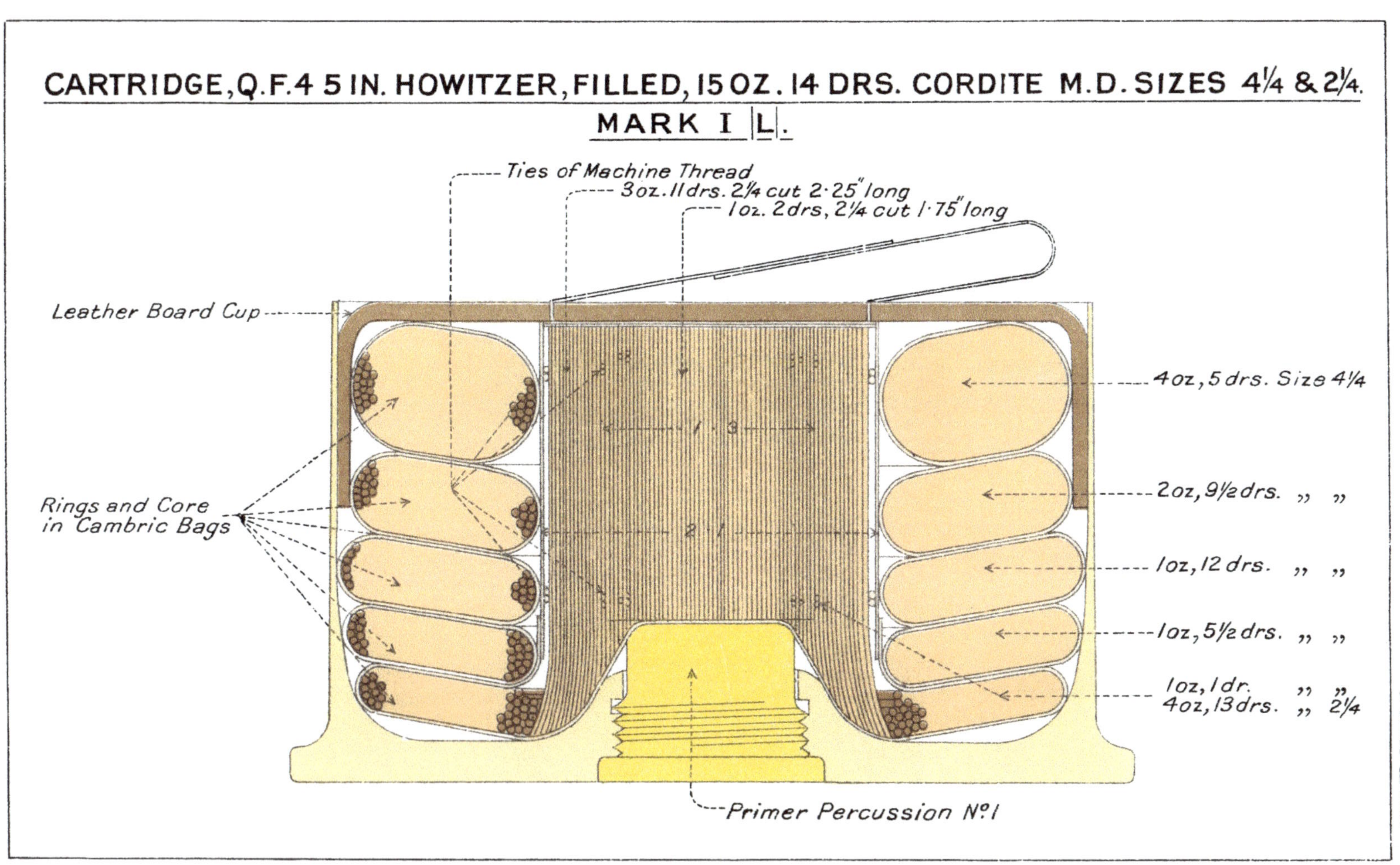

3567.

Malby & Sons, Lith.

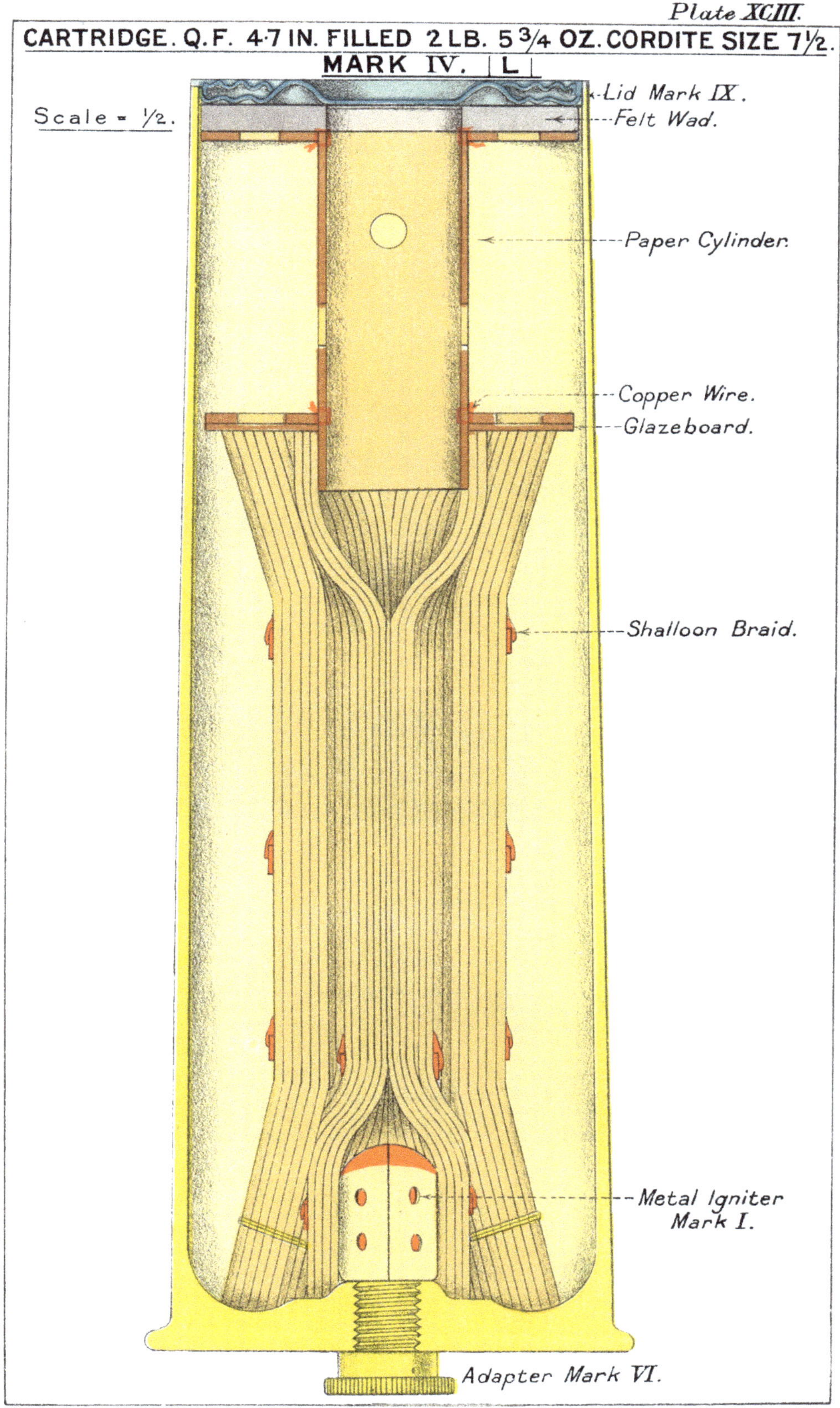

Malby & Sons, Lith.

Q. F. CARTRIDGES FOR PAPER SHOT.

LAND SERVICE.

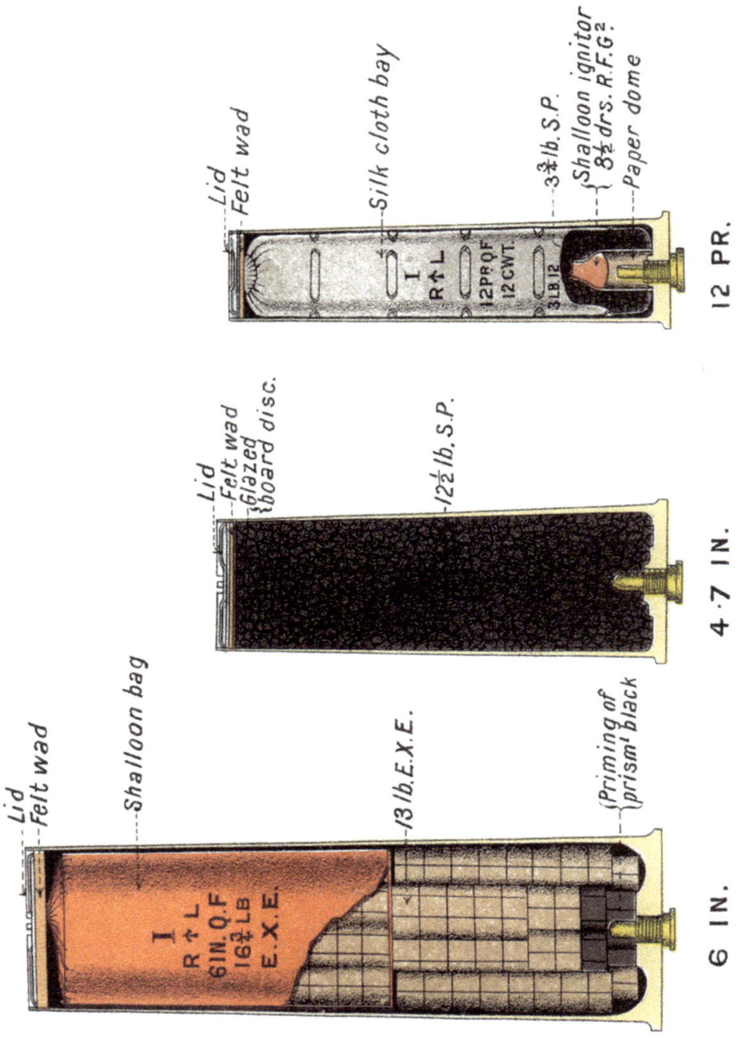

Lid
Felt wad
Shalloon bag
I
R ↑ L
6 IN. Q.F
16¾ LB
E.X.E.
13 lb. E.X.E.
Priming of prism^l black

6 IN.

Lid
Felt wad
Glazed board disc.
12½ lb. S.P.

4·7 IN.

Lid
Felt wad
Silk cloth bay
I
R ↑ L
12 PR. Q.F
12 CWT.
3 LB. 12
3¾ lb. S.P.
Shalloon ignitor 8½ drs. R.F.G.²
Paper dome

12 PR.

3567.

Malby & Sons, Lith.

CARTRIDGE Q.F. BLANK
6 P^{R}. FILLED MARK IV.
15 O^{Z}., BLANK L.G.

CARTRIDGE Q.F. BLANK
3 P^{R}. FILLED MARK IV.
11 O^{Z}., BLANK L.G.

SCALE = ½

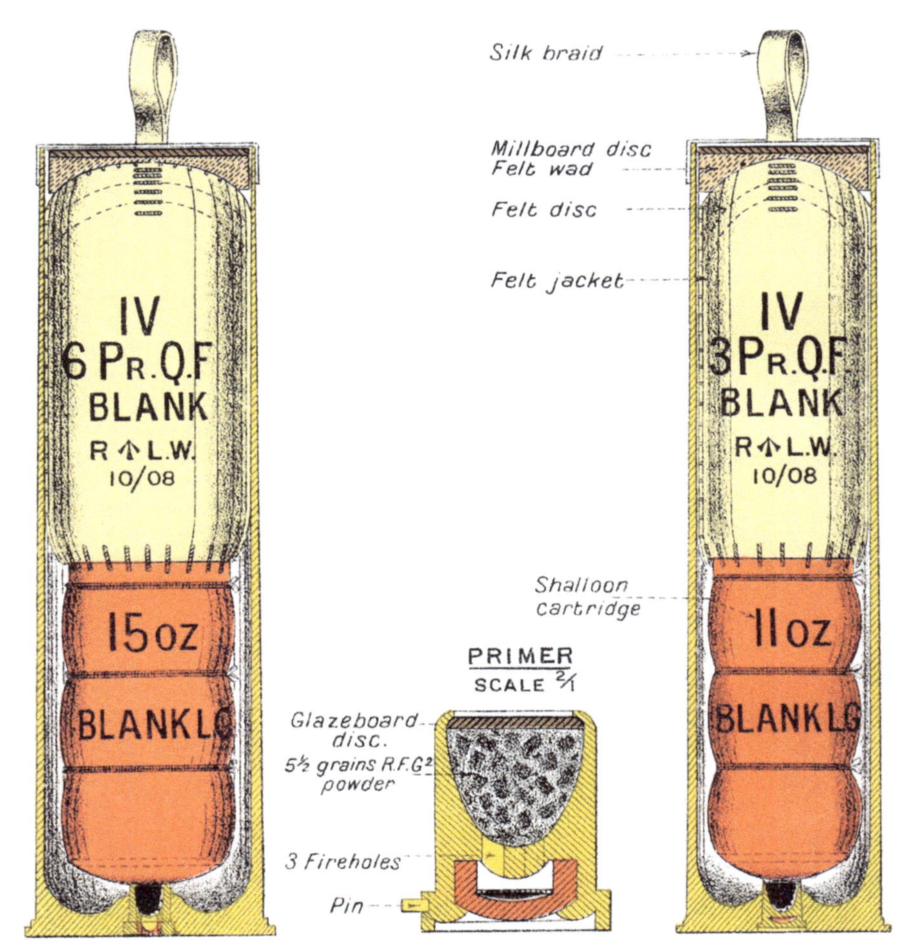

METHOD OF FILLING :—

CARTRIDGE Q.F. BLANK 18-PR., FILLED (MARK II) |L|

BRASS, 1LB. BLANK L.G. IN SILK CLOTH BAG, WITH JACKET, LEATHER BOARD CUP, SPLIT PAPER RING & PERCUSSION PRIMER.

SCALE = 1/2.

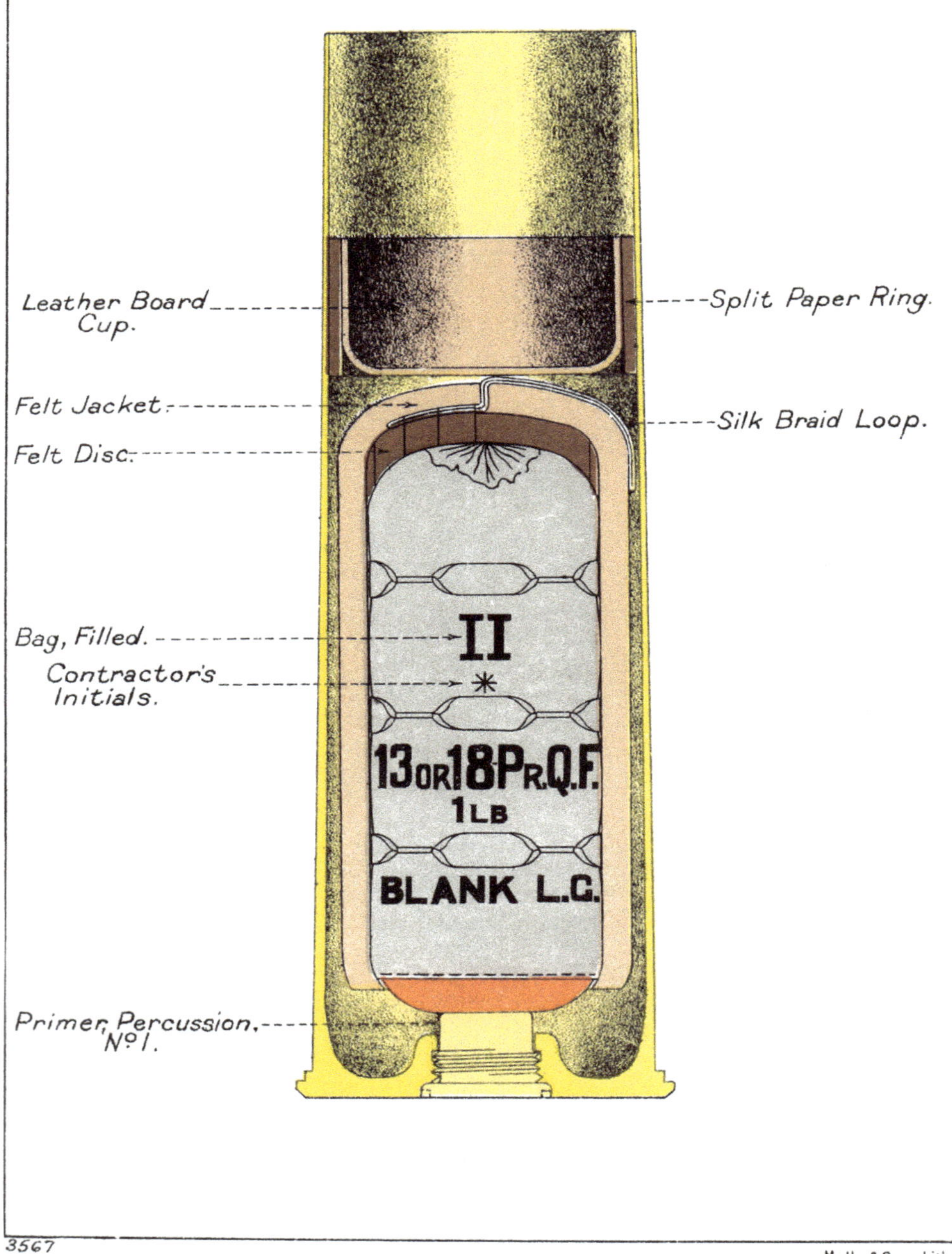

CARTRIDGE. Q.F. 12 PR 8 CWT BLANK.

1½ LBS. BLANK WITH FELT WAD JACKET AND IGNITER.

SCALE = ½.

NAVAL.

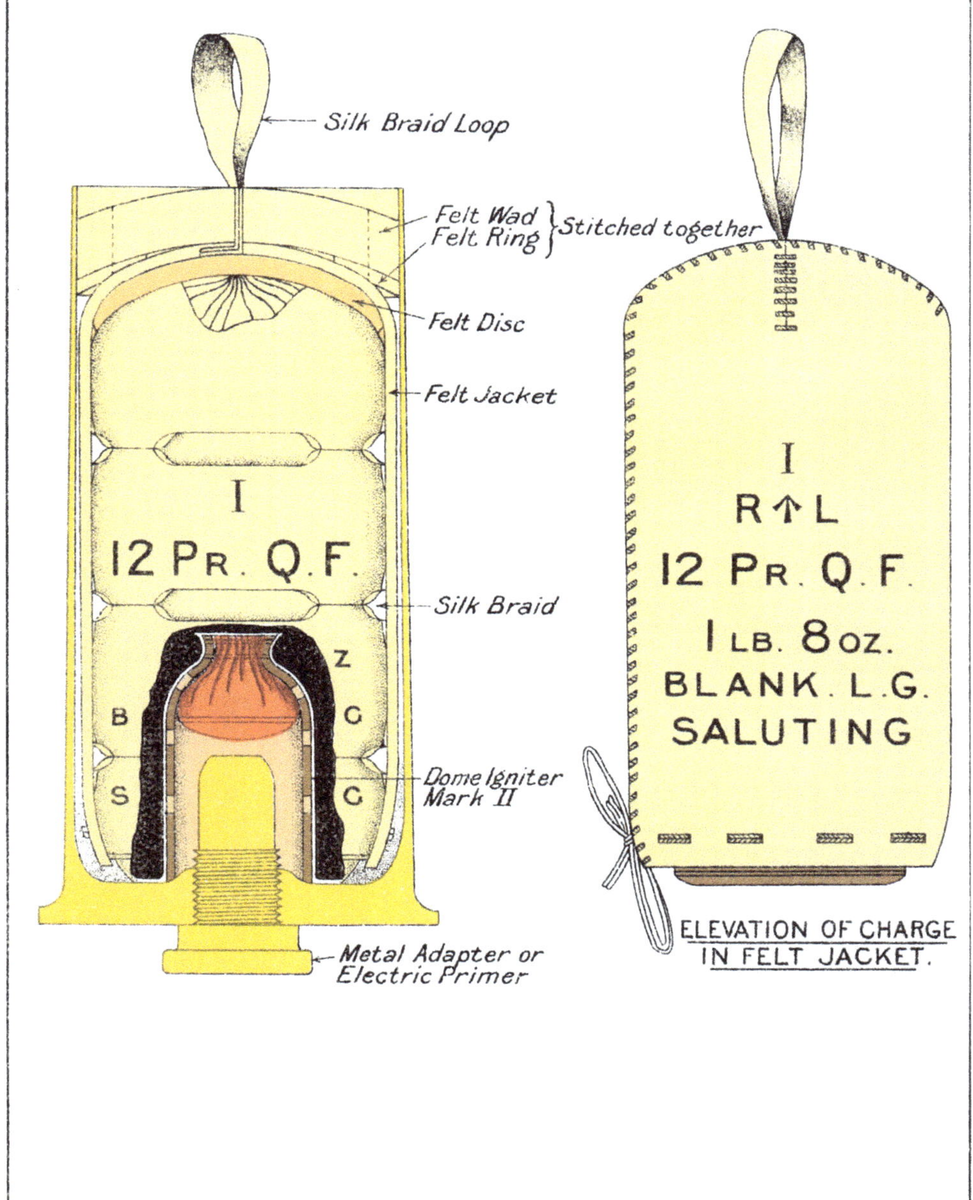

 Malby & Sons, Lith.

CARTRIDGE, Q.F., 12 PR. 12 CWT. BLANK (1½ LBS. L.G.) |C|.

SCALE = ¼

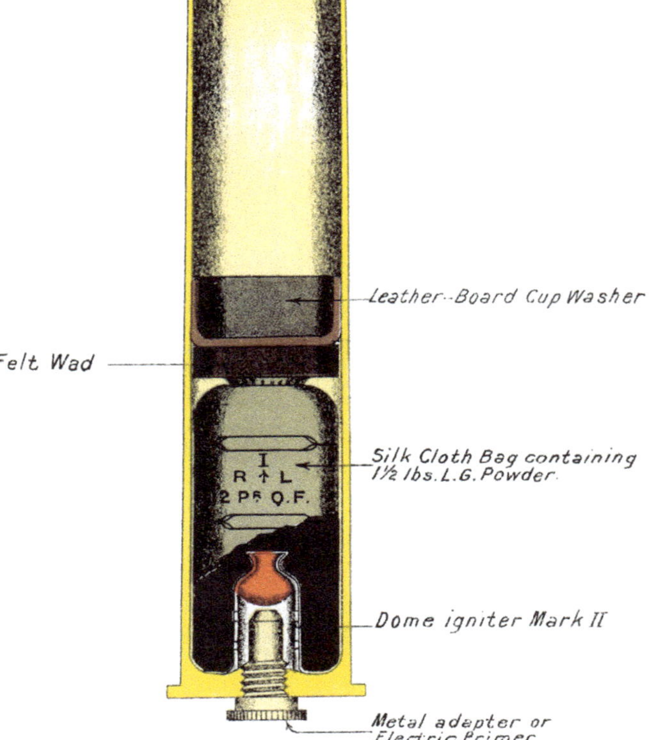

METAL RING FOR INSERTING LEATHER BOARD CUP.

WOOD DRIFT.

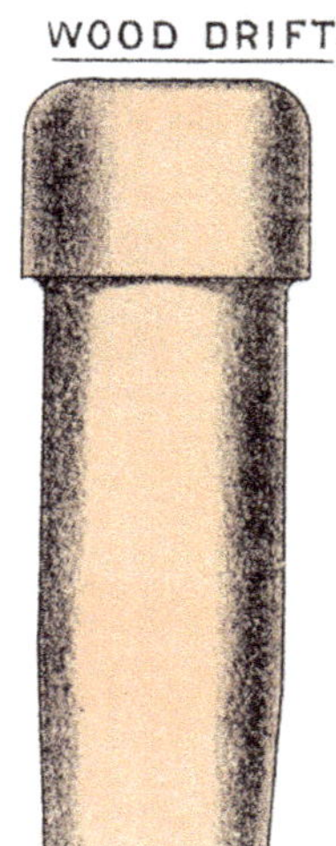

CARTRIDGE, Q.F. 4·7 INCH FILLED BLANK (3 LBS. L.G.) |C|.

SCALE = 1/4

MARK V.

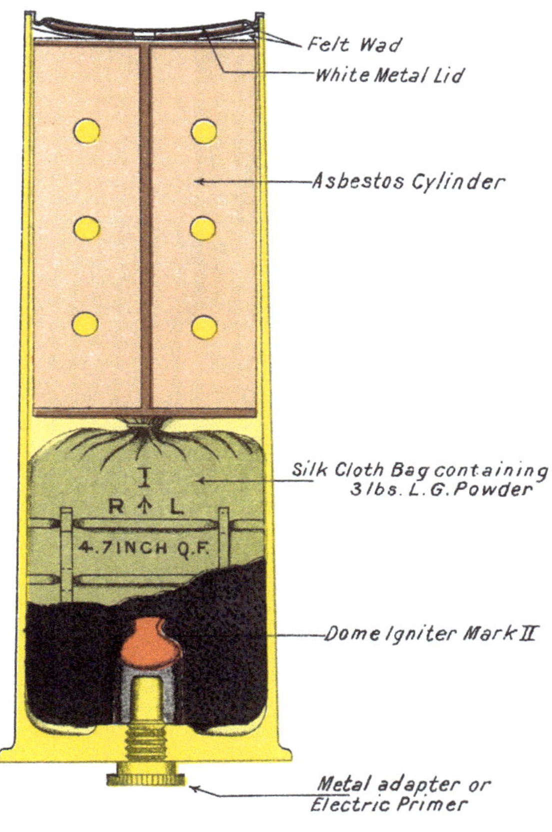

 Malby & Sons, Lith.

To face page 484 *Plate C.*

CARTRIDGE, Q.F. OR Q.F.C., DRILL, 6 INCH SHORT, (MARK IV). (C)
CARTRIDGE, Q.F. DRILL 4·7 INCH MARKS I TO IV* GUNS (MARKS V). (C).
CARTRIDGE, Q.F. DRILL 12 P^R (MARK IV). (C).

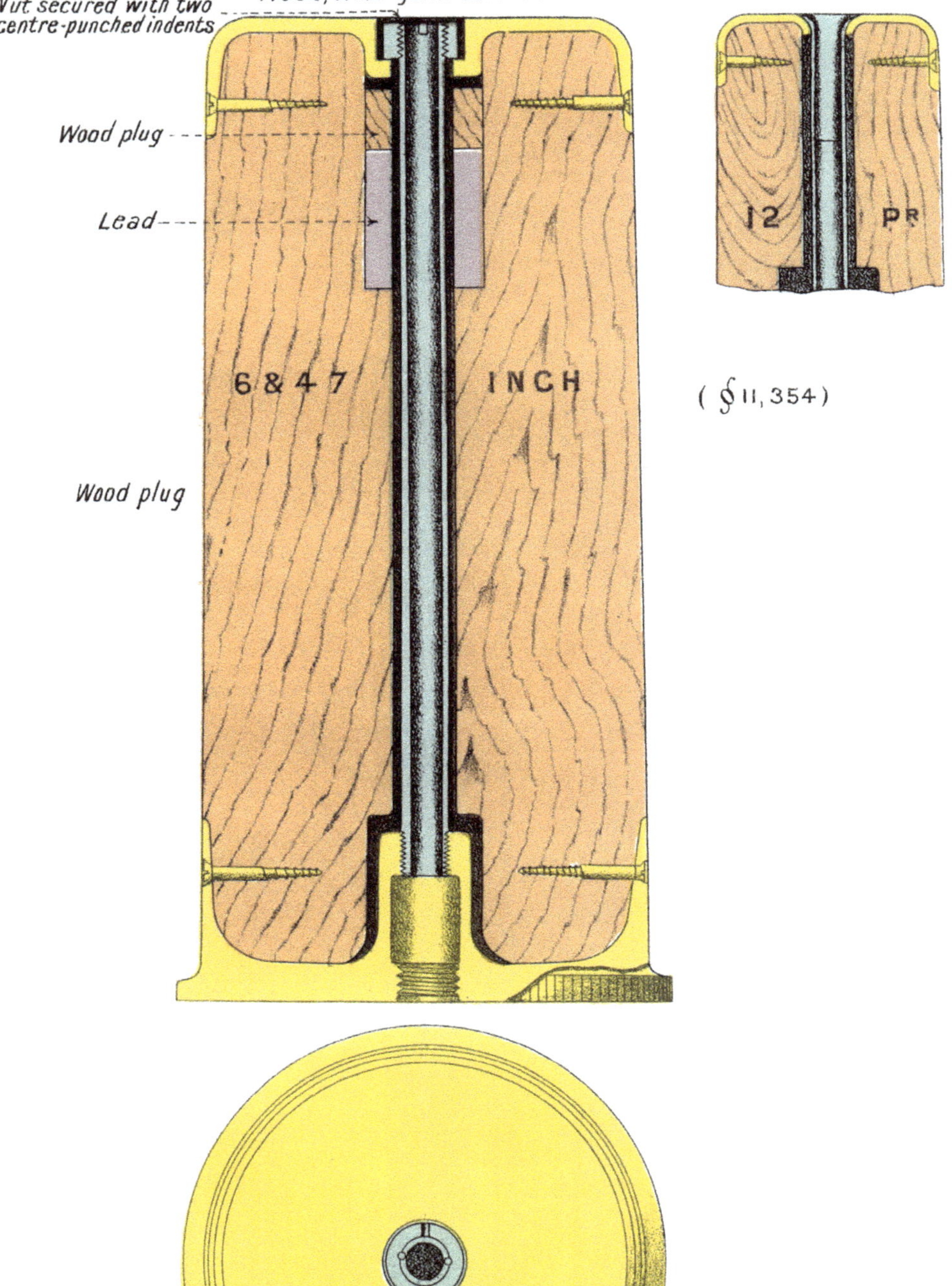

3567.

Malby & Sons, Lith.

To face page 495. Plate CI.

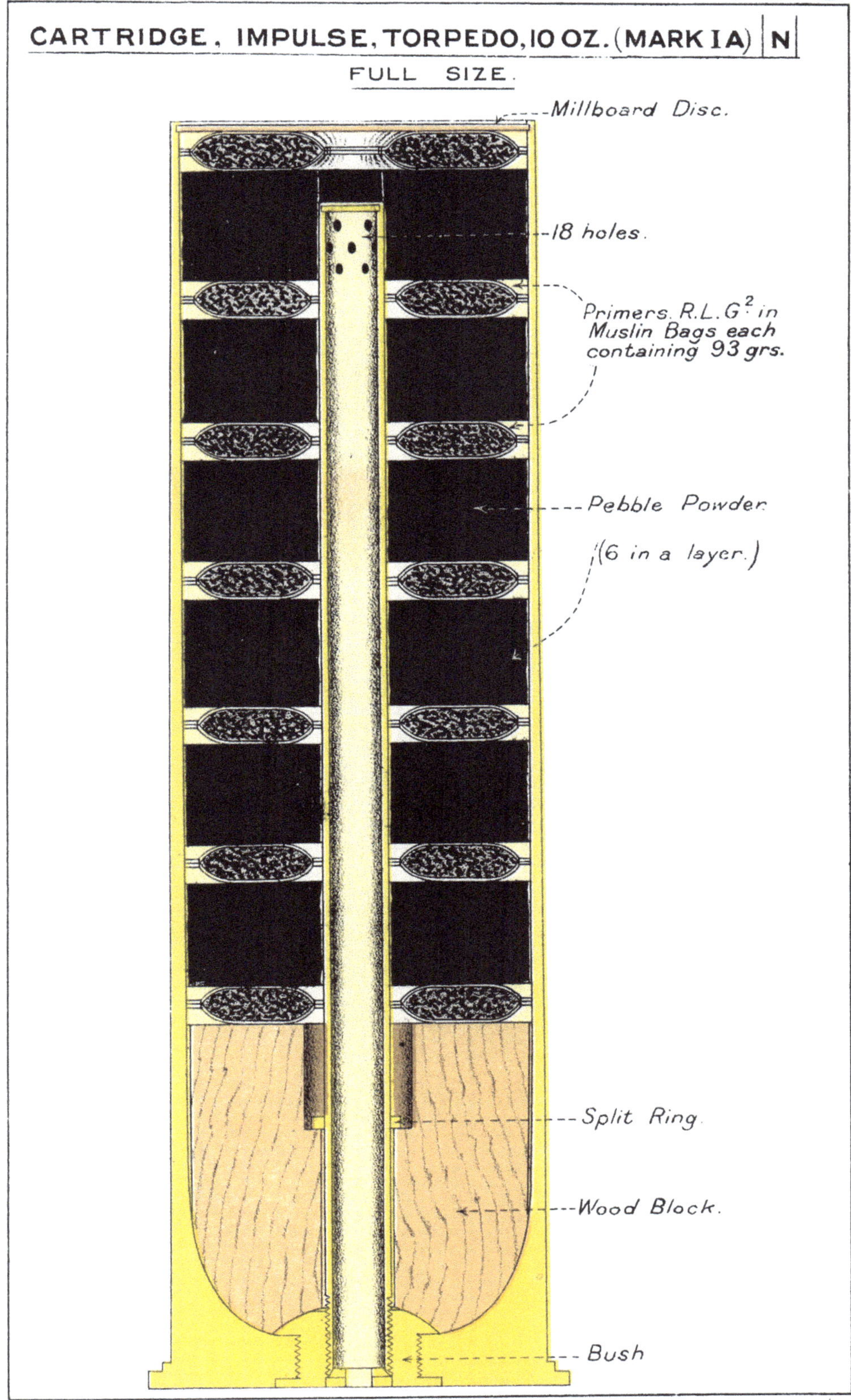

3567

Malby & Sons, Lith.

To face page 516.

Plate CII.

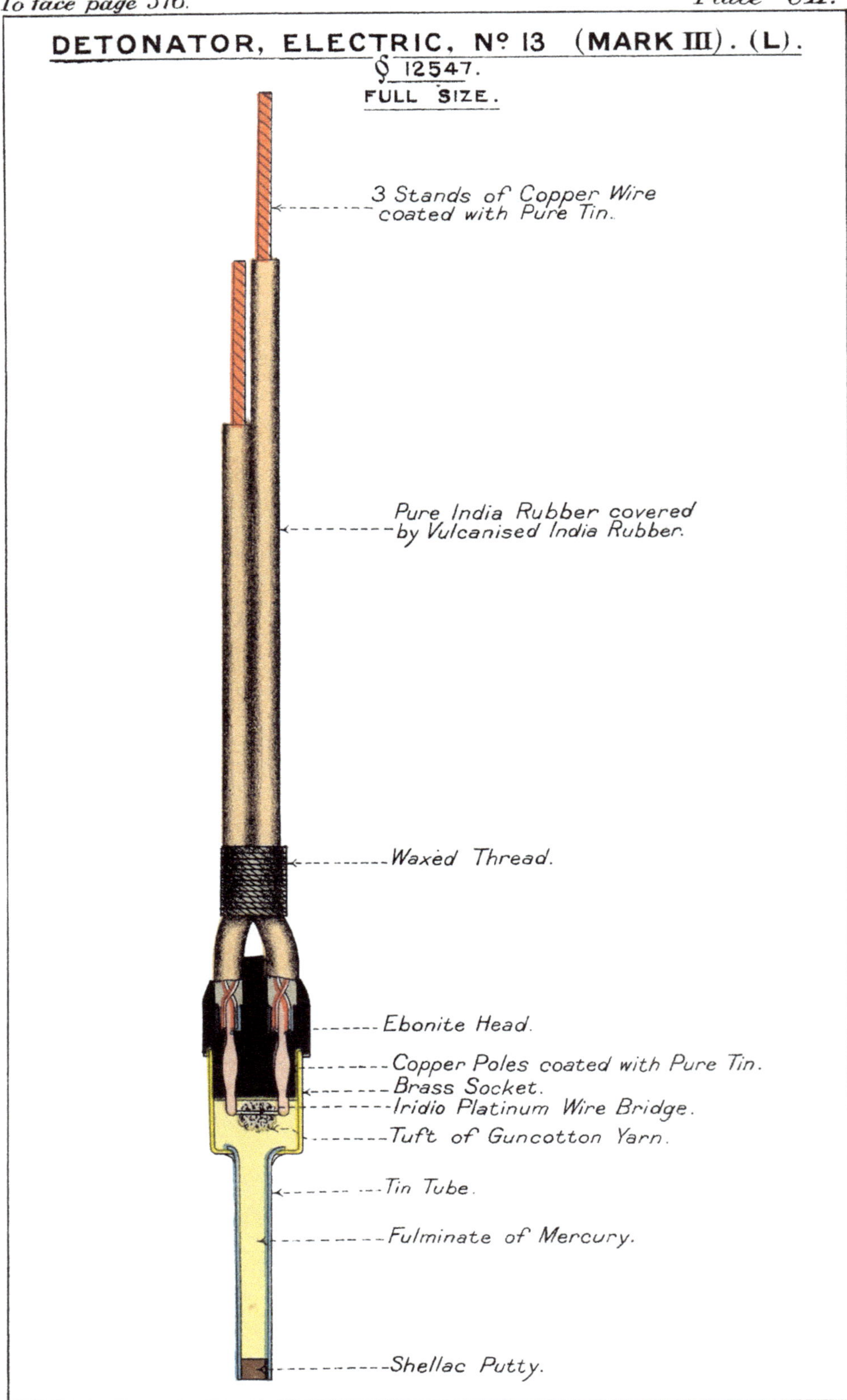

3567. Malby & Sons. Lith

DETONATOR, ELECTRIC, Nº 9. (MARK IV). (N).

FULL SIZE.

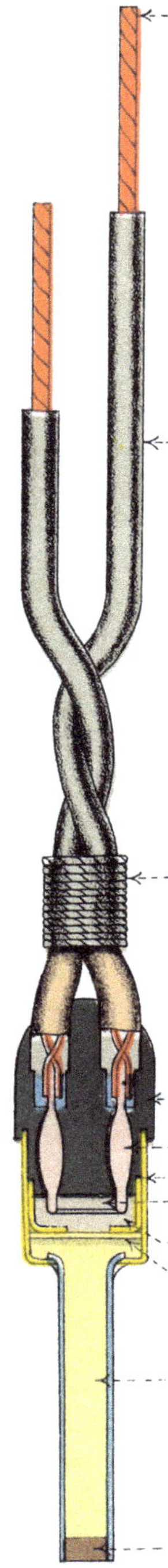

 Malby & Sons Ltd

FUZE, ELECTRIC, No. 14, MARK III

LAND SERVICE.

SCALE 1/1.

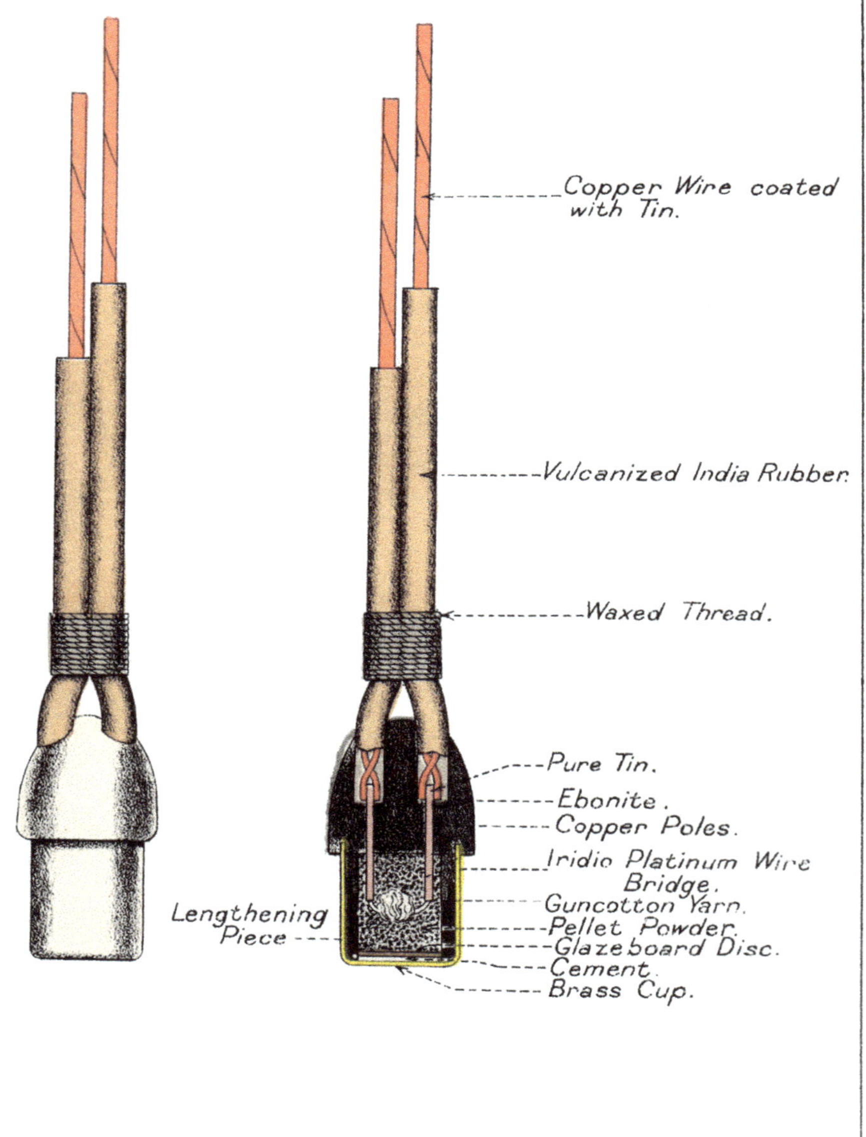

Malby & Sons, Lith.

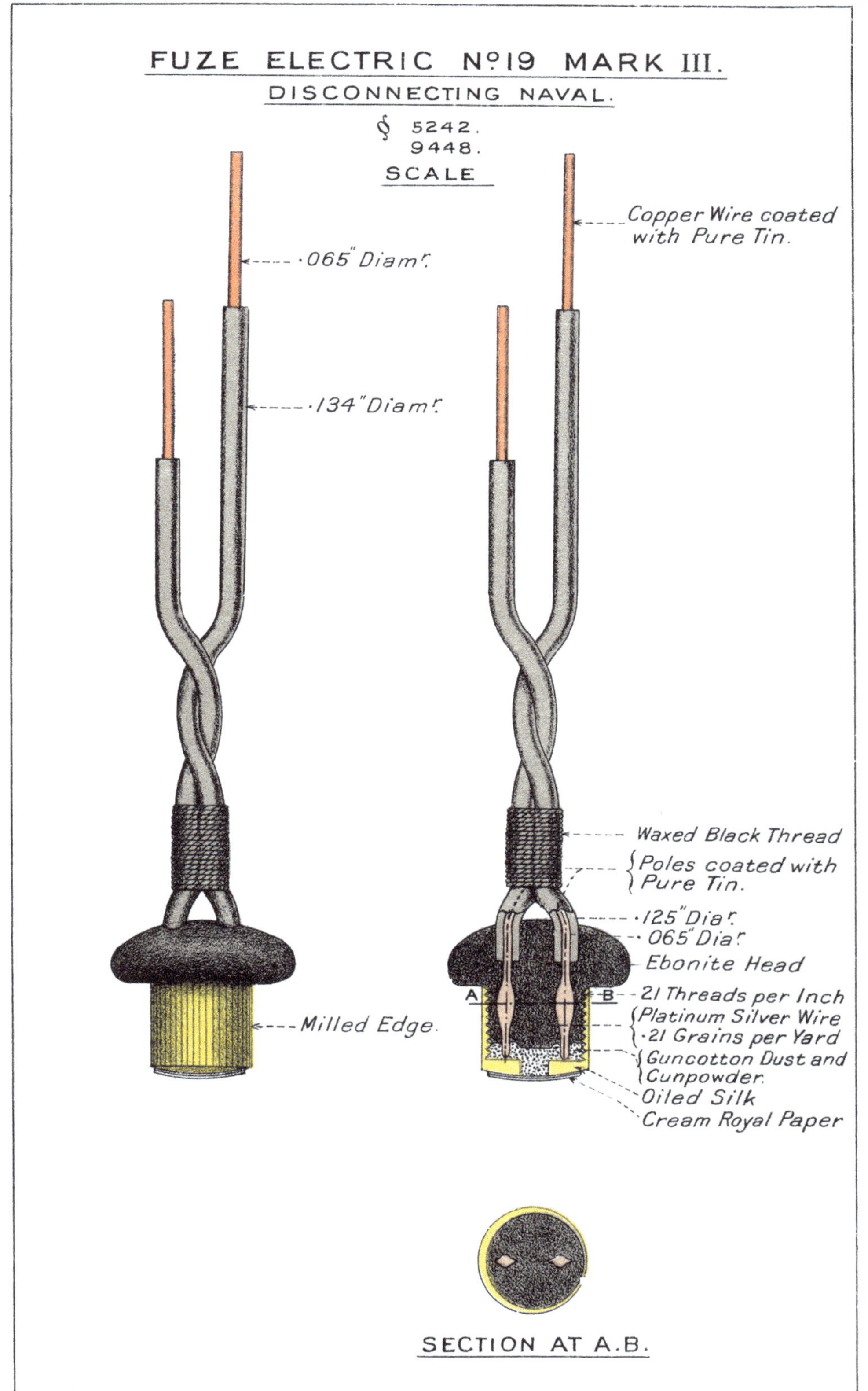

 Malby & Sons, Lith.

CARTRIDGE, AIMING RIFLE, 1 INCH, ELECTRIC, MARK V. M. |C|.

FULL SIZE.

DISTINGUISHING MARK ON BOX.

Paper Patch, two turns, lubricated with pure Beeswax to 8 from Base.

Bullet. (Lead and Tin)

White Card Wad about ·048 thick.

Felt Wad lubricated with Beeswax.

Wad Grease Proof Card.

R.F.G.^{2} Powder about 400 grains.

Guncotton Dust.

Card Wad

{Iridio
{Platinum Wire Bridge.

Brass Contact Pin.

PRIMER, MARK IV "K.N" CARTRIDGE.

SECTION

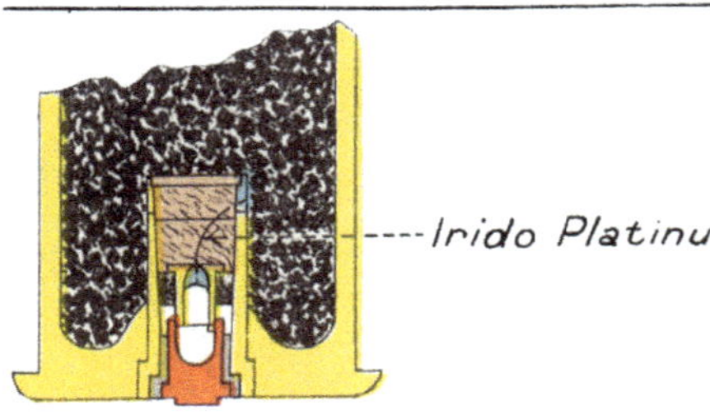

Malby & Sons, Lith.

CARTRIDGE, AIMING RIFLE, 1 INCH ELECTRIC, CORDITE, MARK I C

SCALE = 1/1

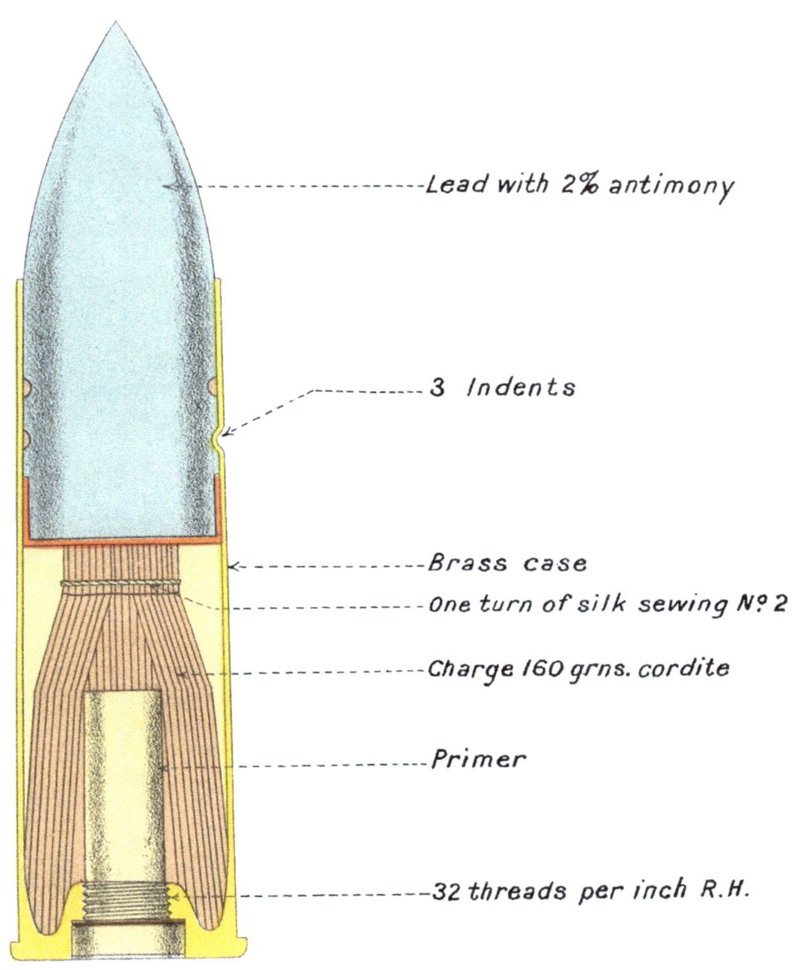

PRIMER.

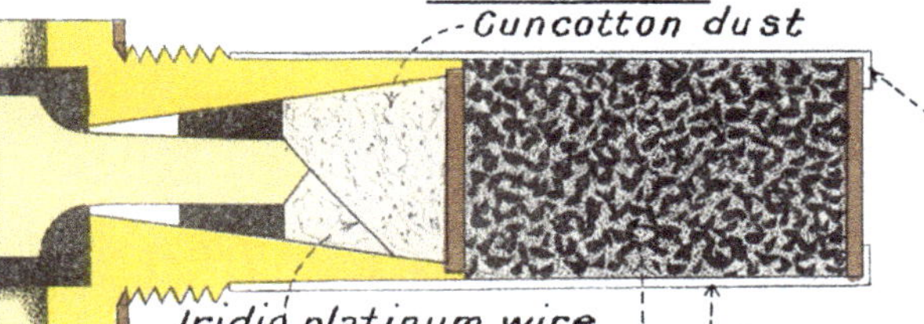

CARTRIDGE, AIMING, RIFLE, 1-INCH PERCUSSION MARK I |L|.

FULL SIZE.

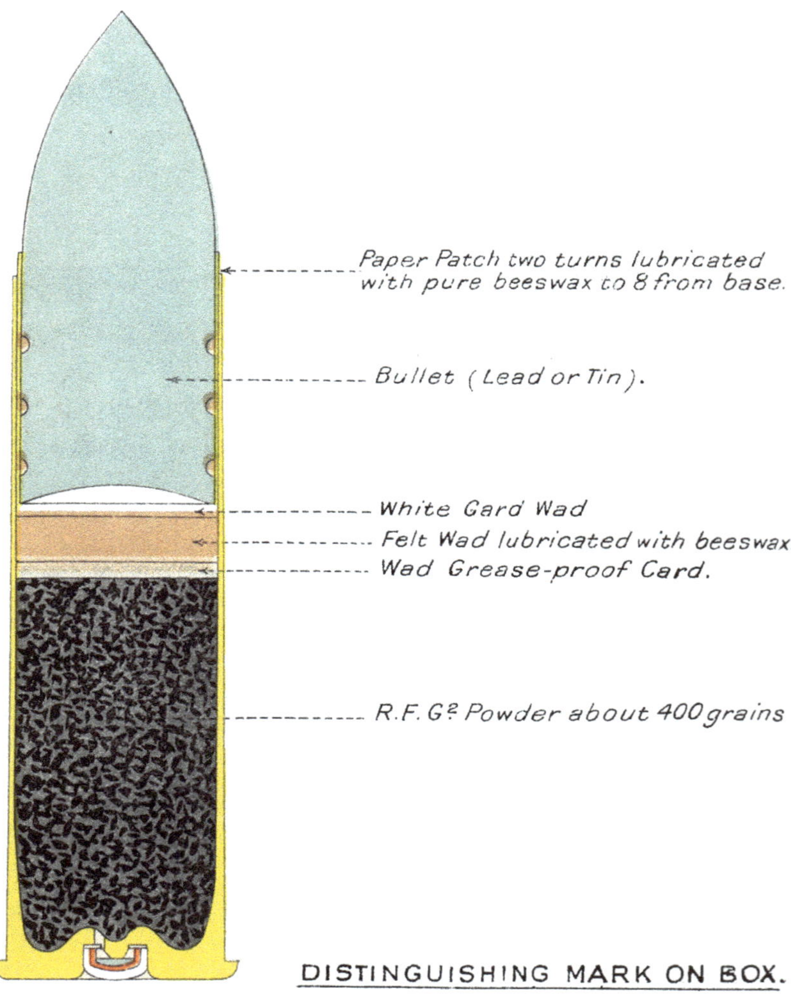

SECTION.

DISTINGUISHING MARK ON BOX.

Malby & Sons, Lith.

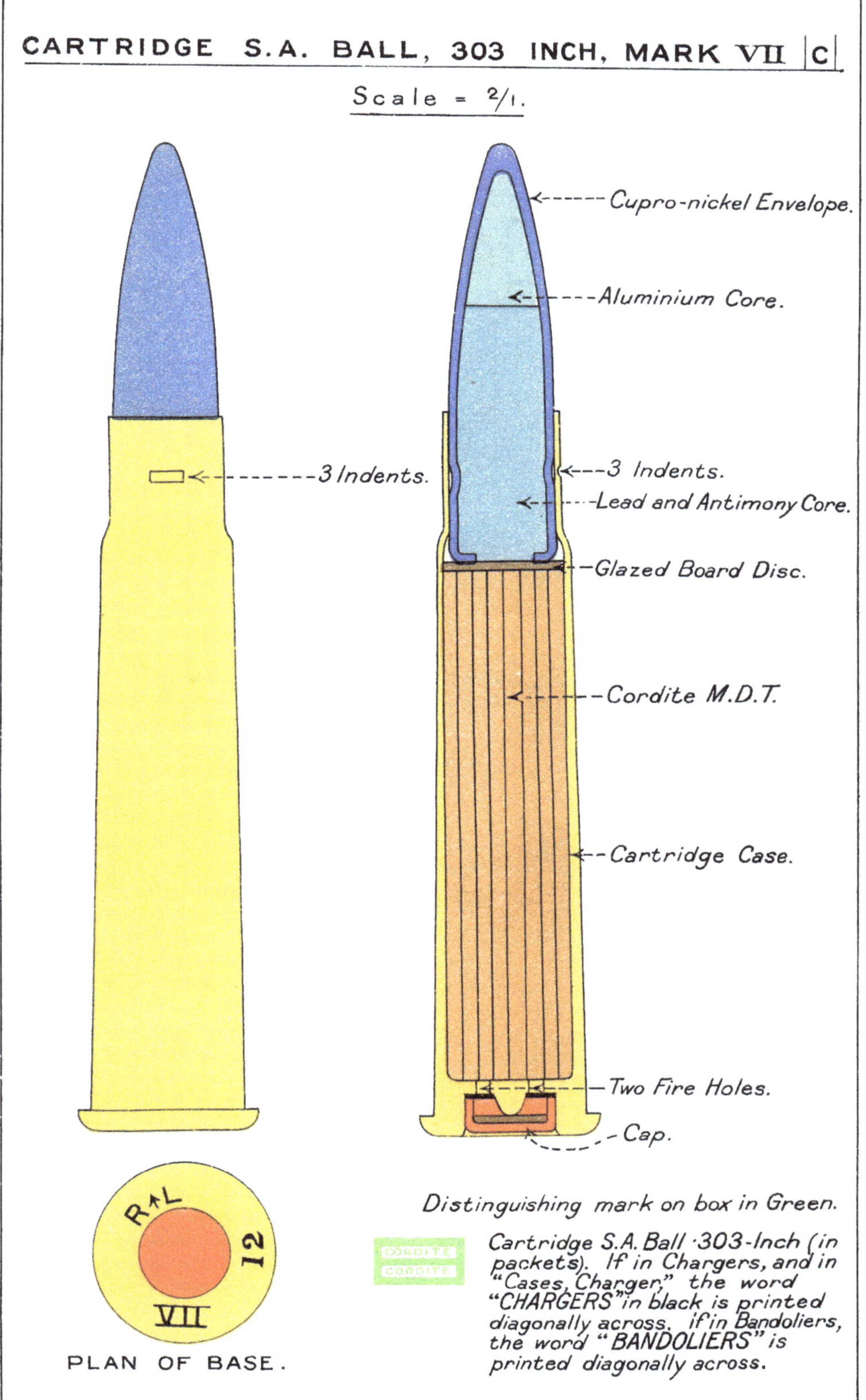

3567 Malby&Sons, Lith.

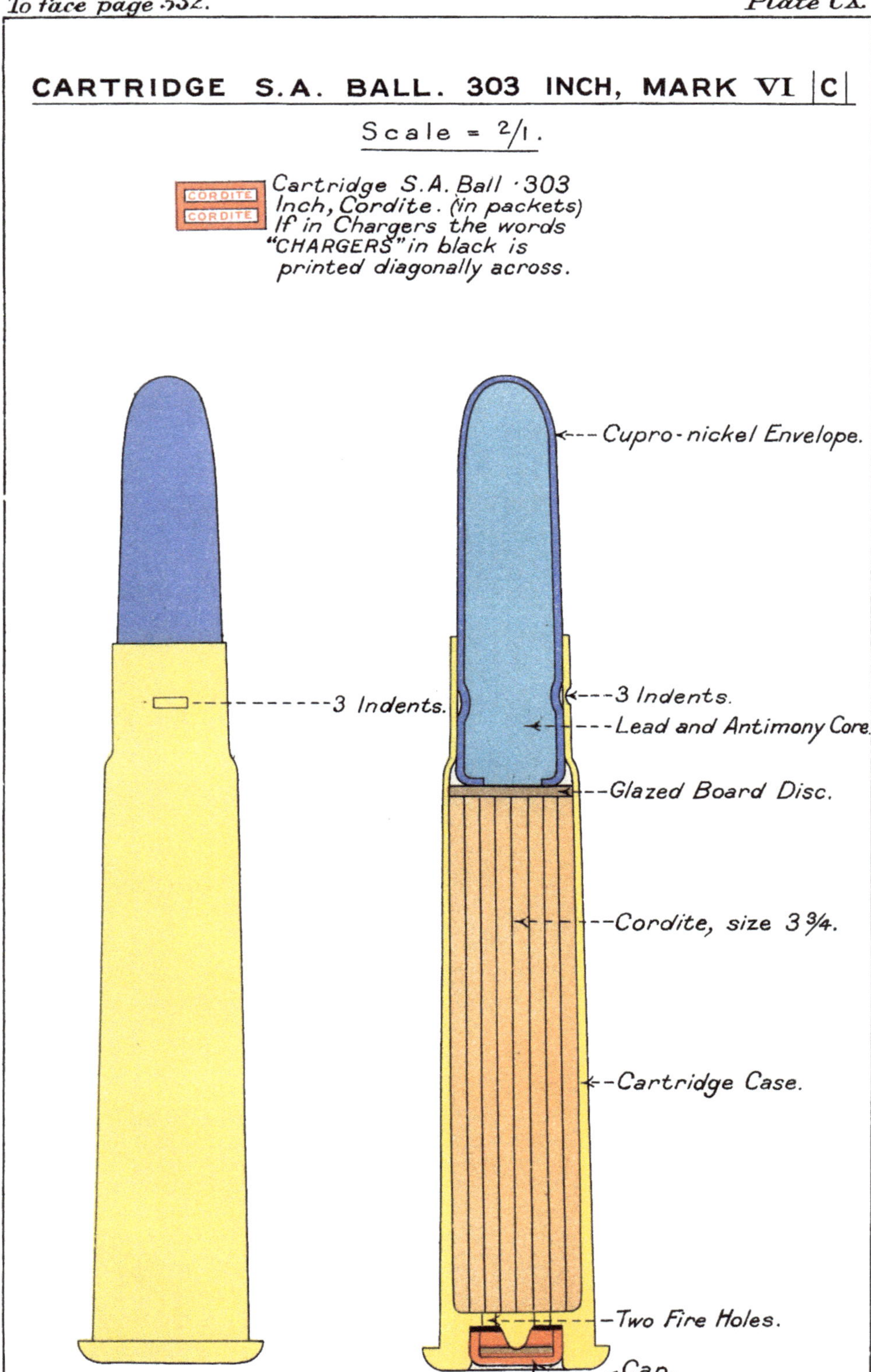

3567.

Malby & Sons, Lith.

CARTRIDGE S.A. BLANK ·303 INCH WITHOUT BULLET MARK V|C|.

SOLID CASE.

FULL SIZE.

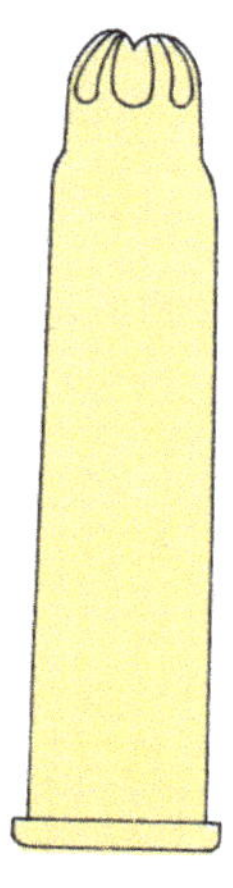

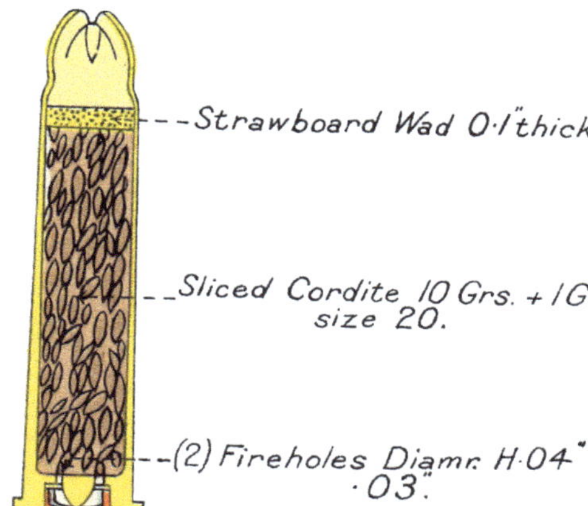

CARTRIDGE, S.A. BALL, PISTOL, WEBLEY, MARK II.

SCALE = 5/1.

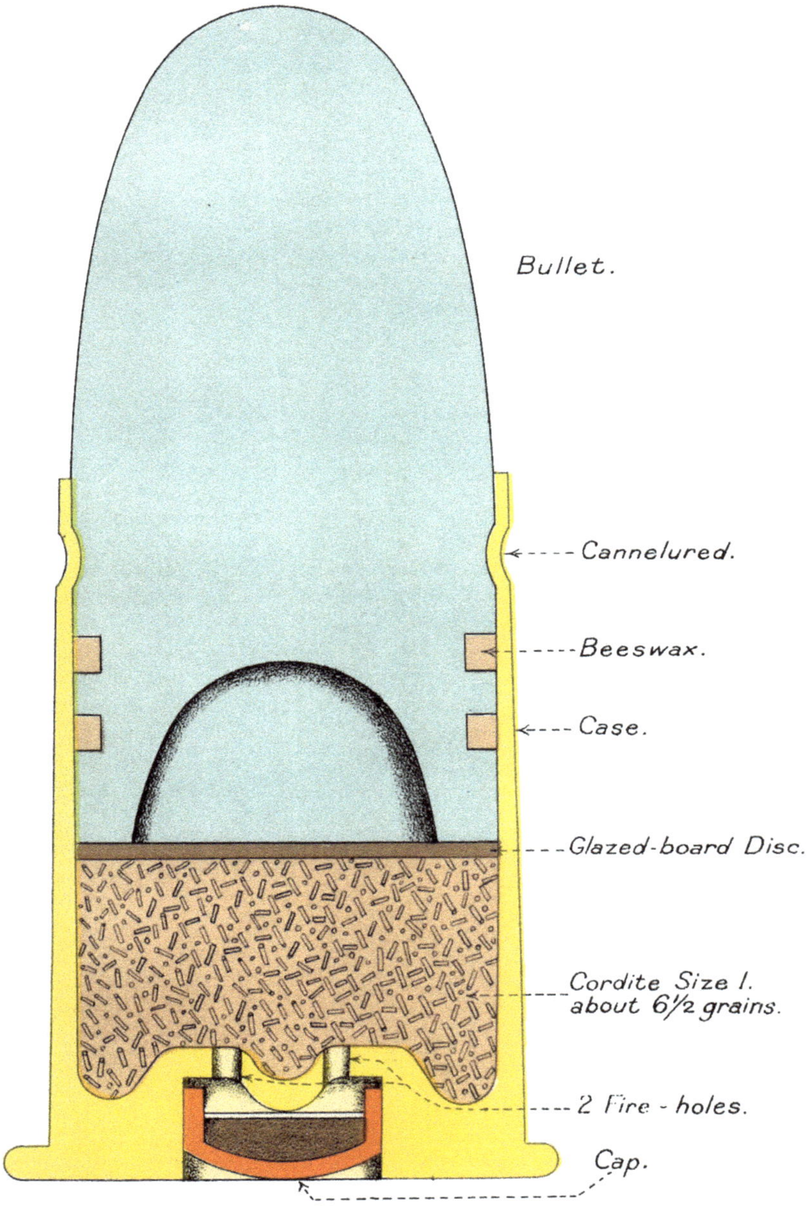

Bullet.
Cannelured.
Beeswax.
Case.
Glazed-board Disc.
Cordite Size I. about 6½ grains.
2 Fire-holes.
Cap.

3567. Malby & Sons, Lith.

CARTRIDGE, S.A., BALL, PISTOL, SELF LOADING, WEBLEY AND SCOTT, 0·455 INCH, MARK I|N|.

SCALE 4/1

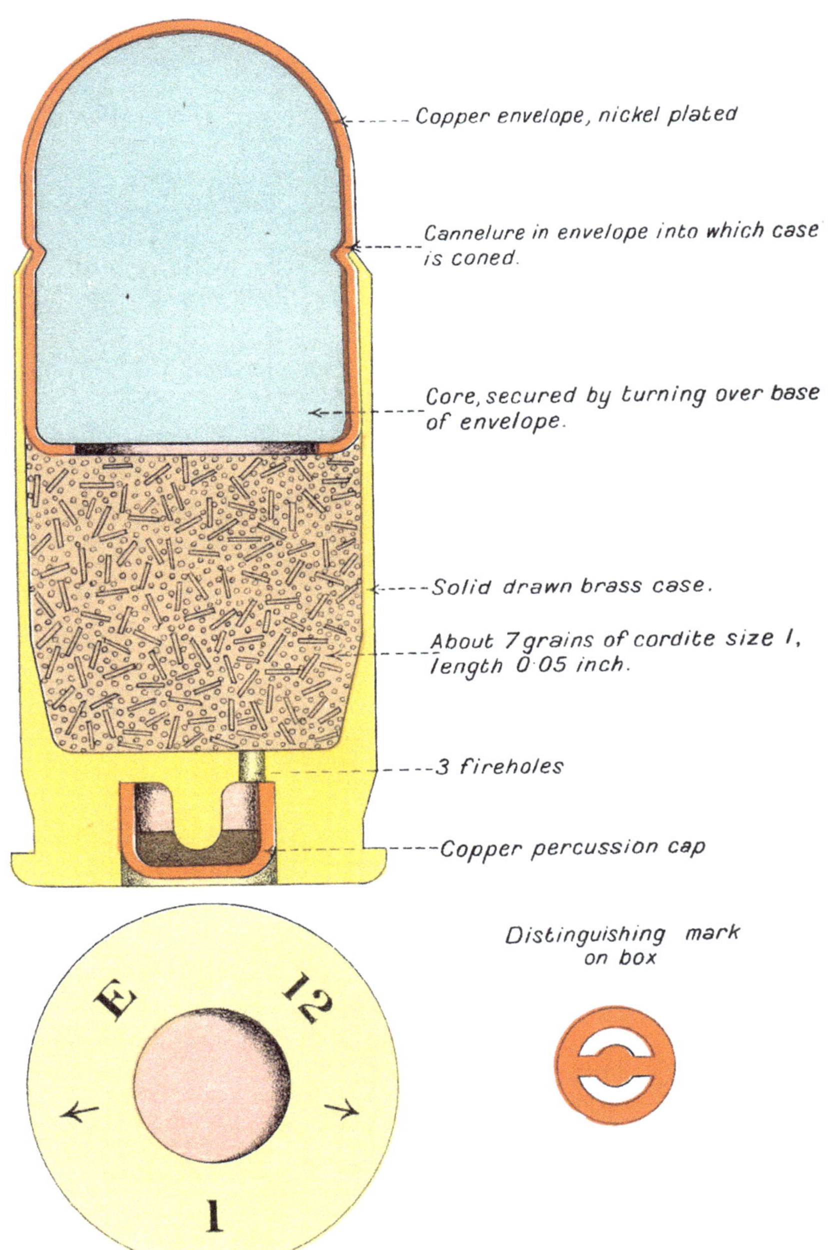

DISTINGUISHING MARKS FOR S.A. AMMUNITION BOXES.

Cartridge, Aiming Rifle. I. Inch Electric. Cordite.

Cartridge, Aiming Rifle. I. Inch Electric. Powder.

Cartridge Aiming Rifle. I. Inch Percussion.

Cartridge Aiming Tube. C.F.

Cartridge Aiming Tube. R.F. No. I.

Cartridge M.G. Ball ·45 Inch.

Cartridge M.G. Ball ·45 Inch. Cordite.

Cartridge, S.A. Dummy Drill Magazine Rifle.

Cartridge S.A. Ball ·303. Inch Mark VII.

Cartridge S.A. Ball ·303. Inch except Mark VII.

NOTE:- If in Chargers, and in "Cases, Charger," the word "CHARGERS" in black is printed diagonally across. If in Bandoliers, the word "BANDOLIERS" is printed diagonally across.

Cartridge S.A. Ball Pistol Webley, Cordite.

Cartridge S.A. Ball Pistol self-loading, Webley and Scott.

Cartridge S.A. Ball, 303 Inch Cordite, Short Range Pratice.

Cartridge S.A. Ball M.H. Carbine, Powder.

Cartridge M.H. Carbine Ball Cordite.

Cartridge S.A. Ball. M.H. Rifle Powder.

Cartridge Ball M.H. Chamber, Cordite.

Malby & Sons, Lith.

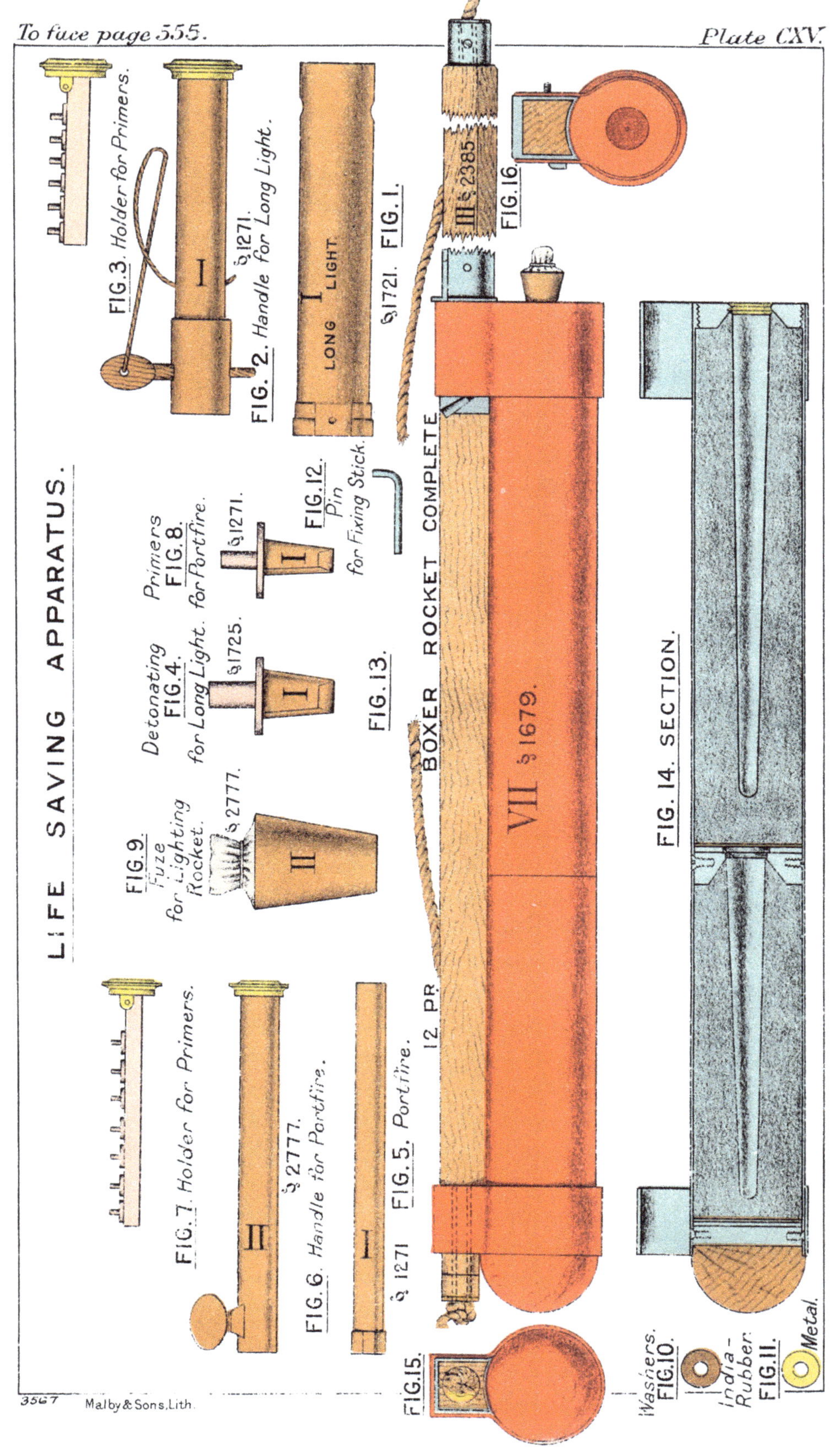

3567 Malby & Sons, Lith.

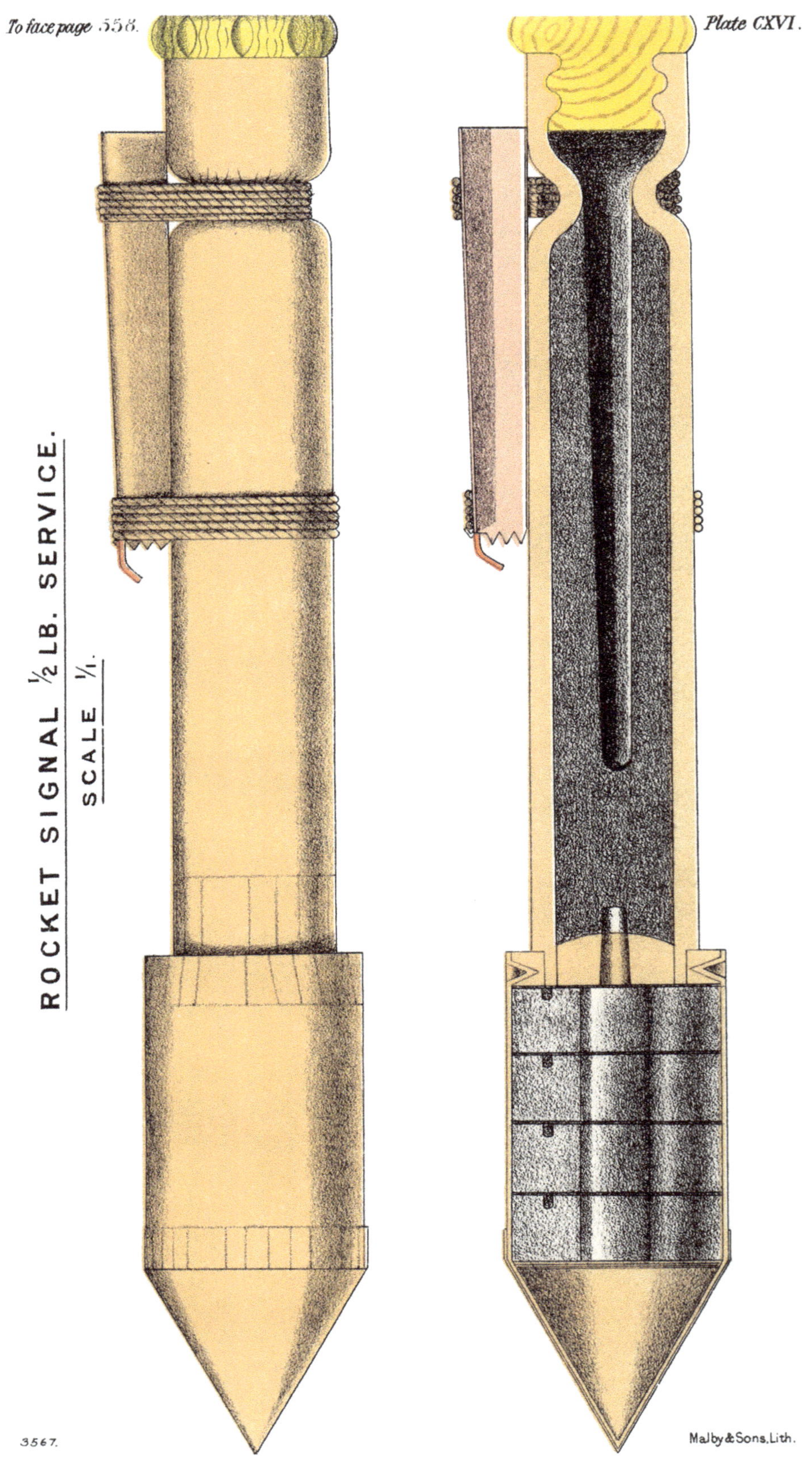

ROCKET SIGNAL ½ LB. SERVICE.

SCALE 1/1.

3567.

Malby & Sons, Lith.

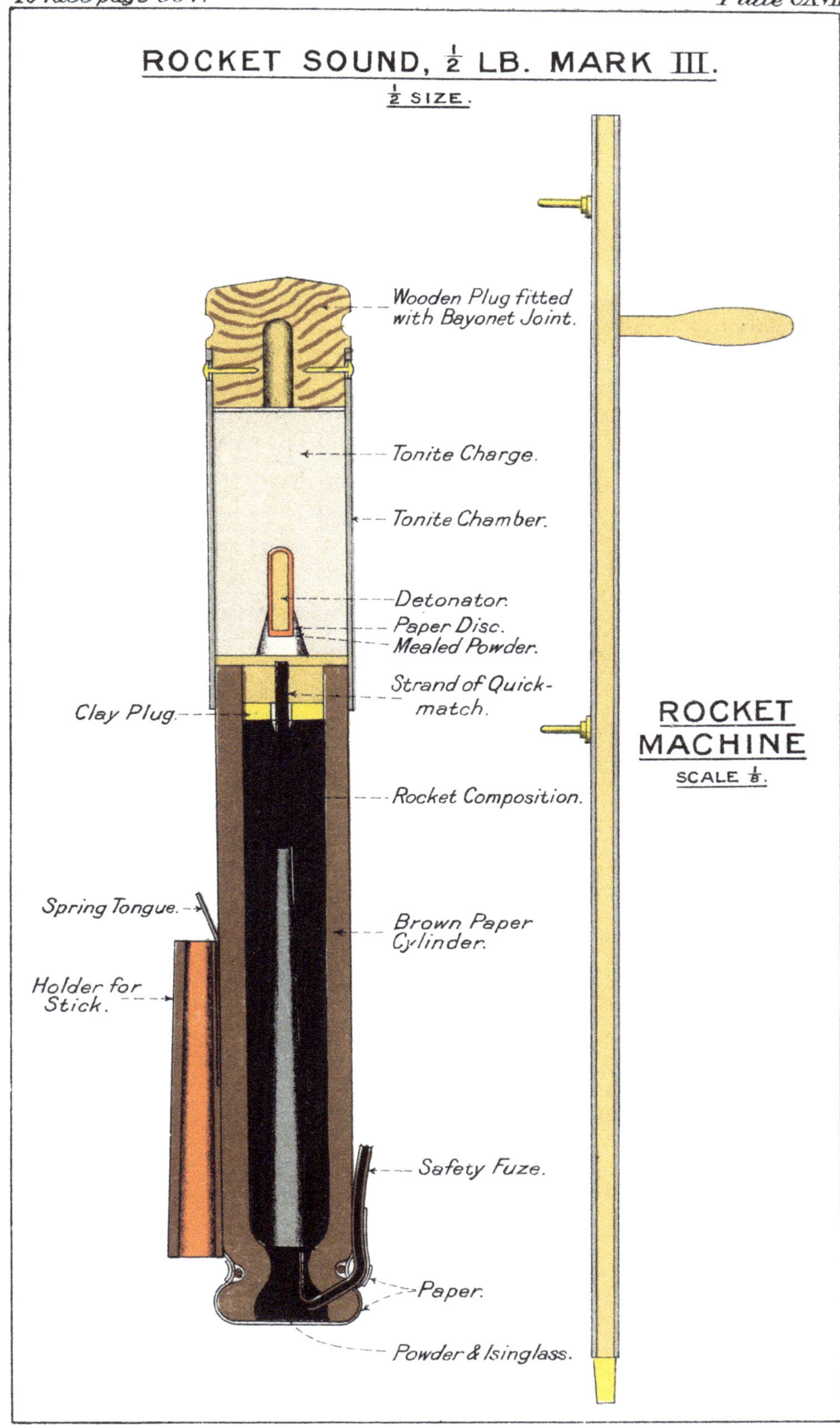
ROCKET SOUND, ½ LB. MARK III.
½ SIZE.
Wooden Plug fitted with Bayonet Joint.
Tonite Charge.
Tonite Chamber.
Detonator.
Paper Disc.
Mealed Powder.
Strand of Quick-match.
Clay Plug.
Rocket Composition.
Spring Tongue.
Brown Paper Cylinder.
Holder for Stick.
Safety Fuze.
Paper.
Powder & Isinglass.
ROCKET MACHINE
SCALE ⅛.

LIST OF TABLES.

LIST OF ABBREVIATIONS.

A.C.	Adjusted Charge.
A.O.	Army Order.
A.O.S.	Regulations for Army Ordnance Services.
A.P.	Armour-Piercing.
A.P.C	Armour-Piercing, Capped.
B.L.	Breech Loading.
B.L.C.	Breech Loading Converted.
| C. |	Common to both Land and Naval Service.
C.E.	Composition Exploding (or Tetryl).
C.S.	Cast Steel.
C.P.	Common Pointed.
C.P.C.	Common Pointed, Capped.
C.I.	Cast Iron.
C.F.	Central Fire.
D.T.	Day Tracer.
F.S.	Field Service or Forged Steel.
F.G.	Fine Grain.
F. of M.	Fulminate of Mercury.
G.G.	Gardner-Gatling.
G.S.	General Service.
fs.	Feet per Second.
H.A.	High Angle.
H.E.	High Explosive.
L.S. or | L. |	Land Service.
L.G.	Large Grain.
M.G.	Machine Gun.
M.H.	Martini-Henry.
M.L.	Metal Lined.
M.D.	Modified Cordite.
M.D.T.	Modified Tubular Cordite.
M.D.S.	Modified Strip Cordite.
M.V.	Muzzle Velocity.
N.S. or | N. |	Naval Service.
N.T.	Night Tracer.
Q.F.	Quick Firing.
Q.F.C.	Quick Firing Converted.
R.F.G.	Rifled Fine Grain.
R.F.	Rim Fire.
R.L.	Royal Laboratory.
S.F. and A.W.	Small Flange and Above Water.
S.V.	Specific Volume.
S.V.	Striking Velocity.
S.F.G.	Sulphurless Fine Grain.
T.N.T.	Tri-nitro-toluene (or Trotyl).
V.S.	Vent Sealing.
W.I.	Wrought Iron.
§	Paragraph in List of Changes.

PREFACE TO TENTH EDITION.

The matter in this edition is CONFIDENTIAL ; attention is directed to the note on the title page.

An attempt has been made in this edition to make the Treatise on Ammunition as complete as possible.

Chapters have been added describing the materials used in the manufacture of ammunition, and technical manufacturing terms are explained.

Most of the text has been rewritten and brought up to date, and the amount of matter increased.

The number of plates has been increased from 94 to 117, and a large number of new wood-cuts added.

Most of the tables and descriptions of packages and "Implements Ammunition" have been placed at the end of the chapters to which they refer, thus avoiding a mass of tables at the end of the Treatise.

This edition has been prepared by the Instructional Staff (Ammunition Branch) of the Ordnance College, who desire to acknowledge with thanks the assistance given by the staffs of the Chief Inspector, Woolwich, and of the Superintendent, Royal Laboratory.

Ammunition Branch,
Ordnance College.
1st *August*, 1914.

7

The information given in this Treatise is *not* to be regarded as an authority for demand, issue, or condemnation of Stores.

(N.B.—This edition is corrected up to 1st August, 1914. As some typographical errors may have occurred in publication, it is requested that, should any be discovered, they may at once be pointed out in writing to the Secretary, War Office, S.W.)

TREATISE ON AMMUNITION.

(NOTE.—*For the regulations as to handling, making up, inspection, packing, and marking of ammunition*, see *Regulations for Army Ordnance Services, Part II. For the regulations affecting the classification and storage of explosives*, see *Regulations for Magazines, &c.*)

CHAPTER I.—GENERAL REMARKS ON EXPLOSIVES.

COMPOSITION; SERVICE EXPLOSIVES; CHARACTERISTICS.

(*For further details, see Text Book of Service Explosives.*)

Explosives can be made by forming intimate *mechanical mixtures* of the combustible and the oxygen supplying substance or by causing oxygen to enter into *chemical combination* with such elements as nitrogen, hydrogen, carbon, &c.

Gunpowder is an example of the mixture type of explosive, guncotton is a chemical compound, and cordite is a mixture of two chemical compounds.

Most of the chemical compound type of explosives are nitro compounds. In these, oxygen is held in unstable chemical combination with nitrogen. The application of heat upsets the stability, and the oxygen set free forms new combinations, usually gaseous, with the other constituents of the explosive.

When this chemical action takes place great heat is evolved and considerable expansion of the products of combustion takes place. The quicker the action and the greater the heat evolved the more violent the explosion.

The explosives commonly met with in the service are gunpowder, cordite, guncotton, lyddite, picric powder, trinitrotoluene, fulminate of mercury and composition exploding.

Explosives may be divided into three classes :—

(1) *Propellants.*
(2) *Disruptives or High Explosives.*
(3) *Detonators.*

(1) This class includes gunpowder, cordite, and all explosives used for the firing of projectiles.

These explosives are intended to act very much more slowly than those in Classes 2 and 3.

The inertia of the projectile has to be overcome gradually, otherwise the projectile or gun may be strained or fractured. The rate of burning can be to some extent regulated by altering the composition, moisture, density, size of grain or cord, &c.

Explosion is very rapid combustion.—Combustion is the oxidation of such elements as carbon, hydrogen, &c., contained in the substance ; if the oxidising process is moderately slow it is called *burning*, if very rapid, explosion.

The rate of combustion of an explosive is affected principally by *pressure* and the amount of area exposed ; the action takes place on the surface only.

The higher the pressure the more rapid the explosion.

For example, a cordite cartridge ignited in the open burns away rapidly but harmlessly, but if it is confined so that the gaseous products formed cause the pressure to rise, explosion takes place.

The greater the area exposed the more rapid is the explosion.

A cartridge made up of small size cordite burns faster than one of equal weight made up of large size sticks, because the igniting flame can act on a greater surface or larger number of particles at the same time.

(2) This class includes guncotton, dynamite, lyddite, &c. These explosives are intended *to detonate,* though most of them can be made to simply *explode*, or even *burn*, but they cannot be used as propellants, whereas some of the latter may be made to act as disruptives. (*See* Cordite.)

Detonation is a different action from explosion, for whereas in the latter combustion is confined to the surface and takes place layer by layer, in the case of detonation the action takes place instantaneously throughout the whole mass of the substance.

Detonation is a wave action, and the detonating wave will pass through space and through solids.

An explosive, when detonated, is considerably more powerful than when simply exploded.

The usual means of securing detonation of a high explosive is by igniting a small quantity of fulminate of mercury in contact with it ; but a blow, friction, or a sudden rise in temperature, will in some cases cause detonation.

The speed of detonation varies with the different disruptives and with the medium in which they lie, but it is approximately 4 miles a second.

Some disruptives can only be completely detonated when compressed and strongly confined.

(3) The commonest service example of this class is fulminate of mercury.

The action of this class is even more rapid than that of the last, and though the heat evolved at the moment of detonation is great, it lasts for so short a time that there is no incendiary effect. This class is used for initiating explosion or detonation in other explosives. It is thought that the brusque explosive blow given by fulminates to surrounding explosives causes direct chemical disintegration, and so starts a detonating wave.

CHAPTER II.—GUNPOWDER.

Composition and Different Forms met with; Manufacture; Properties; Uses; Classification; Packing; Storage.

Composition.

Gunpowder is an intimate mechanical mixture of its constituents, namely, saltpetre, charcoal, and, usually, sulphur.

The gunpowders which are met with are :—

"Black," "Brown," and "Sulphurless."

The composition of black powder is approximately :—

Saltpetre	75 parts.
Charcoal	15 ,,
Sulphur	10 ,,
Moisture	·7 to 1·5 per cent.

Brown powders contain more saltpetre, more charcoal and more moisture, but much less sulphur; they burn more slowly than black powders, and are more difficult to ignite; a cartridge of brown prism powder requires a primer of several black prisms.

Sulphurless gunpowder was introduced because the sulphur in the powder igniters of cartridges was considered to have a bad effect on the stability of the cordite.

It contains about 70 per cent. of saltpetre, and 30 per cent. of charcoal, with a small percentage of moisture. No more is to be made.

Prism or moulded powders are made in regular hexagonal prisms, having a hole through the centre of each, to give even burning. No more will be made.

Pebble powder is in the form of cubes.

Mealed powder is largely used in the manufacture of ammunition: it is gunpowder reduced to an impalpable dust: it is used in fuze compositions, quickmatch, friction tubes, &c.

Other powders commonly met with are :—

Large Grain (L.G.), Rifle Fine Grain (R.F.G.^{2}).

"P" Mixture is a mixture of pebble and fine grain powders.

The charcoal is the only constituent that can be altered in any way to obtain the various rates of burning of gunpowder required for different purposes.

The nature of the wood and the number of hours it is burnt largely affect the resulting charcoal, and subsequently the rate of burning of the finished gunpowder.

Manufacture.

The system of manufacture as carried out at Waltham Abbey is briefly as follows :—

Woods used for charcoal.—The woods used are alder, willow or dogwood, according to the nature of gunpowder required. Dogwood is used for R.F.G.^{2} and is charred for 8 hours. The charcoal is ground fine.

Weighing and mixing.—The ground charcoal, refined saltpetre, and finely ground sulphur are weighed in the correct proportions and thoroughly mixed in a revolving drum. After mixing, this " green charge," as it is termed, is incorporated.

The incorporating mill consists of a flat iron or stone bed on which the iron or stone " runners " revolve on their edges.

The green charge is put on the bed and watered and then milled for some hours. The charge is continually turned over and distributed by " ploughs," which scrape the charge from the edges of the bed to the centre as the runners revolve.

Breaking down.—At the end of the milling operation the charge is mostly in the form of " cake " with some dust. This is taken to the breaking-down machine, which, by means of a series of rollers, reduces the mill cake to the form of " meal " or fine dust.

Pressing.—The meal is pressed between a number of metal plates in a hydraulic press, so forming a number of slabs of " press cake " about ½ inch thick.

This press cake is broken up by hand and put through a *granulating machine*, which crushes it between rollers forming grains which pass over sieves in the machine.

The required size of grain is obtained by passing the granulated powder over sieves of the necessary mesh.

The "grain" then goes through the following finishing operations :—

Dusting in revolving barrel sieves to remove dust.

Glazing in revolving wooden barrels with or without the addition of graphite.

Stoving.—The grain is laid on trays and dried in a steam-heated stove at 100° F.

The time taken depends on the nature of powder and the amount of moisture it contains.

Finishing.—The grain is put in the finishing reel to remove any dust and give a finished glaze.

The finishing reels are skeleton wood reels covered with canvas.

Blending is the process by which a uniform " brand " (corresponding to the " Lot " with cordite) is obtained.

After blending the powder is ready for packing in barrels and issue.

Pebble, which is a cut powder, is made by placing the *press cake* in a machine which cuts it into a number of strips ; these strips are then cut across, so forming a number of cubes, which are " drummed " to take off the sharp corners.

Prism powders were pressed to shape in special machines.

Properties.

Gunpowder ignites at between 550° F. and 600° F. ; it will deteriorate at lower temperatures than that which causes ignition. Sulphur begins to vaporize at about 240° F., and water is given up at a still lower temperature.

Liability of gunpowder to explode by a blow or by friction.—Gunpowder can be exploded by a blow or friction, especially when a thin film of powder-dust is nipped between hard surfaces; a property made use of in the D.A.I. Fuze No. 13.

A glancing blow, even between wooden surfaces, may cause explosion.

This fact forms a clue to most of the precautions laid down to be taken when dealing with gunpowder. When they are carefully observed the risk of an explosion is small.

Gunpowder stands climate well. Pure saltpetre is not deliquescent to any great extent; the glaze and density of powder aid in preserving it from damp, besides affecting the rate of burning; still, in damp climates, or in a damp magazine, it is necessary to keep powder in air-tight metal cases. Also moulded powders, especially brown powders, contain a comparatively high percentage of water, hence these powders must be kept in air-tight metal cases to prevent the loss of moisture, which would otherwise occur in hot and dry magazines.

Keeping gunpowder in a damp atmosphere will tend to separate the ingredients by wetting a portion of the saltpetre, which on a change of atmospheric conditions may effloresce, *i.e.*, be carried to the surface of the grains, and appears in the form of white specks.

The dryness and proper ventilation of powder magazines are therefore points of great importance.

Gunpowder must not be allowed to remain in direct contact with metal, as if there is the least damp present the saltpetre in the powder will corrode the metal, thus damaging both. Again, paper, if of any thickness, is found to cause powder to deteriorate, as it is apt to absorb moisture; this causes part of the saltpetre to be absorbed into the paper, which becomes more or less impregnated with it at the expense of the powder; this difficulty can be overcome by varnishing the paper.

Uses.

Gunpowder is used :—

(*a*) For blank charges.
(*b*) For charges when firing paper shot.
(*c*) For the igniters of most cordite cartridges.
(*d*) For the bursting charges of shell.
(*e*) In fuzes and tubes.
(*f*) In various combustible compositions, such as quickmatch, priming and cap compositions, &c.
(*g*) Occasionally to fire projectiles from some old guns.

Classification. (*See* Regulations for A.O. Services, Part II.)

The main classes of powders are :—

Class I	.. ..	Service.
Class II	.. ..	Blank.
Class III	.. ..	Shell.

Service powders are of the highest grade. Examples of use:—Igniters of cordite cartridges, charges for firing paper shot.

Blank powders are mainly used for blank charges; to give other examples of their use: Blank L.G. may be used for the bursting charge of Armour-Piercing Shell; Blank F.G. may be used with pebble powder for filling certain shell.

Shell powders are mainly used for filling shell; the following table enumerates the various natures of powders employed for this purpose.

TABLE No. 1.

Powders used for Shell Filling.

Nature of Powder.	Nature of Shell.	Remarks.
"P" Mixture	Common, & Common-pointed Shell 12-pr. and upwards.	("P" Mixture is Shell P. and Shell F.G.)
Shell Q.F. or "P" Mixture	Common, & Common-pointed Shell 12-pr. to 6-inch.	(Shell Q.F. is Q.F.[1] reduced to Class III; it is mixed with Shell F.G.)
Blank, L.G., new or converted from Black powder	Common Pointed Shell, B.L. 4-inch H., Mark II.	
Blank, L.G. {New or Converted L.G., R.L.G., R.L.G.,[2] and R.L.G.[4]	Armour Piercing Shell. Uncapped Armour Piercing Shell in the Land Service.	("P" Mixture was formerly used for Uncapped Armour Piercing Shell.)
F.G., R.F.G., R.F.G.[2] Shell F.G. *New* Blank F.G. *New*	Shrapnel Shell.	Q.F. Shell, F.G. and S.F.G.[2] may be used for filling Field Shrapnel in L.S.
R.F.G.[2]	Star Shell.	
Q.F. Shell, F.G.	Q.F. 3 and 6-pr. Shell.	
Shell, Q.F. F.G. converted	Q.F. 3-pr. Shell.	Naval Service only.
Shell L.G.	Practice Shell (iron).	

PACKING.

Gunpowder is usually packed in barrels, the powder being placed in a rubber bag inside the barrel; the barrels hold about 100 lbs.

Powders are, however, sometimes packed in serge bags, containing about 10 lbs., which are then packed in barrels or metal-lined cases for issue to Royal Artillery and in brass cases for issue to H.M. ships.

Moulded (*i.e.*, prism) powders are packed in zinc-lined wooden rectangular cases, known as "Case, powder, 100 lbs."

The zinc is japanned on the inside.

Packages for Gunpowder.

Barrels, powder.—The present pattern is Mark III. There are four sizes: whole, half, quarter, and eighth.

Powder barrels consist of three parts, viz.:—

1. Staves.
2. Heads.
3. Hoops.

The most protuberent part of the barrel is known as the "bilge," and the centre of the bilge is known as the "pitch."

Between the bilge and the end of the barrel is the "quarter."

The extreme end is known as the "chime."

To distinguish one end of the barrel from the other, that at which the barrel is opened is known as the "top end," the other as the "back end." The top end may be known by having the staves bevelled off close to the chime to facilitate heading.

All powder barrels have four copper hoops; the remaining hoops are of ash or hazel. If the copper hoops are not tinned, they must be coated with black tar varnish before being used for high explosives.

Powder barrels are used to transport and store cubical and granulated powders when not made up into cartridges; they are also occasionally used to hold cannon cartridges, and would then be fitted with brown paper linings.

Barrels of gunpowder are not to be rolled.

The whole barrel takes about 100 lbs. of powder, the half 50 lbs. and the quarter 25 lbs.

A rubber bag is always used inside the barrel, except as above.

Case, Powder, 100 *lbs.* | *C* | .

☞ *Case, powder,* 100 *lbs. Mark V,* is for prism powders in bulk; it is a wooden box lined with zinc, which is japanned on the inside; the lid is secured by brass screws; the zinc lining has a recess for luting, and is removable; the case is "khaki" colour for N.S. and "service colour" for L.S. Dimensions, 2 ft. 5¾ in. by 1 ft. 3½ in. by 9¼ in. deep. There is a circular recess in the lid and side, which was for the sealing label. These cases, when used to contain picric acid, have the lining removed and a calico bag substituted.

Cases, Powder, Pentagon.

Cases, Powder, Pentagon, are made in two sizes, whole and half.

The whole case is a five-sided prism made of sheet brass, with the exception of the top and the fittings, which are of cast brass. The body is formed by bending the brass sheet up into the required prism, and rivetting and soldering the junction, and then fitting into the open ends of the prism the top and bottom.

In the top of the case is a circular opening, closed with a bung and a lid with a hinged bolt. There are slots in the rim of the lid, and corresponding projections on the neck of the case; the lid will only open when the slots and projections are in a corresponding position. There is a lever and spanner issued for opening the case.

The half Pentagon is literally a half of the case described above, but has a smaller opening.

The Pentagon Cases were originally issued for B.L. powder cartridges up to 6 in.; they are still employed for packing loose gunpowder in 10 lb. and 15 lb. serge bags, for issue to H.M. ships.

To close the case, the groove round the neck is filled with Mark III luting; the bung is then inserted and tapped down with a wooden mallet, the recess round the bung filled in with luting, smoothed down and wiped clean. The projection on the point of the lever is then inserted in the ring on the curved bolt, and the lever turned from left to right. The spanner is then used to turn the screw and jamb the curved bolt.

To open the case, unscrew the set screw free from the curved bolt with the spanner, turn the lid from right to left with the brass lever until the inclines are clear, then raise the lid and lift out the bung.

Storage.

Gunpowder is stored in Magazines in Group I, Division I^B.

Wetted gunpowder, which is treated as an explosive, is stored in Explosive Stores in Group II, Division II.

CHAPTER III.—NITROGLYCERINE.

Manufacture ; Properties ; Uses.

Nitroglycerine is formed by the action of strong nitric acid on glycerine.

On a manufacturing scale mixed acids, nitric and sulphuric, are used.

The sulphuric acid is to absorb water in the nitric acid, and also any that may be formed during the chemical action of nitration.

Manufacture.

Briefly the process as carried out at Waltham Abbey is as follows :—

The nitration is effected in a large lead vessel in which charges of 1,500 lbs. of glycerine are dealt with. The mixed acids are run in from the bottom, and are cooled by cooling coils in the vessel ; the glycerine charge is then sprayed into the acids by air pressure while the temperature is kept down to about 20° C. (68° F.) ; the contents are kept agitated by jets of air ; the glycerine is immediately converted into nitroglycerine. When the whole charge of glycerine has been nitrated, the nitroglycerine is allowed to separate to the top, where it floats on the acids ; the level of the liquid is then raised by running in waste acid at the bottom, and the nitroglycerine runs off to the prewash tank.

The first two washings are in plain water ; the third in a solution of carbonate of soda. The nitroglycerine is then run down a gutter to the *washing house,* where it is washed in three changes of soda solution.

It is then given a final washing in plain water to rid it of excess of soda.

The nitroglycerine is then run into a tank in the *mixing house.* This tank has a layer of sponges in it, and as the nitroglycerine filters through, the sponges absorb the water adhering.

The nitroglycerine is now finished and is run straight from these filter tanks on to the guncotton for the manufacture of cordite.

The manufacture of nitroglycerine is attended with a certain amount of danger.

Properties.

Nitroglycerine is a clear, oily liquid, nearly colourless ; it has an intensely sweet taste, but no smell ; it is actively poisonous, and when absorbed through the skin, or when the vapour is inhaled, it causes headache and giddiness.

It detonates if given a sharp blow or heated to 257° C.

It freezes at about 45° F. ; thawing it is a dangerous operation.

The direct rays of the sun decompose it and an even more unstable substance is formed, which will detonate if left.

It is too dangerous to use by itself, and is never kept stored in bulk.

Uses.

Nitroglycerine is one of the components of cordite, and is also used in ballistite, dynamite, blasting gelatine and many other explosives.

Directly after manufacture the nitroglycerine is incorporated with other substances, themselves explosives in many cases, and forms safe explosives, some suitable for use as propellants, others for blasting purposes, &c. (*see* Miscellaneous Explosives, page 35).

CHAPTER IV.—GUNCOTTON.

Manufacture; Properties; Uses; Issue; Packing; Storage.

Guncotton is cotton, or cellulose, that has been nitrated by the action of strong nitric acid.

The exact chemical composition of the resulting nitrated cotton depends on the conditions under which nitration took place, *i.e.*, temperature, strength of the acids, &c.

Guncotton is the tri-nitrate of cellulose; the lower nitrates which are more easily formed are known as *soluble or collodion cottons.*

After nitration the cotton increases in weight (about 70 per cent.) and feels harsh, but retains its original colour and appearance.

Manufacture.

It is not possible to obtain on a manufacturing scale pure cotton or nitric acid sufficiently strong for the purpose.

Cotton waste is used and sulphuric acid is mixed with the nitric to absorb the water formed during the process of nitration.

There is no danger attending the manufacture of guncotton.

The method of manufacture carried out at Waltham Abbey is briefly as follows:—

The best cotton waste which has been freed from grease by boiling with benzine and alkalies, and which should contain not more than ·5 per cent. oily matters, is employed. It is picked over to remove foreign substances, passed through a teasing machine, and dried. It is then steeped in the mixed acids, which are prepared to the following analysis:—

70 per cent. sulphuric acid.
21 ,, ,, nitric acid.
9 ,, ,, water.

The cotton (now converted into guncotton) is boiled, to rid it of the free acid, which, if left in, even in minute quantities, would not only be fatal to the keeping qualities of the guncotton, but would make it dangerous to store, owing to its liability to decomposition and spontaneous combustion. After the boiling process, it is pulped and washed, and then alkaline matter is added, except when required for the manufacture of cordite, and in this state it is moulded and pressed into any required form, the pressure employed being 6 tons to the square inch. When finished, it should contain from ·8 to 1·5 per cent. of alkaline matter. In this state it is ordinary wet guncotton as found in the Service in slabs.

The guncotton for making dry primers is subsequently dried till it contains only 2 per cent. of water.

The cylinders of guncotton for making cordite are moulded under a pressure of 34 lbs. to the square inch and then dried.

The operations in the process are as follows :—

Picking.—The cotton is picked over by hand to remove string, wire, wood, &c.

Teasing.—The fibres are separated by means of wooden rollers fitted with iron teeth.

Willowing.—Cotton dust, small wood and grit is removed by this process, which is similar to the teasing process, but there are no teeth fitted to the rollers of the machine.

Drying.—The cotton is passed on an endless band backwards and forwards in a large steam-jacketed stove.

Dipping or Nitrating.—The dipping is done in circular earthenware pans, about 3 feet 6 inches in diameter, and 1 foot deep, which are arranged in rows in a large room. The mixed acids (nitric and sulphuric) are run into the pans from the bottom. About 20 lbs. of cotton are then immersed in the acid in each pan ; nitration takes 2½ hours.

Drawing off the acids.—The acids are run off from the bottom of the pans ; at the same time water is run into the pans from the top, until the acids have been entirely replaced by water.

Removing acids from the guncotton.—The guncotton (as the cotton has now become) is allowed to drain.

It is then boiled in seven changes of water, and each washing is displaced by cold water.

Pulping.—The guncotton is reduced to a wet pulp in a " beater," in which a stream of water carries it between two sets of knives—one set on a revolving iron roller, and the other set on a bedplate.

Pulping reduces the length of the fibres and produces a form of guncotton convenient for moulding.

Removal of mechanical impurities.—The pulp, carried along by water, passes over the poles of a magnet through grit traps, wood traps and over blankets ; this removes wood, sand, small pieces of metal, &c.

Poaching.—The poachers are large tanks provided with paddle wheels to keep the finely divided guncotton agitated in a large volume of water. The pulp receives three washings in this machine ; it is allowed to settle after each washing, and the washing water removed by a skimmer ; this removes impurities lighter than the guncotton.

The poaching process also blends a number of different batches, so as to get a uniform product.

Addition of alkali.—Lime water and enough carbonate of soda to precipitate the lime from the lime water as carbonate of lime is added to all guncotton intended to pass into the service in the form of compressed guncotton. No alkali is added to guncotton intended for the manufacture of cordite.

Washing.—The pulp is given a further washing in the poachers to remove any excess of carbonate of soda.

Moulding.—The pulp, still in a very wet state, is run into moulds, and a great deal of the water drawn off by a vacuum engine.

Pressing (a) For cordite.—The pulp, if intended to be formed into guncotton cylinders for making cordite, is subjected to a pressure of 34 lbs. to the square inch, which is sufficient to give it such a consistency that it may be handled without crumbling; it is subsequently dried.

Pressing (b) For guncotton slabs and primers.—The pulp if intended to form guncotton slabs or primers is wrung in centrifugals to rid it of excess of water. This partly dried pulp is weighed out for each slab or primer as required, moulded, and then compressed at about 6 tons to the square inch. Slabs are then dipped in a solution of carbolic acid. In this state it is ordinary wet guncotton (20 parts of water to 100 parts of dry guncotton).

If intended to pass into the Service as dry primers it is dried till 1·5 to 2 per cent. of water remains.

The finished primers or slabs have the *test number* stamped on them with a special red ink so that the date and details of manufacture can be traced.

General Properties.

Guncotton is very susceptible to explosion by influence. This property was utilised in "countermining" where a large charge of guncotton was fired in a mine field, thus firing all the mines in the vicinity.

Guncotton can be very easily detonated by means of fulminate of mercury and when detonated it is instantaneous in its action, there is therefore no need to tamp a charge as must be done with gunpowder for demolitions.

It is nearly as powerful as dynamite, when detonated, but requires a stronger detonator.

The chief products of the explosion of guncotton are aqueous vapour, carbon monoxide, carbon dioxide and nitrogen, there is therefore no smoke. The carbon monoxide is highly dangerous in confined spaces such as mines, as it is an active poison and very inflammable; for this reason guncotton is unsuitable for mining purposes in the L.S., except under special conditions, when the miners are not required to enter the ground again for some time after the charge has been fired.

A nitrate added to guncotton acts as an oxygen supplier, so that no carbon monoxide can be formed. With this addition (*see* Tonite) guncotton can be used for mining.

Guncotton is unaffected by water and can therefore be stored and used in the wet state, it is then the safest of explosives.

Used for demolitions it is about four times as powerful as gunpowder, but it only has a local shattering effect, whereas gunpowder has a general lifting action.

Properties of Dry Guncotton.

Guncotton when dry *may* ignite at 277° F.; it *must* ignite at 400° F. Its mean igniting point may be taken at about 340° F. The low ignition point of guncotton is taken advantage of by using loosely twisted strands as an igniter in some cordite cartridges. Guncotton is also used in the form of dust, mixed with mealed powder, as priming for electric tubes, &c.

When a small quantity of dry guncotton, perfectly unconfined, is ignited by a flame or by a heated body, it burns very rapidly with a bright yellow flame; if, however, the cotton is ignited when confined in a strong case, even of wood, the action is very different; for it then explodes with great violence, and the strength of the explosion will depend upon the strength of the case; to develop it fully a strong iron case is required.

A large quantity ignited in the open burns for a time till the temperature of the surrounding mass is raised sufficiently high, explosion or detonation then takes place.

It is easily fired by percussion or friction, but the detonation is usually very local and confined to the part struck.

Compressed dry guncotton is easily detonated, even when unconfined, by the action of various explosives, of which fulminate of mercury is found the most suitable.

About 5 grains of fulminate of mercury enclosed in a tube and ignited in close contact with compressed dry guncotton produces detonation, but to guard against any chance of failure, Service detonators contain from 12 to 77 grains. A description of Service detonators will be found in Chapter XX.

To produce detonation, the guncotton must be in a compressed form; loose flocculent guncotton cannot be detonated; it is merely scattered by the detonator.

Ordinary dry guncotton contains about 2 per cent. of moisture. If the moisture amounts to 5 per cent. the guncotton can be easily ignited by a flame, but probably will not detonate with a small detonator.

Uses of Dry Guncotton.

Dry guncotton is used chiefly for the manufacture of cordite, and in the form of primers to detonate charges of wet guncotton.

It is also used in the form of yarn, in the igniters of certain obsolescent cordite cartridges, and in some primers and tubes, &c., also as dust in the priming composition of certain tubes, &c.

There is a special 2-oz. primer used in the ½ lb. Sound Rocket, Mark II.

Forms in which dry guncotton is used.

Dry guncotton primers are generally cylindrical; the perforations, which are slightly tapered, are intended to take the detonators. They are packed in brass, copper, or tin cylinders; a cylinder filled with primers is known as a "charge priming" (except when the primers are to be used separately); the cylinders are packed in wooden boxes for convenience of transport and storage.

The nomenclature of the primers at present in the service is shown below; for detailed particulars *see* table and Plate I.

Guncotton, dry, primers, Field, 1 oz. | C |.
,, ,, ,, Rocket, 2 oz. | C |.
,, ,, ,, Warhead, 1 oz. | N |.
,, ,, ,, Warhead, 2 oz. | N |.
,, ,, ,, Mine, 9 oz. | N | and | C |.
,, ,, ,, Mine, 1½ lbs. | L |.

Guncotton, dry, primers, Field, 1 oz. are conical in form and are coated on the exterior only with a hard gelatine surface by dipping them in acetone to keep them dry and to prevent the surface crumbling ; they are intended for all Field Service with the 15 oz. field slab.

Early issues of the 1 oz. primer were not coned and were paraffined on the surface.

They were known as primers " H."

Guncotton, dry, primers, Rocket, 2 *oz. Mark I* was originally known as primer " F." It is acetoned. It is used in the Mark II Sound Rocket. Early issues were paraffined.

Guncotton, dry, primers, Warhead, 1 *oz., Marks I and II* and *Guncotton, dry, primers, Warhead,* 2 *oz., Marks I and II* were originally known as torpedo primers ; they are not acetoned.

Guncotton, dry, primers, Mine, 9 *oz., Mark I,* | C | are used with the outrigger torpedo, and for submarine mining, &c. They have two perforations to take electric detonators.

Guncotton, dry, primers, Mine, 9 *oz., Marks II and III,* | N | are used in the Mines spherical. The Mark II forming the top primer of the charge priming, is perforated to take the special detonator.

Guncotton, dry, primers, Mine, 1½ *lb., Mark I,* | L | are used in the Countermine, which is now only used in the Colonies.

Each primer has two perforations to take the electric detonators, which fit it to the top primer of the priming charge.

Guncotton, dry, yarn, No. I, is used for filling the tube, torpedo, dropping gear, and a small tuft of it is wrapped round the bridges of *Fuze, Electric, No.* 14 and *Detonator No.* 13. The *No.* 4, which is about fifteen times as thick as the No. 1, is used to form the packing in the *Charge, priming, Warhead,* 1 *lb.* 1 *oz.,* for the igniters of some cordite cartridges, and a small tuft of it is wrapped round the bridge of the *Primer, Electric, Large.*

Issue and Packing of Dry Guncotton.

General remarks.—All dry guncotton primers, except those forming the priming charges for Warheads, are packed in tin cylinders, lacquered black internally and painted black externally. The internal lacquering is to prevent rusting of the tin by the action of moisture which might be deposited from the guncotton at high temperatures.

The lids are secured to the cylinders by bayonet joints, except that of the Charge, priming, 2¼ lbs., Mark III.

Guncotton yarn is packed in similar tin cylinders.

The tin cylinders containing the dry primers or yarn are packed in boxes painted service colour for Land, and khaki colour for Naval Service.

Cylinders.

Cylinder, guncotton, primer, Field, 1 *oz., Mark III* is made of tin, lacquered internally and painted black externally. It holds 10 " Primers, field, 1 oz." ; each primer is inserted into a small conical waterproof paper holder before being placed into the cylinder. A tape lifting becket, passing through the primers and attached to a glazed-board disc, enables them to be withdrawn. The lid of the cylinder is secured by means of a bayonet joint and is attached to the body by a piece of whipcord.

The Mark II cylinders differ from the above in dimensions; the 1 oz. primers packed therein were not fitted with paper cylinders and had no tape lifting becket.

Cylinder, guncotton, primer, Rocket, 2 oz., Mark I is made of tin, black lacquered internally and painted black externally, and the lid is fastened by bayonet joints. The cylinder holds nine 2-oz. primers. A doubled tape with the bight passed below the primers and attached underneath to a glazed board disc enables them to be withdrawn.

Cylinders, guncotton, yarn, are made of tin, black lacquered inside and painted black outside. It holds 2 ozs. of guncotton yarn, and the lid is secured by bayonet joints and attached to the body by a piece of whipcord.

Charges, Priming.

Cylinder, guncotton, charge, priming, Warhead, 6 oz., Mark III is made of copper, and holds six 1-oz. primers. It is screw-threaded on the exterior at one end, and recessed to take the torpedo detonator; the other end is closed by an indiarubber diaphragm between two metal plates, which can be pressed together by means of a screw. To prevent the indiarubber disc adhering to the sides of the copper cylinder the edges are covered with paper.

Cylinder, guncotton, charge, priming, Warhead, 1 lb. 1 oz., Mark III is made of tinned brass, and one end is threaded on the exterior to fit the pistol, and threaded internally to receive a plate for closing it. This plate has a recess in the centre to receive the detonator, closed by a screwed washer; after the eight primers are inserted a small grummet of guncotton yarn is placed on top and the closing plate screwed in; an indiarubber disc is placed above the recess for detonator and secured by the screwed washer.

Cylinder, guncotton, charge, priming, Mine, spherical, $2\frac{1}{4}$ *lbs, Mark I* | *N* | contains four 9-oz. primers, three being unperforated and one perforated to receive the detonator. One end of the cylinder is closed by a lid secured by bayonet joints, and tape band, the other is closed by a gunmetal fitting prepared internally to receive the detonator and screwed externally for attachment to the primer holder of the pistol. The cylinder is lacquered black internally and painted black externally. When filled it forms the "*Guncotton, dry charge, priming, Mine, spherical, Mark I* | *N* | ."

Cylinder, guncotton, charge, priming, Mine, $2\frac{1}{4}$ *lbs., Mark I* contains four 9-oz. primers, forming the "Guncotton, dry, charge, priming mine, $2\frac{1}{4}$ lbs., Mark I | L | ." The primers have two perforations, and are prevented from shaking about by a felt wad placed on top.

Cylinder, guncotton, charge, priming, $2\frac{1}{4}$ *lbs., Mark III* | *C* | contains four 9-oz. primers, each having two perforations. A varnished paper cylinder is used to fill up the space above the primers. Felt wads were used in the Mark II for this purpose. The cylinder is painted black and the lid is secured by a tape band shellaced on. When filled it forms the "*Guncotton, dry charge, priming,* $2\frac{1}{4}$ *lbs., Mark III* | *C* | ."

When firing the "*Guncotton, dry charge, priming, Mine,* $2\frac{1}{4}$ *lbs., Mark I* | *L* | " and the "*Guncotton, dry charge, priming,* $2\frac{1}{4}$ *lbs., Mark III* | *C* | " a mouth-piece is used to close the mouth of the cylinders. The mouth-piece consists of a disc of indiarubber held between two iron plates, four bolts pass through these and are used to tighten the plates and so cause the disc to expand. Electric detonators are used to fire the charges and two holes are bored through the plates and indiarubber disc to admit of the wires passing through. The tightening of the bolts causes the disc to expand and close the holes round the detonator wires in addition to closing the mouth of the cylinder.

Cylinder, guncotton, charge, priming, Mine, $4\frac{1}{2}$ *lbs., Mark I* | *C* | contains three Primers, Mine, $1\frac{1}{2}$ lbs. A felt wad is placed above and below the primers.

The cylinder is lacquered and the lid secured in the usual manner.

Boxes.

Box, guncotton, dry, primers, Field, 1 *oz., Mark III* | *C* | . This box is made of deal, with copper wire handles at each end. It is painted service colour for Land and khaki colour for Naval Service.

It is fitted with wood packing pieces and holds six cylinders Guncotton primers, Field, 1 oz.

The lid of the box is secured by metal screws.

The Mark II box is slightly smaller.

Box, guncotton, dry, primers, Rocket, 2 *oz., Mark II* | *C* | . The above box is similar to the Box, guncotton, dry, primers, field, 1 oz., Mark III, but holds four cylinders, Guncotton primers, Rocket, 2 oz.

The Mark I box is longer, wider and deeper than the Mark II. It holds either 36 primers 2 oz. or 24 detonators No. 8 in four cylinders.

Box, guncotton, dry, charges, priming, Mine, $2\frac{1}{4}$ *lbs.* | *C* | . This box is made of wood, with elm ends and copper wire handles. The lid is secured by metal screws.

It is painted stone colour.

It holds 10 charges.

The other boxes for priming charges are similar in construction.

Box, guncotton, dry, yarn, is made of wood, with copper wire handles. It has divisions to hold eight cylinders.

Storage.

Dry guncotton is stored in Magazines in Group I, Division II.

TABLE NO. 2.—DRY GUNCOTTON.

Table showing Nomenclature and system of packing of Guncotton Primers and Priming Charges in Land and Naval Service in accordance with § 15890, § 14397, 14398, § 14354 and § 13939 *List of Changes.*

<table>
<tr><th>PRIMERS.</th><th>PRIMING CHARGES.
(Note.–A priming charge includes the cylinder shown in the next column.)</th><th>CYLINDERS.</th><th>BOXES.</th></tr>
<tr><td>Guncotton, dry, primers, Field, 1 oz., Mk. I | C |
(This primer is conical and has one perforation.)</td><td>... ...</td><td>Cylinder, guncotton, primers, Field, 1 oz. (Tin, to hold 10). | C |</td><td>Box, guncotton, dry, primers, Field, 1 oz. (to hold 6 cylinders). | C |</td></tr>
<tr><td>Guncotton, dry, primers, Rocket, 2 oz., Mk. I | C |
(One perforation.)</td><td>... ...</td><td>Cylinder, guncotton, primers, Rocket, 2 oz. (Tin, to hold 9). | C |</td><td>Box, guncotton, dry, primers, Rocket, 2 oz. (For 36 primers in 4 cylinders.) | C |</td></tr>
<tr><td>Guncotton, dry, primers, Warhead, 1 oz.
Mk. I (unperforated). | N |
Mk. II (perforated). | N |</td><td>Guncotton, dry, charges, priming, Warhead, 6 oz., Mk. II | N | (i.e., 6 primers in a cylinder).</td><td>Cylinder, guncotton, charge, priming, Warhead, 6 oz., Mk. III | N | (copper with detonator holder.)</td><td>Boxes, guncotton, dry, charges, priming, Warhead, 6 ozs. (There are 2 sizes, one holds 12 and the other 6 charges.) | N |</td></tr>
<tr><td>Guncotton, dry, primers, Warhead, 2 oz.
Mk. I (unperforated). | N |
Mk. II (perforated). | N |</td><td>Guncotton, dry, charges, priming, Warhead, 1 lb. 1 oz., Mk. II | N | (i.e., 8 2-oz. primers in a cylinder).</td><td>Cylinder, guncotton, charge, priming, Warhead, 1 lb. 1 oz., Mk. III | N | (brass, tinned with detonator holder).</td><td>Boxes, guncotton, dry, charges, priming, Warhead, 1 lb. 1 oz. (There are 2 sizes, one holds 12 and the other 6 charges.) | N |</td></tr>
<tr><td>Guncotton, dry, primers, Mine 9 oz.,, Mk. I | C | (two perforations).
,, ,, Mk. II | N | (one perforation).
,, ,, Mk. III | N | (unperforated).</td><td>Guncotton, dry, charges, priming, Mine, 2¼ lbs., Mk. I | L |
Guncotton, dry, charges, priming, 2¼ lbs., Mk. III | C |
Guncotton, dry, charges, priming, Mine, spherical, 2¼ lbs., Mk. I | N |</td><td>Cylinder, guncotton, charge, priming, Mine, 2¼ lbs., Mk. I | L |
Cylinder, guncotton, charge, priming, 2¼ lbs., Mk. III | C |
Cylinder, guncotton, charge, priming, Mine, spherical, 2¼ lbs., Mk. I | N | (Tin.)</td><td>Box, guncotton, dry, charges, priming, Mine, 2¼ lbs. (to hold 10 charges.) | C |</td></tr>
<tr><td>Guncotton, dry, primers, Mine, 1½ lbs., Mk. I | L | (two perforations).</td><td>Guncotton, dry, charges, priming, Mine, 4½ lbs., Mk. I | L | (i.e., 3 primers in a cylinder).</td><td>Cylinder, guncotton, charge, priming, Mine, 4½ lbs., Mk. I | L | (Tin.)</td><td>Box, guncotton, dry, charges, priming, Mine, 4½ lbs. (to hold 4 charges.) | L |</td></tr>
</table>

Guncotton Yarn is issued 2 oz. in a tin cylinder, 8 cylinders in "Box, guncotton, dry, yarn." | C |

Properties of Wet Guncotton.

Wet guncotton contains 20 parts of water to 100 parts of dry guncotton, or about 16·5 per cent. moisture.

The wet guncotton in torpedo warheads contains 22·5 parts water to 100 parts of dry guncotton.

Guncotton in its Service forms cannot be kept for any length of time in a large volume of water because the compressed slabs would disintegrate. The guncotton, however, would suffer no chemical change, and, if re-pressed, would regain all its former properties.

It must be remembered that loose guncotton cannot be detonated by means of the ordinary Service detonators; 16 to 17 per cent. of moisture is sufficient to make guncotton quite safe from ignition, and, at the same time, is not enough to cause the compressed slabs to disintegrate.

A Service slab of guncotton, if put on to a fire, only smoulders away as the outer portions become dry.

A rifle bullet fired into wet guncotton will not fire it.

At the same time it can be used in the wet state for demolitions as it can be readily detonated by suitable means.

These properties make wet guncotton a very safe explosive for military purposes.

Wet guncotton can be detonated by fulminate of mercury, but a considerable amount of the latter must be used.

As it is not desirable to carry or handle large quantities of fulminate the safest and most convenient way is to detonate a small quantity of dry guncotton, in close contact with the wet, by means of a small Service detonator.

The weight of dry guncotton that has to be used depends on the amount of wet guncotton to be detonated.

Wet guncotton, when detonated, is rather more powerful in its effects than dry.

This is because the detonating wave is passed throughout the mass more readily by the water in the interstices than by the air which fills up the spaces in dry guncotton.

Wet guncotton may be dried by exposing it to the air of a dry room till it ceases to lose weight; on no account must artificial heat be used to evaporate the moisture.

Wet guncotton can be sawn to shape or drilled with the utmost safety, care being taken to keep the cotton thoroughly wet during these operations.

Wet guncotton keeps very well now that wood and felt packing pieces, &c., which encouraged fungus and bacilli to form, have been done away with.

The dipping of slabs in carbolic solution is to prevent the formation of fungus, &c.

Uses of Wet Guncotton.

Wet guncotton is used for demolitions in the field and in torpedo warheads and mines.

Forms in which Wet Guncotton is used.

Wet guncotton is issued in the form of slabs. There are five different slabs in the Service, varying in weight from 14 oz. to 2½ lbs.

The nomenclature is shown below; for detailed particulars *see* table and Plate II.

Guncotton, Wet, Slab, Field, 15 oz., Mark I | L | .
,, ,, ,, Mine or Warhead, 2½ lbs., Mark I | C | .
,, ,, ,, Mine, 1½ lbs., Mark I | N | .
,, ,, ,, Mine, 1¼ lbs., Mark I | N | .
,, ,, ,, Mine, 14 oz., Mark I | N | .

Guncotton, wet, slab, Field, is rectangular, 6 in. by 3 in., and has one conical perforation to take the 1-oz. Field primer.

Guncotton, wet, slab, Mine or Warhead, 2½ lbs., is square, 6⅛ in. by 6⅛ in.

Guncotton, wet, slab, Mine, 1½ lbs., is square, 4¾ in. by 4¾ in.

Guncotton, wet, slab, Mine, 1¼ lbs., is similar to the 15-oz. Field slab, but is slightly larger and has no perforation.

Guncotton, wet, slab, Mine, 14 oz., is the 1¼ lbs. slab with a semicircle cut out of one side to fit round the priming charge cylinder in the mine.

Issue and Packing of Wet Guncotton.

Wet guncotton slabs are packed :—

(1) Into *metal* " cases," which are then packed into boxes or into mines.

(2) Direct into " boxes " (metal-lined).

(3) Direct into torpedo warheads.

(4) Into metal-lined wooden cases, known as " Case, powder, metal-lined, half, special."

(There is one instance in which wet guncotton is issued in the form of " primers." *See* Table 3.)

Packages containing wet guncotton are always hermetically sealed, generally by soldering on the lid.

The closing plates of the cases or metal linings and the " doors " of warheads are fitted with a screw plug, to allow the escape of carbon dioxide gas that may have formed, and the re-wetting of the guncotton. It is to be noted that, in all cases where the weight of guncotton is given independently of the package, the weight of that material in a dry state is intended.

The packages are :—

Box, guncotton, wet, Mine or Warhead | C | .
,, ,, ,, Field, Mark II | C | .
,, ,, ,, Charges, Field, 15 oz. | L | .
,, ,, ,, ,, Naval, 16¼ lbs.
,, ,, ,, ,, Warhead | N | .
Case ,, ,, Slab, Field | L | .
Cases, guncotton | N | , Nos. 1 to 5.
Case, powder, metal-lined, half, special.

Box, guncotton, wet, Mine or Warhead, Marks II and III | C |*, holding 20 slabs, was formerly known as the 50-lb. box ; it is of wood, with a tinned copper inner case, black lacquered inside and out ; in the lid is a screw-plug, the removal of which allows access to the plug in the inner case ; on the closing plate of the latter is a label showing the weight of the copper case and its contents, and the limits of increase or decrease in weight (± 2 per cent. of the weight of the wet guncotton) allowed at inspections.

Stencilled on the wooden lid is the nature and the number of slabs in the box, place and date of manufacture, and test number of the guncotton, and the gross weight.

Packing pieces of corrugated tinned copper, black lacquered, are used in the inner case ; they were formerly of wood.

These also serve to keep the heat of the soldering iron from the guncotton.

Fig. 1.

BOX, GUNCOTTON, WET, MINE, OR WARHEAD, MK. III.

28⅝ in. × 12½ in. × 8⅝ in.

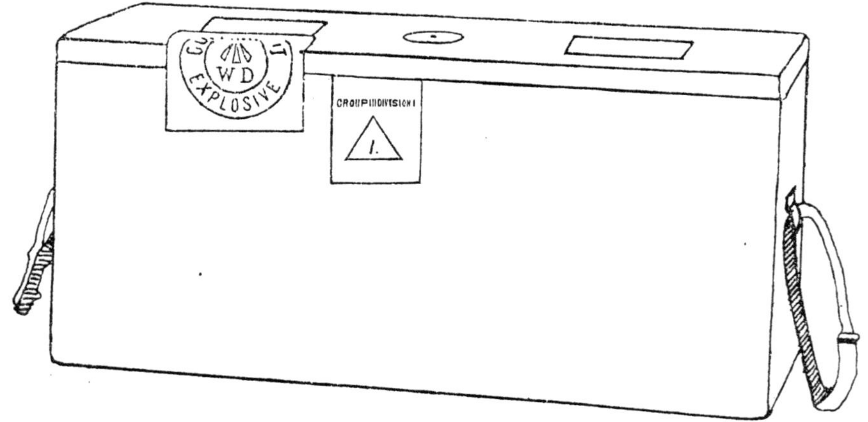

Box, guncotton, wet, Field, Mark II | C | is similar in all respects to the above, except as regards size and details of packing pieces.

It holds 14 slabs, Field, 15 oz.

Box, guncotton, wet charges, Field, 15 *oz., Mark I* | *L* | .—These boxes are made of deal with elm ends, the bottom secured with brass screws. To each end is attached a copper wire handle secured by means of cleats. The interior of the box is divided into 16 partitions, each to hold one " Case, guncotton, wet slab, Field." The top is secured by brass screws.

Case, guncotton, wet slab, Field, Mark III | *L* | is made of tinned copper, to contain one " Slab, Field, 15 oz." The lid of the case is secured by means of a soldered strip, similarly to fuze cylinders. It has a small loop on top for lifting, and a screw plug for re-wetting, and is fitted with a tinned copper packing piece.

Mark II differs from the above in having a wood packing piece, and is $\frac{1}{10}$ in. shorter.

Mark I had no screw re-wetting plug, and was closed with a tape band.

Guncotton, wet charges, Naval, 16¼ *lbs., Mark IV* is used for outrigger torpedoes, and for general demolition purposes.

The present pattern, Mark IV, consists of a tin cylinder about 10⅓ in. high, and the same diameter.

Two holes pass completely through the cylinder—one in the axis, the other parallel to it; the former is for attaching the case to a spar, and the other to contain the dry priming charge. The case is painted black inside, and contains thirty 9-oz. primers, wetted.

They are issued two in "*Box, guncotton, wet, charges, Naval,* 16¼ *lbs., Mark I.*"

Cases, Guncotton (Mark II Spherical Mine), Mark II, Nos. 1 to 5, Naval.

The above cases are used for filling submarine mines. They are made of tinned copper, shaped to fit the interior of the mine and are packed with wet slabs. The lids of the cases are soldered on, and are fitted with screw plugs for re-wetting the guncotton, and handles for the purpose of removing the cases from the mine when necessary. A small tinned copper label is soldered on to the lid giving the number of the case, gross weight, limits of weight allowed on examination, &c.

The No. 5 Case, which forms the centre of the charge when packed in the mine, has a central hole to take the " Cylinder, guncotton, charge, priming, Mine, 2¼ lbs."

Case, powder, metal-lined, half, special.

Some *Cases, powder, metal-lined, half,* have been specially prepared to take guncotton, by making the bottom removable, so that the metal lining may be taken out and weighed when necessary, and by inserting in the bung a screw plug which can be taken out to allow of the escape of gas or to admit of re-wetting. In the wood lid is a small hole to allow of access to this plug without opening the lid. An arrow on the bung and another on the exterior of the case denote when the bung is in the correct position, with the plug under this hole. The interior of the lining is also coated with black paint. These altered cases are still available for packing gunpowder if specially required for that purpose, but care must of course be taken to see that they are thoroughly dry before inserting the powder. After inserting the guncotton, the lining is closed by the metal bung, which is smeared round the edge with thick luting.

Damaged guncotton.—Pieces of guncotton sometimes become detached from discs and slabs in handling, packing and unpacking boxes, and unloading mines. Small fragments and dust also accumulate in cutting and shaping slabs in loading mines and warheads. All these should be carefully collected, and together with the solid pieces cut off the slabs in filling mines should be re-wetted, placed in half metal-lined cases prepared for guncotton, and kept wetted until disposed of.

Guncotton, wet charges, Warhead, are issued to Naval Service in "Box, guncotton, wet charges, Warhead," containing one charge. *See* Chapter VIII.

Storage.

Wet guncotton, except in Mines, is stored in Group III, Division I.

Wet guncotton in Mines in Group II, Division II.

TABLE NO. 3.—§ 14354. WET GUNCOTTON SLABS AND CHARGES WITH THEIR CASES AND BOXES.

(Except Warhead charges and certain slabs which will be obsolete when existing stock is used up.)

<table>
<tr><th>SLABS.</th><th>CHARGES.</th><th>CASES.</th><th>BOXES.</th></tr>
<tr><td rowspan="2">Guncotton, wet,
Slab, Field, 15 ozs., Mk. I | L |
$6'' \times 3'' \times 1\frac{3}{8}''$
(rectangular, one conical perforation).</td><td>Guncotton, wet,
Charges, Field, 15 ozs., Mk. I | L | (consists of one slab in the case shown in next column).</td><td>Case, guncotton, wet,
Slab, Field, Mk. II or III | L | (to contain one 15-oz. Field Slab), (copper, tinned).</td><td>Box, guncotton, wet, charges, Field, 15 ozs., Mk. I | L | (to hold 16 charges).</td></tr>
<tr><td>... ...</td><td>... ...</td><td>Box, guncotton, wet, Field | C | Mk. II (with copper tinned inner case, to hold 14 slabs, Field, 15 ozs.).</td></tr>
<tr><td>*Guncotton, wet,
Slab, Mine, $2\frac{1}{2}$ lbs., Mk. I | C |
$6\frac{1}{8}'' \times 6\frac{1}{8}'' \times 1\frac{3}{4}''$</td><td>... ...</td><td rowspan="4">For filling naval mines, the slabs are packed in Cases, Guncotton | N | Nos. 1 to 5.
(These cases are of tinned copper, with screw plug for re-wetting.)
A number of cases are then packed into the mine.</td><td rowspan="4">Packed in—
Box, guncotton, wet, Mine, Mks. II* and III | C | (with tinned copper inner case: for 20 slabs, Mine, $2\frac{1}{2}$ lbs.).

The following boxes may still be met with, but no more are to be made:—
Box, guncotton, wet, Mine, Mk. II | C | (wire handles).
Box, guncotton, wet, Mine, Mk. II | N | (wire handles).
Box, guncotton, wet, Mine, Mks. II* and III | N | (wire handles).</td></tr>
<tr><td>Guncotton, wet,
Slab, Mine, $1\frac{1}{2}$ lbs., Mk. I | N |
$4\frac{3}{4}'' \times 4\frac{3}{4}'' \times 1\frac{5}{8}''$</td><td>... ...</td></tr>
<tr><td>Guncotton, wet,
Slab, Mine, $1\frac{1}{4}$ lbs., Mk. I | C |
$6\frac{1}{8}'' \times 3\frac{1}{16}'' \times 1\frac{3}{4}''$</td><td>... ...</td></tr>
<tr><td>Guncotton, wet,
Slab, Mine, 14 ozs., Mk. I | N |
$6\frac{1}{8}'' \times 3\frac{1}{16}'' \times 1\frac{3}{4}''$ with semi-circle cut to radius 1·88″.</td><td>... ...</td></tr>
<tr><td>*Guncotton, wet,
Slab, Warhead, $2\frac{1}{2}$ lbs., Mk. I | C |
$6\frac{1}{8}'' \times 6\frac{1}{8}'' \times 1\frac{3}{4}''$</td><td>... ...</td><td>... ...</td><td>The slabs are cut to shape and packed into the Warhead of a torpedo.</td></tr>
<tr><td>... ...</td><td>Guncotton, wet.
Charges, Naval, $16\frac{1}{4}$ lbs., Mk. IV | N | (consisting of 20 whole and 10 cut primers, packed in cylindrical tin case: the primers are the 9-oz. primers, wetted).</td><td>... ...</td><td>Box, guncotton, wet, charges, Naval, $16\frac{1}{4}$ lbs., Mk. I | N | (to hold two "Guncotton, Wet, Charges, Naval").</td></tr>
</table>

* These slabs are identical in size and shape and will in future be designated "Guncotton, Wet Slab, Mine or Warhead, Mark I (C)."

CHAPTER V.—CORDITE.

COMPOSITION; MANUFACTURE; GENERAL PROPERTIES; CORDITE M.D. COMPARED WITH CORDITE, MARK I; CORDITE IN THE GUN; CORDITE IN STORE; ADVANTAGES AND DISADVANTAGES; FORMS IN WHICH ISSUED; NOMENCLATURE; EXAMINATION AND TESTING; PACKING; STORAGE.

Cordite was introduced into the Service in 1893. It consists of nitroglycerine, guncotton, and mineral jelly (vaseline), incorporated together and gelatinized with the aid of a solvent, namely, acetone. The mineral jelly was originally added to act as a lubricant and to waterproof. It acts further as a stabiliser.

There are two natures of cordite in the Service :—

CORDITE.—Commonly known as Cordite, Mark I (though this is not its correct nomenclature).

CORDITE M.D.—Standing for "Modified."

The proportions of the ingredients are :—

——	Mark I.	M.D.
Nitroglycerine	58	30
Guncotton	37	65
Mineral jelly	5	5

MANUFACTURE OF CORDITE.

The system of manufacture, as carried out at Waltham Abbey, is briefly as follows :—

Drying the guncotton.—The guncotton is brought to the drying stoves in the form of cylindrical "primers," 3 in. in diameter and 5 in. long, which contain 48 to 50 per cent. of moisture.

The primers are placed on copper wire racks in the stoves; warm air is blown into the stove, and the temperature kept at about 40° C. (104° F.).

The guncotton is dried down to 0·5 per cent. of moisture, the process lasting about 100 hours.

Weighing.—When dry and cool the guncotton is weighed into bags made of rubber-lined canvas.

Adding the nitroglycerine.—The nitroglycerine is *measured*, not weighed; this method offers great advantages, as great accuracy is obtained with less risk and inconvenience. It is run from the filter tanks into fixed lead burettes, and from the burettes on to the guncotton in the bags; the dry guncotton cylinders absorb the nitroglycerine, but the latter is not evenly distributed through the guncotton.

Mixing the N.G. and guncotton.—The contents of the bags are mixed by rubbing through ½-in. mesh leather sieves into canvas bags.

The mass is now known as " Cordite, or Cordite M.D. paste," according to its composition.

Incorporating.—The incorporating machine is an iron box, open at the top, and with the bottom shaped to form two semi-circular troughs, in each of which a spindle with propeller-shaped blades revolves; the trough has a cooling jacket.

Two men manage the operation as follows:—The spindles start revolving, and one man pours acetone on to them from a can like a watering can (*i.e.*, with a " rose "); the other operator puts the cordite paste into the troughs gradually, the acetone being continually sprayed on; the mixture is being continually squeezed between the blades. When the charge is all in, a metal cover is put on, and the incorporating goes on for 3 hours.

Cordite requires 19 per cent. acetone and Cordite M.D. 37 per cent. and 3 per cent. of water. The water prevents the strands sticking together after pressing.

Addition of mineral jelly.—The mineral jelly (5 per cent.) is then added, and the machine set to work for another 3 hours for cordite M.D. and 2 hours for cordite.

The " dough," as it is now called, is put in bags, and removed to the press houses.

Pressing into cords.—This is done in hydraulic presses; the dough is placed in a cylinder, and rammed tight with a wooden rammer; the cylinder is then put into the press and the cordite squirts through muslin resting on a 100 mesh bronze gauze (which retains small foreign bodies, such as chips of wood from the guncotton) and then through dies; the number of dies is three in the case of size 45, and more in the smaller sizes. The pressure on the dough is about 2 tons per square inch. It is cut to length as it emerges from the press, and is laid in trays for the next process. Some of the smallest sizes are wound on drums.

Drying or stoving.—Almost all the acetone (85 per cent.) is driven off by keeping the cordite in the stoves at 110° F. for a period which varies with the diameter of the cords. Fifty per cent. of the acetone is " recovered " by drawing the air from the stoves through a sodium bisulphate solution. M.D. requires longer than Mark I because, being harder, it retains the solvent more. M.D. size 45 requires 38 days, size 3¾ requires 3 days, M.D.T. 5–2 12 days.

Blending.—As the trays come from the stoves the cordite from at least 6 trays is packed in cases, each holding about 100 lbs.; when about 25 cases have been filled they are blended into another 25 cases, so forming a *blend*. Multiples of these blends are blended together to form a *lot*, which is issued to the Service.

Size 3¾ small-arm cordite is blended as follows:—10 single-strand reels are wound off simultaneously on to a larger reel, and then six of the 10-strand reels are wound off simultaneously on to a drum. When the reeling is completed the ends of the 60 strands are secured by a band of tape wound round the drum, and the drum is packed in a box or barrel for transport.

M.D.T. size 5-2 is blended by running four 11-strand reels on to a drum.

A " lot " of cordite may consist of any quantity up to 30,000 lbs., according to the size of the cordite.

General Properties of Cordite.

Cordite is a "colloid," that is, a substance that has no internal structural form.

It is owing to its " colloid " form that cordite burns comparatively slowly. The hot gases cannot penetrate to the interior of the sticks, but each layer of the substance has to be burnt away successively.

Finished cordite resembles a cord of gutta-percha, and its colour varies from light to dark brown ; when in good condition it is smooth and tough and has little smell.

Finely ground cordite ignites at about 355° F.

Cordite is poisonous.

Cordite ignited in the open burns away without explosion.

Cordite can be fired by percussion, and can be detonated, and is then almost as powerful as guncotton ; it is for this reason that care has to be taken in the ignition of charges of cordite in a gun to avoid all risk of detonation, therefore cartridges of M.D. cordite never had igniters of guncotton yarn.

It may be used on emergency in place of guncotton for demolitions, and when properly detonated is nearly as powerful.

Small size cordite should be used for this purpose.

Cordite is almost smokeless ; on explosion a very thin vapour is produced which is dissipated rapidly. This smokelessness can be understood from the fact that the products of combustion are nearly all non-condensible gases and contain no solid products of combustion which would cause smoke.

The yellowish smoke observed when firing projectiles with cordite charges arises from the vaporising of the copper of the driving bands, and also from the gunpowder in the cartridge igniters.

Cordite suffers if exposed to direct sunlight, but not when enclosed in either shalloon or silk cloth cartridges. It is therefore ordered that when being made up into cartridges it should not be so exposed.

Cordite is not affected in any way by damp or water.

Cordite is difficult to ignite; for this reason all B.L. cordite cartridges have igniters of gunpowder or guncotton yarn attached to them to carry on the flash from the tube.

Sweating.—An oily exudation is sometimes seen on cordite ; this is due to either " *sweating*," which is an exudation of nitroglycerine, *or* to the mineral jelly appearing on the surface.

Sweating is liable to take place when the cordite has been subjected to low temperatures and subsequently warmed : for below 45° F. for M.D. and 32° F. for Mark I, the nitroglycerine freezes and crystallizes out on the surface, and the melting of these crystals forms the sweating.

The exudation of mineral jelly is due to cordite being kept at high temperatures.

Neither sweating nor the exudation of the mineral jelly injuriously affects the cordite.

Stability of cordite.—Cordite is not a thoroughly stable substance. It begins to deteriorate from the day it is made, and, if kept long enough, would eventually ignite spontaneously.

The higher the temperature at which it is stored the more rapid the deterioration.

For these reasons all cordite, except that in small-arm ammunition, is periodically inspected and tested, and, to give the cordite a long life, it should not be stored in magazines, whose temperatures habitually exceed 70° F. ; at the same time it must not be exposed to low temperatures, as this gives rise to " sweating."

Contact with metals, especially iron and wood, affects the stability owing to the oxidising action set up, and lower nitrates are formed which are unstable.

Properties of Cordite M.D. compared with those of Cordite, Mark I.

Cordite M.D. is harder and more brittle.

It is slower burning, therefore a smaller size is used in a given gun to obtain the same ballistics as with Mark I.

A heavier charge has to be used, as the explosion temperatures are lower than with Mark I and therefore the pressures are lower.

Since the explosive temperatures are lower, due to there being less nitroglycerine in M.D., the rifling of a gun may be expected to last some three times as long as when firing Mark I cordite.

M.D. is rather more difficult to ignite.

M.D. does not keep quite so well as Mark I, probably owing to the hardness of the material, which prevents the distribution or escape of any local deterioration.

Cordite in the Gun.

For the same muzzle velocity a smaller charge of cordite is required than of gunpowder, owing to the greater volume of gas produced by it and the high temperature.

Cordite is very slow burning compared with gunpowder, and on this account the maximum pressures are comparatively low and the pressure on the projectile in the bore is well sustained.

The rate of burning of a charge of cordite depends on the diameter of the cords ; with equal weights a large size presents less superficial area than a small size, and so, other things being equal, a charge of large size cordite burns more slowly than one of small. Charges in a gun are difficult to ignite, and an " igniter " of guncotton yarn or fine grain powder is used to extend the flash from the tube.

The effect of *the temperature of the charge before firing* on ballistics varies with every gun ; it varies inversely with the specific volume, and so is more noticeable in Q.F. guns. One degree Fahrenheit rise in temperature of the charge increases the M.V. by 1 to 3·5 fs.

Great erosion of the bore is caused by the extremely high temperature of Cordite, Mark I, due to the large percentage of nitroglycerine which it contains; to minimize this, Cordite M.D. was introduced, containing only 30 per cent. of nitroglycerine.

All cordite has a tendency to cause excessive rusting of the bore; this is due to the high temperature and to the oxides of nitrogen formed on explosion.

Back flash is due to the products of combustion (carbon monoxide and hydrogen) meeting the air at the muzzle, as soon as the projectile leaves, and burning down the bore. As soon as the breech is opened a flame issues.

This is particularly noticeable with heavy guns, especially when firing to windward.

The heavier guns used in the Navy are fitted with an air blast which blows up the bore before the breech is opened. With field guns back flash is not sufficiently serious to call for special orders, but with any heavier guns the detachments are warned against standing in rear of the breech or exposing naked cartridges to the flame.

Cordite in Store.

Sweating.—Neither sweating nor the exudation of the mineral jelly injuriously affects the cordite.

If sweating occurs in made-up cartridges, no action need be taken; if it appears on cordite in bulk, the cordite should not be handled till it recovers its normal state, which it will do by the re-absorption of the nitroglycerine if its temperature is kept above 45° F.

To distinguish the exudation of nitroglycerine from that of mineral jelly, wipe a stick of the cordite with a strip of clean blotting paper, so that the stain from the exudation appears about the centre of the strip; then, in some comparatively dark place, hold the strip in a horizontal position and light it at one end. If the exudation is of nitroglycerine, the flame will travel faster and become distinctly green on reaching the stain.

A waterproof (or non-absorbent) paper lining is placed in all boxes or cases containing cordite in bulk.

Effect of zinc chloride.—The flux of zinc chloride formerly used in soldering cylinders and cases for B.L. cartridges affected the stability of the cordite and caused yellow stains and deterioration of the silk cloth.

The use of this flux has been discontinued.

Effect of cordite on silk cloth of cartridges.—In some cases slight deterioration of cordite has caused discoloration and rotting of the silk cloth.

Effect of gunpowder on cordite.—The sulphur in gunpowder in close contact with cordite was considered to have a bad effect on the stability of the latter, hence the introduction of S.F.G.[2] (sulphurless gunpowder), but on reconsideration it was found that the deleterious effects were exaggerated, and the use of R.F.G.[2] is now reverted to.

Magazines.—Owing to the liability of cordite to spontaneous ignition, it is laid down that cordite in bulk should not be stored in magazines, the temperature of which habitually exceeds 70° F. The temperature of magazines containing cordite *in bulk* before filling into cartridges should not be allowed to fall below 45° F. if it can be prevented.

Examination and testing of cordite (*see* Regulations for A.O. Services, Parts II and III).—Samples of all cordite, except that in small-arm ammunition, are periodically examined and tested.

Advantages and Disadvantages of Cordite.

To sum up, cordite possesses the following advantages :—

It is smokeless.
It is safe, as it will not explode unless confined.
It is not affected by moisture.
It is a better propellant than gunpowder, since a smaller weight of it is required and maximum pressures are lower.

The chief objections to its use are :—

Erosion of the bore.
The bright muzzle flash.
Back flash.
The ballistics of a charge are affected by temperature.
Sweating.
Deterioration.

Forms in which Cordite is Issued.

Cordite, both Mark I and M.D., is generally in the form of sticks or cords, circular in section.

Cordite, M.D.T.—M.D. cordite is also made in the form of tubes, the object being to obtain a more constant burning surface, and therefore more constant evolution of gas. This is known as Cordite M.D.T. At present there are only a few cartridges in the Service made up of M.D.T. cordite.

Cordite cylinders.—Mark I cordite is pressed into cylinders, all of which are 1 in. internal diameter. They are made 4 in. long with walls ·2 in., ·15 in., ·1 in. or ·05 in. thick to support Marks III and IV 1¼-oz. igniters in Q.F. or Q.F.C. cartridges, also 2½ in. long with walls ·05 in. thick to support Marks I and II 10 dram igniters for M.D. cordite charges in the Q.F. 12-pr., 12 and 18 cwt. cartridges, or the Mark I 8-dram igniter in the Q.F. 14-pr. cartridge.

Sliced cordite.—Cordite for ·303 blank ammunition is prepared by slicing size 20 into thin flakes ; it is known as "Sliced Cordite."

Cordite size 1, cut to a length of ·05 in., is issued for pistol cartridges.

Nomenclature.

All lots of cordite from each manufacturer have consecutive numbers, irrespective of size, and one or more initial letters to identify the manufacturer. Some early issues of Waltham Abbey cordite had only the lot number and no initial letter.

To identify cordite completely, it is only necessary to quote the initial letter and lot number, except in the case of lots of Waltham Abbey manufacture numbered 1,757 and below, when the *size* must also be stated.

Former designation.—Cordite used to be distinguished as follows :—

(*a*) When made in specified lengths, by a fraction the numerator of which represented the diameter in hundredths of an inch of the hole in the die through which the cordite is pressed, and the denominator the normal length of the sticks in inches.

With M.D.T. cordite the numerator gave in hundredths of an inch the mean external and internal diameter of the finished cordite.

The nominal length is the mean specification length.

(*b*) When made in indefinite lengths (to be cut as required) by a number representing the diameter of the die as before.

Present designation.—Cordite is now distinguished as follows :—

Cordite, Mark I and M.D., by a number representing in hundredths of an inch the diameter of the hole in the die through which the cordite is pressed, and the length of the sticks is shown in detail appended to designation.

Example :—

Cordite M.D., size 37, length 31 in.
Cordite M.D.T., size 20–10, length 7 in.

Examination and Testing of Cordite.

(*See* Regulations for A.O.D. Services, Parts II and III.)

Samples of all cordite, except that in small-arm ammunition, are periodically examined and tested.

The cordite is examined for general appearance, colour, smell and sweating. Deteriorated cordite may be more brittle, and is usually darker than new cordite, sometimes having a reddish-brown translucent appearance ; it has a sour smell.

Testing.—The tests for stability are the " Abel Heat Test," and the " Waltham Abbey Silvered Vessel Test."

All samples are tested by the heat test, and it is only when this test shows the stability of the cordite to be " doubtful " that the silvered vessel test is applied. The heat test only lasts a few minutes, but the silvered vessel test is much more elaborate, and may take ten days or more to complete, the cordite being under test the whole time—day and night.

M.D. cordite is not silver vessel tested.

Heat test.—In the heat test a small quantity of finely ground cordite is kept at 160° F. in a test tube ; the length of time which the evolved gases take to discolour a sensitive test paper affords a measure of the stability of the sample.

Silvered vessel test.—In the silvered vessel test, a quantity of ground cordite is kept at 80° C. in a vacuum jacketted and silvered flask, till a spontaneous rise of 2° in the temperature indicates decomposition.

From this latter test an estimate as to how many years cordite will remain serviceable can be arrived at.

Packing.

The following cases and drums are used for packing cordite in bulk :—

Case, cordite, 100 *lbs., Mark V* is a wooden box about 40 in. by 17 in. by 9 in.

Case, cordite converted is a 100 lb. prismatic powder case converted, the zinc lining being removed.

Drum, cordite, transport is used for small sizes of cordite when not cut to length.

Case, cordite, drum, is a square wooden box in which the above drum is packed.

Barrel, special transporting cordite is now used for transporting small sizes of cordite on drums to India.

In tropical stations cordite in bulk, or cordite cartridges, are always packed in hermetically sealed packages.

Small irregular quantities are packed in "Cases, powder, cylindrical," or "Cylinder, cartridge," of convenient size.

The cordite is wrapped in non-absorbent paper to prevent contact with metal.

Cases, powder, 100 lbs., Marks IV* and V may be used, the cordite being wrapped in non-absorbent paper.

Case, Cordite, 100 *lbs.*

Case, cordite, 100 *lbs., Mark V* | *C* | is for cordite in bulk ; it is a rectangular wooden box, of greater length than the 100 lbs. powder case ; on the under side of the lid, running round near the edge, a strip of fearnought (a sort of felt) is secured, to make a tight joint when the lid is screwed on.

There is no metal lining, but non-absorbent paper is used to keep the cordite from touching the wood of the box.

The bottom is strengthened by two battens, which also serve to provide an air space between the cases when stacked in tiers.

A number of grooves are cut inside the case, in which movable partitions may be placed to suit different lengths of cordite.

A circular recess at the junction of lid and body is provided for a sealing label.

The box is painted "khaki" colour.

The Naval cases have the letter "N" cut at each end.

The cases are as a rule of deal, but L.S. cases for India and East and West Africa are of teak.

Copper wire handles are fitted at each end.

The earlier Marks of this case differ in dimensions and number of grooves, and had rope handles.

Case, Cordite, converted.

Case, cordite, converted, Mark I | *N* | .—The lining is taken out of the 100 lbs. powder case, the lid prepared with fearnought, the case painted and lined with non-absorbent paper.

Case, cordite, converted, Mark II | *C* | .—This case differs from Mark I in having two sets of grooves cut in the sides.

Drum, Cordite, transport.

Drum, cordite, transport, Mark I, and Case, cordite, drum, Mark I are used for packing and transport of Cordite, Mark I, sizes 5 and 3¾, Cordite M.D., sizes 2¼ and 4¼, and M.D.T. 4–2 and 5–2 when not cut to length.

The drum is a tinned iron cylinder, 8 in. in diameter and about 22¼ in. long. To each end a tinned iron-flanged ring is secured with copper rivets.

Case, Cordite, drum.

The case is a square deal box strengthened by wood battens and provided with cleats and copper wire handles. On the inside of the top and bottom four wood corner pieces are secured, which fit over the flanges of the drum and steady it.

The lid is attached by screws.

Barrel, special, transporting cordite.

Barrel, special, transporting cordite is used for transporting Cordite, sizes 3¾ and 3 on special wooden drums to India. It is a machine-made barrel of deal, length 26 in., diameter of bilge 17 in., diameter of head 15¾ in.

The heads are in three and sometimes in four pieces; the middle piece is of deal and the cants of oak; on the inside of the head is secured, with four iron screws, a conical disc of deal 8⅞ in. diameter; these discs fit in the end of the drums, and prevent shaking during transport.

It is bound with 14 ash hoops—four on each chime, and three on each quarter. There are two dowels in bottom end to support the back head.

STORAGE.

Cordite is stored in Magazines in Group I, Division IA.

CHAPTER VI.—MISCELLANEOUS EXPLOSIVES.

Ballistite.

Ballistite is a mixture of about 50 per cent. nitroglycerine and 50 per cent. soluble nitro-cellulose, with or without the addition of a small percentage of camphor or aniline. It was originally gelatinised by passing the mixture between hot rollers ; it is, however, sometimes incorporated with the aid of a solvent, *i.e.*, benzol.

Ballistite was used in the earliest cartridges for Q.F. 15-pr. and 2·95-inch and B.L. 9·45-inch Howitzers.

It is classed for storage in Group 1, Division IA.

Blasting Gelatine.

Blasting gelatine is made by dissolving finely divided soluble nitro-cellulose in nitroglycerine. The product is a gelatinous mass the colour of honey, and varying in consistency from a tough leather-like substance to a jelly. It contains 93 to 95 per cent. of nitroglycerine, and is a most powerful explosive. It is practically unaffected by water, and can be detonated by a Service detonator. It is made up in cartridges, and is used for blasting purposes only. It is classed for storage in Group I, Division II, and should not be stored in a magazine the temperature of which exceeds 120° F.

Composition Exploding or Tetryl.

Composition exploding is a pale yellow substance, and is used in powder form, also in compressed pellets in Fuzes Nos. 18, 19, 19A, 44 and 45, and in the compressed form in some H.E. shell.

Its composition is secret.

Packing.—It is packed in the same manner as picric powder, and the barrels are marked in the same way, but in red.

For storage it is classified in Group I, Division I.

Composition Priming.

Composition priming consists of 60 per cent. mealed R.F.G.[2] powder and 40 per cent. guncotton dust.

It is used in some electric tubes, fuzes and primers to form an easily ignitable priming round the electric bridge.

It is stored in Group II, Division IA.

Dynamite.

Dynamite is a brown pasty mass consisting of a siliceous earth, called "Kieselguhr," impregnated with about 75 per cent. of nitroglycerine. In actual contact with water the nitroglycerine at once separates from the earth. Like most other nitroglycerine mixtures it freezes at about 40° F., and, in its frozen state, is, under ordinary circumstances, less liable to explosion from detonation or percussion than when thawed, but more susceptible to explosion by simple ignition; should any of the nitroglycerine have exuded, the dynamite cartridges are much more sensitive to explosion by a blow. Dynamite must not be exposed to the rays of the tropical sun. It can be detonated by a Service detonator. For storage it is classed in Group I, Division II, and should not be stored in magazines the temperature of which exceeds 120° F.

Fulminate of Mercury.

Fulminate of mercury.—There are two varieties—white and brown—the former is the purer of the two.

It is prepared by the action of alcohol on a solution of mercury in nitric acid.

In the dry state it detonates readily and with great violence when rubbed, struck, or heated to about 190° C. (375° F.).

Contact with sulphuric or nitric acids also causes its detonation.

It is too dangerous a substance in the dry state to store in bulk.

In explosive factories it is kept under water, and is then practically harmless. Small quantities are carefully dried as required for use.

It is possible to *ignite* small quantities if quite unconfined; it then burns with a very bright flash.

Its sensitiveness to friction and percussion makes it a suitable ingredient for cap composition, &c.

The white fulminate is used alone, in most of the detonators for guncotton, for lyddite in hand grenades, and in the detonators of Fuzes Nos. 18, 19, 19A, 44 and 45.

The brown is used mixed with other substances in—caps, friction tubes, &c.

The flash given by fulminate of mercury by itself has not a sufficiently high temperature, and is too sudden to ignite the powder in igniters; it is therefore mixed with other substances which prolong the flash.

Fulminate of mercury is not stored in bulk in the Service.

Fuze, Instantaneous.

Fuze, instantaneous, Mark III | *L* |, consists of two or more strands of quick-match, enclosed in a tube of waterproof tape, round which cotton is twisted, the whole being contained in a gutta-percha covering. The gutta-percha covering is braided with yellow worsted and varnished orange on the outside. This readily distinguishes it from the safety fuze. It burns at the rate of about 30 yards per second, so is practically instantaneous. It can easily be ignited by a portfire or vesuvian, but it is unsafe to hold it in the hand when lighting it as may be done in the case of safety fuze.

Mark I, which differed only in minor details, may be known by its being unpainted. Mark II is obsolete.

It is issued in 100-yards lengths on a wood or tin reel, packed in a zinc box.

Instantaneous fuze is used for hasty demolitions with No. 8 detonator.

It is stored in Group I, Division IB.

Fuze, Safety.

Fuze, safety, No. 9, Mark II | *C* | , consists of a train of F.G. powder enclosed in jute yarn, contained in a flexible tube of waterproof composition, covered by waterproof tape. It burns 90 $\pm$ 15 seconds per yard. The rate of burning is shown on the label of the cylinder. It is easily ignited by a portfire or vesuvian, but not always by a lighted piece of paper. To prepare the fuze the gutta-percha must be removed by an oblique cut, and the powder laid bare, both at the end in contact with the charge and at that which is to be ignited.

Colour, black.

Old fuze should be tested before use, as it sometimes deteriorates, and instances have occurred where it has burned much too rapidly.

It is issued in tin cylinders containing 8 fathoms each.

It is used for hasty demolitions, with Nos. 8 and 15 detonators, in rockets, &c.

Stored in Group II, Division IA.

Gelignite.

Gelignite, or gelatine dynamite, is blasting gelatine with the addition of an absorbing powder, consisting of saltpetre and wood meal.

It is made up in the form of cartridges and is used for blasting purposes only.

Gelignite, which is less violent in its action than blasting gelatine, is now largely used in commerce. It is classed for storage in Group I, Division II, under similar conditions to dynamite.

Lyddite.

Lyddite is picric acid that has been fused at about 280° F. and allowed to solidify; it is thus brought to a much denser form, with long interlacing crystals.

In this state it can be more easily detonated.

Lyddite can be detonated by the direct action of fulminate of mercury, but some substances chemically related to it—picric powder or T.N.T.—is more effective, and a space is always left in lyddite shell to take an exploder of one of the above substances.

Detonation of lyddite is uncertain when the exploder is fired by the flash from gunpowder only, the usual method now employed being to fire a small quantity of F. of M. in the fuze; this detonates the C.E. in the fuze which passes on the detonating wave to the exploder and to the lyddite.

Lyddite is dark yellow in colour.

It is used as the burster in practically all small common and in some A.P. shell, and in the hand grenade.

It is not affected by moisture.

Picric acid, or *Tri-nitro-phenol,* is a pale yellow crystalline substance, formed by the action of nitric acid on phenol (or carbolic acid).

It was used in the dyeing industry many years before being adopted as a Service explosive. When closely confined or brought to a dense state by fusion it can be detonated. (*See* Lyddite.)

It can be fired by percussion.

With some bases it forms *picrates,* most of which are more sensitive to percussion or friction than itself.

Lead picrate is especially sensitive.

Picrates will act as detonators to any picric acid within reach of their explosive influence, and therefore great care must be taken to prevent picric acid or lyddite coming in contact with certain metals, metallic oxides, lime, rust, verdigris, &c., as picrates may be formed. There is no action between picric acid and tin or aluminium.

Packing and storage.—Picric acid is packed in rubber bags in powder barrels.

The barrels have a band of yellow ochre, 1 in. wide, painted round the bilge, and are stencilled on the top with the words, " To be used for picric acid only," also in yellow ochre.

Stored in Group I, Division III.

Picric Powder.

Picric powder is a mixture of 43 parts ammonium picrate and 57 parts of saltpetre.

Ammonium picrate is formed by the action of ammonia on picric acid. It is very stable and quite insensitive to shock, and when ignited, unconfined, only burns, but when mixed with saltpetre and confined can be readily detonated.

Picric powder is used to form the detonator, or, as it is technically termed, the exploder, in most lyddite shell.

The ingredients are thoroughly mixed together in the dry state ; first issues of picric powder were granulated in the same way as gunpowder.

Picric powder readily picks up moisture ; when damp it is useless as a detonator.

Packing and storage.—It is packed in the same way as picric acid, but the barrel has a blue band painted round it.

Stored in Group I, Division II.

Quick-Match.

Quick-match is made of cotton wick boiled with a solution of mealed powder and gum, and afterwards dusted over with mealed powder before it is quite dry.

Unenclosed it burns at the rate of about 1 yard in 13 seconds ; when enclosed in a tube of any kind it burns much more rapidly, the pressure causing the gas to rush forward and fire the whole length practically simultaneously. Quick-match is made up in paper or calico tubes when this rapid action is required, and when so made up is termed a "leader."

The proportions of powder, &c., vary with the number of threads in the wick. Quick-match is largely used for priming in rockets, detonators, &c.

It is demanded by weight and issued in hanks in metal-lined cases.

It is stored in Group I, Division IB.

Slow-Match.

Slow-match is made up of pure hemp slightly twisted and boiled in a ley of water and wood ashes in the proportion of water 50 gallons, wood ashes 1 bushel; this serves for 100 lb. of yarn. It burns at the rate of 1 yard in 8 hours; it is used for lighting portfires, &c. Slow-match may be equally well made by boiling in a solution of 8 oz. saltpetre to 1 gallon of water.

It is issued in skeins or parts of skeins, placed in a case with other stores. When large quantities are demanded it is issued in bales or casks. It should be demanded by weight; about 4 yards go to 1 lb.

Stored in Group II, Division IA.

Smoke Composition.

Smoke composition is used in the practice shell for the Q.F. 4·5-inch Howitzer.

It is composed of a mixture of :—

Lead sulphate.
Tin oxide.
Aluminium powder.
Naphthaline.

When the gunpowder burster of the shell is fired the ingredients of the smoke composition are vaporized, and then condense, forming a dense white smoke.

The particular substances in the mixture are used because of their smoke-producing qualities, and because the mixture formed is safe against prematures caused by the shock of discharge.

Smoke composition is not stored in bulk.

Sonite.

Sonite is a mixture of guncotton and soluble nitro-cotton, with the addition of 1 per cent. of mineral jelly.

It is made in tubular form, and has been tried for smokeless blank cartridges. For storage it is classed in Group I, Division II.

Tonite.

Tonite, or *nitrated guncotton*, is guncotton mixed with about an equal weight of barium nitrate.

This mixture is compressed into cylinders weighing about 3½ ozs., with a recess at one end to take the detonator.

The cylinders are wrapped in paraffined paper and then dipped in molten paraffin wax.

The nitrate supplies the extra oxygen needed for the complete detonation of guncotton.

Therefore tonite can be used for mining purposes, since no poisonous carbon monoxide is formed.

Tonite is a safer explosive than guncotton; it burns very slowly and without the least danger of explosion.

It is waterproof.

It is used in the Service in rockets :—

Rocket, Sound, 1½ lb., Mark III, and
Rocket, Light and Sound, 1 lb., Mark I.

Packing.—Tonite is packed, 50 cartridges in a quarter M.L. case, with a rectifier in a leather pocket.

Stored in Group I, Division II, and should not be stored in magazines the temperature of which exceeds 120° F.

Tri-nitro-toluene, or Trotyl.

Tri-nitro-toluene, or T.N.T., is used as an exploder for certain lyddite shell, both in the form of a brown powder and in a compressed condition.

It is formed by the action of nitric acid on toluene (C_7H_8), which is like benzine, and is obtained from coal-tar. It is neither difficult nor dangerous to make; it behaves as a very stable substance when exposed to the air under varying conditions of temperature (— 10° to + 50° C.) for several months. It melts at about 180° F.

It cannot be exploded by flame, nor by heating it in an open vessel. It is only slightly decomposed by strong percussion on an anvil.

It does not form metallic salts as picric acid does.

A fulminate detonator produces the best explosive effect, but it must be in the compressed form and strongly enclosed to be completely detonated.

It is slightly soluble in hot water; it is decomposed by dilute alkalies and alkaline carbonates.

It does not pick up moisture.

Trotyl is packed in barrels in the same way as picric powder; and the barrels are marked in a similar manner, but in *light green.*

Stored in Group I, Division III.

Substances used in the Manufacture of Explosives.

Acetone is a volatile, colourless liquid. It is obtained by heating calcium acetate.

Aluminium in a fine state of division mixed with a provider of oxygen burns with a bright flash and produces a dense white smoke.

Barium nitrate or *nitrate of baryta* is used as a source of oxygen.

Its rate of combustion is comparatively slow.

It is much less hygroscopic than saltpetre.

Barium salts impart a green colour to flames.

Charcoal carbon.—Carbon is the combustible element of most explosive mixtures, and is present, combined with hydrogen, in most explosive compounds.

When it combines with oxygen, the heat resulting causes increased volume of the gases produced.

Chlorate of potash is also used as a source of oxygen. As, however, it has the property of detonating on being rubbed or struck when mixed with sulphur or sulphide of antimony, it cannot with safety be used in the ordinary burning compositions. It would manifestly be highly dangerous to use it instead of saltpetre—in gunpowder, for instance. This property, however, renders it most useful in detonating compositions, in all of which (except when fulminate of mercury alone is used) it is to be found. In mixing these compositions, great care is necessary to avoid accidents from detonation.

Carbonate of copper, basic, commonly known as *verdigris,* gives a blue colour to flames, and is used in blue " fires " and " lights."

Carbonate of strontia gives a bright crimson colour to flames, and is used in red " fires " and " lights."

Calomel or *mercurous chloride* is used in the *blue lights,* &c. It increases the blue colour given to flames by copper salts.

Magnesium, in the finely divided state, mixed with a good provider of oxygen, burns with an intensely brilliant white flame and produces considerable smoke.

Naphthaline is a transparent crystalline solid derived from coal-tar. It is inflammable, and burns with a smoky flame.

Powdered glass is used to increase the sensitiveness of cap compositions that contain no fulminate.

Red orpiment or *sulphide of arsenic,* when mixed with saltpetre, burns with a brilliant white flame.

Saltpetre, or *nitrate of potash,* or *nitre,* is used as a source of oxygen. One cubic inch of saltpetre contains about 207 grains weight of available oxygen, equivalent to that contained in about 3,000 cubic inches of air. As most of the laboratory compositions burn in a more or less confined space, some source of oxygen is indispensable. This need is supplied by saltpetre, which is found in most non-detonating compositions.

Sulphide of antimony is readily ignited by the very sudden flash of fulminate of mercury.

When it burns it forms solid red-hot particles which are driven on to the explosive in the cartridge, so igniting it.

Mixed with chlorate of potash it explodes violently when heated. These properties make it useful in compositions intended to ignite other bodies at a little distance ; thus it is used in caps, friction tubes, &c.

Sulphur ignites at a low temperature (500° F.), and burns with a long flame.

This is useful in enabling the other ingredients of the various compositions, of which sulphur forms a part (*e.g.*, gunpowder), to ignite in the first place, and the heat given out by the sulphur, combining with the oxygen of the saltpetre, increases the rapidity and power of action of the whole. This substance is found in all the burning compositions, and, owing to its property of detonating with chlorate of potash, in most of the detonating compositions.

PACKAGES FOR MISCELLANEOUS EXPLOSIVES.

Cylinders, Ammunition.

Cylinders, ammunition, half and quarter barrels.—These are iron cylinders, used for the conveyance of almost all sorts of explosive stores.

Dimensions :—Half-size, $19\frac{1}{8}$ in. by 17 in.
Quarter-size, 19 in. by $14\frac{3}{4}$ in.

When explosives sent by rail are packed in these cylinders they are not subjected to such strict regulations as if sent in their ordinary packages.

Fig. 2.
METALLIC CYLINDER AMMUNITION QUARTER BARREL.
19 in. × $14\frac{3}{4}$ in.

Case, Transport, Explosives.

Cases, transport, explosives, Marks I and II*, are made of sheet steel, with angle pieces of metal, and a metal framework riveted to the top. The lid is secured by 12 screws and the joint is made by asbestos attached to the lid. Two metal hinged handles are riveted to the ends of the case. It is painted red.

Dimensions :—33⅜ in. by 18½ in. by 11 in.

Mark I cases are similar but have galvanized-iron wire handles.

Use.—The use of these cases is the same as that of the cylinders above mentioned.

Fig. 3.

METALLIC CASE TRANSPORT EXPLOSIVES.

33⅜ in. × 18½ in. × 11 in.

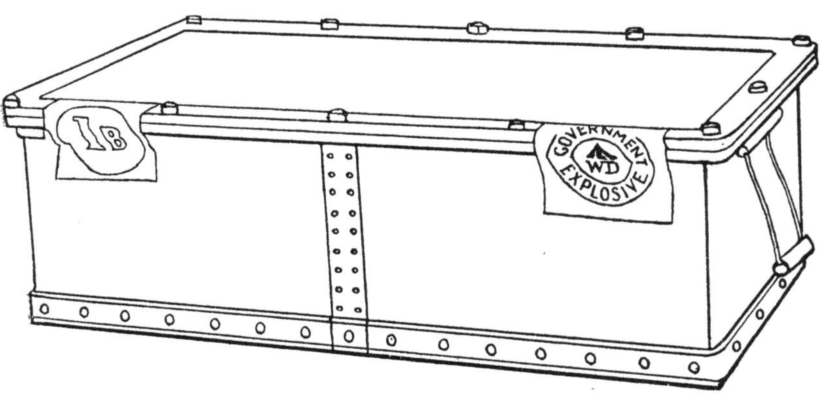

Table 4.

Combustible Compositions for Laboratory Stores, &c.

Friction Tubes.

Copper.	lb.	oz.	*"T" Tubes.*	lb.	oz.
Potash, chlorate of ...	0	6	Potash, chlorate of ...	0	6
Antimony, sulphide of ...	0	6	Antimony, sulphide of ...	0	6
Sulphur, ground	0	0½	Sulphur, ground	0	0½
			Powder, mealed	0	0½
			Glass, ground	0	0½

Quill.	lb.	oz.	*"T" (Push).*	Parts.
Potash, chlorate of ...	0	6	Potash, chlorate of ...	16
Antimony, sulphide of ...	0	6	Antimony, sulphide of ...	12
Sulphur, ground	0	0½	Glass, ground	3
Powder, mealed	0	0½	Sulphur, ground	1
Glass, ground	0	0½	Powder, mealed	1

TABLE 4—*continued*.

PERCUSSION PRIMERS.

Nos. 1, 2 *and* 4.

	Parts.
Potash, chlorate of ...	12
Antimony, sulphide of ...	18
Powder, mealed	1
Sulphur, ground	1
Glass, ground	1

No. 3.

	Parts.
Potash, chlorate of ...	14
Antimony, sulphide of ...	18
Fulminate of mercury ...	8
Sulphur, ground	1
Powder, mealed	1

PERCUSSION CAPS.

S.A., M.G. Ammunition (*Cordite*) *and Impulse Torpedo Tubes.*

	Parts.
Potash, chlorate of ...	14
Antimony, sulphide of ...	18
Fulminate of mercury ...	8
Sulphur, ground	1
Powder, mealed	1

S.A., M.G., 1-inch A.R. Ammunition (*Powder*).

	Parts.
Fulminate of mercury ...	6
Potash, chlorate of ...	6
Antimony, sulphide of ...	4

Q.F. 1-pr., 3-pr., 6-pr., 2·95-inch.

	Parts.
Potash, chlorate of ...	12
Antimony, sulphide of ...	18
Powder, mealed	1
Sulphur, ground	1
Glass, ground	1

Igniters, Torpedo.

	Parts.
Fulminate of mercury ...	33
Potash, chlorate of ...	42
Antimony, sulphide of ...	25

R.L. CAPS, AND DETONATORS USED IN FUZES.

Fuzes filled with C.E., Nos. 18, 19, 19A, 44 *and* 45.

4 grains of fulminate of mercury.

D.A. Impact, No. 13.

1 grain of mealed powder.

R.L. Caps, &c., used in most of Percussion, Time, and T. and P. Fuzes.

	Parts.
Fulminate of mercury ...	6
Potash, chlorate of ...	6
Antimony, sulphide of ...	4

(*Note.—The proportions of the above ingredients vary in some of the latest fuzes.*)

ROCKETS.

War.

	24-pr. lb.	9-pr. lb.
Saltpetre, ground ...	64·75	8·75
Sulphur, ground ...	14·75	2·0
Charcoal, alder ...	20·5	2·5

Life Saving. (Quick-burning Pellet.)

	lb.	oz.
Saltpetre, ground	8	9½
Charcoal, alder, ground ...	2	1
Sulphur, ground	1	13½

Signal.

	Parts.
Saltpetre, ground	8
Charcoal	3
Sulphur, ground	2

Life Saving.

	lb.	oz.
Saltpetre, ground	7	0
Charcoal, alder, ground ...	3	4
Sulphur, ground	2	0

TABLE 4—*continued.*

STARS FOR SIGNAL ROCKETS.

Signal, Service.	lb. oz.	*Blue.*	Parts.
Saltpetre, ground	8 0	Chlorate of potash ...	50
Sulphur, ground	2 0	Basic, carbonate of copper	8
Antimony, sulphide of ...	2 0	Gum shellac	10
Isinglass	0 3½	Calomel	32
Vinegar	1 quart		

Green.	Parts.	*Red.*	Parts.
Chlorate of baryta ...	5	Chlorate of potash ...	6
Chlorate of potash ...	3	Carbonate of strontia ...	1
Gum shellac	1	Shellac	1

Magnesium Light (for ½-lb. Light Rocket).	lb. oz.	*Red Star (for 1-lb. Light and Sound Rocket).*	Parts.
Magnesium powder ...	1 8	Chlorate of potash ...	5
Chlorate of potash ...	1 2	Carbonate of strontia ...	3
Nitrate of baryta	1 9½	Shellac	2
Oil, linseed, boiled ...	0 3		

STARS USED WITH SHELL.

Stars for all Star Shell.	lb. oz.	*Stars, Incendiary.*	lb. oz.
Nitrate of baryta	1 4¼	Indiarubber solution ...	0 2
Nitrate of potash	0 13½	Powder, mealed	0 5
Magnesium powder coated with paraffin	1 2	Saltpetre, ground	0 1
Add 3 per cent. boiled linseed oil.		Paraffin wax	0 0¼
		Coal-tar	0 1

LIGHTS.

Long, and Short, G.S. and Coastguard.	lb. oz.	*Blue.*	Parts.
Saltpetre	17 8	Chlorate of potash ...	50
Sulphur	4 6	Gum shellac	10
Red orpiment	1 4	Basic, carbonate of copper	8
		Calomel	32

Green.	Parts.	*Red.*	Parts.
Chlorate of baryta ...	5	Chlorate of potash ...	6
Chlorate of potash ...	3	Carbonate of strontia ...	1½
Gum shellac	1	Shellac, ground	1

TABLE 4—*continued.*

LIGHTS—*continued.*

Illuminating Wreck.

	lb.	oz.
Saltpetre	6	0
Sulphur	1	12
Red orpiment	0	9

Igniting Composition for G.S. and Coloured Lights.

	Parts.
Chlorate of potash ...	8
Gum shellac	6
Bone charcoal	1
Carbonate of potash ...	1
Charcoal, alder	1

"COMPOSITION PRIMING."

	Parts, by weight.
Gunpowder, R.F.G.[2], mealed	3
Guncotton dust	2

SMOKE COMPOSITION.

	Parts.
Lead sulphate	300
Tin oxide	1,080
Aluminium powder ...	240
Naphthalene	180

QUICK MATCH.

	2 threads.		3 threads.		4 threads.		6 threads.		10 threads.	
	lb.	oz.	lb.	oz.	lb.	oz.	lb.	oz.	lb.	oz.
Cotton wick ...	2	0	3	0	1	10	2	2	2	7
Gum arabic ...	0	10½	1	0	0	11	0	12	0	13
Powder,mealed,L.G.	18	0	26	0	20	0	20	0	24	0
Water, distilled ...	8 pints		12 pints		8 pints		9 pints		10 pints	

SLOW MATCH.

Hemp yarn, pure, Russian	lb. 100
Ashes, wood ...	bushel 1
Water	gals. 50

PORTFIRES.

	lb.	oz.
Saltpetre, ground	6	0
Sulphur, ground	2	0
Powder, mealed	1	4

NIGHT TRACERS.

FRICTION COMPOSITION.

	Parts.
Sulphur	135
Sulphide of antimony ...	345
Ground glass	300
Lamp black	187
Chlorate of potash ...	720

The above well mixed with gum water: Water, 1 pint; gum arabic, 1 oz.

PRIMING COMPOSITION.

	lb.	oz.
Mealed powder	4	0
Saltpetre	6	0
Magneisum (fine)	4	0
Linseed oil	0	4

ILLUMINATING COMPOSITION.

	lb.	oz.
Saltpetre	7	0
Sulphide of antimony	3	8
Sulphur	3	12
Magnesium (coarse)	4	4
Linseed oil	2	4¼
Resin	1	0

CHAPTER VII.

(A) Metals used in the Manufacture of Ammunition; (B) Textile and Paper Goods, &c.; (C) Terms used in the Manufacture of Ammunition.

(A) Metals used in the Manufacture of Ammunition.

Steel is an alloy of iron, cast in the fluid state into a malleable ingot. Carbon has a greater effect on steel than any other beneficial ingredient.

The composition of steel for the manufacture of shell differs according to the nature of the shell. Generally speaking steel for shell contains about ·35 per cent to ·7 per cent. carbon, together with small percentages of nickel, manganese and silicon.

Both cast and forged steel are used for the manufacture of shell; cast steel shell cannot be made with as thin walls as those of forged steel, because the material is not so good, and there is also always the risk of blow holes and porous metal being present.

The process of forging improves the material, it increases its tensile strength and minimises the chance of porous metal remaining.

Although the term *forged steel* shell is still used the process of forging shell under the hammer has been discontinued.

Forged shell are made by punching the solid cast ingots to form the cavity, and the exterior is *drawn* to shape by being forced through dies. Practically all shell go through the *heat treatment,* which is an annealing process; they are heated to a fairly high temperature and allowed to cool slowly. This not only relieves strains in the material but also improves the quality.

Certain shell are *oil toughened* by dipping in rape oil whilst red-hot.

This process improves the elasticity and tensile strength.

Shell for attack of armour are made of a higher percentage carbon alloy steel and the points are chilled to give a glass hard surface to them.

Cast iron contains about 2 per cent. carbon. It is brittle and hard, with no elasticity and practically no ductility.

The use of cast iron for projectiles is now almost entirely confined to practice shot.

Alloys.

Strictly speaking, all the brasses are alloys of copper and zinc, and the true bronzes alloys of copper and tin, but the term "bronze" is now wrongly applied to some of the brasses.

Brass is used principally for making Q.F., S.A.A. cases, tubes and various small parts used in tubes, fuzes, &c. It is an alloy of 70 parts copper and 30 parts zinc.

Brass is a soft material, makes good castings, and is easily worked in the lathe.

The term " metal " is used to denote the group of alloys known as

Class A, B and C metals.

These metals are supplied in the form of rolled and drawn bars.

The contractor is not tied down to any exact composition, and the metal is accepted provided it complies with the specification limits of *tenacity*.

They are all alloys of 60 per cent. copper with 40 per cent. zinc. Small percentages of iron, manganese, tin, &c., may be present.

Class A metal is 1½ times the strength of *Class B*.

Class C metal is lower in strength and quality, and is practically the ordinary commercial *brass*.

Aluminium bronze is an alloy of copper with about 10 per cent. aluminium.

It is hard, forgeable, and practically non-corrodible.

Manganese bronze is a 60—40 alloy of copper and zinc, with the addition of 1 per cent. manganese and iron. It possesses mechanical properties similar to those of mild steel, but is relatively non-corrodible.

Phosphor bronze is a true bronze, and is an alloy of copper and tin, with a small percentage of phosphorus.

It is hard and somewhat brittle.

It cannot be forged.

Forgeable alloys are known as No. 1 and No. 2. They are both 60—40 alloys of copper and zinc. *Forgeable No.* 1 must not contain more than ·03 per cent. lead, and is sometimes called *Leadless alloy*.

Forgeable No. 2 contains about 1 per cent. lead.

Gunmetal is an alloy of copper and tin; the actual proportions are varied according to the qualities required. It is about half as strong as mild steel.

It is unforgeable.

Crown metal is an alloy of copper, nickel and zinc. It is hard and does not tarnish.

It is practically the same as *German silver*.

White metal is a very brittle alloy of 4 parts of tin and 1 of zinc.

Iridio-platinum is an alloy composed of 10 per cent. iridium and 90 per cent. platinum by weight.

Platinum silver is an alloy of 1 part platinum and 2 parts silver by weight.

(B) Textile and Paper Goods, &c.

The undermentioned materials, when used in connection with Service ammunition, must all be of the best quality and free from acid.

Textile Goods.

Batiste is cambric, waterproofed by being coated with rubber on both sides.

It is of extremely fine texture.

Cambric is a species of fine white linen.

Cotton is made from the down of the cotton tree.

Dowlas is made of flax, in the form of a closely-woven canvas.

Fearnought is a woven woollen fabric, similar in appearance to felt.

Felt is made of a mixture of wool, hair, and vegetable fibre.

Flax is a plant, the stalks of which yield a fibre which is used for making thread and cloth, such as linen, cambric, lawn, lace, &c.

Jean is a cotton fabric of medium texture.

Jute is the fibre of the inner bark of an Indian plant from which coarse fabrics such as bags, mats, yarn, &c., are made.

Lasting cloth is made of all wool. It is closely woven, and must possess a smooth and even surface; it is fireproof under certain tests laid down.

Linen tape and *linen thread* are made of half-bleached flax.

Shalloon cloth and *shalloon braid* are made from all worsted.

Silk cloth and *silk braid* are made from pure silk obtained from the outside of the cocoons. This silk is often too short and coarse for the manufacture of fine strong material.

Silk webbing and silk sewing (or silk thread) are made from long draft, spun silk obtained from the interior of the cocoons. They are of superior quality and strength.

Tape, white, is made of bleached cotton.

Vulcanized cashmere is a woollen cashmere material, vulcanized.

Worsted is usually made of the coarser samples of wool.

Paper Goods.

Cardboard is formed by *pasting* together as many sheets of fine paper as will give the required thickness; so that both "middle" and "facings" are of the same quality.

Leatherboard is made of a number of sheets of leather paper pressed together.

It must be pliable and tough.

Millboard is produced to any required thickness by successive dips into a vat of pulp; hence millboards are solid single boards.

Papier-mâché is formed by *glueing* together sheets of strong brown paper, the upper surface being sometimes coated with hard enamel. There is also a papier-mâché made by running pulp into moulds and subsequently coating it.

Pasteboard is made like cardboard, but the "middles" are of an inferior quality of paper.

Strawboard is a thick board made of straw-pulp, run into moulds and pressed.

Vegetable paper.—This is a thin, transparent, waterproof paper; it is made entirely of vegetable substances and is used to waterproof the fuze composition in the time rings of all modern Time and T. and P. fuzes.

Vulcanized paper.—This is a thick, transparent paper which has been specially treated so as to make it a perfect insulator; it is used to insulate the contact piece from the body in the Tube, V.S. Electric, "S," Mark I.

Ebonite is indiarubber, 5 parts, mixed with 2 to 3 parts of sulphur, and cured at 167° F., under pressure.

It is a good electrical insulator, but is brittle and deteriorates with age.

Vulcanized fibre is made either of wood-pulp or paper, and vulcanized with the addition of sulphur and French chalk.

TABLE 5.

Non-Combustible Compositions for Laboratory Stores.

Luting, Mark I.

Tallow } Equal parts.
Beeswax }

Luting, Mark III.

	Parts.
Whiting	80
Mineral jelly	20
Castor oil	1

Transparent Lacquer for Brass Work.

	lb.	oz.
Seed lac	5	0
Turmeric	2	8
Spirits, methylated ...	gals.	5

Black Lacquer for Q.F. Cases.

Methylated spirits... ...	gals.	3
Seed lac	lbs.	3
Turps	gal.	0½
Vegetable black	lbs.	2

Lacquer for Inside of Shells.

	lb.	oz.
Resin	12	0
Brown, Spanish	2	0
Plaster of Paris	1	0
Turpentine, spirits ...	pint	0½

Velvril Paint for Inside of Shell.

	Parts.
Zinc oxide	24·0
Ochre, yellow	3·5
Red oxide of iron ...	0·5
Nitrated castor oil ...	15·0
Nitro-cellulose (of a very low degree of nitration)...	7·5
Acetone oil	60·0

Lubricant for Lids, Q.F. Cartridges.

Tallow, Russian } Equal parts.
Beeswax ... }

Beeswax, Composition.

	lb.	oz.
Pitch, Swedish	16	8
Tallow, Russian	4	8
Beeswax	44	0
Resin	36	0

Varnish for Inside of Capped Common Pointed and A.P. Shell.

	lb.	oz.
Shellac	3	0
Methylated spirits... ...	pints	4½
Terebene	pint	0¾

Varnish for Lyddite Shell.

Copal varnish, free from all metallic impurity in any form.

Waterproof Varnish for Percussion Caps.

	lb.	oz.
Shellac, gum	2	2
Spirits, methylated ...	gal.	1

Pettman's Cement for Waterproofing Purposes.

	lb.	oz.
Shellac gum	7	8
Methylated spirits (1 gal.)	8	0
Tar, Stockholm (½ gal.) ...	5	0
Red, Venetian	20	12

Shellac Putty.

	lb.	oz.
Whiting	6	0
Shellac, gum	2	0
Spirits, methylated ...	qrt.	1

Paste.

Flour	lbs.	6
Water	gals.	2
Glue	ozs.	2
Carbolic acid ...	fluid ozs.	3

Day Tracer Cement.

	Parts.
Venetian red	19
Gum arabic	3
Glycerine	1

(C) Terms used in the Manufacture of Ammunition.

Annealing.

Annealing is the subjection to a long continued heat of metals that are naturally brittle, or rendered so in the processes of manufacture.

The material is then allowed to cool slowly, and becomes tougher and is better able to withstand shock in consequence.

Tubes, S.A. and Q.F. brass cases are annealed to take the strains out of the metal set up during manufacture or on firing.

Shell are annealed to relieve the metal of strains set up during manufacture. This annealing is technically termed "*heat treatment.*"

Blending.

Blending is the process by which a number of small batches of an explosive, which may vary slightly owing to small differences in the raw materials, conditions of manufacture, &c., are mixed up together to form a uniform lot for issue to the Service. (*See* Gunpowder and Cordite.)

Burring.

Burring is the bending over of an upstanding edge of metal on an object to retain another part in position. The ends of V.S. tubes are *burred* over to keep the closing cork discs in position.

Canneluring.

Canneluring is a process employed in some instances to attach the bullet to its brass case.

The case is pressed all round into one of the cannelures on the bullet as in the Webley pistol cartidges.

Choking.

Choking is the usual method of closing bags containing explosives, in powder form, namely, blank cartridges, exploders, &c.

It is also used to reduce the opening in signal rockets, &c. The neck of the bag to be closed is drawn together into several pleats, and then tied round with silk sewing.

Coning.

Coning of the brass case is sometimes used to hold the projectile to its case. The mouth of the brass case is pressed all round into a groove formed either on the projectile or on its driving band, or is pressed on to the parallel portion.

In the latter instance the case is usually also *indented.*

Frapping.

Frapping is the drawing together of two or more ropes or wires. It is used in the electric detonators and fuzes to draw together the insulated wires issuing from the head. A length of thread is passed several times round the two wires, drawn tight and tied.

Hooping.

Hooping is used to strengthen a cartridge when made up and enable it to retain its shape.

A silk braid is passed round the cartridge, being threaded through slits in the silk cloth, and is then tied in a knot. This is termed a *hoop*.

The number of hoops on a cartridge varies with its weight and length.

Most of the latest cordite cartridges are not hooped.

Indenting.

Indenting is used to hold the projectile to its brass case. In some instances three or more impressions are formed in the case, so as to grip the projectile in one of its cannelures.

Indenting is also used to hold together the two portions of some detonators.

Milling.

Milling is the shaping of metals by means of slowly revolving circular cutters.

Milling is quicker and more accurate than *planing* and *shaping*.

Milled Head.

Milled head.—The edges of screws, &c., are sometimes milled into a succession of ridges to enable the fingers to obtain a hold without slipping.

Riveting.

Riveting means either the turning over of the heads of rivets, to clamp two or more thicknesses of metal together, or the hammering over of the edge of one metal article to make a tight joint with another which is driven or screwed into the former.

Spinning.

Spinning is the moulding of circular articles of thin sheet metal by pressure applied during rotation in a lathe, *i.e.*, aluminium containers for A.P. shell are *spun*.

Spinning is also used to turn down the upstanding edges of various articles to hold other parts in place, *i.e.*, the brass disc over the steel plug in Fuzes, D.A., Impact No. 18 is *spun* in.

Stabbing.

Stabbing is used to prevent a screwed part working loose.

Three or more centre punch " dabs " are given to the metal where the fixed and movable parts meet.

Stemming.

Stemming is the process used to press the C.E. into the various channels of some of the fuzes. It is done by hand with a wooden rod or *drift*.

Rockets, lights, &c., are *driven* or *stemmed* with the rocket composition in a similar manner, but the drift is used with a mallet.

Sweating.

Sweating is the soldering of metallic surfaces without the use of a soldering iron.

The surfaces are cleaned, heated and covered with a film of solder. The articles to be united are then brought together and heated till the solder flows and unites.

CHAPTER VIII.—MINES AND WARHEADS.

Mine, Spherical, Mark II.

Plate III.

The mine case is made of two thin steel hemispheres riveted together to form a sphere, with openings at the top and bottom. The lower opening is just sufficiently large to take a bronze spindle that passes through it. The top opening is large enough to enable the wood platform and cases of guncotton to be passed through when fitting up and filling the mine. The bronze spindle screws into the lower crosshead of the safety mooring arrangement in the interior of the mine; its outer end is flattened and has a hole drilled in it. About 17 feet of steel chain is attached to the spindle, and to this chain is fastened the steel cable of the *sinker*.

At the bottom of the mine case about 50 lb. of lead, cast in segments, is secured in position by lead run in.

The interior of the mine is fitted with a wood platform, with four sides and four corner pieces.

The cases of guncotton are placed in position on the platform, a felt pad placed on top, and on this a wood cover, held in position by two battens, secured by bolts and nuts. The mouth of the mine is closed by a dermatine washer and steel plate tightly nutted down.

The interior of the mine and the wood fittings are painted red and the exterior grey. When filled, mines have the following information stencilled on the side:—

Monogram of filling station.
Serial number.
Mark.
Date of filling.
Actual and total weights.

The mine contains about 260 lb. of guncotton, contained in 11 tinned copper cases, and weighs, filled, about 860 lb.

The firing, safety and sinker arrangements are confidential.

The Mark I mine differs from the Mark II in having the guncotton packed in a greater number of smaller cases.

A number of R.E. mines, which held 100 lb. of guncotton, have been converted to hold 260 lb., and are known as:—

Mines, Spherical, Mark I (or II), Converted.

These mines are slightly larger than the Mark II described above. The guncotton is not contained in copper cases, but is packed direct into the box formed by the deck and the side pieces. These mines have no safety mooring device.

A large number of mines have been filled with lyddite. The molten picric acid is poured into cases similar to those for guncotton, and these are packed into the mine in the usual manner. Only seven cases are used. Mines so filled are distinguished by a yellow band painted round the neck.

Torpedo Warheads.

A warhead carries the explosive charge (guncotton or trotyl) of a torpedo, and is attached to the front of the latter by a number of screws.

Warheads are made of thin sheet phosphor bronze to the form shown in Fig. 4, with a flanged metal ring brazed to the end. To this ring is attached, by a number of bolts and nuts, a dished metal closing plate or "*door*," which is made a watertight fit by means of a rubber washer. These doors, for guncotton filling, are fitted with the usual re-wetting plugs; for trotyl filling the plugs are not, of course, required.

Inside, along the lower side of the head, is attached a lead ballast, and fitting into the nose is a metal plate, bored and threaded to receive the firing *pistol*.

Brazed to this nose plate is a brass tube to take the *charge priming*.

Along the upper side of the head is fitted a "*deck*," which is intended to form an air space to give the necessary buoyancy.

The decks are made of sheet copper, and for guncotton filling are in the form of an open framework; for trotyl they are entirely closed in.

The interior of warheads and the fittings are painted black; for trotyl they are varnished. The exterior is left bright with the exception of the doors, which are painted black or, when filled with trotyl, yellow with a green band across. On the door is stencilled the following information:—

Number of the head.
Size, type and Mark of head.
Actual weight.
Low weight (for guncotton only).
Date and station of filling.
Guncotton yarn (if fitted).
Weight and lot number of trotyl (if so filled).
Weight of water to dry guncotton (if so filled).
"N."

Stamped on the nose:—

↑

Mark of head.

Type of head.

Number of head.

Contractor's initials.

Date of manufacture.

Trotyl in red (if so filled).

S.N. or S.D. stamped on the nose denotes strengthened nose or deck.

To fill the heads with guncotton the slabs are cut to the required shape to conform to the interior, and are packed straight in in layers starting at the nose.

Each layer of slabs is numbered consecutively from the nose, and every slab in a layer has its layer number stencilled on it. A 21-in. Mark II head requires 123 slabs, packed in 17 layers to fill it, *i.e.*, 280 lb. wet Guncotton.

Over the top layer is coiled a flat spiral of No. 4 guncotton yarn, wetted, to form a packing.

To fill with trotyl the substance is melted and simply poured in and allowed to set.

Early Marks of warheads had wooden decks and felt packing pieces.

Warheads are packed and issued filled, one in a wood case.

In the lid of the case is a small hinged door to give ready access to the re-wetting plugs.

For details of warheads, *see* Table 6.

Fig. 4.

WARHEAD, 21-INCH TORPEDO, MARK II, FOR TROTYL.

Scale $\frac{1}{16}$.

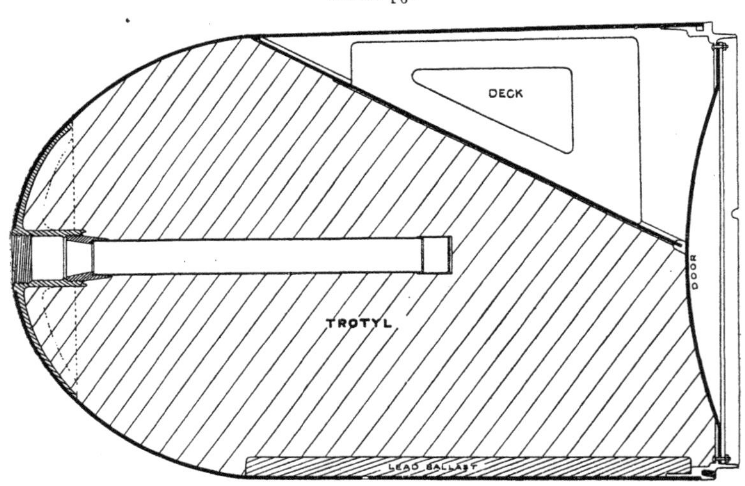

TABLE 6.—*Warheads for Torpedoes.*

Type of Torpedo.	Guncotton filled Heads ($22\frac{1}{2}$ parts water).					Trotyl filled Heads.			Method of Securing Head to Air Vessel.
	Warhead empty with door and Ballast.	Guncotton Charge, including $22\frac{1}{2}$ per cent. of water.	Deck (Sheet, Copper).	Guncotton Yarn or Felt (between Door and Charge).	Warhead complete with Deck and Guncotton Yarn, but without Pistol.	Warhead empty, with Fixed Deck, Felt Strip, Door and Ballast.	Trotyl Charge.	Warhead complete, but without Pistol.	
	lb. oz.	lb. oz.	lb. oz.	lb. oz.	lb. oz.	lb. oz.	lb. oz.	lb. oz.	
14-inch, R.G.F., Mark IX	30 $6\frac{1}{2}$	94 4	3 4	0 $2\frac{1}{2}$	128 1	...	...	...	Male flange on head, 1·2 inch deep; V pieces; secured by 12 diagonal screws.
,, ,, ,, X & X*	37 $5\frac{1}{2}$	94 4	3 12	0 $2\frac{1}{2}$	135 8	...	...	...	
,, ,, ,, XI	47 7	94 4	2 $11\frac{1}{2}$	0 $2\frac{1}{2}$	144 9	...	...	...	
,, Weymouth, Mark I	57 0	94 4	2 $13\frac{1}{2}$	0 $2\frac{1}{2}$	154 4	...	...	...	Male flange on head $\frac{13}{16}$ inch deep; secured by 12 diagonal screws; one stud.
18-inch, R.G.F., Mark I*	83 10	230 5	7 4	1 6	322 9	...	...	...	Male flange on head, 1·5 inch deep; V pieces; secured by 36 screws placed zig-zag.

NOTE.—When guncotton and trotyl warheads are shown against same torpedo, some heads are filled with one and some with the other explosive.

TABLE 6.—*Warheads for Torpedoes*—continued.

Type of Torpedo.	Guncotton filled Heads (22½ parts water).					Trotyl filled Heads.			Method of Securing Head to Air Vessel.
	Warhead empty with door and Ballast.	Guncotton charge, including 22½ per cent. of water.	Deck (Sheet Copper).	Guncotton Yarn or Felt (between Door and Charge).	Warhead complete with Deck and Guncotton Yarn, but without Pistol.	Warhead empty, with Fixed Deck, Felt Strip, Door and Ballast.	Trotyl Charge.	Warhead complete, but without Pistol.	
	lb. oz.	lb. oz.	lb. oz.	lb. oz.	lb. oz.	lb. oz.	lb. oz.	lb. oz.	
18-inch, R.G.F., Mark II*	83 10	230 5	7 4	1 6	322 9	...	...	...	Male flange on head 1·5 inch deep; V pieces; secured by 18 diagonal screws.
,, ,, ,, III & IV	91 7¼	209 6½	7 5	0 3¾	308 6½	...	...	...	
,, ,, ,, V & V*	75 15	209 6½	6 9½	0 3½	292 2½	...	...	...	
,, ,, ,, VI–VI*** H ...	87 8½	200 0	5 12	0 3½	293 8	...	...	...	Male flange on head 1·2 inch deep; U pieces; secured by 18 diagonal screws.
,, ,, ,, VII–VII*** ...	87 4	200 0	6 0	0 4	293 8	98 0	240 0	338 0	
,, R.N.T.F., Mark VIII	...	...	...	...	...	93 12	320 0	413 12	Male flange on head 1·2 inch deep; U pieces; secured by 24 diagonal screws.
,, Fiume, Mark II & II*	75 8	228 10	7 2	2 0	313 4	...	...	...	Male flange on head ·625 inch deep; guide on top; secured by 12 diagonal steel screws.

,, ,, ,, III	68 12	200 0	13 0	0 4	282 0	...	...	...	Male flange on head ·826 inch deep; V pieces; secured by 18 diagonal screws.
,, ,, ,, III* & III* H ...	76 10	200 0	10 1	0 3	286 14	...	...	...	Male flange on head ·826 inch deep; stud in lieu of V pieces; secured by 18 diagonal screws.
,, Weymouth, Mark I	70 4	198 0	6 2	0 4	274 10	...	...	...	Male flange on head ·826 inch deep; V pieces; secured by 18 diagonal screws.
,, ,, ,, I*	72 12	198 0	6 2	0 4	277 2	...	...	...	
21-inch, R.G.F., Mark I & I*	119 15	225 0	8 0½	0 4½	353 4	...	...	...	Male flange on head 1·2 inch deep; U pieces; secured by 18 diagonal screws.
,, ,, ,, II–II***	135 5	280 0	9 6½	0 4½	425 0	149 8	400 0	549 8	
,, ,, ,, II*** V.B.	...	...	...	...	...	145 8	302 0	447 8	
,, Weymouth, Mark I	132 7	330 0	6 12	0 4	469 7	...	...	...	Male flange on head 1·125 inch deep; U pieces; secured by 18 diagonal screws.

Note.—When guncotton and trotyl warheads are shown against same torpedo, some heads are filled with one and some with the other explosive.

CHAPTER IX.—CARTRIDGES FOR B.L. AND B.L.C. GUNS AND HOWITZERS.

(A) General Remarks on B.L. Cartridges; (B) Description of Cordite Cartridges for B.L. Guns; (C) Description of Cordite Cartridges for B.L. Howitzers; (D) Marking on B.L. Cordite Cartridges; (E) Cordite Cartridges for R.M.L. Guns; (F) Powder Cartridges for B.L. Guns and Howitzers; (G) Drill Cartridges and Cartridges for Instruction.

SECTION (A). GENERAL REMARKS ON B.L. CARTRIDGES.

Materials used and requisite properties of:—Classes of silk cloth and shalloon, silk and shalloon braid, silk webbing, silk sewing, various charges used, cartridges made up into fractions, adjusted charges, density of loading, specific volume, igniters, making up and issue.

For safety, convenience, and rapidity in loading, the powder or cordite for charges for B.L. guns and howitzers is placed in a bag, which is called an "empty cartridge."

Requisite properties of material:—

1. It should be strong enough to bear reasonable knocking about when filled, and to stand the wear and tear of travelling.
2. If it is to contain powder, it should be so close in texture that the powder, even if slightly dusty, will not readily work its way through.
3. It must be permeable to the flash of the tube.
4. It should not deteriorate in store from the chemical action of the explosive upon it, or from any other cause.
5. It should be entirely consumed in the gun when fired; no smouldering fragments or sparks should be left in the bore.

If much residue is left, the vent is apt to get choked; if sparks remain, and the gun be re-loaded almost immediately, a serious accident will probably occur.

Several accidents have thus occurred, generally when using blank cartridges.

Materials used.—These conditions are very well fulfilled by the materials in the service, namely, silk cloth and shalloon.

The materials are examined with a magnifying glass for closeness of texture, and tested chemically for purity; any admixture of cotton would be most objectionable, as it tends to hold fire.

Silk cloth.—Silk cloth is made of the refuse silk from the outside of the cocoons. It is strong and of a close texture, and has little tendency to "hold fire" or smoulder. The cloth is quite free from all acids, and no substance is to be added to the silk while in the course of manufacture or after the cloth is finished.

The first issues of silk cloth were steeped in a cold saturated solution of boracic acid and at a temperature not exceeding 120° F., to prevent fungoid growths forming on the cartridge, but this was found to be unnecessary and was discontinued. Silk cloth was originally introduced on the score of safety for blank cartridges, as in firing these there is not so much heat and pressure as when shotted rounds are fired, and therefore less chance of the cartridge being completely consumed.

Classes of silk cloth.—Silk cloth is divided into three classes—Nos. 1, 2 and 3, according to its strength ; No. 3 is the strongest.

Test for silk cloth.—To test silk cloth, a test piece free from holes for braids should be cut from the cartridge in the direction of either warp or weft, 10 in. long and 1 in. wide (when lengths of 10 in. cannot be obtained, shorter pieces must be used). The test piece is passed through the ring of a weight made up to the necessary amount, which it must support when lifted by the two ends.

The following table gives the minimum weights the three classes of silk cloth should lift (when tested as above) to be considered serviceable :—

No. 1 Class, minimum for use	.. ..	56 lb.
,, 2 ,, ,, ,,	.. ..	70 ,,
,, 3 ,, ,, ,,	.. ..	84 ,,

Cordite cartridges made of silk cloth.—Class No. 1 is used for all cartridges for guns 30-pr. up to and including 6-inch.

Class No. 2 is used for all cartridges for 7·5-inch guns and upwards, except half charges for 12-inch and 13·5-inch, and all charges for 15-inch.

Class No. 3 is used for half charges for 12-inch and 13·5-inch guns, and all charges for 15-inch.

All powder cartridges, with the exception of the B.L. 5-inch howitzer blank, are made of silk cloth.

Class I silk cloth is used for powder charges up to 14 lbs. in weight inclusive.

Class II silk cloth is used for powder charges from 14 lbs. to 85 lbs. in weight.

Class III silk cloth is used for powder charges above 85 lbs., and for all prism powder cartridges ; this is on account of the sharp edges and the hardnesss of the prisms.

Shalloon.—Shalloon was originally introduced because being thinner than silk cloth it formed less obstruction to the flash from the tube where the cartridge is choked.

The choke was liable to come underneath the vent when a very small cartridge in radial vented guns was used.

It is made entirely of "long" wool and is woven twilled ; it is dyed red for service use, but issues have been made of undyed shalloon.

There are two classes of shalloon, namely :—

No. 1 for laboratory work.

No. 2 for necks of burster bags.

Shalloon, though not so strong as silk cloth, is more permeable to the flash of a tube ; it is used for the smaller cartridges, *i.e.*, for B.L. 2·75-inch, 10-pr., 12-pr. of 6 cwt., 15-pr., and for all B.L. howitzers.

It is also used for the igniters of all B.L. cartridges.

Silk braid.—Silk braid for hooping is made in two widths—0·35 in. and 0·65 in. These braids must support weights of 28 and 85 lb. respectively. For beckets the silk braid is also made in two widths—1·5 in. and 1 in.—which must support weights of 250 and 160 lb.

The 0·35 in. braid was also used in securing the outer layer of cordite sticks in the building up of the heavier type of cartridges.

Silk webbing.—Silk webbing is made from pure, long draft spun silk; it is supplied 0·35 in. in width, and is used in securing the outer layer of cordite sticks in the building up of the heavier type of cartridges instead of the 0·35 in. silk braid when the existing stocks of the latter are used up.

Silk webbing is much stronger than the silk braid. The breaking-load of a piece of silk webbing, fixed in a testing machine, the clamps being 20 in. apart, must not be less than 80 lbs.

Shalloon braid.—Shalloon braid, undyed, for hooping, is made in three widths, 0·35, 0·65, and 1 in.

The 0·35 in. braid is also used in making up cordite charges for Q.F. guns.

Silk sewing.—Silk sewing is used for sewing up cartridge bags, and igniters, and for hooping and choking the smaller natures of blank cartridges; it is also used in building up the cordite charges for B.L. guns and howitzers. The following table gives the minimum weights the various natures of silk sewing must support:—

Silk sewing:—

No. 1 should support	.. .. ..	13 lbs.
No. 2 „ „	.. .. ..	$5\frac{1}{2}$ „
No. 20–3 „ „	.. .. ..	$5\frac{1}{2}$ „
No. 30–3 „ „	.. .. ..	4 „
No. 40–3 „ „	.. .. ..	$2\frac{3}{4}$ „
No. 50–3 „ „	.. .. ..	$2\frac{1}{4}$ „
No. 60–3 „ „	.. .. ..	$1\frac{3}{4}$ „

Silk sewing used in making up cordite cartridges is first greased with vaseline to prevent rotting.

Tape, linen.—Linen tape 1 in. wide is now used for lifting beckets for heavy B.L. cordite cartridges, *i.e.*, 7·5-inch and up; a linen tape becket is also used with certain B.L. 6-inch cartridges when packed in rectangular cases for Naval Service.

The various charges are as follows:—

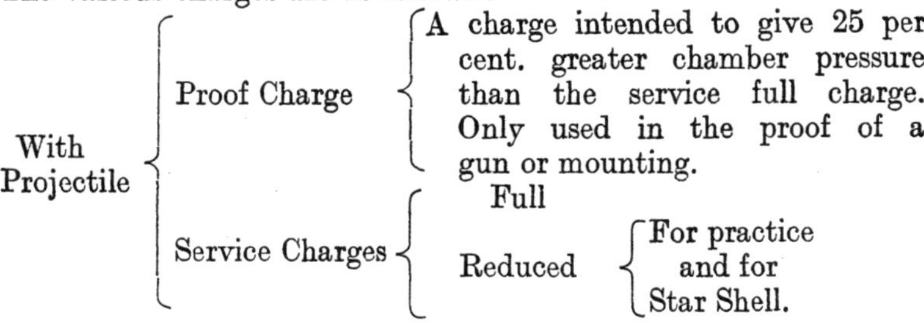

Without Projectile } Blank Charge.

Note.—There are in addition certain charges which are fired with paper shot.

Blank charges.—Gunpowder (Blank L.G.) is generally used for blank charges ; cordite is unsuitable, as, in the absence of a projectile, there is not enough pressure to cause cordite to burn with the necessary rapidity. Smokeless blank cartridges have been introduced for certain field guns.

Charges made up in fractions.—Charges, both powder and cordite, for the heavier guns are made up in fractions for convenience in handling, and to enable reduced charges to be fired :—

<table>
<tr><td>B.L.C. gun,</td><td>6-inch .. ..</td><td>½ charges.</td></tr>
<tr><td>B.L. gun,</td><td>6-inch, Marks VII and VIII ..</td><td>½, ⅓ and ⅔ charges.</td></tr>
<tr><td>,,</td><td>6-inch, Mark XI .. ..</td><td>⅓ and ⅔ ,,</td></tr>
<tr><td>,,</td><td>7·5-inch .. ..</td><td rowspan="4">½ and ¼ charges.</td></tr>
<tr><td>,,</td><td>9·2-inch (except on H.A. mountings)</td></tr>
<tr><td>,,</td><td>10-inch .. ..</td></tr>
<tr><td>,,</td><td>12-inch .. ..</td></tr>
<tr><td>,,</td><td>13·5-inch .. .. ..</td><td rowspan="2">¼ charges.</td></tr>
<tr><td>,,</td><td>15-inch .. ..</td></tr>
</table>

The 9·2-inch gun on H.A. mountings fires various charges.

Adjusted charges.—With a view to eliminating one of the sources of error in shooting, B.L. and Q.F. cartridges 12-pr. and upwards, will, where this is possible, have the weight of the cordite charge adjusted, so that, so far as the charge is concerned, the gun will shoot up to its normal velocity. The charges will continue to be designated by their normal weight, but will be distinguished by having the letters A.C. (denoting adjusted charge) appended to the Lot No. of the cordite as follows :—

B.L. cartridges.—Stencilled or stamped on back of bag.

Q.F. cartridges. *Separate loading* (*except* 4·5-*inch howitzer*).—Written or stamped on label affixed to lid.

Q.F. cartridges. *Fixed ammunition* (*and* 4·5-*inch howitzer*).—Stencilled on base of case.

The first issue of cartridges with adjusted charges for Naval Service had the actual weight of cordite (plus or minus) necessary for the adjustment appended to the Lot No., instead of the letters A.C.

Specific volume (S.V.).—The specific volume of a charge in the chamber of a gun is the number of cubic inches allotted to each lb. in the charge ; it is obtained by dividing the capacity of the chamber in cubic inches by the weight of the charge in lbs.

Example :—
$$\frac{\text{Chamber capacity} = 5{,}546 \text{ cubic in.}}{\text{Weight of charge} = 100 \text{ lbs.}} = 55{\cdot}46 = \text{S.V.}$$

Density of loading.—The *density of loading* of a charge in the chamber of a gun is the ratio of its weight to the weight of water which will fill the space behind the projectile in the gun.

To find out the *density of loading* it is necessary to know the space occupied by 1 lb. of water, *i.e.*, 27·73 cubic inches.

Density of loading is then found by dividing the number of cubic inches occupied by 1 lb. of water (27·73) by the "specific volume" of the charge.

$$\text{In this case } \frac{27{\cdot}73}{55{\cdot}46} = {\cdot}5 = \text{Density of loading.}$$

Igniters.—Every B.L. cordite cartridge requires an igniter, which is attached to, and forms part of, the cartridge.

In the case of small natures, where the cartridge is made up as a full charge only, an igniter is attached to each end; in the case of heavier natures, 7·5-inch and up, the cartridge is fitted with an igniter at one end only.

Cartridges for the B.L. 6-inch guns, Mark VII to XII, 9·2-inch H.A. and the B.L.C. 6-inch are issued, laced together, to form a "full charge," and have then an igniter at each end.

All cartridges made up with M.D. cordite, and most of those of Mark I cordite, have their igniters filled with R.F.G.[2] gunpowder, but in certain marks of the latter cartridges, below 6-inch, guncotton yarn (waterproofed by being dipped in indiarubber solution) has been used.

Blank F.G. and S.F.G.[2] gunpowder has also been used for filling igniters of cordite cartridges.

The use of Blank F.G. as an alternative to R.F.G.[2] for filling igniters is now discontinued, and the existing stock of S.F.G.[2] will be used up for filling igniters of certain Mark I cordite cartridges for Land Service.

Igniters for B.L. gun cartridges up to the 5-inch inclusive (except 2·75-inch, 10-pr., 12-pr. and 15-pr.) are cross-stitched radially to form four compartments; 6-inch and up are now fitted with standardized igniters.

These igniters are suitable for different charges for the same gun; they are stitched across to form five parallel compartments in the case of igniters for 6-inch to 13·5-inch, and six compartments in the case of 15-inch igniters.

Most of the above guns now have two different sizes of standardized igniters for making up their cordite charges, *i.e.*—

Igniter "A" and Igniter "B."

They differ only in dimensions as shown in the accompanying table.

TABLE 7.

Nature of Igniter.	Number of Compartments.	Weight of Powder.	Maximum Diameter.	Remarks.
		ozs.	inches.	
15-inch "A" ...	6	16	14·0	—
13·5-inch "A" ...	5	16	12·25	—
12-inch "A" ...	5	16	12·25	—
12-inch "B" ...	5	10	11·5	—
10-inch "A" ...	5	8	11·0	—
10-inch "B" ...	5	12	11·25	—
9·2-inch "A" ...	5	8	11·0	—
9·2-inch "B" ...	5	8	11·0	Central hole for stick.
9·2-inch "C" ...	5	8	10·0	Central hole for stick.
9·2-inch "D" ...	5	6	10·25	—
7·5-inch "A" ...	5	6	8·9	—
6-inch "A" ...	5	2	7·4	4 felt studs.
6-inch "B" ...	5	2	7·1	4 felt studs.

Earlier marks of B.L. cartridges 6-inch and up were fitted with igniters cross-stitched radially to form four compartments in the case of 6-inch and 7·5-inch, and eight compartments in the case of heavier natures.

Fig. 5.

SPECIMENS OF OLD AND NEW TYPE OF IGNITERS FOR B.L. GUN CARTRIDGES 6 INCH AND UP.

New type. (Standardized Igniters.)

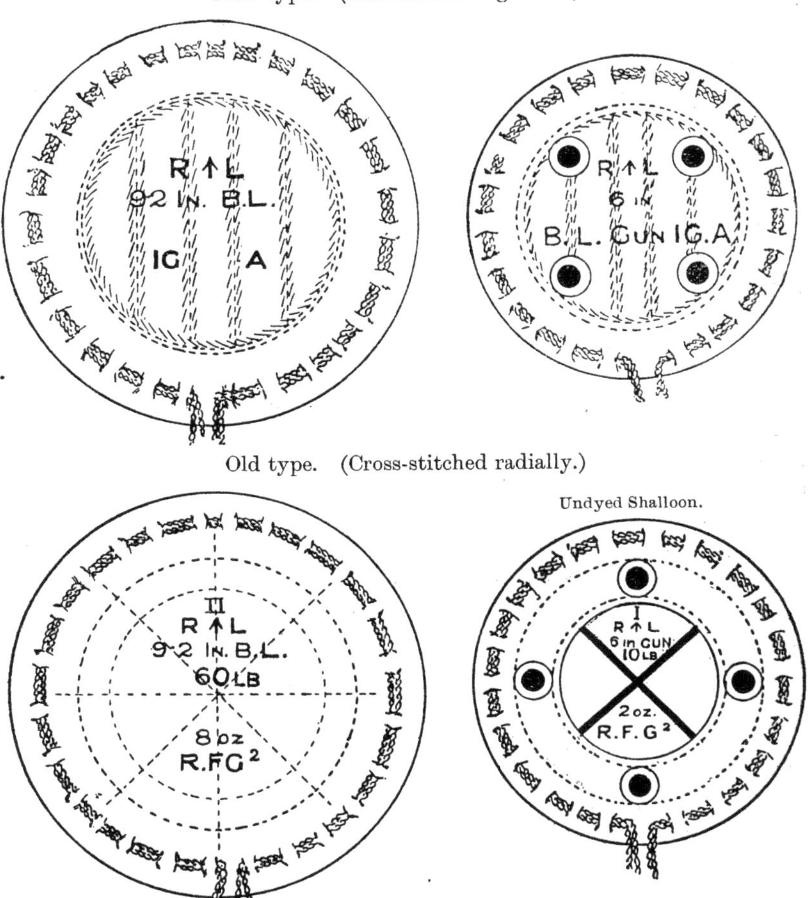

The object of cross-stitching the igniters so as to form a number of compartments is to ensure that some of the gunpowder shall always be opposite the vent when the cartridge is loaded in the gun.

Note.—The first issues of cordite cartridges for the 6-inch B.L. Mark VII and VIII guns were fitted with igniters divided into two compartments only, and some 9·2-inch to 13·5 inch cartridges may be met with, having igniters cross-stitched to form four compartments only.

Shalloon *dyed red* is now used for all B.L. cartridge igniters. Numbers of cartridges were at one time issued for Land Service with igniters of undyed shalloon, and on these was stamped a *black cross* (for 6 inch and upwards) to distinguish the igniter end of the cartridge.

Marking.—After being cut out to shape the cartridges are marked with printer's ink as shown.

Paint must not be used for marking, as it holds fire.

Making up, &c.—Cordite cartridges are issued filled. In case they should have to be made up locally, a sample cartridge or drawing, and detailed description of method of manufacture, must be procured from the Royal Arsenal, Woolwich.

For cutting cordite to the required length for making up cartridges, the "Apparatus, cutting cordite" was introduced; this acts very well for Mark I cordite, but M.D., being brittle, is liable to split, so a special "Machine, cutting cordite" has been introduced for the larger sizes of M.D. and M.D.T.; it is also suitable for cutting Mark I cordite, but the "Apparatus, cutting cordite" will generally be used for this purpose.

Apparatus, cutting cordite consists of an oblong mahogany board, about 9½ in. by 40½ in., having a gunmetal knife working vertically in a fixed pivot near one end; it is provided with a scale and an adjustable end, which may be fixed at any required distance from the knife.

Machine, cutting cordite.—In the "Machine, cutting cordite," the sticks are pressed between a revolving drum and a concave knife-block; from the concave surface of the knife-block projects an aluminium bronze rib, which acts as a knife.

Record of cordite used.—A record of the cordite used in all cartridges, with lot letter and number, and date of filling, is kept in a book for reference at all stations where cartridges are made.

SECTION (B). DESCRIPTION OF CORDITE CARTRIDGES FOR B.L. GUNS.

Cordite cartridges for B.L. guns are here divided into types as illustrated on Plates No. IV to XII.

Plate IV illustrates the method of making up the cartridge, B.L. 2·75-inch, 7 oz. 9 drs., cordite, size 7½, Mark I | L |.

The latest Marks of the following cartridges are made up in the same manner as the above, differing only in weight and dimensions:—

Cartridge, B.L. 10-pr., 6 oz. 14 drs. Cordite, Size 5, Mark II | L |.
Cartridge, B.L. 10-pr., 3 oz. 9 drs. Cordite, Size 3¾, Mark II | L |.
Cartridge, B.L. 12-pr., 12 oz. 7 drs. Cordite, Size 5, Mark III | L |.
Cartridge, B.L. or B.L.C. 15-pr., 1 lb. 1 oz. 11 drs. Cordite, M.D., Size 4¼, Mark I | L |.

The charge consists of a cylindrical bundle of cordite (all sticks cut to the same length) tied in several places with silk sewing.

The charge is inserted into a shalloon cartridge, each end of which is closed by an igniter.

Each igniter consists of two discs of shalloon sewn together round the edges so as to form a flat circular bag.

The igniters are filled with R.F.G.^{2} or S.F.G.^{2} gunpowder.

Earlier Marks of 10- and 12-pr. cartridges.—The earlier Marks of these cartridges had igniters filled with guncotton yarn (waterproofed with indiarubber solution).

Mark I Cordite Cartridges for B.L. or B.L.C. 15-Pr.

Cartridge, B.L. or B.L.C. 15-*pr.*, 15¾ *oz. cordite, size* 5, *Mark I* | *L* | . —This cartridge is of a special type; it consists of a bundle of cordite (all the sticks being the same length) secured in several places with silk sewing. Near each end of the charge is fitted an igniter consisting of a skein of guncotton yarn wound round the cordite sticks;

Fig. 6.

CARTRIDGE, B.L. OR B.L.C. 15-PR., 15¾ OZ. CORDITE, SIZE 5 MARK I.

Scale ½.

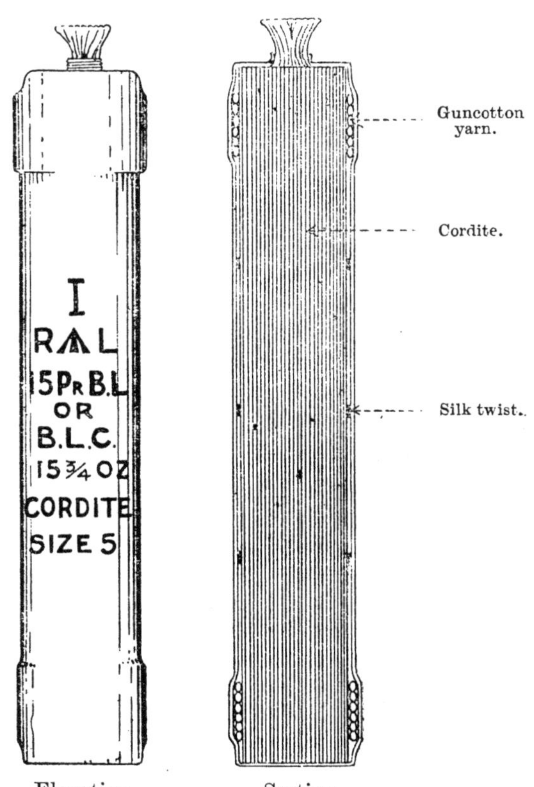

Elevation. Section.

the centre of the guncotton yarn is one inch from the end of the charge. The whole is then inserted into a shalloon cartridge choked at one end with silk sewing.

The cartridge for the B.L. and B.L.C. 15-pr., for star shell, is a reduced charge; it consists of 4 oz. of cordite, size 5, bundled round a paper cylinder, with an igniter at each end of guncotton yarn, and is similar in other respects to the full charge described above.

Cartridge, B.L. 4-inch, 3 lb. 7 oz. Cordite M.D., Size 4¼, Mark I (S.I.).

The charge consists of a cylindrical bundle of M.D. cordite sticks cut to the required length, and tied in three places with silk or shalloon braid.

The charge is inserted into a No. 1 silk cloth cartridge, each end being closed by a shalloon igniter.

The igniter is made up of two discs of shalloon sewn together round the edges and cross-stitched to form four compartments; the pockets thus formed are filled with R.F.G.² gunpowder.

The igniters are secured to the ends of the cartridge by silk sewing.

The Cartridge, B.L. 4-inch, 3 lb. Cordite, Size 5, Mark III, is made up similar to the above, differing only in weight and dimensions, and in the outer layer of cordite sticks being tied with silk sewing instead of shalloon braid.

The Mark II cartridge for the B.L. 30-pr., and the Mark II cartridge, B.L. 4-inch, 3 lb. 1 oz., are fitted with igniters filled with waterproofed guncotton yarn and are not cross-stitched.

Plate V illustrates a slightly different method of building up the cartridge used for larger charges. Of this type are the following :—

Cartridge, B.L. 60-pr., 9 lb. 12 oz. Cordite M.D., Size 16, Mark II | L | .

Cartridge, B.L. 4-inch, 9 lb. 5 oz. 15 drs. Cordite M.D., Size 16, Mark II | N | .

Cartridge, B.L. 4-inch, 5 lb. 6 oz. Cordite M.D., Size 16, Mark II | N | .

Cartridge, B.L. 5-inch, 4 lb. 11¼ oz. Cordite M.D., Size 4¼, Mark I (S.I.).

The cartridge for the B.L. 60-pr. is here described in detail; the remainder are similar in construction, differing only in dimensions, &c. The cartridge for B.L. 5-inch is hooped with silk braid.

For particulars of the above cartridges *see* Table 8, page 78.

Cartridge, B.L. 60-pr., 9 lb. 12 oz. Cordite M.D., Size 16, Mark II | L | .

The charge consists of a bundle of cordite secured in several places with silk sewing; the centre sticks are shorter than the outer ones, thus forming a circular recess at each end for the igniter. The charge is inserted into a silk cloth cartridge provided with silk braid hoops, and is closed at each end by an igniter.

The igniter consists of two discs of shalloon and a ring of silk cloth sewn together; the shalloon discs are stitched across the centre to form four compartments for the R.F.G.² gunpowder.

One of the igniters is sewn to the bottom of the cartridge prior to the insertion of the charge; in the other igniter the silk cloth ring carries a silk draw-string; it is placed over the end of charge, after the latter has been inserted into the cartridge, and secured by the draw-string and by being stitched to the cartridge.

The Mark I cartridges for the B.L. 60-pr. and B.L. 4-inch guns differ from the Mark II, in being hooped with silk braid and the sticks of cordite used in building up the charge being cut to a different length.

Certain issues of the above Marks of cartridges were fitted with igniters containing S.F.G.[2] instead of R.F.G.[2] gunpowder.

Plate VI illustrates a type of B.L. cartridge, the main features of which are :—The igniters are fitted with four perforated felt studs and the cartridges are made up in two fractions (two ½ charges, or a ⅓ and ⅔ charge) laced together. Of this type are the following cartridges :—

Cartridge, B.L. 6-inch gun, 23 lb. Cordite M.D., Size 16, Mark I to III | C | .

Cartridge, B.L. 6-inch gun, 23 lb. Cordite M.C., Size 16, Mark I and II | C | .

Cartridge, B.L. 6-inch gun, 20 lb. Cordite Mark I, Size 10, Mark I to III | C | .

Cartridge, B.L. 6-inch gun, 28 lb. 10 oz. Cordite M.D., Size 26, Mark II | N | .

Cartridge, B.L. 6-inch gun, 32 lb. 1½ oz. Cordite M.D., Size 26, Mark II | N | .

Cartridge, B.L. 6-inch gun, 27 lb. 2 oz. Cordite M.D., Size 19, Mark I | N | .

Cartridge, B.L.C. 6-inch gun, 20 lb. 15 oz. Cordite M.D., Size 16, Mark II | L | .

Use of the above 6-inch B.L. Cartridges.

The 23 *lb. M.D. cartridge* is made up in two ½ charges, and is for use with B.L. 6-inch, Mark VII and VII^{v} guns in Land Service, and Mark VII to VIII guns in Naval Service, other than those guns on Mark II twin mountings.

The 20 *lb. cartridge* is also made up in two ½ charges ; it is for use with all Mark VII and VIII guns.

The 28 *lb.* 10 *oz. M.D.* cartridge is made up in ⅓ and ⅔ charges ; it is for use with Mark VII guns other than those on twin mountings and unstrengthened P IV mountings for shell with cupro-nickel driving bands in Naval Service.

The 32 *lb.* 1½ *oz.* M.D. cartridge, made up in ⅓ and ⅔ charges, is used with the Mark XI and XI* guns in Naval Service.

The 27 *lb.* 2 *oz. M.D. cartridge,* made up in ⅓ and ⅔ charges, is used with the Mark XII gun in Naval Service.

The latest Mark of the 23 lb. Cordite M.D. cartridge is here described in detail.

Cartridge, B.L. 6-inch gun, 23 lb. Cordite M.D., Size 16, Mark III | C | .

The cartridge is made up of two fractions (½ charges) laced together.

Each fraction consists of a cylindrical bundle of cordite sticks tied together with silk sewing. The cordite sticks in the charge are all the same length; no recess being left for the igniter. The charge is enclosed in a silk cloth cartridge hooped with silk braid. The mouth of the cartridge is closed by a standardized "A" igniter.

The igniter consists of two discs of shalloon, with a strengthening disc of silk cloth on the underside, sewn together round the circumference, and divided into five parallel compartments, each filled with R.F.G.[2] gunpowder.

The outer shalloon disc is fitted with a draw-string of silk sewing and four perforated felt studs. The object of the felt studs is to prevent the heated axial vent from pressing against the powder igniter, and so causing a premature.

The igniter is placed over the end of the charge and is held in position by the draw-string, and by being stitched to the mouth of the silk cloth cartridge.

Near the base a strengthening band of silk cloth is stitched round the cartridge; this prevents the silk cloth of the cartridge from being torn when the two fractions are laced together.

The Cartridge, B.L. 6-inch gun, 23 lb. Cordite M.C., Size 16, Mark II, is made up in the same way as the above.

Earlier Marks of 6-inch B.L. gun Cartridges.

The following cartridges for 6-inch B.L. guns differ from the above design in the following particulars:—

(*a*) A recess is formed at the end of the cordite charge for the igniter.

(*b*) An earlier type of igniter is used, which is cross-stitched radially to form four compartments, and is fitted on the outside with a silk cloth ring which carries the felt studs and draw-string.

The Mark I and II, 23 lb. Cordite M.D., Cartridge.
The Mark I, 23 lb. Cordite M.C., Cartridge.
The Mark I, 20 lb. Cordite Cartridge.
The Mark I, 28 lb. 10 oz. Cordite M.D., Cartridge.
The Mark I, 32 lb. 1½ oz. Cordite M.D., Cartridge.

The Marks I and II 23 lb. charge is described below, and is illustrated on Plate VII.

Cartridge, B.L. 6-inch gun, 23 lb. Cordite M.D., Size 16, Mark II.

The full cartridge is made up into two fractions (½ charges), laced together. Each fraction consists of a bundle of cordite sticks tied with silk sewing. The centre sticks are slightly shorter than the outer layers, forming a circular recess at one end only. The bundle of cordite sticks is placed into a silk cloth cartridge hooped with silk braid.

The mouth of the cartridge is closed by the igniter. This consists of two discs of shalloon, and one of silk cloth sewn together round the edges. The shalloon discs are sub-divided into four compartments to take the R.F.G.[2] gunpowder. To the outside of the shalloon discs is stitched a ring of silk cloth with a draw-string ; four perforated felt studs are stitched to the silk cloth ring.

The igniter is placed over the end of the charge, the silk cloth disc next the cordite, the four compartments of S.F.G.[2] powder fitting into the recess, and is held in position by the draw-string and by being stitched to the silk cloth cartridge. A strengthening band of silk cloth stitched round the bottom prevents the cartridge from tearing when the two fractions are laced together.

Changes in design.

The following are the improvements that have been adopted in the design of the above type of cartridge. The improvements extend over a long period, and did not advance the numeral of the cartridges. Many of the older designs will still be met with:—

(*a*) Felt studs on the igniter reduced from six to four, to minimise residue.
(*b*) Igniter divided into four instead of two compartments.
(*c*) Strengthening disc of silk cloth sewn to underside of the igniter to prevent the sharp edges of the cordite sticks injuring the shalloon.
(*d*) A single ring of silk cloth sewn to outside discs of the igniter, instead of having two rings.

The introduction of the "Standardized igniter" for this type of cartridge in all cases advanced the numeral of the filled cartridge.

B.L. 6-inch, Mark III (chase-hooped), IV and VI Guns.

There are two cartridges for the above-mentioned old type of B.L. 6-inch guns, namely :—

A 14 lb. 12 oz. charge of Mark I cordite, which is made up like the cartridges for the heavier guns, and a 16 lb. 12 oz. charge of M.D. cordite made up in a full charge as described for the Mark II 11½ lb. ½ charge for the Mark VII gun, but it has no strengthening band of silk cloth at the base. It has an igniter with felt studs at one end.

Cordite Cartridges for Heavy B.L. Guns.

Cordite cartridges for B.L. guns, 7·5-inch to 15-inch inclusive, with a few exceptions to be detailed later, are made up on one model, the only differences being in external shape, in the hoops, and in the arrangements of the lifting beckets. They are made up in ½ and ¼ charges, with igniter and mill-board protecting disc or cover at *one* end only.

For particulars of weight, mark, nature and size of cordite, &c., *see* Table 8.

A description of the Mark III B.L. 9·2-inch, 60 lb. M.D. cartridge is given below ; it is typical of the latest heavy B.L. cartridges.

Typical Heavy B.L. Gun Cartridge with Standardized Igniter.

Cartridge, B.L. 9·2-inch, 60 lb. Cordite M.D., Size 37, Mark III | C | ½ Charge.

The charge.—The charge consists of a bundle of M.D. cordite, the sticks all cut to the same length, and tied in several places with silk webbing. The charge is inserted into a silk cloth cartridge bag, which has no hoops; the mouth of the cartridge is closed by the igniter.

Igniter.—The "A" igniter consists of two outer discs of shalloon and an inner disc of silk cloth, sewn together round the circumference, and then divided into five parallel compartments, each filled with R.F.G.[2] gunpowder.

The outer disc of shalloon is provided with a draw-string of silk sewing.

The igniter is placed over the end of the charge and is secured by the draw-string and by being sewn to the mouth of the cartridge bag.

Cover for igniter.—The protecting cover consists of a disc of millboard, to the outside of which is stitched and glued a larger disc of silk cloth.

This silk cloth disc is painted with a red cross, and is fitted with a draw-string of linen tape, and a small loop of linen tape on which the words "*Tear off*" are printed.

The cover is placed over the igniter, and is secured by means of the linen tape draw-string, which is drawn tightly round the top of the cartridge and tied with a slip-knot, the running end being fastened to the "Tear off" loop, so that one operation will remove the cover from the igniter.

The cover is also held from slipping off by being loosely secured at each side to the cartridge bag by a single stitch of silk sewing.

Lifting becket.—Three loops of braid sewn to the cartridge—one on the bottom and one on each side near the top—form "fairleads" to keep in position a linen tape lifting becket (1 in. wide) which passes round the cartridge. The ends of this lifting becket are tied on top of the cover by a slip-knot.

Before loading, the becket must be slipped and unrove from the fairleads and thrown to one side; the "cover" can then be removed.

A becket of silk braid is also sewn to the base to form a grip for convenience in loading.

Earlier Type of Heavy Cartridge for B.L. Guns.

The earlier type of heavy B.L. cartridges were fitted with igniters, cross-stitched radially to form a number of compartments—four in the 7·5-inch and eight in 9·2-inch to 13·5-inch.

Some 9·2-inch to 13·5-inch igniters may be met with, however, divided into four compartments.

In the earlier type of cartridge a recess was also formed at the end of the cordite charge for the igniter to fit into.

Typical Heavy B.L. Cartridge with Old Type of Igniter.

Cartridge, B.L. 9·2-inch, 60 lb. Cordite M.D., Size 37, Mark II | C | ½ Charge.

The Cartridge, B.L. 9·2-inch, 60 lb. M.D., size 37, Mark II | C | ½ charge, for Marks IX and X guns, consists of a half charge, a silk cloth cartridge, an igniter, protecting disc, and tape becket. (*See* Plate IX.)

Half charge.—The charge is cylindrical in shape and is made up as follows :—A central bundle of cordite sticks is tied in several places with silk sewing. Round this central bundle is placed the remainder of the charge, consisting of a layer of cordite sticks, somewhat longer than the central bundle, also secured with silk sewing; a circular recess is thus formed at one end for the igniter.

Cartridge.—The charge is inserted into a silk cloth cartridge which is hooped with silk braid, and has a silk braid becket sewn to the bottom. Three loops of braid sewn to the cartridge—one on the bottom, and one on each side near the top—form "fairleads" to keep a linen tape lifting becket in position.

The mouth of the cartridge is closed by the igniter.

Igniter.—The igniter consists of two discs of shalloon which are strengthened on the side next the cordite by a disc of silk cloth. These discs are sewn together round the edges and centre, and are stitched radially to form eight compartments which are filled with R.F.G.² powder. This ensures a portion of the powder being always opposite the vent. The outer shalloon disc is provided with a draw-string of silk sewing. The igniter fits over the end of the charge, the compartments filled with powder fitting into the recess; it is secured by the draw-string and by being sewn to the mouth of the cartridge.

Protecting disc.—A millboard protecting disc, to the outside of which a disc of silk cloth is attached, is placed over the igniter, and lightly attached by four stitches. The disc is painted with a red cross, and is provided with a loop of linen tape on which the words "Tear off" are printed.

Lifting becket.—A length of 1-in. linen tape is passed completely round the cartridge from top to bottom; it passes through the three guide loops above mentioned, and is tied on top of the protecting disc with a bow so that it can easily be removed.

A becket of silk braid is also sewn to the base to form a grip for convenience in loading.

Improvements, Alterations, &c., in the Design of Heavy B.L. Cordite Cartridges.

Charges.

(1) Charges first issued had the cordite sticks in the outer layers and in the centre of the charge cut slightly longer than the intermediate sticks, thus forming a recess in the shape of a ring at one end for the igniter.

Some of the above charges were cylindrical in shape and some conical.

(2) The long sticks in the centre of the charge were discontinued.

(3) In the latest design of cartridges fitted with standardized igniters and covers, the sticks in the charge are all cut to the same length, no recess being left for an igniter.

(4) Charges first issued had the outer layer of cordite sticks secured with ties of silk sewing ; this was replaced by silk or shalloon braid.

Silk webbing is now used for this purpose in making up the B.L. 15-inch cartridges, and will supersede the silk braid for all B.L. 7·5-inch and up when existing stock of silk braid is used up.

The latest type of cartridge for the B.L. 7·5-inch and up have no hoops ; the maintenance of the proper shape of the cartridge depends almost entirely upon the silk webbing or silk braids.

Beckets.

(1) *Central silk cord becket.*—The first becket used was one of silk cord. This cord ran up on the double through the centre of the charge and passed through a hole in the centre of the igniter and protecting disc, the bight of the cord forming a loop at the top ; the ends of the cord were attached to a silk braid loop on the inside of the base of the cartridge.

To take the extra strain the base of the cartridge was strengthened by having two small discs of silk cloth stitched to it—one on the inside and the other on the outside.

(2) *Removable silk braid becket.*—Two loops of silk braid were sewn to the outside of the cartridge near the top, and another piece of braid (forming a lifting becket) was passed through these loops and tied on top of the protecting disc with a slip-knot, so that it could easily be removed.

This form of lifting becket was found to be unsatisfactory, owing to the silk braid loops being unable to stand the strain of lifting the cartridge out of the cylinder.

(3) *All round lifting becket of silk braid.*—This lifting becket passed completely round the cartridge and was kept in position by three "fairleads"—one on the base and one at each side of the cartridge near the top.

(4) *All round becket of linen tape.*—This becket was attached in the same way as above. It was introduced owing to the cordite having a rotting effect on the silk braid becket.

Igniters.

(1) All the igniters were formerly stitched radially to form four compartments ; for later 9·2-inch and up they were cross-stitched to form eight compartments. The latest Marks of cartridges are now fitted with "Standardized igniters" stitched across to form five parallel compartments ; in the 15-inch cartridge the igniter is stitched across to form six parallel compartments.

(2) The first igniters consisted of an inner disc of silk cloth and an outer disc of shalloon. The silk cloth rotted and allowed the powder

to fall among the sticks. The gunpowder priming is now contained between two discs of shalloon, with a strengthening disc of silk cloth on the underside.

(3) The central hole in "igniter" and "tear off" disc was discontinued when the central silk cord becket was superseded by the removable lifting becket.

(4) A certain number of cartridges have been issued with igniters of undyed shalloon, which is now discontinued. Most cartridges that are fitted with an igniter at one end only, if the igniter is of undyed shalloon, have a black cross extending across the outer disc to make the igniter of the cartridge more conspicuous.

(5) S.F.G.² gunpowder introduced for filling igniters followed by the re-introduction of R.F.G.² instead of S.F.G.², stock of S.F.G.² being used up for igniters of cartridges made up with Mark I cordite in the Land Service only.

(6) Protecting disc for igniter fitted with a larger silk cloth disc, and called "Cover for igniter."

This cover for igniter is secured by a draw-string of linen tape in addition to being secured by single stitches of silk sewing.

Composite Cartridges.

(*See* Plate XII.)

A different system of making up cartridges with *Mark I cordite* was adopted for the heaviest guns; it is now discontinued, but many of the above type of cartridges will still be met with, until the existing stock is used up.

Guns which have composite cartridges.—In this type of cartridge two sizes of cordite are used—a large size and a small; such cartridges are known as "Composite cartridges." They were introduced for the B.L. 9·2-inch, and early Marks of 12-inch and 13·5-inch guns.

In external appearance they do not differ from other cartridges; some are cylindrical and some are enlarged at one end.

After tying the central bundle of large-sized cordite with silk sewing, a bundle or rope of cordite, consisting of 60 or 120 cords of very small size ($3\frac{3}{4}$), is wound spirally round the core from one end to the other, and secured with silk sewing; the outer layers of large-size cordite are then built up as usual.

The Mark I composite cartridge has the central silk cord becket (*see* Plate XII). The Mark II has the all-round removable lifting becket.

Advantages claimed for composite cartridge.—This method of making up enables a larger charge to be fired, increasing the muzzle velocity, but without any increase in the chamber pressure. It has also been found that with this design of cartridge ballistics vary less with change of temperature, and are better maintained in worn guns.

Cartridges for 9·2-inch B.L. Guns on H.A. Mountings.

(*Special type.*)

The cartridges for 9·2-inch B.L. guns on high angle mountings are laced together in the same way as those for the B.L. 6-inch Marks VII to XII guns.

Fig. 7.

CARTRIDGE, B.L. 9·2-INCH, 16 LB. 1 OZ. CORDITE M.D., SIZE 8, CONSISTING OF 12 LB. 6 OZ. AND 3 LB. 11 OZ. CHARGES, FOR GUNS ON H.A. MOUNTINGS.

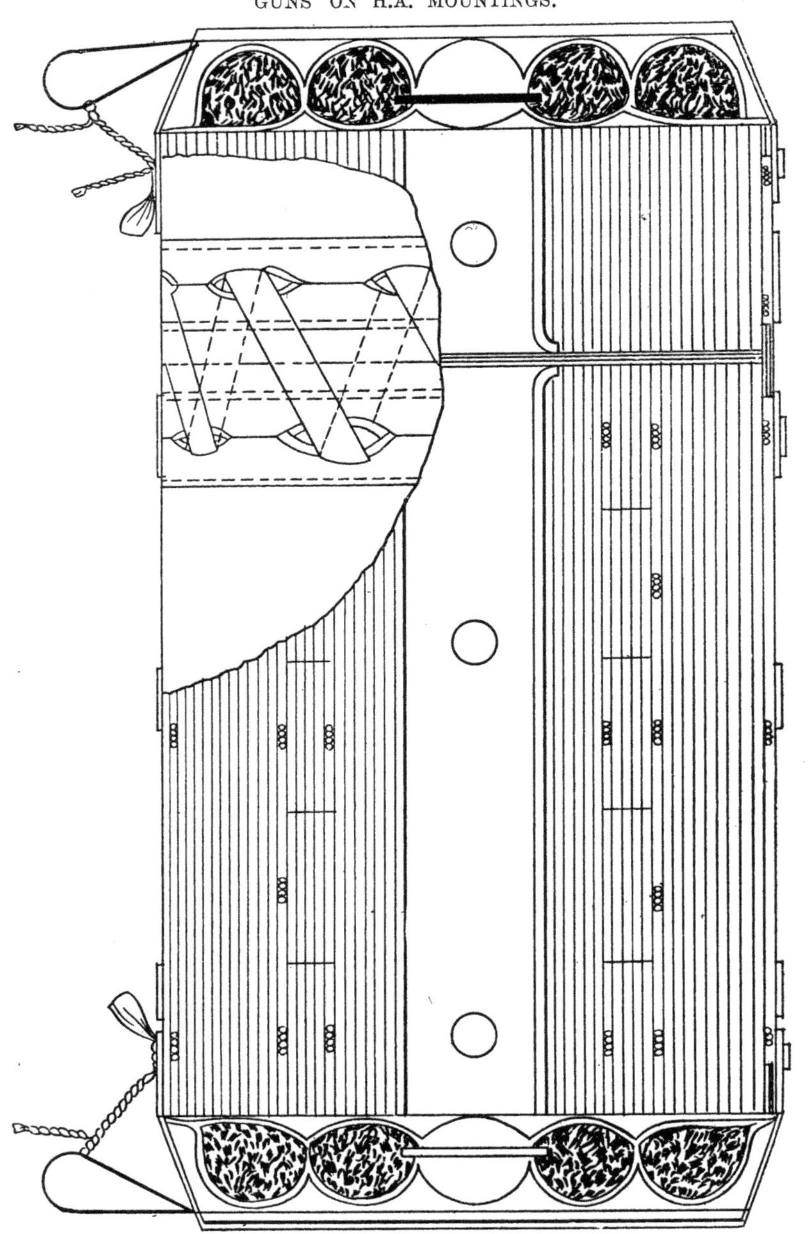

A peculiar feature of these cartridges is a perforated paper cylinder built up in the centre, to take a wooden stick intended to prevent the projectile slipping back when the gun is elevated.

There are four cartridges, as follows :—

29 lb. M.D. cordite, size 16 15 lb. 12 oz. M.D. cordite, size 16	Laced together to form a 44 lb. 12 oz. (full charge).
12 lb. 6 oz. M.D. cordite, size 8 3 lb. 11 oz. M.D. cordite, size 8	Laced together to form a 16 lb. 1 oz. (full charge).

The above-mentioned 44 lb. 12 oz. and 16 lb. 1 oz. (full charges) are made up in the same manner.

Each charge consists of a bundle of cordite tied together with silk sewing. In the centre of the charge is a perforated paper cylinder.

The charge is inserted into a silk cloth cartridge hooped with silk braid ; it has a hole in the bottom for a stick.

The " igniter " and " cover for igniter " are similar to those already described on page 72, but have central holes to allow the stick to pass through each end of the cartridge.

The heavier portion of each laced-up cartridge is provided with a removable lifting becket, tied over the " cover for igniter."

The stick is of beech, 1·25 in. in diameter, in two parts which screw together—one part 3 in. in length (exclusive of screwed part), the other part 39·4 in. in length—the end being strengthened by a brass ferrule.

The stick is made up in two parts so that it can be easily lengthened by partly unscrewing the two portions, so as to arrange for over-ramming in worn guns.

The Mark I cartridges for the 9·2-inch H.A. guns differ from the Mark II described above in the following particulars :—

(*a*) An annular recess is formed at one end of the charge for the igniter.

(*b*) The igniter is cross-stitched radially to form eight compartments.

(*c*) The igniter is protected by a " tear off " disc secured by four single stitches instead of a " cover " secured by a linen tape draw-string.

TABLE 8.—*Cordite Cartridges for B.L. and B.L.C. Guns.*

Para. in List of Changes.	Nature of Gun.	Mark of Cartridge.	Service.	Weight.	Nature of Cordite.	Size of Cordite.	Igniter.	Diameter in inches.	Length in inches.	Number in Package.	Package.	Lifting Becket.	Remarks.
16298 16535	2·75 - inch B.L.	I	L	7 ozs. 9 drams (full charge)	I	$7\frac{1}{2}$	3 drams S.F.G.[2] at each end	2	4·9	 240 115 46 120	Case, Powder, M.L.: Whole Half Quarter Case, M.L., Field	Nil	—
11021 11915	10-pr. B.L.	I	L	6 ozs. 14 drams (full charge)	I	5	1 dram G.C. yarn at each end	2	4·4	 256 115 48	Case, Powder, M.L.: Whole Half Quarter	Nil	—
	Do.	II	L	Do.	I	5	3 drams S.F.G.[2] at each end	2	4·4	Do.	Do.	Nil	Gunpowder replaces G.C. yarn for Igniter.
11022	10-pr. B.L.	I	L	3 ozs. 9 drams (for Star Shell)	I	$3\frac{3}{4}$	1 dram G.C. yarn at each end	1·8	3·1	 445 220 87	Case, Powder, M.L.: Whole Half Quarter	Nil	—
—	Do.	II	L	Do.	I	$3\frac{3}{4}$	3 drams S.F.G.[2] at each end	1·8	3·1	Do.	Do.	Nil	Gunpowder replaces G.C. yarn for Igniter.

8137	12-pr. B.L. (6 cwt.)	I	L	12 ozs. 7 drams (full charge)	I	5	4 drams R.F.G.[2] at each end	2·2	6·2	110	Case, Powder, M.L.: Whole	—	—
11978 10580	Do.	I* } II }	L	Do.	I	5	1 dram G.C. yarn at each end	2·2	6·2	145	Do.	—	G.C. yarn replaces gunpowder for Igniter.
—	Do.	III	L	Do.	I	5	4 drams S.F.G.[2] at each end	2·2	6·2	145	Do.	—	Gunpowder replaces G.C. yarn for Igniter.
7597 8314	15-pr. B.L. or B.L.C.	I	L	15¾ ozs. (full charge)	I	5	2 drams G.C. yarn ½ inch from each end	1·9	11·5	75	Do.	—	Originally for the 12-pr. B.L. 7-cwt. gun. Nomenclature altered to 15-pr. or 12-pr. 7 cwt.
8499 14214	—	—	—	—	—	—	—	—	—	100	Do.	—	Nomenclature altered to 15-pr. B.L. or B.L.C.
16397	15-pr. B.L. or B.L.C.	I	L	1 lb. 1 oz. 11 drams (full charge)	M D	4¼	4 drams R.F.G.[2] at each end	1·9	11·5	100	Do.	—	M.D. Cordite charge introduced.
13344	Do.	I	L	4 ozs. (for Star Shell)	I	5	2 drams G.C. yarn ½ inch from each end	1·9	11·5	100	Do.	—	Has a paper cylinder in the centre.
—	30-pr. B.L.	I	S I	2 lbs. 6 ozs. (full charge)	I	10	½ oz. R.F.G.[2] at each end	2·75	11·5	35	Do.	Nil	—
—	Do.	II	S I	Do.	I	10	3 drams G.C. yarn at each end	2·75	11·5	35	Do.	Nil	G.C. yarn replaces gunpowder for Igniter.
—	Do.	III	S I	Do.	I	10	½ oz. S.F.G.[2] at each end	2·75	11·5	35	Do.	Nil	S.F.G.[2] replaces G.C. yarn for Igniter.

TABLE 8.—*Cordite Cartridges for B.L. and B.L.C. Guns*—continued.

Para. in List of Changes.	Nature of Gun.	Mark of Cartridge.	Service.	Weight.	Nature of Cordite.	Size of Cordite.	Igniter.	Diameter in inches.	Length in inches.	Number in Package.	Package.	Lifting Becket.	Remarks.
7596	4-inch B.L. (except Marks VII to VIII* Guns)	I	L	3 lbs. 1 oz. (full charge)	I	5	½ oz. R.F.G.[2] at each end	3·2	11·5	37 20	Case, Powder, M.L.: Whole Half	Nil	—
11919	Do.	II	L	Do.	I	5	2 drams G.C. yarn at each end	3·2	11·5	Do.	Do.	Nil	G.C. yarn replaces gunpowder for Igniter.
—	Do.	III	L	Do.	I	5	½ oz. S.F.G.[2] at each end	3·2	11·5	Do.	Do.	Nil	Gunpowder replaces G.C. yarn for Igniter.
—	4-inch B.L. (Mark V Gun)	I	S I	3 lbs. 7 ozs. (full charge)	M D	4¼	½ oz. R.F.G.[2] at each end	3·4	11·5	Do.	Do.	Nil	—
14411	4-inch B.L. (Mark VII Gun)	I	N	9 lbs. 15 ozs. (full charge)	M D	19	1 oz. R.F.G.[2] at each end	4·3	18·5	6 12	Rect. "R" Case Rect. "A" Case	Nil	Superseded for future manufacture by 9 lbs. 5 ozs. 15 drams, Size 16.
15948 15947	Do.	I	N	9 lbs. 5 ozs. 15 drams (full charge)	M D	16	Do.	4·3	17·2	6 12	Rect. "R" Case Rect. "A" Case	Nil	Cut from 26-inch lengths. No hoops.

15931	Do.	II	N	Do.	M D	16	Do.	4·3	17·2	 12	Rect. "R" Case Rect. "A" Case	Nil	Cut from 33-inch lengths. No hoops.
15949 15948	4-inch B.L. (Mark VIII Gun)	I	N	5 lbs. 6 ozs. (full charge)	M D	16	1 oz. R.F.G.[2] at each end	3·84	13·15	32 7	Rect. "O" Case Rect. "R" Case	Nil	Was originally 5 lbs. 4 ozs. §§13848 and 15084. Cut from 26-inch lengths.
15931	Do.	II	N	Do.	M D	16	Do.	3·84	13·15	32 7	Rect. "O" Case Rect. "R" Case	Nil	Cut from 33-inch lengths. No hoops.
—	4-inch, jointed	I	S I	2 lbs. 14 ozs. (full charge)	I	5	$12\frac{1}{2}$ drs. R.F.G.[2] in each circumferential igniter	3·3	11·5	37 20 7	Case, Powder, M.L. : Whole Half Quarter	Nil	—
—	5-inch B.L.	I	S I	4 lbs. $11\frac{4}{16}$ ozs. (full charge)	M D	$4\frac{1}{4}$	1 oz. R.F.G.[2] at each end	4·1	12·5	37 20 7	Whole Half Quarter	Nil	—
—	Do.	I	L	8 lbs. (full charge)	I	15	$1\frac{1}{2}$ oz. R.F.G.[2] at each end	4·4	14·75	37 20 7	Whole Half Quarter	Nil	For Victoria.
13169 15933	60-pr. B.L.	I	L	9 lbs. 12 ozs. (full charge)	M D	16	1 oz. R.F.G.[2] at each end	4·7	17	12	Case, Powder, M.L. : Whole	Nil	Was originally 9 lbs. 7 ozs. Cut from 26-inch lengths.
15931	Do.	II	L	Do.	M D	16	Do.	4·7	17	12	Do.	Nil	No hoops. Cut from 33-inch lengths.
9770	6-inch B.L. Gun (Marks III chase-hooped IV and VI)	II	L	14 lbs. 12 ozs. (full charge)	I	20	2 ozs. S.F.G.[2] at one end	5·5	17·5	7	Do.	Central silk cord	—

TABLE 8.—*Cordite Cartridges for B.L. and B.L.C. Guns*—continued.

Para. in List of Changes.	Nature of Gun.	Mark of Cartridge.	Service.	Weight.	Nature of Cordite.	Size of Cordite.	Igniter.	Diameter in inches.	Length in inches.	Number in Package.	Package.	Lifting Becket.	Remarks.
11584 12749	6-inch B.L. Gun (Marks III (chase-hooped), IV and VI	I	L	16 lbs. 12 ozs. (full charge)	M D	16	2 ozs. R.F.G.² at one end	5·75	17·5	6 1	Whole Cylinder No. 39 Case, Powder, M.L. :	Nil	Cut from 26-inch lengths.
15931 16189	Do.	II	L	Do.	M D	16	Do.	5·75	17·5	7 1	Whole Cylinder No. 39 Case, Powder, M.L. :	Nil	No hoops. Cut from 33-inch lengths.
–	6-inch B.L. Gun (Armstrong), 4 tons	I	L	9½ lbs. (full charge)	I	5	2 ozs. S.F.G.² at one end	5·75	11·5	10 2	Whole Cylinder No. 34	Nil	For Australia.
10408 13676	6-inch B.L. Gun(Marks VII to VIII)	I	C	10 lbs. (½ charge)	I	20	Do.	5·75	11·75	14 24 2	Rect. "B" Case Rect. "L" Case Cylinder No. 34	Nil	When issued two half-charges are laced together.
—	Do.	II	C	Do.	I	20	6-inch "A" (2 ozs. S.F.G.²)	5·75	11·75	14 24 2	Rect. "B" Case Rect. "L" Case Cylinder No. 34	Nil	Standardised Igniters. Cut from 17-inch lengths.
—	Do.	III	C	Do.	I	20	Do.	5·75	11·75	14 24 2	Rect. "B" Case Rect. "L" Case Cylinder No. 34	Nil	Cut from 15-inch lengths.

11356 13676	6-inch B.L. Gun(Marks VII toVIII except those on twin mountings)	I	C	11½ lbs. (½ charge)	M D	16	2 ozs. R.F.G.^{2} at one end	5·75	12·7	14 2	Rect. "B" Case Cylinder Nos. 34, 38 or 38A	Nil	Cut from 26-inch lengths.
11635 15931	Do.	II	—	Do.	M D	16	Do.	5·75	12·7	14 2	Rect. "B" Case Cylinder Nos. 34, 38 or 38A	Nil	Cut from 33-inch lengths.
—	Do.	III	—	Do.	M D	16	6-inch "A" (2 ozs. R.F.G.^{2})	5·75	12·75	14 2	Rect. "B" Case Cylinder Nos. 34, 38 or 38A	Nil	Standardised igniter. Cut from 33-inch lengths.
—	Do.	I	C	Do.	M C	16	2 ozs. R.F.G.^{2} at one end	5·75	12·7	14 2	Rect. "B" Case Cylinder Nos. 34, 38 or 38A	Nil	—
—	Do.	[illegible]	C	Do.	M C	16	6-inch "A" (2 ozs. R.F.G.^{2})	5·75	12·75	14 2	Rect. "B" Case Cylinder Nos. 34, 38 or 38A	Nil	Standardised igniter.
—	6-inch B.L. Gun (Mark XII)	I	N	27 lbs. 2 ozs. (full charge)	M D	19	6-inch "A," 2 ozs. R.F.G.^{2} at each end	5·25	36·2	4	Rect. "W" Case	All round 1-inch linen tape	⅔ charge = 18 lbs. 1⅓ ozs. ⅓ charge = 9 lbs. 0⅔ ozs. } Laced together.
—	Do.	II	N	Do.	M D	19	6-inch "B," 2 ozs. R.F.G.^{2} at each end	5·25	36·2	4	Do.	Do.	Do.

TABLE 8.—*Cordite Cartridges for B.L. and B.L.C. Guns*—continued.

Para. in List of Changes.	Nature of Gun.	Mark of Cartridge.	Service.	Weight.	Nature of Cordite.	Size of Cordite.	Igniter.	Diameter in inches.	Length in inches.	Number in Package.	Package.	Lifting Becket.	Remarks.
12948 15946	6-inch B.L. Gun (Mark VII other than those on twin and unstrengthened P IV mountings)	I	N	28 lbs. 10 ozs. (full charge)	M D	26	2 ozs. R.F.G.[2] at each end	6·1	28·5	8 8	Rect. "D" Case Rect. "F" Case	All round 1-inch linen tape	For cupro-nickel driving bands. Formerly a 29-lb. charge. $\frac{2}{3}$ charge = 19 lbs. $1\frac{1}{3}$ ozs. } Laced together. $\frac{1}{3}$ charge = 9 lbs. $8\frac{2}{3}$ ozs. }
—	Do.	II	N	Do.	M D	26	6-inch "A" (2 ozs. R.F.G.[2])	6·1	28·5	4 1	Rect. "T" Case Cylinder No. 41	Do.	Standardised Igniter.
13019 13750 15946	6-inch B.L. Gun (Marks XI and XI*	I	N	32 lbs. $1\frac{1}{2}$ ozs. (full charge)	M D	26	2 ozs. R.F.G.[2] at each end	6·2	31	4 8 1	Rect. "T" Case Rect. "F" Case Cylinder No. 41	Do.	First a 32 lbs 10 ozs. charge. Then a 33-lb. charge. $\frac{2}{3}$ charge = 21 lbs. $6\frac{1}{3}$ ozs. } Laced together. $\frac{1}{3}$ charge = 10 lbs. $11\frac{1}{6}$ ozs. }
—	Do.	II	N	Do.	M D	26	6-inch "A" (2 ozs. R.F.G.[2])	6·2	31	1 8 1	Rect. "T" Case Rect. "F" Case Cylinder No. 41	Do.	Standardised Igniter.
11357	6-inch B.L.C. Guns	I	L	10 lbs. $7\frac{1}{2}$ ozs. ($\frac{1}{2}$ charge)	M D	16	2 ozs. R.F.G.[2] at one end	5·75	11·75	2	Cylinder No. 34	Nil	Cut from 26-inch lengths.
15931	Do.	II	L	Do.	M D	16	6-inch "A" (2 ozs. R.F.G.[2])	5·75	11·85	2	Do.	Nil	Standardised Igniter. Cut from 33-inch lengths.

—	Do.	I	L	Do.	M C	16	6-inch "A" (2 ozs. R.F.G.[2])	5·75	11·85	2	Do.	Nil	Cut from 33-inch lengths.
13771 12758	7·5-inch B.L. (Marks II and II*)	I	S I	31 lbs. 6 ozs. (½ charge)	M D	26	6 ozs. R.F.G.[2] at one end	7	22·5	—	—	—	—
14093 14607	Do.	II	S I	Do.	M D	26	7·5-inch "A" (6 ozs. R.F.G.[2])	7	22·5	—	—	—	Standardised Igniter.
12758 13771	Do.	I	S I	15 lbs. 11 ozs. (¼ charge)	M D	26	6 ozs. R.F.G.[2] at one end	7	11·25	—	—	—	—
14093 14607	Do.	II	S I	Do.	M D	26	7·5-inch "A" (6 ozs. R.F.G.[2])	7	11·25	—	—	—	Standardised Igniter.
14093 14607 15945	7·5-inch B.L. (Marks I to II* and V)	I	N	30 lbs. 8 ozs. (½ charge)	M D	26	6 ozs. R.F.G.[2] at one end	7	22·5	4 4	Rect. "O" Case Box, Cartridge, 12-inch	All round	—
—	Do.	II	N	Do.	M D	26	7·5-inch "A" (6 ozs. R.F.G.[2])	7	22·5	4	Do.	Do.	Standardised Igniter.
14093 14607 15945	Do.	I	N	15 lbs. 4 ozs. (¼ charge)	M D	26	6 ozs. R.F.G.[2] at one end	7	11·25	8 8	Rect. "O" Case Box, Cartridge, 12-inch	Do.	—
—	Do.	II	N	Do.	M D	26	7·5-inch "A" (6 ozs. R.F.G.[2])	7	11·35	8 8	Rect. "O" Case Box, Cartridge, 12-inch	Do.	Standardised Igniter.
12819	7·5-inch B.L. (Marks III to IV*)	I	N	27 lbs. 2 ozs. (½ charge)	M D	26	6 ozs. R.F.G.[2] at one end	6·7	22	8	Rect. "L" Case	Central silk cord	—
12819	Do.	II	N	Do.	M D	26	6 ozs. R.F.G.[2] at one end	6·7	22	3	Box, Cartridge, Q.F. 6-inch, Naval Transport	Linen tape	—
12819	Do.	III	N	Do.	M D	26	6 ozs. R.F.G.[2] at one end	6·7	22	3	Do.	Do.	—
13502	Do.	I	N	13 lbs. 9 ozs. (¼ charge)	M D	26	6 ozs. R.F.G.[2] at one end	6·7	11	—	Rect. "L" Case	Do.	—

TABLE 8.—*Cordite Cartridges for B.L. and B.L.C. Guns*—continued.

Para. in List of Changes.	Nature of Gun.	Mark of Cartridge.	Service.	Weight.	Nature of Cordite.	Size of Cordite.	Igniter.	Diameter in inches.	Length in inches.	Number in Package.	Package.	Lifting Becket.	Remarks.
—	7·5-inch B.L. (Marks III to IV*)	II	N	13 lbs. 9 ozs. (¼ charge)	M D	26	6 ozs. R.F.G.[2] at one end	—	—	3	Box, Cartridge, Q.F. 6-inch, Naval Transport	Linen tape	—
11766	9·2-inch B.L. (Mark VIII)	I	N	33 lbs. (½ charge)	I	44 & 3¾	Do.	7·5	21·5	6	Rect. "I" Case	—	—
11766	Do.	I	N	16 lbs. 8 ozs. (¼ charge)	I	44 & 3¾	Do.	7·5	12·5	9	Do.	—	—
—	9·2-inch B.L. (Mark IX)	II	L	50 lbs. (½ charge)	I	44	8 ozs. R.F.G.[2] at one end	8·25	33·0	1	Cylinder No. 36	—	—
—	Do.	II	L	25 lbs. (¼ charge)	I	44	Do.	8·25	16·5	2	Do.	—	—
10233 15143	9·2-inch B.L. (Marks IX, X, X^v and X*)	I	C	51 lbs. 8 ozs. (½ charge)	I	44 & 3¾	Do.	8·25	32·5	4 1 2	Rect. "N" Case Cylinder No. 36 Rect. "S" Case	Central silk cord	—
12838	Do.	II	C	Do.	I	44 & 3¾	Do.	8·25	32·5	4 1 2	Rect. "N" Case Cylinder No. 36 Rect. "S" Case	All round	Differs from Mark I in lifting beckets.
9766	9·2-inch B.L. (Marks III to VII)	I	C	26 lbs. 12 ozs. (½ charge)	I	30	Do.	7·5	19·5	6 1	Rect. "I" Case Cylinder No. 22	Central silk cord	—
9766	Do.	II	C	Do.	I	30	Do.	7·5	19·5	6 1	Rect. "I" Case Cylinder No. 22	Do.	—

—	9·2 - inch B.L. (Marks IV, IV^A, and VI to VI^C)	III	L	Do.	I	30	Do.	7·5	19·5	6 1	Rect. "I" Case Cylinder No. 22	All round	—
9766	Do.	I	N	13 lbs. 6 ozs. (¼ charge)	I	30	Do.	7·5	9·75	12	Rect. "I" Case	Do.	
9766	9·2 - inch B.L. (Mark VIII)	II	N	31 lbs. 8 ozs. (½ charge)	I	40	Do.	7·5	21·5	5	Do.	Central silk cord	—
—	Do.	II	N	15 lbs. 12 ozs. (¼ charge)	I	40	Do.	7·5	12·5	9 2	Do. Cylinder No. 27	Do.	—
10233 15143	9·2 - inch B.L. (Marks IX, X, X^V and X*)	I	C	25 lbs. 12 ozs. (¼ charge)	I	44 & 3¾	Do.	8·25	16·25	8 2 4	Rect. "N" Case Cylinder No. 36 Rect. "S" Case	Do.	—
12838	Do.	II	C	Do.	I	44 & 3¾	Do.	8·25	16·25	8 2 4	Rect. "N" Case Cylinder No. 36 Rect. "S" Case	All round	Differs from Mark I in lifting beckets.
12212	Do.	I	C	60 lbs. (½ charge)	M D	37	Do.	8	32·5	2 5	Box, Cartridge, 9·2-inch Rect. "G" Case	Central silk cord	—
12839 13500	Do.	II	C	Do.	M D	37	Do.	8	32·5	4 2	Rect. "N" Case Rect. "S" Case	All round	Differs from Mark I in lifting beckets.
13985	Do.	III	C	Do.	M D	37	9·2-inch "A," 8 ozs. R.F.G.[2]	8	32·5	1	Cylinder No. 36	Do.	Standardised Igniter.
12212 12839	Do.	I	C	30 lbs. (¼ charge)	M D	37	8 ozs. R.F.G.[2] at one end	8	16·25	4 10	Box, Cartridge, 9·2-inch Rect. "G" Case	Central silk cord	—
13500 13985	Do.	II	C	Do.	M D	37	Do.	8	16·25	8 4	Rect. "N" Case Rect. "S" Case	All round	Differs from Mark I in lifting beckets.
—	Do.	III	C	Do.	M D	37	9·2-inch "A," 8 ozs. R.F.G.[2]	8	16·3	2	Cylinder No. 36	Do.	Standardised Igniter.

TABLE 8.—*Cordite Cartridges for B.L. and B.L.C. Guns*—continued.

Para. in List of Changes.	Nature of Gun.	Mark of Cartridge.	Service.	Weight.	Nature of Cordite.	Size of Cordite.	Igniter.	Diameter in inches.	Length in inches.	Number in Package.	Package.	Lifting Becket.	Remarks.
—	9·2 - inch B.L.(Marks X, X[v] and X*)	I	L	53½ lbs. (½ charge)	M D	26	9·2-inch "A," 8 ozs. R.F.G.[2]	7·75	30·25	1	Cylinder No. 36	All round	—
—	Do.	I	L	26 lbs. 12 ozs. (¼ charge)	M D	26	Do.	7·75	15·5	2	Do.	Do.	—
13518	9·2 - inch B.L.(Marks XI to XI*)	I	N	64 lbs. 4 ozs. (½ charge)	M D	37	8 ozs. R.F.G.[2] at one end	8·25	32·5	2 4 2	Rect. "S" Case Rect. "N" Case Box, Cartridge, 9·2-inch	Do.	Formerly 65 lbs. 4 ozs.
13858	Do.	II	N	Do.	M D	37	9·2-inch "A" (8 ozs. R.F.G.[2])	8·25	32·5	2 4 2	Rect. "S" Case Rect. "N" Case Box, Cartridge, 9·2-inch	Do.	Standardised Igniter.
13518 13858	Do.	I	N	32 lbs. 2 ozs. (¼ charge)	M D	37	8 ozs. R.F.G.[2] at one end	8·25	16·25	4 8 4	Rect. "S" Case Rect. "N" Case Box, Cartridge, 9·2-inch	Do.	Formerly 32 lbs. 10 ozs.
—	Do.	II	N	Do.	M D	37	9·2-inch "A" (8 ozs. R.F.G.[2])	8·25	16·25	4 8 4	Rect. "S" Case Rect. "N" Case Box, Cartridge, 9·2-inch	Do.	Standardised Igniter.

13501 13981	9·2 - inch B.L. (Marks B VI and VI^c on H.A. mountings)	I	L	44 lbs. 12 ozs. (full charge)	M D	16	8 ozs. R.F.G.[2] at each end	9·375	20·3	1	Cylinder No. 16	Do.	A 29-lb. fraction and A 15 lb. 12 oz. fraction } Laced together.
15931	Do.	II	L	Do,	M D	16	Do.	9·375	20·3	1	Do.	Do.	Differs from Mark I in being cut from 33-inch instead of 26-inch lengths.
—	Do.	III	L	Do.	M D	16	9·2-inch "B" (8 ozs. R.F.G.[2])	9·375	20·3	1	Do.	Do.	Standardised Igniter. Cut from 33-inch lengths.
13501 13981	Do.	I	L	16 lbs. 1 oz. (full charge)	M D	8	8 ozs. R.F.G.[2] at each end	7·375	13·9	2	Cylinder No. 28	Do.	A 12 lbs. 6 ozs. fraction and A 3 lbs. 11 ozs. fraction } Laced together.
	Do.	II	L	Do.	M D	8	Do.	7·375	13·9	2	Do.	Do.	Standardised Igniter.
9671 9719	10-inch B.L. (Marks I to IV*)	III	C	38 lbs. (½ charge)	I	30	8 ozs. R.F.G.[2] at one end	8·25	23·25	4	Rect. "J" Case	Central silk cord	Marks I and II were brought up to Mark III pattern locally.
9645 9671 13461	Do.	II	N	19 lbs. (¼ charge)	I	30	Do.	8·25	13·5	6	Do.	Do.	Mark I was brought up to Mark II pattern locally.
12876	Do.	I	L	40 lbs. (½ charge)	M D	16	Do.	8·25	26	1	Cylinder No. 15 or 15A	All round	Cut from 26-inch lengths.
15931	Do.	II	L	Do.	M D	16	Do.	8·25	26	1	Do.	Do.	Cut from 33-inch lengths.
12819 15857	10-inch B.L. (Marks VI to VII)	I	N	36 lbs. 11 ozs. (¼ charge)	M D	45	Do.	9	15	4 2	Rect. "U" Case Box, Cartridge, 4·7-inch Naval outfit	Central silk cord	—

TABLE 8.—*Cordite Cartridges for B.L. and B.L.C. Guns*—continued.

Para. in List of Changes.	Nature of Gun.	Mark of Cartridge.	Service.	Weight.	Nature of Cordite.	Size of Cordite.	Igniter.	Diameter in inches.	Length in inches.	Number in Package.	Package.	Lifting Becket.	Remarks.
12819	10-inch B.L. (Marks VI to VII)	II	N	36 lbs. 11 ozs. (¼ charge)	M D	45	8 ozs. R.F.G.[2] at one end	9	15	1	Cylinder No. 17	All round	—
—	Do.	III	N	Do.	M D	45	10-inch "A" (8 ozs. R.F.G.[2])	9	15	1	Do.	Do.	Standardised Igniter. Cut from 35-inch lengths.
—	Do.	IV	N	Do.	M D	45	Do.	9	15	1	Do.	Do.	Cut from 23-inch lengths.
10205	12-inch B.L. (Marks I to VII)	II	L	22 lbs. 2 ozs. (¼ charge)	I	30	8 ozs. R.F.G.[2] at one end	9·2	11·5	2	Cylinders Nos. 15, 15A and 16	Central silk cord	—
9739	12-inch B.L. (MarkVIII)	I	N	83 lbs. 12 ozs. (½ charge)	I	50	Do.	9·5	32·5	 1 1 1	Cylindrical Cases: "R" Large "R" Small Cylinder No. 33	Do.	—
9739	Do.	III	N	41 lbs. 14 ozs. (¼ charge)	I	50	Do.	9·5	16·25	 2 2 1 2	Cylindrical Cases: "R" Large "R" Small Cylinder No. 17 Cylinder No. 33	Do.	—

9927 12838	Do.	II	N	87 lbs. (½ charge)	I	50 & 3¾	12 ozs. R.F.G.[2] at one end	9·5	32·5	1 1 1	Cylindrical Cases : "R" Large "R" Small Cylinder No. 33	All round	—
9927 12838	Do.	II	N	43½ lbs. (¼ charge)	I	50 & 3¾	Do.	9·5	16·25	2 2 2 1	Cylindrical Cases : "R" Large "R" Small Cylinder No. 33 Cylinder No. 17	Do.	—
13230	Do.	I	N	100 lbs. (½ charge)	M D	45	8 ozs. R.F.G.[2] at one end	10	34·5	1	Cylindrical Case: "R" Large	Do.	Cut from 37 - inch lengths.
15392	Do.	II	N	Do.	M D	45	Do.	10	34·5	1	Do.	Do.	Cut from 24 - inch lengths.
—	Do.	III	N	Do.	M D	45	12-inch "B" (10 ozs. R.F.G.[2])	10	34·5	1	Do.	Do.	Standardised Igniter. Cut from 35 - inch lengths.
—	Do.	IV	N	Do.	M D	45	Do.	10	34·5	1	Do	Do.	Cut from 23 - inch lengths.
13230	Do.	I	N	50 lbs. (¼ charge)	M D	45	8 ozs. R.F.G.[2] at one end	10	17·5	2 1	"R" Large Cylinder No. 26	Do.	Cut from 37 - inch lengths.
15392	Do.	II	N	Do.	M D	45	Do.	10	17·5	2 1	"R" Large Cylinder No. 26	Do.	Cut from 24 - inch lengths.
—	Do.	III	N	Do.	M D	45	12-inch "B" (10 ozs. R.F.G.[2])	10	17·25	2 1	"R" Large Cylinder No. 26	Do.	Cut from 23 - inch lengths.
—	Do.	IV	N	Do.	M D	45	Do.	10	17·25	2 1	"R" Large Cylinder No. 26	Do.	Cut from 35 - inch lengths.
10367 10388 12838	12-inch B.L. (Mark IX)	II	N	52 lbs. 12 ozs. (¼ charge)	I	50 & 3¾	12 ozs. R.F.G.[2] at one end	10·25	16·75	2	Cylindrical Cases : { "Q" { "N"	Do.	Composite.

TABLE 8.—*Cordite Cartridges for B.L. and B.L.C. Guns*—continued.

Para. in List of Changes.	Nature of Gun.	Mark of Cartridge.	Service.	Weight.	Nature of Cordite.	Size of Cordite.	Igniter.	Diameter in inches.	Length in inches.	Number in Package.	Package.	Lifting Becket.	Remarks.
13517 13985	12-inch B.L. (Mark IX)	I	N	61½ lbs. (¼ charge)	M D	45	16 ozs. R.F.G.[2] at one end	10·4	18·4	2	Cylindrical Cases: "Q" "N" Rect. "I" Case Box, Cartridge, 12-inch	All round	For "Duncan" and "Formidable" Class. Cut from 37-inch lengths.
15392	Do.	II	N	Do.	M D	45	Do.	10·4	18·4			Do.	Cut from 24-inch lengths.
—	Do.	III	N	Do.	M D	45	12-inch "A" (16 ozs. R.F.G.[2])	10·4	18·4	1	Cylinder No. 40	Do.	Standardised Igniter. Cut from 23-inch lengths.
—	Do.	IV	N	Do.	M D	45	Do.	10·4	18·4	1	Do.	Do.	Cut from 35-inch lengths.
13517 13985	Do.	I	N	63½ lbs. (¼ charge)	M D	45	16 ozs. R.F.G.[2] at one end	10·4	18·8	2	Cylindrical Cases: "Q" "N" Rect. "I" Case Box, Cartridge, 12-inch	Do.	"King Edward VII" Class. Cut from 37-inch lengths.
15392	Do.	II	N	Do.	M D	45	Do.	10·4	18·8			Do.	Cut from 24-inch lengths.
—	Do.	III	N	Do.	M D	45	12-inch "A" (16 ozs. R.F.G.[2])	10·4	18·8	1	Cylinder No. 40	Do.	Standardised Igniter. Cut from 23-inch lengths.
—	Do.	IV	N	Do.	M D	45	Do.	10·4	18·8	1	Do.	Do.	Cut from 35-inch lengths.

13554 13857 13985	12-inch B.L. (Marks X and X*)	I	N	64½ lbs. (¼ charge)	M D	45	16 ozs. R.F.G.² at one end	10·5	18·8	2	Cylindrical Cases : "Q" "N" Rect. "I" Case	Do.	Cut from 37-inch lengths.
15392	Do.	II	N	Do.	M D	45	Do.	10·5	18·8		Box, Cartridge, 12-inch	Do.	Cut from 24-inch lengths.
—	Do.	III	N	Do.	M D	45	12-inch "A" (16 ozs. R.F.G.²)	10·5	18·8	1	Cylinder No. 40	Do.	Standardised Igniter. Cut from 23-inch lengths.
—	Do.	IV	N	Do.	M D	45	Do.	10·5	18·8	1	Do.	Do.	Cut from 35-inch lengths.
—	Do.	V	N	Do.	M D	45	Do.	10·5	18·8	1	Do.	Do.	Cut from 25-inch lengths.
14762	12-inch B.L. (Marks XI to XII)	I	N	76 lbs. 12 ozs. (¼ charge)	M D	45	16 ozs. R.F.G.² at one end	10·4	24	2	Cylindrical Case "L"	Do.	Cut from 24-inch lengths.
15341	Do.	II	N	Do.	M D	45	Do.	10·4	24	2	Do.	Do.	Cut from 37-inch lengths.
—	Do.	III	N	Do.	M D	45	12-inch "A" (16 ozs. R.F.G.²)	10·4	24	2	Do.	Do.	Cut from 23-inch lengths.
—	Do.	IV	N	Do.	M D	45	Do.	10·4	24	2	Do.	Do.	Cut from 35-inch lengths.
12853 15340	13·5-inch B.L. (Marks I to IV)	II	N	93 lbs. 12 ozs. (½ charge)	I	44 & 3¾	12 ozs. R.F.G.² at one end	11·8	23	1	Cylindrical Case "P"	Do.	Composite.
12853	Do.	II	N	46 lbs. 14 ozs. (¼ charge)	I	44 & 3¾	Do.	11·8	11·5	2	Do.	Do.	Do.
15340	13·5-inch B.L. (Mark V)	I	N	73 lbs. 4 ozs. (¼ charge)	M D	45	16 ozs. R.F.G.² at one end	10·25	22·5	2	Cylindrical Case "L"	Do.	Cut from 37-inch lengths.
15340	Do.	II	N	Do.	M D	45	Do.	10·25	22·5	2	Do.	Do.	Cut from 24-inch lengths.

TABLE 8.—*Cordite Cartridges for B.L. and B.L.C. Guns*—continued.

Para. in List of Changes.	Nature of Gun.	Mark of Cartridge.	Service.	Weight.	Nature of Cordite.	Size of Cordite.	Igniter.	Diameter in inches.	Length in inches.	Number in Package.	Package.	Lifting Becket.	Remarks.
—	13·5-inch B.L. (Mark V)	III	—	73 lb. 4 ozs. (¼ charge)	M D	45	13·5-inch "A" (16 ozs. R.F.G.²)	10·25	22·5	2	Cylindrical Case "L"	All round	Standardised Igniter. Cut from 23-inch lengths.
—	Do.	IV	N	Do.	M D	45	Do.	10·25	22·5	2	Do.	Do.	Cut from 35-inch lengths.
16405	Do.	I	N.	74 lbs. 4 ozs. (¼ charge)	M D	45	16 ozs. R.F.G.² at one end	10·25	22·5	2	Do.	Do.	Cut from 23-inch lengths.
16787	Do.	II	N	Do.	M D	45	Do.	10·25	22·5	2	Do.	Do.	No recess in end of charge for Igniter. Cut from 23-inch lengths.
—	Do.	III	N	Do.	M D	45	13·5-inch "A" (16 ozs. R.F.G.²)	10·25	22·5	2	Do.	Do.	Standardised Igniter. Cut from 35-inch lengths.
—	Do.	IV	N	Do.	M D	45	Do.	10·25	22·5	2	Do.	Do.	Cut from 35-inch lengths.
—	15-inch B.L.	I	N	107 lbs. (¼ charge)	M D	45	15-inch "A" (16 ozs. R.F.G.²)	11·7	26	2	Cylindrical Case "M"	Do.	Charge tied with silk webbing. Cut from 24·75-inch lengths.
—	Do.	II	N	Do.	M D	45	Do.	11·7	26	2	Do.	1-inch silk braid	Cut from 35-inch lengths.

SECTION (C).—DESCRIPTION OF CORDITE CARTRIDGES FOR B.L. HOWITZERS.

With the exception of the B.L. 9·45-inch, all howitzer cartridges are made up in the same way, but in a manner entirely different from that employed for B.L. gun cartridges for the following reasons:—

(1) A howitzer is intended for "high angle" fire so as to obtain a steep angle of descent, hence the charge is much lighter compared with that of a gun of the same calibre, viz.:—

The B.L. 6-inch gun has a charge of 23 lbs.

The B.L. 6-inch howitzer has a charge of 2 lbs. 8½ oz. (full).

(2) A howitzer is a comparatively short piece of ordnance, hence the charge must be quick burning (small size of cordite sticks), so that the whole of the charge may be consumed before the projectile has left the muzzle.

The B.L. 6-inch gun cartridge takes Cordite M.D., size 16.

The B.L. 6-inch howitzer cartridge takes Cordite M.D., size 4¼.

(3) With a howitzer, large angles of descent must be obtained at short as well as long ranges. This is done by altering the weight of the charge; therefore a cartridge for a howitzer is made up in such a manner that its weight can be readily reduced.

Howitzer cartridges are made in the form of a mushroom-shaped core of cordite, upon the stalk of which three or more rings of cordite are placed. The rings are all removable, so that the core alone, or the core plus one or more rings, may be fired.

Cartridges for 6-inch 30-cwt. B.L. Howitzer.

(*See* Plate XIII.)

The 6-inch 30-cwt. Howitzer has 3 charges, namely:—

1 lb. 12 oz. of Mark I Cordite (core and 3 rings).	For Mark I Howitzer only.
1 lb. 15½ oz. of Mark I Cordite (core and 4 rings).	For Mark I* Howitzer with Heavy Shell.
2 lb. 8½ oz. of M.D. Cordite (core and 3 rings).	For Mark I* Howitzer with Light Shell.

The M.D. Cordite Charge.

Cartridge, B.L. 6-inch 30 *cwt. Howitzer,* 2 *lb.* 8½ *oz. Cordite M.D., Size* 4¼, *Mark I* | *L* | is for use with the 100 lb. "light" shell in howitzers having enlarged chambers (*i.e.* Mark I* howitzers).

The cartridge consists of a core, three rings, a shalloon bag for the core, and the shalloon coverings for the rings; the shalloon bag for the core also carries the igniter.

The core is composed of a stalk and a mushroom head; the stalk consists of a bundle of cordite secured in places with silk sewing. Round one end of this is placed a ring of cordite, also secured with sewing, which forms the mushroom head. The whole of the core is placed in a shalloon bag, which is then sewn up. The rings, weighing 3½ oz., 8½ oz. and 13 oz. respectively, are each enclosed in a shalloon covering, the ends of each covering being brought together, and sewn with silk sewing. The rings fit over the stalk and are secured to the core by two pieces of shalloon braid, the bights of which are stitched to the shalloon bag. One end of each braid is brought up inside the rings and tied to the other end, which is outside them.

The igniter in the base of the core is divided into four compartments filled with R.F.G.[2] powder; some early issues have been made with S.F.G.[2].

Cartridge for Mark I Howitzer.

The 1 lb. 12 oz. charge of Cordite, Mark I, is for use with the Mark I Howitzer.

The Mark III cartridge is generally similar in construction to the cartridge above described, but differs in having the shalloon bag for the core in two parts—one part for the stem, the other for the mushroom head; this latter part is secured by a draw-string. The igniter is 2½ drms. of guncotton yarn.

Cartridges fitted with igniters containing S.F.G.[2] powder are known as Mark IV.

Mark I Howitzer Charge of Mark I Cordite.*

For Mark I* howitzers when firing the "heavy" shell, 122 lbs., the above-mentioned 1 lb. 12 oz. charge, with an additional 3½ oz. ring, is used; cartridges having the additional ring are designated "1 lb. 15½ oz., Cordite, Size 5."

Charges for 6-inch Howitzer Star Shell.

The charges for use with 6-inch Howitzer Star Shell are as follows:—

In Mark I Howitzer.—The core and 2-oz. ring of the 1 lb. 12 oz. cartridge.

In Mark I* Howitzer.—The core and 3½-oz. ring of the 2 lb. 8½ oz. M.D. cartridge.

5-INCH B.L. HOWITZER CARTRIDGES.

(*See* Plate XIV.)

There are two different cartridges for this howitzer, namely:—

11$\frac{7}{16}$ oz. of Mark I Cordite (core and 3 rings).

14$\frac{5}{16}$ oz. of M.D. Cordite (core and 3 rings).

(*This M.D. charge is special for India.*)

The Mark V, 11$\frac{7}{16}$ oz. Cartridge.

Cartridge, B.L. 5-inch Howitzer, 11$\frac{7}{16}$ oz. Cordite, Size 3¾, Mark V | L | consists of a core, three rings, and a shalloon bag, the latter carrying the igniter. (*See* Plate XIV.)

Core.—The core is mushroom-shaped and consists of a bundle of cordite, size 3¾, about 7 ft. long, twisted into shape, and secured in places with silk sewing. The mushroom-head portion is covered by the shalloon bag, the rest of the cordite is bare.

The shalloon bag is formed of two discs of shalloon; the inner disc is smaller than the outer, and is stitched to it round the edge. The bag thus formed is sub-divided into four compartments filled with S.F.G.[2] gunpowder.

The outer shalloon disc is fitted with a silk draw-string; it is placed over the mushroom head, the draw-string tightened round the stalk part of the core, and tied with a reef knot.

Each ring consists of $2\frac{9}{16}$ oz. of cordite tied in places with silk sewing; the three rings of bare cordite fit over the stalk and are secured to the core by two pieces of shalloon braid, the bights of which are stitched to the shalloon bag. One end of each braid is brought up inside the rings and tied to the other end, which is outside them.

Becket.—A small becket of silk sewing is attached to the base of the shalloon bag to facilitate withdrawing the cartridge from its silk cloth cover.

Earlier Marks of the $11\frac{7}{16}$ oz. Cartridge.

Mark IV Cartridge.—The Mark IV cartridge had 2 drams of waterproofed guncotton yarn instead of S.F.G.[2] powder; the igniter was not cross-stitched.

Mark III Cartridge.—The Mark III cartridge differs from Mark IV in having the core and rings entirely covered with shalloon, and the core is made up similar to that for B.L. 6-inch howitzer.

Charge for Star Shell.

The charge used with star shell is :—

"Core and one ring."

M.D. Cordite charge, 5-inch Howitzer.

The M.D. cordite charge consists of $14\frac{5}{16}$ oz. of size $4\frac{1}{4}$.

The core is made up in the same way as the 6-inch howitzer cartridge (*see* Plate XIII); the core is entirely covered with shalloon.

The S.F.G.[2] powder igniter is in the form of a ring.

Covers, cartridges, dowlas, are issued for protecting 5-inch B.L. howitzer $11\frac{7}{16}$ oz. cartridges, Marks I to III, when packed in boxes, in ammunition wagons or in limbers in Mark I* or II equipments.

Covers, cartridges, silk cloth, B.L. 5-inch howitzer, are used with the $11\frac{7}{16}$ oz. cartridges, Marks IV and V, and the $14\frac{5}{16}$ oz. M.D. charge. The covers are provided with a silk braid draw-string at the mouth, and a becket at the base.

Cartridges for B.L. 6-inch 25 cwt. Howitzer and B.L. 5·4-inch Howitzer.

The above guns are special for India.

The latest Marks of cartridges for the 6-inch howitzer of 25 cwt. are :—

"Cartridge, B.L. 6-inch Howitzer (25 cwt.), 2 lbs. 7 oz. Cordite, M.D., Size $4\frac{1}{4}$, Mark I. (Core and five rings.)

"Cartridge, B.L. 6-inch Howitzer (25 cwt.), 2 lbs. 1 oz. Cordite, Size $3\frac{3}{4}$, Mark VI." (Core and three rings.)

The latest cartridge for the B.L. 5·4-inch howitzer is :—

"Cartridge, B.L. 5·4-inch Howitzer, $13\frac{1}{2}$ oz. Cordite, Size $3\frac{3}{4}$, Mark V." (Core and three rings.)

The above-mentioned cartridges are made up in the same way as the 2 lbs. 8½ ozs. cartridge for the 6-inch howitzer of 30 cwts. described on page 95.

Cartridge for 9·45-inch B.L. Howitzer.

The cartridges for the 9·45-inch B.L. howitzer are made up on an entirely different model from the other howitzer cartridges. (*See* Fig. 8.)

The charge is of cordite M.D.T.

Cartridge, B.L. 9·45-inch Howitzer, 5 lbs. 8 ozs. 4 drams Cordite M.D.T., Size 18–10, Mark II | L | .

The charge is made up in two layers of cordite.

The base portion consists of one layer containing 2 lbs. 8 ozs. 4 drams made up in the form of a disc. The cordite M.D.T. is cut to a length of about 1·3 in., and is tied with silk sewing. It is then placed in a paper ring. The whole is then enclosed in a shalloon bag having an igniter of 1 oz. of R.F.G.[2] gunpowder in five parallel compartments.

The top layer consists of six wedge-shaped sections, two containing 1 lb. each and four containing 4 ozs. each of cordite M.D.T. cut to a length of about 1·6 in. and tied with silk sewing.

Each section is supported by a perforated glazed-board wall and enclosed in a shalloon bag. The sections are lightly stitched to the edge and centre of the base portion.

Each portion is marked with the nature and numeral of the cartridge and the weight and Lot No. of the cordite M.D.T. contained in it.

The weight of the full charge and the station and date of filling are also shown on the base of the cartridge.

Fig. 8.

Cartridge, B.L. 9·45-inch Howitzer, 5 lbs. 8 ozs. 4 drams Cordite M.D.T., Size 18–10, Mark II | L | .

Scale ⅓.

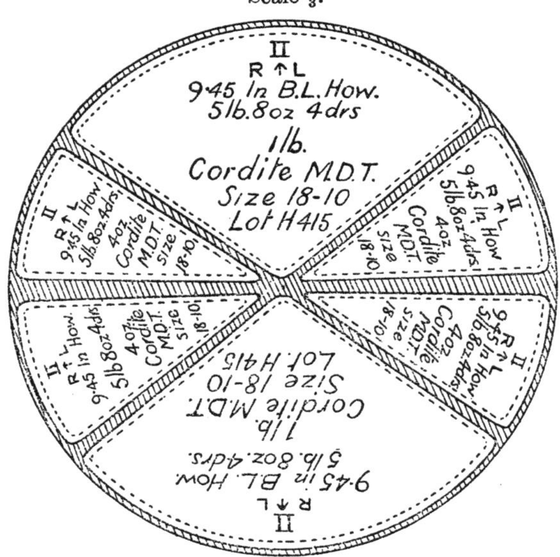

CARTRIDGE, B.L. 9·45-INCH HOWITZER, 5 LBS. 8 OZS. 4 DRAMS CORDITE M.D.T., SIZE 18–10. MARK I | L | .

The cartridge consists of a core and six sections, the whole being secured by shalloon braid to a shallow mill-board cup. On loading, this cup with its cordite charge is inserted in the obturating cup of the howitzer. The core of M.D.T. cordite tubes about 3 ins. long is contained in a cylindrical shalloon bag sewn with silk sewing; this bag also carries the igniter.

The igniter consists of S.F.G.[2] powder in a pocket formed by a disc of shalloon sewn to the bag and stitched across the centre to form four compartments.

The six sections are built up as follows:—

Two are arc shaped; each of these consists of 1 lb. of M.D.T. cordite (tied loosely in five bundles) enclosed in a shalloon bag; each of the remaining four sections consists of a bundle of 4 ozs. of M.D.T. cordite in a shalloon bag.

The six sections are placed round the core and secured in position in the mill-board cup by shalloon braid as shown in the figure.

Fig. 9.

CARTRIDGE, B.L. 9·45-INCH HOWITZER, 5 LBS. $8\frac{1}{4}$ OZS. CORDITE M.D.T. SIZE 18–10, MARK I | L | .

Scale $\frac{1}{3}$.

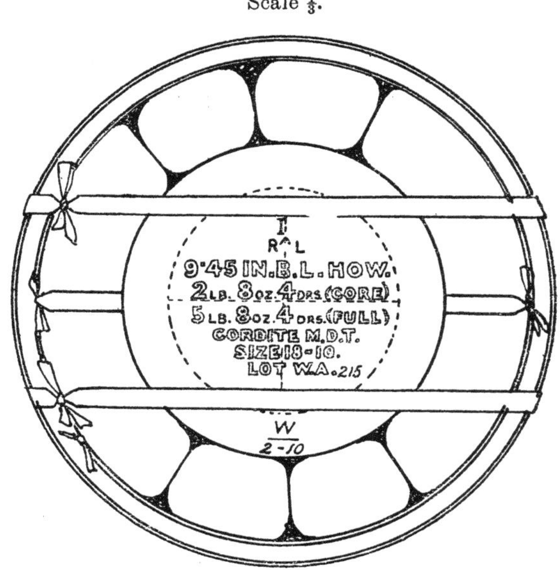

TABLE NO. 9.—*Cordite Cartridges for B.L. Howitzers.*

Para. in List of Changes.	Nature of Howitzer.	Mark of Cartridge.	Weight.			Cordite.		Igniter.	Length.	Diameter.	Number packed.	Package.	Remarks.
			Full.	Core.	Rings.	Nature.	Size.		Inches.				
8224	5-inch	I	$11\frac{7}{16}$ ozs.	$3\frac{12}{16}$ ozs.	Three $2\frac{9}{16}$ ozs. each	I	$3\frac{3}{4}$	8 drams R.F.G.^{2} in 4 compartments	3·2	3·8	100	Case, powder, metal-lined :— Whole	
9263	Do.	II	Do.	Do.	Do.	I	$3\frac{3}{4}$	8 drams R.F.G.^{2} in a ring	3·2	3·8	100	Do.	Core and rings covered with shalloon. (For 50-lb. shell.)
10579	Do.	III	Do.	Do.	Do.	I	$3\frac{3}{4}$	2 drams G.C. yarn	3·2	3·8	100	Do.	
12453	Do.	IV	Do.	Do.	Do.	I	$3\frac{3}{4}$	Do.	3·2	3·8	100	Do.	Base portion only, covered with shalloon. Each cartridge enclosed in a "Cover, cartridge, silk cloth." (For 50-lb. shell.)
	Do.	V	Do.	Do.	Do.	I	$3\frac{3}{4}$	8 drams S.F.G.^{2} in 4 compartments	2·8	3·4	120 60 24	Case, powder, metal-lined :— Whole Half Quarter	
	Do.	I	$14\frac{5}{16}$ ozs.	$5\frac{9}{16}$ ozs.	Two $2\frac{3}{4}$ ozs., one $3\frac{1}{4}$ ozs.	MD	$4\frac{1}{4}$	8 drams R.F.G.^{2} in 4 compartments	3·2	3·84	—	—	Core entirely covered with shalloon. Special for India. (For 50-lb. shell.)

For Star shell the core ($3\frac{12}{16}$ ozs.) and one ring ($2\frac{9}{16}$ ozs.) would be used.

5·4-inch	I	$13\frac{1}{2}$ ozs.	$4\frac{1}{4}$ ozs.	Three 3 ozs. each	I	$3\frac{3}{4}$	8 drams R.F.G.[2]	3·2	3·75	100	Case, powder, metal-lined :— Whole	
Do.	II	Do.	Do.	Do.	I	$3\frac{3}{4}$	Do.	3·2	3·75	100	Do.	Special for India.
Do.	III	Do.	Do.	Do.	I	$3\frac{3}{4}$	2 drams G.C. yarn	3·2	3·75	100	Do.	
Do.	IV	Do.	Do.	Do.	I	$3\frac{3}{4}$	Do.	3·2	3·75	100	Do.	
Do.	V	Do.	Do.	Do.	I	$3\frac{3}{4}$	8 drams R.F.G.[2]	3·2	3·75	100	Do.	
6-inch, 25 cwts.	I	2 lbs. 1 oz.	15 ozs.	Three— 11 ozs. 5 ozs. 2 ozs.	I	$3\frac{3}{4}$	12 drams R.F.G.[2]	10	4·3	30 16 5	Case, powder, metal-lined :— Whole Half Quarter	
Do.	II	Do.	Do.	Do.	I	$3\frac{3}{4}$	Do.	10	4·75	Do.	Do.	
Do.	III	Do.	Do.	Do.	I	$3\frac{3}{4}$	Do.	10	4·75	Do.	Do.	
Do.	IV	Do.	Do.	Do.	I	$3\frac{3}{4}$	$2\frac{1}{2}$ drams G.C. yarn	10	4·75	Do.	Do.	Special for India.
Do.	V	Do.	Do.	Do.	I	$3\frac{3}{4}$	Do.	10	4·75	Do.	Do.	
Do.	VI	Do.	Do.	Do.	I	$3\frac{3}{4}$	12 drams R.F.G.[2]	10	4·75	Do.	Do.	
Do.	I	2 lbs. 7 ozs.	1 lb. $2\frac{12}{16}$ ozs.	Five— 1, 13 ozs. 1, $3\frac{1}{2}$ ozs. 3, $1\frac{1}{4}$ ozs.	M D	$4\frac{1}{4}$	Do.	10	5			
Do.	II	Do.	Do.	Do.	M D	$4\frac{1}{4}$	Do.	10	5	Do.		Rings without shalloon covering.

TABLE NO. 9.—*Cordite Cartridges for B.L. Howitzers*—continued.

Para. in List of Changes.	Nature of Howitzer.	Mark of Cartridge.	Weight.			Cordite.		Igniter.	Length.	Diameter.	Number packed.	Package.	Remarks.
			Full.	Core.	Rings.	Nature.	Size.		Inches.				
9022	6-inch, 30 cwts. Mark I	I	1 lb. 12 ozs.	14 ozs.	Three— 1, 8 ozs. 1, 4 ozs. 1, 2 ozs.	I	5	12 drams R.F.G.[2]	6·5	4·4	50	Case, powder, metal-lined :— Whole	
9263	Do.	II	Do.	Do.	Do.	I	5	Do.	6·5	4·4	50	Do.	
11919	Do.	III	Do.	Do.	Do.	I	5	2½ drams G.C. yarn	6·5	4·4	50	Do.	
	Do.	IV	Do.	Do.	Do.	I	5	12 drams R.F.G.[2]	6·5	4·4	50	Do.	
13084	Do.	Charge for Star shell	1 lb.	Do.	2 ozs.	I	5	According to mark of cartridge					
12297	6-inch, 30 cwts. Mark I*	I	1 lb. 15½ ozs.	Do.	Four— 1, 8 ozs. 1, 4 ozs. 1, 2 ozs. 1, 3½ ozs.	I	5	2½ drams G.C. yarn	—	—	48	Case, powder, metal-lined :— Whole	For heavy shell.

	Do.	II	Do.	Do.	Do.	I	5	12 drams R.F.G.[2]	—	—	48	Do.	Do.
12550	Do.	I	2 lbs. 8½ ozs.	15½ ozs.	Three— 1, 13 ozs. 1, 8½ ozs. 1, 3½ ozs.	M D	4¼	Do.	6	5	32	Do.	For light shell.
1308	Do.	Charge for Star shell	1 lb. 3 oz.	Do.	1, 3½ ozs.	M D	4¼	Do.					
13259 14228	9·45-inch	I	5 lbs. 8¼ ozs.	2 lbs. 8¼ ozs.	6 sections— 2, 1 lb. 4, 4 ozs.	MDT	18–10	1 oz. R.F.G.[2]	3·3	9·8	7	Cylinder, No. 16	
15885	Do.	II	Do.	Do.	Do.	MDT	18–10	Do.	3·3	9·8	7	Do.	Built up in two tiers.

SECTION (D). MARKING ON B.L. CORDITE CARTRIDGES.

Gun Cartridges.

B.L. cordite gun cartridges are marked on one side with the following information .—

(*a*) Mark of cartridge.
(*b*) Manufacturer's initials.
(*c*) Calibre of gun.
(*d*) Weight of charge.
(*e*) Nature of cordite (cordite, or cordite M.D.).
(*f*) Size of cordite.
(*g*) Fraction of charge (if it is a fraction).

On the opposite side of the cartridge is marked the following :—

(*a*) Lot letter and number of the cordite.
(*b*) The letters "A.C." when charge is adjusted ; earlier issues in Naval Service had the actual weight ± of the adjustment shown.
(*c*) " N " (if for Naval Service).
(*d*) Monogram of station where filled.
(*e*) Date of filling (month and year).

Marking on Igniters.

The igniters of all B.L. gun cartridges, 2·75-inch to 5-inch, have the following information printed on them :—

(1) Mark of cartridge.
(2) Maker's initials.
(3) Calibre of gun.
(4) Weight of charge.
(5) Weight of powder in igniter.

The above information will also be found on the old type of B.L. cartridge, 6-inch to 13·5-inch, viz., those not fitted with " Standardized igniters."

" Standardized igniters " (that is, igniters which are suitable for making up different cartridges for the same gun) are now used with the latest Marks of cordite cartridges, 6-inch and up.

They are marked as follows :—

(1) Maker's initials.
(2) Calibre of gun.
(3) Igniter " A " or igniter " B," etc.

Marking on " Protecting Disc " or " Cover for Igniter."

The protecting disc on the old type of cartridge is marked with :

(1) Mark of the cartridge.
(2) Calibre of the gun.
(3) Weight of the charge.
(4) A red cross.

The " Cover for igniter " on the new type of cartridge is marked with the calibre of the gun and the letter of the igniter with which used.

Example :—Cover for 9·2-inch.
"A" igniter.

MARKINGS ON B.L. HOWITZER CARTRIDGES.

B.L. howitzer cartridges are marked with similar information.

Howitzer cartridge.—The monogram of station and the date of filling are found on the stalk of the core (in the case of 5-inch howitzer cartridges where the core is not covered with shalloon this is found on the mushroom head).

Each ring is marked with :—

(*a*) Its own weight.
(*b*) Calibre of howitzer, and, in the case of the 6-inch, "25 cwt." or "30 cwt."
(*c*) Mark of cartridge.
(*d*) Size of cordite.
(*e*) Lot letter and number of cordite.

SPECIAL MARKINGS ON B.L. CARTRIDGES AND THEIR PACKAGES FOR THE COLONIES.

Land Service.	D↑ for Australia.
	C↑ for Canada.
	U↑ for South Africa.
	N ↑ Z for New Zealand.
Naval Service.	A.N. for Australia.
	C.N. for Canada.
	N.Z. for New Zealand.

SECTION (E).—CORDITE CARTRIDGES FOR R.M.L. GUNS.

The following cordite cartridges have been sealed for the 10-inch R.M.L. Mark III guns on H.A. mountings :—

15 lb. 3 oz.	cordite	M.D.,	size 8.
11 lb. 3 oz.	,,	,,	,, 8.
8 lb. 4 oz.	,,	,,	,, 8.
4 lb. 13 oz.	,,	,,	,, 4¼.

The cartridges are all built up in the same manner as follows :—

The cordite charge is cylindrical in shape; it is bundled round a wooden stick, tied in places with silk sewing, and enclosed in a silk cloth cartridge open at each end. The ends of the cartridge are choked with silk braid into grooves at each end of the stick. Two rings of silk cloth, each containing R.F.G.² powder, are sewn around the cartridge, one at each end, and divided into a number of compartments by cross stitching.

The stick is of varnished deal 2·5 inch square, except at the ends which are grooved for choking the cartridge.

Fig. 10.

CARTRIDGE, R.M.L. 10-INCH, 15 LB. 3 OZ. CORDITE M.D., SIZE 8, FOR MARK III H.A. GUNS.

Scale $\frac{1}{3}$.

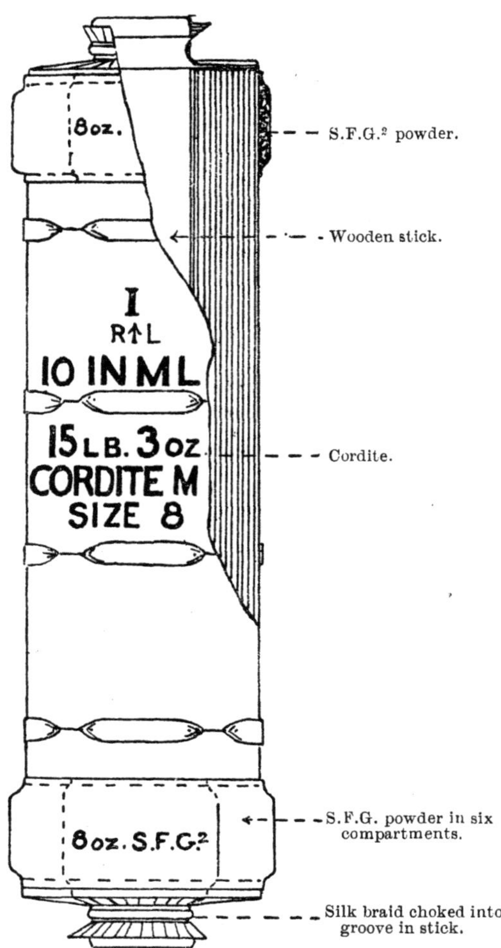

Cartridge for Case Shot.

A cordite charge of 20 lb. 6 oz. has been issued for use with 10-inch R.M.L. guns with special case shot.

It is built up in the same way as the cartridge above described, but the rings of silk cloth containing the R.F.G.[2] powder are sewn to the cartridge a short distance apart near the centre. This is to ensure the igniter being immediately below the vent no matter which end of the cartridge is loaded first.

SECTION (F).—POWDER CARTRIDGES FOR B.L. GUNS AND HOWITZERS.

There are still a few "Service" charges of gunpowder for B.L. guns in existence, but gunpowder is now mainly used for "blank" charges and for charges which are fired with paper shot.

Materials.

All powder-filled cartridges, with the exception of the B.L. 5-inch howitzer blank, are made of silk cloth and are hooped with silk braid. The 5-inch howitzer blank cartridge is made of shalloon.

Blank Cartridges for B.L. Ordnance.

Blank charges, in the Land Service, are fired from B.L. guns and howitzers up to and including the B.L. 6-inch.

All blank cartridges are made up in the same way.

(For particulars of weights, &c., *see* Table 11.)

With the exception of the 1½-lb. blank charge for the B.L. 15-pr. Mark I gun, which is pear-shaped (to prevent it being pushed past the radial vent), all blank cartridges are cylindrical in shape.

Cartridge, B.L. or B.L.C., 15 or 12-pr., 1 lb. 4 oz., Blank, Mark I.

The charge consists of 1 lb. 4 oz. Blank L.G. powder enclosed in a silk cloth cartridge bag; the cartridge is hooped with silk braid and choked with silk sewing; any superfluous choke is cut off.

Fig. 11.

CARTRIDGE, B.L. OR B.L.C., 15 OR 12-PR., 1 LB. 4 OZ. BLANK (MARK I)

Scale ½.

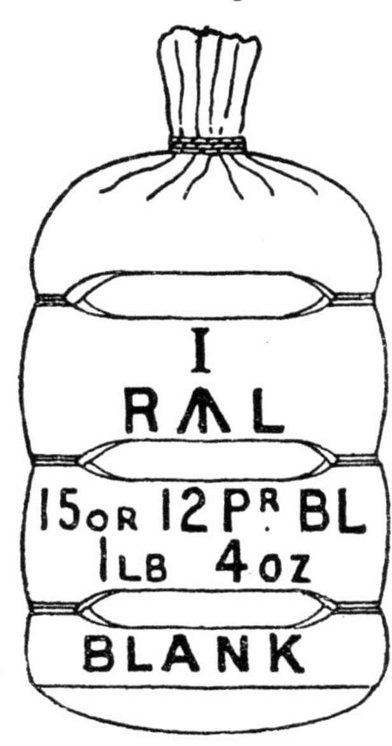

Blank Cartridge for B.L. 5-inch Howitzers.

The cartridge, B.L. 5-inch howitzer, blank, consists of 1 lb. blank L.G. gunpowder enclosed in a shalloon bag which is hooped and choked with silk sewing.

Covers, cartridge paper, are brown paper bags on which are marked the contents and numeral of the cartridge which they cover.

They are for the blank charges for the 2·75-inch, 10-pr., 12-pr. and 15-pr. B.L. guns :—

Cartridges for use with Paper Shot in B.L. Guns.

The following cartridges are used in the Land Service for firing paper shot from B.L. guns.

Cartridge B.L. 10 inch, 63 lb., Prism I brown		.. ¼ charge.
,, ,, 9·2-inch, 54 lb., ,, ,,		.. ¼ charge Mark X gun.
,, ,, ,, 41 lb., ,, ,,		.. ¼ charge Mark III to VII guns.
,, ,, 7·5-inch, 31½ lb., ,, ,,		.. ¼ charge for Mark II gun.
,, ,, 6-inch, 22 lb. E.X.E. (large prisms)		.. ½ charge for Mark VII guns.
,, B.L.C. 6-inch, or B.L. 6-inch, Mark VII guns in examination batteries	12 lbs. E.X.E.	¼ charge for 6-inch B.L.C. gun. Full charge for 6-inch, Mark VII.
,, B.L. 6-inch, Marks III, IV and VI guns.. ..	12 lbs. E.X.E.	¼ charge.

The shape of the above cartridges is prismatic, and there is no choke; the 31½ lb. charge for the 7·5-inch is fitted with an igniter containing 8 ozs. of R.F.G.[2] powder.

Cartridge, B.L. 9·2-inch, 54 lb. Prism I Brown, Silk Cloth, ¼ Charge, for Paper Shot, Mark X Gun.

The empty cartridge is made of No. 3 silk cloth with ten silk braid hoops, and is provided with two lifting beckets at each end.

The charge is built up in 16 tiers (15 tiers of 37 prisms each, and 1 tier of such convenient number as will ensure the correct weight). Seven black prisms being inserted in each end to act as an igniter. Each end of the cartridge has a central hole, covered with silk netting to allow the flash from the tube readily to ignite the prisms. To prevent the escape of powder dust, each hole is covered with a shalloon disc (secured with shellac).

The shalloon disc of the last cartridge placed in the bore in loading must be removed.

Fig. 12.
Scale $\frac{1}{4}$.

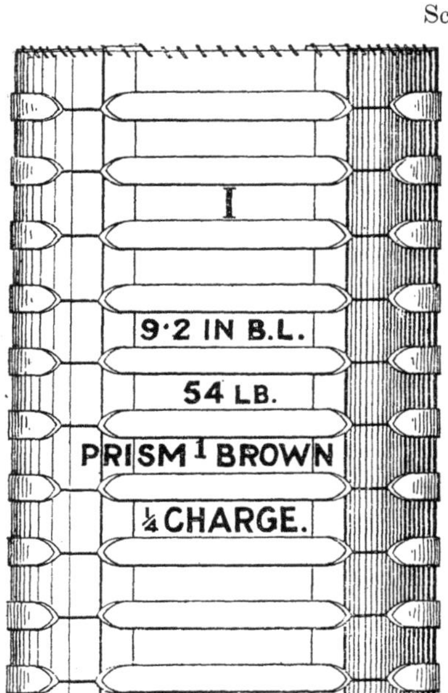

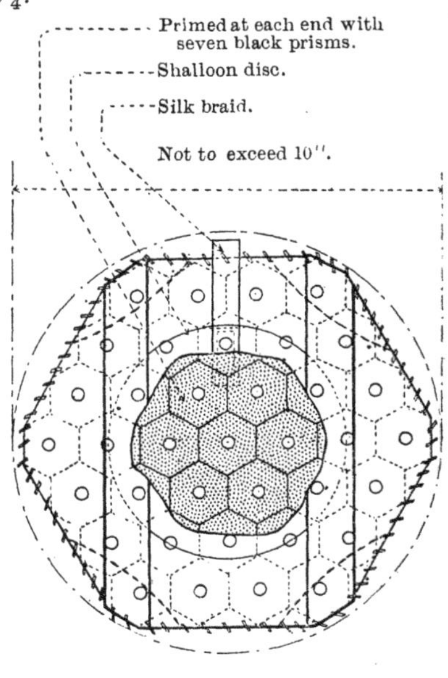

Plan of top
with part of top removed.

CARTRIDGE, B.L. 7·5-INCH, 31½ LB. PRISM I BROWN, SILK CLOTH, ¼ CHARGE, FOR PAPER SHOT, MARK II GUN.

The charge is built up in 13 tiers of prisms, 25 prisms in each tier, enclosed in a No. 3 silk cloth cartridge bag provided with 8 silk braid hoops.

The mouth of the cartridge bag is closed by an igniter. The igniter consists of two outer discs of shalloon and an inner disc of silk cloth, sewn together and stitched across to form five parallel compartments in the same way as the " Standardized igniters " for the heavy type of cordite cartridges.

The igniter contains 8 oz. of R.F.G.[2] powder and is top sewn to the mouth of the cartridge with double silk sewing No. 1.

CARTRIDGE, B.L. AND B.L.C. 6-INCH GUNS, 12-LB. E.X.E., MARK I | L | .

This is a ¼ charge for 6-inch B.L.C. guns with paper shot (120 lb.), and a full charge for B.L. 6-inch, Mark VII guns in Examination Batteries with a paper shot weighted up to 50 lb. only.

This charge is built up in 9 tiers: 8 tiers of 14 prisms each, and one of such convenient number of prisms (not less than 10) as will bring the total weight of the powder up to 12 lb.

There is also another *12 lb. E.X.E. charge, Mark I | C |* built up in a similar manner in 5 layers of 22 prisms each, and one of not less than 16 prisms; this is a ¼ charge for the B.L. 6-inch, Marks III, IV, and VI guns.

SECTION (G).—DRILL CARTRIDGES.

Cartridges, drill, are made to the same shape, weight, and dimensions as the Service cartridges they represent. They consist of wooden cylinders built up in segments, usually containing a cast-iron cylinder to give the necessary weight, and covered with raw hide. Cartridges, drill, B.L., representing prismatic cartridges, 8-inch to 13·5-inch, are weighted with lead and are made with polygonal sides, and have rope handles at each end so fitted that they do not project.

Drill representatives of cordite cartridges are issued for B.L. guns and howitzers except the 10-pr., 12-pr. of 6 cwt. and 15-pr. B.L. guns and the 5-inch B.L. howitzer. They are made of wood covered with raw hide, and conform to the general shape, weight, and dimensions of the Service cartridges. The representatives of cartridges, 6-inch and upwards, however, have no disc to be torn off, and the igniter end of the cartridge is represented by being painted white with a red cross. It has also a small becket projecting from its centre. These cartridges are stamped with the usual marking to be found on Service cartridges.

Fig. 13.

CARTRIDGE, B.L. DRILL, 9·2-INCH, 31 LB. 8 OZ., MARK II | N | .

Raw hide; ½ charge for wire guns.

Scale ⅙.

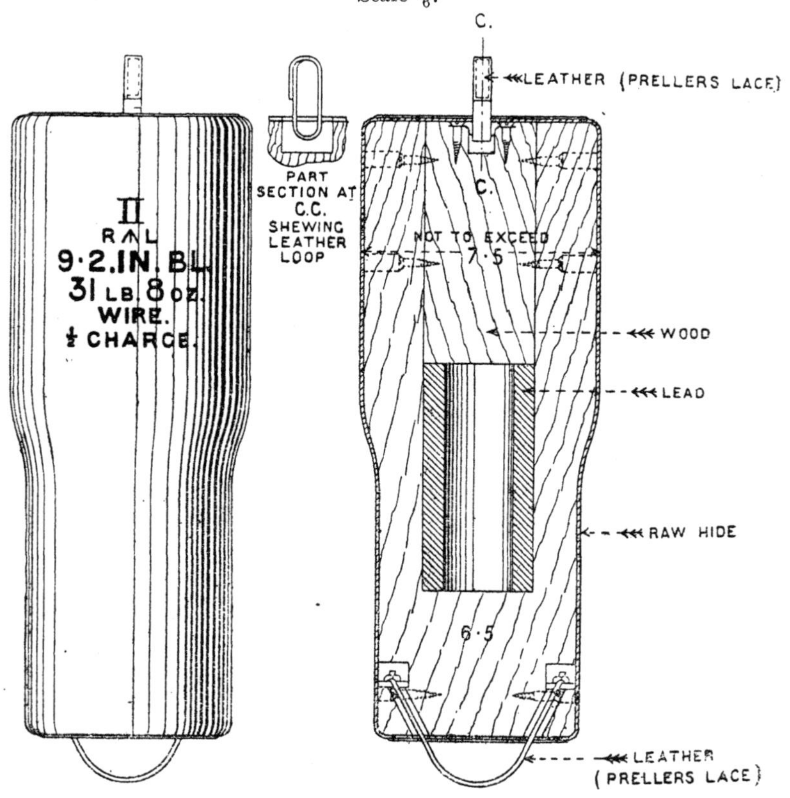

Drill cartridges have been issued, built up of 1-inch rope, and weighted with one layer of lead sticks, a felt disc being placed at each end, and the whole covered with stout canvas; this type of drill cartridge has been declared obsolete as soon as existing stock is used up.

Cartridge, B.L., for instruction, are issued for the B.L. howitzers. They consist of a core of wood and three or more rings of twine covered with leather. The rings are secured to the core by two leather strips, and the cartridge is marked in a similar manner to the Service one. They are used for instruction and not for drill.

Cartridges for instruction for 5-inch B.L. howitzers are now made of twine, and consist of a core and three rings, the base portion of the core only being covered with leather.

TABLE No. 10.—*Powder Cartridges for use with B.L. or B.L.C. Guns when Firing Paper Shot in the Land Service.*

Para. in List of Changes.	Nature of Gun.	Mark of Cartridge.	Weight and Nature of Powder.	Length.	Diameter.	No. in Package.	Package.	Remarks.
				Inches.				
			B.L. or B.L.C.					
12924	6-inch B.L. or B.L.C. ...	I	12 lbs. E.X.E.	9·2	6·6	4	Cylinders {No. 22, No. 30}	¼ charge, for paper shot.
12924	6-inch B.L., Mark VII ...	I	Do.	9·2	6·6	4	Do.	Full charge for 6-inch Mark VII guns in Examination Batteries when firing Paper Shot weighted up to 50 lbs.
10959	6-inch B.L., Marks VII and VII^{v}	I	22 lbs. E.X.E.	16	6·6	5	Case, powder, metal-lined whole	½ charge.
—	9·2-inch B.L., Marks III to VII	I	41 lbs. Prism' Brown	9·25	11·7	2	Cylinders {No. 26, No. 7}	¼ charge.
11754	9·2-inch B.L., Marks X and X^{v} guns on Mark V Barbette Mountings	I	54 lbs., Prism, Brown	16·25	10	1	Cylinder No. 13 ...	¼ charge.
—	10-inch B.L.	I	63 lbs., Prism' Brown	12·25	12	1	Cylinders {No. 5, No. 5A}	¼ charge.

TABLE NO. 11.—*Blank Cartridges for B.L. or B.L.C. Guns and Howitzers.*

Para. in List of Changes.	Nature of Gun or Howitzer.	Mark of Cartridge.	Service	Weight.	Nature of Explosive.	Length. Inches.	Diameter. Inches.	No. Packed.	Package.	Remarks.
									Case, powder, metal-lined :—	
11068	10-pr. and 2·75-inch ...	I	L	1 lb.	Blank, L.G.	5·25	2·7	102	Whole.	
								40	Half.	
								18	Quarter.	
13458	12-pr. 6 cwt.	I	L	6 ozs.	Sonite 12—10	6·5	3·2	66	Whole	Primed at one end with 6 drams of S.F.G.[2]
13458	15-pr., Mark I gun ...	I	L	7 ozs.	Do.	6·5	3·6	50	Do.	
13458	15-pr., Mark IV gun ...	I	L	7 ozs.	Do.	8·0	3·2	50	Do.	
10541	15-pr., Mark I gun ...	II	L	1½ lbs.	Blank, L.G.	6·25	3·0	64	Do.	Pear-shaped.
10427 14214	B.L. or B.L.C., 15 or 12-pr., except B.L. 15-pr., Mark I	I	L	1¼ lbs.	Do.	5·5	3·0	90	Do.	
								40	Half.	
								18	Quarter.	
13734	4-inch or 5-inch gun and 60-pr.	I	C	3 lbs.	Do.	6·75	4·3	36	Whole	For Land Service.
								17	Half	
								7	Quarter	
								36	Rect. "A" Case... ...	For Naval Service.
									Case, powder, metal-lined :—	
—	5-inch Howitzer	I	L	1 lb.	Do.	—	—	120	Whole	Bag, all shalloon.
								48	Half	
								18	Quarter	
11310	6-inch Howitzer	I	L	5 lbs.	Do.	6·5	6·0	32	Whole.	
8458	6-inch gun	I	C	7 lbs.	Do.	7·8	6·0	16	Rect. "A" Case... ...	
12924	6-inch gun, Mark VII in Examination Batteries	I	L	12 lbs.	E.X.E.	9·2	6·6	4	Cylinder No. 22 ... ... / Cylinder No. 30	With paper shot weighted up to 50 lbs.

CHAPTER X.—PACKAGES FOR B.L. CARTRIDGES, &c.

General Remarks.

The use of iron or steel is forbidden in the construction of packages intended for explosives which are placed in magazines; any exception to this rule must be specially authorized.

The interior of all packages must be examined to ensure their being dry before being used for explosives.

All packages for Naval Service containing explosives or explosive stores have two red bands painted round them.

Air-tightness.

Packages for B.L. *Cartridges* are made air-tight by the use of—

Luting, or
A dermatine washer, or
Both luting and dermatine washer.

Luting.—Luting is a mixture of whiting, mineral jelly and castor oil. It is placed as a rule in a groove in the metal of the package, into which the metal lid fits.

Mark III luting.—The luting for securing the lids of packages is Mark III, which consists of 80 parts by weight of whiting, 20 parts of mineral jelly (vaseline), and one part of castor oil. It is issued from Woolwich ready mixed in tin cyliners, each containing 1 lb.

The luting, before use, is to be beaten up with a wooden mallet till it is of the required consistency.

Marks I and II luting.—Mark I luting (equal parts of beeswax and tallow) is for naval mining and torpedo services. With this exception Marks I and II luting are not to be used in future.

Dermatine washer.—Dermatine is a rubber composition. The washer is placed in a groove in the underside of the lid of the package; when the lid is in position the metal rim of the package compresses the washer, thus forming an air-tight joint.

Testing Air-tightness.

After closing the lid of any cylinder or case, except metal-lined cases, the joint may be tested by immersion in warm water, about 20° F. above the temperature of the atmosphere. If air bubbles escape *at the joint* the case or cylinder must be re-closed and re-tested; if elsewhere, the case needs repair. *This test is only applied to empty packages.*

At depots at home and abroad, such naval powder cases as are fitted with testing plugs will be tested for air-tightness by the Pump, air, testing powder cases:—

1. On receipt into store.
2. After packing.

The test will only be considered satisfactory when the indicator remains stationary (under compression or exhaust) for not less than 1½ minutes at 1½ lb. per square inch.

Pump, air, testing powder cases, Mark I, for the above test is a single-acting pump fitted with a gauge for indicating compression or exhaust, which will read 3 lb. either way. Connection with the powder case is made by an india-rubber pipe having a nozzle which screws into the hole in the lid of the case.

Packages for B.L. Cartridges.

The following are the packages in which B.L. cartridges (both gunpowder and cordite) are transported and stored :—

Land Service.	Naval Service.
Cylinders, cartridge. Cases, powder, metal-lined. Cases, metal-lined, field.	Cylinders, cartridge. Cases, powder, metal-lined. Brass cases :— Cases, powder, rectangular. ,, ,, cylindrical. Boxes, cartridge.

NOTE.—Powder barrels are occasionally used (R.A.O.S., Part II., para. 133).

Packages for B.L. Cartridges in the Land Service.

Cylinders, Cartridge.

Cylinders, cartridge, are used for packing cartridges for B.L. guns, 6-inch to 12-inch.

In the Land Service they are used to transport, store, and bring up to the gun, such cartridges.

In the Naval Service they are only used for transport purposes and storage ashore. Cartridges are transferred from them to the Naval (brass) cases for storage in ships' magazines.

They are also used for packing small irregular quantities of cordite in tropical climates, the cordite being first wrapped in non-absorbent paper.

The cylinders are known by their numbers, which are stamped on them. Table 12, page 134, shows the cartridges that may be packed in them.

Description of " Cylinders, Cartridge."

There are two different patterns of cylinders, known as :—

" B " pattern, with handles on top of lid.

" C " pattern, with handles on the side.

The above cylinders have tape bands for sealing the lid.

The latest " C " pattern cylinders manufactured or repaired since June, 1906, have a luting joint instead of a tape band.

Latest "C" Pattern Cylinder.

Fig. 14.

CYLINDER, CARTRIDGE.

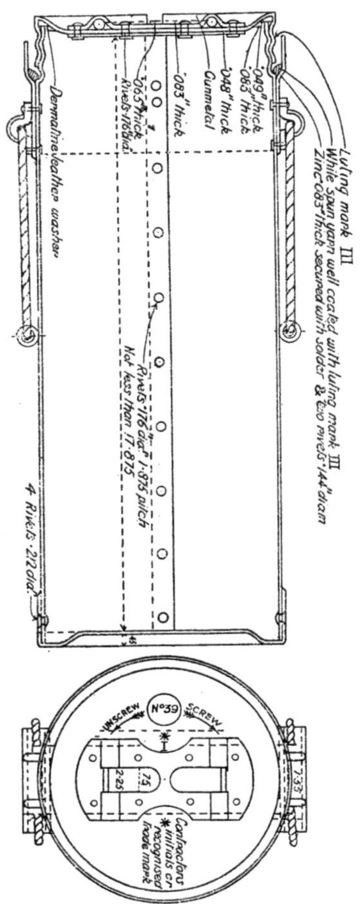

The body of the cylinder is made of zinc. It is lap-jointed, riveted and soldered. The top is screwed to receive the lid. A ring of zinc is soldered around the inside at the top, thus doubling the thickness of zinc at that point. A cupped ring of zinc is riveted and soldered around the outside near the top, to receive the lower part of the lid.

Two galvanized wrought-iron or mild-steel brackets are riveted to the body, and the joints are soldered.

The brackets are each fitted with a handle of galvanized-iron wire on which is threaded a piece of galvanized wrought-iron or mild-steel tubing, the ends of the wire being spliced together.

The bottom of the cylinder is made of zinc ; it is secured to the body by rivets and solder.

The lid is flanged and screwed to fit the top of the cylinder.

To the top is fitted a piece of zinc, having two eyes formed in it, into which metal lugs are hinged.

A zinc ring is soldered on the underside of the lid to form a groove, into which a thick dermatine washer is secured by rubber solution.

Wooden disc.—A painted wooden disc is placed in the bottom of each cylinder; except in certain cases, where one cylinder is used for packing cartridges of different lengths, when a "packing-piece" is used for the shorter cartridges instead of the disc.

Formerly these discs and packing pieces were varnished and may still be met with.

Earlier "C" Pattern Cylinders.

The "C" pattern cylinders, manufactured prior to June, 1906, differ from the latest pattern in being without the cupped ring of zinc intended to receive the lower part of the lid.

They are sealed with a tape band shellaced on.

"B" Pattern Cylinders.

These have two handles on the top of the lid instead of on the side.

They are sealed with a tape band shellaced on.

Paint.—New zinc cylinders are not painted; *repaired* cylinders are painted stone colour.

Wooden cases for protecting cylinders.—When zinc cylinders are issued containing filled cartridges they are protected by wooden skeleton cases which are made up locally as required, brass screws being used. Serviceable zinc cylinders, when issued empty, are similarly protected.

Flux used in Soldering Cartridge Cylinders and Cases, Powder.

The flux used in soldering zinc cylinders and Naval powder cases is that known as "Flux, soldering." (Commercial oleic acid—Fatty acid.)

The joints are carefully cleaned after soldering and the above-mentioned flux leaves no "free acid."

Prior to May, 1908, zinc chloride was used as a flux.

All zinc cylinders manufactured prior to May, 1908, on being emptied, are boiled out with a solution of sodium carbonate sufficiently strong to neutralize any free acid that may be present. After this treatment the cylinders have a circle "O" *stamped* on them above one of the handle brackets.

They also have a blue disc stencilled on them under one of the handles.

Bearer, cartridge, cylinder, Mark III.—This is an ash stave, 4 ft long, 1 in. thick, 2½ in. broad at the centre, tapering off to each end. It has three grooves cut in the side near the centre to take the handles of the cartridge cylinders. The bearer is also used as a lever in unscrewing and screwing home the lids of the cylinders.

CASES, POWDER, METAL-LINED :—WHOLE, MARK IV | C | ; HALF, MARK IV | C | ; QUARTER, MARK IV | C | .

Cases, Powder, Metal-lined.

Cases, powder, metal-lined, Mark IV, are of three sizes—whole, half and quarter; they are rectangular cases of deal, strengthened by oak corners; the cleats are of ash; the sides and ends are secured by dovetailing, and the top and bottom by brass screws. They are lined with tinned copper. A square lid opens on copper wire hinges on top of the case; it is screwed down by two metal bolts which enter nuts in the top of the case. This lid covers a circular opening in the lining, which is closed by a bung of tinned copper.

Originally the cases had rope lifting handles if for Land Service, and copper wire if for Naval Service. In future, copper wire handles will be used in both Services.

Fig. 15.

CASE, POWDER, METAL-LINED, "WHOLE" (typical also of "half" and "quarter").

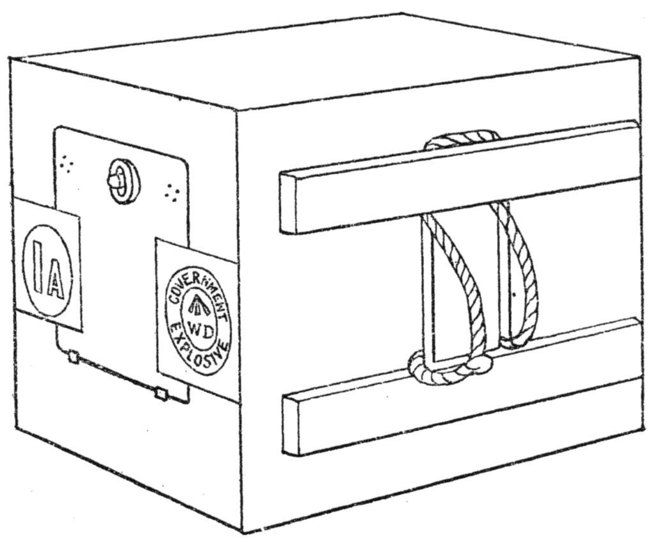

Dimensions for Stowage.

Whole	.. ..	17·625 × 17 × 21·625 in.
Half	.. ..	14·25 × 13·6 × 17·7 in.
Quarter	.. ..	11·125 × 10·4 × 14·7 in.

Use.—Cases, powder, metal-lined, are used in both Land and Naval Services for the storage and transport of the smaller cartridges, powder and cordite, and for small combustible stores generally, the half size also for wet guncotton, when specially fitted as described on page 23.

The whole size will take cartridges, B.L., up to 6-inch Mark VI gun, inclusive.

Painting.—Land Service cases, new or repaired, if for use abroad, are treated with " Oil, mineral, preserving wood." For use at home stations, they are painted " service colour," except as stated below.

Naval Service cases, are painted " stone colour," except as stated below.

Both Land and Naval Service cases, are painted " red " for blank cartridges and " yellow " if containing explosives for lyddite shell.

Case, Powder, Metal-Lined, Special, Half, Mark II | *N* | .

This case is similar to the ordinary " Case, powder, metal-lined, half," but differs slightly in dimensions.

It is specially suitable for storage in H.M. Ships " Swiftsure " and " Triumph."

The Mark I case was slightly smaller.

Case, Metal-Lined, Field, Mark II.

This case is of wood, about the same length as a S.A.A. box, provided with a sliding lid with pin, and with a cleat and rope handle at each end. Dimensions :—

$21\frac{7}{8}$ in. × $11\frac{5}{8}$ in. × 13 in. Weight, empty—28 lbs.

The interior is fitted with a removable tinned copper lining, with a groove for luting at the top, and the lid of the lining is flanged to fit into the groove. The case is for " Field Service," B.L. reserve cartridges, and is only used for the carriage of cordite cartridges which are packed without their covers on, in wagons in ammunition columns.

Note.—The Mark I case is obsolete for manufacture ; it is known as " Case, metal-lined, Mark I, Special."

CASES, POWDER, RECTANGULAR.

These cases are placed in a lettered series, many of which are now obsolete for future manufacture, but will still be met with until existing stock is used up.

For particulars of use, packing, &c., *see* Table 14, page 137.

The rectangular cases that are now being manufactured are :—

Cases, Powder, Rectangular, " O," Mark III.
,, ,, ,, " R," Mark III.
,, ,, ,, " S," Mark III.
,, ,, ,, " T," Mark II.
,, ,, ,, " U," Mark I.
,, ,, ,, " W," Mark I.

All the above-mentioned cases are fitted with the latest pattern lid, and all, except the " S " case, are termed " weakened cases," *i.e.*, specially constructed to be strong enough when filled to stand the strain of transit between naval depots and H.M. Ships, but when finally placed in the magazine they can be weakened, so that, should the cordite ignite spontaneously, the top of the case will be blown off before the gas evolved attains sufficient pressure to cause a violent explosion and so cause the cordite in adjacent cases to explode.

A description of the " R," Mark III case follows.

The " O," " T," " U " and " W " cases differ only in dimensions, and in the bottom of the " W " case being dished.

Case, Powder, Rectangular, "R," Mark III.

Fig. 16.

CASE, POWDER, RECTANGULAR, "R," MARK III.

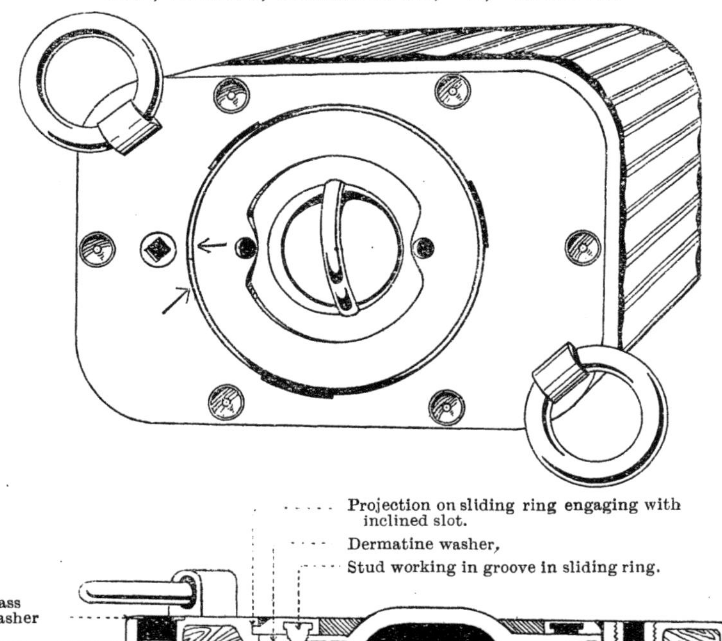

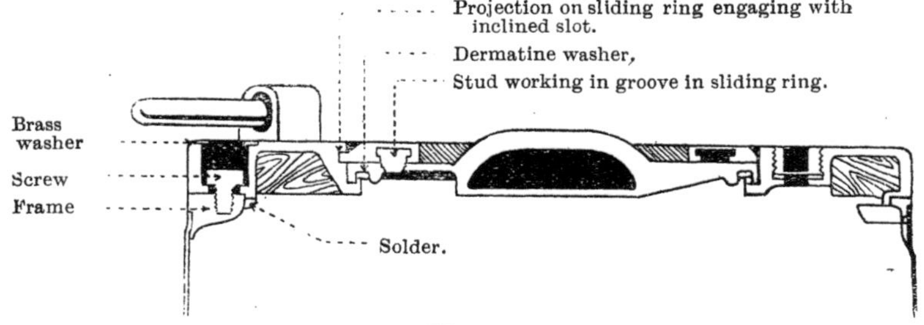

Fig. 17.

SECTION SHOWING HOW SCREW SECURES TOP OF "R," MARK III, CASE TO FRAME.

Scale full size.

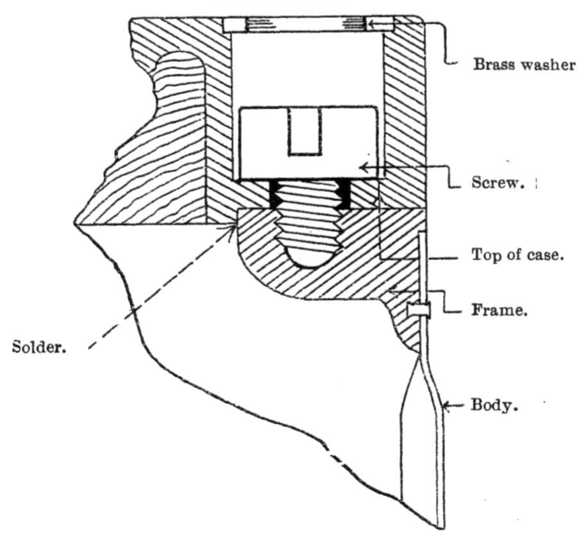

Case, Powder, Rectangular, "R," Mark III.

The body of the case is made of corrugated sheet brass, lap-jointed, riveted and soldered.

The bottom is riveted and soldered to the body.

The frame has six lugs on the inside, which are bored out and screw-threaded to take the transport screws ; it is riveted and soldered to the body.

The top of the case is lined with wood to prevent the cartridges being cut by the lower ends of the top fittings ; it is lightly soldered to the frame on the inner edges.

This light solder attachment is the only means of holding the top on the case when in store.

To make the case stronger when in transit, "transport screws" are provided ; these screws enter recesses in the top of the lid and screw into the lugs of the frame.

Small brass washers fitted into the upper portion of these recesses prevent the "transport screws" from falling out when they are unscrewed from the frame.

The lid is circular and provided with a handle on the top.

It is recessed on the underside to receive a dermatine washer, intended to bear against a raised lip formed in the mouth of the case.

A sliding ring, having an undercut groove on the underside, is held to the lower part of the lid by three studs.

The studs have enlarged heads which fit in the undercut groove, and are screwed and riveted to the lid.

A hole is drilled in the top of the ring to admit of the insertion of the studs, and, after insertion, this hole is closed by a brass disc soldered in. Three projections are formed on the outer circumference of the sliding ring ; these engage with three inclined grooves in the mouth of the case to secure the lid in position.

Two key-holes are drilled through the sliding ring to receive corresponding projections on the key for turning it.

The case is fitted with two copper wire lifting handles covered with leather.

Testing plug.—A hole is drilled through the top of the case and screw-threaded to receive a screw plug with a leather washer under it.

This hole is used to connect the "*Pump, air testing, powder cases.*"

Paint.—The case is painted stone colour inside and out.

Case, Powder, Rectangular, "S," Mark III.

The "S," Mark III case differs from the "R," Mark III, above described, in dimensions and in having the top riveted and soldered direct on to the body.

The bottom of the case is dished.

Case, Powder, Rectangular, "S," Mark II.

The "S," Mark II case differs from the "S," Mark III, in having a flat bottom.

Dished bottoms will be fitted to Mark II cases as they pass through Ordnance Factories for repair, and a star will be added to the numeral.

Cases, Powder, Rectangular, "O," and "R," Mark II.

Fig. 18.

CASE, POWDER, RECTANGULAR, "R," MARK II.

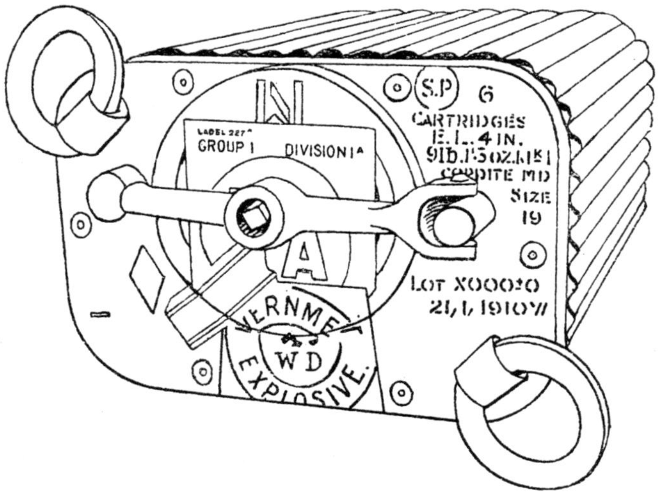

Fig. 19.

SECTION SHOWING HOW NUT SECURES TOP OF CASE TO FRAME.

Scale Full Size.

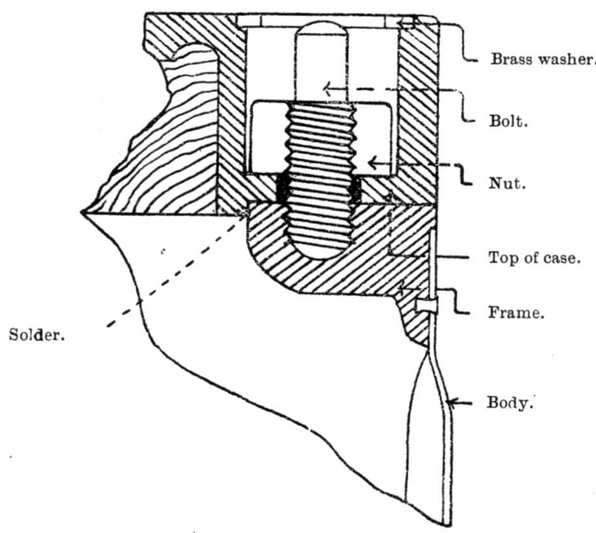

The above-mentioned cases are weakened in the same way as the "O" and "R," Mark III, but instead of having transport screws they are fitted with screw bolts.

These bolts are fitted with metal nuts which screw tightly down into the recesses in the lid.

In store these nuts are unscrewed from the bolts by means of a "forked bit."

The cases are also fitted with a different pattern of lid, which is secured by a crossbar working on a pivot at one end, and fitting under a projection at the other.

A central screw bolt bears upon the lid; when this is tightened, all is fixed in position; when unscrewed, the bar can be turned on one side and the lid lifted out.

The bolts and nuts in the above rectangular cases will be replaced by transport screws as the cases pass through Ordnance Factories for repair.

Case, Powder, Rectangular, "R," Mark I.

This is of a different design to that described above; it is fitted with two steel travelling bands as shown in the woodcut.

Fig. 20.

CASE, POWDER, RECTANGULAR, "R," MARK I.

Scale $\frac{1}{5}$.

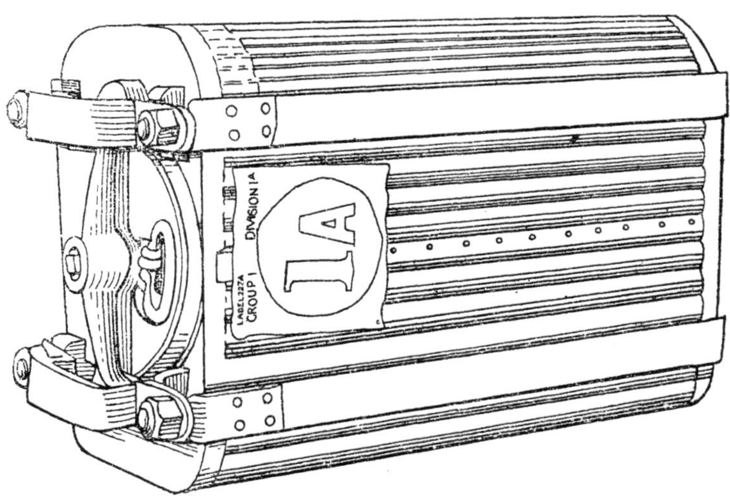

The body of the case is of corrugated sheet brass to which the gunmetal top is lightly soldered; neither rivets nor securing bolts are used.

It is not provided with lifting handles, nor fitted with a wood lining, but has four projections or cradles cast on it to keep the travelling bands in position.

The travelling bands are made of steel; the set consisting of two straps and two bars. The straps fit round the two sides and the bottom of the case and are provided with a bolt and nut at each end for securing the top bars, which lie in the cradles on the top of the case.

Fig. 21.

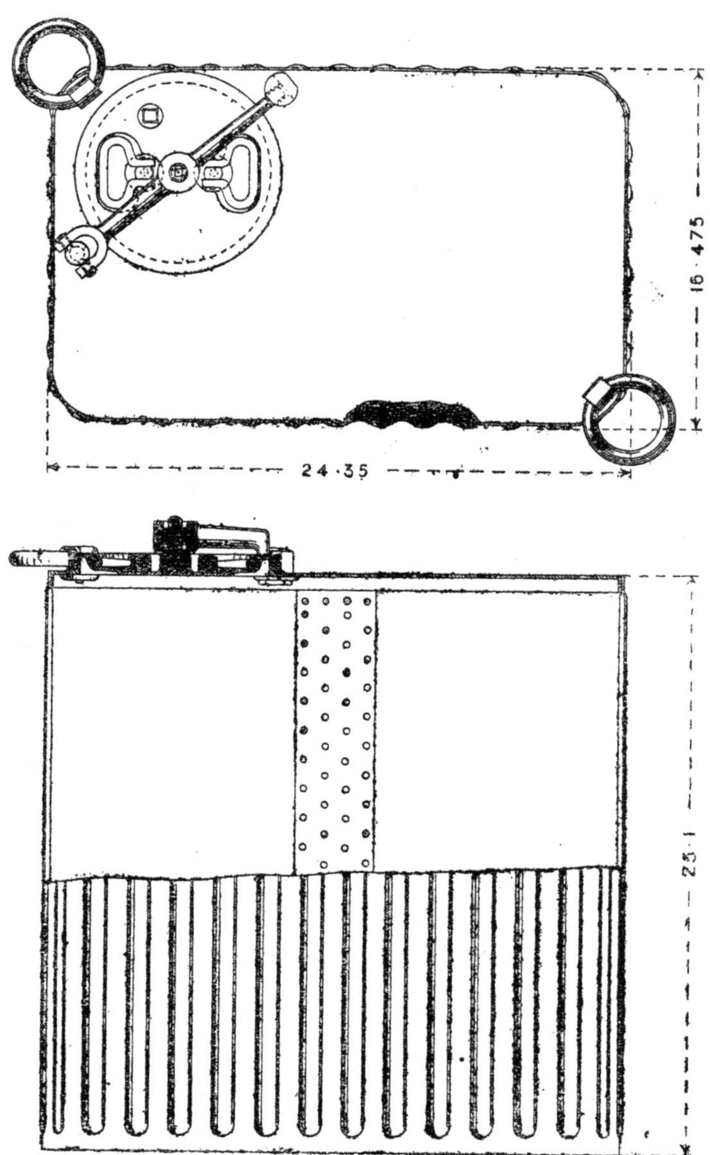

The top bars are bridged to form handles for lifting purposes. This was the first type of weakened case issued, but no more will be manufactured and existing stock used up.

Case, Powder, Rectangular, " T," Mark I.

This is another design of weakened case; it is made of corrugated sheet brass, the body joined with a lap-joint, riveted and soldered. Each side is indented near the top, to receive a brass bracket which is riveted to the body. Holes are made through the top of the case and each bracket to receive screws for the purpose of securing the top of the case to the body in addition to the solder attachment. These screws are to be removed when the case is not in transit.

The lid is similar to that on the " O " and " R," Mark II cases (Fig. 18), and is secured in the same way.

Cases, Powder, Rectangular, not weakened at top when in store.

These are of the early pattern of rectangular cases. They are all made of corrugated sheet brass, double riveted, with strips of brass on the vertical joints at the sides. The top of the case is riveted and soldered to the body, and is fitted with a lid and handles similar to those already described. (*See* Fig. 21.)

Special Marking on Cases where Resin Oil was used as a Flux.

Where resin oil has been used as the flux in soldering, a ◊ 1 in. in width is to be stencilled in blue paint, and a ◊ $\frac{3}{8}$ in. long stamped in corner remote from bung. (Cases stamped with date of manufacture subsequent to 1.1.10 will not be so marked.)

Fig. 22.

KEYS, CASE, POWDER, RECTANGULAR, NO. 1, MARK V | N |.

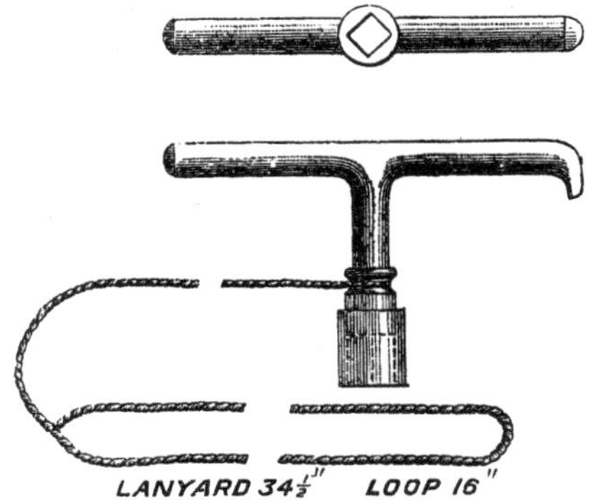

Keys for Cases, Powder, Rectangular.

Key, case, powder, rectangular, No. 1, *Mark V* | *N* |, is used for " Cases, powder, rectangular," with cross bar for securing lid; also for Q.F. or Q.F.C. Naval cartridge boxes, 6-inch to 12-pr., except transport.

It is made of aluminium bronze with cross handle, one end of which is formed into a toe for use as a lever.

The key is fitted with a lanyard. (*See* Fig. 22.)

Keys, Case, Powder, Rectangular | *N* | .

No. 2, Mark I .. For " U," Mark I case.
No. 3, Mark I .. For " O," Mark III and " S," Mark II cases.
No. 4, Mark I .. For " R," Mark III and " T," Mark II cases.
No. 5, Mark I .. For " W," Mark I case.

The above keys are used for locking or unlocking the lids of the latest patterns of rectangular cases. They are made of metal to the form shown in the accompanying drawing and are similar in design, differing only in dimensions.

Fig. 23.

KEY, CASES, POWDER, RECTANGULAR, NO. 2, MARK I | N | .

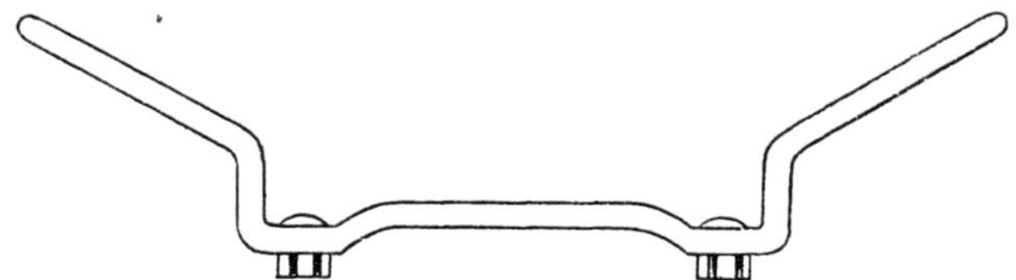

CASES, POWDER, CYLINDRICAL.

Cylindrical powder cases are used for B.L. gun cartridges, 12-inch to 15-inch.

They are also used to pack small irregular quantities of cordite at tropical stations.

When so used the cordite is first wrapped in non-absorbent paper.

The cases are placed in a lettered series ; many of them, however, were made to hold powder charges, and are now obsolete.

All cylindrical powder cases are made air-tight by a dermatine washer only.

(For cartridges packed in each case, *see* Table 13, page 136.)

The cases at present in the Naval Service may be classified as " *New Pattern* " and " *Old Pattern*."

Those now in use are :—

New Pattern, " *N,*" " *L* " *and* " *M.*"
Old Pattern, " *P,*" " *Q* " *and* " *R.*"

The *new pattern* cases are constructed with both ends removable, and the ends can be placed in a weakened position when in the magazine.

The *old pattern* cases have only one end removable, and have no arrangement for weakening when in the magazine.

Case, Powder, Cylindrical, " *L,*" *Mark III.*

The Case, powder, cylindrical, " L," Mark III, | N | , consists of a body, two intermediate rings, lifting plate with handle, two end rings and two lids.

The body of the case is made of sheet brass and may be either solid drawn or lap-jointed, riveted and soldered.

Fig. 24.
CASE, POWDER, CYLINDRICAL, "L," MARK III.

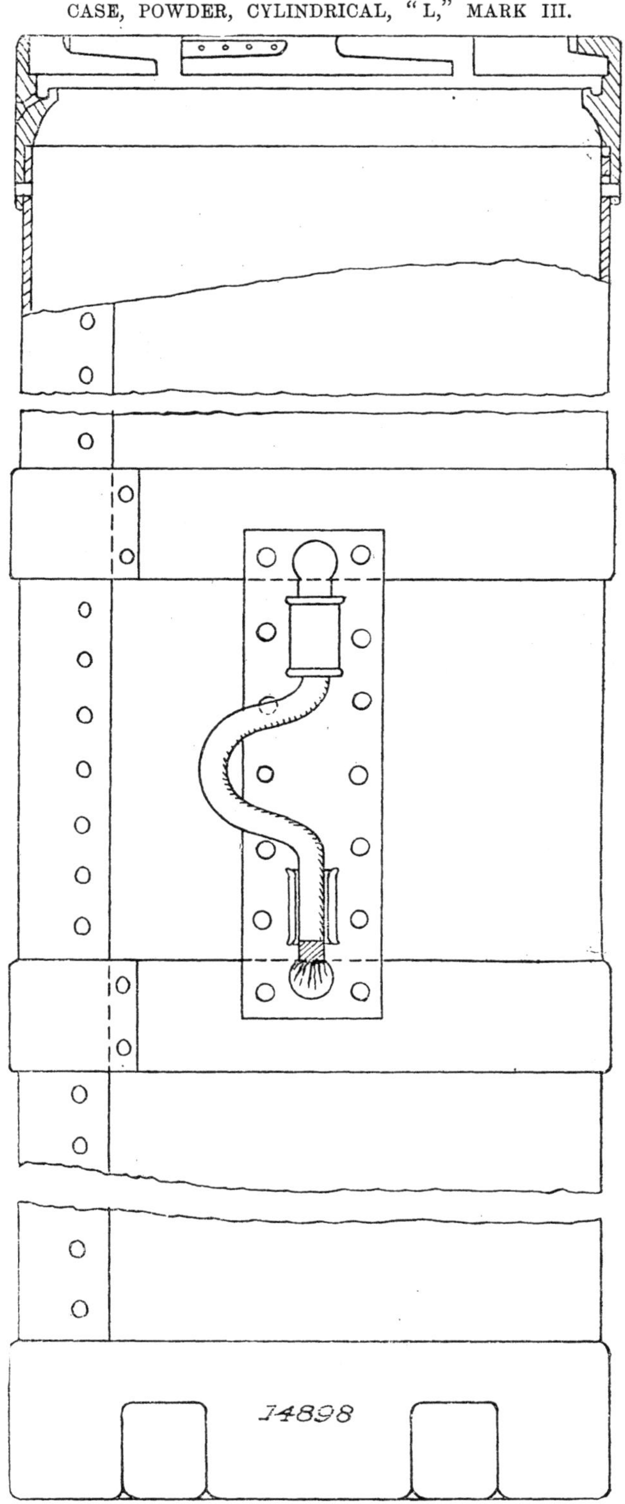

The intermediate rings are tinned all over and are secured to the body with solder.

The lifting plate is cast with two lugs ; holes are bored through the lugs to receive a copper wire lifting handle which is covered with leather.

A knob consisting of 7 parts of lead to 1 part of antimony is cast on each end of the wire handle, to retain it in position in the lugs of the lifting plate.

The lifting plate is tinned all over and riveted and soldered to the body of the case.

The end rings are cast with four bearing pieces on the outside of each to give stability in stowage. Six inclined grooves are milled on the inside to take the projections on the locking rings.

Metal pads.—Three metal pads are lightly riveted and soldered for three of the projections on the locking ring to engage with when the cases are in the magazine.

The rings are tinned all over and are riveted and soldered to the body.

The lids are dome shaped on top and have two pockets to facilitate lifting. A hole is drilled and screw-threaded in the centre of the lid to receive the connection for air testing. This hole is closed by a screw plug, having a leather washer under it. A recess is turned on the underside and tinned to receive a dermatine washer. The underside of the lid is lined with wood to prevent the sharp edges from cutting the igniters of the cartridge.

Fig. 24A.

CASE, POWDER, CYLINDRICAL, "L," MARK III.

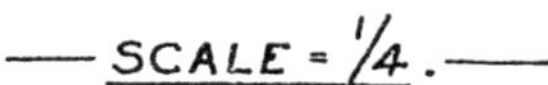

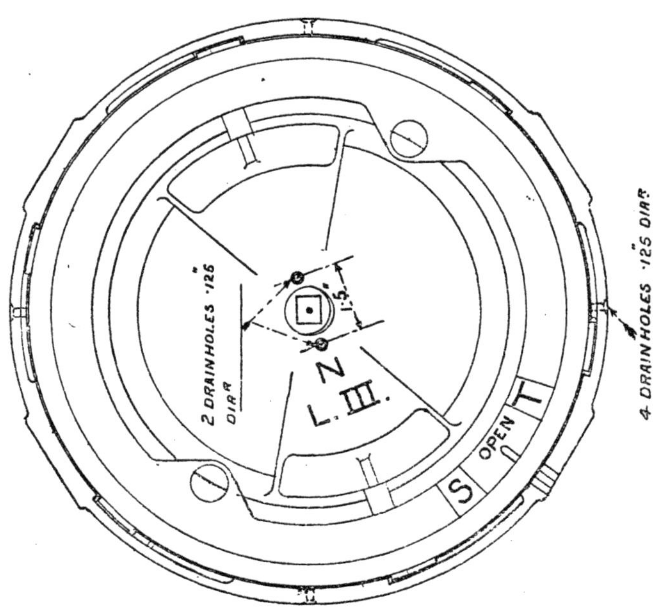

The locking rings are cast with six projections around the outside and are secured to the lid by three screws, the heads of which work in undercut grooves in the locking rings.

Two keyholes are drilled in each ring.

An index line is cut on each locking ring and the word " Open " and the letters " S " and " T " are shown.

An index line is also cut on the top of the end rings and in conjunction with the marks on the locking rings show when the latter are in the position to "*Insert*" or "*Remove,*" or for "*Storage*" or "*Transport.*" Holes and slots are cut in the end rings and the lid to receive a tape for sealing.

Fig. 25.

CASE, POWDER, CYLINDRICAL, MARK I.

Scale $\frac{1}{8}$.

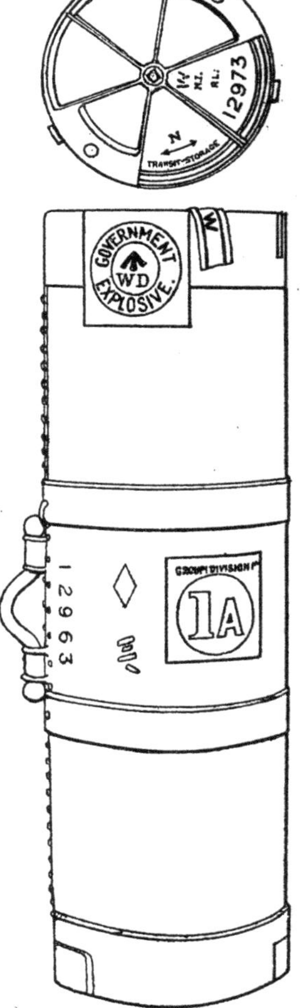

Case, Powder, Cylindrical, " L," Mark II.

A small number of the above-mentioned cases have been issued.

The body is solid drawn.

The end rings have no inclined planes. They are grooved to take the ends of the locking bolts and have a recess formed on the inside into which a dermatine washer is fitted.

Each lid consists of an annular top plate screwed to a base plate and recessed to suit the flange of the cam plate. It has three single and three double locking bolts, on each of which a projection is formed to take an anti-friction roller which runs in one of the grooves in the underside of the cam plate. The bolts are actuated by turning the cam plate with a " Key, case, powder, cylindrical, ' L,' No. 3."

Case, Powder, Cylindrical, " L," Mark I.
Case, Powder, Cylindrical, " N," Mark I.

These cases differ from the " L," Mark III, in the lid, which has only three projections formed around it to engage with three inclined slots in the end rings.

It has no locking ring, the lid itself having to be turned to the desired position.

Case, Powder, Cylindrical, "M," Mark I.

The Case, powder, cylindrical, "M," Mark I | N | , differs from the "L," Mark III, in dimensions and in having four intermediate rings instead of two.

Old Pattern Cylindrical Cases.

Cases, Powder, Cylindrical, "P," "Q" and "R."

The cases are made of stout sheet brass, butt jointed; on the exterior two metal bands are attached to the body with solder. Between these bands, over the joint, is secured, with rivets and solder. a bar with two loops; through the loops fits a copper wire lifting handle, covered with leather. At each end of the case there is an end ring, also of metal, secured to the body with rivets and solder. Each ring has four bearing surfaces, which give stability to the case in stowage; the top ring forms the mouth of the case, inside of which a groove is formed, and in this groove the locking cams on the lid work.

The lid is made of metal, having a flange on the upper surface. Locking cams, worked by a key, are fitted in it at right angles and secured from the underside of the lid by means of a nut held by a set-screw. These cams engage in short grooves inside the top of the case, thus securing it. A recess is formed on the underside of the lid to receive a *dermatine washer,* and a wood packing piece, painted stone colour, is also secured to its underside by four metal screws.

It is fitted with the usual plug for the air test.

Fig. 26.

CASE, POWDER, CYLINDRICAL, "R."

Scale $\frac{1}{6}$.

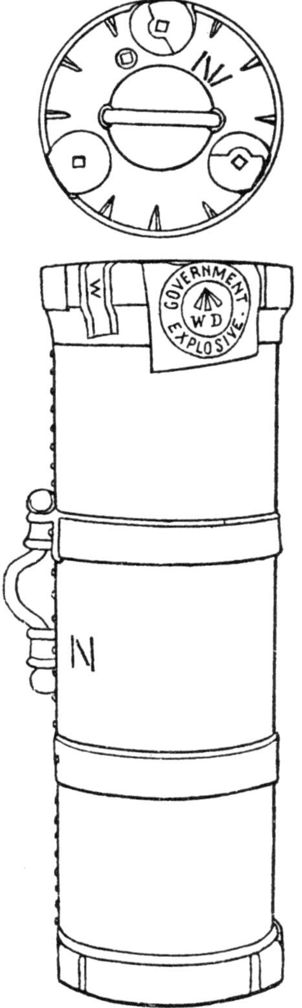

Special Marking on Cases where Resin Oil was used as a Flux.

Where resin oil has been used as the flux in soldering, a ◊ 1 in. in width is to be stencilled in blue paint, and a ◊ $\frac{3}{8}$ in. long stamped on the body near monogram of station.

(Cases stamped with date of manufacture subsequent to 1.1.10 will not be so marked.)

KEYS, FOR CASES, POWDER, CYLINDRICAL.

The under-mentioned keys are used for the *New Pattern* cases :—

Keys, Case, Powder, Cylindrical, " L " and " N " :—

No. 1. Mark I. Single handed ; for Mark I cases.
No. 2. Mark I. Double handed ; for Mark I cases.

Key, Case, Powder, Cylindrical, " L " :—

No. 3. Mark I. For Mark II case.

Key, Case, Powder, Cylindrical, " L " and " M " :—

No. 4. Mark I. Double handed ; for " L," Mark III and " M," Mark I cases.

Fig. 27.

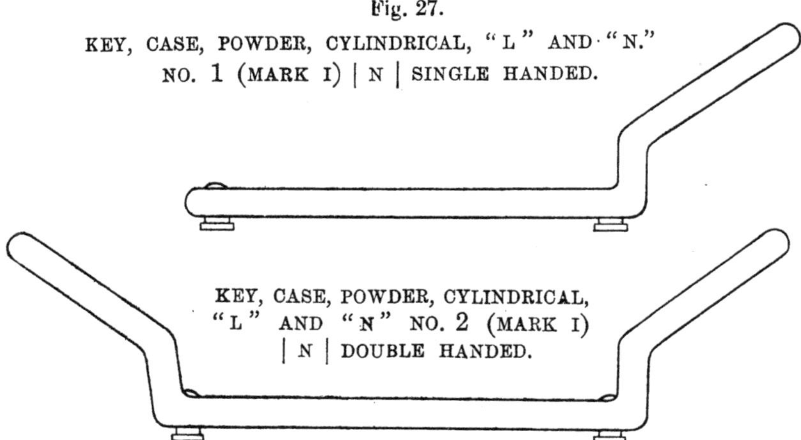

The keys are of metal. Nos. 1 and 2 are illustrated in Fig. 27.

The No. 1 key is single handed, and the Nos. 2, 3 and 4 keys, double handed.

The Nos. 1, 2 and 4 keys have two manganese bronze pins, to suit the keyholes in the lids of the cases, screwed in and riveted.

The No. 3 key has one end cranked inwards, and is provided with two projections which engage with grooves in the lid of the Case, powder, cylindrical, " L," Mark II. It has a spring stop which is worked by a finger grip, and which fits into a slot in the lid on the case.

The No. 4 key is similar to the No. 2, but the pins are further apart.

The under-mentioned key is used with the *Old Pattern* cases.

Key " P " to " R," Mark II.

This is a metal key with a cross-handle. It has a square projection which fits into the recesses in the locking cams, and is fitted with a lanyard of white line.

It is also used with all cylindrical and rectangular cases for the air-testing plug.

Fig. 28.
KEY "P" TO "R," MARK II.
Scale ½.

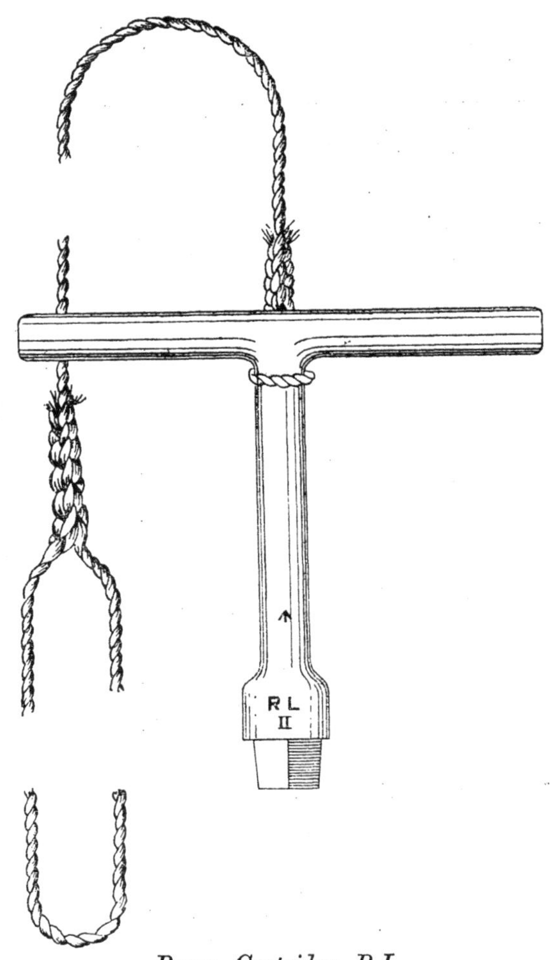

Boxes, Cartridge, B.L.

Boxes, Cartridge, B.L. have been made for 12-in., 9·2-in., 7·5-in., and 6-in. B.L. cartridges for Naval Transport purposes.

They are strong deal boxes, rectangular in shape, with elm ends, hard wood cleats, and copper wire handles covered with leather. The lid is hinged and is detachable, the hinge pins being removable. The lid is secured by two metal hasps secured by leather strips and a piece of whipcord. They are zinc-lined, the lid of the lining being secured to the lid of the box; a flange formed on the upper edge of the lining forms a groove for luting.

The boxes are not painted; the contents are stencilled in black paint on the front of the body.

They were introduced in lieu of Cylinders, Cartridge, but were found to be unsuitable. No more will be made. Existing stock will be used up.

Table No. 12.

Cylinders, Cartridge.

Para. in List of Changes	Number of Cylinder.	Mark.	Service	Cartridges packed.	No. packed.	Remarks.
9233 13215	5 or 5A	II to IV	L	B.L. 10-inch 63 lbs. Prism' Brown ...	1	Charge for paper shot.
9233 13215	7	I to III	L	B.L. 9·2-inch 41 lbs. Prism' Brown ...	2	Charges for paper shot. For Marks III to VII guns.
9233 13215	13	I to III	L	B.L. 9·2-inch 54 lbs. Prism' Brown ...	1	Charge for paper shot. For Marks X and X^{v} guns.
9233 12876 13215	15 or 15A	III to VII	L	B.L. 10-inch 40 lbs. Cordite M.D. ... B.L. 12-inch 22 lbs. 2 ozs. Cordite ...	1 2	
9233 13302 13981	16	I and III	L	B.L. 9·2-inch 44 lbs. 12 ozs. Cordite M.D. ... ... B.L. 9·45-inch Howitzer 5 lbs. 8¼ ozs. Cordite... B.L. 12-inch 22 lbs. 2 ozs. Cordite ...	1 7 2	For guns on H.A. mountings.
9233 13215	17	I and II	N	B.L. 12-inch— 43½ lbs. Cordite ... 41 lbs. 14 ozs. Cordite ... ... B.L. 10-inch— 36 lbs. 11 ozs. Cordite M.D. ...	 1 1 1	H.M. ships "Swiftsure" and "Triumph."
9233 13302 13215 14129	20 or 20A	III to VII	L	B.L. 10-inch— 38 lbs. Cordite ... B.L. 9·2-inch— 34½ lbs. Cordite ...	 1 1	Practice charge.
9233 13302 13215 14129	22	I to IV	L	B.L. 6-inch— 12 lbs. E.X.E. ... B.L. 9·2-inch— 26¾ lbs. Cordite ... 34½ lbs. Cordite ...	 4 1 1	Practice charge.
9233 13215 13302	26	I and II	L	B.L. 9·2-inch— 41 lbs. Prism' I, Brown ... ... B.L. 12-inch— 50 lbs. Cordite M.D.	 2 1	Charge for paper shot. For Marks I and II guns.

TABLE 12—*Cylinders, Cartridge*—continued.

Para. in List of Changes	Number of Cylinder.	Mark.	Service	Cartridges packed.	No. packed.	Remarks.
9233 13215	27	I	L	B.L. 9·2-inch— 15 lbs. 12 ozs. Cordite ... ... B.L. 10-inch— 38 lbs. Cordite ... 19 lbs. Cordite ...	 2 1 2	
9233 13202 16443 13461 13981	28 and 28A	I and I*	L	B.L. 9·2-inch— 16 lbs. 1 oz. Cordite M.D. ... B.L. 10-inch— 19 lbs. Cordite ...	 2 2	For guns on H.A. mountings.
9233 13215 13302	30	I and II	L	B.L. 6-inch— 12 lbs. E.X.E. ...	 4	
13215 9852	33	I	N	B.L. 12-inch— 41 lbs. 14 ozs. Cordite ... ... 83 lbs. 12 ozs. Cordite ... ... 43½ lbs. Cordite ... 87 lbs. Cordite ...	 2 1 2 1	
9389 13215 13302	34	I and IA	L	B.L. 6-inch— 10 lbs. Cordite ... B.L.C. 6-inch— 10 lbs. 7½ ozs. Cordite M.D. ...	 2 2	
		I* to III	L	B.L. 6-inch— 10 lbs. Cordite ... 11½ lbs. Cordite M.D.	 2 2	With packing piece.
9507 9852	36	I and II	C	B.L. 9·2-inch— 51 lbs. 8 ozs. Cordite ... ... 25 lbs. 12 ozs. Cordite ... ... 50 lbs. Cordite ... 25 lbs. Cordite ... 60 lbs. Cordite M.D. 30 lbs. Cordite M.D. 53½ lbs. Cordite M.D. ... ... 26¾ lbs. Cordite M.D.	 1 2 1 2 1 2 1 2	
11635 13215 13302	38 and 38A	I and I*	L	B.L. 6-inch— 11½ lbs. Cordite M.D.	 2	

TABLE 12.—*Cylinders, Cartridge*—continued.

Para. in List of Changes	Number of Cylinder.	Mark.	Service.	Cartridges packed.	No. packed.	Remarks.
12749	39	I	L	B.L. 6-inch— 16 lbs. 12 ozs. Cordite M.D. ...	1	For conveying cartridges from magazine to gun.
13320 13517	40	I	N	B.L. 12-inch— 63½ lbs. Cordite M.D.	1	Mark IX guns: King Edward VII class.
				61½ lbs. Cordite M.D.	1	Duncan and Formidable class.
				65 lbs. Cordite M.D.	1	Mark X and X* guns.
13320 13321	41	I	N	B.L. 6-inch— 28 lbs. 10 ozs. Cordite M.D. ...	1	
				32 lbs. 1½ ozs. Cordite M.D. ...	1	

TABLE No. 13.

Cases, Powder, Cylindrical | *N* | .

Para. in List of Changes.	Designation.	Mark.	Cartridges packed.	No. packed.	Remarks.
9830 12989	"P"	I II	B.L. 13·5-inch—		
			93 lbs. 12 ozs. Cordite	1	½ charge } Marks I to IV guns.
			46 lbs. 14 ozs. Cordite	2	¼ charges } Marks I to IV guns.
9584 12989	"Q"	I II	B.L. 12-inch—		
			52 lbs. 12 ozs. Cordite	2	Mark IX gun.
			61½ lbs. Cordite M.D.	2	Duncan and Formidable class.
			63½ lbs. Cordite M.D.	2	King Edward VII class.
			65 lbs. Cordite M.D....	2	Marks X and X* guns.
8843	"R"	I and II	(Altered to Large Mark II below.)		
8852		III	(Altered to Small Mark III below.)		

TABLE 13.—*Cases, Powder, Cylindrical* | *N* | —continued.

Para. in List of Changes.	Designation.	Mark.	Cartridges packed.	No. packed.	Remarks.
10521	"R"	II*	B.L. 12-inch—		
12989	Large	and	83 lbs. 12 ozs. Cordite	1	Mark VIII gun.
		IV	41 lbs. 14 ozs. Cordite	2	
			87 lbs. Cordite ...	1	
			43½ lbs. Cordite ...	2	
			100 lbs. Cordite M.D.	1	
			50 lbs. Cordite M.D....	2	
10521	"R"	III*	B.L. 12-inch—		
12989	Small	and	83 lbs. 12 ozs. Cordite	1	Mark VIII gun.
		V	41 lbs. 14 ozs. Cordite	2	
			87 lbs. Cordite ...	1	
			43½ lbs. Cordite ...	2	
15403	"N"	I	B.L. 12-inch—		
			52 lbs. 12 ozs. Cordite	2	Mark IX gun.
			61½ lbs. Cordite M.D.	2	Duncan and Formidable class.
			63½ lbs. Cordite M.D.	2	King Edward VII class.
			65 lbs. Cordite M.D....	2	Marks X and X* guns. (Top and bottom weakened.)
14837	"L"	I	B.L. 12-inch—		
16407		to III	76 lbs. 12 ozs. Cordite M.D.	2	Marks XI and XII guns. (Top and bottom weakened.)
15340			B.L. 13·5-inch—		
			73 lbs. 4 ozs. Cordite M.D.	2	Mark V gun.
			74 lbs. 4 ozs. Cordite M.D.	2	Mark V gun.
—	"M"	I	B.L. 15-inch—		
			107 lbs. Cordite M.D.	2	

TABLE NO. 14.

Cases, Powder, Rectangular | *N* |.

Para. in List of Changes.	Designation.	Mark.	Cartridges packed.	No. packed.	Remarks.
1369	"A"	I	B.L. 6-inch—		
1402		II	7 lbs. blank L.G. ...	16	
5714		III	B.L. 4-inch—		
9317		IV	3 lbs. blank L.G. ...	36	

TABLE 14.—*Cases, Powder, Rectangular* | *N* | —continued.

Para. in List of Changes.	Designation.	Mark.	Cartridges packed.	No. packed.	Remarks.
			B.L. 4-inch—		
			9 lbs. 15 ozs. Cordite M.D.	12	
			9 lbs. 5 ozs. 15 drs. Cordite M.D. ...	12	
9317	"B"	III	B.L. 6-inch—		
			10 lbs. Cordite ...	14	
			11½ lbs. Cordite M.D.	14	
9317	"D"	III	B.L. 6-inch—		
			28 lbs. 10 ozs. Cordite M.D.	8	
9317	"F"	III	B.L. 6-inch—		
			28 lbs. 10 ozs. Cordite M.D.	8	
			32 lbs. 1½ ozs. Cordite M.D.	8	
9317	"G"	III	B.L. 9·2-inch—		
			60 lbs. Cordite M.D....	5	
12839			30 lbs. Cordite M.D....	10	
9317	"I"	II	B.L. 12-inch—		
			61½ lbs. Cordite M.D.	2	
			63½ lbs. Cordite M.D.	2	
			65 lbs. Cordite M.D....	2	
			B.L. 9·2-inch—		
			26 lbs. 12 ozs. Cordite	6	
			13 lbs. 6 ozs. Cordite...	12	
			33 lbs. Cordite ...	6	
			16½ lbs. Cordite ...	9	
			31½ lbs. Cordite ...	5	
			15 lbs. 12 ozs. Cordite	9	
9317	"J"	II	B.L. 10-inch—		
			38 lbs. Cordite ...	4	
			19 lbs. Cordite ...	6	
			20 lbs. Cordite M.D....	8	
9317	"L"	IV	B.L. 7·5-inch—		
			27 lbs. 2 ozs. Cordite M.D.	8	
			13 lbs. 9 ozs. Cordite M.D.	16	
			B.L. 6-inch—		
			20 lbs. Cordite ...	12	
			10 lbs. Cordite ...	24	
10219	"N"	I	B.L. 9·2-inch—		
10233			51 lbs. 8 ozs. Cordite	4	
			25 lbs. 12 ozs. Cordite	8	

TABLE No. 14.—*Cases, Powder, Rectangular* | *N* | —continued.

Para. in List of Changes.	Designation.	Mark.	Cartridges packed.	No. packed.	Remarks.
			B.L. 9·2-inch—*contd.*		
12212			60 lbs. Cordite M.D....	4	
			30 lbs. Cordite M.D....	8	
13858			64 lbs. 4 ozs. Cordite M.D.	4	
			32 lbs. 2 ozs. Cordite M.D.	8	
12761	"O"	I	B.L. 7·5-inch—		
15044		to III	30 lbs. 8 ozs. Cordite M.D.	4	"O," Mark II is a weakened case. It has moveable nuts, which are loosened when in store.
			15 lbs. 4 ozs. Cordite M.D.	8	
			B.L. 4-inch—		
			5 lbs. 6 ozs. Cordite M.D.	32	
14716	"R"	I to III	B.L. 4-inch—		Weakened:—
14912 15084			9 lbs. 15 ozs. Cordite M.D.	6	Travelling bands (Mark I). Movable nuts (Mark II).
			9 lbs. 5 ozs. 15 drs. Cordite M.D. ...	6	
			5 lbs. 6 ozs. Cordite M.D.	7	
14763	"S"	I	B.L. 9·2-inch—		
		to III	60 lbs. Cordite M.D....	2	
			30 lbs. Cordite M.D....	4	
			64 lbs. 4 ozs. Cordite M.D.	2	
			32 lbs. 2 ozs. Cordite M.D.	4	
			51 lbs. 8 ozs. Cordite...	2	
			25 lbs. 12 ozs. Cordite	4	
14717	"T"	I	B.L. 6-inch—		
14838		and II	28 lbs. 10 ozs. Cordite		
15155			M.D.	4	
15856			32 lbs. 1½ ozs. Cordite M.D.	4	
15085	"U"	I	B.L. 10-inch—		
15857			36 lbs. 11 ozs. Cordite M.D.	4	H.M. ships "Swiftsure" and "Triumph."
	"W"	I	B.L. 6-inch—		
			27 lbs. 2 ozs. Cordite M.D.	4	

CHAPTER XI.—PROJECTILES FOR B.L., B.L.C., Q.F. AND Q.F.C. GUNS, AND B.L. AND Q.F. HOWITZERS.

(A) General Remarks; (B) Driving Bands, &c.; (C) Powder-Filled Shell; (D) Lyddite and H.E. Shell; (E) Shot, Various; (F) Practice Projectiles; (G) Markings on Projectiles; (H) Packing, Transport and Storage; (I) Drill Shell; (J) Blinds and Prematures; (K) Stores used in Connection with Shell.

SECTION (A). GENERAL REMARKS ON PROJECTILES.

Types of projectiles fired; Length and weight of projectiles; Diameter; Cast or forged with bands; Radius of head; Fuze-hole gauges; Capped shell; Projectiles suitable for gun or howitzer; Heavy and light shell; Tracers.

Projectiles Fired from Modern Guns and Howitzers.

Shell:

- *Common*
- *Common Pointed.* (*C.P.*)
- *Common Pointed with Cap.* (*C.P.C.*)
- *Armour Piercing.* (*A.P.*)
- *Armour Piercing with Cap.* (*A.P.C.*)
- *Lyddite Common*
- *High Explosive.* (*H.E.*)
- *Shrapnel*
- *Star*

Shot:

- *Armour Piercing*
- *Practice*
- *Case*
- *Paper*
- *Proof*

Of the above the following are becoming obsolete:—Armour-piercing shot, case shot, and the common shell of cast steel; common shell of cast iron, for *practice only*, are still made for certain guns and howitzers. (*See* page 234.)

Weight of Projectiles.

There are certain considerations which govern the weight of projectiles.

If two projectiles of different weight be fired, under similar conditions, from identical guns, they will, at the muzzle (unless the difference in weight be excessive), have equal energies, or approximately so.

The energy of both being the same, it follows that the lighter projectile must have the higher velocity, the velocities varying inversely as the square roots of their weights; but owing to the resistance of the air the heavier projectile will part more slowly with its muzzle energy than the light one.

The weight is, to a certain extent, limited by the length of the projectile, and also by the strain admissible on the gun and mounting.

At moderate ranges light projectiles will have flatter trajectories than heavy ones; but at long ranges the reverse is the case.

Ratio of Weight to Calibre.

The approximate relation between weight and calibre can be expressed as follows; where D equals diameter in inches and W equals weight in lbs.:—

$$\frac{D^3}{2} = W.$$

Example:—B.L., 9·2-inch—

$$= \frac{9{\cdot}2^3}{2} = 389.$$

Actual dead weight, 380 lbs.

Note.—With the latest projectiles for Naval Service for B.L. 13·5-inch and B.L. 15-inch guns, the ratio of weight to calibre is expressed by the formula:—

$$\frac{D^3}{1{\cdot}75} = W.$$

Example:—B.L., 13·5-inch—

$$\frac{13{\cdot}5^3}{1{\cdot}75} = 1{,}405.$$

Actual dead weight, 1,400 lbs.

Projectiles Suitable for both Gun and Howitzer of same Calibre.

The chamber pressure in a gun is much greater than that in a howitzer of the same calibre; *e.g.*, it may not be safe to fire from a 6-inch gun, a projectile designed for the 6-inch howitzer.

Another consideration which affects the question of a projectile suitable for both gun and howitzer is the length of the projectile; a long projectile requires a quicker twist of rifling than a short one to keep it travelling point foremost.

[The fact that the gun has a higher M.V. than the howitzer does not affect the question, as the twist of rifling necessary is the same for both high and low velocities. (*See* Text Book of Gunnery, 1914.)]

The driving band should be such that, if the projectile is loaded into a howitzer, it will not slip back at high angles of elevation.

In spite of the above difficulties, there *are* Shrapnel Shell suitable for both gun and howitzer; such projectiles are stamped on the base "Gun or How."; this information is also painted on the shoulder.

Heavy and Light Projectiles.

Certain guns and howitzers have been issued with "heavy" and "light" projectiles.

They are easily distinguished by the letters "H" and "L" stamped on the base, and stencilled on the shoulder, after the numeral of the shell.

For guns and howitzers issued with heavy and light projectiles, *see* Table 15.

Table No. 15.

Ordnance Firing Heavy and Light Projectiles.

Calibre.	Weight of Projectile.		Remarks.
	Heavy.	Light.	
	Lbs.	Lbs.	
B.L. 13·5-inch, Marks I to IV.	—	1,250	
B.L. 13·5-inch, Mark V...	1,400	1,250	
B.L. 12-inch, Marks I, IA, III to VII.	—	714	
B.L. 12-inch (wire) Marks VIII to XII.	850	—	(Naval Service.)
B.L. 9·2-inch, I to XI ...	380	—	
B.L. 9·2-inch, IV and VI (on H.A. mountings).	—	290	(Lyddite shell and practice shot only.)
B.L. 4-inch, I to VI ...	—	25	
B.L. 4-inch, VII and VIII	31	—	(Naval Service.)
6-inch Howitzer, 30 cwt., I	122 $\frac{9}{16}$	—	With 1 lb. 12 ozs. Cordite, size 5
6-inch Howitzer,30 cwt., I*			With 1 lb. 15½ ozs. Cordite, size 5.
6-inch Howitzer,30 cwt., I*	—	100	With 2 lbs. 8½ ozs. Cordite, M.D. 4¼.
5-inch Howitzer	50	—	With 11 $\frac{7}{16}$ ozs. Cordite, size 5. / 14 $\frac{5}{16}$ ozs. Cordite, M.D. 4¼.
5-inch Howitzer	—	40	With 14 ozs. 13 drams Cordite, size 5.

Capped Shell.

Caps of mild steel have been introduced for armour piercing and common pointed shell; they are fixed over and firmly attached to the point of the projectiles in the various ways detailed below.

The latest caps are made hollow, so that there shall be less metal over the point for the shell to penetrate on impact.

Hadfield's.—Fixed by indenting portions of the lower edge of the cap into indents on the shoulder of the shell.

Firth's.—Shallow grooves turned on the shoulder of the shell. Similar grooves are formed on the inside of the cap; the lower portion of the cap round the grooves is then soldered to the shell.

R.L.—Cap held on by two cotter pins—one at each side—which pass through the cap and into grooves formed on each side of the shoulder of the shell.

Advantages claimed for capped shell.—The cap with a high striking velocity, at normal impact, adds about $\frac{1}{6}$th to the penetrating power against a cemented plate.

No advantage is gained if the S.V. is below 1,800 f.s.

The cap holds the head together and diminishes the sudden violence of the blow tending to shatter the extreme point of the projectile the moment it meets the hard face of the plate.

The cap also transfers the blow on impact from the extreme point to the shoulder of the shell. There is therefore less tendency for the point itself to become broken or crushed to powder.

Radius of Head.

The advantages obtained by increasing the radius of head from 2 to 4 calibres are :—

(1) Increased range and striking velocity.
(2) Flatter trajectory, therefore a large increase in the length of danger zone, especially at long ranges.

Fuze-hole Gauges.

Projectiles taking nose fuzes :—

- Common.
- Common lyddite.
- H.E. shell.
- Shrapnel.
- Star.

Projectiles taking base fuzes :—

- Common Pointed and C.P.C.
- Armour Piercing and A.P.C.
- Double.

All holes in the nose of shell are threaded right-handed ; all holes in the base are threaded left-handed so as to prevent any tendency for an adapter, plug, or fuze to unscrew during flight.

(A) *Nose-fuzed Shell.*

Gauge	Shell
G.S. gauge (1 inch)	Common shell. Common lyddite, 12-pr. and up. H.E. shell. Star shell. Old type of shrapnel.
2-inch gauge	Shrapnel shell for 2·75-inch, 13, 18 and 60-pr., 3-inch, Mark V shrapnel for 4·7-inch, 6-inch, and 4·5-inch Howitzer, and all latest Marks of Naval shrapnel. 4·5-inch practice shell filled "Smoke Composition."

Special gauge.—3 and 6-pr. lyddite.

1-pr. gauge.—For 1, $1\frac{1}{2}$ and 2-pr. shell.

(B) *Base-fuzed Shell.*

Large, base, fuze gauge	Common-pointed and C.P.C. shell, armour-piercing and A.P.C. shell, 6-inch and up.
Medium, base, fuze gauge	Common-pointed and armour-piercing shell, 12-pr. to 5-inch. Also double shell, 2·95-inch Q.F.
Hotchkiss gauge ..	Powder-filled shell, 3 and 6-pr.
Armstrong gauge	Some old 4·7-inch shell taking No. 9 fuze.
Special gauge	Common pointed shell for 9·45-inch howitzer.

Inserting Plugs and Fuzes.

(*a*) *Projectiles taking nose fuzes.*—If to be fired immediately, the plug or fuze may be inserted without lubrication, but if not required for immediate use the threads of the plug or fuze must be lubricated with luting. (*See* page 114.)

(*b*) *Projectiles taking base fuzes.*—The plug or fuze must always be coated with thin luting on the threads, and unthinned luting should be placed under the flange, immediately before screwing them home into the shell. The luting acts as lubrication, as a waterproofing arrangement, and prevents the powder gas getting into the shell.

BANDS.

Some projectiles are made larger in diameter at the shoulder and base than at the centre of the body; these parts of greater diameter are known as the front and rear bands respectively; these names are apt to be confused with that of the driving band. Bands obviate a lot of machining, as the central part need not be brought accurately to gauge.

Fig. 29.

With Bands. *Without Bands.*

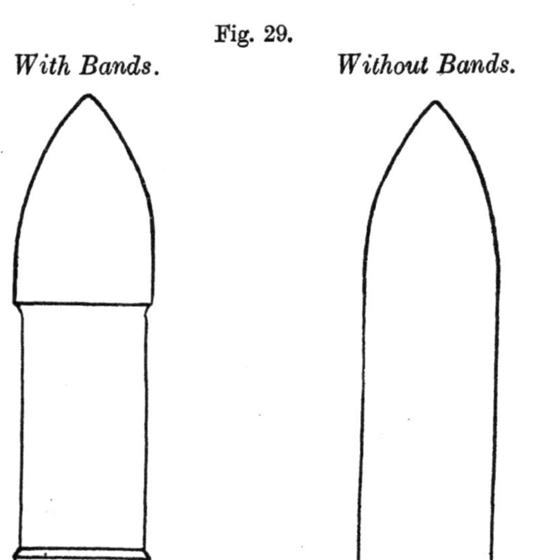

Radius of Head.

The form of head for Service projectiles is ogival and, prior to 1909, the heads of nearly all projectiles were struck with a radius of 2 calibres.

Projectiles are now being manufactured with the heads struck at approximately 4 calibres, and in order to distinguish projectiles of concurrent marks, which differ only in the radius of head, the following procedure of nomenclature has been adopted.

The letter " A " following the Roman numeral will denote those projectiles which have a radius of approximately 4 calibres.

Shell that may be introduced with a radius of between 4 and 6 calibres will have the letter " B " added after the numeral.

Table No. 16.

Dead weight limits of Shell.

Nature.	Nominal weight.	Dead weight limits.			
	Lbs.		Lbs.	ozs.	
15-inch	1,920	H	1,921	14½	
		L	1,918	1½	
13·5-inch H	1,400	H	1,401	6½	
		L	1,398	9½	
13·5-inch L	1,250	H	1,251	4	
		L	1,248	12	
12-inch	850	H	850	13½	
		L	849	2½	
12-inch L	714	H	714	11½	
		L	713	4½	
10-inch	500	H	500	8	
		L	499	8	
9·2-inch H	380	H	380	6	
		L	379	10	
9·2-inch L	290	H	290	4½	
		L	289	11½	
7·5-inch	200	H	200	3¼	
		L	199	12¾	
6-inch	100	H	100	1½	
		L	99	14½	
			Lbs.	ozs.	drs.
60-pr.	60	H	60	0	15
		L	59	15	1
5-inch	50	H	50	0	13
		L	49	15	3
4·5-inch	35	H	35	0	9
		L	34	15	7
4·7-inch	45	H	45	0	11½
		L	44	15	4½
4-inch H	31	H	31	0	8
		L	30	15	8
4-inch L	25	H	25	0	6½
		L	24	15	9½
18-pr.	18½	H	18	8	5
		L	18	7	11

TABLE NO. 16.—*Dead weight limits of Shell*—continued.

Nature.	Nominal weight.			Dead weight limits.			
	Lbs.				Lbs.	ozs.	drs.
15-pr.	14			H	14	0	3½
				L	13	15	12½
2·95-inch	12½			H	12	8	3¼
				L	12	7	12¾
13-pr. ... ... 12-pr. and 14-pr. ... ... 2·75-inch	12½			H L	12 12	8 7	3 13
10-pr.	10			H	10	0	2½
				L	9	15	13½
6-pr.	6			H	6	0	1½
				L	5	15	14½
	Lbs.	ozs.	drs.				
3-pr.	3	4	15½	H	3	5	0½
				L	3	4	14½
3-pr. with tracer	3	8	0½	H	3	8	1½
				L	3	7	15½
2·95-inch (double)	18	2	8	H	18	2	12
				L	18	2	4

All projectiles for the same gun (except star) will in future be filled to dead weight, *i.e.*, to within a limit of ± 0·1 per cent. of the nominal weight.

See Table 16 for nominal weights and dead weight limits of Service projectiles.

Length of shell.—As the weights of all natures of projectile (except star) for a given gun is the same, the length must vary. The lowest admissible length for accurate shooting is found to be 2 calibres.

On the other hand, a very long projectile necessitates a sharper twist of rifling and therefore a greater strain on the gun.

The length of projectiles does not exceed 4½ calibres, the majority being 3½ to 4 calibres long.

Diameter.—The diameter of B.L. and Q.F. projectiles across the bands, or across the body of those which are cast without bands, is less than the bore of the gun across the lands, by an amount varying from about ·05-inch with heavier natures to 0·008-inch with the smallest.

The clearance between the projectile and the bore is known as windage.

Paint and marking.—All projectiles, except case shot, are painted.

Lyddite and H.E. shell are painted yellow; all others black, or slate colour; the driving band is never painted.

For markings on projectiles, *see* page 236.

TRACERS.

The latest Marks of certain projectiles are now fitted with "Tracers."
There are two types of tracers, namely :—

"*Day Tracer*" *and* "*Night Tracer.*"

Day tracer.—In a shell with a "Day tracer" a cavity is formed in the base; this cavity is filled with a black liquid (turpentine and aniline dye), which, during flight, escapes through a small hole in a steel plug ("Plug, day tracer, vent").

This hole is closed by a brass disc sweated in. On firing, the pressure blows this disc in, so unmasking the hole.

Fig. 30.

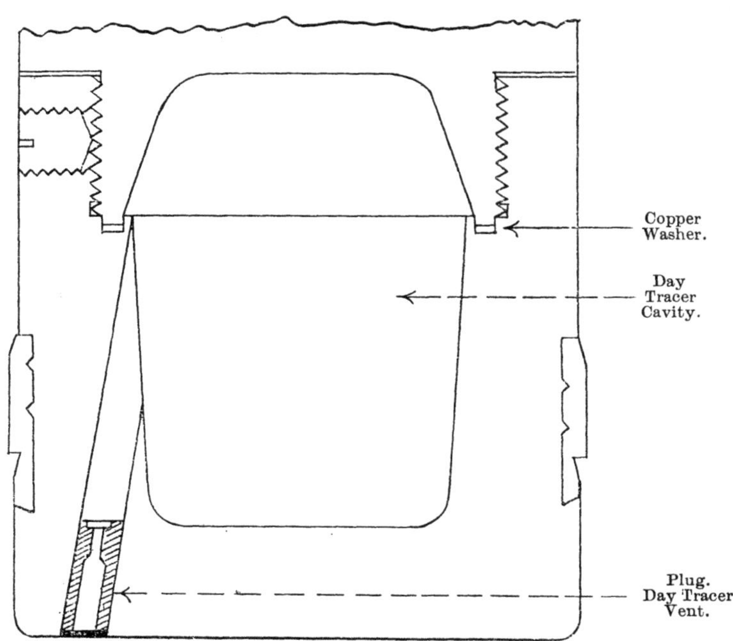

Night tracer.—There are two patterns of "Night tracer," namely, "Internal" and "External."

They are similar in internal construction, and differ only in the method of attachment to the base of the projectile.

A detailed description of the "Internal tracer" is given below. (*See* Plate XV.)

TRACER, SHELL, NIGHT, INTERNAL (MARK I) | C | OR SHOT.

The Mark I internal tracer consists of the following parts:—Body: illuminating, igniting and friction compositions; friction bar with brass cup and nut; two lead washers; a washer of press board; two steel washers; and cover plate.

The body is made of bronze about 2·25 in. long; the head is enlarged, below which it is screw-threaded for a short distance with a left-handed screw-thread to screw into the base of the shell. It is bored out and tapped to a depth of about 1·25 in. with a right-handed screw-thread and is partly filled with illuminating composition, firmly pressed or stemmed in, so as to grip the interior screw-thread.

A conical hole is bored into the top of this for a small charge of igniting composition, a small recess being left in the centre of the latter for the cup of the friction bar.

The friction bar is of hard brass wire, the outer end screw-threaded for the securing nut, the inner end flattened and serrated.

The serrated end of the bar fits into a small cup made of annealed brass, filled with friction composition. This cup fits into a recess in the igniting composition, over which is pressed a thin lead washer.

A small groove is turned inside the body above this for a thick lead washer, which is pressed home and expanded into it.

A small press-board washer dipped in paraffin and resin, and two steel washers, are pressed down on top, as shown in the plate.

The outer end of the tracer is closed by a steel cover-plate through which the outer end of the friction bar passes. This plate is secured by a brass nut.

The cover-plate is pierced at the side with a small hole, about 0·05 in., called a "gas-port"; this hole leads into the recess in the head of the tracer.

The first issues of Mark I internal night tracer were fitted with a celluloid disc over the cover plate. This disc occasionally masked the gas-port and caused blinds; they have been removed locally.

Marking.—The tracer is marked with manufacturer's initials, numeral, and date of filling.

Action.—When a projectile fitted with a tracer is fired from a gun, the gas from the cordite charge passes through the gas-port in the cover-plate and fills the gas chamber with gas under high pressure.

As soon as the projectile leaves the bore of the gun, the pressure outside the tracer falls to less than atmosphere pressure. The gas in the chamber being unable to escape quick enough through the gas-port, blows out the cover-plate, carrying with it the friction bar, firing the friction composition in the brass cup and so ignites the illuminating composition, which, burning for about 15 seconds, traces the flight of the shell.

External tracer.—The external tracer differs from the above in not having an enlarged head, and is screw-threaded at the opposite end.

Packing.—Internal tracers are packed ten in a tin cylinder, No. 115.

External tracers are packed ten in a tin cylinder, No. 116.

Key No. 33 *(Mark I), Night Tracer* | *C* |.

This is a steel key and is used for inserting and removing night tracers, or "Plugs, base, shell, Nos. 7 and 8."

It is T-shaped, the lower end of the upright part being fitted with projections to fit into the key holes in the internal night tracer.

One end of the cross-piece is formed into a wrench, with teeth to fit over and grip the body of the external night tracer, whilst the other end is formed like a screwdriver, to fit into the slots in the heads of the "Plugs, base, shell, Nos. 7 and 8."

Fig. 31.

Scale ½.

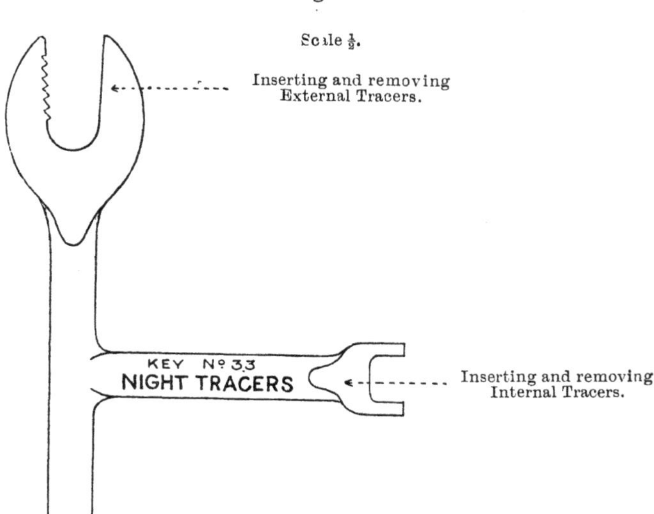

SECTION (B). DRIVING BANDS.

Points governing design; Description of various types; Augmenting strips; Augmenting rings; Augmenting bands.

Projectiles for B.L., B.L.C., Q.F. and Q.F.C. guns, and B.L. and Q.F. howitzers, are rotated by means of a band, generally of copper, but in a few cases of cupro-nickel, attached to them near the base. This band is larger in diameter than the bore of the gun, and consequently when the charge is fired the soft metal is compressed into the grooves and cut into by the lands, overflows to some extent into cannelures or towards the base, and at the same time receives from the grooves a motion of rotation which it imparts to the projectile.

Points Governing Design.

Material.

The material for a driving band should possess the following qualities :—

1. It should be soft enough readily to take the rifling.
2. It should not be so hard as to throw an excessive strain on the base of the shell when it takes the rifling.

3. It should have a high melting point.
4. It should not be so soft as to strip.
5. It should not leave a deposit in the bore.
6. It should not cause smoke.

Copper is generally used in the Service, but is not entirely satisfactory. It vaporises to some extent, giving rise to a brown smoke, and also leaves a deposit in the bore.

Other materials are being experimented with.

The hardness of two samples of the same copper differently prepared will vary, and it is found in practice that the best condition of the metal for this purpose is that in which it is cut into the shape of a ring from drawn tubing, which is afterwards annealed and then forced into the groove round the circumference of the shell by a powerful hydraulic or other press. Electro-deposited copper is also used.

Cupro-nickel (95 parts of copper and 5 parts of nickel) has been introduced as the material for driving bands for use with the latest 6-inch B.L. guns.

Fig. 32.

UNDERCUT GROOVES FOR DRIVING BANDS, SHOWING DIMENSIONS OF UNDERCUT.

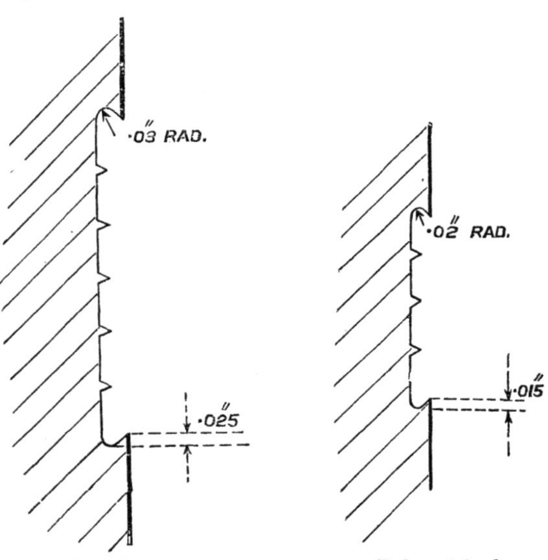

4-inch and upwards. Below 4-inch.

Position of Driving Band.

The driving band should be as near the base of the projectile as possible, it being generally found that the more rearward position of the band gives the most accurate shooting. In practice, however, this is limited by the thickness of metal behind the band which is necessary to support the great strain thrown upon the shell when the band is forced through the grooves, tending to tear off the base. In fixed ammunition the driving band must be far enough forward to allow the projectile to be firmly secured in the case.

Attachment of Band.

(a) *Undercut groove.*—The band must be firmly attached to the shell so that it cannot be torn out of its groove when it takes the rifling.

Firmness of attachment is secured by undercutting the groove in the shell as shown in Fig. 32.

The band is forced into the groove by a powerful press and is afterwards turned to shape.

Fig. 33.

OLD METHOD OF ATTACHING DRIVING BANDS.

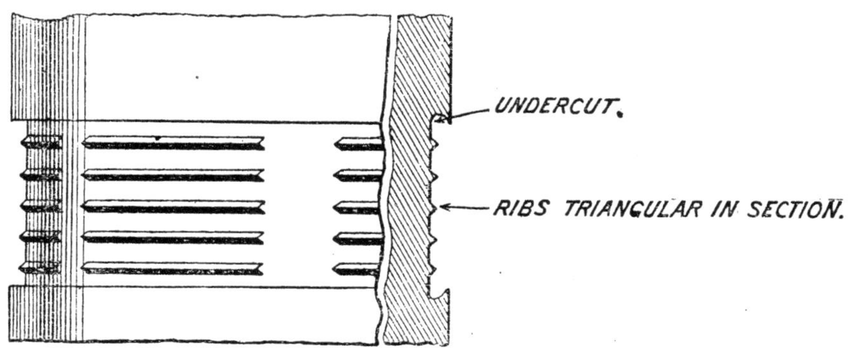

NEW METHOD OF ATTACHING DRIVING BANDS.

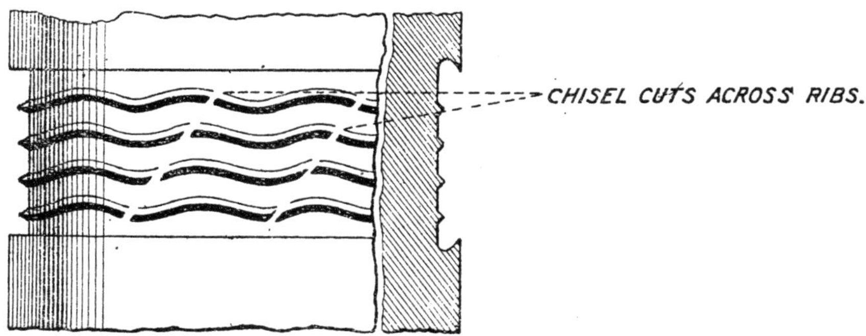

(b) *Waved ribs.*—Originally, at the bottom of the groove in the shell, there were a number of raised ribs which were *straight, i.e.*, running circumferentially round the shell; they were cut away at intervals to prevent the band slipping. (*See* Fig. 33.)

Since June, 1901, *waved* ribs have been employed instead of the above (except in 3 and 6-pr. shell which have straight ribs); these waved ribs are not cut away, but have chisel cuts across them to allow the air in the channels between the ribs to escape when the driving band is being pressed on.

Shape of the Band.

The first portion of the band should be so shaped as to prevent "over ramming" of the projectile.

The band should be designed to entirely seal windage.

The rear of the band should be shaped to prevent "fringing" or "fanning," that is, the surplus copper is dragged back by the lands of the gun as the projectile passes down the bore and forms a sort of fringe behind the band. When the shell leaves the gun this fringe is no longer supported, and the pressure of the gas behind it turns it up at various angles to the axis of the projectile, thus forming variable resistances and causing irregularity of flight.

DESCRIPTION OF VARIOUS TYPES OF DRIVING BANDS.

Plate XVI shows sectional views of the principal driving bands.

The numbers assigned to the bands are for reference in connection with the Table of Projectiles.

No. 1.—The Narrow Vavasseur.

This band was used with the early Marks of projectiles. It was made of copper; the front part bevelled off at a slope of 7 degrees to fit the cone seating between the bore of the gun and the powder chamber.

The objection to this narrow band of large diameter is excessive fringing causing the shell to be unsteady in flight.

The band used with projectiles for the B.L. 9·45-inch howitzer is very similar to the narrow Vavasseur band.

No. 2.—Broad Vavasseur.

To overcome the above defect the broad Vavasseur band was introduced. It differs from the narrow band in the exterior diameter being reduced, and the width considerably increased, and a number of cannelures were cut round the band to receive the surplus metal forced back by the lands of the rifling. The front of the band was bevelled off at a slope of 7 degrees, and in the first issues was not serrated as shown in the plate.

To prevent projectiles fitted with the broad Vavasseur band from being rammed too far home in worn guns, narrow strips of copper, termed "*Augmenting Strips,*" were introduced to fit into and fill up the cannelures in the band.

For description of Augmenting Strips, *see* page 155.

A special steel chisel was issued for the purpose of preparing the cannelures in the band to take the augmenting strip.

By means of this chisel and a hammer, V-shaped grooves were cut in the bottom angles of the cannelures as shown by the dotted lines, *a, a,* in the sketch.

Fig. 34.

Scale $\frac{1}{1}$.

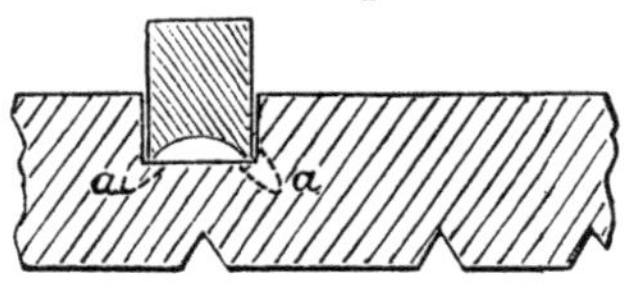

No. 3.—*Broad Vavasseur Band.* (With cannelures undercut for augmenting strips.)

To prevent the necessity of preparing the cannelures in the broad Vavasseur band for augmenting strips locally, No. 3, the broad Vavasseurs with undercut cannelures, was introduced for all natures except the 30-pr., 5-inch, and 5·4-inch howitzer, 4·7-inch Q.F. In this type the bottom angles of the cannelures (except the rear bottom angle of the rear cannelure) are undercut, and have the letter "U" stamped on the band between the first and second cannelures.

No. 4.—*Broad Vavasseur Band.* (With groove in shell undercut.)

This is the same band as No. 3, but the groove in the shell is undercut (introduced April, 1899). This type of driving band is still being fitted to the latest Marks of projectiles for Q.F., 3- and 6-pr., 6-inch howitzer (heavy shell), and the Q.F. 4·5-inch howitzer.

No. 5.—*Broad Vavasseur Band with Gas-check (Original).*

Owing to the great erosion of the gun caused by cordite charges, this type of band was introduced for all projectiles fired with *cordite* charges from B.L. and B.L.C. guns, 6-inch and up, but not for smaller natures. The band differs from the "Broad Vavasseur band" in having a second slope formed near the centre of the band, and the metal cut away behind, thus forming a lip or gas-check. The apex of this lip is considerably higher than the remainder of the band, thus more efficiently sealing the bore. In addition the gas pressure on the under side of the lip tends to force it outwards against the bore of the gun. The band has cannelures of the ordinary type in the cylindrical portion in front of the lip; the slope of the lip is serrated to hold the projectile in the chamber. The cannelures are not undercut, since augmenting strips are not required with this form of band, as the gas-check lip prevents over-ramming in worn guns.

Modifications have been made in this driving band since it was first introduced, principally with regard to the extent and depth of the serrations on the gas-check lip. These are now made deep and sharply pointed in order the better to ensure the projectile remaining fast in the chamber and not slipping back.

The original broad Vavasseur band with gas-check is still used for banding B.L. or Q.F. 6-inch, but with B.L. 9·2-inch to 13·5-inch this type of band was found to have the same defect as the narrow Vavasseur band, namely, fringing, which greatly affected the shooting of the gun. To overcome this defect the shape of the rear portion of the band was altered for B.L. 9·2-inch to 13·5-inch projectiles as shown in type No. 6.

No. 6.—*Broad Vavasseur with Gas-check (Modified).*

In this type of band the size of the "grave" (the space behind the gas-check slope) is so arranged that the metal of the lip when jammed back by the passage of the band through the gun just fills it, and the rear portion is so sloped away that any tendency to fringe is eradicated.

With projectiles for the 10-inch to 15-inch, both the front slope and the gas-check slope are serrated; in the 9·2-inch the front slope is not serrated.

Projectiles which originally had the broad Vavasseur band, when re-banded with the gas-check band (No. 5), had a (*) added to their numeral; when re-banded with the gas-check band (No. 6) they have two stars (**) added to their numeral.

No. 7.—Cupro-Nickel Band | *N* | .

The cupro-nickel band is now used with the latest Marks of 6-inch B.L. projectiles, N.S. only. (The 29 lb. cordite M.D. charge was introduced for use with this driving-band.)

The band is made of an alloy, 95 per cent. of copper and 5 per cent. of nickel. It has no cannelures, and the metal in rear of the second slope is not grooved out to form a gas-check lip. Projectiles banded with cupro-nickel are distinguished by having a white band ½ in. wide painted round the body immediately above the driving-band.

No. 8.—Broad Copper Band (" Hump ").

Projectiles fired from B.L. 7·5-inch, Marks I to II** and V guns, and the latest Marks of B.L. 9·2-inch are now fitted with this new type of driving-band. It differs principally from the gas-check band in having no cannelures, the front slopes are not serrated, the metal in rear of the second slope is not grooved out to form a gas-check lip, and the rear portion of the band is turned down to the same diameter as the base of the shell.

No. 9.—Broad Copper Band (" Hump ") (for Practice Shot, 7·5-inch and 9·2-inch).

This band differs from the above in having the cylindrical portion, which is behind the rear slope in No. 8 type of band, placed in front of the front slope. This change in design is necessary with cast-iron projectiles so as to leave sufficient breadth of metal in rear of the band to support the strain when the band takes the rifling.

The diameter of the projectile in the rear of the band is also reduced $\frac{2}{10}$ of an inch, leaving more space for the jamming back of the metal, so as to prevent the base of the projectile from being torn off.

No. 10.—Plain Band.

This type was introduced to supersede the broad Vavasseur band for all natures below the 6-inch calibre. It is of much larger diameter in the centre than the broad Vavasseur band (No. 3), hence it has no cannelures to take augmenting strips. It slopes from a short cylindrical portion to the front and rear. The front slope is serrated. This plain band is still used with the latest marks of projectiles for the B.L. 30-pr., 4-inch^{L}; 60-pr. Q.F. 4·7-inch and 6-inch 30 cwt. howitzer (^{L}).

No. 11.—E.O.C. Band.

This band was introduced with projectiles for the 7·5-inch and 10-inch B.L. guns in H.M.S. " Swiftsure " and " Triumph." It has now been adopted for use with projectiles for the B.L. 4-inch, Mark VII and VIII guns. The band differs from the plain type in having a gas-check.

No. 12.—Special Narrow Band.

No. 12 shows the band at present used with the 2·75-inch, 10-pr., 12 and 14-pr. and 15-pr. It is a narrow band, having a short steep slope in front, and it is also sloped off in rear.

No. 13.—Old Special Narrow Band for 13 *and* 18-*pr. Q.F.*

The driving band used with the early Marks of 13 and 18-pr. projectiles differs from the special narrow band used with the B.L. projectiles in having a very narrow lip formed on the rear of the band. This lip is intended to prevent the shell from being inserted too far into the cartridge case.

No. 14.—New Special Narrow Band for 13 *and* 18-*pr. Q.F. prepared for attachment of case.*

The latest driving band for 13 and 18-pr. projectile is prolonged to the rear, and is prepared for the case to be attached to it by coning at the mouth.

No. 15.—New Type of Band for 3-*inch Q.F.*

This band is similar to the narrow Vavasseur band, but the rear portion is grooved out for the coning of the mouth of the brass cartridge case.

AUGMENTING STRIPS.

Strips, augmenting, B.L., Mark I., are for use with certain projectiles having broad Vavasseur driving bands, and fired under certain conditions. (*See* below.)

Fig. 35.

AUGMENTING STRIP.

Scale $\frac{1}{1}$.

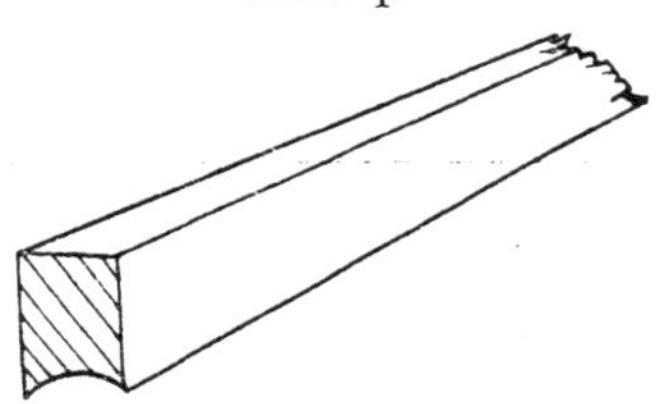

The strips consist of pure copper, of even section throughout, and grooved on one side (as shown in the sketch). The lengths of the strips vary with the calibre, and they are marked for the nature of the gun with which they are intended to be used.

Mode of Use.

The augmenting strip is placed in the cannelure grooved side downward, and hammered round the shell until the two ends meet.

The augmenting strip was originally introduced in the days of gunpowder charges for use in worn guns, to correct the ramming and to hold the projectile till the charge was well ignited; it was placed in the front cannelure.

A few years after the introduction of augmenting strips, cordite charges came into use. Owing to the erosion caused by cordite, the gas-check form of driving-band was introduced for guns 6-inch and upwards firing cordite. A number of projectiles for these guns still exist, however, which are not fitted with the gas-check band, and to use these up, they may, except as *stated below*, be fired with cordite charges *at practice*, provided an augmenting strip is used in the *rear* cannelure; the strip in this case is intended to act as a gas-check. The order on the subject states that: "Projectiles for guns 6-inch and above, not fitted with gas-check driving-bands, may be fired with cordite charges at practice, provided that an augmenting strip is used with the existing broad Vavasseur band, except when firing full charges from 9·2-inch B.L., Marks IX, X and X^{v} guns, in which case projectiles with gas-check driving-bands must invariably be used."

RINGS, AUGMENTING, B.L. 9·2-INCH, 380 LB. PROJECTILES.

No. 1, Mark I, for projectiles with No. 6 type of band } *See*
No. 2, ,, I ,, ,, No. 5 ,, ,, } Plate XVI.

These rings are fitted to all 9·2-inch 380 lb. projectiles forming part of the equipment of guns in approved armaments. The rings are made of copper, and are intended to fit into and fill up the groove in rear of the gas-check portion of the driving-band, as shown in Fig. 31. The rings are split so that they can be increased in diameter to slip over the base of the projectile.

Fig. 36.

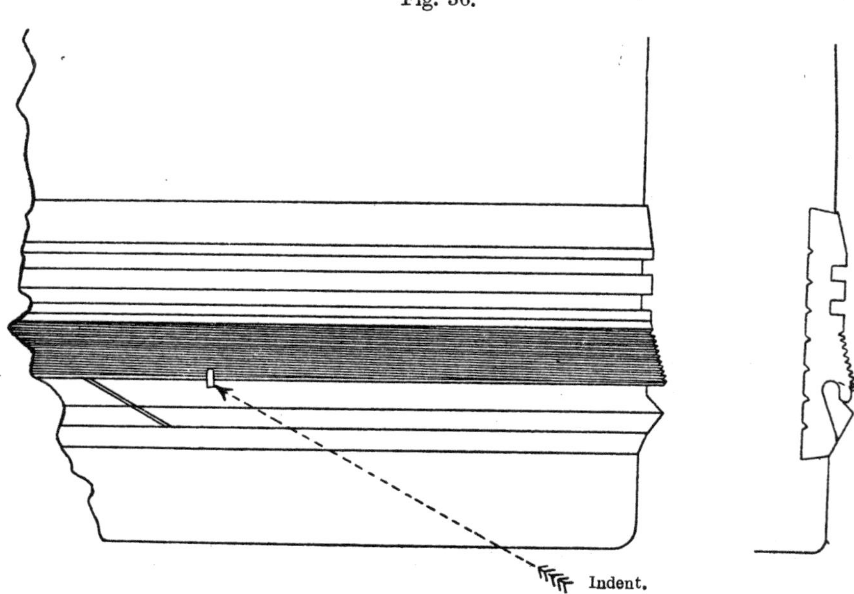

Instructions for fitting Augmenting Rings.

1. Remove the grummet from the projectile and place the latter on its side with the base slightly raised; for Land Service the projectile should be placed in a block No. 3. (*See* Fig. 152.)

Should it appear that the gas-check has been pressed down so that the ring will not fit into its groove, the ring must be removed, and the groove rectified with the "*Tool, rectifying, gas-check,* 9·2-*inch.*"

2. Spring the ring over the base of the projectile into the groove in rear of the gas-check portion of the driving band.

3. With a blunt chisel (the edge held parallel to the axis of the projectile) and a mallet, strike the lip of the gas-check in 5 places, so as to turn a small portion of the metal over, thus securing the augmenting ring firmly. Care must be taken to damage the gas-check portion of the band as little as possible.

The position of the five "indents" or "turnovers," and the order in which they are to be made are as follows:—

First.—Diametrically opposite to the split.

Second and third.—On each side, midway between the first turnover and the split.

Fourth and fifth.—Close on each side of the split.

4. Replace the grummet on the projectile and, where single ring grummets are used, lace them together in three places, equidistant from each other, with spun yarn threaded through the grummet.

Projectiles fitted with augmenting rings should not be fired at practice from guns bearing on crowded waterways, or over houses, &c.

RINGS, AUGMENTING, B.L. 13·5-INCH PROJECTILES, MARK I | N |.

These are similar to the "Rings, augmenting, No, 1," for B.L. 9·2-inch projectiles.

They are for use with the broad Vavasseur with gas-check (modified) driving bands.

They are supplied in accordance with arrangements made by the Admiralty, and are fitted in the same way as those for the 9·2-inch projectiles.

AUGMENTING BANDS.

Bands, augmenting, B.L. 13·5-inch projectiles are made of copper, and will be used only in guns so much worn that there is a tendency for the projectile to slip back after loading. They are never to be used in a new or comparatively new gun, and when necessary they can easily be removed from any projectile to which they have been fitted. A special steel tappet ring is supplied for fixing these augmenting bands in position.

The band is placed over the point of the projectile and tapped home with the tappet ring until the augmenting band is in the position shown below.

Fig. 37.

BANDS, AUGMENTING, B.L. 13·5-INCH PROJECTILES | N | .

See § 15425, "Band, augmenting, B.L. 13·5-inch projectiles. (Mark I.)"

Scale full size.

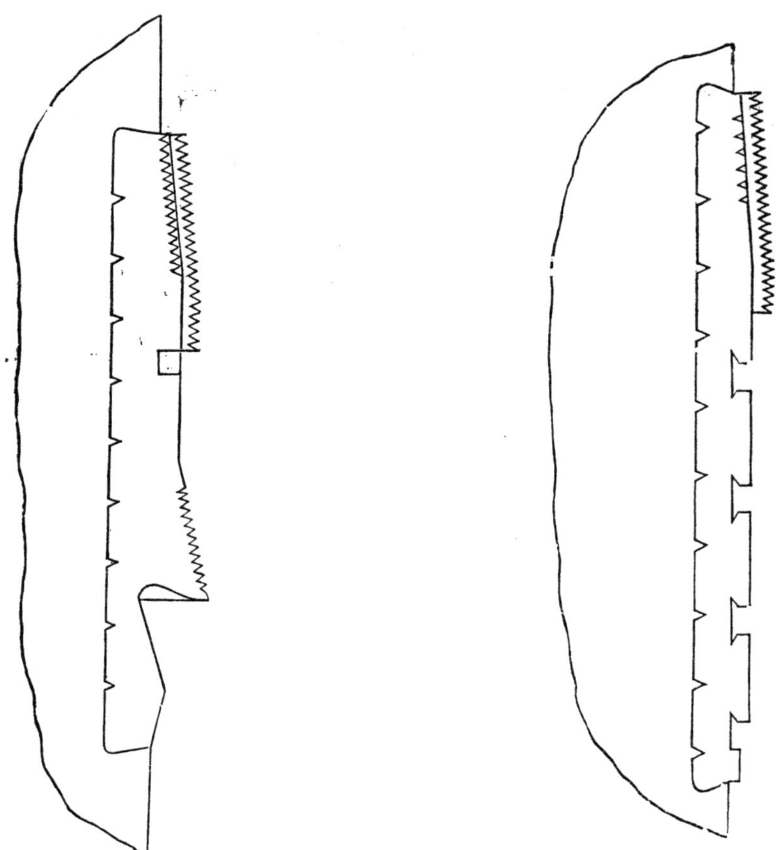

Copper Band in position on gascheck driving band.

Steel Band in position on old pattern driving band without gascheck.

SECTION C. POWDER-FILLED SHELL.

Points governing design and description of Common Shell; Common-Pointed and C.P.C. Shell; Armour-Piercing and A.P.C. Shell; Shrapnel and Star.

COMMON SHELL.

The type of shell known as "Common" has a hole in the nose for a fuze, and a very large cavity, which is completely filled with an explosive.

If this explosive is lyddite the shell is known as "Common lyddite," if Trotyl or C.E. it is known as H.E. shell, but if the explosive is gunpowder, the shell is known as "Common shell."

[*Exceptions.*—There are some projectiles which, although known as common shell, have no hole in the nose, and consequently take a base fuze; their points are blunted, *e.g.*, the double common shell for Q.F. 2·95-inch.]

"Common shell" (*i.e.*, common shell with gunpowder bursters) are used mainly for practice; they are almost obsolete for war, though at one time they were universal.

Considerations governing design.—A common shell is not designed for penetration of armour, but is intended to do damage by the force of its burst, by flying fragments, and by setting fire to material; it is designed to hold as large a bursting charge as possible, but must be strong enough to stand the shock of discharge without deformation, and the shock of impact without breaking up; the walls are consequently made as thin as possible, consistent with the necessary strength.

They have been made of cast iron, cast steel, and forged steel but forged shell have the disadvantage of breaking up into only a small number of fragments with a powder bursting charge; the stronger the material, the thinner the walls can be made, and hence the larger the bursting charge. This varies from 5 per cent. to 16 per cent. of the weight of the filled projectile.

Use—

(*a*) *For active service.*—Common shell will still be met with in some coast batteries; as to active service in the field, broadly speaking, common shell are not used in our service; the Q.F. 2·95-inch, however, fires a common shell (also the B.L. 10-pr. shell special for India).

(*b*) *For practice.*—Common shell of cast iron are fired from siege guns and howitzers, and from field howitzers, and heavy field guns, *e.g.*, B.L. 60-pr. and Q.F. 4·7-inch travelling; they are not fired from light field guns.

General construction.—They are about 3·5 calibres long, with ogival heads struck with a radius of two diameters; the nose is fitted, with a gunmetal bush threaded internally to take the fuze or plug, to the G.S. gauge.

The interior of the shell is lacquered, varnished, or velvrilled so as to give a smooth surface.

From the 9·2-inch upwards they have a hole bored in the side opposite the centre of gravity, and threaded to take an eye-bolt for lifting purposes.

Common shell 6-inch and upwards are of cast steel in the later patterns; below 6-inch the latest patterns are of forged steel for service, and of cast iron for practice.

Bursting Charges of Common Shell 4-inch and up.

The bursting charge for 4-inch shell and upwards is enclosed in a bag made of dowlas (a material made from flax) as a precaution against the premature explosion of the charge from the shock of discharge and friction against the walls of the shell. The portion of the bag nearest the fuze is made of shalloon, which is more permeable to a flash than dowlas.

Primers.—Several 7-dram primers are placed next the fuze; they consist of small bags of shalloon filled with fine grain powder.

Bursting Charges of Common Shell below 4-inch.

The bursting charges of common shell below the 4-inch are not enclosed in a bag, and so 7-dram primers are not used; the lacquer is sufficient precaution against prematures from friction.

Powder.—The bursting charge of a common shell usually consists of P mixture (a mixture of shell P. and shell F.G. powders).

Cast-iron common shell for practice are filled with shell L.G. powder.

Closing the Bases of Common Shell.

Cast-iron common shell below 6-inch have solid bases as a rule; cast-iron common shell 6-inch and above have a hole in the base closed by a gunmetal screw-plug; a recess left over the latter is sealed by a lead *disc* hammered in; in some cases a lead *ring* seals the junction between plug and shell, in place of the lead disc (*e.g.*, 6-inch B.L. Howitzer, Mark III).

Cast-steel common up to 5-inch have solid bases.

Cast-steel common above 5-inch have their bases closed in one of the following ways:—

(*a*) A solid gunmetal plug; the joint between plug and body sealed by a lead ring.

(*b*) A large base plug screwing into an adapter; the joint between adapter and shell is closed by a lead ring; the joint between plug and adapter by luting (formerly a lead ring was used under the flange of the base plug instead of the luting).

(*c*) The large base plug only, without adapter.

(*d*) Some 6-inch take a plug and lead *disc* similar to the cast-iron common shell.

Forged-steel common were almost invariably made with solid bases.

Lead discs.—As mentioned above, *discs, base plug,* are employed with B.L. cast-iron common shell, 6-inch and upwards, and 6-inch cast steel, Marks IV and V, to seal the joint when the base plug is screwed in.

They are made of lead, slightly concave, and have a square projection which fits into the hole in the base plug. When hammered flat they fit tightly into the slightly undercut recess in the base of the shell.

Fig. 38.

DISC, BASE PLUG.

Full size.

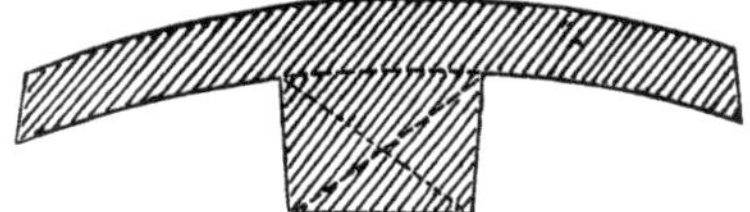

Common-Pointed Shell.

Considerations governing design.—*See* Common Shell; the point is not, as a rule, especially designed to penetrate armour; the main consideration is that such shell should contain a large bursting charge; a pointed shell, however, has an advantage over an ordinary common shell in the attack of shipping, and it is for such purposes that they have been made; common-pointed shell are not used in the field, with the exception of B.L. 9·45-inch howitzer, which fires a common-pointed shell.

Material.—They are made of cast steel, with the exception of the Q.F. 3 and 6-pr., which are made of forged steel.

Cast-iron common-pointed shell, filled with salt cake for *practice only,* have been made.

They are obsolete for future manufacture.

General construction.—Common-pointed shell are about 3·5 calibres long, ogival heads struck with a radius of 2 calibres, the walls being about $\frac{1}{7}$ the diameter in thickness.

7·5-inch and up were cast with bands, and 9·2-inch and up have a hole at the C.G. for the "Bolt, eye, lifting."

They are velvrilled internally (earlier shell lacquered).

With the exception of the Q.F. 12-pr. they are fitted with a dowlas burster bag, the neck of which is primed with a number of 7-dram primers.

They are filled with "P mixture," with the following exceptions:—

12-pr. to 6-inch may be filled with "Shell, Q.F. and shell F.G.," instead of "P mixture."

4-inch heavy Mark II is filled with Blank L.G.

Closing the Bases of Common-pointed Shell.

12-pr. to 5-inch ..		Medium base fuze (No. 12). or Plug, base, shell, No. 2.
6-inch		Large base fuze (No. 11). or Plug, base, shell, No. 1.
7·5-inch and up ..	Steel bush and	Large base fuze (No. 11). or Plug, base, shell, No. 1.

Exceptions—

12-pr. Mark I shell is closed by Plug, base, shell, No. 3 or Hotchkiss base fuze.

4·7-inch (Marks I, II and III Shell) { By Armstrong base plug or fuze.

9·45-inch B.L. Howitzer { By special plug or fuze.

Early marks of shell 9·2-inch and up were fitted with a gunmetal adapter instead of a steel bush: the joint between adapter and shell was closed by a lead ring.

Colour.

All common-pointed shell are painted black, with the exception of those suitable for the Q.F. 12-pr. of 12 and 8 cwts. guns only | N |, which are painted lead colour to distinguish them from shell suitable for *any* 12-pr. or 14-pr. gun. For markings on the shell *see* page 236.

Description of Common-Pointed Shell for B.L. 9·2-inch.

(*Plate* XVIII.)

Shell, B.L. common-pointed, 9·2-inch, Mark VI | N |.—The shell is made of cast steel 3·8 calibres long; the head struck with a radius of 2 calibres. It is cast with bands and eye-bolt, lifting hole. It is rotated by means of a "Broad Vavasseur driving band with gascheck" (No. 6 type), pressed into an undercut groove with waved ribs, near the base.

This band is now fitted with an augmenting ring. The interior of the shell is coated with velvrill, and is fitted with a dowlas burster bag filled with "P mixture."

The neck of the bag is made of shalloon and is choked with twine; a number of 7-dram primers (generally six) are inserted on top of the choke, so as to be close to the base fuze.

The hole in the base of the shell is closed by a steel bush screwed in and riveted up; this steel bush takes the Plug, base shell, No. 1, or Fuze, percussion, base, large, No. 11.

The shell is painted black, with a white ring denoting steel and a red ring denoting filled.

For markings, *see* page 236.

Description of Common-Pointed Shell for Q.F. 12-pr.

(*Plate* XIX.)

Shell, Q.F. common-pointed, 12-pr., 12 and 8 cwt., Mark VI | C |. —The shell is made of cast steel, 4 calibres long; the head struck with a 2-calibre radius.

It is rotated by means of the "Special Narrow" type of driving band (No. 12) pressed into an undercut groove with waved ribs.

Internally the shell is velvrilled, but has no burster bag; it is filled with "P mixture," or with "Shell, Q.F." and "Shell, F.G."

No 7-dram primers are used.

The hole in the base is closed by Plug, base, shell, No. 2, or Fuze, percussion, base, medium, No. 12.

The shell is painted black for Land Service and lead colour for Naval Service.

The shell, Q.F., common-pointed, 12-pr. of 18 cwt., Mark I | N |, and the shell Q.F., common-pointed, 12- and 14-pr., Mark II | C |, are stronger in design than the above and may be fired from any 12- or 14-pr. gun.

They are painted black for both Land and Naval Service.

9·45-*inch Howitzer Shell.*

The bursting charge is composed of a number of compressed pellets of ammonium powder. Each pellet has a central hole in which is placed a primer of F.G. powder in a muslin bag. The pellets are wrapped in thin paper and tinfoil. Two of these pellets are then placed together and enclosed in a calico bag and pushed into the shell.

Capped Common-Pointed Shell. (*C.P.C.*).

(*Plate* XX.)

Capped common-pointed shell have been introduced for B.L. 6-inch and up for Naval Service only.

This type of shell has excellent penetrating powers against armour, and, at the same time, carries a large bursting charge; it may be classified as a " Semi-armour-piercing projectile."

Main features of the C.P.C. shell.—(*a*) The points are specially hardened for the attack of armour.

(*b*) Owing to the above hardening, spontaneous cracks are possible. to avoid danger of the bursting charge being fired from this cause, a copper container is placed in the shell before the bag and bursting charge.

(*c*) The provision of the container necessitates a removable base to the shell.

(*d*) Though designed to attack armour, the bursting charge is much larger than in the case of armour-piercing shell.

(*e*) They are of either cast or forged steel, and have neither bands nor hole for hoisting purposes.

They take the Fuze, percussion, base, large, bronze, No. 15.

TABLE NO. 17.—*Shell, B.L., B.L.C., Q.F. or Q.F.C. Common-Pointed and Common-Pointed with Cap.*

Para. in List of Changes.	Nature of Gun.	Mark of Shell.	Service.	Type No. of Band.	Length in Inches.	Diameter. Bands or Body.	Diameter. Driving Band.	Approximate weight empty, lbs.	Weight filled.	Remarks.
7726	12-pr., 8 and 12 cwts.	I	C	4	12·55	2·97	3·085	10½	12½	
8106	Do.	II	C	4	12·0	2·97	3·085	10	12½	
9957	Do.	III	C	4	12·0	2·97	3·085	10	12½	
10301	Do.	IV	C	10	12·0	2·97	3·095	10	12½	
11234	Do.	V	C	10	12·0	2·97	3·095	10	12½	
12033	Do.	VI	C	12	12·0	2·97	3·095	10	12½	
12596	12-pr. of 18 cwts.	I	N	12	12·15	2·98	3·095	10	12½	
13103 } 14554 }	12 and 14-pr.	II	C	12	11·84	2·98	3·095	10	12½	
8142	4-inch B.L., Q.F., or Q.F.C. Light (except B.L. Marks VII to VIII* and Q.F. Marks IV and V guns)	I	C	3	13·48	3·96	4·11	21½	25	
9957	Do. do.	II	N	4	13·48	3·96	4·11	21½	25	
11193	Do. do.	III	N	10	13·48	3·96	4·1	21½	25	
15071 } 16445 }	Do. do.	IV	N	10	13·48	3·96	4·1	21½	25	
14461 15071	4-inch, B.L., Marks VII to VIII* guns and Q.F. Mark IV gun	I	N	11	16·42	3·97	4·23	27¼	31	

16445 16454	4-inch, B.L., Marks VII to VIII* guns and Q.F. Mark IV gun, also for Q.F. Mark V gun	II	N	11	15·63	3·97	4·23	$27\frac{10}{16}$	31	Filled Blank, L.G.
8104	4·7-inch Q.F.	IV	C	3	17·36	4·7	4·819	$39\frac{1}{2}$	45	
9957	Do.	VI	N	4	17·36	4·7	4·819	$39\frac{1}{2}$	45	
10412 11234 }	Do.	VII	N	4	17·36	4·7	4·819	$39\frac{1}{2}$	45	
12036	Do.	VIII	N	7	17·36	4·7	4·819	$39\frac{1}{2}$	45	
8142	6-inch Q.F. gun	I	C	4	22·5	5·96	6·115	88	100	
9272	Do.	II	C	5	22·78	5·97	6·33	88	100	
9957	Do.	III	C	5	22·78	5·97	6·33	88	100	
10412	Do.	IV	N	5	22·53	5·97	6·33	$88\frac{1}{2}$	100	
11234	6-inch B.L., Marks VII and VIII, or Q.F. guns	V	N	5	22·53	5·97	6·33	$88\frac{1}{2}$	100	
12267	6-inch B.L., Marks VII, VIII, XI, XI*, or Q.F. guns	VI	N	7	22·53	5·97	6·33	$88\frac{1}{2}$	100	
14977	6-inch B.L., Marks VII, XI, XI*, or Q.F. guns	VIIA	N	7	23·5	5·97	6·33	$88\frac{1}{2}$	100	With cap and copper container.
	6-inch B.L., Marks VII, and XI to XII guns	VIIIAQ	N	8	23·5	5·97	6·33	90	100	Do.
12387	7·5-inch, except Marks III and IV	I	N	6	27·88	7·465	7·99	$181\frac{1}{4}$	200	
		I*	N	6	27·88	7·465	7·99	$181\frac{1}{4}$	200	
14518	Do. do.	II	N	6	27·88	7·465	7·99	$181\frac{1}{4}$	200	
14574 14874 }	Do. do.	IIIA	N	8	28·9	7·465	7·99	$181\frac{1}{4}$	200	Do.
12855	7·5-inch, Marks III and IV ...	I	N	11	28·45	7·465	7·88	181	200	
		II	N	11	27·88	7·465	7·88	$181\frac{1}{4}$	200	

TABLE No. 17.—*Shell, B.L., B.L.C., Q.F. or Q.F.C. Common-Pointed and Common-Pointed with Cap*—continued.

Para. in List of Changes	Nature of Gun.	Mark of Shell.	Service.	Type No. of Band.	Length in Inches.	Diameter.		Approximate weight empty, lbs.	Weight filled.	Remarks.
						Bands or Body.	Driving Band.			
9963	9·2-inch	IV	N	6	35·1	9·165	9·71	347½	380	
10412 10508	Do.	V	N	6	35·4	9·165	9·71	347½	380	
11234	Do.	VI	N	6	35·4	9·165	9·71	347½	380	
14703	Do.	VII	N	8	35·4	9·165	9·71	347½	380	
13088	9·45-inch Howitzer	I	L	1	28·31	9·425	9·59	227	280	
		II	L	1	28·31	9·425	9·59	233	280	
8105 9957	10-inch, Mark I to IV* ...	I	C	4	37·85	9·95	10·145	460¼	500	
		II	C	4	37·85	9·95	10·145	460¼	500	
9963	Do.	III	C	4	37·85	9·95	10·145	460¼	500	
9963	Do.	IV	C	6	38·32	9·965	10·61	460¼	500	
11399	Do.	V	C	6	38·32	9·965	10·61	460¼	500	
12855	10-inch, Marks VI and VII ...	I	N	11	38·0	9·965	10·52	460½	500	
		II	N	11	38·32	9·965	10·52	460¼	500	

9963	12-inch, Mark I to VII	III	N	4	42·93	11·965	12·71	634½	714	
10177	Do.	IV	N	5	42·93	11·965	12·71	634½	714	
11234, 12119	Do.	V	N	5	42·93	11·965	12·71	634½	714	
9963	12-inch, Mark VIII to XII ...	III	N	4	48·01	11·965	12·51	766½	850½	
9963	Do. do. ...	IV	N	4	48·01	11·965	12·51	764¼	850	
10187	Do. do. ...	V	N	5	48·01	11·965	12·76	764¼	850	
11234	Do. do. ...	VI	N	6	48·1	11·965	12·76	764¼	850	
14178	Do. do. ...	VIIA	N	6	49·2	11·965	12·76	736¼	850	Cap and copper container.
8105, 16410	13·5-inch (Light)	I	N	3	50·34	13·45	13·65	1,163	1,250	
9957	Do.	II	N	4	50·34	13·45	13·65	1,163	1,250	
9963	Do.	III	N	5	50·96	13·465	14·26	1,163	1,250	
10252	Do.	IV	N	6	50·96	13·465	14·26	1,163	1,250	
11234	Do.	V	N	6	50·96	13·465	14·26	1,163	1,250	Waved ribs.
16661	Do.	VA	N	6	60	13·465	14·26	1,129	1,250	Cap and copper container.
16660	13·5-inch (Heavy)	IA	N	6	65	13·465	14·26	1,280	1,400	Do.
	15-inch	IA	N	6	67	14·965	15·76	1,770	1,920	Do.
	Do.	IIA	N	6	67	14·965	15·76	1,787½	1,920	Do.

ARMOUR-PIERCING SHELL.

In these shells, the desirability of having a large bursting charge has to give way to the necessity for great strength in the shell; hence the *comparatively small bursting charges used.*

The material is either cast or forged steel.

The points are made extremely hard, while the bodies are softer, so as to give greater tenacity and enable the shells to hold together when they strike hard-faced armour.

There is great thickness of metal in the head, and the walls are thicker than in other shell; cases have been known of the walls of the shell resisting the burster, and the fuze being blown out.

No holes for lifting eye-bolts are provided in A.P. projectiles, because of their weakening effect.

They are shorter than common-pointed shell; the exact length, however, is left to the contractor.

Spontaneous splits.—The heads of A.P. shell are liable to split spontaneously from the strains set up during the hardening processes; should such a split extend to the cavity in a filled shell, the sudden fracture might cause the explosion of the bursting charge; for this reason A.P. shell (and also capped common-pointed) are stored for three months before filling to allow latent cracks to develop.

Burster bags.—Armour-piercing shell are now *varnished* internally and fitted with burster bags of a closely-woven material called "Lasting cloth."

A large number of old A.P. shell will still be met with in the Service; they are lacquered internally and have dowlas burster bags.

Lasting cloth, being closer in texture than dowlas, prevents the powder bursting charge, if dusty, working through the bag, therefore there is less chance of the bursting charge being fired should the head of the shell split spontaneously.

Bursting charges.—All modern A.P. shell are fitted with *Blank L.G.* new, or *Blank L.G. converted* powder; L.G., R.L.G., R.L.G.^{2} and R.L.G.^{4} powders may also be used.

Old A.P. shell with dowlas burster bags are filled with "P mixture."

A.P. shell (powder filled) are not fitted with containers, owing to their comparatively small bursting charges.

Closing the bases.—All A.P. shell, 7·5-inch and up, manufactured since 1899 have the hole in the base closed by means of a steel bush, screwed in and riveted over; this bush takes the Plug, base, shell, No. 1, or the Large base fuze.

With 6-inch shell the use of this steel bush is optional; the base of the shell itself may be threaded to take the plug or fuze.

Old A.P. shell manufactured prior to 1899, had the hole in the base closed with a gun-metal bush and lead ring instead of a steel bush.

Paint.—The shell are painted black and have two white rings round the head denoting A.P. shell; when filled, a red ring is stencilled between the two white rings.

For markings, *see* page 237.

Armour-Piercing Shell with Cap (A.P.C. Shell). (Powder Filled.)

Capped A.P. shell are similar to uncapped A.P. shell, but are stronger in design, having thicker walls and more metal in the head, hence smaller bursting charges.

The 9·2-inch A.P. shell takes an 18 lb. bursting charge.

The 9·2-inch A.P.C. shell takes a 7¾ lb. bursting charge.

They are made of cast or forged steel, and may be fitted with any approved design of cap.

They are varnished internally, fitted with lasting cloth burster bags, and are filled with L.G. powder.

For future manufacture all A.P.C. shell will have the base fitted with an adapter large enough to admit the use of a container if necessary.

A.P.C. shell, when filled with powder, are not fitted with containers.

A.P.C. shell, when filled with lyddite, are always fitted with aluminium containers. (*See* page 209.)

Fuzes used.—All capped armour-piercing shell filled powder take the No. 15 base fuze.

Uncapped armour-piercing shell 6-inch and up, take the No. 11 base fuze, but the No. 15 base fuze can also be used for shell without cap, for Naval Service.

Shell, B.L. Armour-Piercing, with Cap, 9·2-inch, Mark V.

The shell is made of either cast or forged steel about 3 calibres in length ; the head is struck at a radius of 2 calibres, hardened for the attack of armour, and fitted with a mild steel cap.

The interior is varnished and takes a bursting charge of L.G. powder contained in a lasting cloth burster bag ; the choke of the burster bag is primed with a number of 7-dram primers.

The base of the shell is closed with a steel bush screwed in and riveted over ; this bush is bored out and threaded to take the Plug, base, shell, No. 1, or the Large base percussion fuze, No. 15.

The shell is rotated by a " Hump " pattern driving band pressed into an undercut groove with waved ribs near the base.

The Mark IV differs from the above in having a broad Vavasseur driving band with gas-check.

Note.—For A.P. shell filled with lyddite, *see* page 209.

TABLE NO. 18.—*Shell, B.L. or Q.F., Armour-Piercing, and A.P. with Cap.*

Para. in List of Changes	Nature of Gun.	Mark of Shell.	Service.	Type No. of Band.	Length in Inches.	Diameter.		Approximate weight empty, lbs.	Weight filled.	Remarks.
						Bands or Body.	Driving Band.			
7379 5966	4·7-inch Q.F.	III	L	3	14·3	4·694	4·808	$41\frac{6}{16}$	45	Armstrong fuze, No. 9.
8460	Do.	IV	L	3	14·53	4·7	4·819	$41\frac{15}{16}$	45	Base medium fuze, No. 12.
9529	6-inch B.L. or Q.F.	I	N	5	19·24	5·97	6·33	$93\frac{1}{4}$	100	
10027	Do.	II	C	5	19·24	5·97	6·33	92	100	
11234	Do.	III	C	5	19·24	5·97	6·33	92	100	
11267	6-inch B.L.	IV	N	7	19·24	5·97	6·33	92	100	
12807	Do.	V	C	7	(Left to Contractor)	5·97	6·33	92	100	Capped.
16552	Do.	VI	N	7	Do.	5·97	6·33	$95\frac{1}{4}$	100	Do.
	Do.	VII	L	5	17·29	5·97	6·33	94	100	
12387	7·5-inch except Marks III to IV* guns	I	C	7	Do.	7·465	7·99	$193\frac{1}{2}$	200	Do.
12925	Do. do.	I*	C	8	Do.	7·465	7·99	$193\frac{1}{2}$	200	Do.
14518	Do. do.	II	C	8	Do.	7·465	7·99	$193\frac{1}{2}$	200	Do.
		IIIA	N	8	28·9	7·465	7·99	$193\frac{1}{2}$	200	Do.
16532	Do. do.	III	N	8	28·9	7·465	7·99	$193\frac{2}{16}$	200	Do.

12855	7·5-inch Marks III to IV* guns ...	I	N	11	(Left to Contractor)	7·465	7·88	$188\frac{3}{4}$	200	Do.
12855	Do. do. ...	II	N	11	Do.	7·465	7·88	$193\frac{1}{2}$	200	Do.
16552	Do. do. ...	III	N	11	Do.	7·465	7·88	$193\frac{2}{16}$	220	Do.
10307	9·2-inch	I	C	5	Do.	9·165	9·71	$359\frac{1}{2}$	380	
10508	Do.	II	C	5	Do.	9·165	9·71	$359\frac{1}{2}$	380	
10508	Do.	III	C	6	Do.	9·165	9·71	$359\frac{1}{2}$	380	
11234	Do.	IV	C	6	Do.	9·165	9·71	$359\frac{1}{2}$	380	Do.
	Do.	V	C	8	Do.	9·165	9·71	$359\frac{1}{2}$	380	Do.
16552	Do.	VI	N	8	Do.	9·165	9·71	$369\frac{1}{2}$	380	Do.
14705	Do.	VIA	C	8	Do.	9·165	9·71	$369\frac{1}{2}$	380	Do.
16253	Do.	VIIA	L	8	Do.	9·165	9·71	365	380	Capped with aluminium container (No. 16 Fuze).
16552	Do.	VII	C	6	Do.	9·165	9·71	$369\frac{1}{2}$	380	Capped.
10508 11234 11399	10-inch, except III to IV guns ...	I II	C	6	Do.	9·965	10·61	$472\frac{1}{2}$	500	
12855	10-inch, Mark VI and VII ...	II	N	11	Do.	9·965	10·52	$475\frac{1}{2}$	500	Do.
11234	12-inch, Marks I to VII	II	C	6	Do.	11·965	12·66	$676\frac{1}{2}$	714	
12119	Do. do.	III	C	6	Do.	11·965	12·66	$676\frac{1}{2}$	714	
10307	12-inch, Marks VIII to XII ...	I	N	6	Do.	11·965	12·76	805	850	
11234	Do. do. ...	II	N	6	Do.	11·965	12·76	805	850	
12821	Do. do. ...	III	N	6	Do.	11·965	12·76	805	850	Do.
16552	Do. do. ...	IV	N	6	Do.	11·965	12·76	805	850	Do.

TABLE 18.—*Shell, B.L. or Q.F. Armour-Piercing, and A.P. with Cap*—continued.

Para. in List of Changes	Nature of Gun.	Mark of Shell.	Service.	Type No. of Band.	Length in Inches.	Diameter.		Approximate weight empty, lbs.	Weight filled.	Remarks.
						Bands or Body.	Driving Band.			
14528 14874	12-inch, Marks VIII to XII—*contd.*	IVA	N	6	49·2	11·955	12·76	805	850	Capped.
14528 14874 16658	Do. do. ...	VA	N	6	49·2	11·965	12·76	805	850	Do.
16659	Do. do. ...	VIA	N	6	49·2	11·965	12·76	805	$859\frac{7}{16}$	Capped with aluminium container (No. 16 Fuze).
10307	13·5-inch (Light)	I	N	6	(Left to Contractor)	13·465	14·0	1,185	1,250	
11234	Do.	II	N	6	Do.	13·465	14·0	1,185	1,250	
16661	Do.	IIA	N	6	60	13·465	14·26	1,224	$1,266\frac{1}{2}$	Do.
16660	13·5-inch (Heavy)	IA	N	6	65	13·465	14·26	1,350	1,400	Do.
	15-inch	IA	N	6	67	14·965	15·76	$1,856\frac{1}{2}$	1,920	Do.

SHRAPNEL SHELL.

General remarks.—Shrapnel shell are designed to hold as many bullets as possible.

With Field shrapnel the bullets are the only useful part, the weight of the rest of the shell is reduced to a minimum consistent with the necessary strength to prevent it setting up in the bore or breaking up during flight.

In the Q.F. 18-pr. shrapnel the bullets form 48 per cent. of the total weight.

Bursting charge.—The body of shrapnel is not intended to break up when the bursting charge is fired.

The bursting charge is usually of F.G. powder contained in a tin cup; it is intended to blow off the head and drive out the bullets, which are carried forward with the remaining velocity of the shell.

Certain shrapnel are now fitted with a compressed pellet of powder instead of loose F.G.; this tends to give the bullets additional velocity.

Shrapnel have the bursting charge in the base, except the 4-inch (light) Marks IV, V and VI, which have the burster in the head.

Tin cup.—This, though known as a *tin* cup, is made of tinned sheet iron, and is consequently liable to rust. Its object is to prevent prematures which might occur from friction.

Once a tin cup has got rusty it defeats its own object.

The cup as a rule is coned at the top to facilitate emptying.

Head.—Most shrapnel have the head made in a piece separate from the body; this separate head is filled up with a block of wood and is lightly secured to the body by a row of rivets, and a row of twisting pins, the head being slotted so that the twisting pins offer no resistance to the head being blown off.

A band of solder round the outside of the shell at the junction of the head and body covers the rivets and twisting pins.

The B.L. 7·5-inch, Mark III, Q.F. 4·7-inch, Mark V and B.L. 2·95-inch, Mark II shrapnel have the head lightly screwed on to the body instead of being attached by rivets and twisting pins.

Some small modern shrapnel have no separate head, *e.g.*, Q.F. 13- and 18-pr., 2·75-inch, 3-inch and 12-pr., Mark IX.

Socket.—All shrapnel have in the head a socket for the fuze to screw into; this socket is secured to the head by screwing in, or by solder, or both.

All modern shrapnel have a 2-inch fuze-hole gauge; earlier types have the G.S. gauge (1-inch).

Size of bullets.—The bullets are made as small as possible so as to get the maximum number into the shell, but the bullets must be effective. It is considered that, for a Field shrapnel bullet to be effective against personel it should have a striking energy of at least 60 ft.-lb.

The striking energy of the bullet depends on its weight and striking velocity; the higher the striking velocity, the lighter it may be. It is found that for light Field guns, bullets 41 to the lb., and heavy Field guns and howitzers, bullets 35 to the lb., satisfy requirements.

Some old shell, however, have bullets as heavy as 14 to the lb.

Material of bullets.—A shrapnel bullet should be made of the heaviest available metal, as a heavy bullet keeps up its velocity better than a light one. Lead, however, is too soft, therefore a lead alloy is used called "*Mixed metal,*" *i.e.*, 7 parts lead and 1 part antimony. Earlier Marks of shrapnel, 7·5-inch to 13·5-inch, had bullets made of cast iron called "Sand shot."

The shell B.L., shrapnel, 12-inch (heavy), Mark I | N |, has 12-oz. steel balls.

In the latest Naval shrapnel, 6-inch to 15-inch, mixed metal bullets 27 to the lb. are now used.

For particulars re size of bullets, &c., *see* Table 19.

Angle of opening.—When the shell bursts, the bullets fly forward in a "cone of dispersion" due to the combined effect of the spin of the shell and the forward velocity; the angle of opening is the angle between the lines of flight of the outer bullets. It is varied in different shell to suit tactical requirements by various means; *e.g.*, in the B.L. 10-pr. shell the bullets are in a perforated tin cage to *decrease* the angle. In the Q.F. 18-pr. shell, the central tube is filled with pellets of powder to ensure a *larger* angle of opening.

Effect of range on the angle of opening.—The longer the range, the larger becomes this angle, for the forward velocity of the shell falls off, but its spin does not decrease much.

Shoulder and protecting disc.—To protect the bursting charge from the set-back of the bullets, a steel or wrought-iron disc is placed over the tin cup. This disc rests on a shoulder inside the shell near the base. In shrapnel which have a compressed burster a bronze disc is used; it rests on the top of the screwed base.

Brown paper lining.—The bodies of those shell which have a separate head are lined with brown paper; the brown paper lining prevents the resin adhering to the walls of the shell, and thus less resistance is offered to the bursting charge blowing out the bullets.

Bullets in resin.—Melted resin is poured into the shell between the bullets; when it sets it prevents them moving about during flight.

Central tube.—A central tube conveys the flash from the fuze to the bursting charge; the bottom of the tube screws into the disc and the top is as a rule merely a mechanical fit with the bottom of the fuze socket; but with light shrapnel, having 2-inch gauge, the top of the tube is soldered to the fuze-hole bush to prevent resin working up into the recess for the fuze.

In most shell this central tube is empty, but in some small modern shrapnel it is fitted with perforated pellets of powder.

Primer, h apnel shell.—In all shrapnel having the G.S. fuze-hole gauge (with the exception of the 4·7-inch, Mark IV shrapnel) a metal primer is screwed into the bottom of the fuze-socket or into the top of the central tube to convey the flash from the fuze down the tube to the burster; it also prevents the loose powder of the burster working up into the fuze-hole socket.

For description of Primer, shrapnel shell, *see* page 252.

A primer is never used in shrapnel shell which have the fuze-hole of the 2-inch gauge, because fuzes of the 2-inch gauge have a much larger magazine, and give a more powerful flash than fuzes of the G.S. gauge.

Means of preventing powder getting into the threads of the fuze-hole.—In shell having a 2-inch fuze-hole, in some cases a shalloon disc is shellaced into the fuze-socket to prevent the powder working up; in the latest heavy shrapnel for N.S. a "Wad fuze-hole, 2-inch" is used instead of the shalloon disc.

The Q.F. 4·7-inch, Mark IV, shrapnel, has no primer; a "Wad, fuze-hole, G.S." is used; it fits into a recess in the bottom of the fuze socket.

Paint.—

Shrapnel shell are painted black, except:—

(*a*) those for the B.L. or B.L.C. 15-pr., which are lead-colour to distinguish them from the B.L. 12-pr. shrapnel.

(*b*) the Mark II and III shrapnel for Q.F. 13- and 18-pr., which are lead-colour to distinguish them from the Mark I shrapnel.

(*c*) 12-pr. shrapnel, suitable for Q.F. 12-pr., of 8 and 12 cwt. guns only (Naval Service).

All shrapnel have the nose painted red in addition to the red ring denoting filled. (*See* page 236.)

Action of Shrapnel.

The flash from the fuze passes down through the central tube, firing the primer, or perforated powder pellets in the shell which have them, and ignites the bursting charge of the shell, the explosion of which blows off the head (or where there is no separate head, blows out the socket and fuze) and so liberates the bullets.

Types of Shrapnel Shell. Description of.

The various shrapnel shell are here divided into types, illustrated in Plates XXIII to XXX.

In addition there are some of exceptional construction.

TYPE SHOWN ON PLATE XXIII.

The shrapnel shell for the B.L. 10-pr., B.L. or Q.F. 12-pr., B.L., B.L.C., or Q.F. 15-pr., Marks I to VI, and the B.L. 30-pr., are alike in construction. The main feature of this type is a perforated tin cylinder to contain the bullets, intended to decrease the angle of opening.

Plate XXIII illustrates the shrapnel shell for the B.L., B.L.C., or Q.F. 15-pr., a short description of which is given below.

SHELL, B.L., B.L.C., OR Q.F. SHRAPNEL, 15-PR. (MARK VI) | L |.

The body is made of forged steel. The bursting charge is contained in a tin cup, above which rests a steel disc, into which the metal tube screws. This tube is in two parts screwed together, the upper part enlarged, and threaded internally for the primer, and externally for a metal nut. The shell is lined with brown paper and fitted with a perforated tin cylinder which rests upon the steel disc. This cylinder is filled with mixed metal bullets; on top of the bullets is placed a flanged tin ring, and the top of the cylinder, which is fringed, is bent over and soldered to this ring. The space between the bullets is filled in with molten resin.

A short gunmetal socket screws into the head and is further secured by solder. This socket is threaded inside to the G.S. taper and pitch, and a flange formed on the inside at the bottom fits round the top of the central tube, to which it is secured by a locking nut screwed on the top of the tube, and prevented from working loose by four centre punch stabs.

The head is attached by means of brass screws and steel twisting pins, which are covered with solder. The twisting pins are in the top row and fasten the tin cylinder to the head.

TYPE SHOWN ON PLATE XXIV.

The shrapnel shell for the Q.F. 13- and 18-pr., and the Mark VII shrapnel for the B.L., B.L.C. or Q.F. 15-pr., are similar in construction, differing only in dimensions and type of driving band.

The main features are: They have no separate head nor shrapnel primer; they have the 2-inch fuze-hole, and the central pipe is filled up with perforated powder pellets for the purpose of increasing the angle of opening.

Shell, Q.F., Shrapnel, 13-pr., Mark III | L |.

The body of the shell is of forged steel, oil-hardened, but the front portion is subsequently annealed, and is thereby enabled to open out slightly when the burster acts so as to allow the bullets to escape.

Its length is about 2·6 calibres; the walls near the base are thickened, forming a shoulder on which rests a steel disc; below the disc is placed a tin cup for the bursting charge. A hole is bored through the centre of this disc, and screwed into it is the lower end of a brass central pipe, which passes through the disc and projects into the mouth of the tin cup. The shell contains about 236 mixed metal bullets (7 parts lead, 1 part of antimony), 41 to the lb., the spaces between the bullets being filled up with resin. The front end of the shell is closed by a flanged gunmetal bush, or "Fuze socket," screwed in. The fuze socket is screw-threaded in the interior to the 2-inch gauge to take the T. and P., No. 80 fuze, a hole being bored through the bottom of the socket for the top of the central pipe. To prevent the resin working up into the fuze socket the top of the central pipe is soldered to the latter.

The shell is rotated by means of a copper driving band pressed into an undercut groove with waved ribs near the base. The driving band is of the special narrow type (No. 14. Plate XVI).

A projecting rim and a coning groove are formed on the band for the mouth of the cartridge case.

The bursting charge is $1\frac{1}{4}$ oz. of F.G. powder which fills the tin cup, and the central pipe is filled with six perforated powder pellets; the bottom ends of the two lower powder pellets are covered with discs of muslin, and a disc of shalloon shellaced in the bottom of the fuze-hole socket prevents the F.G. powder working out of the tin cup.

The Plug, fuze-hole, 2-inch, No. 1, is used.

TYPE SHOWN ON PLATE XXV.

This is a new type of shrapnel; its main features are:—

1. It has no separate head.
2. It has a separate base which screws on.
3. It has the 2-inch fuze-hole gauge.
4. The bursting charge is in the form of compressed pellets of gunpowder.
5. It has no primer, the central tube being filled with powder pellets.

The B.L. 2·75-inch, and the Q.F. 12- and 14-pr., Mark IX shrapnel are manufactured to the above design.

The shrapnel shell for the B.L. 4-inch (heavy), is similar, but has a separate head attached to the body by rivets and twisting pins.

SHELL, B.L. SHRAPNEL, 2·75-INCH, MARK I | L |.

The body and base are of forged steel; two holes are formed in the base to take the key when screwing it home, and a small steel screw passing through the wall of the shell prevents the base from unscrewing. The tin cup fits into the front of the base; it is in two parts, and contains two flat powder pellets, with a central perforation in each; the central hole in the pellets is fitted with a small powder pellet, also perforated.

On the top of the pellets rests a tin disc with central hole and disc of paper shellaced to the under side.

On top of the tin cup, and resting on the base, is a bronze disc into which the central tube screws.

A small hole in the centre of the disc allows the flash to pass from the tube to the bursting charge. A disc of paper is shellaced to the underside of the disc.

Over the bronze disc is placed a flanged tin collar, intended to prevent the resin from working through into the base.

The shell is not lined with brown paper; it contains about 253 mixed metal bullets, 41 to the lb.

The fuze-socket is threaded to the 2-inch fuze-hole gauge and is soldered to the top of the central tube; it is fitted with a steel set-screw for securing the fuze in position.

The central tube contains eight perforated powder pellets; a disc of shalloon shellaced to the bottom of the fuze socket retains the powder pellets in position.

TYPE SHOWN ON PLATE XXVI.

This is a new type of shrapnel fitted with tracers, intended for the attack of aircraft.

TYPE OF SHRAPNEL SHOWN ON PLATE XXVII.

The shrapnel shell for the B.L. 60-pr., the Q.F. 4·5-inch howitzer, the Marks XI, XIA, XIII, XIIIA and XVAQ shell for the B.L. 6-inch gun, the Mark XVI shell for the B.L. or Q.F. 6-inch gun and howitzer, and the 6-inch howitzer (light) shrapnel are generally similar in construction.

They have the 2-inch fuze-hole gauge; the central tube has no primer, nor perforated powder pellets.

The 6-inch howitzer shrapnel has a wood block surrounding the central tube. (*See* Fig. 39.)

SHELL, B.L. SHRAPNEL, 60-PR., MARK I | L |.

The body is made of forged steel and has a recess in the base to contain a bursting charge of 4½ ozs. of F.G. powder contained in a tin cup.

The bursting charge is covered with a steel disc resting on a shoulder in the interior of the shell. A hole is bored through the centre of this disc and screwed into it is the lower end of a brass central tube.

The shell is lined with brown paper and contains about 990 mixed metal bullets (35 to 37 to the lb.), the space between the bullets being filled up with molten resin.

The head is of steel and contains a wood block, below which is placed a felt wad; it is attached to the body by means of screws and twisting pins covered with solder.

The metal fuze-socket is screwed and soldered to the head; it is threaded to the 2-inch fuze-hole gauge and has a steel fixing screw for securing the No. 83 fuze, or the adapter 2-inch, when a No. 62 fuze is used.

SHELL, B.L. SHRAPNEL, 6-INCH HOWITZER (LIGHT), MARK I | L |.

The 6-inch howitzer shrapnel differs from the above in dimensions, and in having a wood block surrounding the central tube so as to reduce the weight but not the length of the shell.

First issue of Shrapnel Shell for Q.F. 4·5-inch Howitzer.

The shrapnel shell that were first issued for the Q.F. 4·5-inch howitzer (§ 15434), differed from that shown on Plate XXVII as follows:—

The tin cup has no lid; the steel disc covering the bursting charge is tinned, and has soldered to its underside the flanged top of the tin cup as shown in Fig. 40.

Fig. 39.

SHELL, B.L. SHRAPNEL, 6-INCH HOWITZER (LIGHT), MARK I | L | .

Scale ⅓.

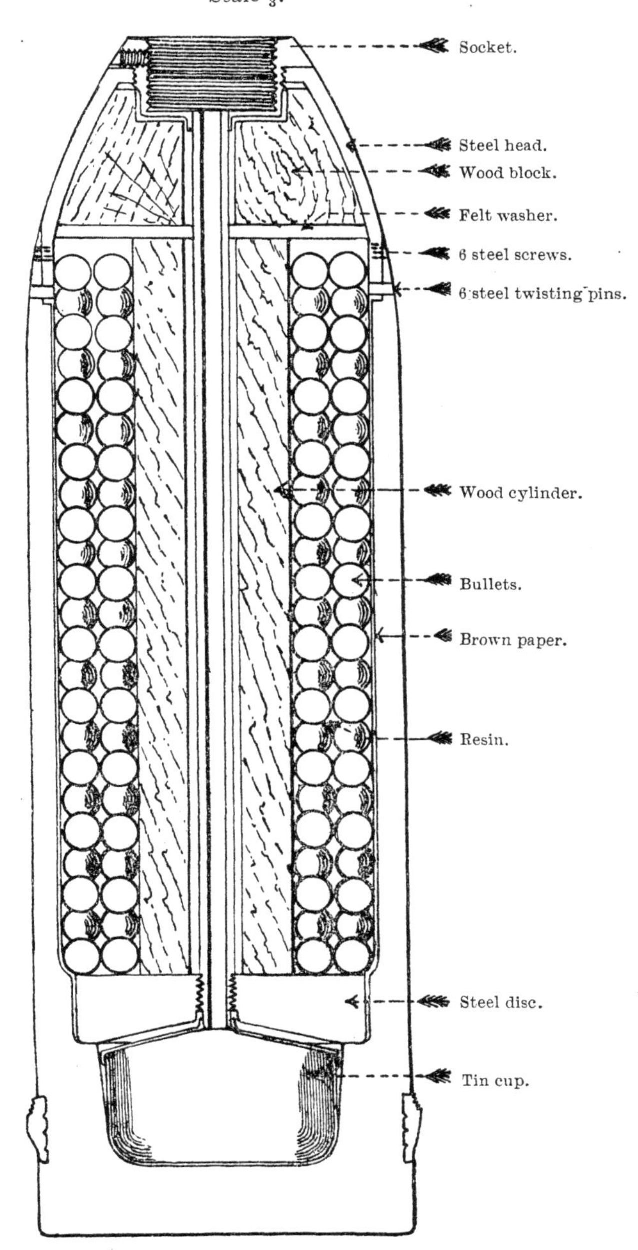

Fig. 40.

SHELL, Q.F. SHRAPNEL, 4·5-INCH HOWITZER, MARK I.,

OLD TYPE.

Scale ½.

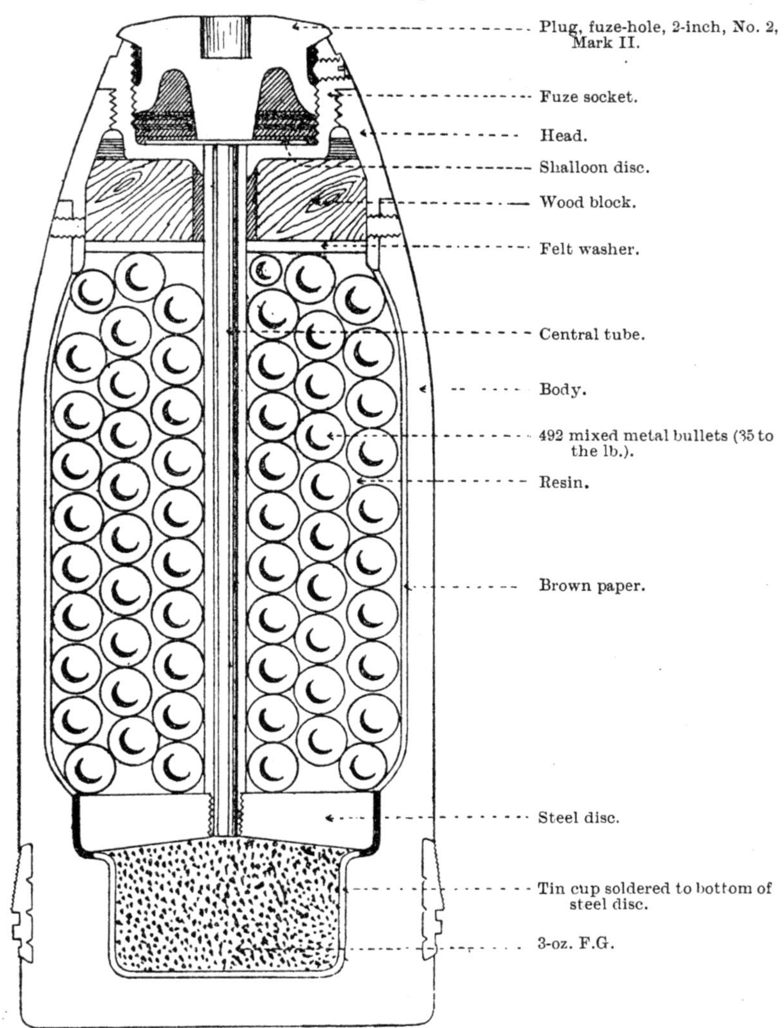

TYPE OF SHRAPNEL SHOWN ON PLATE XXVIII.

This class of shrapnel is generally known as the "*Composite socket type.*"

The shrapnel shell, Marks I to III, for the Q.F. 4-inch, Mark VI to VIII for the B.L. 4-inch; Marks III and IV for the B.L. 5-inch gun; and Marks IV to IX and Mark XII shrapnel, for the B.L. or Q.F. 6-inch gun and howitzer are of this type.

Plate XXVIII illustrates the Mark IX shrapnel shell for the B.L. or Q.F. 6-inch gun and howitzer.

Shell, B.L. and Q.F. Shrapnel, 6-inch Gun and Howitzer, Mark IX | C |.

The body of the shell is of cast steel ; the bursting charge is contained in a tin cup ; the disc is of steel, but the central tube is of brass with a comparatively narrow orifice ; the bullets are of mixed metal (viz., 4 parts lead to 1 part antimony), and the socket is of different pattern from that for the heavier shell. The top of the brass tube is tapped to take the metal primer, and to the exterior is attached by solder a small tin socket which fits over a gunmetal socket attached to the interior of the head of the shell. The head has a wood and the body a paper lining, and a felt washer covers the bullets and prevents the resin from working into the socket, where it might cause a blind, as well as allow the bullets to get loose.

Shrapnel of Exceptional Construction.

Latest marks of 5-inch gun shrapnel.—The latest marks of shrapnel for the B.L. 5-inch gun (*i.e.*, Marks V and VI) are different from the type shown above in that they are not fitted with the composite socket ; the fuze socket is longer and the lower part fits inside the top of the central tube ; the socket is screwed to take the shrapnel primer.

Mark IV Shrapnel for Q.F., 4·7-inch.

The Mark IV shrapnel shell for the Q.F. 4·7-inch is similar in design to the latest marks of shrapnel for the B.L. 5-inch above referred to, but has no shrapnel primer ; a *wad fuze-hole* is used to prevent the powder working out into the thread of the fuze-hole. (*See* Fig. 41 on p. 182.)

Fig. 41.

SHELL, Q.F. SHRAPNEL, 4·7-INCH, MARK IV.

Scale ½.

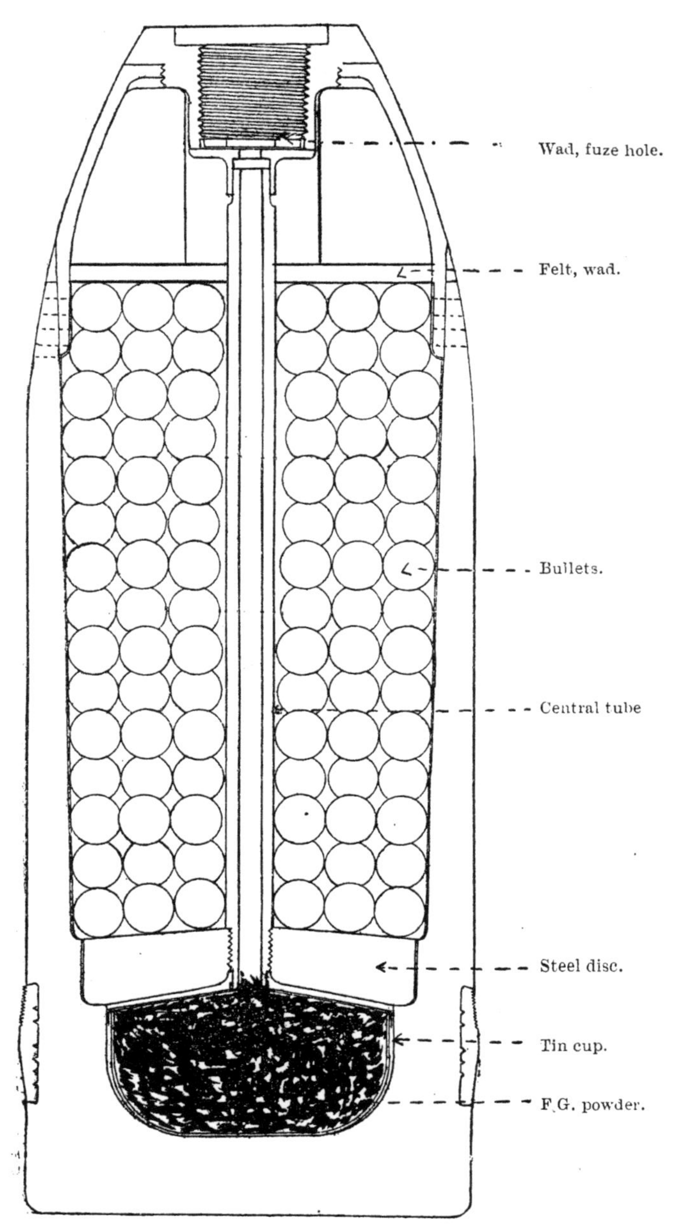

Shell, Q.F. Shrapnel, 4·7-inch, Mark V | L |.

Mark V shrapnel for Q.F. 4·7-inch.—The Mark V shrapnel differs from the Mark IV, in having the head screwed on instead of being attached by screws and twisting pins; the gunmetal fuze-hole bush is threaded to the 2-inch gauge. It contains 712 mixed metal bullets (35 to the lb.).

The first issue of the Mark V shrapnel had the steel disc covering the bursting charge tinned, and had soldered to it the tin cup, as shown on Fig. 40.

Shell, B.L. Shrapnel, 7·5-inch, Mark IIIa | N |.

This is a new type of shrapnel. The head is screwed on and secured by four steel screws, and is weakened by six saw-cuts, each ·04 inch wide, filled in with solder. The fuze-hole bush fits over the top of the central tube and is threaded internally to the 2-inch gauge.

There is no wood lining to the head; the bullets come right up to the bottom of the fuze-hole bush. The shell contains 407 2-oz. sand shot.

Fig. 42.

Shell, B.L. Shrapnel, 7·5-inch, Mark IIIa | N |.

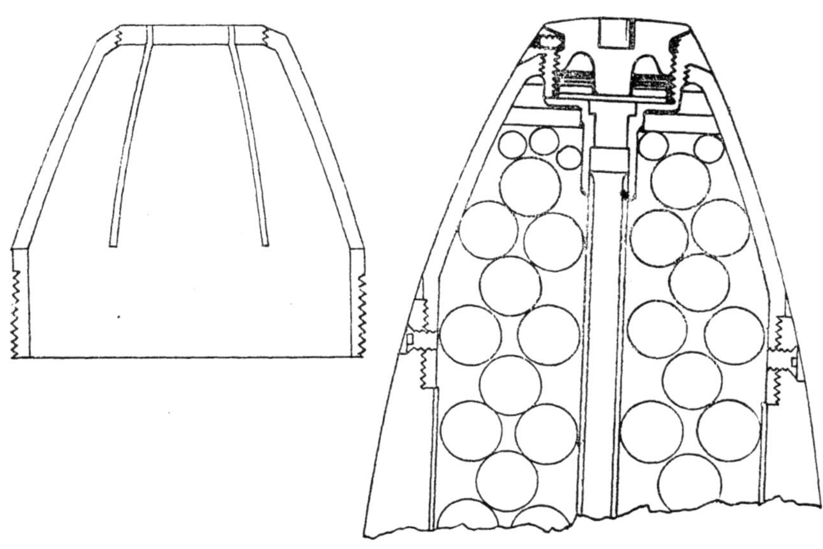

Fig. 43.

SHELL, B.L., Q.F. OR Q.F.C. SHRAPNEL, 4-INCH, FORGED STEEL, MARK V.

Scale ½.

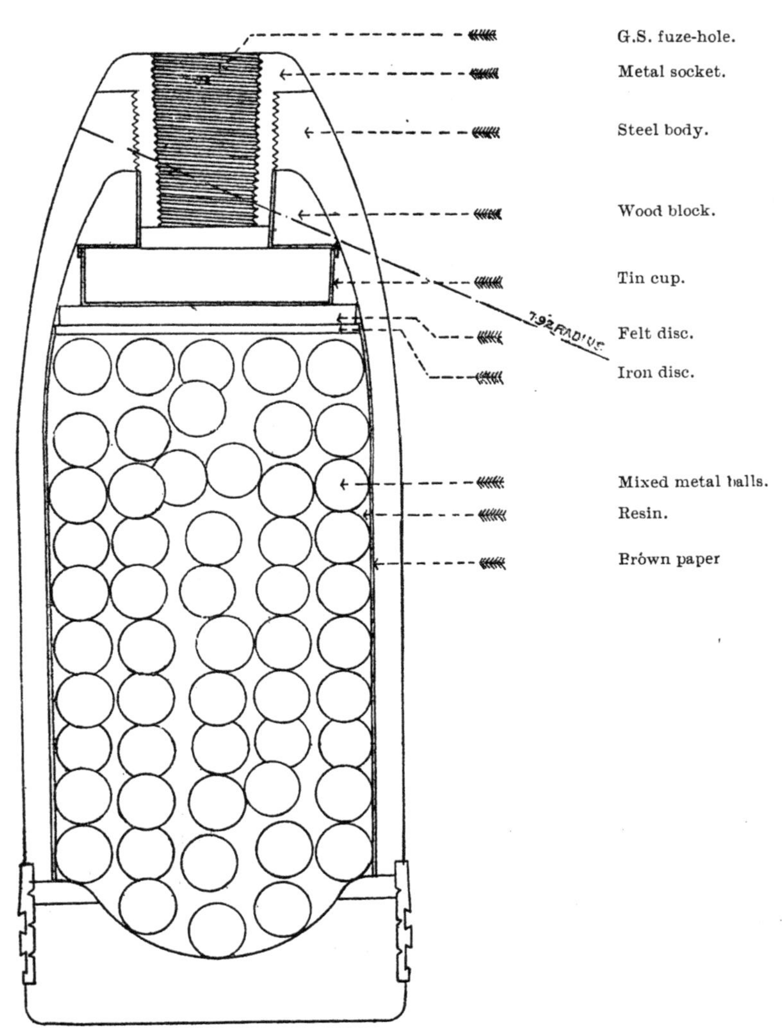

SHELL, B.L. OR Q.F. SHRAPNEL, 4-INCH (HEAVY), MARK II | N |.

This shell has a compressed pellet burster and is similar in design to type shown on Plate XXV, but differs in having a separate head, with wood lining and felt washer, attached to the body by rivets and twisting pins.

It contains about 500 mixed metal bullets (35 to the lb.).

The central tube contains 9 perforated powder pellets.

The Mark I shell differs from the Mark II in having a slightly longer head, which is weakened by four saw-cuts.

SHRAPNEL SHELL WITH BURSTING CHARGE IN THE HEAD.

The Shell, B.L. shrapnel, 4-inch, Marks IV, V and VI, are made of forged steel, and have the bursting charge contained in the head.

This shell has a body of forged steel tubing, head and body in one piece, and malleable cast iron or mild steel base. A flanged gun-metal bush screws into the nose, and is tapped to the G.S. pitch and taper, being made long enough to take the short T. and P. fuze over the G.S. wad. A tin cup is placed inside the head to contain the bursting charge, the neck of the tin cup fitting over the lower portion of the bush, and the space between the projecting portion of the latter and the tin cup is filled by a wood block. A felt disc is placed next the tin cup, then a wrought-iron one, which presses on top of the bullets and prevents the resin from working up. The body is lined with brown paper, and the bullets inserted from the base, molten resin being afterwards run in amongst them. The base is then lightly attached by six steel shearing pins and two steel keys, the latter fitting into undercut slots in the body and base. They prevent the base being twisted off, but do not oppose any resistance to a direct blow from the front.

Action.—The explosion of the bursting charge slightly checks the velocity of the bullets, and acting through them blows off the base of the shell. The body then slides over them like a glove, the bullets continuing their course with a velocity slightly reduced.

TYPE OF SHRAPNEL SHOWN ON PLATE XXIX.

Plate XXIX illustrates the old type of heavy shrapnel introduced for B.L. guns, 7·5-inch to 13·5-inch.

The main features are :—

(1) They are made of cast steel.

(2) The central tube is of iron, instead of brass or gunmetal, and is of larger diameter.

(3) The fuze socket is lengthened so as to extend into the top of the central pipe, the upper part threaded to G.S. gauge, the lower part threaded to take the primer.

(4) Sand shot (cast iron) are used instead of mixed metal bullets—2-oz. sand shot in the 7·5-inch and 9·2-inch, and 4-oz. in the 10-inch, 12-inch and 13·5-inch.

Shell, B.L. Shrapnel, 9·2-inch, Mark VIII | L |.

The shell is made of cast steel about 3·5 calibres in length; the walls near the base are thickened, so as to form a shoulder, on which rests a steel diaphragm, beneath which is the powder chamber.

The bursting charge is contained in a tin cup, which is coned at the top to facilitate unloading, and the neck of which fits on to the bottom of the wrought-iron pipe. The use of this cup is to guard against possible prematures from the roughness of the interior of the shell, and to prevent loss of powder. The bottom of the pipe has asbestos wrapped round it and is covered with Pettman cement in order to make a tight joint with the tin cup and so prevent resin working in among the powder. The diaphragm, which rests upon the shoulder of the powder chamber, is of steel, and into it is screwed a piece of 1-inch gas pipe, smoothed and lacquered inside. The interior of the shell is lined with brown paper, and contains 638 2-oz. sand shot; the interstices are filled up with melted resin and covered with a felt washer.

The head is of steel, fitted with a wood block, and is attached to the body by a row of rivets and a row of twisting pins covered with solder.

The head is fitted with a metal fuze-socket; the upper part of this socket is screw-threaded to the "G.S. gauge," the lower part threaded to receive the "Primer, shrapnel, shell."

There is no attachment between the gunmetal socket and the wrought-iron tube; but the end of the latter is slightly recessed to receive the socket, which is carefully turned to fit it.

The shell is rotated by means of the "Hump" pattern of driving band which is pressed into an undercut groove near the base

TYPE OF SHRAPNEL SHOWN ON PLATE XXX.

Plate XXX illustrates the new type of heavy shrapnel introduced for B.L., 7·5-inch to 15-inch, for Naval Service.

They differ from the old type of heavy shrapnel in the following particulars:—

(1) The fuze-socket is threaded to the 2-inch gauge.

(2) No shrapnel primer is used.

(3) A "Wad, fuze-hole, 2-inch," is used to prevent the powder working up into the threads of the fuze-hole.

(4) Latest marks are now filled with mixed metal bullets, 27 to the lb.; the earlier marks were filled with sand shot; the B.L. 12-inch, Mark I^{H} shrapnel, had 12-oz. steel balls.

(5) The 12-inch, 13·5-inch and 15-inch shell have "stepped walls," to give greater capacity for bullets.

For particulars of number and size of bullets, weight of burster and other details of all Service shrapnel, *see* Table 19 overleaf.

TABLE No. 19.—*Shell, B.L., B.L.C., Q.F. or Q.F.C. Shrapnel.*

Para. in List of Changes	Nature of Gun.	Mark of Shell.	Service.	Type No. of Band.	Length in Inches.	Diameter. Band or Body.	Diameter. Driving Band.	Number and Nature of Bullets.	Weight empty. Lbs.	Weight filled. Lbs.	Fuze Hole. Gauge.	Remarks.
11021	10-pr. B.L.	I	L	12	7·915	2·73	2·855	163 (48 per lb.)	$9\frac{2}{16}$	10	G.S.	Tin cylinder.
11140 } 11696 }	Do.	II	L	12	7·915	2·73	2·855	Do.	$9\frac{2}{16}$	10	G.S.	Do.
16298	2·75-inch B.L.	I	L	12	9·65	2·73	2·845	253 (41 per lb.)	$11\frac{11}{16}$	$12\frac{1}{2}$	2-inch	Pellet burster.
10898	2·95-inch Q.F.	I	L	3	8·78	2·941	3·017	175 (41 per lb.)	$11\frac{1}{2}$	$12\frac{1}{2}$	G.S.	
11792	Do.	II	L	4	9·075	2·941	3·017	203 (41 per lb.)	$11\frac{1}{2}$	$12\frac{1}{2}$	G.S.	
	3-inch Q.F.	I	L	15	8·92	2·99	3·095	79 (41 per lb.)	$11\frac{2}{16}$	$12\frac{1}{2}$	2-inch	Day and night tracer.
	Do.	II	N	15	8·36	2·99	3·095	83 (41 per lb.)				
	Do.	III										
8233	12-pr., 12, 8, and 6 cwt....	II	C	4	8·349	2·98	3·085	156 (35 per lb.)	$11\frac{9}{16}$	$12\frac{1}{2}$	G.S.	Tin cylinder.
8502	Do. ...	III	C	4	8·349	2·98	3·09	Do.	$11\frac{9}{16}$	$12\frac{1}{2}$	G.S.	Do.
9957	Do. ...	IV	C	4	8·349	2·98	3·09	Do.	$11\frac{9}{16}$	$12\frac{1}{2}$	G.S.	Do.
10301	Do. ...	V	C	10	8·349	2·98	3·095	Do.	$11\frac{9}{16}$	$12\frac{1}{2}$	G.S.	Do.
11235	12-pr. Q.F. only ...	VI	N	10	8·349	2·98	3·095	182 (41 per lb.)	$11\frac{9}{16}$	$12\frac{1}{2}$	G.S.	Do.
11235	12-pr. B.L., 6 cwt. ...	VII	L	12	8·349	2·98	3·095	Do.	$11\frac{9}{16}$	$12\frac{1}{2}$	G.S.	Do.
12033	12-pr. Q.F. only ...	VIII	N	12	8·349	2·98	3·095	Do.	$11\frac{9}{16}$	$12\frac{1}{2}$	G.S.	
14476	12 and 14-pr.	IX	N	12	8·12	2·98	3·095	227 (41 per lb.)	$11\frac{3}{4}$	$12\frac{1}{2}$	2-inch	Pellet burster.

12775	13-pr. Q.F.	I	L	13	8·0	2·99	3·09	236 (41 per lb.)	11$\frac{3}{4}$	12$\frac{1}{2}$	2-inch	Head struck 1$\frac{1}{4}$ calibres.
13497	Do.	II	L	13	7·988	2·99	3·09	234 (41 per lb.)	11$\frac{3}{4}$	12$\frac{1}{2}$	2-inch	Head struck 2 calibres.
	Do.	III	L	14	7·988	2·99	3·09	Do.	11$\frac{3}{4}$	12$\frac{1}{2}$	2-inch	No cannelure.
8502	15-pr. B.L.	III	L	4	9·075	2·98	3·09	192 (35 per lb.)	13$\frac{2}{16}$	14	G.S.	Tin cylinder.
9957	Do.	IV	L	4	9·075	2·98	3·09	Do.	13$\frac{2}{16}$	14	G.S.	Do.
10301 12835 11235 12835	15-pr. B.L., B.L.C. or Q.F.	V	L	10	9·075	2·98	3·095	Do.	13$\frac{2}{16}$	14	G.S.	Do.
		VI	L	12	9·075	2·98	3·095	230 (41 per lb.)	13$\frac{2}{16}$	14	G.S.	Do.
	Do. do.	VII	L	12	8·94	2·98	3·095	264 (41 per lb.)	13$\frac{2}{16}$	14	2-inch	Pellets in central tube.
12774	18-pr. Q.F.	I	L	13	9·88	3·29	3·39	364 (41 per lb.)	17$\frac{3}{4}$	18$\frac{1}{2}$	2-inch	Head struck 1$\frac{1}{2}$ calibres.
13497	Do.	II	L	13	9·86	3·29	3·39	375 (41 per lb.)	17$\frac{3}{4}$	18$\frac{1}{2}$	2-inch	Head struck 2 calibres.
	Do.	III	L	14	9·86	3·29	3·39	Do.	17$\frac{3}{4}$	18$\frac{1}{2}$	2-inch	No cannelure.
5515	4-inch B.L. Q.F., or Q.F.C. Light (except for B.L., Marks VII to VIII*, and Q.F., Marks IV and V)	V	C	4	10·3	3·97	4·115	230 (22$\frac{1}{2}$ per lb.)	24$\frac{3}{4}$	25	G.S.	Head burster.
9957	Do. do.	VI	C	4	10·3	3·97	4·115	Do.	24$\frac{3}{4}$	25	G.S.	Do.
10543	Do. do.	VII	C	4	10·64	3·97	4·115	169 (22$\frac{1}{2}$ per lb.)	24	25	G.S.	Base burster.
14161 11527 11103	Do. do.	VIII	C	10	10·64	3·97	4·105	270 (35 per lb.)	24	25	G.S.	Do.

TABLE No. 19.—*Shell, B.L., B.L.C., Q.F. or Q.F.C. Shrapnel*—continued.

Para. in List of Changes	Nature of Gun.	Mark of Shell.	Service.	Type No. of Band	Length in Inches.	Diameter.		Number and Nature of Bullets.	Weight empty.	Weight filled.	Fuze Hole.	Remarks.
						Band or Body.	Driving Band.		Lbs.	Lbs.	Gauge.	
14161 15071 16445	4-inch B.L., or Q.F. Heavy (for B.L., Marks VII and VIII*, and Q.F., Mark IV)	I	N	11	11·85	3·97	4·23	498 (35 per lb.)	$30\frac{2}{16}$	31	2-inch	Pellet burster. (Saw-cuts in head, pellets in tube.)
16456	Do. do. (Also for Q.F., Mark V)	II	N	11	11·85	3·97	4·23	500 (35 per lb.)	$29\frac{3}{4}$	31	2-inch	Pellet burster. (Pellets in tube.)
	30-pr. B.L.	II	S I	4	12·45	3·97	4·115	313 (27 per lb.)	29	30	G.S.	Tin cylinder.
	Do.	III	S I	4	12·45	3·97	4·115	Do.	29	30	G.S.	Do.
	Do.	IV	S I	4	12·55	3·97	4·115	Do.	29	30	G.S.	Do.
9957	4·7-inch Q.F.	II	C	4	13·0	4·7	4·809	242 (14 per lb.)	$43\frac{14}{16}$	45	G.S.	
10546	Do.	III	C	4	13·6	4·7	4·814	205 (14 per lb.)	$43\frac{14}{16}$	45	G.S.	
12037	Do.	IV	C	10	12·96	4·7	4·814	580 (35 per lb.)	$43\frac{14}{16}$	45	G.S.	No primer.
15487	Do.	V	L	10	12·7	4·7	4·815	712 (35 per lb.)	$42\frac{14}{16}$	45	2-inch	Pellets in central tube.
15434	4·5-inch Howitzer, Q.F....	I	L	3	11·03	4·48	4·62	492 (35 per lb.)	$33\frac{10}{16}$	35	2-inch	Total weight includes fuze.

·	5-inch B.L. Howitzer "H"	V	L	4	13·1	4·97	5·115	623 (35 per lb.)	48 15/16	50	G.S.	Gun or howitzer.
12076	Do. do.	VII	L	10	13·15	4·97	5·115	625 (35 per lb.)	48 15/16	50	G.S.	
	5·4-inch Howitzer ...	II	S I	4	13·6	5·36	5·51	397 (16 per lb.)	58 6/16	60	G.S.	
	Do. ...	III	S I	4	13·6	5·36	5·51	Do.	58 6/16	60	G.S.	
13169	60-pr.	I	L	10	15·28	4·97	5·125	990 (35 per lb.)	58¼	60	2-inch	No primer.
5621	6-inch B.L. or Q.F. Gun and Howitzer	IV	C	4	18·925	5·96	6·115	536 (14 per lb.)	98 9/16	100½	G.S.	
6465	Do. do.	V	C	4	18·925	5·96	6·115	518 (14 per lb.)	98 9/16	100½	G.S.	
9272	Do. do.	VI	C	5	18·925	5·97	6·33	Do.	98 9/16	100½	G.S.	
9957	Do. do.	VII	C	5	18·925	5·97	6·33	Do.	98 9/16	100½	G.S.	
10097	Do. do.	VIII	C	5	18·582	5·97	6·33	453 (14 per lb.)	98 9/16	100½	G.S.	
11234	Do. do.	IX	C	5	18·582	5·97	6·33	Do.	98 9/16	100½	G.S.	
12267												
15030	6-inch B.L., Marks VII to VIII and XI to XI*	X	N	7	18·582	5·97	6·33	Do.	98 9/16	100½	G.S.	
	Do. do.	XI	N	7	18·15	5·97	6·33	445 (14 per lb.)	97 12/16	100	2-inch	
	Do. do.	XIA	N	7	19·21	5·97	6·33	441 (14 per lb.)	97 12/16	100	2-inch	
	6-inch B.L. or Q.F. Gun and Howitzer	XII	C	5	18·58	5·97	6·33	444 (14 per lb.)	98	100	G.S.	
	6-inch B.L., Marks VII to VIII and XI to XI*	XIII	N	7	19·21	5·97	6·33	441 (14 per lb.)	97 12/16	100	2-inch	
	Do. do.	XIIIA	N	7	19·21	5·97	6·33	439 (14 per lb.)	97 12/16	100	2-inch	
	6-inch B.L. or Q.F. Gun and Howitzer	XIV	L	5	17·59	5·97	6·33	1,317 (35 per lb.)	98 3/16	100	2-inch	
	6-inch B.L., Marks VII and XI to XII	XVAQ	N	7	18·15	5·97	6·33	874 (27 per lb.)	98 6/16	100	2-inch	

TABLE No. 19.—*Shell, B.L., B.L.C., Q.F. or Q.F.C., Shrapnel*—continued.

Para. in List of Changes	Nature of Gun.	Mark of Shell.	Service.	Type No. of Band	Length in Inches.	Diameter.		Number and Nature of Bullets.	Weight empty.	Weight filled.	Fuze Hole.	Remarks.
						Band or Body.	Driving Band.		Lbs.	Lbs.	Gauge.	
3089	6-inch Howitzer (Light)...	I	L	10	19·62	5·97	6·12	905 (27 per lb.)	$97\frac{14}{16}$	100	2-inch	A wood block round central tube.
12387, 14683	7·5-inch B.L., except Marks III to IV*	I	N	7	25·465	7·465	7·99	368 (2-oz. shot)	$197\frac{11}{16}$	200	G.S.	
14684	Do. do.	II	N	8	25·12	7·465	7·99	Do.	$197\frac{9}{16}$	200	2-inch	
15346	Do. do.	IIIA	N	8	25·34	7·465	7·99	407 (2-oz. shot)	$197\frac{9}{16}$	200	2-inch	
15083	7·5-inch B.L., Marks III to IV*	I	N	11	25·12	7·465	7·88	368 (2-oz. sand shot)	$197\frac{5}{16}$	200	2-inch	
8695	9·2-inch B.L.	V	L	4	32·7	9·15	9·7	638 (2-oz. shot)	377	380	G.S.	
9957	Do.	VI	L	4	32·7	9·15	9·7	Do.	377	380	G.S.	
111234	Do.	VII	L	6	32·97	9·165	9·7	Do.	377	380	G.S.	
	Do.	VIII	L	6	32·97	9·165	9·7	Do.	$376\frac{1}{2}$	380	G.S.	
14840	Do.	IX	N	8	32·97	9·165	9·7	630 (2-oz. shot)	$376\frac{1}{4}$	380	2-inch	
9271	10-inch B.L.	II	L	6	36·59	9·965	10·4	466 (4-oz. shot)	$497\frac{10}{16}$	500	G.S.	
11234	Do.	III	L	6	36·59	9·965	10·4	Do.	$497\frac{10}{16}$	500	G.S.	

5011	12-inch B.L. (Light) ...	III	L	4	42·0	11·95	12·13	1,120 (4-oz. shot)	$710\frac{4}{16}$	714	G.S.
6401	Do. ...	IV	L	4	40·5	11·97	12·13	1,000 (4-oz. shot)	$709\frac{11}{16}$	714	G.S.
8649	Do. ...	V	L	6	40·83	11·965	12·71	Do.	$709\frac{11}{16}$	714	G.S.
12119	Do. ...	VI	L	6	40·83	11·965	12·66	Do.	$709\frac{11}{16}$	714	G.S.
14877	12-inch B.L. (Heavy) ...	I	N	6	44·12	11·965	12·76	341 (12-oz. steel shot)	$844\frac{14}{16}$	850	2-inch
14878	Do. ...	IA	N	6	45·15	11·965	12·76	Do.	$844\frac{14}{16}$	850	2-inch
		IIA	N	6	42·28	11·965	12·76	7,766 (27 per lb.)	$844\frac{14}{16}$	850	2-inch
5349 5620	13·5-inch B.L. (Light) ...	I	N	4	49·0	13·45	13·65	1,348 (4-oz. shot)	1,244	1,250	G.S.
11234	Do. ...	II	N	6	49·0	13·45	14·0	Do.	1,244	1,250	G.S.
	Do. ...	IIA	N	6	47·88	13·46	14·25	10,280 (27 per lb.)	$1,242\frac{1}{2}$	1,250	2-inch
	13·5-inch B.L. (Heavy)...	IA	N	6	52·13	13·46	14·25	11,907 (27 per lb.)	$1,392\frac{1}{2}$	1,400	2-inch
	15-inch B.L.	IA	N	6	56·92	14·965	15·76	13,770 (27 per lb.)	1,912	1,950	2-inch

STAR SHELL.

Star shell are intended to illuminate an enemy's position; they should burst high in the air, the burst igniting and scattering the stars, which fall slowly to the ground, giving out a bright light.

Star shell are much lighter than other projectiles for the same gun, and are fired with special small charges, the use of which necessitates special fuzes.

They are all made to the same general design, and only differ from each other in dimensions and details.

They resemble shrapnel as regards the head and body; the body is of forged or cast steel; the head is of mild steel or iron, and is attached as in shrapnel, but brass instead of steel rivets are used as they offer less resistance to the bursting charge blowing off the head; the central tube is larger in diameter and is perforated.

The stars, which are cylinders of light-giving composition, occupy the same place in a star shell as bullets do in a shrapnel. Like the majority of shrapnel, star shell have a wrought-iron or mild-steel disc to protect the bursting charge from the set-back of the interior of the shell on discharge, but the following points may be noted :—

(1) The bursting charge is smaller than that of a shrapnel, and consequently the recess in the base is smaller.

(2) The R.F.G.[2] powder burster is contained in a flat shalloon bag, and is primed with quickmatch; no tin cup is used.

(3) To allow the small burster to open the shell, the head is not so securely attached as in shrapnel; it is not soldered to the body, but the joint is waterproofed with cement, and in the later marks a lead ring is inserted at the junction for the same purpose.

The stars are placed in the shell in two tiers, separated by an iron disc.

Arrangements have to be made—

(*a*) To prevent the stars from being crushed, which might occur on the shock of discharge, or when the shell bursts.

(*b*) To ensure that the flash from the fuze penetrates all over the interior to ignite the stars.

(*c*) To prevent the stars and the other internal parts of the shell from rotating independently of the body of the shell.

The object mentioned in (*a*) above is secured by placing supports in between the stars; wooden supports, triangular in section, are used, except in the shell for the 6-inch howitzer, which has corrugated steel supports.

With reference to (*b*) above, the socket, central tube, and central disc, are all perforated, and the stars are primed with quickmatch. In the 6-inch howitzer star shell, and the earlier marks of the B.L. 10-pr., a primer of gunpowder in a shalloon bag is fixed halfway down the central tube.

(*c*) To prevent the internal parts from rotating independently of the body, a pin at the bottom of the body engages with a slot in the steel disc ; further, pins (in the 6-inch howitzer shell) projecting from the steel discs in the centre and base of the shell engage with the corrugated steel supports. In the star shell for the 13 and 18-pr. Q.F., a further safeguard against the internal parts turning is provided by two projections on the central disc engaging with slots in the interior of the body.

Wood block.—The head is in all cases fitted with a wood block.

Socket.—The socket of all star shell is threaded to the G.S. gauge ; it is secured to the head by solder, and in some cases by being screwed in as well.

Felt washers.—To prevent injury to the stars, felt washers are placed under the disc, and at each end of the stars.

Fig. 44.

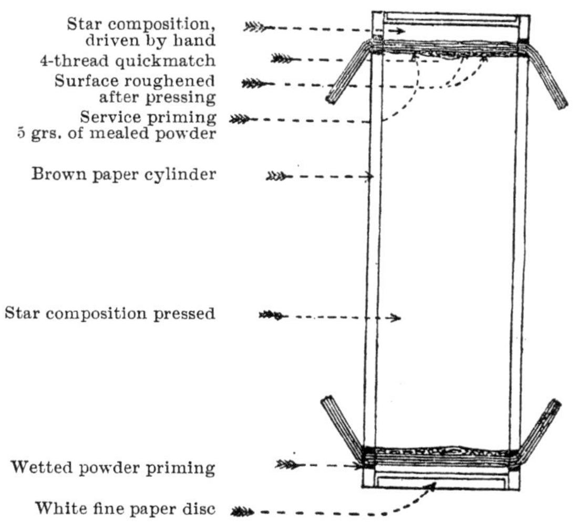

Star composition, driven by hand
4-thread quickmatch
Surface roughened after pressing
Service priming 5 grs. of mealed powder
Brown paper cylinder
Star composition pressed
Wetted powder priming
White fine paper disc

Interior to be dry.—Care is taken that all the internal parts of the shell are thoroughly dry before the shell is assembled.

Lining.—The interior of the body is lacquered or painted, and lined with brown paper.

Construction of the stars.—The stars consist of brown paper, or sheet brass and brown paper, rolled to form a cylinder ; the star composition is pressed in to within a short distance of each end ; a priming of saltpetre, sulphur, and mealed powder is added, then the quickmatch is threaded through the holes at each end ; finally, this is covered with star composition and the ends of the cylinder closed with paper discs. For star composition, *see* page 45.

Fuzes.—A fuze designed to "arm" on the shock of discharge of a full charge will not do so when fired with the small charge used with star shell, so a special time fuze (No. 25), which has a weak stirrup spring, is used. It is of the G.S. gauge, as all star shell have their fuze holes threaded to that gauge. For particulars of this fuze, *see* page 309.

Shell, B.L. Star, 10-pr., Mark II | L |.

(*Plate XXXI.*)

This shell illustrates the points referred to in the above general remarks on star shell, *e.g.*, the head is secured by four brass screws and four steel twisting pins; the socket, central tube, and central disc, are all perforated; the supports for the stars are pieces of wood, triangular in section; the disc on top of the burster is prevented from turning by a steel pin in the body; the burster is in a shalloon bag and is primed with six strands of quickmatch threaded through it; the socket, as in other star shell, is soldered to the head, but in this particular shell it is further secured by being screwed in; the head is lined with a wood block; the stars are in two tiers, separated by the centre disc, and felt washers are placed at the ends of each tier. In the central tube is a small primer of gunpowder in a shalloon bag; it rests on a bridge in the central tube, and is kept in position by two copper wires passing through holes in the tube.

This shell, though for a 10-pr. gun, only weighs about 7 lbs.

The Mark III star shell for the B.L. 10-pr. differs from the Mark II in having no primer in the central tube.

Shell, B.L. Star, 6-inch Howitzer, Mark I | L |.

(*Plate XXXII.*)

The special feature of this shell is the supports for the stars which are of corrugated steel, semi-circular in shape, two in each tier.

The body is made of steel, and has a recess in the base for the reception of a bursting charge of 10 drams of R.F.G.[2] powder in a shalloon bag threaded with quickmatch.

The head contains a wood block and is fitted with a perforated metal fuze-hole socket, screwed to the G.S. gauge.

The head is attached to the body by six brass screws and six steel twisting pins.

The fuze socket fits into the upper end of a wrought-iron central tube, which is in two parts, connected by a screwed gunmetal junction piece, containing a shalloon primer of 70 grains of R.F.G.[2] powder, which is kept in position by a piece of copper wire, passed through a central tube.

The tube is pierced with fire holes, and is screwed into a disc of wrought iron covering the bursting charge.

The shell contains twelve stars in two tiers, six in a tier.

Each tier is supported by a corrugated steel support.

A perforated iron disc is placed between the two tiers.

The figure shows the interior arrangement of the shell with one of the upper corrugated steel supports removed, showing two of the stars in position; the pins projecting upwards from the bottom disc, and from both sides of the central disc, lock the corrugated steel supports and prevent them from turning.

The shell weighs about 58 lbs.

Fig. 45.

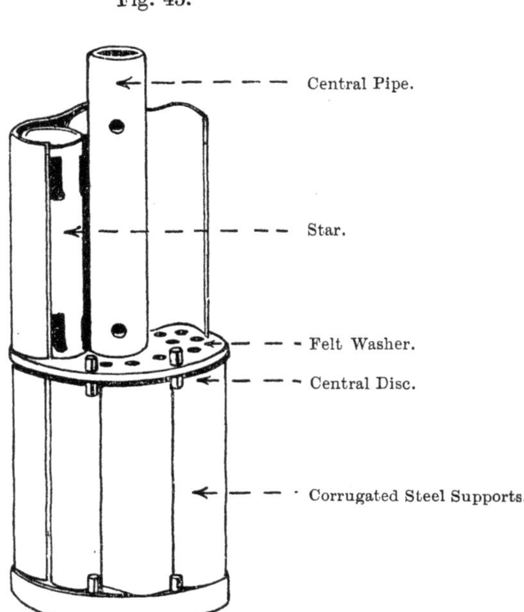

STAR SHELL FOR B.L. 5-INCH AND 5·4-INCH HOWITZERS.

The star shell for the B.L. 5·4-inch and 5-inch howitzers differ from the above in having wood wedges instead of corrugated steel supports, and the central tube is in one piece and has no powder primer.

The 5-inch has 8 stars, the 5·4-inch 10 stars.

SHELL, Q.F. STAR, 13-PR., MARK II | L |.

The shell is made of steel recessed in the base to receive a bursting charge of 3¼ drams of R.F.G.[2] powder contained in a shalloon bag primed with quickmatch. A metal central pipe perforated with 12 holes is screwed into a flat covered wrought-iron disc resting over the bursting charge.

The interior of the shell is velvrilled and lined with brown paper; it contains 10 stars in two tiers of five; a perforated iron disc covered with felt separates the tiers. The iron disc is supported by wood wedges placed between the stars, and is prevented from turning by means of two projections or feathers fitting into two featherways cut down the inside of the shell.

The head is lined with wood, and is attached to the body by four brass screws and four steel twisting pins. A felt wad is placed between the wood block and the top tier of stars. A gunmetal fuze-hole bush is fitted to the head, threaded internally to the G.S. gauge to take the No. 25 time fuze.

Star Shell for Q.F., 18-pr.

The star shell for the Q.F., 18-pr., is similar to the Q.F., 13-pr. star shell described above, differing only in weight and dimensions. (*See* Table 20.)

Star Shell for Q.F., 2·95-inch.

Star shell.—The Mark II star shell is made of forged steel, and is similar to the 13-pr., Q.F. star shell, but has a bursting charge of 3 drams of R.F.G.² powder and 5 per cent. oxide of iron, and is fitted with the plain type of driving band.

The Mark I star shell differed from Mark II in having a gun-metal head attached by brass screws and steel twisting pins; the tiers of stars were not separated by a perforated iron disc, and the bursting charge was 1 dram of R.F.G.² powder.

TABLE No. 20.—*Shell, B.L. and Q.F. Star.*

Para. in List of Changes.	Nature of Gun or Howitzer.	Mark of Shell.	Type of Band.	Length in Inches.	Diameter.		Number of Stars.	Burster R.F.G.[2]	Total Weight.	Remarks.
					Band or Body.	Driving Band.		Drams.	Lbs.	
11025	10-pr. B.L.	I	10	8·318	2·73	2·815	10	1	6½	Wood wedges; 15 grain primer in central tube.
11278, 12804	Do.	II	12	8·318	2·73	2·845	10	3	6½	(as above)
13812	Do.	III	12	8·318	2·73	2·845	10	3¼	6½	Wood wedges; no primer in central tube.
13811	13-pr. Q.F.	I	13	8·44	2·99	3·09	10	3¼	7 10/16	Wood wedges.
	Do.	II	14	8·44	2·99	3·09	10	3¼	7 10/16	No cannelure around shell.
13811	18-pr. Q.F.	I	13	10·32	3·28	3·39	10	3	10¼	Wood wedges.
	Do.	II	14	10·32	3·28	3·39	10	3	10¼	No cannelure around shell.
13350	15-pr. B.L., B.L.C., or Q.F.	I	12	9·075	2·98	3·095	10	3	9 6/16	Wood wedges.
10898, 11979	2·95-inch Q.F. ...	I	4	8·78	2·94	3·017	10	1	8¼	Do
13384	Do. ...	II	4	8·78	2·94	3·017	10	3	8¼	Do.
11318	5-inch Howitzer ...	II	4	14·87	4·95	5·115	8	2	30	Do.
	Do. ...	III	10	14·87	4·95	5·115	8	2	30	Do.
	5·4-inch Howitzer (India)	II	4	16·33	5·35	5·115	10	2	39½	Do.
10410, 11071	6-inch Howitzer ...	I	4	17·69	5·97	6·12	12	10	58¼	With corrugated steel supports in lieu of wood wedges; 70-grain primer in central tube.

NOTE.—Star Composition: Nitrate of Baryta, 108 parts; Nitrate of Potash, 72 parts; Magnesium Powder, 96 parts.

SECTION (D). LYDDITE AND H.E. SHELL.

General remarks on Common Lyddite; *Methods of Filling and Marking; Exploders Picric Powder and Trotyl; Description of types of Filled Lyddite; A.P. Lyddite and H.E. Shell.*

Lyddite shell are intended to detonate.

If detonation takes place the shell is torn into a large number of comparatively small fragments.

These fragments are projected over a comparatively large area in all directions and do not all go forward with the remaining velocity of the shell, as do the fragments of a powder-filled shell.

The fragments would be very effective against personnel, but as they have little energy, they are not very effective against material except in the vicinity of the detonation, where an intense shattering effect is obtained.

There is no incendiary effect, and the effect of the fumes is small.

A lyddite shell is vastly superior to a gunpowder-filled shell for the attack of earthworks, casements, &c., being about four times as powerful.

To obtain the maximum effect, lyddite must be detonated well inside, and not on the surface of the target, hence a delay action is desirable.

When used against " Heavily armoured ships," or against " Earthworks," a fuze giving a short delay is best.

Against " Torpedo craft," instantaneous action is essential, otherwise the shell will have passed through before detonating.

Detonation is indicated by the bursting shell spreading its fragments over a large area, giving an all-round effect, and by the smoke being black to grey, or even nearly white.

The latter appearance is due to the steam produced, which shows up more clearly under certain atmospheric conditions.

Yellow smoke denotes simple explosion, and the effect is not so great as regards the spread of fragments.

The proportion of yellow smoke to that of black, grey or white, may therefore be taken as a guide to the nature of the explosion.

Common Lyddite.

The premature explosion of a common lyddite shell in the gun would be so disastrous that, in designing these shell, every precaution is taken to prevent such an accident, hence :—

(1) They are made of forged steel.

(2) As these shell take a large bursting charge, it is not considered desirable to have any opening in the base ; they are therefore forged with solid bases to prevent any chance of gas getting into the shell.

(3) To prevent gas getting through a possible flaw, a recess is turned in the base, so that the interior metal may be examined for piping. This recess is filled in with a plate steel disc screwed in and riveted. Flaws are most likely to exist in the centre of a forging.

For future manufacture this steel plate will be of larger diameter more effectively to cover the base of the shell.

Early issues had no steel plate, and this type will still be met with.

(4) As picric acid easily forms dangerous picrates with metallic bases, the interior of the shell is coated with copal varnish to prevent the formation of picrate of iron; the exterior, painted with a special yellow paint (containing no lead) to prevent formation of picrates; the fuze-hole bush, fuze-hole plug, &c., are made of a leadless alloy.

Steel base plates.—The introduction of the steel base plate advanced the numeral of lyddite shell for all natures.

Earlier marks of shell, when passing through Ordnance Factories, will be fitted with a steel base plate, and will have a star added to their numeral.

The introduction of the large steel base plate again advanced the numeral of the shell.

Coned walls.—Lyddite shell first issued had "parallel walls" (*i.e.*, walls of uniform thickness). The walls in nearly all natures are now tapered, being thicker at the base and thinner near the shoulder. This design of shell is stronger, and has a greater capacity.

For example :—

		Lyddite.
6-inch, Mark VI shell (parallel walls)	..	10 lb. 4 oz.
6-inch, Mark VII shell (tapered walls)	..	13 lb. 6½ oz.
Increase	..	3 lb. 2½ oz.

Tracer shell.—The base of common lyddite fitted for "Internal night tracer," is thicker, and has a boss formed in the interior as shown in Plate XXXVI.

Method of Filling.

There are three methods :—

(1) "*Long central cavity filling.*"—A long central cavity is left in the lyddite for the long exploder of "Picric powder."

The original method for all natures up to B.L. 10-inch is still retained for B.L. 60-pr., 5-inch, 5·4-inch and Q.F. 4·5-inch howitzers.

(2) *Solid filling.*—There is no central cavity, but a small place is left on the top of the lyddite in the nose of the shell.

The first issues, B.L. 6-inch and up, had this space filled with "7-dram exploders" of picric powder. Later issues with "Exploder pellets"; 12-pr. to 4·7-inch have the space in the nose filled with exploders of trotyl.

(Solid filling with exploder pellets is the latest method of filling for B.L. 2·75-inch and 9·2-inch and up.)

(3) *Solid filling with short cavity in the top of lyddite.*—Q.F. 3-pr.

Notes on the Methods of Filling Lyddite Common.

The first lyddite shells were filled up to the bottom of the fuze-hole bush and had a long central cavity; the exploder consisted of granulated picric powder in a shalloon bag.

Various changes have been made and details introduced; they took place in the following order :—

(1) A waterproof-paper cylinder was introduced to take the above-mentioned shalloon bag. Shell fitted with exploders enclosed in this waterproof cylinder are marked with a rectangle stencilled on the side of the shell. The cylinder had a paper cap.

(2) Dry mixed picric powder was next introduced; this was indicated by the letters **D.M.** stencilled in the rectangle.

D.M.

(3) An aluminium cap was introduced instead of the paper cap; the letter **A** is marked above the rectangle. The aluminium cap is perforated with a number of small holes; these are closed by a disc of paper shellaced to the inside of the cap.

A
D.M.

(4) An asbestos paper tube, closed at the bottom, was introduced; it is placed into the lyddite before it solidifies, and the lyddite sets round it and sticks to it. At the same time as the above asbestos tube was adopted, the method of filling was changed as follows :—

The shell is not filled right up to the fuze-hole, as had been the procedure up to that time; a space is thus left beneath the fuze-hole, which space is left empty. To prevent lyddite dust getting into the fuze-hole threads, the surface of the lyddite is covered with a thin layer ($\frac{1}{10}$-inch thick) of "composition beeswax," originally known as "kit composition."

A
D.M.
K.C.

The asbestos tube prevents friction between the metal "former" (the tool used to form the cavity) and the lyddite; the tube being left in the shell, prevents the walls of the cavity from breaking away.

The **K.C.** stencilled below the rectangle indicates both asbestos tube and "composition beeswax."

(5) "Solid filling" was introduced in November, 1904, for shell 6-inch and upwards.

In this method the space above the lyddite is filled with 7-dram exploders; these consist of dry mixed picric powder in small seamless bags; no "composition beeswax" is used.

(6) The shalloon bag in the long exploder was done away with and a batiste bag introduced; such exploders are known as Mark II.

Shell that were first issued with Mark II exploders were stencilled on the body with a rectangle and the numeral "II" inside the rectangle, the markings "A" and K.C. "D.M." being omitted.

(7) Shell that were originally filled up to the bottom of the fuze-hole bush with lyddite, when refitted with a Mark II exploder, had the letters "K.C." omitted, so as to distinguish them from shell with an asbestos tube and "composition beeswax."

The letters "K.C." were also omitted on lyddite shell for B.L. 60-pr. and Q.F. 4·5-inch howitzer, as all shell for the above natures have "composition beeswax" on top of the lyddite.

(8) Solid filling was introduced in November, 1909, for B.L. and Q.F. guns, 12-pr. to 4·7-inch, and the shell fitted with small exploder bags of batiste containing *tri-nitro-toluene* (or trotyl) instead of picric powder.

Shell so filled are marked as follows :—

"EXPLODERS, T.N.T."

← T.N.T. in bags.

(9) Existing shell for B.L. and Q.F. guns, 12-pr. to 4·7-inch, are converted as follows :—

The picric powder exploder is removed, the cylindrical cavity filled up with compressed pellets of T.N.T. and the cavity in the nose of the shell filled up with exploders of T.N.T. in bags.

Shell so converted are marked as follows :—

EXPLDRS T.N.T.

$5\frac{3}{4}$ *OZ* & ← Weight of pellets.

← T.N.T. in bags.

(10) B.L., Q.F. or Q.F.C. 4-inch (light) and Q.F. 4·7-inch shell originally filled up to bottom of fuze-hole bush and fitted with a long exploder, when converted, will have the cavity filled up with pellets of trotyl only, there being no space for T.N.T. in bags in the nose.

(11) Exploder pellets introduced for solid-filled shell, 6-inch and up, instead of 7-dram exploders. (Introduced 30–5–10.)

Compressed pellets of picric powder are inserted into a batiste bag resting on the top of the lyddite. One or more 7-dram exploders are used to fill up the space in the nose of the shell after the bag has been choked.

Shell so filled have stencilled on the body :—

"EXPLODER PELLETS."

(12) *New marking on lyddite shell having central cavity.*—The rectangle denoting waterproof paper cylinder is now dispensed with, and all shell when filled on the above method are stencilled as under :—

Front.		Reverse.	
W.	= Monogram of filling station.		
27.2.12	= Date of filling.		
$4\frac{1}{4}$ oz.	= Weight of exploder.	N.	= If Naval Service.
Expldr. = II =	Mark of exploder. If Mark I exploder is used the letters D.M. will be appended to numeral if the picric powder is dry mixed.	K.C. =	If an asbestos tube and "composition beeswax" are present.
Lot 24	= Lot No. of picric acid.		

(13) *Method of filling for shell taking No.* 19A *D.A. Impact fuzes, with composition exploding* (*i.e.*, 3-pr.). In this method the shell is filled to within a short distance of the fuze-hole bush.

A short asbestos paper cylinder is then inserted by means of a "former" into the lyddite before its solidifies; this cylinder takes an exploder of T.N.T. in a batiste bag.

The space in the nose of the shell is filled with "composition beeswax."

Shell so filled are marked as follows :—

"EXPLODER, T.N.T. (or TROTYL)."

| 3 DRS. |

EXPLODERS FOR LYDDITE SHELL.

Long cylindrical exploder.—The latest Mark is Mark III ; it consists of a cylindrical bag of vulcanized cashmere filled with dry mixed picric powder; the mouth is choked with silk sewing.

It is inserted, choke-end down, into a waterproof paper cylinder, closed with a perforated aluminium cap secured with shellac. The paper cylinder is fitted with a lifting becket of silk sewing.

The Mark II exploder was made of "batiste" (a waterproof cotton fabric) which had a tendency to rot.

The Mark I was made of shalloon ; the first issues were filled with granulated picric powder.

On the left of fig. 46 is shown the batiste, or vulcanized cashmere bag filled with picric powder ; in the centre is the exploder complete ; on the right is the asbestos paper tube (open at the top) which is inserted into the lyddite before it has solidified, and into which the exploder is placed.

Stencilled on the bag, and also on the waterproof paper cylinder, are particulars of the mark, weight and length of the exploder, also the lot No. of P.P. used.

Long exploders fitted with powder primer.—With shell above 5-inch, issued prior to November, 1904, filled with a central cavity, a primer of 8 drams black gunpowder (R.F.G.[2]) is included in the exploder

Fig. 46.

Scale $\frac{1}{3}$.

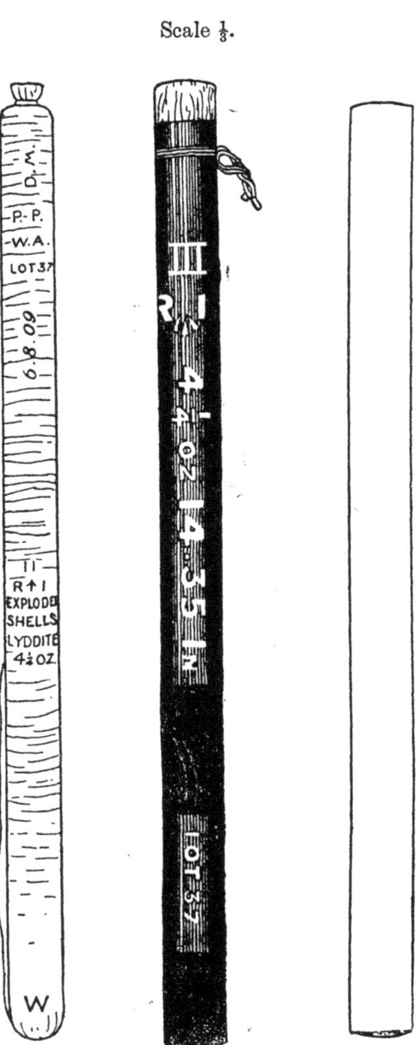

on top of the picric powder, and in a separate bag of shalloon (*see* Fig. 47); its presence is indicated by a black disc " ● " stencilled on the opposite side of the shell to the rectangle, and also on the waterproof paper cylinder.

Exploders for central cavity.—The lengths of these varies of course with the length of the shell in which they are used.

Fig. 47.

EXPLODER COMPLETE. EXPLODER WITH CAP REMOVED.

Scale $\frac{1}{3}$.

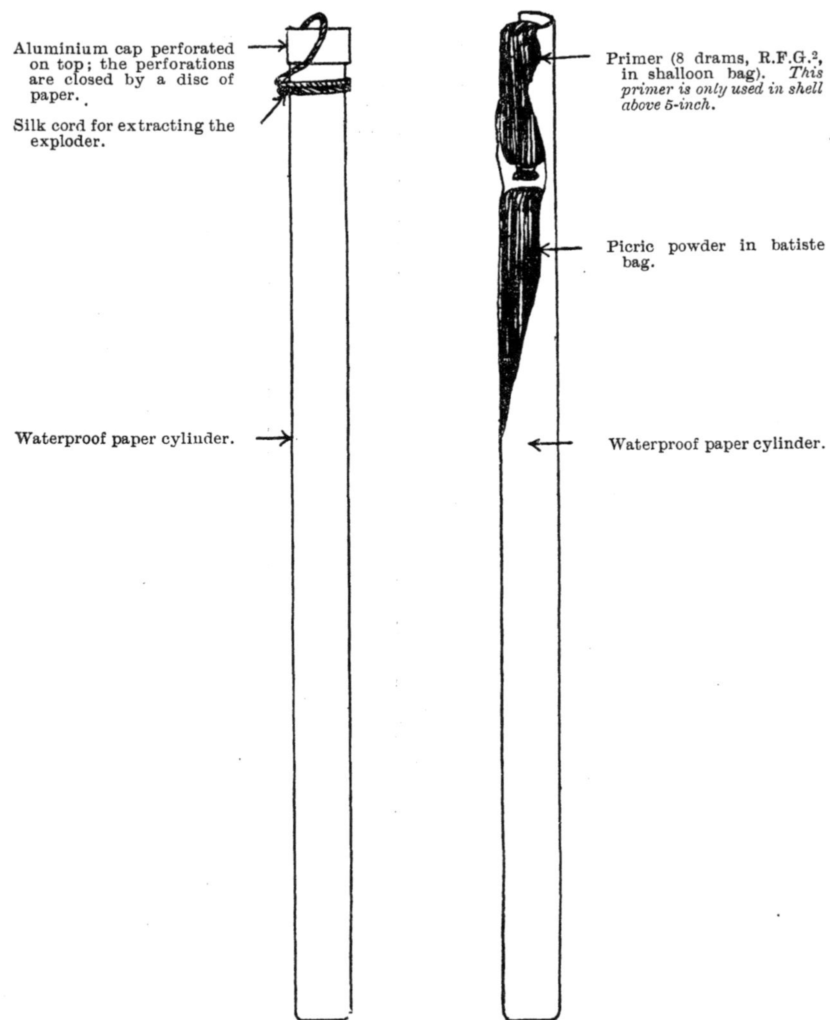

NOTE.—The 8-dram primer is not used with shell 5-inch and below.

The following table gives the various long exploders in the Service :—

TABLE No. 21.

Exploders, Lyddite Shell.	—	In Vulcanized Cashmere Bag and Paper Cylinder.
5 ozs., with primer, 17·5-inch. (Mark III)	C	B.L. or Q.F. 6-inch shells, formerly taking 5¾ ozs. exploder.
4¼ ozs., with primer, 16·1-inch. (Mark III)	C	B.L. howitzers, 5·4-inch and above ; B.L. or Q.F. gun shells, 6-inch and above ; and R.M.L. 10-inch (except shells formerly taking 5¾ ozs. or 4 ozs. exploders).
4¼ ozs., without primer, 14·35-inch. (Mark III)	L	B.L. 5-inch howitzer, and 60-pr. Marks II to III shells.
4 ozs., 13·35-inch. (Mark III)	C	Q.F. 4·7-inch, Marks I to II** shells.
3¾ ozs., 12·5-inch. (Mark III)	C	B.L. or B.L.C. 5-inch gun ; and Q.F. 4·7-inch, Marks III to IV* shells.
3½ ozs., 11·4-inch. (Mark III)	C	B.L. 9·2-inch ; B.L. or Q.F. 6-inch ; Q.F. 4·7-inch, Mark I ; B.L. 6-inch howitzer, 5-inch howitzer, and R.M.L. 10-inch shells formerly taking 4 ozs. exploder.
3¼ ozs., 10·75-inch. (Mark III)	L	B.L. 60-pr., Marks I and I*, and 30-pr. shells.
3 ozs., 9-inch. (Mark III)...	N	B.L., Q.F., or Q.F.C. 4-inch shell.
2 ozs., 6·25-inch. (Mark II)	N	Q.F. 12 and 14-pr. shell. (In batiste bag.)

7-DRAM EXPLODERS.

Fig. 48.

Scale ½.

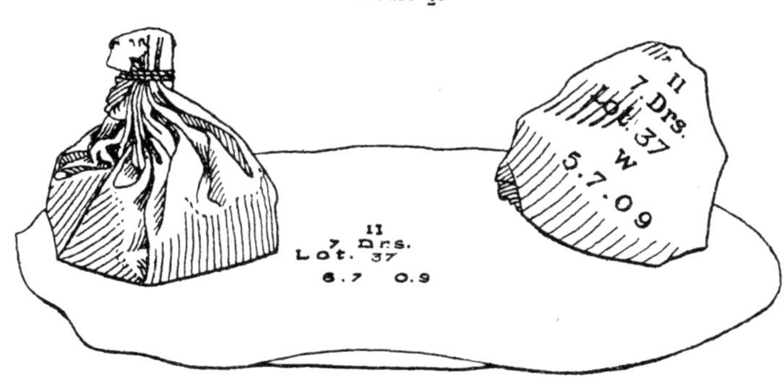

7-DRAM EXPLODERS.

The above figure shows " 7-dram exploders," Mark III, one empty and two filled.

The exploder is merely a small seamless bag of vulcanized cashmere filled with dry mixed picric powder and tied with silk sewing. The bag is made seamless because picric powder is very fine, and if there were any seams in the bag, it would work out and so get into the threads of the fuze-hole.

The Mark II, 7-dram exploder, was made of batiste. The Mark I exploders were made of oiled silk, which were apt to stick together and get torn when removed for inspection.

Bags, Pellet Exploders, Lyddite Shell | C |.

No. 1. *Mark II | C |*.—Vulcanized cashmere; for shell with exploders of picric powder pellets, except B.L. 9·2-inch (heavy), Marks IV to V* and 2·75-inch.

No. 2. *Mark II | C |*.—Vulcanized cashmere; for shell with exploders of picric powder pellets, B.L. 9·2-inch (heavy), Marks IV to V* and all shell formerly fitted with 7-dram exploders.

No. 3. *Mark II | L |*.—Vulcanized cashmere; for shell with exploders of picric powder pellets, B.L. 2·75-inch.

The Mark I bags differ from the Mark II, in being made of batiste.

Exploders for A.P. Lyddite Shell.

(See *Page* 210.)

Exploder bag, 4-*oz., Mark II*.—The exploder bag is made of vulcanized cashmere and contains 4 ozs. of D.M. picric powder; the mouth of the bag is choked with silk sewing.

It is fitted with a silk sewing loop.

The Mark I was made of batiste.

Table No. 22.

Exploders, T.N.T. (or Trotyl).

5¾ ozs. T.N.T. (Mark I) ...	L	Q.F. 4·7-inch, Marks II and II* shell, formerly taking 4 ozs. picric powder exploder.
5½ ozs. T.N.T. (Mark I) ...	L	Q.F. 4·7-inch, Marks III to IV shell, formerly taking 3¾ ozs. picric powder exploder.
3¾ ozs. T.N.T. (Mark I) ...	N	B.L., Q.F. or Q.F.C. 4-inch shell, formerly taking 3 ozs. picric powder exploder.
2½ ozs. T.N.T. (Mark I) ...	N	Q.F. 12 and 14-pr. shell, formerly taking 2 ozs. picric powder exploder.
3 drams T.N.T. (Mark I)...	C	In waterproof bag; Q.F. 4·7-inch, B.L., Q.F. or Q.F.C. 4-inch, and Q.F. 12 and 14-pr., and Q.F. 3-pr.
Bag, T.N.T. ... ... (Various weights.)	C	In waterproof bag; Q.F. 4·7-inch, B.L., Q.F. or Q.F.C. 4-inch, Q.F. 12 and 14-pr., and Q.F. 3-pr.

The exploders are made up, in the case of the first four named, in the form of compressed pellets of T.N.T., enclosed in white fine paper wrappers, and fitted with a lifting becket of silk sewing.

The 3-drams T.N.T. is made up similarly to the 7-dram picric powder exploder (*see* Fig. 48), but contains T.N.T. in crystalline form.

The " Exploders, bag, T.N.T." are made up in different sizes to suit the various natures of shell, and consist of T.N.T. in crystalline form, enclosed in small bags of vulcanized cashmere or waterproofed batiste.

Shells filled prior to 6/03 will not require the two last-mentioned exploders when converted to T.N.T. filling as described on page 201, as all these shells are filled up to the bottom of the fuze-hole bush.

For special fuze-hole plugs used with lyddite shell, *see* page 254.

Packages for Exploders, Lyddite Shell.

Up to 16·1 inches in length they can be packed in " *Box, exploders, lyddite.*"

Above 16·1 inches in length they can be packed in " *Case, powder, metal-lined, whole* " (*see* page 118), or " *Cylinder No.* 6."

Box, Exploders, Lyddite.

Box, exploders, lyddite, Mark I | *C* | .—The box is of deal with elm ends, painted yellow ; the lid is secured by brass screws working in nuts let into the top of the box. It is fitted with a tinned copper lining which has a luting groove formed round the top. Into this the flange of a tinned copper lid is pressed so as to make an air-tight joint ; cleats with handles are attached to each end of the box. The box is provided with packing pieces of varnished wood for use with different size exploders.

Mark II differs in dimensions and the nuts for screws securing lid ; a felt wad over the exploders is also used.

Cylinder No. 6, *Mark I* | *C* | .

The cylinder is made of tin and is issued for the use of Inspecting Ordnance Officers. The cylinder, which is painted yellow, is provided with a lid having a groove on the inside, which is filled with luting and secured by a tape band and shellac ; it is intended for the carriage of " exploders " when a small number is required for replacing those taken for moisture test. At forts, &c., where no laboratory exists, the cylinder may be used for packing the exploders removed from shell for moisture test.

For transport purposes, three of these cylinders are packed in a " Cylinder, ammunition, half-barrel."

Types of Filled Lyddite Shells, Description of.

(Type shown on Plate XXXIII.)

Shell, B.L. 60-pr. Common Lyddite, Mark III | L | .

The shell is made of F.S. nearly 4 calibres long ; the head is struck with a radius of 2 calibres, the point being truncated and fitted with a gunmetal bush which is tapped to the G.S. fuze-hole gauge. A steel plate is screwed into the centre of the base and the shell is rotated by the plain type of driving band.

The interior of the shell is varnished with copal varnish and is filled nearly to the top with melted picric acid, a long central cavity being left in the lyddite for the exploder; this cavity is lined with an asbestos paper tube.

The top surface of the lyddite is covered with a thin layer of "composition beeswax" $\frac{1}{10}$-inch thick, as shown in the plate.

The exploder is described on page 202; before it is inserted into the asbestos tube one or more glazed board discs may be inserted so as to adjust the depth of the cavity.

The Plug, fuze-hole, special, No. 1, Mark III, with a leather washer under the flange, is screwed into the fuze-hole bush and is prevented from unscrewing by the raised lip on the plug being stabbed into the bush of the shell in three places.

Type of Lyddite shown on Plate XXXIV.

Plate XXXIV. illustrates a "solid-filled" shell (without a steel base plate) fitted with "7-dram exploders."

B.L. and Q.F. 6-inch to 13·5-inch have been filled with "7-dram exploders."

In this type of filling no cylindrical cavity is left in the lyddite, but a slight space is left at the top. The surface of the lyddite is *not* covered with "composition beeswax."

The space in the nose of the shell is then filled with exploders, each consisting of 7 drams of picric powder (*see* page 205); the 6-inch generally taking 12 to 15, the 7·5-inch and up generally 8 to 10, but these numbers may be increased if necessary.

Type of Lyddite shown on Plate XXXV.

This is the latest method of filling for B.L. 2·75 and 9·2-inch and up.

B.L. and Q.F. 6- and 7·5-inch shell have also been filled on this system.

In this type the shell is filled with lyddite in the same manner as above, but the space in the nose is filled with a number of compressed pellets of picric powder (about 20 in the B.L. 2·75-inch, 60 in all other natures).

The pellets are inserted into a batiste or vulcanized cashmere bag resting on top of the lyddite, one or two "7-dram exploders" being used to fill up the slight space in the nose of the shell after the bag has been choked.

For description of "Bags, pellets, exploders," *see* page 206.

Shell, B.L. Common Lyddite, 6-inch Gun, Mark XII. A.Q.N.T. | N |.

This is a 4-calibre head shell, with tapered walls, large base-plate, and is prepared to take a "night tracer."

The letter "Q" after the numeral denotes that the shell is of stronger design, introduced for B.L. 6-inch, Mark XII gun.

Shell, Q.F. Common Lyddite, 12- and 14-pr., Mark IV | C |.

In Plate XXXVII the latest mark of shell for the Q.F. 12- and 14-pr. is shown. Owing to the increased thickness of the base for the tracer this shell takes a much smaller charge of lyddite than earlier marks.

Type of Lyddite shown on Plate XXXVIII.

Plate XXXVIII illustrates the method of filling Q.F. 3-pr. shell.

The shell is nearly filled with molten picric acid, which is then allowed to solidify. A small quantity of molten picric acid is then poured in and a short "former," carrying an asbestos cylinder, is inserted.

This forms a short cylindrical cavity and displaces sufficient acid to fill the shell nearly to the fuze-hole bush. After the acid has solidified the former is removed and the space in the nose filled with "composition beeswax."

The exploder, 3 drams of T.N.T. in a vulcanized cashmere bag, is inserted into the asbestos cylinder, choke down.

Fuze, D.A. Impact, No. 19A, or Plug, fuze-hole, special, No. 4A, is used with this shell.

Type of Lyddite shown on Plate XXXIX.

Plate XXXIX illustrates an earlier method of filling for Q.F. 3-pr., 12- and 14-pr., 4-inch B.L. or Q.F. (heavy); B.L., Q.F. or Q.F.C. 4-inch (light), and Q.F. 4·7-inch.

The above shell are filled on the *solid system*, the space in the nose of the shell being filled with exploders of T.N.T., instead of picric powder.

Conversion of shell filled with cylindrical cavity to take pellets and exploders of T.N.T.—The first issues of the above lyddite shell, *i.e.*, those filled with a cylindrical cavity, are being converted as follows :—The cylindrical exploder of picric powder is removed, the cavity inside the asbestos tube is then filled with pellets of compressed T.N.T., covered with three thicknesses of fine white paper, discs of fine cardboard being placed on top and underneath the pellets. The space in the nose of the shell above the kit composition is filled with bags of batiste or vulcanized cashmere containing T.N.T.

Armour-Piercing Shell. (Lyddite Filled.)

Armour-piercing shell filled with lyddite have been introduced for B.L. guns, 6-inch to 15-inch.

They may be either cast or forged.

They are fitted with the latest type of cap. The cavity is lined with a container of aluminium ; the base is closed by a large adapter, and is threaded to take a special gas-sealing arrangement (plate cover) to prevent the gas from the charge getting into the shell.

Shell, B.L. Armour-Piercing, with Cap, 9·2-inch, Mark VIIa | L |.

(With Aluminium Container.)

(Plate XL.)

The shell is generally similar to the Mark V shell described on page 169. The head is fitted with the latest type of cap, struck with a radius of nearly 4 calibres.

The interior of the shell is varnished and fitted with an aluminium container, also varnished internally.

The base is closed by a large adapter screwed in; this adapter is bored out and threaded to take the No. 16 large base fuze and screw-threaded on its outer edge to take the "Plate cover."

The shell is filled nearly to the base with lyddite, in which a short cavity is formed; this cavity is lined with an asbestos paper tube, fitted with an exploder bag, containing 4 ozs. of picric powder, inserted choke end first. The space between the lyddite and the adapter is filled with "composition beeswax."

After insertion of the fuze, the key slots in the adapter are filled in with steel plugs, and the key slots in the fuze are partially filled with luting and metal plugs.

A "Plate, gas-check, copper," is placed over the base of the fuze and adapter, and is held in position by the "Plate cover."

The plate cover is in two parts, *i.e.*, a perforated steel plate and a steel locking ring; the latter screws on to the base of the adapter and is prevented from unscrewing by a steel set-screw.

The shell is painted yellow with 2 white rings denoting A.P., and, when fuzed, the "Plate-cover" is painted red.

A.P. shell filled with lyddite are issued *fuzed* to both Land and Naval Services.

If these shell are plugged, the "Plug, base, shell, No. 6," is used.

Plates, Gas-check, Copper.

These are issued in three sizes, viz.:—

Large, for B.L. 9·2-inch.
Medium, for B.L. 12-inch to 15-inch.
Small, for B.L. 6-inch.

They are flat discs of pure copper, shaped to fit into the "Pressure, plate recess," and the partially filled key slots in the No. 16 base fuze.

With A.P. lyddite shell taking the medium and small "*Plate, gas-check, copper,*" a smaller "*Plate cover*" is used, and the locking screw is inserted vertically, instead of radially, in the shell.

High Explosive Shell.

Shell which take a bursting charge of "Composition exploding" are painted yellow in the same way as lyddite shell; they are distinguished by a red band round the centre, and are known as "High explosive shell."

This nature of projectile is used with the Q.F. 3-inch, anti-airship gun.

Shell, Q.F., High Explosive and Tracer, 3-inch, Mark II | C |.

The shell is made of forged steel, 3 calibres in length; the head struck with a radius of 2 calibres.

In the centre of the base is screwed a tracer socket with fixing screw for the night tracer.

The lower part of the body, which forms the day tracer, is filled with a black liquid (turpentine and aniline dye).

A small conical hole is bored through the base of the shell, and is fitted with a steel plug. (Plug, day tracer, vent.)

This plug is pierced with a central hole closed on the inside by a brass disc soldered on. On firing the pressure blows this disc in, so unmasking the hole.

The front of the shell is threaded internally for a "gunmetal container," a small groove being formed for a copper washer intended to seal the joint between the container and the body.

The front of the container is closed by a gunmetal fuze-socket secured by a steel locking screw.

The fuze-socket is bored out and threaded to the G.S. gauge to take the No. 44 D.A. fuze.

The shell is rotated by a special type of band, prepared so that the cartridge case can be attached to it by coning.

TABLE NO. 23.—*Shell, B.L., B.L.C., Q.F., or Q.F.C. Common Lyddite.*

Para. in List of Changes	Nature of Gun.	Mark of Shell.	Ser-vice.	Type No. of Band.	Length in Inches.	Diameter.		Bursting charge, including Exploders.		Nature and weight of Exploders.	Weight filled.	Remarks.
						Band or Body.	Driving Band.	Lbs.	ozs.		Lbs.	
15566	3-pr. Q.F.	I	N	4	6·59	1·842	1·902	0	$4\frac{14}{16}$	T.N.T., $\frac{1}{2}$ or $\frac{3}{4}$ dram and bags	$3\frac{5}{16}$	For No. 19 fuze.
16480	Do.	II	N	4	6·59	1·842	1·902	0	$4\frac{14}{16}$	T.N.T., 3 drams	$3\frac{5}{16}$	For No. 19A fuze.
16480	Do.	IIINT	N	4	*6·59	1·842	1·902	0	$4\frac{14}{16}$	Do.	$3\frac{1}{2}$	* Over tracer holder = 7·35 inches. Fitted for external night tracer. For No. 19A fuze.

	Do.	IV	N	4	6·59	1·842	1·902	0	$4\frac{14}{16}$	Do.	$3\frac{5}{16}$	Same as Mark II except } That these Marks are fitted with a larger base plate.
	Do.	V_{NT}	N	4	6·59	1·842	1·902	0	$4\frac{14}{16}$	Do.	$3\frac{1}{2}$	Same as Mark III except } That these Marks are fitted with a larger base plate.
16298	2·75-inch B.L. ...	I	L	12	11·53	2·73	2·845	1	$6\frac{1}{2}$	Exploder pellets	$12\frac{1}{2}$	
13652	12 and 14-pr. Q.F. ...	I	N	12	9·57	2·98	3·095	1	$1\frac{1}{2}$	2 ozs., long, picric powder	$12\frac{1}{2}$	
15443	Do. ...	II	N	12	9·57	2·98	3·095	1	$1\frac{1}{2}$	T.N.T., 3 drams and bags	$12\frac{1}{2}$	Base plate.
15450	Do. ...	III	C	12	9·57	2·98	3·095	1	5	Do.	$12\frac{1}{2}$	Base plate and tapered cavity.
	Do. ...	IV_{NT}	C	12	8·92	2·98	3·095	0	$12\frac{1}{2}$	T.N.T., 5 drams	$12\frac{1}{2}$	For internal night tracer.
	Do. ...	V	C	12	9·57	2·98	3·095	1	5	T.N.T., 3 drams and bags	$12\frac{1}{2}$	Same as Mark III, but has large base plate

TABLE No. 23.—*Shell, B.L., B.L.C., Q.F., or Q.F.C., Common Lyddite*—continued.

Para. in List of Changes	Nature of Gun.	Mark of Shell.	Service.	Type No. of Band.	Length in Inches.	Diameter.		Bursting charge, including Exploders.	Nature and Weight of Exploders.	Weight filled.	Remarks.
						Band or Body.	Driving Band.	Lbs. ozs.		Lbs.	
	30-pr. B.L.	I	S I	4	14·22	3·96	4·11	3 14¼	3¼ ozs., long, picric powder	30	
	Do.	II	S I	4	14·22	3·96	4·11	3 14¼	Do.	30	
	Do.	III	S I	10	14·22	3·96	4·1	3 14¼	Do.	30	
15443	Do.	IV	S I	10	14·22	3·96	4·1	3 14¼	Do.	30	Base plate.

8784	4-inch B.L., Q.F., or Q.F.C. Light (except for B.L., Marks VII to VIII* and Q.F. Marks IV and V guns)	I	N	3	12·3	3·97	4·115	3	$3\frac{1}{2}$	3 ozs., long, picric powder	25	
9957	Do.	II	N	10	12·3	3·97	4·105	3	$3\frac{1}{2}$	Do.	25	Groove for driving band undercut.
11193	Do.	III	C	10	12·45	3·97	4·105	3	$3\frac{1}{2}$	Do.	25	Waved ribs in groove for driving band.
15073	Do.	IV	C	10	12·25	3·97	4·105	3	$2\frac{2}{16}$	T.N.T., 3 drams and bags	25	Tapered cavity. Base plate.
16446	Do.	V_{NT}	C	10	11·93	3·97	4·105	3	0	Do.	25	For internal night tracer.
	Do.	VI	C	·10	12·25	3·97	4·105	3	$2\frac{2}{16}$	Do.	25	Large base plate. Otherwise same as Mark IV.
16445 14461	4-inch B.L. or Q.F. Heavy (for B.L., Marks VII to VIII*, and Q.F., Mark IV guns)	I	N	11	14·25	3·97	4·23	3	2	3 ozs., long, picric powder	31	
16445 15443	Do.	II	N	11	14·25	3·97	4·23	3	2	T.N.T., 3 drams and bags	31	Base plate.
16453	Do.	III	N	11	16·19	3·97	4·23	5	0	Do.	31	Tapered cavity.
16455	Do. (Also for Q.F., Mark V gun)	IV_{NT}	N	11	15·16	3·97	4·23	4	3	Do.	31	For internal night tracer.
	B.L. Marks VII to VIII*, and Q.F., Mark IV guns	V	N	11	16·19	3·97	4·23	5	0	Do.	31	Large base plate; otherwise the same as Mark III.
	B.L., Marks VII to VIII*, and Q.F., Marks IV and V guns	VI_{NT}	N	11	14·96	3·97	4·23	3	$15\frac{1}{2}$	Do.	31	For internal night tracer. Strengthened walls.

TABLE NO. 23.—*Shell, B.L., B.L.C., Q.F., or Q.F.C. Common Lyddite*—continued.

Para. in List of Changes	Nature of Gun.	Mark of Shell.	Service.	Type No. of Band.	Length in Inches.	Diameter.		Bursting charge, including Exploders.		Nature and Weight of Exploders.	Weight filled.	Remarks.
						Band or Body.	Driving Band.	Lbs.	ozs.			
15434	4·5-inch Q.F. Howitzer	I	L	4	15·12	4·46	4·62	6	2	$3\frac{1}{4}$ ozs. long, picric powder	35	Base plate.
	Do.	II	L	4	15·12	4·46	4·62	6	1	Do.	35	Thicker walls than Mark I.
	Do.	III	L	4	15·12	4·46	4·62	6	1	Do.	35	Large base plate.

8479	4·7-inch Q.F. ...	I	N	3	17·06	4·7	4·819	7	5	4 ozs., long, picric powder	$46\frac{9}{16}$	
9957	Do. ...	II	C	4	17·06	4·7	4·819	7	5	Do.	$46\frac{9}{16}$	Groove for driving band undercut.
11193	Do. ...	III	C	4	16·365	4·7	4·819	6	10	$3\frac{3}{4}$ ozs., long, picric powder	$46\frac{9}{16}$	Waved ribs in groove for driving bands. Thicker walls.
12036	Do. ...	IV	C	10	16·365	4·7	4·815	6	10	Do.	$46\frac{9}{16}$	Plain driving band introduced.
15449	Do. ...	V	C	10	16·36	4·7	4·815	6	$13\frac{6}{16}$	T.N.T., 3 drams and bags	45	Tapered cavity. Base plate.
16446	Do. ...	VI INT	C	10	15·81	4·7	4·815	6	$5\frac{12}{16}$	Do.	45	For internal night tracer.
	Do. ...	VII	C	10	16·36	4·7	4·815	6	$13\frac{6}{16}$	Do.	45	Large base plate; otherwise same as Mark V.
8463	5-inch B.L. Howitzer	I	L	2	15·00	4·96	5·11	4	14	$5\frac{1}{4}$ ozs., long, picric powder, without water-proof paper cylinder	$50\frac{10}{16}$	This shell was originally filled with gunpowder § 8231.
9347	Do.	II	L	2	18·225	4·97	5·115	9	15	Do.	50	Longer shell. Thinner walls.
9957	Do.	III	L	4	18·225	4·97	5·115	9	15	$4\frac{1}{4}$ ozs., long, picric powder	50	Groove for driving band undercut.
11318	Do.	IV	L	10	18·225	4·97	5·115	9	15	Do.	50	Plain driving band introduced.
	Do.	V	L	10	18·2	4·97	5·115	9	15	Do.	50	Base plate.

TABLE NO. 23.—*Shell, B.L., B.L.C., Q.F., or Q.F.C. Common Lyddite*—continued.

Para. in List of Changes	Nature of Gun.	Mark of Shell.	Service.	Type No. of Band.	Length in Inches.	Diameter.		Bursting charge, including Exploders.		Nature and Weight of Exploders.	Weight filled.	Remarks.
						Band or Body.	Driving Band.	Lbs.	ozs.		Lbs.	
	5·4-inch B.L. Howitzer	I	S I	4	19·44	5·36	5·515	12	10	$4\frac{1}{4}$ ozs., long, picric powder	60	
	Do.	II	S I	4	19·44	5·36	5·515	12	10	Do.	60	
15443	Do.	III	S I	4	19·44	5·36	5·515	12	10	Do.	60	Base plate.
13169	60-pr. B.L.	I	L	10	15·48	4·97	5·125	4	0	$3\frac{1}{4}$ ozs., long, picric powder	60	

13169	Do.	II	L	10	19·05	4·97	5·125	8	0	$4\frac{1}{4}$ ozs., long, picric powder	60	Longer shell. Thinner walls. Greater capacity for burster.
15443	Do.	III	L	10	19·05	4·97	5·125	8	0	Do.	60	Base plate.
8479	6-inch B.L. or Q.F....	I	N	5	21·81	5·97	6·33	13	12	$5\frac{3}{4}$ ozs., long, picric powder without water-proof paper cylinder	$102\frac{1}{4}$	
9957	Do.	II	C	5	21·81	5·97	6·33	13	12	$4\frac{1}{4}$ ozs., long, picric powder	$102\frac{1}{4}$	Groove for driving band undercut.
9960	Do.	III	C	5	20·57	5·97	6·33	10	6	Do.	$101\frac{2}{16}$	Shorter and lighter.
11234	Do.	IV	C	5	20·57	5·97	6·33	10	6	Do.	$101\frac{2}{16}$	Waved ribs in groove for driving band.
12267	Do.	V	N	7	20·57	5·97	6·33	10	6	7 drams, changing to exploder pellets	$101\frac{2}{16}$	Cupro - nickel driving band.
15443	Do.	VI	L	5	20·57	5·97	6·33	10	6	Do.	$101\frac{2}{16}$	Base plate.
15458	6-inch B.L., Marks VII, VIII, XI and XI* guns	VII	N	7	21·53	5·97	6·33	13	$6\frac{1}{2}$	Do.	100	Cupro - nickel driving band. Base plate. Tapered cavity.
15008	Do.	VIIA	N	7	22·89	5·97	6·33	13	5	Exploder pellets	100	Base plate. Tapered cavity.
15448	6-inch B.L. or Q.F. guns	VIII	L	5	21·53	5·97	6·33	13	$6\frac{1}{2}$	Do.	100.	Differs from Mark VII only in the driving band.
	6-inch B.L., Marks VII and XI to XII guns	IXAQNT	N	7	21·4	5·97	6·33	10	15	Do.	100	For internal night tracer.

TABLE No. 23.—*Shell, B.L., B.L.C., Q.F., or Q.F.C. Common Lyddite*—continued.

Para. in List of Changes	Nature of Gun.	Mark of Shell.	Service.	Type No. of Band.	Length in Inches.	Diameter.		Bursting charge, including Exploders.	Nature and weight of Exploders.	Weight filled.	Remarks.
						Band or Body.	Driving Band.	Lbs. ozs.		Lbs.	
	6-inch B.L., Marks VII, VIII, XI and XI* guns	X	N	7	21·53	5·97	6·33	13 6½	5 drams T.N.T.	100	Large base plate; otherwise the same as Mark VII.
	Do.	XA	N	7	22·89	5·97	6·33	13 5	Do.	100	Large base plate; otherwise the same as Mark VIIA.
	6-inch B.L. or Q.F. guns	XI	L	5	21·53	5·97	6·33	13 6½	Do.	100	Large base plate; otherwise the same as Mark VIII.
	6-inch B.L., Marks VII and XI to XII guns	XIIAQNT	N	7	21·4	5·97	6·33	10 15	Do.	100	For internal night tracer. Large base plate; otherwise the same as Mark IXAQNT.

12298	6-inch B.L. Howitzer (Light)	I	L	10	22·05	5·97	6·12	14	4	7 drams, changing to exploder pellets	100	
15443	Do.	II	L	10	22·05	5·97	6·12	14	4	Do.	100	Base plate.
9007	6-inch B.L. Howitzer (Heavy)	I	L	2	27·225	5·97	6·12	18	14	$4\frac{1}{4}$ ozs., long, picric powder	$122\frac{9}{16}$	
9957 } 12552 }	Do.	II	L	4	27·225	5·97	6·12	18	14	7 drams picric powder	$122\frac{9}{16}$	Groove for driving band undercut.
11234 } 12298 }	Do.	III	L	4	27·225	5·97	6·12	18	14	Do.	$122\frac{9}{16}$	Waved ribs in groove for driving band.
12925 } 12387 }	7·5-inch B.L., Marks I to II** and V	I	C	7	25·58	7·465	7·99	19	7	$4\frac{1}{4}$ ozs., long, picric powder	200	Cupro-nickel driving band.
14518	Do.	II	C	8	25·58	7·465	7·99	19	7	7 drams picric powder, changing to exploder pellets	200	Copper driving band.
15072	Do.	III	C	8	27·9	7·465	7·99	26	3	Do.	200	Tapered cavity. Base plate.
14879	Do.	IIIA	N	8	28·5	7·465	7·99	24	14	Do.	200	Do.
	Do.	IV	C	8	27·9	7·465	7·99	26	3	5 drams T.N.T.	200	Large base plate; otherwise the same as Mark III.
	Do.	IVA	N	8	28·5	7·465	7·99	24	14	Exploder pellets	200	Large base plate; otherwise the same as Mark IIIA.

TABLE NO. 23.—*Shell, B.L., B.L.C., Q.F., or Q.F.C. Common Lyddite*—continued.

Para. in List of Changes	Nature of Gun.	Mark of Shell.	Service.	Type No. of Band.	Length in Inches.	Diameter.		Bursting charge, including Exploders.		Nature and Weight of Exploders.	Weight filled.	Remarks.
						Band or Body.	Driving Band.	Lbs.	ozs.		Lbs.	
12855	7·5-inch B.L., Marks III to IV* guns	I	N	11	25·58	7·465	7·88	19	7	7 drams picric powder	200	
	Do.	II	N	11	26·98	7·465	7·88	23	5	5 drams T.N.T.	200	Tapered cavity. Base plate.
12507	9·2-inch B.L. (Light)	I	L	6	27·83	9·165	9·71	37	7	7 drams picric powder	290	
15443	Do.	II	L	6	27·83	9·165	9·71	37	7	Exploder pellets	290	Base plate.
9636 9957	9·2-inch B.L. (Heavy)	I	L	5	33·86	9·165	9·71	40	0	4¼ ozs., long, picric powder	380	Groove for driving band undercut.
10508	Do.	II	L	6	33·86	9·165	9·71	40	0	Do.	380	Fitted with "Broad Vavasseur with gas-check, modified" driving band.
11234	Do.	III	C	6	33·86	9·165	9·71	40	0	Do.	380	Waved ribs in groove for driving band.

12752	Do.	IV	C	6	30·24	9·165	9·71	25	10	7 drams picric powder, changing to exploder pellets	380	Shorter shell and stronger in design.
14703	Do.	V	C	8	30·24	9·165	9·71	25	10	Do.	380	Broad copper driving band.
15447	Do.	VI	C	8	32·11	9·165	9·71	33	4	Do.	380	Tapered cavity. Base plate.
16191	Do.	VIIA	L	8	33·7	9·165	9·71	33	0	Exploder pellets	380	Differs from Mark VI in length, capacity and radius of head, and has no base plate.
	Do.	VIIIA	L	8	33·7	9·165	9·71	33	0	Do.	380	Base plate.
	Do.	VIII	L	8	32·11	9·165	9·71	33	$10\frac{3}{16}$	Do.	380	Large base plate; otherwise the same as Mark VI. (For Canada only.)
12355 } 15443 }	9·45-inch B.L. Howitzer	I	L	1	28·646	9·415	9·605	53	2	Do.	280	Waved ribs in groove for driving band. Base plate.
10508 } 10510 }	10-inch B.L. (except Marks VI to VII guns)	I	L	6	36·55	9·965	10·61	46	0	$4\frac{1}{4}$ ozs., long, picric powder	500	

TABLE No. 23.—*Shell, B.L., B.L.C., Q.F., or Q.F.C. Common Lyddite*—continued.

Para. in List of Changes	Nature of Gun.	Mark of Shell.	Ser-vice.	Type No. of Band.	Length in Inches.	Diameter.		Bursting charge, including Exploders.		Nature and Weight of Exploders	Weight filled.	Remarks.
						Band or Body.	Driving Band.	Lbs.	ozs.		Lbs.	
11399 / 11459	10-inch B.L. (except Marks VI to VII guns)	II	C	6	36·55	9·965	10·4	46	0	$4\frac{1}{4}$ ozs. long, picric powder.	500	Maximum diameter of driving band reduced.
15443	Do.	III	C	6	36·55	9·965	10·4	46	0	Exploder pellets	500	Base plate.
15007	10-inch B.L., Marks VI to VII guns	I	N	11	37·5	9·962	10·52	52	1	Do.	500	Tapered cavity. Base plate.
14875	12-inch B.L. (Heavy)	I	N	6	46·5	11·965	12·76	113	4	7 drams picric powder, changing to exploder pellets	850	Do.
14876	Do.	IA	N	6	48·25	11·965	12·76	112	3	Do.	850	Differs from Mark I in radius of head, length and capacity.

15006	13·5-inch B.L.(Light)	I	N	6	48·8	13·465	14·26	108 12	Do.	1,250	Tapered cavity. Base plate.
15505	Do.	IA	N	6	57·42	13·465	14·26	170 11	Do.	1,250	Differs from Mark I in radius of head, length and capacity.
	Do.	IIA	N	6	58·58	13·465	14·26	183 13	Do.	1,250	Differs from Mark IA in length and capacity.
	Do.	II*A	N	6	58·58	13·465	14·26	181 3	Do.	1,259$\frac{13}{16}$	With adapter to R.L., Des. 20369c.
	13·5-inch B L.(Heavy)	IA	N	6	63·42	13·465	14·26	183 10	Exploder pellets	1,400	Base plate.
	Do.	I*A	N	6	63·42	13·465	14·26	181 9	Do.	1,407$\frac{13}{16}$	With adapter to R.L., Des. 20369c.
	15-inch B.L.... ...	IANT	N	6	67	14·965	15·76	223 12	Do.	1,920	Base plate. Fitted for internal night tracer.

SECTION ("E") SHOT.

Armour-Piercing Shot.

Armour-piercing shot are no longer manufactured; those which exist are being used up for practice.

They have a small cavity, and their weight can be adjusted by inserting dust shot and sawdust; the hole in the base is closed with a large steel plug, screwed and soldered in.

Armour-piercing shot have been issued fitted with caps, which were painted white to distinguish them from capped *shell.*

Paint.—For marking, *see* page 236.

Practice Shot.

Practice shot are used for practice over sea ranges; they are of solid cast iron, fitted with the service driving band for whatever nature of gun they are intended for, except 7·5-inch and 9·2-inch, *see* Type 9, Plate XVI. The 9·2-inch and 10-inch have a hole for a lifting eye-bolt.

The letter "P" denoting practice is stamped on the base.

The latest marks of practice shot are fitted with internal night tracers.

Paint.—For marking, *see* page 236.

Case Shot.

Case shot have been made for guns and howitzers of almost all calibres. (*See* Table 24.)

The principal exceptions are the Q.F., 13- and 18-pr., 4·7-inch and 4·5-inch howitzer; B.L., 2·75-inch, 60-pr. and 6-inch howitzer.

They are becoming obsolete, however, more especially in the case of those made for large and medium guns.

In the British Service they will be met with as follows :—

Case shot for B.L.C., 15-pr.	In the equipment for the Territorial Force.	
Q.F., 15-pr.	,, ,,	In movable armament.
B.L., 12-inch	,, ,,	Fort armaments.

In the Indian Service they are included in the equipment as follows :—

(1) B.L., 4-inch and 5-inch	Fort armaments.
(2) B.L., 12- and 15-pr. ,, 30-pr. ,, 5·4-inch howitzer	Field.
(3) B.L., 10-pr.	Mountain.

A case shot is essentially a close quarter projectile; case shot from field guns is effective up to about 300 yards.

Construction.—The shot consists of an envelope containing sand shot (cast iron) or "mixed metal" bullets, packed in clay and sand, and is of the same weight as a long-range projectile.

The envelope must be weak enough to break up at the muzzle of the gun and release the bullets, but it must also be strong enough not to set up on discharge, as this would cause it to take the rifling; should a case shot leave the bore with the spin of other projectiles, the dispersion of the bullets laterally would be very great and their range to the front very small.

As it is essential that case shot should not rotate, they have no driving band. Some case shot, such as that for the 15-pr. described below, have at the base a copper band, which is intended to act as a gas-sealer, and which takes the rifling like a driving band, but such projectiles are specially designed so that this band does not cause the projectiles themselves to rotate.

The following case shot are made on the same principle, illustrated on Plate XLI and explained below in the description of the shot for the 15- and 12-pr. :—

Mark V case shot for 15-pr. and 12-pr., B.L. and Q.F.
Mark I ,, ,, 10-pr. B.L.
Mark II ,, ,, 5-inch howitzer.
Mark I ,, ,, 5·4-inch howitzer.
Mark II ,, ,, 2·95-inch Q.F.

Note.—The bodies of case shot are not painted; they must be oiled to prevent rust.

Case Shot, 15- and 12-pr.

The *Shot, B.L., B.L.C., or Q.F. Case*, 15 *and* 12-*pr., Mark V*, is for use with the following guns :—

B.L., 12-pr. B.L. and B.L.C., 15-pr. Q.F., 15-pr.	Land Service.
Q.F., 12-pr.	Naval Service.

This shot has a body of tinned sheet iron, lap-jointed and soldered; the base is of forged steel and is secured to the body by the bottom of the latter being pressed into a recess for, and being held by, a copper band. The bottom of the recess is milled. The base is recessed and is fitted with a straight wrought-iron handle. A small hole is bored through the bottom of the recess and is closed on the inside by a thin wrought-iron disc soldered into a shallow recess. Inside the body are two steel segments, and in rear of them, resting loosely upon the base, is a wrought-iron or mild-steel disc. The body contains mixed metal balls (34 per lb.), the interstices being filled with clay and sand. The top is closed by a mild steel disc, over which the body is spun and lightly soldered.

Action.—On firing, the band seals the bore; the gas rushes through the hole in the base, forcing out the soldered disc and, acting upon the loose disc and the bullets, breaks the body away from the base, thus ensuring the release of the bullets before the shot leaves the bore of the gun.

Case shot, B.L., 30-pr. and B.L., 4-inch.—The **Mark I** case shot for the B.L., 30-pr., is similar to the Mark I shot for the B.L., 4-inch (*see* Fig. 49).

Shot, B.L. Case, 4-inch.

The body is of tin, in three longitudinal pieces, lap-jointed and soldered; the bottom is of tin, fastened in the same way, and is partly covered by a copper ring of the form shown, which is attached by iron rivets. Six studs of soft metal are fixed round the exterior in front of the copper ring to act as stops in loading. The top is of sheet iron, tinned, fitted with one handle attached by

Fig. 49.

SHOT, B.L. CASE, 4-INCH, MARK I.

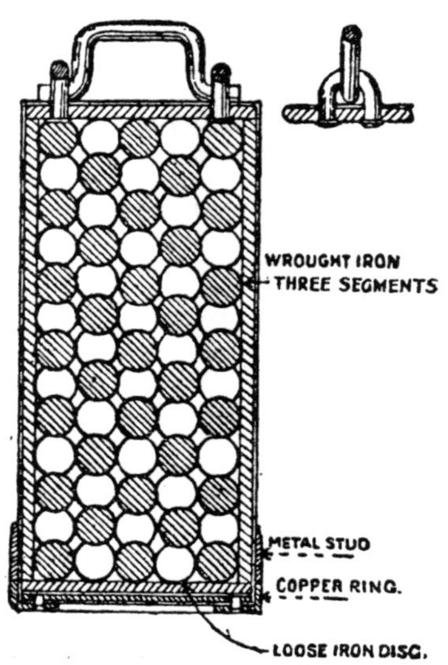

two staples, and is secured by the fringed end being bent over on it and soldered.

Inside the case there is a loose disc, three segments, and mixed metal bullets.

Case Shot, 6-inch and 5-inch.

The case shot for the B.L., 6-inch and 5-inch guns are similar in construction (*see* Fig. 50).

In the above case shot there are metal studs round the base, but no copper ring.

Case Shot for B.L., 9·2-inch.

The case shot for B.L., 9·2-inch is of the same type as those for the 5-inch and 6-inch. It has, however, a central wrought-iron stay-bolt, two handles at the top and a hole for the lifting eyebolt.

Fig. 50.

CASE SHOT, MARK I, FOR B.L., 5-INCH.

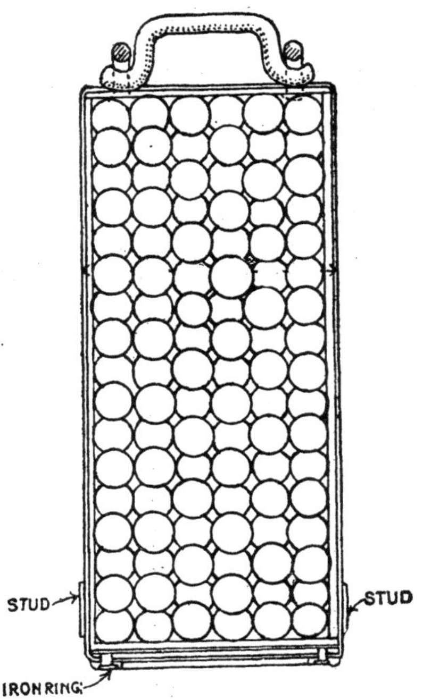

Paper Shot.

Paper shot are intended for use in the Land Service, to be fired from guns which cannot, owing to their position, fire Service projectiles in time of peace. They are used when specially ordered, to test the mounting, &c. They are designed to cause the same amount of recoil as a Service projectile, and to break up in the bore. The heavier shot are made in fractional portions.

They are to be fired with *powder charges only* (except when special permission is given for firing with certain cordite charges), and have the words "*not to be fired with cordite*" stencilled with white paint on the body.

The body is made of wood pulp sufficiently strong to stand loading and slinging when filled, the ends being strongly secured by oak pins and glue; the shot is coated with black japan lacquer, and is filled with small shot and sawdust. Earlier marks were made of papier-mâché.

Fig. 51.

SHOT, B.L. OR Q.F., PAPER, EMPTY, 6-INCH, WITH BUNG, WEIGHT, FILLED, 120-LB.

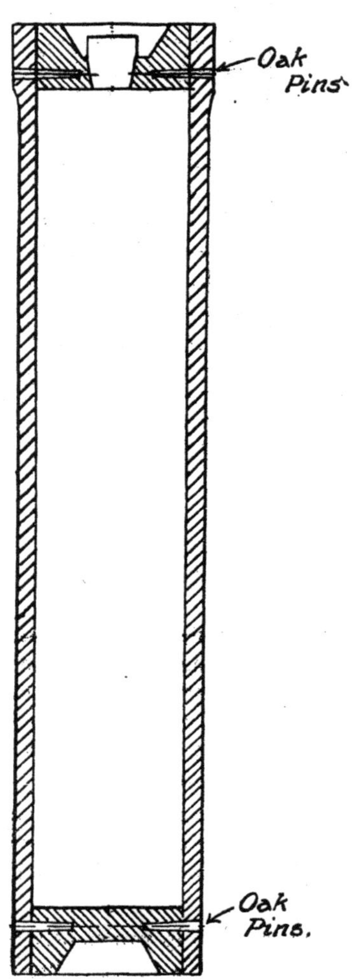

TABLE NO. 24.—*Shot, B.L. and Q.F. Case.*

Calibre.	Mark.	Para. in List of Changes.	Diameter, Body.	Length over Handles.	Contents. Balls. No.	Contents. Balls. Nature.	Contents. No. of Segments.	No. of Handles.	No. of Metal Studs at Base.	Approximate Weight of Balls.	Weight.	Weight. ±	Remarks.
			Ins.	Ins.						Lbs. ozs.	Lbs. ozs.	Lbs. ozs.	
12-inch (Light) ...	I	4574	11·88	35·2	828	8-oz. sand shot	6	2	6	414 0	714 0	21 0	Staybolt up the centre, and fitted with hole for eyebolt.
10-inch	I	7980	9·93	34·5	528	Do.	6	2	6	254 8	500 0	15 0	
9·2-inch	I	4653	9·08	33·15	414	Do.	6	2	6	207 0	380 0	11 0	
8-inch ...	I	4523	7·89	23·5	486	4-oz. sand shot	6	2	6	117 0	210 0	6 0	No hole for eyebolt.
	II	4792	7·89										Fitted with hole for eyebolt.
6-inch	I	4226	5·94	20·1	207	Do.	6	1	3	51 9	100 0	3 0	
5·4-inch Howitzer...	I	5/9/98	—	—	245	2-oz. sand shot	—	1	—	30 10	60 0	—	Copper ring at base.
5-inch	I	4376	4·95	12·3	450	Mixed metal balls (14 per lb.)	4	1	6	32 3½	50 0	1 8	Do.
5-inch Howitzer ...	I	8234	4·95	13·91	433	Do.	4	1	—	30 15	50 0	1 8	Do.
Do. ...	II	9515	4·97	14·9	185	2-oz. sand shot	3	-	—	23 2	50 0	—	Do.
30-pr.	I	—	3·95	11·45	300	Mixed metal balls (14 per lb.)	3	1	6	1 9	30 0	3 p.c.	Do.
4-inch	I	4265	3·95	9·45	245	Do. (16½ per lb.)	3	1	6	15 5	25 0	0 12	Do.
12-pr.	I	5146 8108	2·96	8·5	314	Do. (34 per lb.)	3	1	3	9 4	12 15	0 6	Copper ring at base fitted with handle becomes Mark I*.

TABLE NO. 24.—*Shot, B.L. and Q.F. Case*—continued.

Calibre.	Mark.	Para. in List of Changes.	Diameter, Body.	Length over Handles.	Contents.			No. of Handles.	No. of Metal Studs at Base.	Approximate Weight of Balls.	Weight.		Remarks.
					Balls.		No. of Segments.					±	
					No.	Nature.							
12-pr.	II	7570				Same as Mark I							Copper ring at base, fitted with handle.
15-pr. or 12-pr. ...	III	8235	2·9	8·9	300	Mixed metal balls (34 per lb.)	3	1	—	8 13	12 8	0 6	Corrugated in three pieces. For use with cordite.
Do. ...	IV	8736	2·97	9	290	Do. do.	2	1	—	—	12 14	—	Base screwed to body.
Do. ...	IV*	10283	This is Mark IV with two partial perforations in base for N.S.										
Do. ...	V	9456, 9815, 11027	2·96	8·9	290	Do. (34 per lb.)	2	1	—	8 9	13 4	—	Base pressed into recess in body.
10-pr.	—	11021	2·72	8·3	211	Do. do.	2	1	—	6 3	10 0	—	
2·95-inch Q.F. ...	I	10898, 15804	2·941	11·95	355	Mixed metal— 140 (28 per lb.) 170 (34 per lb.) 45 (36 per lb.)	3	-	—	11 4	15 0	—	
Do. ...	II	11979	2·941	10·45	404	(41 per lb.)	6	-	—	9 12	15 0	—	
Do. ...	III	16009	2·941	9·23	330	(41 per lb.)	6	-	—	8 0	13 0	—	

Proof Shot.

Proof shot are of forged or cast steel, cylindrical in shape, so that they shall not penetrate too far into the butt.

They have the same driving band, and are of the same weight as the Service projectiles.

Shot, Flat-Headed | N | .

Shot, flat-headed, are similar to proof shot, but are made of *cast iron.*

They are used for firing full elevation rounds at gun trials on ships.

They are issued for all Naval guns, 3-*inch to* 15-*inch*.

SECTION ("F.") PRACTICE PROJECTILES.

Practice Over Sea Ranges.

For practice over sea ranges the following projectiles are used :—

(1) Practice *shot* as a rule.
(2) Common-pointed practice *shell* of cast iron filled with salt, until existing stock is used up.
(3) Old Palliser shot .. .. }
(4) Old A.P. shot .. .. } No longer required and sentenced for practice.
(5) Old common shell filled with salt .. .. .. }
(6) Service shell, emptied and filled with salt, or with bursting charge wetted.

Filled Service shell, except Q.F., 12-pr., 4-inch and 4·7-inch, when fired at practice are not to be emptied, but the powder will be wetted, not more than four days and not less than two days, before the shell are required for firing.

After the bursting charge has been drowned, the shell must be placed on their bases, with the plugs removed, for a few hours, to allow any surplus water to drain off.

Sufficient wetted powder should then be removed to bring the shell to correct weight.

The fuze-hole thread should be carefully cleaned and the plug lubricated before being replaced.

If the plug cannot be removed, shell may be fired at a drifting target, but not at a towed one.

Q.F., 12-pr., 4-inch and 4·7-inch powder-filled shell to be fired at practice from "Fixed armaments" will be emptied and brought up to weight with salt cake, dust shot, iron turnings, &c.

Practice projectiles for Q.F. guns.—All practice projectiles having base plugs will, before being fired from Q.F. guns, 12-pr. and up, in both Land and Naval Service, have the key hole fitted with a wooden plug to prevent the lid of the cartridge from taking a seating in the plug.

Practice over Land Ranges.

Field guns fire their Service shrapnel shell for practice.

B.L. howitzers, B.L., 60-pr., and the 4·7-inch Q.F., on travelling carriage, fire a cast-iron common shell for practice. These shell are velvrilled internally and take a bursting charge of shell L.G. powder which is enclosed in a dowlas bag; 7-dram primers are used. (*See* Fig. 52, which shows an empty cast-iron shell for the B.L., 60-pr.)

Fig. 52.

SHELL, B.L. COMMON, 60-PR., MARK II, CAST IRON.
(FOR PRACTICE.)

Scale $\frac{1}{5}$.

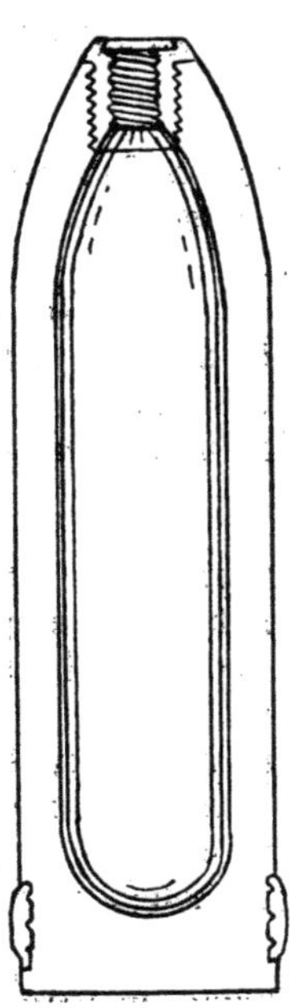

Shell, Filled Pitch and Powder.

The 5-inch and 6-inch howitzers fire old iron common shell, or old lyddite shell, partially filled with pitch; on top of the pitch is a blowing charge of L.G. powder in a shalloon bag; 7-dram primers are placed next the fuze-hole. They are for use in temperate climates only.

Fig. 53.

PRACTICE SHELL FILLED WITH PITCH AND POWDER.

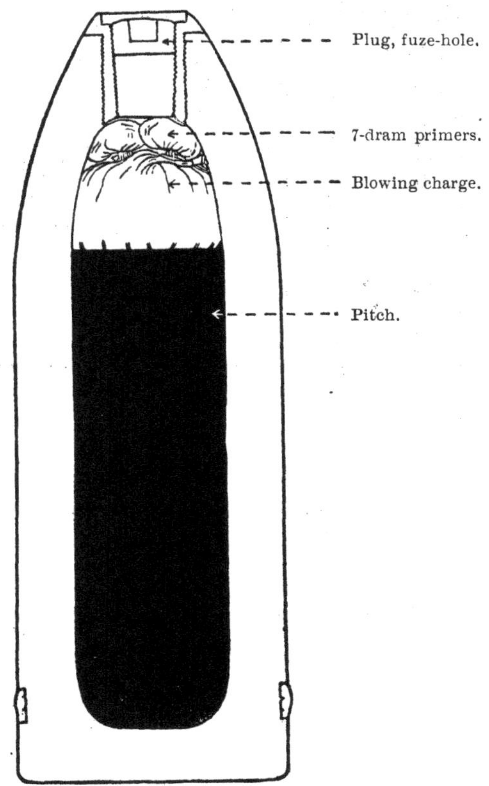

Shell Filled with Smoke-producing Composition.

A new design of practice shell has been introduced for the Q.F., 4·5-inch howitzer.

The shell is intended to represent shrapnel.

Shell, Q.F. Common, 4·5-inch Howitzer, Mark I | L | .

The shell is made of cast iron. It is prepared to take fuzes or plugs of the 2-inch gauge, and is filled with smoke-producing composition stemmed in around a central burster of R.F.G.[2] powder contained in a batiste bag, which is enclosed in a paper cylinder and is removable.

The shell is issued plugged, and is painted black with two yellow bands around the body. The "Plug, fuze-hole, 2-inch, No. 1, Mark II," is used. The fuze used is the T. and P. No. 82.

Fig. 54.

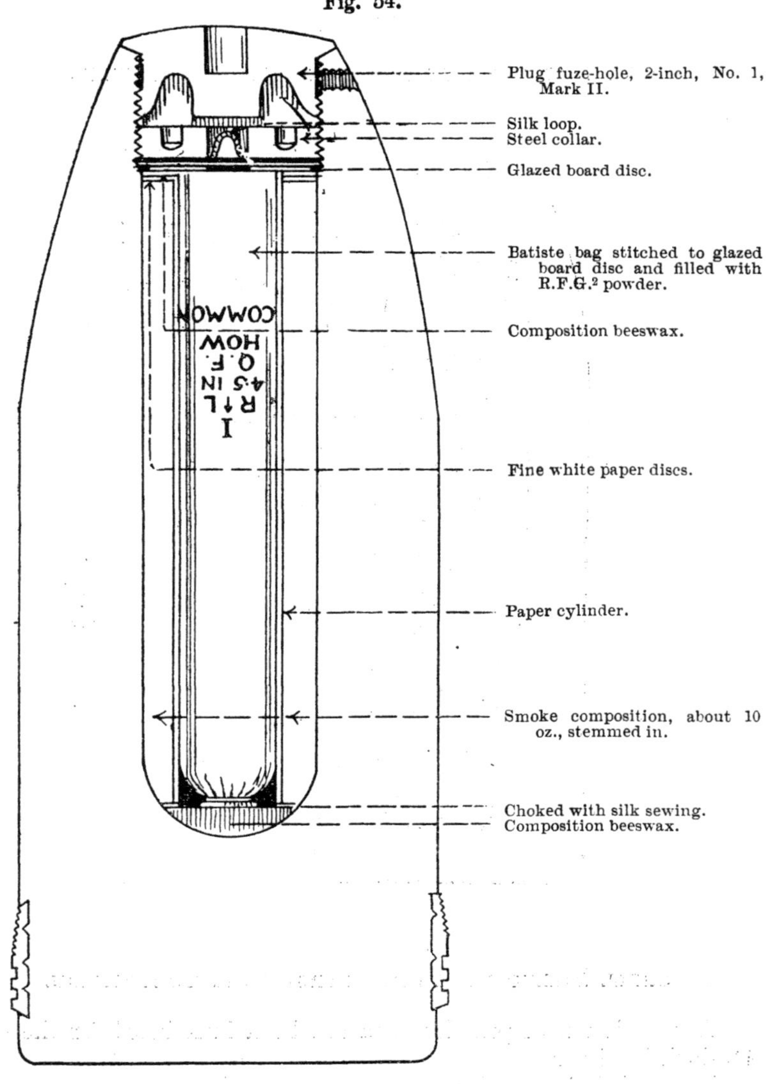

SECTION (G). MARKINGS ON PROJECTILES.

(See also *Regulations for Army Ordnance Services.*)

A. Paint Marks.

(1) *Tips :—*

Shot (except case)—To have white tips. Capped shot to have the cap painted white.

Common shell (except lyddite)—To have black tips.

Shrapnel shell—To have red tips.

(2) *Ends :—*

Case shot containing chilled-iron balls—To have a white disc painted on the end.

(3) *Bands* :—

Steel projectiles (except star, shrapnel, high explosive and lyddite)—To have a white band round the head.

Armour-piercing shell—Two white bands round the head.

All filled shell—To have a red band round the head.

High explosive shell to have a red band round the body.

All projectiles with cupro-nickel driving bands—To have a white band round the body immediately above the driving-band.

Practice projectiles—To have a yellow band round the body.

Pitch and powder and smoke composition practice shell—To have two yellow bands round the body.

Lyddite shell, empty, for drill—To have a black band round the body.

(4) *Bodies* :—

All lyddite and high explosive shells to be yellow; B.L., 15-pr. shrapnel, Q.F., 13- and 18-pr., Marks II and III, shrapnel shell, and 12-pr.; projectiles (except lyddite) suitable for 12- and 8-cwt. guns only, to be painted lead colour; all other projectiles to be black, excepting case shot, which will not in future be painted.

(5) *Star* :—

Star shells to have a star in red, on a white disc, on the shoulder.

(6) *Numeral and calibre* to be stencilled on all projectiles, both Naval and Land Service above 6-pr. (except fixed ammunition).

Four-calibre head projectiles have the letter "A" after their numeral.

B.L., 6-inch projectiles of strengthened design introduced for Mark XII gun to have letter "Q" after numeral.

Shell prepared to take night tracer to have "N.T." after numeral.

Description.	Additional Markings.
13·5-inch	"H" or "L."
12-inch...	
9·2-inch lyddite and practice shot... ...	
6-inch howitzer	
4-inch	
6-inch	"Gun" or "Howitzer."
5-inch	
10-inch, Marks VI to VII guns	"Special."
7·5-inch, Marks III to IV* guns	
12-pr. (Land Service)	12 or 6 cwt., as the case may be.
12 and 14-pr. in Naval Service, suitable only for 12-pr., 12 and 8 cwt. guns	"I" / 12-pr., 12 and 8 cwt.
12 and 14-pr., suitable for all 12-prs., and 12-pr. guns	"12- and 14-pr."
6- and 3-pr. annealed shell	"A."

(7) The word "sand" or "salt" to be stencilled on the body in white paint if a shell is filled with sand or salt for practice.

(8) The following additional markings will be shown on all shell, excepting 6-pr. and 3-pr. Q.F. :—

(*a*) The word "fuzed" if the shell is fuzed.

(*b*) The monogram of the filling station. (*See* para. 42, Regulations for Magazines, &c.)

(*c*) The date of filling.

(*d*) The nature of powder with which filled, *e.g.*, P., Q.F., L.G. (except shrapnel shell).

(*e*) The marking "*Dead weight*" or "actual weight," except shrapnel, star, practice, and fixed ammunition.

(*f*) The letter "N" on the bodies of all Naval projectiles to be stencilled over the "N" which has been stamped on.

(*g*) If the shell is fuzed with a base fuze, the head of the fuze, or where an Armstrong base fuze is used, the lead cap covering the fuze will be painted *red*.

Fuzed A.P. lyddite will have the "plate cover" painted *red*.

(*h*) Projectiles for B.L., 12-inch, 13·5-inch and 15-inch will be stencilled ⊖ in three places equally spaced around the projectiles to indicate the centre of gravity when filled.

(9) The following markings will be stencilled on projectiles designed to take day tracers :—

Distinguishing mark.

D	When a projectile is fitted for day tracer but is not filled with tracer liquid.
D (with T)	When a projectile is fitted for day tracer and filled with tracer liquid.

(10) The following markings will be stencilled on projectiles designed to take night tracers :—

T.	When fitted with night tracer.
	When fitted for, but not with, night tracer.
T.F.	When fitted with combined tracer fuze.
P.	When fitted for, but not with, combined tracer fuze.

(11) If for Colonial Service :—

" A.N." for Australia. (Naval.)
" D↑." ,, ,, (Land.)
" N.Z." ,, New Zealand. (Naval.)
" N.↑Z." ,, ,, ,, (Land.)
" C.N." ,, Canada. (Naval.)
" C↑." ,, ,, (Land.)
" U↑." ,, South Africa.

(12) For special distinguishing marks stencilled on lyddite shell, *see* page 200.

(13) Shell which have been emptied will be marked on the head with the letter " E " in white paint, and also the monogram of the station.

(14) Shell which have been examined as to condition of bursting charge, &c., will be stencilled " Exd," date of examination and the monogram of the station.

(15) Shell to which external repairs (including re-painting) have been carried out are to be stencilled " Repd," date, and monogram of station.

B. Markings Stamped on Projectiles.

The markings are generally stamped on the base, but in the case of fixed ammunition the markings will be found on the body above the driving band :—

(1) Calibre and numeral.

(2) Manufacturers " initial " or " Trade Mark."

(3) Date of completion of manufacture.

(4) " Gun " or " Howitzer."

(5) " H " or " L " = denoting Heavy or Light Shell.

(6) " C.S." or " F.S." = denoting Cast or Forged Steel.

(7) " A.P." = denoting Armour-Piercing.

(8) " P " = denoting Practice Shot.

(9) " A " after numeral = denoting 4-calibre head.

(10) " Q " after numeral = denoting Shell specially made to stand pressure of 6-inch, Mark XII gun.

(11) B.L., 7·5-inch and B.L., 10-inch projectiles for H.M.S. " Swiftsure " and " Triumph " are stamped as follows :—

7·5-inch = " Marks III and IV guns."
10-inch = " Marks VI and VII guns."

(12) $\bar{\text{A}}$ on 3- and 6-pr. annealed shell.

The letter " N " is stamped on the body of all Naval projectiles.

All projectiles are stamped with broad arrow " ↑ " and various work marks in definite positions by the inspectors.

The work marks generally consist of one or more letters.

Fig. 55.

MARKING STAMPED AND STENCILLED ON COMMON-POINTED SHELL.

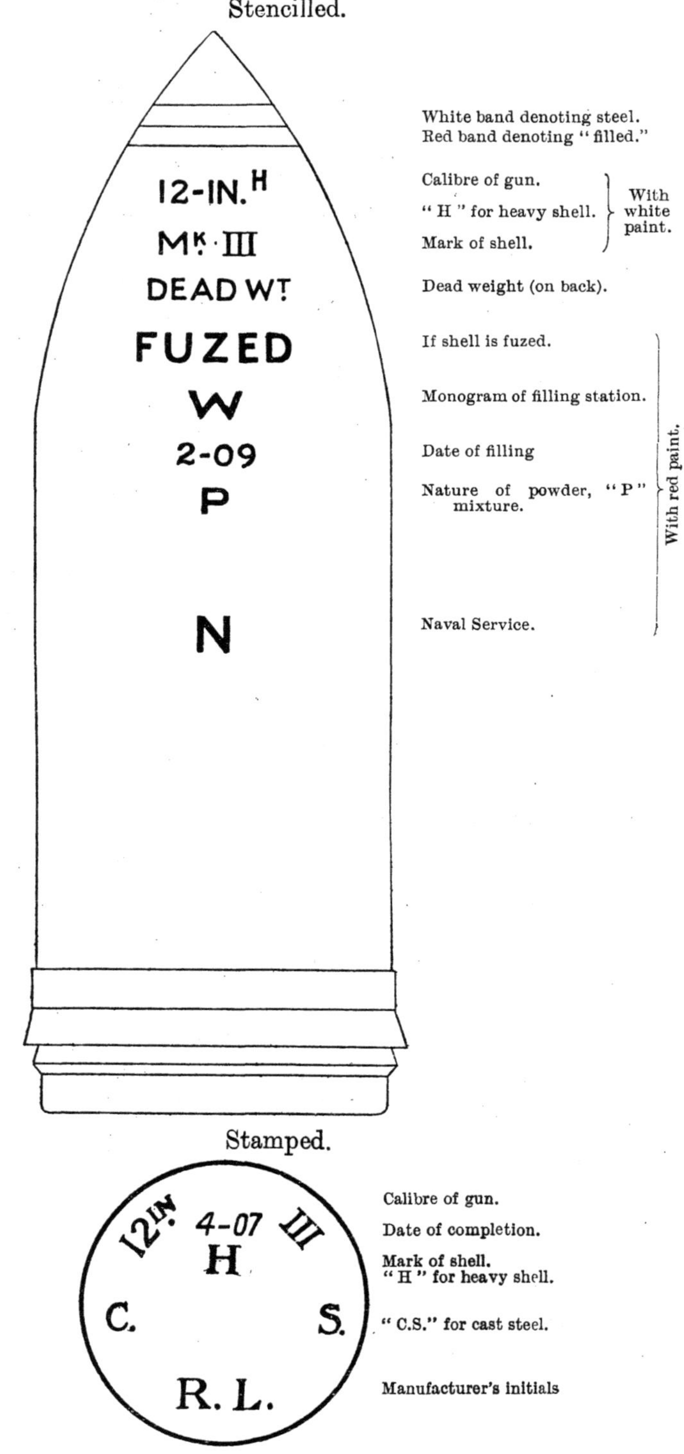

Fig. 56.

MARKING STAMPED AND STENCILLED ON CAPPED ARMOUR-PIERCING SHELL.

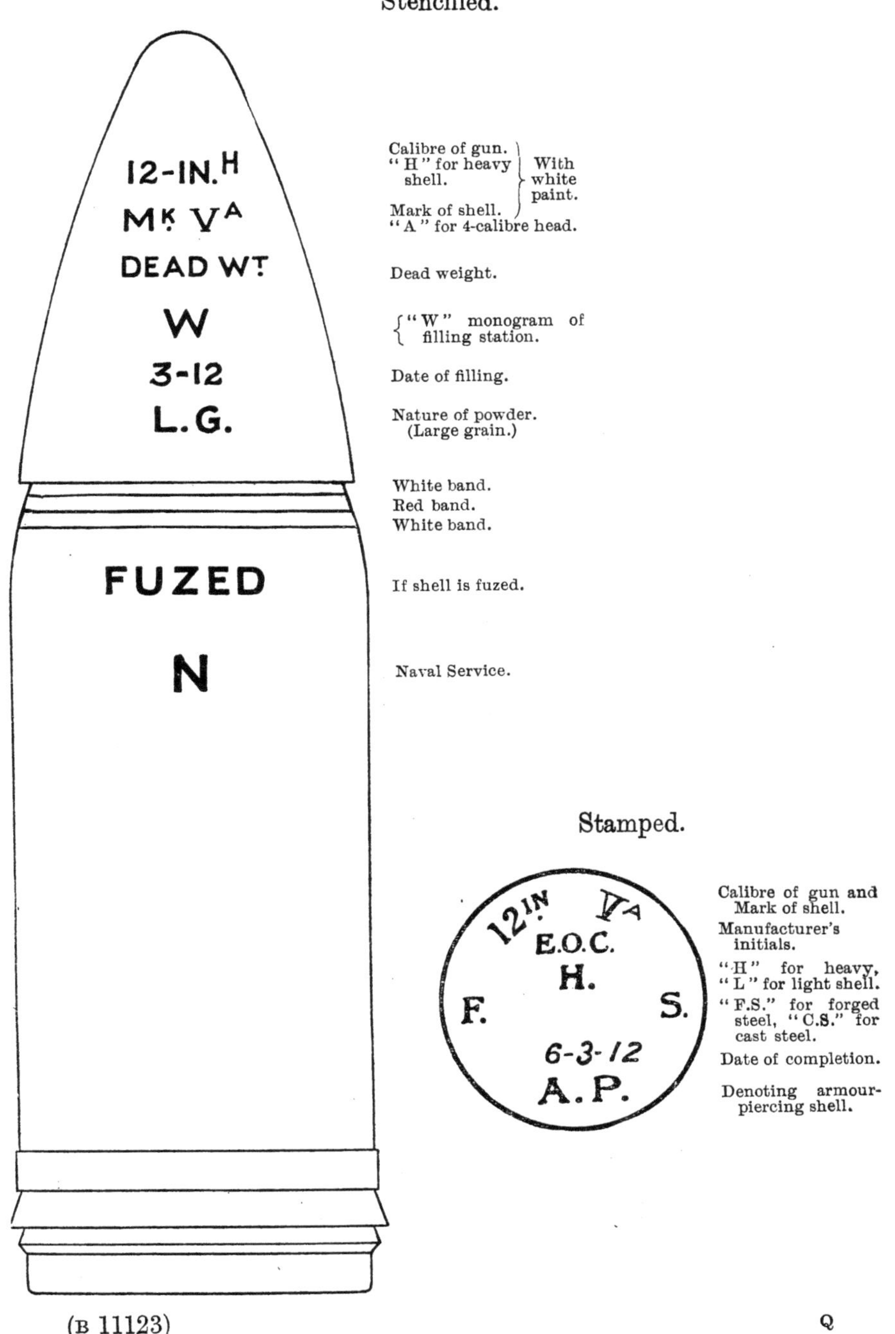

Fig. 57.

MARKING STAMPED AND STENCILLED ON SHRAPNEL SHELL.

Stencilled.

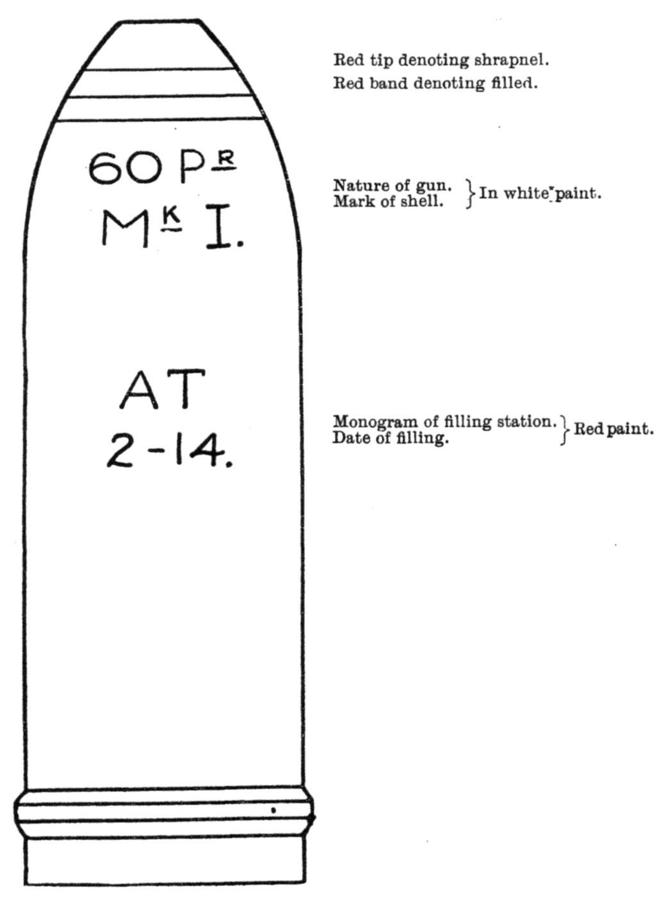

Stamped.

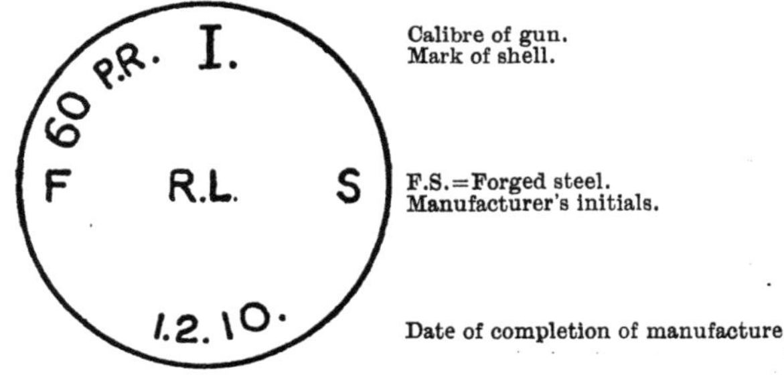

Fig. 58.

MARKING STENCILLED ON LYDDITE SHELL.

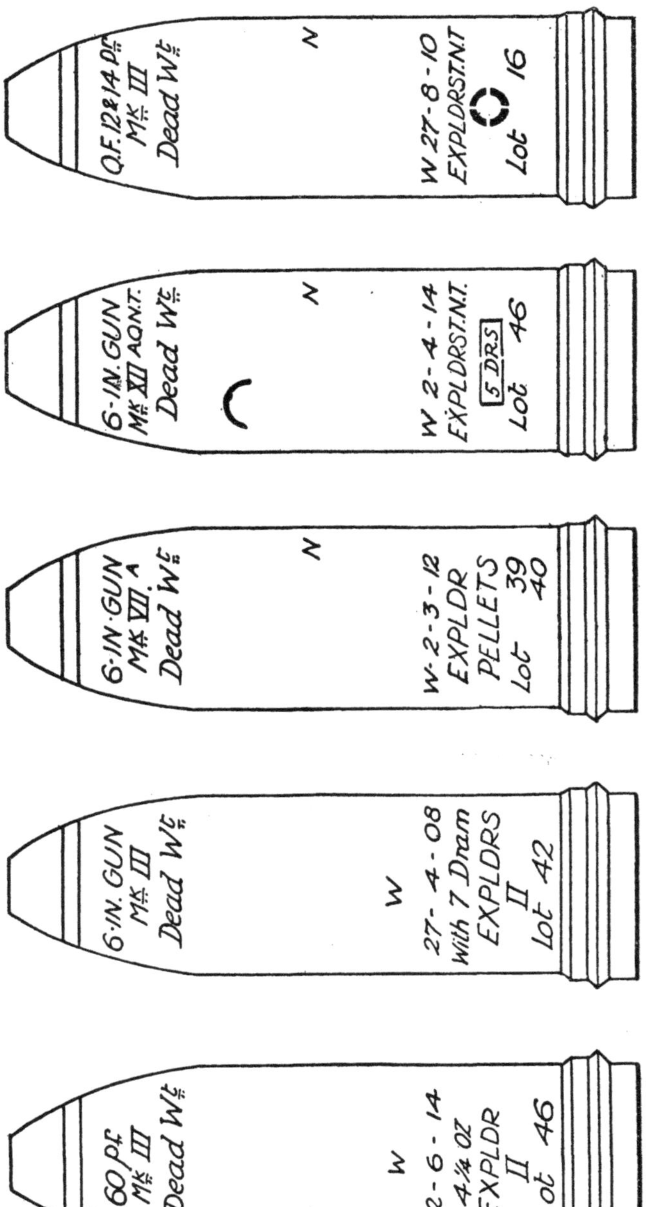

60 Pr.
Mk. II
Dead Wt.
W
12-6-14
4¼ OZ
EXPLDR
II
Lot 46

6·IN. GUN
Mk. II
Dead Wt.
W
27-4-08
With 7 Dram
EXPLDRS
II
Lot 42

6·IN·GUN
Mk. VII.A
Dead Wt.
N
W-2-3-12
EXPLDR
PELLETS
Lot 39
40

6·IN. GUN
Mk. XII AQN.T.
Dead Wt.
N
W 2-4-14
EXPLDRST.N.T.
5 DRS
Lot 46

Q.F. 12 & 14 Pr.
Mk. III
Dead Wt.
N
W 27-8-10
EXPLDRST.N.T
Lot 16

Some Former Markings on Projectiles.

The markings are changed from time to time; the following old markings will be met with :—

(1) The weight of a shell *empty* stamped on the driving band.

(2) The weight of the shell, filled, or "dead weight," stencilled close above the driving band, instead of on the shoulder.

(3) The markings R ↑ L stencilled on the bodies of the filled shell; this denoted "Royal Laboratory."

Projectiles *filled* in the R.L. are now merely stencilled "W" (Woolwich).

(4) The word "Bag" (meaning that the bursting charge was contained in a bag) and a red "disc" (meaning that the shell contained 7-dram gunpowder primers) were formerly stencilled on common, common pointed, and A.P. shell.

SECTION (H).—PACKING OF B.L., B.L.C., Q.F. AND Q.F.C. PROJECTILES FOR TRANSPORT AND STORAGE.

TABLE 25.

Nature of Ordnance.	Nature of Projectile.	How packed for	
		Transport.	Storage.
(a) Naval and Land Service.			
(i) All natures ...	"Fixed" ammunition	Boxed	Boxed.
(ii) B.L., Q.F. or Q.F.C., 5-inch and under ...	Common pointed Armour piercing	Boxed	Boxed or in bulk.
(iii) All natures ...	Star shell ...	Boxed	Stored in bulk except 8-inch spherical which are boxed.
(b) Naval Service.			
Q.F., 4·7-inch, B.L. and Q.F., 4-inch Q.F., 12-pr. and 14-pr. ...	Lyddite common, if *fuzed* Shrapnel ...	Boxed	Boxed.
(c) Land Service.			
(i) All natures in Horse, Field and Heavy Batteries, except B.L., 60-pr.	All natures ...	Boxed	Boxed.
(ii) B.L., 6-inch Howitzer in Siege and Territorial Artillery	All natures ...	Boxed	Boxed.
(iii) B.L., 5-inch, 4-inch, 10-pr. and 2·75-inch, Q.F., 4·7-inch and 2·95-inch guns in Movable Armament, Mountain and Territorial Artillery	All natures ...	Boxed	Boxed.
(iv) B.L., B.L.C. and Q.F., 15-pr., and B.L., 12-pr.	All natures ...	Boxed	Boxed.

TABLE 25—*continued.*

Nature of Ordnance.	Nature of Projectile.	How packed for	
		Transport.	Storage.
	(c) Land Service—continued.		
(v) B.L., 60-pr., in Heavy Artillery	All natures ...	In slings or boxes	In slings or boxes.
(vi) B.L.C., 6-inch, in Siege Artillery	All natures ...	In slings or boxes	In slings or boxes.
(vii) B.L., 5-inch Howitzer in Territorial Artillery	All natures ...	In slings or boxes	Shells which on mobilization will be carried in limbers and wagons (and are not so stored during peace) will be stored in bulk. Those which will be carried in hired wagons will be stored in slings or boxes. Of the shells which are stored by A.O.D., and are not included in the above, 50 per cent. will be stored in slings or boxes, and the remainder in bulk.
(viii) Q.F., 4·7-inch and under in Fixed Armaments	Lyddite common plugged	Boxed	Boxed or in bulk when stored by A.O.D. In bulk when in R.A. charge.

PROTECTION OF PROJECTILES WHICH ARE TRANSPORTED AND STORED IN BULK.

The following projectiles which are transported and stored in bulk will be protected as follows :—

Naval Service.

(*a*) All shell with fuzes, base percussion, large, with steel protecting plate, need not be protected. For the protection of large base percussion fuzes without steel protecting plate, a fuze protector, which consists of a dished iron disc with two projections which engage in the wrench holes of the fuze, should be used. When these fuze protectors are not available " Protectors, projectile " should be used for transport, these being removed before the shell is stowed.

(*b*) Uncapped A.P. shell will, in transport, be fitted with " Protectors, projectile."

(*c*) All B.L. and Q.F. projectiles with the driving band with gas-check will have one grummet on each side of the gas-check portion of the band to be kept on in store and where possible in the shell rooms.

(*d*) Filled and fuzed A.P. lyddite shell have the fuze protected by the "Plate cover."

(*e*) ***Transport and storage of shells fuzed with base fuzes.***—Shells fuzed with base fuzes No. 11, Marks I and II, without "Fuze protectors," must never be placed, either temporarily or permanently, with the point of one against the base of another, and the fuze protectors should never be removed during transport so long as there is the least risk of the point of a projectile coming into contact with the base of a shell having a base fuze. Special precaution will be taken to prevent movement of base fuzed shells during transport.

SECTION (I).—DRILL SHELL.

Drill shell are issued for B.L. guns, 30-pr. to 15-inch, and for 5·4-inch and 6-inch B.L. howitzers.

No more drill shell for the B.L., 9·2-inch will be made. Where drill shell are not already provided for B.L., 9·2-inch guns, one Service cast-iron, common, nose fused shell per gun will be set apart for drill purposes and marked "Drill" in one-inch white letters round the nose, so as to ensure the same shell always being used. This shell is to be rammed home at the first round only, and local arrangements made for extraction at conclusion of drill (for this purpose a light spar may be required).

Drill shell are not issued with the equipments of the B.L., 2·75-inch, B.L., 12-pr. of 6 cwts., B.L. and B.L.C., 15-pr., nor for Q.F., 4·5-inch and 5-inch B.L. howitzers.

The drill shell for the *B.L., 30-pr., for the B.L., 4-inch, and for the 5·4-inch howitzer* are made of cast iron, fitted with gunmetal bands to prevent injury to the rifling in loading and unloading. The nose is bushed similar to the common shell and the base is fitted with a gunmetal plug hollowed out to form a bridge for the hook of the extractor to engage with. A groove is formed between the front end of this lug and the body of the shell to take a rope grummet which prevents the shell being rammed too far home.

Shell, B.L. drill, 6-inch howitzer, Mark II, are similar to the above, but are fitted with a steel plug instead of one of gunmetal.

The shell, B.L. drill, 7·5-inch, Mark I, and the Shell, B.L., or Q.F. drill, 6-inch, Mark V | L | (*see also* Drill shell for Q.F. 6-inch), are similar to the drill shell already described, but the groove formed between the gunmetal base and the lower metal band is bound with spunyarn as shown in Fig. 59.

The binding of spunyarn should be well smeared with Russian tallow before the shell is taken into use, and again when necessary to prevent difficulty in extraction.

Should the spunyarn binding require renewal it may be carried out as described in para. 12553, L. of C. in War Material.

The Mark IV drill shell for the B.L., 6-inch gun had a rope grummet instead of the frapping of spunyarn.

Existing B.L., 6-inch, Mark IV, drill shell will be bound with spunyarn locally and a | * | added to the numeral.

Shell, B.L. drill, 6-inch, Naval, Mark III, is made of gunmetal and has no rope grummet or frapping of spunyarn. The hollow in the base has a bayonet joint recess to receive the extractor when

Fig. 59.

SHELL, B.L. DRILL, 7·5-INCH, MARK I.

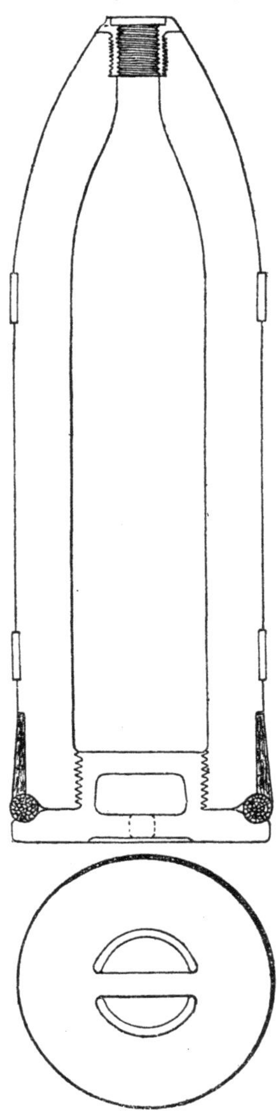

the latter is employed in withdrawing the projectile. It is brought up to the weight of a Service projectile with lead ash.

The drill shell for the B.L., 9·2-inch, 10-inch, 12-inch, and 13·5-inch are made of cast iron of about the same dimensions and weight as the Service common shell; they are filled with salt and the hole in the base closed by a large gunmetal screw plug which is recessed, a

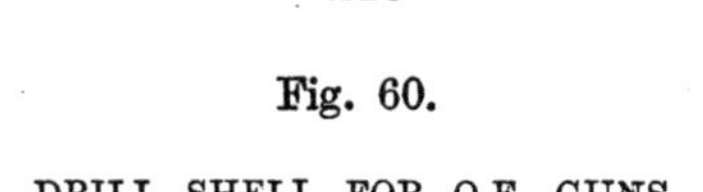

Fig. 60.

DRILL SHELL FOR Q.F. GUNS.

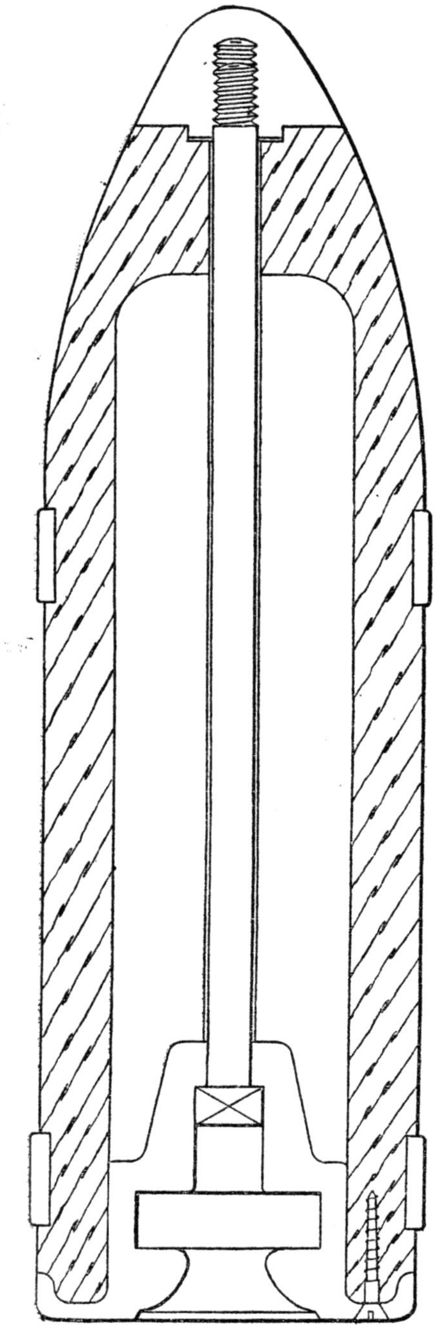

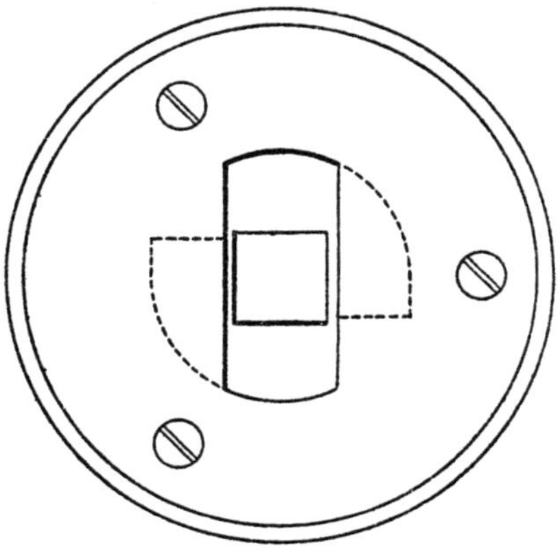

PLAN OF BASE.

bridge being formed across the recess for the extractor to engage with when withdrawing the shell. They are fitted with two gunmetal bands—one near the base, the other at the shoulder. The base end of the shell is turned down and screwed to take a gunmetal ring which projects over the side and prevents injury to the gun in loading. A groove is formed round the junction of this gunmetal ring and the body to take a rope grummet which prevents the shell jamming when rammed home.

The shell are fitted with a hole in the side at the centre of gravity to take the "Bolt, eye, lifting."

In future manufacture of drill shell for Land Service the eye bolt lifting hole will be dispensed with.

A common-pointed shell weighted with salt, and stencilled near the point with the word "DRILL" is used for drill purposes with the B.L., 9·45-inch howitzer.

Drill shell for the 4·7-inch, 4-inch and 6-inch Q.F. guns (*see also* Shell, B.L. or Q.F. drill, 6-inch), also the 5-inch B.L.C. are made of wood, brought up to weight of the Service projectile by a lead core. They are fitted with a bolt which passes through the centre of the shell; the front end of this bolt screws into a gunmetal nut which forms the point of the shell.

The base is of gunmetal, having a recess in which is formed a slot to receive the "T" shaped projection of the extractor. Two bands of copper are fitted to the shell; the rear band is sufficiently large to prevent the shell being rammed home too far.

Shell, Q.F. Drill, 12-pr., Mark II.

The drill shell for the 12-pr., Q.F., is made of gunmetal of the same diameter and weight as the Service projectile, except that instead of being fitted with a driving-band the diameter is increased near the base to prevent the shell being rammed home too far. The base of the shell is closed by a plug recessed to receive the hook of the extractor.

SECTION (J).—CAUSES OF BLINDS AND PREMATURES.

BLINDS.

The fact of a fuzed shell failing to burst may be generally attributed to some fault in the manufacture of the fuze, or in its preparation, but it may also be due to the shell not being properly filled, a damp bursting charge, damp exploder, or a dirty shrapnel primer. "Graze" percussion fuzes will cause blinds if the shell is not sufficiently checked. The most trying conditions for a fuze of this description are a heavy shell with a high velocity and a small angle of incidence, since a heavy body moving at a high rate of speed is more difficult to check than a light one.

To sum up, we may divide the probable causes of blinds into two classes :—

Due to Shell.	*Due to Fuzes.*
(1) Empty, or shell not properly filled. (2) Damp bursting charge. (3) 7-dram primers omitted. (4) With shrapnel, a dirty or defective shrapnel primer. (5) With lyddite, damp or defective exploders.	(1) Fuze not properly prepared. (2) Fuze used in a gun for which it was never intended. (3) Defective fuze. (4) With graze fuzes, angle of descent too slight, and shell not sufficiently checked on graze.

Prematures.

Premature explosions are even more serious than blind shell, and it is difficult to over-rate the importance of getting rid of them, especially with high explosive shell.

A premature may be due to a weak shell breaking up in the gun, or to the penetration of gas into the shell through a flaw. It may also be due to the ignition of the bursting charge due to the distortion of the shell, or from friction or shock of discharge.

Distortion may be due to (*a*) oil in the bore ; (*b*) weakness of the shell walls :—

(*a*) The oil collects in front of the driving band as the projectile travels up the bore and, as it cannot escape, forces in the walls of the shell.

(*b*) The walls of the shell being unable on firing to withstand the pressure acting on the base, bulge outwards.

These two forms of distortion are termed " set in " and " set down " respectively.

It appears to be conclusively proved that prematures may occur from shell not being properly filled, and also from the great friction due to the rotation of the shell and the powder setting back on shock of discharge.

For this reason the insides of shell are either velvrilled or varnished, to give them a smooth surface.

As an additional precaution, burster bags to contain the charge are used with practically all powder-filled common and armour-piercing shell, except the smallest. The necessity for these bags has been demonstrated on several occasions when shell have been fired without them.

With time fuzes, apart from defects of manufacture, a premature may occur in metal fuzes from the time ring not being clamped down, when the flash of the burning composition could travel round the underside of the ring.

With graze percussion fuzes a rebound action may be set up by the shock of discharge, which might throw the detonator against the needle. To check this action, a spiral spring is placed in all modern fuzes between the pellet and the body of the fuze.

There is also the fact that a very light shell may be checked by an irregularity in the bore of the gun, sufficiently to set the fuze in action.

With direct action fuzes prematures are less likely, and the cause, if one does occur, is more obscure. Placing a disc of tin foil under the detonating composition was found to have a great effect in preventing prematures.

Base fuzes have introduced additional causes for prematures from failures due to a weak fuze, or from the penetration of gas into the shell between the fuze and the body of the shell, or between the pressure plate and the fuze body.

To sum up, the probable causes of prematures may be divided into two classes :—

Due to Shell.	*Due to Fuze.*
(1) Bad lacquer.	(1) Fuze not properly prepared.
(2) Grit in shell.	(2) Fuze used in a gun for which it was never intended.
(3) Burster bag omitted.	(3) Defective fuze.
(4) In shrapnel, a rusty tin cup.	(4) Penetration of gas through or over the fuze.
(5) Weak or defective shell.	
(6) Shell not properly rammed home.	
(7) Obstruction in the bore.	
(8) Oil in the bore.	

SECTION (K).—STORES USED IN CONNECTION WITH SHELL.

BAGS, BURSTER.

Bags, burster, are used with common, common-pointed, and A.P. shell, 4-inch and above.

For *common shell* they are made of dowlas. Common shell filled through the base have the tip, while those filled through the nose have the neck and shoulder of the bag made of shalloon. Shalloon is used to lessen the resistance to the penetration of the flash from the fuze.

With heavy natures, " collar cloth " is employed at the shoulder of the bag to form a cushion for the powder as it sets back on the shock of discharge.

The burster bags for *common-pointed shell* are also made of dowlas, with a neck and shoulder of shalloon.

For *armour-piercing shell* the burster bags are made entirely of " lasting cloth," a material much closer in texture than dowlas, the object being to prevent the bursting charge, if dusty, from working through the bag.

The burster bags for A.P. shell were formerly made of dowlas, and such bags will still be met with.

Burster bags for shell filled with " pitch and powder " are made of shalloon.

Burster bags for iron or steel shells or different marks of shell of the same calibre are not interchangeable, as the capacities of shells of different material may vary considerably. The bags have the numeral, nature of shell, and contractor's mark on them.

The following shows the marking on burster bags :—

R ↑ L	R ↑ L
I	II
4-in.	9·2-in.
B.L. or Q.F.	B.L.
Common	Armour-
Pointed.	Piercing.
Heavy.	Marks I to
	III Shell.

Bags, Primer, 7 Drams.

Bags, primer, 7 drams, Mark I, are small shalloon bags containing 7 drams of F.G. powder. Two or more of these primers are inserted in all common, common-pointed and armour-piercing shells filled with bursting charges of powder in bags, to facilitate igniting the bursting charge and to fill up the shell.

Primer, Shrapnel Shell, Mark III.

This primer is made of brass, solid at the top, with a conical cup-shaped recess ; the bottom of the cup is perforated with three small

Fig. 61.

PRIMER, SHRAPNEL SHELL, MARK III.

Full Size.

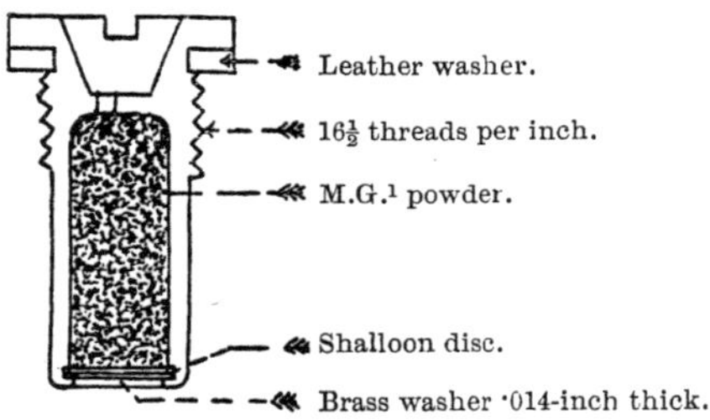

holes, communicating with loose powder, with which the body of the primer is filled. The bottom is closed by a thin brass washer covered with shalloon.

There is a slot in the head for the screwdriver, and a leather washer is placed under the shoulder.

The Mark II had a smaller head, and had no leather washer under the shoulder.

Wads, Fuze-Hole.

Wad, fuze-hole, G.S., Mark III is made of papier-mâché, and has a hole in the centre covered by a thin disc of black shalloon cemented to one side. It is used with 4-inch B.L. shrapnel, Marks V and VI, and the 4·7-inch Q.F., Mark IV shrapnel to prevent the powder from working up into the fuze-hole of the shell. The side covered with shalloon is placed downwards in the shell.

Wad, fuze-hole, 2-inch, Mark I | *N* | is made of millboard similar to the Wad, fuze-hole, G.S., Mark III, but larger in diameter. It is used with Naval shrapnel having the 2-inch fuze-hole.

FUZE PLUGS.

Plugs used with Nose-Fuzed Shell.

Plug, fuze-hole, G.S., with loop, Mark I.	For all shell with G.S. gauge (except lyddite), carried in limbers, &c.
Plug, fuze-hole, G.S., without loop, Mark I.	For all shell with G.S. gauge (except lyddite), not carried in limbers, &c.
Plug, fuze-hole, G.S., without loop, Naval.	G.S. plug with flange and washer for old common shell in N.S.
Plug, fuze-hole, Special, No. 1, Marks I to III.	For lyddite shell, 2·75-inch, 4·5-inch howitzer and above, except those taking Nos. 18 and 44 fuzes.
Plug, fuze-hole, Special, No. 2, Mark I.	Steel plug, used with empty lyddite shell to protect nose. (Special to India.)
Plug, fuze-hole, Special, No. 3, Mark I.	For all lyddite shell taking Nos. 18 and 45 fuzes.
Plug, fuze-hole, Special, No. 4, Mark I.	For Mark I lyddite shell for Q.F., 3-pr.
Plug, fuze-hole, Special, No. 4A, Mark I.	For lyddite shell, Q.F., 3-pr., except 3-pr., Mark I shell.
Plug, fuze-hole, 2-inch, No. 1, Marks I and II (Mark I white metal; Mark II, forgeable alloy).	For shrapnel for B.L., 2·75-inch, Q.F., 13- and 18-pr., 4·5-inch howitzer. Common shell, 4·5-inch howitzer.
Plug, fuze-hole, 2-inch, No. 2, Marks I and II.	For all shrapnel having 2-inch gauge; except B.L., 2·75-inch, Q.F., 13- and 18-pr., and Q.F., 4·5-inch howitzer.

Plug, fuze-hole, G.S., with loop, Mark I | *L* | is a conical plug without shoulder, having a square hole in the head to take the G.S. key, and a loop of tarred white line about 1¼ inch long; the loop facilitates extracting shell from limbers.

Plug, fuze-hole, G.S., without loop, Mark I | *C* | differs from the above in having no loop.

Plug, fuze-hole, G.S., without loop, Naval, Mark II | *N* | has a flange ·19 inches wide, under which is a leather washer, soaked in ozokerine, to make a water-tight joint.

Plugs for Lyddite Shell.

Plugs, fuze-hole, Special, No. 1.—The latest mark is Mark III; it is secured after having been screwed home, by being stabbed in three places at the top of the flange. The Mark II plug is identical in dimensions with Mark III, but the metal of which the plug was manufactured was not free from traces of lead.

The Mark I plug is flat on top, and so cannot be secured by stabbing. The Mark I plugs have now been altered (to enable them to be stabbed) by turning a groove round their upper surface; such plugs are known as Mark I.*

It was the introduction of stabbing the plugs that rendered kit plasters unnecessary.

Fig. 62.

PLUGS, FUZE-HOLE, SPECIAL, NO. 1.

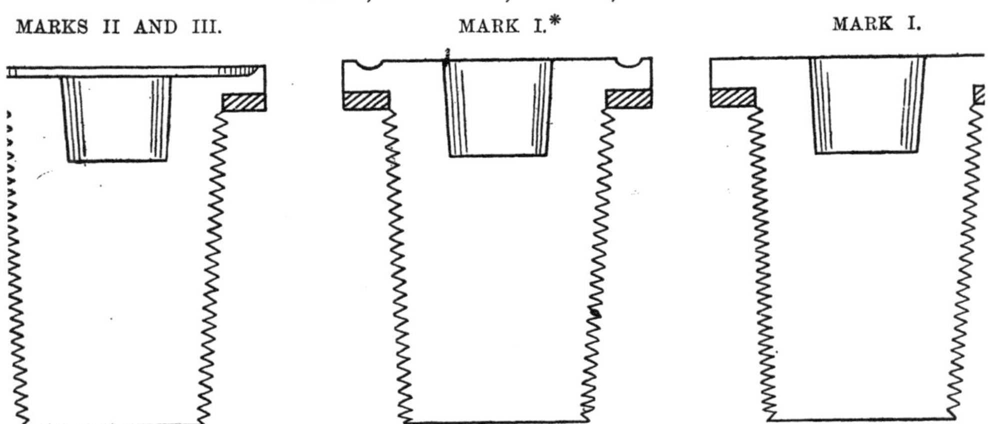

Plug, fuze-hole, Special, No. 2, *Mark I* | *S.I.* | is made of steel, blackened, and fitted with a leather washer. It is used for the protection of the nose of empty lyddite shell, and is specially made for India. (*See* Fig. 63.)

Fig. 63.

PLUGS, FUZE-HOLE, SPECIAL, NO. 2 | S.I. |.

Scale full size.

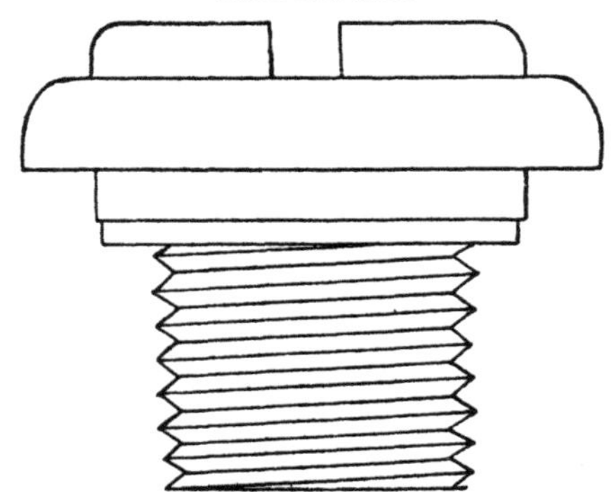

Plug, fuze-hole, Special, No. 3, *Mark I* | *C* | is made of the same metal as the Plug, fuze-hole, Special, No. 1, Mark III (free from any trace of lead); it is used with lyddite shell that are intended to take Fuzes, percussion, D.A. Impact, Nos. 18 and 45.

Fig. 64.

PLUG, FUZE-HOLE, SPECIAL, NO. 3, MARK I.

Scale Full Size.

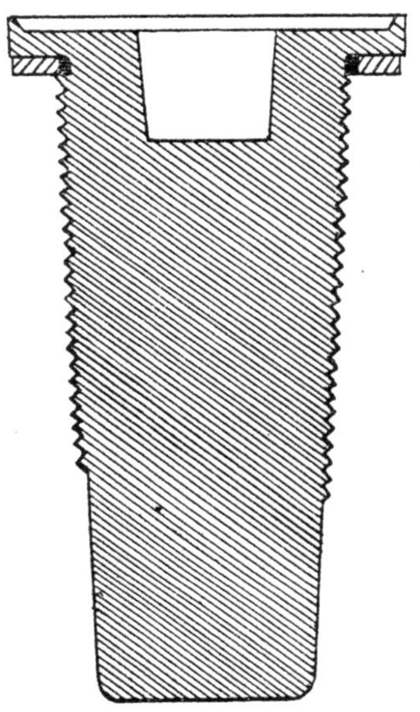

Plug, fuze-hole, Special, No. 4, *Mark I* | *N* | .—The No. 4 Plug, fuze-hole, special, is used with the Mark I, 3-pr. Q.F. lyddite shell. This shell takes the Fuze, percussion, D.A. Impact, No. 19. It is similar to the No. 3 plug illustrated above, but smaller, and the body is not tapered.

Plug, fuze-hole, Special, No. 4A, *Mark I* | *N* | .—The No. 4A plug, fuze-hole, special, differs from the above in being tapered; it is used with shell taking the Fuze, percussion, D.A. Impact, No. 19A.

Plug, fuze-hole, 2-inch, No. 1, *Mark II* | *L* | .—This plug is made of "forgeable alloy" to the form shown in Fig. 65.

Fig. 65.

PLUG, FUZE-HOLE, 2-INCH, NO. 1, MARK II | L | .

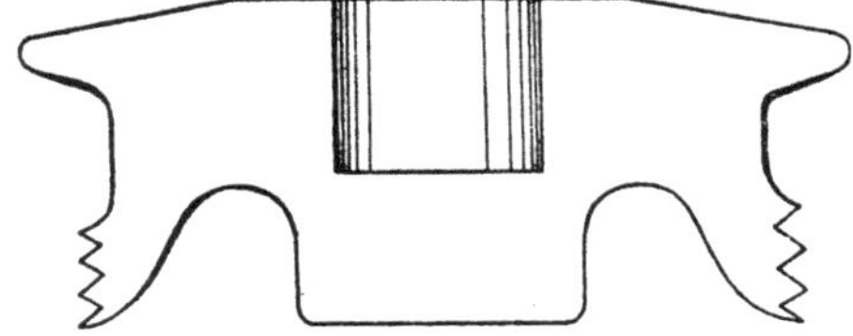

Plug, fuze-hole, 2-inch, No. 1, *Mark I* | *L* | is made of white metal. The top is convex, and projects over the screwed portion and forms a protecting flange for the fuze-hole of the shell, and it has a square key-hole in the top.

Plug, fuze-hole, 2-inch, No. 2, *Mark II* | *C* | is made of " forgeable alloy " to the form shown in Fig. 66. It has a longer stem than the No. 1, Mark II plug ; it is used with shrapnel shell having the deep fuze-socket.

Fig. 66.

PLUG, FUZE-HOLE, 2-INCH, NO. 2, MARK II | C |.

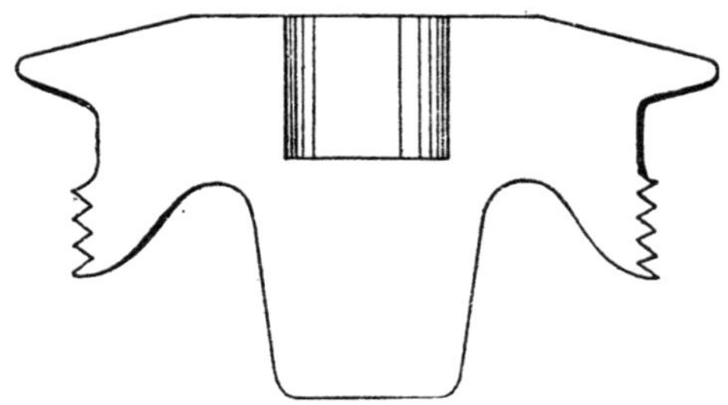

Plug, fuze-hole, 2-inch, No. 2, *Mark I* | *C* |.—The plug, fuze-hole, 2-inch, No. 2, Mark I is similar to the Mark II plug, but is screw-threaded to the shoulder ; the set-screw in the nose of the shell, when screwed home, has a tendency to damage the thread of this plug.

Plugs used with Base-Fuzed Shell.

Plug, base, shell, No. 1, Marks I and II.	For filled shell taking the large base fuze, except No. 16 ; 6-inch, and above.
Plug, base, shell, No. 2, Marks I to IV.	For filled shell taking the medium base fuze, 12-pr. to 5-inch (2·95-inch double shell).
Plug, base, shell, No. 3, Mark I.	For Q.F., 3- and 6-pr., practice plugged shell.
Plug, base, shell, No. 4, Mark I.	For empty shell (C.P. and A.P., 6-inch and above), also old C.I. practice C.P. shell.
Plug, base, shell, No. 5, Mark I.	For empty shell, C.P., 12-pr. to 5-inch (also old C.I. practice C.P. shell).
Plug, base, shell, No. 6, Mark I.	For shell taking No. 16 base fuze.
Plug, base, shell, No. 7, Mark I.	Or shot (metal), preserving night tracer recess for all shell, except Q.F., 3-pr. and 3-inch.
Plug, base, shell, No. 8, Mark I.	Or shot (metal), preserving night tracer recess ; Q.F., 3-pr. and 3-inch.

Plugs, base, shell, No. 1, *Marks I and II*, are made of gunmetal of the same gauge as the large base fuze ; they have a recess in the head to take the " Key No. 8."

The Mark II plug is longer and heavier than the Mark I; it was introduced to conform with the new design of base fuzes (*i.e.*, No. 11, Mark V and No. 15, Marks II and III).

Fig. 67.

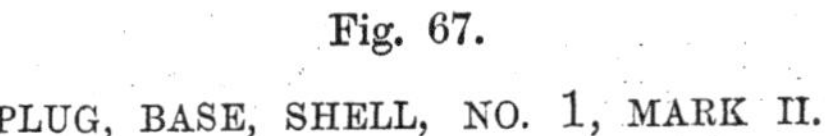

PLUG, BASE, SHELL, NO. 1, MARK II.

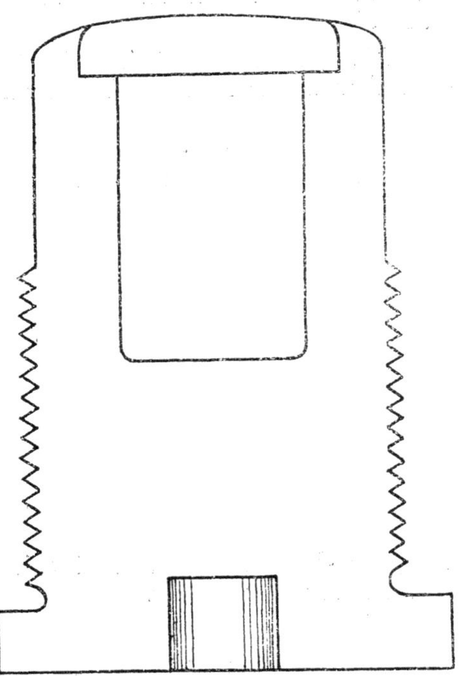

Plug, Base, Shell, No. 2, Marks I to IV.

The plugs are generally similar to the No. 1 base plug, but smaller, and are of the same gauge as the medium base fuze.

The Mark II plug is longer than Mark I; it was introduced with the Mark V No. 12 medium base fuze.

The Mark III plug is $\frac{1}{10}$ of an inch longer than Mark II; it was introduced with the Mark VI No. 12 medium base fuze.

The Mark IV plug differs from the Mark III in having a shorter length of screw thread in order that the threads may not project into the cavity of the shell.

Unserviceable Nos. 11 and 12 Base Fuzes to be used as Plugs.

(*a*) *In salt filled or empty shell.*

The fuzes when used for this purpose will be prepared by removing the magazine plug and detonator, breaking the needle and stamping the word "EMPTY" on the base.

(*b*) *For use in filled shell for Land Service.*

The above-mentioned fuzes when ordered to be used as plugs in filled shell will be prepared by burning out the fuze, and in order to avoid the possibility of a live fuze getting into a shell in mistake for a plug, the magazine and detonator plug of the fuze will be replaced by a wooden plug of the same size and dimensions.

Fuzes are to be stamped on the base with the word "EMPTY."

Plug, base, shell, No. 3, *Mark I* | *C* | is made of gunmetal to the same external dimensions as the Hotchkiss base fuze.

Fig. 68.

PLUG, BASE, SHELL, NO. 3.

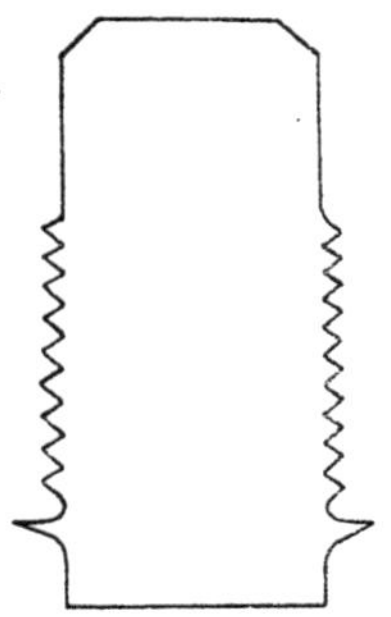

Plugs, preserving, fuze-hole, Hotchkiss, Mark I is made of white metal and is used for store purposes only.

Fig. 69.

PLUG, PRESERVING, FUZE-HOLE, HOTCHKISS, MARK I.

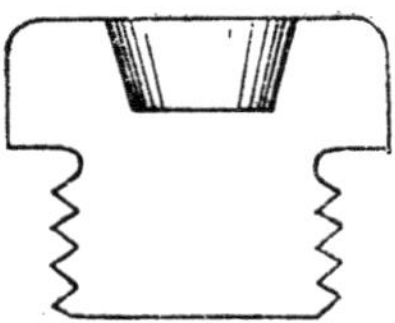

Plug, base, shell, No. 4, *Mark I* | *C* | is made of gunmetal and resembles the Plug, base, shell, No. 1, but is much shorter, the inner end being cut off at the termination of the screw thread.

Plug, base, shell, No. 5, *Mark I* | *C* | is similar to the last named, but is of the same gauge as the No. 2 base plug.

Plug, base, shell, No. 6, *Mark I* | *C* | is similar to No. 1, Mark II, but is 0·17 inch shorter. It must be *free from lead.*

Plug, base, shell, No. 7, *Mark I* | *C* | is of metal, and is used to preserve the tracer recess in shells taking the *internal* tracer.

Plug, base, shell, No. 8, *Mark I* | *C* | is of metal, and is used to preserve the tracer recess in shells taking the *external* tracer.

Note.—All Plugs, base, shell, are stamped on the base with their number and numeral, and the letter " P."

Plug, base, shell, Nos. 1, 2 and 5, supplied by the trade are examined and weighed.

Those found correct to weight are stamped with " W."

Those of incorrect weight will be stamped with " C."

Those marked " W " are suitable for storage or for firing in filled shell, whether filled with powder or salt.

Those marked " C " are suitable for storage, but must be used in salt-filled shell only.

SUNDRY STORES USED IN CONNECTION WITH SHELL.

Adapter, 2-inch fuze-hole, Mark I | *C* |, is made of aluminium, screw-threaded externally below the shoulder to suit the 2-inch fuze-hole gauge, and screw-threaded internally to the G.S. gauge. It is for use with shrapnel shell having the 2-inch gauge, to adapt them to take fuzes having the G.S. taper and pitch.

The adapter has a slot cut in the shoulder to take the key for inserting, or removing, and is fitted with a steel set-screw for securing the fuze when screwed home.

Fig. 70.

ADAPTER, 2-INCH FUZE-HOLE.

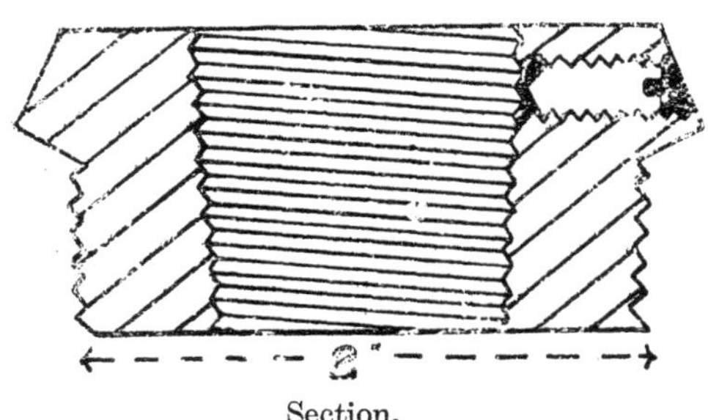

Section.

Adapter, drill, 2-inch fuze-hole, Mark I | *C* |, is similar to the above in dimensions but is made of brass or other hard metal instead of aluminium. It is blackened externally, and is stamped " Drill."

Bolt, eye, lifting projectiles, small, Mark III, is made of wrought iron; it is used for raising to the loading position the projectiles for the 9·2-inch B.L. to 13·5-inch guns. The Mark II bolt, eye, lifting, had two eyes for the hook of hoisting tackle, and three grooves were cut across the screw-thread to allow the escape of any tallow, &c., with which the bolt-hole may be filled. (*See* Fig. 71.)

Fig. 71.

BOLT, EYE, LIFTING PROJECTILES, SMALL, MARKS II AND III.

Mark III. Mark II.

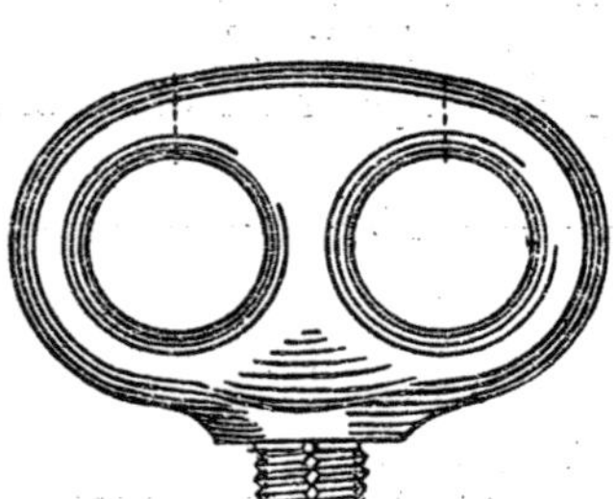

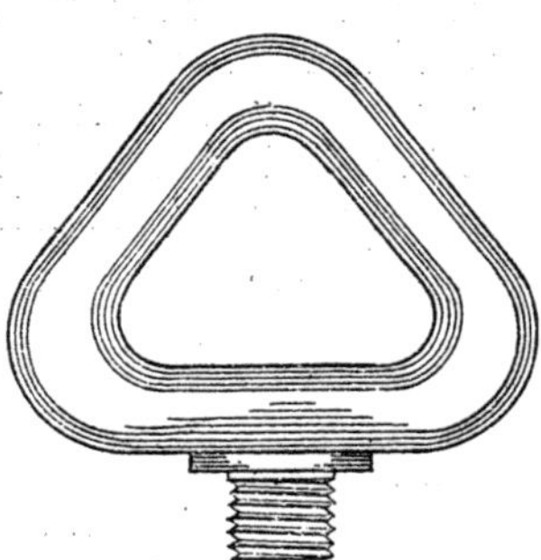

Discs, base plug, are employed with B.L. cast-iron common shell 6-inch and upwards, and the 6-inch cast-steel common shell, Marks IV and V, to seal the joint when the base plug is screwed home.

They are made of lead in two sizes, having a square projection which fits into the hole in the base plug. When hammered home they fit tightly into the undercut recess in the base of the shell.

For the size required for any particular shell, *see* Regulations for Army Ordnance Services, Part II.

Grummets, projectile, sets.—Projectiles having the gas-check form of driving band are fitted with rope grummets to protect the gas-check portion in transport and store.

The latest grummets are made of two pieces of rope—one to go on either side of the gas-check, each having an eyesplice at both ends. The ropes are connected together at intervals by tarred line, and, when fitted to the shell, the ends are secured by a lanyard of tarred line.

The older pattern grummets were spliced together and had to be forced on and off—one from the nose, and the other from the base of the shell.

When in position they are connected by tarred line.

The size of rope used varies from 1½ inch to 3 inch, and, in the case of the heavier natures, the rope in rear of the gas-check portion of the driving band is larger than that in front, in order to give better protection to the gas-check lip.

Protectors, projectile, were placed in a numbered series from 1 to 8. No. 1 is now obsolete. The uses of Nos. 2 to 8 are shown below.

No. 2, Mark II to IV	N	B.L., 13·5-inch.
No. 3, Mark I to III	N	B.L., 12-inch heavy.
No. 4, Mark II to IV	C	B.L., 12-inch light.
No. 5, Mark II to IV	C	B.L., 10-inch.
No. 6, Mark II to IV	C	B.L., 9·2-inch.
No. 7, Mark II to IV	C	B.L., 8-inch.
No. 8, Mark II to IV	C	B.L., B.L.C., Q.F., or Q.F.C., 6-inch.

The protector consists of an elm block recessed to receive the point of the projectile and fitted with a band of hoop iron in one piece, bent, and riveted at the bottom to a dish-shaped plate which fits over the base of the projectile. The top of the protector is secured to the band by two strap bolts, each fitted with studs, which engage

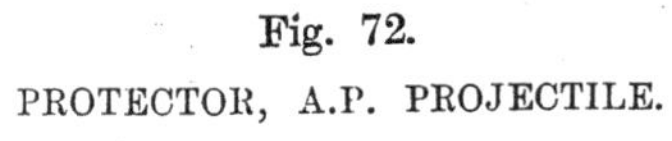

Fig. 72.

PROTECTOR, A.P. PROJECTILE.

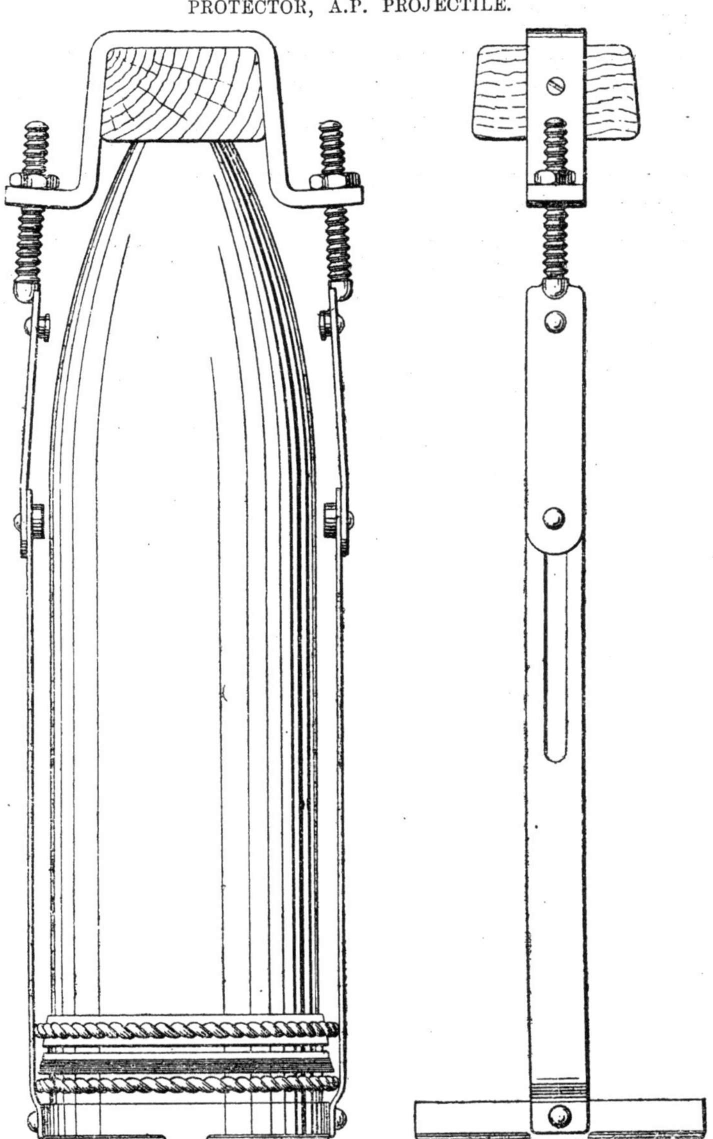

in slots at the top of the band. By engaging any two opposite studs in the slots, the protector can be lengthened or shortened as required, being clamped on to the projectile by the nuts on the strap bolts; in earlier Marks the band was in two pieces.

Certain issues of protectors (B.L. or Q.F., 6-inch to 13·5-inch) differed from the above in being provided with only two studs on the strap bolts.

Protectors, projectile are issued to N.S. with uncapped A.P., and uncapped common-pointed shell. In the L.S. they are issued with uncapped A.P. shell only.

Protectors, armour-piercing projectiles | *C* | .—There is also an older pattern of protector designed for uncapped A.P. shot only. It is not extensible and has no tray for the base, but the strap passing under the base has two studs which fit into the holes of the base plug, and so retain the protector. They have been superseded by the protector, projectile.

Rings, lead, base plug, large and small. They are made of lead ·3-inch in thickness, and hollowed out on the under surface; they are intended to be hammered into the recess between the metal adapter and the shell to seal the joint.

They are used with the earlier marks of cast-steel common and common-pointed shell.

For the size required for any particular shell, *see* Regulations for Army Ordnance Services, Part II.

Screws, preserving, eye-bolt, holes, small, Mark I | *C* | are used for projectiles having eye-bolt holes in the side. Their use is to prevent dirt, &c., filling up the hole.

Slings, lifting, projectiles, B.L., Mark I | *L* | are used for slinging A.P. projectiles. They consist of a band of spring steel, 4 inches wide, having at each end a lifting eye, one being made smaller than the other. The band fits round the centre of the projectile, the smaller eye being passed through the larger one, and the lifting gear attached to the former.

Fig. 73.

SLING, PROJECTILE, B.L., 60-PR.

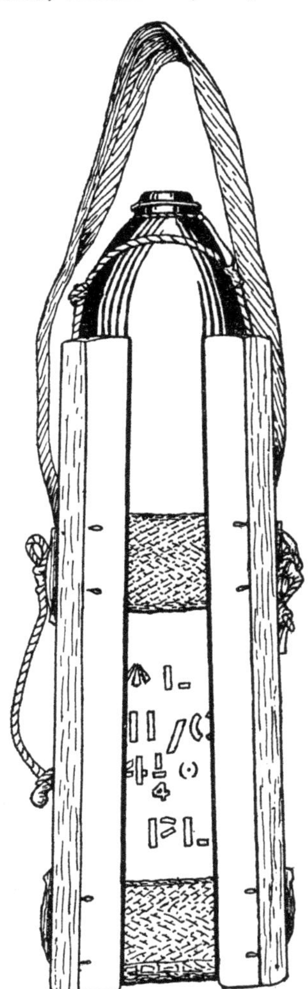

Slings, projectiles, B.L., are used with B.L., 60-pr., 5-inch howitzer, and 6-inch siege guns. The sling consists of four wood battens braced together by two transverse bands of webbing. A handle and base are formed by webbing fixed to the two transverse bands. A short length of white line is provided to secure the projectile in the sling. The sling, which will carry one projectile, is for use in place of boxes in transporting shell in wagons.

Straps, carrying, projectiles, B.L., 5-inch howitzer, Mark I.—The strap consists of two stout leather straps with iron runners attached to an iron handle covered with leather.

There is also a similar strap issued for carrying B.L., 5·4-inch howitzer projectiles.

Selvagees are used in the L.S. for slinging projectiles when loading. They are made of spunyarn and are issued in five sizes as shown below :—

Designation.	Service.	Detail.		—
		Length (inside when stretched straight).	No. of Strands.	
Selvagees—		(3-thread yarn.) in. in.		For projectiles, B.L., or R.M.L.—
29-inch ...	L	29 to 30	6	B.L., 6-inch.
36-inch ...	L	36 to 37	12	B.L., 9·2-inch, R.M.L., 9-inch.
43-inch ...	L	43 to 44	18	B.L., 10-inch, R.M.L., 10-inch.
48-inch ...	L	48 to 49	27	For General Service.
53-inch ...	L	53 to 54	26	B.L., 5-inch howitzer.

CHAPTER XII.—HAND GRENADES.

(*Plate* XLIII.)

Grenade, hand, No. 1, *Mark I* | *L* | consists of a brass body with hook, cap with needle and safety pin, a cast-iron ring, a cane handle with wood block and silk braid tail, detonator holder with detonator, and bursting charge.

Body.—The body is made of solid drawn brass, cylindrical in shape; soldered to it is a hook for attaching the grenade to a waist belt.

Around the body near the centre is a cast-iron ring, secured by a projection on the body and solder. This ring is intended to break up on the detonation of the charge. This ring, by its weight, ensures that the grenade shall fall on the end which carries the cap; the needle is thus driven on to the detonator.

Bursting charge.—Near the centre of the body is a brass disc soldered in. The upper portion of the body above this disc is lined with asbestos and filled with a charge of 4 oz. 2 dr. of lyddite. A cavity is formed in the lyddite into which is placed a hollow pellet of compressed picric acid, lined inside with a paper cylinder. Over the junction of the lyddite and picric acid is placed a paper ring covered with a layer of kit composition.

Detonator holder.—In the centre of the picric acid pellet is placed a brass flanged detonator holder; the flange fits over the kit composition, and closes the end of the body, and is attached by screws fitting into three brass sockets soldered to the body.

Detonator.—Inside the holder fits a flanged detonator of solid drawn copper, filled at the lower end with about 30 grains of F. of M. secured in position by a brass cup, the body being indented to prevent the cup from moving. To secure the detonator, insert it and give it a slight turn; its flange engages under two pins on the holder; a small spring holds it in this position. (*See* Fig. 171, page 522.)

Cap.—Over the end of the body is placed a brass cap, which carries a steel needle riveted to the centre of it; the upper part of the cap is filled with lead.

The cap is secured to the body by a double bayonet joint, and can be placed in three positions, which are marked as follows:—"Remove," "Travel," and "Fire."

Safety pin.—A phosphor bronze safety pin passes through the cap and the needle, and prevents the needle from being crushed on to the detonator; the pin is secured by a becket and leather tie.

Handle.—The handle is made of cane, about 16 inches in length. About 6 inches from the end of the handle is attached a silk braid tail one yard long. The handle fits into a block of beech wood, and is secured by two wood pins, glued, and driven in.

The block is covered by a brass cup, and fits into the lower end of the body and is secured by three brass screws.

The body of the grenade is stamped with the manufacturer's initials and date of manufacture. When filled it has a red band stencilled on. The cast-iron ring is painted yellow.

To Prepare the Grenade for Use.

(1) Turn the cap to the right until the word "remove" is opposite the arrow painted on the body; the cap is then taken off.
(2) Place the detonator into its recess, care being taken that the two grooves on its flange coincide with the two projecting studs; press the detonator gently home, and then turn it to the left, passing the flange under the heads of the studs until the brass plate-spring is released, thus locking the detonator in position.
(3) Replace the cap with the word "remove" opposite the arrows on the body; press down into position and turn ⅛ of a turn to the left, until the word "travel" is opposite the indicating arrows on the body.

To Throw the Grenade.

(1) Unwind the tail and allow it to hang loose at full length.
(2) Turn the cap from the "travel" to the "fire" position.
(3) Remove the safety pin.
(4) Grasp the grooved end of the cane handle and throw the grenade in the required direction, either under or overhand, care being taken that the tail cannot entangle itself with the thrower or with any object near him.

Note.—The grenade should be thrown well upwards at not less than an angle of about 35 degrees. This, besides assisting to increase the range to which the grenade can be thrown, renders its action more certain by causing it to strike the ground nearly vertical. This is especially important when throwing with a following wind.

The circumstances under which "hand grenades" will be employed in war are detailed in Field Service Regulations, Part I.

Grenades for Instruction and Practice.

Grenade, hand, for instruction, Mark I dummy is similar to the Service one, but has a wood block in place of the charge, and is fitted with a dummy detonator, charged with clay instead of F. of M.

It has the word "dummy" painted on it in red, and both the grenade and detonator have holes bored through them.

Grenade, hand, practice, Mark I | L |.—The practice grenade has a beechwood body, fitted with a fixed wrought-iron cap, and has no safety pin. It is fitted with an iron ring, and weighted with lead to the weight of the Service grenade.

Two spare handles are issued with each of the practice grenades.

CHAPTER XIII.—FUZES.

(A) General Remarks; (B) D.A. and D.A. Delay Fuzes; (C) D.A. Impact Fuzes; (D) Graze Fuzes; (E) Time Fuzes; (F) Time and Percussion Fuzes; (G) Keys, &c., Used in connection with Fuzes.

(A.) GENERAL REMARKS ON FUZES.

Classes of Fuzes; *Influences affecting the rate of burning of Time and T. and P. Fuzes; Fuze-hole Gauges; Waterproofing; Marking; Packing; Storage.*

Classes of Fuzes.

The bursting charge of a shell is ignited by means of a fuze designed so as to act at any particular moment during its flight, or upon or after impact.

Fuzes may be divided into *three* classes :—

(1) Percussion Fuzes. { Direct Action. Direct Action Impact. Graze Action. Direct Action Delay.

(2) Time Fuzes.

(3) Time and Percussion Fuzes.

Direct action (D.A.) fuzes.—The direct action fuzes require a heavy blow to make them act. There are no loose or movable parts inside the fuze to move forward when the shell strikes. The head of the fuze is fitted with a copper disc supporting a steel needle over a detonator; this needle must be crushed in to explode the fuze. Safety in transport is ensured by a cap or screw plug covering the top of the fuze which is removed just before loading.

The following percussion fuzes are known as "*Direct Action*" :—

Nos. 1, 3, 17 and 44.

Direct action impact (D.A.I.) fuzes.—D. A. Impact fuzes are used with common lyddite shell; they require a still heavier blow than the D.A. fuzes to make them act.

Instead of a copper disc with steel needle, they have a steel hammer suspended over a detonator by a strong steel shearing pin.

The head of the fuze must be crushed in and the shearing pin broken to make the fuze act.

The following fuzes are known as "*Direct Action Impact*" :—

Nos. 13, 18, 19, 19A and 45.

Graze fuzes.—Graze fuzes are very sensitive; their action depends upon a movable pellet inside the fuze usually carrying a detonator, which, on the shell being checked in flight by grazing an object, moves forward carrying the detonator on to a needle.

Usually these fuzes are required to act very quickly, before the shell has time to rise after glancing off the ground.

In graze action fuzes it is necessary to make special arrangements to guard against premature action which may occur :—

(1) During transport or handling, from jolts, accidental dropping, jars in loading, &c.

(2) After discharge, while the shell is still in the bore of the gun.

(3) During the flight of the shell, before it strikes or grazes.

To guard against (1), there is generally more than one arrangement, and, in the majority of nose fuzes, " safety pins " are employed. These are stout wires so placed that they support some portion of the fuze in such a manner that the detonator or needle pellet cannot move forward until they are pulled out.

Safety pins cannot be used with base fuzes, so, in these, other special safety arrangements are made, which will be understood from the description of the fuzes.

After the safety pins are removed it is necessary that the pellet should still be prevented from moving until after the shock of discharge. This is done in various ways, for instance, by some portion which locks the pellet, " setting back " on the shock of discharge ; or by " centrifugal bolts," which lock the pellet to the fuze body and are caused to withdraw by centrifugal force due to the rotation of the shell.

Very often both of these arrangements are to be found in the same fuze.

It will be understood, therefore, that the pellet is only free to move when the shell is in flight.

Rebound action of pellet.—As regards (2), on discharge, the shell receives a violent shock which has a tendency to cause a rebound action of any loose parts inside the fuze. This rebound action might cause the detonator pellet to jerk forward on to the needle. This action can be overcome to a great extent by making some of the parts of soft metal such as lead, as is done in the case of the Hotchkiss fuze, though this is not necessary in many other types.

There is always a spiral spring placed between the pellet and the fuze body, which must be compressed before the fuze can fire. These springs prevent rebound action ; they are made of various degrees of strength, according to the degree of sensitiveness required from the fuze.

Creeping action of pellet.—These spiral springs also prevent the premature action of the fuze in (3). While the shell is in flight, there is a tendency for the pellet to move forward, because the velocity of the shell is being checked by the resistance of the air, and this resistance cannot affect the interior portions of the fuze, which consequently would move onward with the muzzle velocity of the shell, unless checked by some means. This action is known as " Creeping," and is prevented by the spiral spring.

Boring action of pellet.—If the pellet is allowed to rotate inside the fuze body it has a tendency to gradually screw its way forward until the needle pierces the detonator tending to cause either a blind or a premature. This action is prevented in most cases by a longitudinal slot in the pellet, in which a pin projecting from the inside of the body works. This prevents rotation of the pellet independently of the body of the fuze, while allowing it to move forward on graze or impact.

Direct action delay fuzes.—Some direct action fuzes are designed to have a *delay action*; a percussion arrangement ignites a column of composition on impact, which burns for a fraction of a second before igniting the bursting charge, thus giving the shell time to penetrate.

Example.—The D.A. Delay No. 10 Fuze.

Time fuzes.—Time fuzes are constructed to act at the expiration of an interval of time by the burning of a length of slow-burning composition, this time being regulated by the setting of the fuze previous to loading. The composition is pressed hard into a groove in a ring which is movable, and the setting of a fuze merely means that a longer or shorter portion of the composition is allowed to burn before it explodes the shell. The composition is ignited either by detonator being fired on shock or discharge, or by the rotary motion of the shell causing a detonator pellet to move on to a needle.

The following Time fuzes will be met with :—

Fuzes, time, Nos. 24, 25 and 30.

Time and Percussion (T. and P.) Fuzes.

In addition to the time arrangement described above, many fuzes contain a graze percussion action in the same body, and are then known as "combined" or T. and P. fuzes.

The object of this combination is not so much that the fuze may be used indifferently for time or percussion shrapnel, as that the bursting of the shell may be secured if it should graze before the time arrangement acts.

"Double-Banked" T. and P. Fuzes.

The composition in the time rings is made to burn slowly, but there is a limit to this ; if it is made to burn too slowly the regularity of the fuze is impaired. Therefore, when a fuze to burn for a long time is required, it is necessary to increase the length of the composition. This can be done by making the fuze larger, or by having two time rings instead of one as in some time and percussion fuzes. Fuzes with two time rings are known as "double-banked" fuzes ; in these, the length of the composition is almost doubled, while the size of the fuze is not increased. Usually the top ring is fixed, and the lower one is movable for setting. The fuze is arranged so that the top ring lights first, and after an interval, depending upon the setting of the fuze, it lights the lower one, which burns back in the opposite direction the same length as the top ring, and then fires the bursting charge of the shell.

Tension fuzes.—With the older types of fuzes it was necessary to unclamp the time rings to set the fuze. This was done by loosening the top nut which was tightened up again after setting. To save the time thus taken up, *tension fuzes* were introduced. When these fuzes are assembled the top cap is screwed down till a weight of 24 ozs. at the end of a 6-inch lever will just turn the ring. To set the fuze it is only necessary to turn the ring to the required setting. All modern time and T. and P. fuzes are "*Tension fuzes.*"

Gas-escape hole.—It is essential that there should be a good escape for the gas from the burning composition, otherwise the fuze will explode instead of burning regularly.

The first metal T. and P. fuzes had escape holes in the head of the fuze. The gases escaping from these holes met the resistance of the air direct when the shell was in flight, and so caused the fuze to burn irregularly.

The later fuzes had the head entirely closed and a side channel for the escape of the gases.

This side channel was bored from the interior slightly towards the rear of the fuze, and slanted away from the direction of rotation of the shell.

In all the latest time and T. and P. fuzes, a gas-escape hole is bored into each time ring at the beginning of the composition. These holes are closed by means of thin brass or aluminium discs.

Perforated powder pellets are generally placed in the gas-escape holes to ensure the closing discs being blown out when the composition is ignited.

Influences affecting the rate of burning.—The rate of burning of a time fuze is influenced by its age, the climate in which it has been kept, and the atmospheric pressure. An old fuze burns slower and therefore longer than a new one. Fuzes kept in a very dry or very damp climate will burn irregularly, the effect of extreme dryness being to make them burn faster. Increased atmospheric pressure causes fuzes to burn more rapidly, while diminishing the pressure causes them to burn more slowly.

The reason for the retardation of the time of burning, due to diminished pressure, is briefly this. Each layer of the fuze composition must be raised to the temperature necessary for combustion by heat transmitted from the burning layer above it. When the pressure is diminished the incandescent gases can expand more freely and consequently transmit less of their heat to the layer beneath them, the contact being less close ; moreover the cooling effect due to rapid expansion is well known.

Each diminution of atmospheric pressure of one inch of the barometer increases the *time* of burning of T. and P. No. 64, 65, 66, 67, 68, 80, 81, 82, 83 and 84, and Time No. 25 and 30 fuzes by $\frac{1}{44}$th, and all other T. and P. fuzes by $\frac{1}{30}$th.

The barometer falls about one inch for an increase of 1,000 feet in elevation ; thus at 5,000 feet elevation the time of burning of a No. 80 fuze would be increased by $\frac{5}{44}$ and would therefore burn :—

$$22 + \left(\tfrac{5}{44} \text{ of } \tfrac{22}{1}\right) = 24\cdot5 \text{ seconds.}$$

The effect of varying pressures on the rate of burning explains to some extent the important fact that fuzes burn at different rates when fired out of different guns; as a rule, they are found to burn quicker in large than in small guns, probably because the projectiles from the former keep up their velocity better.

Gauge of Fuzes.

Fuzes are designed to screw into the nose or the base of a shell.

Graze fuzes may be designed for either the nose or base, but all Direct Action, D.A. Impact, Time, and Time and Percussion fuzes screw into the nose.

There are now four sizes of nose fuze-holes in the Service, the (G.S.) General Service gauge (1-inch), and the 2-inch gauge; also the special gauge for the 3-pr. Q.F., lyddite, and the Q.F., 1-pr, 1½-pr. and 2-pr.

With the exception of the Fuze, Time, No. 30, and the Fuze, D.A. Impact No. 19 and 19A, all nose fuzes below No. 80 have the G.S. gauge, and all T. and P. fuzes No. 80 and above and Time No. 30 have the 2-inch gauge; except 1-pr., 1½-pr. and 2-pr.

Fuzes filled with "Composition exploding."—The following D.A. and D.A. Impact fuzes are fitted with special detonators containing fulminate of mercury instead of detonating composition, and are filled with "Composition exploding," instead of F.G. powder:—

D.A., No. 44, D.A. Impact, Nos. 18, 19, 19A and 45.

The above fuzes are used with lyddite shell fitted with T.N.T. exploders.

Marking stamped on fuzes.—All fuzes are now stamped with their mark, lot number, maker's initials, date, and distinguishing number.

Lot number of a fuze.—T. and P. fuzes are manufactured in "lots" of 2,000, percussion fuzes in lots of 1,000.

Waterproofing of Fuzes.

All openings in percussion fuzes are coated with Pettman's cement or waterproofing composition to prevent the ingress of damp.

Waterproofing Time and T. and P. fuzes.—In all the latest Time and T. and P. fuzes, the spaces between the cap, time rings and body, also the set-screw recess of cap and gas-escape hole discs in the time rings, are now waterproofed with a composition composed of bees-wax, mineral jelly and French chalk.

The base plugs are waterproofed by having the threads coated with Pettman's cement before screwing in, and then the whole of the base is coated with varnish.

The flash hole of the base plug is also coated with a special collodion varnish.

The No. 80 Marks IV and V fuzes are further waterproofed by means of a brass "fuze cover," which fits over and is soldered to the body. (*See* page 324.)

Fuze Cylinders.

Fuzes must be carefully protected from damp or they will quickly deteriorate, so they are packed in tin cylinders, the lids of which are soldered on. Each cylinder has a label on the lid showing the number, nature, mark, lot number, date of packing, and manufacturer's initials. All Time and Time and Percussion fuzes have also marked on the label the "*mean time of burning*" at rest when set full. The following is the label for a cylinder containing a "Fuze, T. and P., No. 80."

Fuze cylinders containing fuzes having the 2-inch gauge are painted *green.*

Fuze cylinders containing fuzes filled with C.E. are painted *yellow.*

All other fuze cylinders are painted black.

Issue.

Fuzes are generally issued one in a cylinder, 25 or 50 cylinders in a packing case.

(B). DIRECT ACTION FUZES.

Fuze, Percussion, Direct Action, with Cap, No. 1, Mark III | L | .

(*Plate* XLIV.)

This fuze is used in the Land Service with—

B.L., 60-pr. and 4-inch (jointed).
B.L.C., 6-inch (siege).
Q.F., 4·7-inch (on travelling carriage).
And all B.L. and Q.F. howitzers.

Description.—The fuze consists of the following parts, viz.:—Body, safety cap, screw collar, needle disc with steel needle, screw plug for needle disc, detonator plug with R.L. cap, and bottom plug.

All the parts are made of an alloy resembling gunmetal, with the exception of the steel needle, and the needle disc, which is of copper.

Body.—The body is threaded on the exterior to the G.S. gauge, the upper portion being left plain to receive the safety cap. The lower part of the fuze is bored out to form a small chamber which takes a blowing charge of about 63 grains of F.G. powder, and the bottom is closed by a bottom plug screwed in, having a central fire hole closed on the upper side by a disc of white paper and one of shalloon.

The fuze body is also bored out from the top, and screw-threaded to take the percussion arrangement; below this the centre of the body is recessed and screwed to receive a detonator plug; under this recess a small central fire hole is bored, communicating with the charge in the magazine of the fuze. A disc of paper prevents the powder working up into this fire-hole.

R.L. cap.—The detonator is known as the R.L. cap. It consists of a cap of copper, the top of which is cut out, and the small opening closed by a thin brass disc. The cap contains about 3½ grains of cap composition pressed in, and then varnished, and covered by a disc of tinfoil. The bottom of the cap is then closed by a disc of copper pierced with 4 fire-holes, secured in position by six lugs on the rim of the cap bent down on to it. (*See* Fig. 74.)

Fig. 74.

Scale $\frac{4}{1}$.

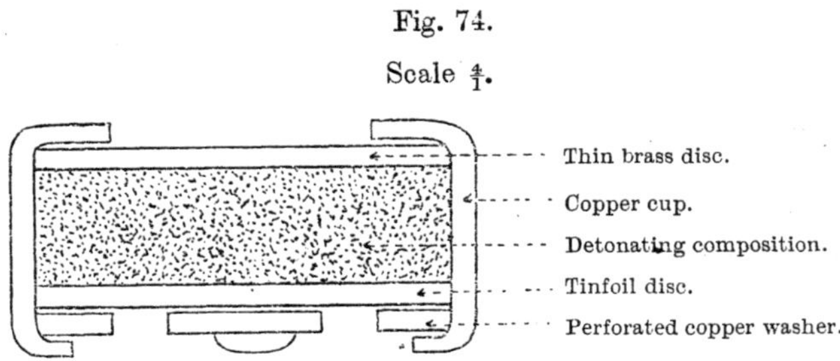

The R.L. cap is retained in position in the recess in the bottom of the percussion chamber by means of the detonator plug.

Screw plug.—The screw plug for the needle disc is threaded so as to screw into the bottom of the percussion chamber; it is recessed for the point of the needle. Two holes are drilled in it, for screwing the plug into the fuze.

Needle disc and needle.—The needle disc, with a steel needle having a single point snapped on to the centre of it, rests on a small shoulder made in the top of the screw plug, and is thus suspended with the point of the needle directly over the R.L. cap.

Screw collar.—The screw collar retains the needle disc in position. It screws into the body over the screw plug; it has a slot cut in its upper edge to take a screwdriver.

Safety cap.—The safety cap fits over the head of the fuze, and has a milled edge. On each side a "T" shaped slot is cut, to fit over two brass pins in the body, and by which the cap is secured to the fuze. A square recess is cut in the top of the cap to take the flat arm of the fuze-key, when screwing the fuze home in the shell.

Waterproofing.—The exterior of the fuze is lacquered, and the fuze is carefully waterproofed by putting a little Pettman's cement on the threads of the plug for needle disc, edge of needle disc, screw collar and bottom plug before screwing them in, and finally painting the top of the fuze below the safety cap with the same cement, so as to completely cover the needle disc ; and also painting the bottom of the fuze and bottom plug. The needle is soldered to the needle disc to improve the water tightness of the fuze.

Action.—The fuze, being prepared by simply removing the safety cap, is quiescent in all its parts till direct impact takes place, or graze at such an angle (10° or over) that the nose of the shell enters the ground. When the needle disc is crushed in, the needle pierces the detonator and fires it, the flash from the exploding detonator passes through the central flash hole and fires the magazine of the fuze, which in turn fires the bursting charge of the shell.

Mark II fuze.—The Mark II fuze differs from Mark III in having nine conical fire holes filled with mealed powder instead of a single fire hole. A recess in the bottom of percussion chamber contains 3½ grains of cap composition pressed in and covered with a brass disc held in position by a copper washer, over which the metal is spun, instead of an R.L. cap. The needle has four points instead of one, and the magazine is slightly larger and contains more powder.

Marks I*, I** and II are identical in construction, but the former are conversions from Mark I, while the latter is a new fuze. Mark I is obsolete.

Issue.—The fuzes are wrapped in brown paper, and issued five in a tin cylinder ; for 5-inch howitzer and 60-pr. equipment, only one in a cylinder.

FUZE, PERCUSSION, DIRECT ACTION, WITH PLUG, No. 3, MARK IV | N |.

This fuze is now only used in the Naval Service ; it is for common shell with—

B.L., 13·5-inch, 10-inch, 9·2-inch and 6-inch and Q.F., 6-inch.

The fuze in general construction is similar to the No. 1 with cap, Mark III. The body, however, is screw-threaded throughout its entire length, and is fitted with a safety plug screwing into the top of the fuze, instead of a safety cap.

This safety plug must be removed at the moment of loading; for this purpose a slot is cut across the upper surface of it to take the fuze key, and it is marked with an arrow, and the word " Unscrew," showing the direction to turn.

Issue.—The fuzes are wrapped in brown paper and packed five in a tin cylinder.

The Mark III differs from Mark IV in having nine conical fire holes filled with mealed powder instead of a single fire hole. A recess in the bottom of percussion chamber containing 3½ grains of cap composition pressed in and covered with a brass disc held in position by a copper washer, over which the metal is spun, instead of an R.L. cap; the needle has four points instead of one, and the magazine is slightly larger and contains more powder.

Marks I and II are obsolete.

Fig. 75.

FUZE, PERCUSSION, DIRECT ACTION, WITH PLUG, NO. 3, MARK IV.

Full Size.

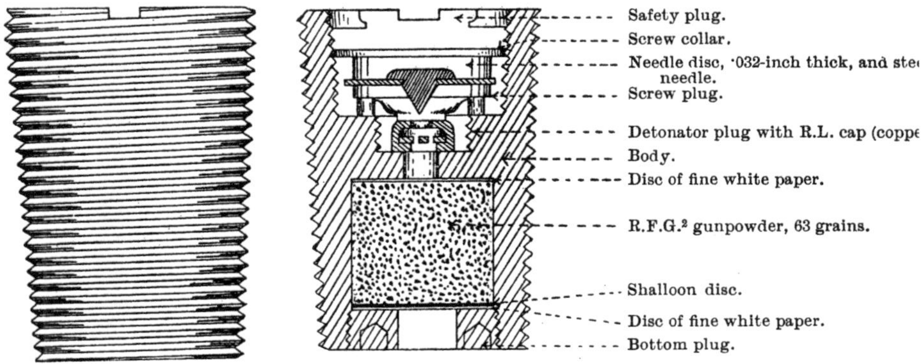

Fuze, Percussion, Direct Action, Delay, No. 10, Mark III | L | .

This fuze is used with the B.L., 6-inch howitzers of 25 and 30 cwt., when specially ordered. (*See* Table 27.)

The construction of the percussion portion of this fuze is very nearly identical with that of the "Fuze, percussion, D.A., with cap, No. 1, Mark II."

There are a few minor differences, namely:—The recess in the bottom of the percussion chamber only contains 1 grain of detonating composition and the brass disc covering this composition is retained in position by a small screw collar. There are five cylindrical fire-holes under the detonator which are not filled with powder. The body of the fuze is longer than the D.A., with cap, No. 1, but generally resembles it in shape.

Two holes are bored in the lower part of the body, parallel to its axis, the smaller forming the delay chamber, and the larger is divided into two compartments by a diaphragm plug, the upper forming a chamber for the gas escaping from the burning composition, while the lower compartment forms the magazine of the fuze.

The delay chamber communicates with the top of the air chamber by a fire hole which is primed with quick match to carry the flash from the detonator to the pressed mealed powder with which the delay chamber is filled.

The bottom of the delay chamber communicates by means of a small fire hole with the magazine of the fuze, which is filled with about 20 grains of M.G. powder, retained in position by the bottom plug, which is screwed into the bottom of the fuze body. The bottom plug has a central fire hole, closed on the lower side by a brass disc spun in.

Fig. 76.

FUZE, PERCUSSION, DIRECT ACTION, DELAY, NO. 10, MARK III, METAL: 1 IN A TIN CYLINDER.

Full Size.

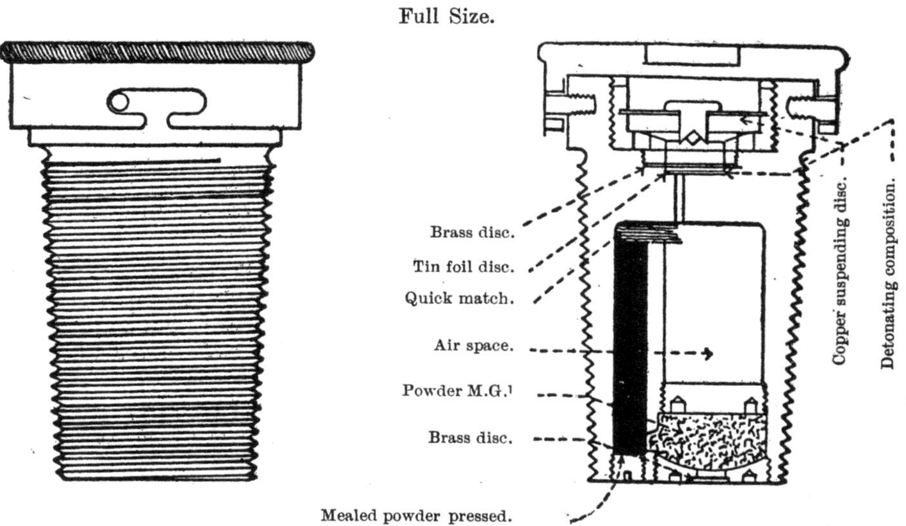

Action.—The safety cap having been removed at the moment of loading, the percussion arrangement, on impact, ignites the quick-match priming and the column of mealed powder in the delay chamber, which burns for about half a second, thus giving the shell time to penetrate well before bursting.

Issued, wrapped in brown paper, one in a tin cylinder.

FUZE, PERCUSSION, DIRECT ACTION, WITH CAP, NO. 17, MARK III | L |.

(*Plate* XLV.)

This fuze is used with all field and siege guns and howitzers when firing lyddite shell, *i.e.* :—

B.L., 2·75-inch and 60-pr.
B.L.C., 6-inch (siege).
Q.F., 4·7-inch (travelling carriage).
All B.L. and Q.F. howitzers.

Parts.—It consists of the following parts :—Body, safety cap, safety pin, needle disc with steel needle, screw collar for needle disc, detonator plug with R.L. cap, steel pellet with screw collar and shearing wire, centrifugal bolt with spiral spring and closing plug, screw plug with brass shield, and bottom plug.

Body.—The body is made of a metal resembling gunmetal; externally, it is similar to the D.A., No. 1, with cap, but longer, and is fitted with the same pattern of safety cap. The upper part of the fuze is bored out to take the needle disc and screw collar, and a small central recess is screw-threaded for the detonator plug and R.L. cap.

Needle disc.—The needle disc is of copper; the rim is flanged and fits into a circular groove in the body. A steel needle snapped and soldered to the centre of the disc is suspended over the detonator. The needle disc is retained in position by a screw collar which screws into the top of the fuze body. A small hole closed at the outer end by a brass disc is bored through the side of the body into the space below the needle disc; this hole is intended to allow the flash from the detonator to escape if the detonator is prematurely fired, and so to prevent the shell from exploding in the bore.

A central fire channel is bored through the body leading from the detonator to the magazine in the base of the fuze. The upper part of this channel is enlarged and screw-threaded to take a small screw collar containing a steel pellet; this pellet is suspended by means of a copper shearing wire passing through it and the collar.

Centrifugal bolt.—The centrifugal bolt is placed into a radial hole in the body; the small end of this bolt projects into, and masks the central fire channel; a small spiral spring bearing against the heavy end of the bolt retains the latter in position until acted upon by centrifugal force.

Safety cap.—The safety cap fits over the head of the fuze; it has two "T" shaped slots to engage with brass pins in the head of the fuze.

Safety pin.—The safety pin is of bronze, and passes in a diagonal direction through the side of the safety cap and fuze body and in front of the enlarged head of the centrifugal bolt, and so retains the cap, and locks the bolt in the safe position.

Brass shield.—Screwed into the upper part of the magazine recess is a hollow plug; a brass cup or shield placed over the front end of the plug seals the lower end of the central fire channel and so prevents the flash from the detonator passing through to the magazine until the brass shield has been perforated by the steel pellet.

The magazine contains about 70 grains of F.G. powder retained in position by the bottom plug. A central flash hole in the bottom plug is closed by means of a thin brass disc spun in.

The fuze is lacquered inside and out, and is stamped with the manufacturer's initials, mark, lot number, and date of manufacture.

Action.—At the last moment of loading the safety pin and safety cap are removed. On the shock of discharge nothing takes place, but on rotation the centrifugal motion of the shell causes the bolt to spin outwards, compressing its spiral spring and opening the fire-channel in the fuze body. On impact the needle disc is crushed in, carrying its needle on to the detonator. The explosion of the detonator forces the steel pellet through the fire channel; it shears its suspending wire and pierces the brass shield, and thus allows the flash from the detonator to fire the magazine of the fuze, which in turn explodes the bursting charge of the shell. If the detonator is fired prematurely before the centrifugal bolt has been spun out by the rotation of the shell, the steel pellet would be unable to pass the small end of the bolt. The flash from the exploding detonator would in this case blow out the brass disc in the small side hole in the upper part of the body, and so would escape to the exterior. The result would be a *blind shell* instead of a *premature* in the bore.

No. 17, *Mark II.*

The Fuze, percussion, D.A., with cap No. 17, Mark II, differs from Mark III in having two instead of one centrifugal bolt; a slightly different form of needle disc and the base of the fuze is fitted with a magazine plug containing a charge of F.G. powder, on the front end of which is placed the brass shield.

No. 17, *Mark I.*

The Mark I differs from Mark II in having a shorter needle, and the lower end of the safety pin hole is not closed with a brass screw plug.

(C). D.A. IMPACT FUZES.

Fuze, Percussion, D.A. Impact, No. 13, Mark V | C |.

(*Plate* XLVI.)

This fuze is used with heavy common lyddite shell, *i.e.*, B.L., 9·2-in. and above.

It may also be used with common lyddite shell with picric powder exploders for the following guns :—

B.L., 30-pr., 4-inch, 6-inch and 7·5-inch.

Q.F., 12- and 14-pr., 4-inch, 4·7-inch and 6-inch.

Description.—The fuze consists of the following principal parts :—Body, safety cap with securing pin, steel hammer with screw collar and shearing pin, detonator plug with mealed powder detonator and brass collar.

The body externally is similar to the Fuze, percussion, D.A., No. 1, with cap, but longer.

In the centre of the body is screwed a steel plug, recessed on the upper side to take one grain of mealed powder, enclosed between two discs of tinfoil, held in position by a small screw collar.

Leading from this recess are four conical fire holes filled with pistol powder, and closed on the underside by a disc of paper. The lower portion of the body is filled with pellet powder, retained in position by a brass disc spun in.

In the upper part of the fuze, suspended by means of a steel shearing pin, is a steel hammer, the lower part of which is reduced in diameter and is directly over the recess in the plug.

The top of the hammer is also reduced in diameter ; over it fits a screw collar ; a brass disc placed on top of the collar, and spun into position, prevents the ingress of damp. The safety cap is of manganese bronze, and has two T-shaped slots to engage with brass pins in the head of the fuze, and is further secured by two securing pins which fit into vertical holes in the cap and body. The pins are connected by copper wire, which is held to the cap by a brass strip bent over it.

Action.—Before loading, the securing pins and safety cap are removed.

Fig. 77.

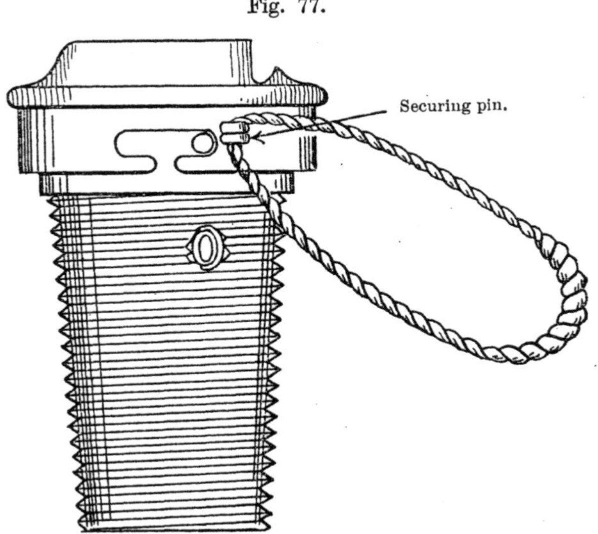

On impact, the hammer is driven in, shearing the pin, firing the mealed powder detonator, the flash passing through the detonator plug into the magazine.

Mark IV fuze.—The Mark IV fuze differs from the above in having a securing pin which passed horizontally through the sides of the cap and body. (*See* Fig. 77.)

In the later issues of this mark a slot was cut across the head of the cap and then painted red to show a straight lead for the withdrawal of the pin.

The Mark III fuze.—In the Mark III, the head of the fuze was entirely covered with a thin brass disc, instead of only the top of the steel hammer.

The Mark II fuze.—This mark of fuze differs from Mark III in having a thicker brass disc.

The Mark I fuze had the top covered with a paper instead of a brass disc.

The Marks I, I**, II*,* and *III** fuzes are those which have been altered to conform with Mark IV.

Fuze, Percussion, D.A. Impact, No. 18, Mark II | C | .

(*Plate* XLVII.)

This fuze was introduced for use with common lyddite shell, in the Land Service, for Coast Defence guns, Q.F., 12-pr. to 4·7-inch inclusive. In the Naval Service with the B.L., 4-inch, Q.F., 12- and 14-pr., 4-inch and 4·7-inch.

This fuze, as existing stock is used up, will be superseded by the D.A. Impact, No. 45, fuze.

Description.—The fuze consists of the following parts:—Body, bottom plug, percussion holder with detonator, screw collar, steel plug and shearing pin, safety cap and securing pin.

Body.—The body is generally similar to the Fuze, percussion, D.A. Impact, No. 13, Mark IV, but longer, and the lower part of the body is not screw-threaded. It is bored out from the bottom to form a magazine and screw-threaded for the bottom plug. The fuze body is also bored out from the top to take the percussion holder, a central channel forming a communication between the percussion chamber and the magazine of the fuze; this channel is filled with loose "composition exploding," stemmed in, and retained in position by a disc of thin white paper secured with shellac.

A disc of tinfoil is shellaced into a recess in the bottom of the percussion chamber.

The magazine of the fuze contains a pellet of "composition exploding," compressed under a pressure of 2 tons per square inch; the bottom is closed by a solid screw plug, covered on the inside with a disc of shalloon and coated on the underside with Pettman's cement.

Percussion holder.—The percussion holder screws into the top of the fuze. Suspended in this holder by means of a steel shearing pin is a steel plug with a needle point; this needle point is directly over a detonator.

The detonator contains four grains of fulminate of mercury; it fits into a brass collar screwed into the underside of the percussion holder.

A brass disc placed on top of the holder, and spun into position, prevents the ingress of damp into the fuze.

Safety cap.—The safety cap is of manganese bronze, and has two "T" shaped slots cut in the flange to engage with two brass pins in the head of the fuze.

Securing pin.—The safety cap is retained in position by means of a securing pin passing through it and the side of the head of the fuze.

The later issues had a slot cut across the head of the cap and then painted red to show the direction in which the securing pin must be withdrawn.

Action.—At the last moment of loading the securing pin and safety cap are removed. The fuze is quiescent in all its parts till direct impact takes place, when the steel plug is crushed in, shearing its steel pin, and carrying its needle point on to the detonator. The explosion of the detonator fires the loose "composition exploding" in the central channel, which in turn fires the magazine of the fuze and the bursting charge of the shell.

Mark I fuze.—The Mark I differs from the above in the following particulars :—

(1) The fuze is not fitted with a percussion holder.
(2) The steel plug is suspended by its shearing wire in the fuze body.
(3) The brass collar for the detonator screws into the fuze body instead of the bottom of the percussion holder.

Fuze, Percussion, D.A. Impact, No. 19a, Mark I | N | .

This fuze is used with Q.F., 3-pr., lyddite shell (except Mark I).

It is similar in design to the Fuze, percussion, D.A. Impact, No. 18, Mark II, but is smaller.

Fuze, Percussion, D.A. Impact, No. 19, Marks I and II | N | .

Fuze, percussion, D.A. Impact, No. 19, Marks I and II, is used with the Mark I, Q.F., 3-pr. lyddite shell. It is similar to the Mark I, No. 18, D.A. Impact fuze, but smaller, and the screwed portion is parallel instead of being tapered.

The Mark II differs from the Mark I in having a smaller shearing pin and a solid securing pin.

(D.) GRAZE ACTION FUZES.

(1) *Nose fuzes.* (2) *Base fuzes.*

(1) Nose Fuzes.

Fuze, Percussion, Q.F., 1-pr., Mark I.

Description.—The fuze consists of a metal body screw-threaded on the outside to fit into the nose of the shell. The head is flanged, to fit over the nose of the shell, and cut away to take the fuze key. The body is bored out from the top, and the bottom is pierced by a fire hole. In it is placed a detonator pellet, carrying the detonator and having a channel driven with powder. The composition in the detonator contains no fulminate of mercury, and is the same as in the cap of the cartridge.

Encircling the front end of the pellet is a split collar of brass, which prevents the pellet moving forward against the needle.

The top of the fuze is closed by a screw plug, carrying a steel needle and creep spring on its underside ; the plug is secured by a set screw.

Action.—On shock of discharge, the split collar sets back over the pellet, thus unmasking the detonator ; on impact the pellet and collar are thrown violently forward and the needle pierces the detonator, thus firing it. The flash from detonator and powder in the pellet passes through the fire-hole and ignites the bursting charge in the shell.

Fig. 78.

FUZE, PERCUSSION, Q.F., 1-PR.

Scale = $\frac{2}{1}$.

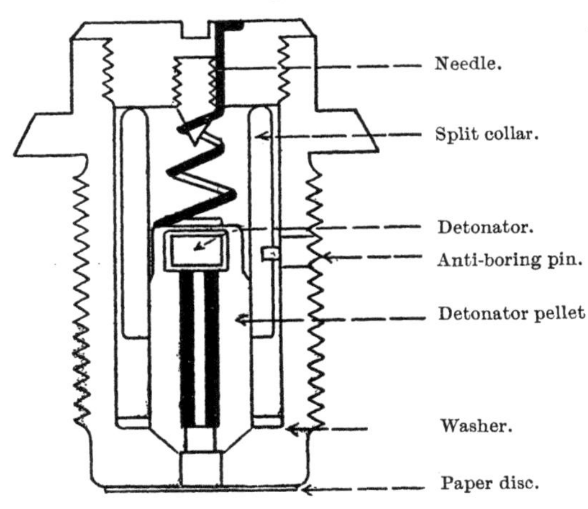

(2) Base Fuzes.

(*See Table of Base Fuzes, page* 295.)

Fuze, Percussion, Base, Large, No. 11, Mark V | C | .

(*Plate* XLVIII.)

Use (*see* page 303).—The Base Fuze, No. 11, is used in the *Land Service* with B.L., 12-inch to 6-inch; B.L.C., 6-inch; Q.F., 6-inch, for armour-piercing shells without cap, and common-pointed shell without cap.

In the *Naval Service*, with uncapped armour-piercing, and uncapped common-pointed shell, B.L., 13·5-inch to 6-inch, and Q.F., 6-inch.

Parts.—The principal parts of the fuze are:—Body, steel protecting plate, pressure plate and spindle, detonator pellet, centrifugal bolt, small retaining bolt, locking pellet with spiral spring, detonator plug with detonator, brass ball with retaining bolt and spiral spring, phosphor-bronze creep spring, and screw cap with needle.

Body.—The body is made of Class "A" metal, and is screw-threaded (left-handed) nine threads to the inch for a length of 1·75 inches. Below the threaded portion a flange is formed, which is coated with Mark III luting before being screwed home into the shell; this is to make a gas-tight joint.

The body is bored out to receive the percussion arrangement, and screw-threaded to take the screw cap; the percussion chamber is coned towards the bottom, and terminates in a small seating, sealed by a brass ball. A flash hole, bored radially through the body, forms a communication between the underside of the ball and the bottom of a vertical channel driven with five perforated powder pellets leading to a circular groove formed in the top face of the body. This groove contains a circular pellet of powder weighing about 110 grains. A hole is bored into the body a little above this flash hole into which is placed a small retaining bolt with spiral spring. This bolt is pressed inwards by the action of its spring, and, bearing against the brass ball, keeps the latter in its seating until spun out by the rotary motion of the shell. In the base of the fuze, to one side and clear of the percussion chamber, a vertical hole is bored for the pressure plate spindle; the lower part of this hole is enlarged to take the gas check pressure plate, and below this, screw-threaded (left-handed) for the steel protecting plate.

Detonator pellet.—The detonator pellet is made of metal, cylindrical in shape, the lower part being coned to fit the bottom of the percussion chamber. A circular flange on the bottom of the pellet, fitting into

Fig. 79.

NEW PATTERN OF R.L. CAP.

(For old pattern, *see* page 272.)

Scale $\frac{4}{1}$.

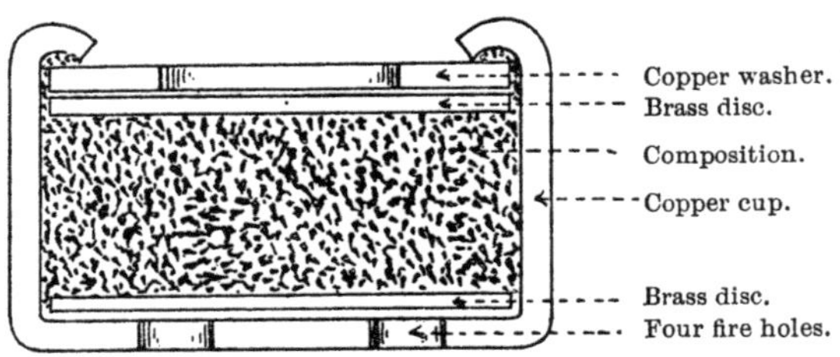

a circular groove in the body, assists the brass ball in sealing the flash hole from the detonator until the pellet is moved forward on graze. A groove is formed round the front end of the pellet to support a phosphor-bronze spring, and in the centre a recess is formed and screw-threaded to take the detonator plug.

Phosphor-bronze spring.—This spring, resting in grooves formed in the front face of the pellet and the underside of the screw-cap, prevents the pellet moving on to the needle until graze or impact.

Small pin.—The tendency of the pellet to turn round during flight is prevented by a small screw pin projecting from the body, engaging with a longitudinal groove cut down the side of the pellet.

Detonator plug.—The detonator plug is made of metal, threaded externally to screw into the pellet, and recessed on the underside to contain the detonator.

Detonator.—The detonator consists of a copper cap pierced with four fire holes; in the bottom of the cap and covering the fire holes is placed a brass disc, on the top of which is pressed $3\frac{1}{2}$ grains of detonating composition. The detonating composition is covered with a brass disc and a copper washer, retained in position by six lugs on the top of the cap being pressed over on to the top of the washer. The detonator plug, when screwed home, is prevented from unscrewing by being stabbed into the face of the pellet in two places. Leading from the underside of the detonator plug a central fire channel is bored first vertically, then diagonally through the pellet. This fire channel leads out into the lower part of the percussion chamber a little above the circular flange on the bottom of the pellet.

Centrifugal bolt.—Passing through the percussion pellet, at right angles to its axis, is a centrifugal bolt, one end of which is enlarged, while the other end projects into a recess in the fuze body. The centrifugal bolt in this position masks the fire channel in the pellet, but when the bolt is spun out of its recess by the rotary motion of the shell, a vertical flash hole in the bolt is brought directly in line with the fire channel in the pellet.

A small brass pin, projecting from the head of the bolt and fitting into a hole in the pellet, prevents the former from turning, and so ensures that the flash hole always remains in a vertical position.

Small retaining bolt.—Placed into a small hole bored through the body, communicating between the upper part of the recess for the pressure plate spindle and the percussion chamber, is a small retaining bolt. This bolt bears against the enlarged head of the centrifugal bolt in the pellet; its outer end is grooved, so as to admit of a slight outward movement when the small diameter part of the spindle has been forced inwards on firing.

Pressure plate and spindle.—The pressure plate and spindle are made of copper in one piece.

The plate is formed with a gas-check lip which fits into a recess in the body. The spindle is cylindrical in shape, about 1·75 inches in length; near its upper end it is reduced in diameter so as to give clearance for the small retaining bolt when the spindle has been pushed home.

When placed in position, the spindle does not extend the whole length of its recess, and the upper part of the spindle bearing against the small retaining bolt keeps the small end of the centrifugal bolt in its recess in the body, and so prevents the forward movement of the detonator pellet.

Protecting plate.—The protecting plate is made of steel, screw-threaded and recessed so as to fit over the pressure plate. It is pierced with four fire holes, intended to allow the pressure of gas on discharge to crush in the pressure plate. When screwed home it is prevented from unscrewing by being stabbed in four places.

Small locking bolt.—To prevent, on graze or impact, the detonator pellet being blown back after it has been carried on to the needle, and so masking the flash from the exploding detonator, a small locking bolt with spiral spring is inserted into a hole in the pellet. This locking bolt, when the pellet is carried forward, is pushed out by its spring, and engages with the recess in the body vacated by the small end of the centrifugal bolt.

Screw cap.—The screw cap is of bronze, flanged to fit over the end of the fuze body. A boss is formed on the underside of the cap threaded to screw into the top of the percussion chamber, and has snapped to the centre a steel needle, which comes directly over the detonator in the pellet.

The flange of the screw cap is pierced with six holes to allow the flash from the powder pellet in the circular recess to pass into the shell. These holes are closed by a washer of paper secured to the underside of the flange.

To prevent the cap from unscrewing, a set-screw passes vertically through it and into the body.

In the base of the fuze there are two elongated holes to take the key for screwing it home into shell. The base is stamped with an arrow showing the direction to turn when screwing the fuze home, and with the lot number, contractor's initials, date of manufacture, No. 11, and Mark of fuze.

Action of the fuze.—This fuze is specially designed to avoid premature action. If the detonator is accidentally fired on the shock of discharge, the flash is masked by the centrifugal bolt, by the flange on the bottom of the pellet fitting into the groove in the body, and by the brass ball closing the fire channel in the base.

On discharge, the gas, acting through the holes in the protecting plate, crushes in the pressure plate, carrying forward the spindle, thus bringing the reduced diameter of the spindle opposite the small retaining bolt. The rotation of the shell causes the bolts to move outwards; the grooved end of the small retaining bolt fitting round the reduced part of the spindle allows the centrifugal bolt to withdraw its projecting end from the recess in the body, and to bring its vertical flash hole in line with the fire channel in the pellet. At the same time the retaining bolt in the base of the fuze, acted upon by centrifugal force, moves outwards, compressing its spiral spring, thus allowing the brass ball to move out of its seating and open the fire channel in the base.

The phosphor-bronze spring prevents any rebound action of the pellet, and the small screw pin prevents it turning round during flight. On graze or impact, the pellet moves forward, carrying its detonator on to the needle, at the same time withdrawing the flange round its base from the groove in the bottom of the percussion chamber.

The locking bolt, engaging with the recess in the body, retains the pellet in the forward position and thus allows the flash from the detonator to pass down through the channel in the centrifugal bolt and pellet, through the hole left open by the brass ball, to the perforated powder pellets in the vertical channel, thus firing the magazine in the front end of the fuze, the flash from which, passing through the six fire holes in the screw cap, explodes the bursting charge of the shell.

The time taken for the pellets to burn gives this fuze a slight delay action.

Fuze, Percussion, Base, Large, No. 11, Mark IV.

The Mark IV fuze is entirely different in design from the Mark V. (*See* Plate XLIX.)

Parts.—The fuze consists of the following principal parts :—Body, needle pellet, centrifugal bolt, pressure plate with spindle and nut, steel protecting plate, screw cap with detonator, screw plug with magazine, phosphor-bronze spring, and four screws.

Body.—The body is threaded externally in a similar way to the Mark V; the interior is bored out to receive the needle pellet and threaded at the top to take the screw cap.

In the centre of the base a hole is bored, through which passes the pressure plate spindle, and a recess is made in the base into which fit the pressure plate and steel protecting plate. A hole is bored through the side of the body to allow of a small recess being formed in the opposite side, in which the small end of the centrifugal bolt engages. This hole is closed by a small screw-plug, the end of which is reduced in diameter, and fitted with a small spiral spring which bears against the heavy end of the centrifugal bolt.

Needle pellet.—The needle pellet is cylindrical and rests on the bottom inside the body ; it is reduced in diameter at the front, forming a shoulder, over which fits a phosphor-bronze spring. A hole is bored through the pellet at right angles to its axis for the centrifugal bolt, and another along its axis in which works the nut of the pressure plate spindle ; the upper part of this hole is threaded to receive the needle plug ; after the needle plug is screwed home the metal of the pellet is spun over it to prevent it from unscrewing. There is a small longitudinal groove in the side of the pellet, into which a screw projects from the side of the body ; this prevents the pellet from turning round in flight.

Centrifugal bolt.—The centrifugal bolt is also cylindrical and fits in the hole in the pellet ; one end is reduced in diameter to fit into the recess in the body. An elongated hole is bored through it from top to bottom, for the spindle ; the upper part of the bolt has a slot formed, into which the flange on the underside of the nut fits ; this locks the bolt, and makes the fuze safe till set in action.

Pressure plate.—The pressure plate is made of copper and has a boss on one side, into which screws the spindle ; it fits in the under-cut recess made near the base of the fuze, and when in position the metal of the body is spun over it.

Spindle.—The pressure plate spindle is threaded at both ends ; one end screws into the boss on the pressure plate, and the other receives the nut on top of the centrifugal bolt ; the end of the spindle is riveted over the nut when screwed home.

Protecting plate.—The protecting plate is made of steel, perforated with eight holes; it fits into a slightly undercut recess in the base, the metal of the body being spun over it. It is intended to protect the pressure plate from accidental blows.

Screw cap.—The top of the fuze is closed by the screw cap and screw plug, screwed together and containing a pressed pellet of R.F.G.[2] powder pierced with a central hole. A recess is made in the underside of the screw cap to receive the detonator (an R.L. cap), and there are six fire holes to convey the flash from it to the powder pellet; the metal is spun over the detonator to keep it in position. After the screw cap has been screwed home it is secured by a locking screw from the side of the body.

Screw plug.—The screw plug is pierced with four fire holes to allow the flash to pass into the shell; it is covered on the inside with a disc of muslin and fits over the powder pellet. When screwed home it is prevented from unscrewing by a locking screw.

Action.—On discharge, the gas acting through the holes in the protecting plate crushes in the pressure plate, carrying forward the spindle and nut, thus releasing the centrifugal bolt. The rotation of the shell causes the heavy end of the bolt to spin out, withdrawing its small end from the recess in the body. The phosphor-bronze spring prevents any rebound action of the pellet, and the small screw fitting into the longitudinal groove prevents the pellet turning round in flight. On graze or impact the pellet is carried forward, compressing the spiral spring and carrying its needle on to the detonator. The flash from the exploding detonator passing through the six holes in the screw cap fires the pellet of powder, which in turn fires the bursting charge of the shell.

No. 11, *Mark III.*

A few Mark III fuzes have been made; they differ from Mark IV in the small end of the centrifugal bolt having less protrusion into the recess in the body.

No. 11, *Mark II.*

The Mark II fuze differs from the Mark III in not having a steel protecting plate for the pressure plate.

This Mark of fuze is more sensitive than Mark I, and is still retained with common-pointed shell for 9-inch R.M.L. high angle guns.

Fig. 80.

FUZE, PERCUSSION, BASE, LARGE, NO. 11, MARK II.

Full Size.

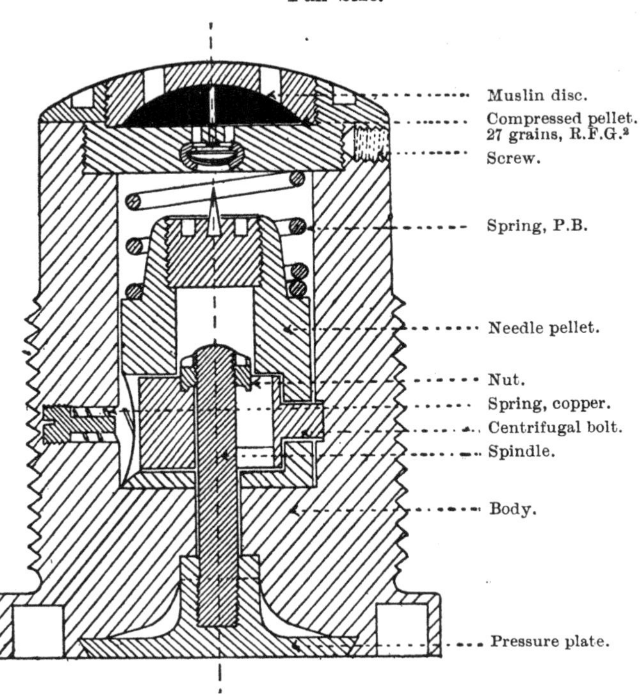

Fig. 81.

FUZE, PERCUSSION, BASE, LARGE, NO. 11, MARK I.

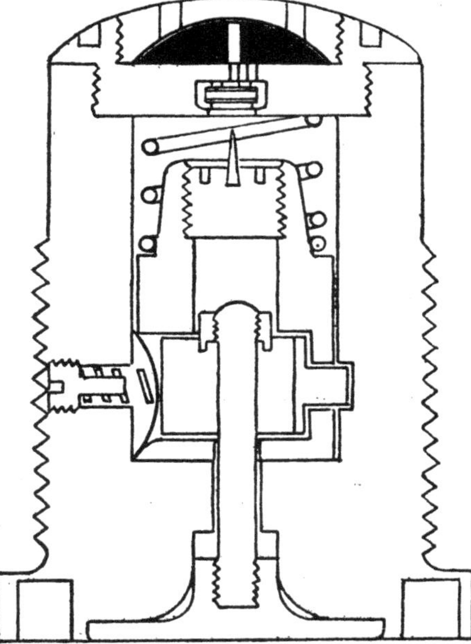

No. 11, *Mark I.*

The Mark I fuze differed slightly from the Mark II in the form of the recess for the pressure plate, the shoulder of which is not so much cut away; it is therefore not so sensitive, since the pressure plate offers greater assistance. (*See* Fig. 81.)

CONVERTED LARGE BASE FUZES.

Most existing Marks I and II large base fuzes have been converted to the Mark IV pattern, by having protecting plates fitted. When altered the fuzes will have a (*) added to their numeral.

In these converted fuzes, the protecting plate is screwed in, instead of fitting into an undercut recess with the metal spun over it.

Fuze, Percussion, Base, Large, Bronze, No. 15 | C | .

There are three marks of this fuze in the Service. They are made of aluminium bronze, and are used in the Land Service with all *capped* armour-piercing shell, in the Naval Service with *capped* armour-piercing and *capped* common-pointed shell, and may be used with uncapped armour-piercing, and uncapped common-pointed shell. With these fuzes, the bodies of which are of stronger material than No. 11, there is less chance of the shell failing to burst through the fuze being blown out.

Mark III.—The latest mark of the No. 15 base fuze is Mark III; it is identical in construction with the latest mark of the No. 11 base fuze, namely, Mark V. (*See* Plate XLVIII.)

Fig. 82.

FUZE, PERCUSSION, BASE, LARGE, BRONZE, NO. 15, MARK II.

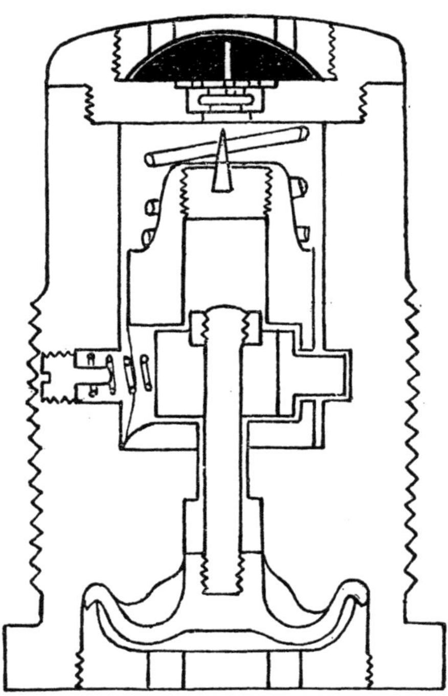

Mark II.—The Mark II fuze differs from the Mark I in being longer, and the pressure plate has a gas-check lip which fits into a recess formed in the body. The steel protecting plate is of a different shape and screws into the base, and is pierced with four instead of eight holes. The locking-nut on the top of the spindle secures the centrifugal bolt more firmly in position, and the detonator is inserted into the screw cap from the top, and is covered with a perforated brass disc, over which the metal is spun.

Mark I.—The Mark I is identical in construction with the No. 11, Mark IV fuze. (*See* Plate XLIX.)

Fuze, Percussion, Base, Large, Bronze, No. 16, Mark I | C | .

(*Plate* L.)

This fuze is used with capped armour-piercing shell filled with lyddite.

It differs from the No. 15, Mark III Base fuze in the following particulars :—

(*a*) It is fitted with a screwed steel collar instead of a steel perforated plate over the copper pressure plate.

(*b*) The copper pressure plate is of a slightly different shape so as to fit inside the steel collar.

(*c*) The screwed cap closing the front of the fuze is longer, has more screw thread, and is bored out in the interior to form a coned seating.

(*d*) The front portion of the detonator pellet is tapered ; the object being that when the pellet moves forward on to the needle on graze, it will wedge itself into the coned seating in the screwed cap at the same time as the needle fires the detonator, thus assisting the locking bolt in preventing the pellet being thrown back and masking the flash hole in the base of the fuze.

A " *Plate gas-check* " with steel " *Cover plate* " is used with all shell taking this fuze. (*See* page 210.)

Fuze, Percussion, Base, Medium, No. 12 | C | .

Use.—There are seven marks of this fuze, all of which will still be met with. They are used in the Land Service with common pointed and A.P. shell for the B.L., 4-inch and 30-pr. ; Q.F., 4-inch and 12-pr., and with the double shell for the Q.F., 2·95-inch.

In the Naval Service with common-pointed shell for the B.L., 4-inch ; and Q.F., 12-pr. and 4·7-inch.

In material, construction, and action, the medium base fuzes are, with one exception (the Mark VI), similar to the No. 11 large base fuzes, but are smaller, and are screw-threaded twelve threads to the inch instead of nine.

No. 12, *Mark VII.*

Mark VII.—Fuze, Percussion, Base, Medium, No. 12, *Mark VII | C |* .—The Mark VII medium base fuze is identical in construction with the large base fuze, No. 11, Mark V, described on page 281.

No. 12, *Mark VI.*

Fuze, Percussion, Base, Medium, No. 12, *Mark VI* | *C* | .—As this mark of fuze differs in construction from the other base fuzes a detailed description follows :—

Parts.—The principal parts of Mark VI fuze are:—Body, detonator pellet, detonator plug with R.L. cap, centrifugal bolt, steel protecting plate, pressure plate with spindle and nut, phosphor bronze spring, needle plate with steel needle, two safety bolts, screw cap, and screw plug with magazine.

Fig. 83.

FUZE, PERCUSSION, BASE, MEDIUM, NO. 12, MARK VI.

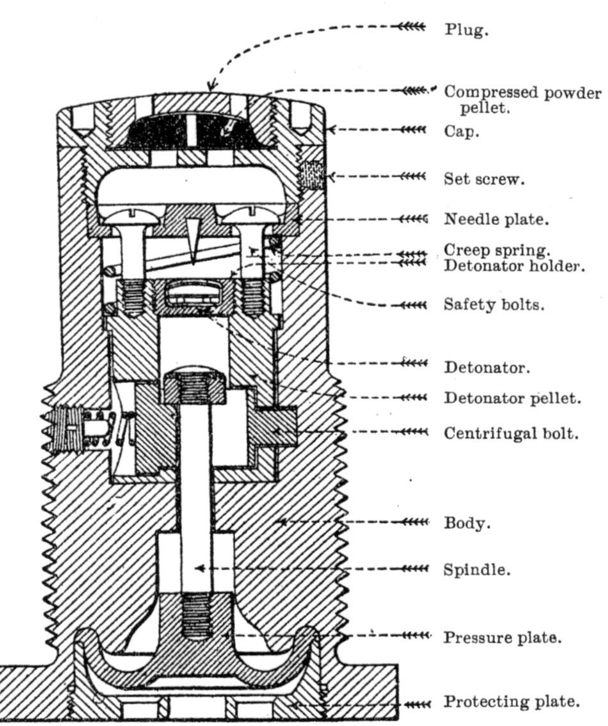

Body.—The body is threaded twelve threads to the inch for a length of 1·5 inches ; the upper part of the fuze is not threaded, while a flange is formed at the base which, as in all base fuzes, is coated with luting so as to make a tight joint when screwed home into the shell. It is bored out to receive the percussion arrangements, but the recess is slightly enlarged at the upper end to form a shoulder on which rests the needle plate and which is threaded to receive the screw cap.

A recess is formed in the base into which fits the pressure plate; the lower part of this recess is screw-threaded for the steel protecting plate. A hole, intended for the spindle of the pressure plate, is bored through the lower portion of the body and forms a communication between the recess in the base and the percussion chamber. A small hole is bored through the side of the body to allow of a recess being formed in the opposite side, in which the small end of the centrifugal bolt engages. This hole is closed by a small screw plug, the end of which is reduced in diameter and fitted with a small spiral spring, which bears against the heavy end of the centrifugal bolt.

Detonator pellet.—The detonator pellet is cylindrical in form and rests on the bottom of the percussion chamber; a small projection is formed around the upper part of the pellet, forming a shoulder on which rests a phosphor bronze spring, the object of which is to prevent any rebound or creeping action.

A hole is bored through the pellet at right angles to its axis, into which fits the centrifugal retaining bolt, and another is bored along its axis for the pressure plate spindle and locking nut; the upper part of this hole is threaded for the detonator plug. Two small holes are bored vertically in the top of the pellet; these holes are threaded for the safety bolts. Down the side of the pellet is cut a longitudinal groove into which projects a screw from the side of the body; this prevents the pellet from turning round in flight.

Centrifugal bolt.—The centrifugal bolt is cylindrical in form, and fits into the hole in the detonator pellet; one end is enlarged, while the other is reduced in diameter to fit into the recess in the body. An elongated hole is bored through the bolt, the upper part being recessed for the locking nut on the pressure plate spindle.

Pressure plate.—The pressure plate is made of copper, formed with a gas-check lip which fits into an undercut recess formed in the base of the fuze; a boss is formed on the inside of the plate into which screws the spindle.

Spindle.—The spindle is threaded at both ends, one end screwing into the boss on the pressure plate; the other receives the locking nut, which secures the centrifugal bolt in the safe position.

Protecting plate.—The protecting plate is made of steel, recessed to fit over the pressure plate; it is pierced with five holes intended to allow the gas to act on the pressure plate on discharge.

This plate screws home into position and is prevented from working loose by four punch stabs.

Detonator plug.—The detonator plug is a small cylinder of metal, threaded externally and recessed to take an R.L. cap, which is retained in position by the metal of the plug being spun over it.

It screws into the top of the detonator pellet after the locking nut has been screwed home.

Needle plate.—The needle plate is of metal with a steel needle snapped into its underside; two holes are bored through it, and countersunk for the enlarged heads of the safety bolts. The plate rests on the shoulder inside the fuze body, and is prevented from turning by a set screw projecting into a slot cut in its side. In this position the needle is directly over the detonator in the pellet.

Safety bolts.—The safety bolts are fitted with large flanged heads; they project through the holes in the needle plate, screwing tightly home into the two holes in the front of the pellet, the flanged heads forming a gas-tight joint in the countersunk holes in the needle plate.

A quantity of Pettman's cement is placed over the heads of the bolts to assist in sealing the joint.

The phosphor bronze creep spring is placed between the front of the pellet and the needle plate.

Screw cap and screw plug.—The top of the fuze is closed by the screw cap and screw plug screwed together and prevented from unscrewing by a locking screw. They form a small magazine containing a pressed pellet of R.F.G.[2] powder pierced with a central hole. A disc of muslin is shellaced to the underside of the screw plug to prevent the powder working out through the four fire holes in the plug through which the flash passes to the bursting charge of the shell.

A recess formed on the underside of the screw cap is intended for the safety bolts to be driven into by the action of the pellet on graze.

There are four holes bored through the cap to allow the flash from the detonator to fire the magazine. A disc of muslin prevents the powder working into the interior of the fuze through these holes.

The screw cap is prevented from unscrewing from the body by a small set screw passing through the body and into the cap.

Action.—On discharge, the gas, acting through the holes in the protecting plate, crushes in the pressure plate, carrying forward the spindle and nut and releasing the centrifugal bolt. The rotation of the shell causes the bolt to move outwards, withdrawing its small end from the recess in the body. On graze or impact the detonating pellet compresses the spring in front of it, and moves forward, carrying the heads of the safety bolts clear of the holes in the needle plate; at the same time the detonator is carried forward on to the needle and fired, the flash passing through the holes in the needle plate, past the heads of the safety bolts, to the magazine of the fuze.

This fuze was designed to prevent premature action on firing. The heads of the safety bolts seal the communication between the detonator and the magazine, and are only released simultaneously with the firing of the detonator on graze or impact.

Note.—Very few of the Mark VI have been issued.

The Mark V Fuze.

The Mark V fuze, externally, is similar to the Mark VI already described, and is fitted with the same pattern of protecting plate, pressure plate, spindle and nut. The pellet, however, is entirely different. Instead of having a detonator plug with R.L. cap, it is fitted with a needle plug and steel needle. There is no needle plate with safety bolts, and the screw cap is not recessed on the underside, nor pierced with fire holes, but has a central recess formed on the upper side to hold the R.L. cap, which is kept in position by a perforated brass disc spun into position over it.

The Mark IV Fuze.

The Mark IV fuze differs from Mark V in the following particulars :— The fuze is shorter ; the pressure plate has no gas-check lip ; the steel plate is pierced with eight instead of five holes, and is not screwed but spun into position ; the locking nut has a flange on its underside fitting into a small slot on the top of the centrifugal bolt ; the

Fig. 84.

FUZE, PERCUSSION, BASE, MEDIUM, NO. 12, MARK V.

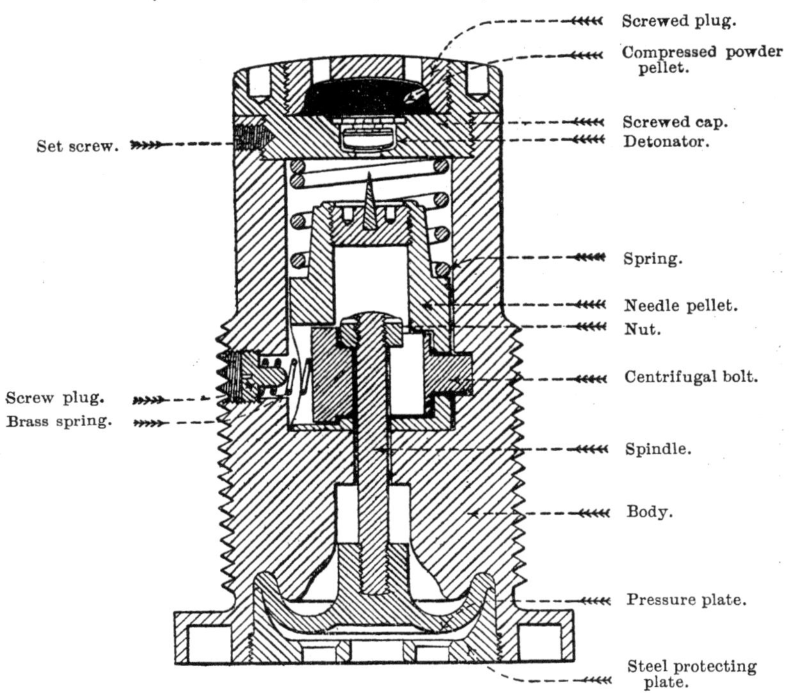

detonator is inserted into a recess formed on the underside of the screw cap, the top of which is pierced with six fire holes to allow the flash to pass to the magazine.

This Mark IV, No. 12, Medium Base Fuze, is similar to the Mark IV, No. 11, large base fuze, but the shape of the centrifugal bolt is slightly different, having the heavy end enlarged, and the pellet recessed to receive it.

The Mark III Fuze.

The Mark III fuze only differs from the Mark IV in the shape of the centrifugal bolt, which is without the increased diameter at the heavy end. Only a very few of this Mark of the No. 12 fuze were made, owing to the centrifugal bolt not withdrawing its small end from the body during flight. Mark IV was introduced to remedy this defect.

The Mark II Fuze.

Fuze, Percussion, Base, Medium, No. 12, *Mark II* differs from the Mark III in the small end of the centrifugal bolt having less protrusion into the recess in the body. It is generally similar to the Mark III, No. 11, base fuze.

The Mark I Fuze.

This fuze is similar to the No. 11, Mark I, fuze. It has no steel protecting plate.

Alteration of Position of Key Slots in Large and Medium Base Fuzes.

In future manufacture the key slots in the base of Nos. 11, 12 and 15 base fuzes will be moved nearer to the centre and will be the same distance apart for all three fuzes. This change involves no advance in numeral of fuze.

Protectors, Large Base Percussion Fuzes, Marks I and II | N | .

These protectors have been issued for use with the earlier Marks (Marks I and II) of No. 11 base fuzes (those without steel protecting plate).

The Mark II protector consists of a dished steel disc blackened all over and provided with two glove-button studs which fit into the key-holes in the base of the fuze. A certain number of Mark I protectors have been issued which differ from the Mark II in having elliptical spring clips instead of the glove-button studs.

These protectors are intended to protect the pressure plate during transport.

TABLE No. 26.—*Table of Base Fuzes, Nos.* 11, 12, 15, and 16.

List of Change	*Fuze, Percussion, Base, Large, No.* 11. Use :—Common-pointed and uncapped A.P. shell, 6-inch and up.	List of Change	*Fuze, Percussion, Base, Medium, No.* 12. Use :—Common-pointed and A.P. shell, 12-pr. to 5-inch, and 2·95-inch double shell.	List of Change	*Fuze, Percussion, Base, Large, Bronze, No.* 15. Use :—Capped A.P. shell, filled gunpowder and capped common-pointed shell.	List of Change	*Fuze, Percussion, Base, Large, Bronze, No.* 16. Use :—Capped A.P. shell, filled with Lyddite.
§ 8099 8315 8788	*Mark I.* Has no protecting plate.	§ 8100	*Mark I.* Has no protecting plate.				
§ 8652 8788 9674	*Mark II.* Differs from Mark I in the recess for the pressure plate being more cut away, making the fuze more sensitive. *It has no protecting plate.*		No change.				
§12114	*Mark III.* Differs from Mark II in having a perforated steel disc spun in position in the base to protect the pressure plate.	§12114	*Mark II.* Differs from Mark I in having a perforated steel disc spun in position in the base, to protect the pressure plate.				
§12114	*Mark IV.* Differs from Mark III in having increased protrusion of small end of centrifugal bolt.	§12114	*Mark III.* Differs from Mark II in having increased protrusion of small end of centrifugal bolt.	§12834	*Mark I.* Same as No. 11, Mark IV, but made of aluminium bronze.		

TABLE No. 26.—*Table of Base Fuzes, Nos.* 11, 12, 15 and 16—continued.

List of Change	*Fuze, Percussion, Base, Large, No.* 11. Use :—Common-pointed and uncapped A.P. shell, 6-inch and up.	List of Change	*Fuze, Percussion, Base, Medium, No.* 12. Use :—Common-pointed and A.P. shell, 12-pr. to 5-inch, and 2·95-inch double shell.	List of Change	*Fuze, Percussion, Base, Large, Bronze, No.* 15. Use :—Capped A.P. shell, filled gunpowder and capped common-pointed shell.	List of Change	*Fuze, Percussion, Base, Large, Bronze, No.* 16. Use :—Capped A.P. shell, filled with Lyddite.
	No change.	§12249	*Mark IV.* The Mark IV differs from the Mark III in having an enlarged head on heavy end of centrifugal bolt, the pellet being recessed to take the new bolt.		No change.		
13126	This pattern was introduced as Mark V, but none were issued for service; the sealed pattern was declared obsolete.	§13126	*Mark V.* The Mark V fuze is longer; the pressure plate has a gas-check lip. The steel protecting plate screws into base, and has 5 instead of 8 holes. Different design of locking nut. The R.L. cap, inserted from top, and covered with perforated brass disc spun in.		*Mark II.* The Mark II fuze is longer; the pressure plate has a gas-check lip. The steel protecting plate screws into base, and has 4 instead of 8 holes. The centrifugal bolt has an enlarged head on heavy end. New design of locking nut. R.L. cap inserted from top and covered with perforated brass disc spun in.		

§14463	This pattern was introduced as Mark V, but sealed pattern was declared obsolete before any were issued to service.	§14463	*Mark VI.* The Mark VI differs from Mark V as follows:—A percussion pellet with detonator, a needle plate with needle between the front of the pellet and screw cap, two safety bolts passing through the needle plate, and screwing into front of pellet, different form of screw cap and creep spring.	§14463	This pattern was introduced as Mark III, but sealed pattern was declared obsolete before any were issued to service.		
§14560	*Mark V.* The Mark V fuze is different from previous marks, and is designed to prevent premature action. The pressure plate and spindle are in one piece and much smaller, and are placed to one side of the centre. The pellet is fitted with a different form of detonator, and has two centrifugal bolts. The large bolt engages with a recess in the body, and masks the fire channel in the pellet. There is a safety flange on the bottom of the pellet, and a brass ball with retaining bolt in the base which seals the flash from the detonator until the ball is displaced.	§14560	*Mark VII.* The Mark VII is similar to No. 11, Mark V.	§14560	*Mark III.* The Mark III is similar to No. 11, Mark V.	§15503 §16255	*Mark I.* This fuze is similar to the No. 11, Mark V, but differs from it in having a screwed collar instead of a steel plate under the pressure plate, which is of a different design. The front end of the detonator pellet and the under side of the screwed cap are coned

Fuze, Percussion, Base, Hotchkiss, Mark IV | C | .

Use.—This fuze is used with steel, and common iron shell, filled with gunpowder for the Q.F. 3- and 6-pr. guns.

Parts.—It consists of the following parts:—Body, percussion pellet, spiral creep spring, screw cap, screw plug, and detonator.

Body.—The body is of copper alloy, Class A, screwed externally with a left-handed screw (12 threads to the inch), and the base is formed with a flange to act as a gas-check, and a projection to take the key by which it is screwed into the shell. (*See* Fig. 85.)

Fig. 85.

FUZE, PERCUSSION, BASE, HOTCHKISS, MARK IV.

Scale $\frac{2}{1}$.

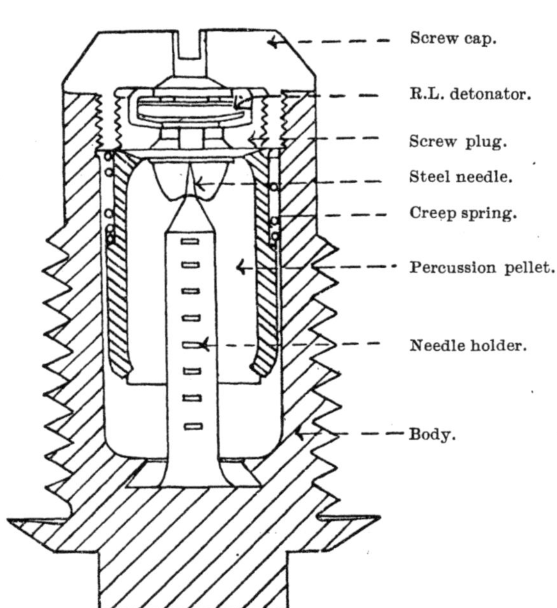

The body is bored out from the front to take the percussion pellet, the interior at the top screwed to receive the screwed cap, and has an undercut recess, the top edge of which is slightly rounded, formed at the bottom.

Percussion pellet.—The percussion pellet consists of a brass casing, filled with an alloy (12 parts lead, 1 part tin), into which a roughened needle holder of hard brass wire, carrying a steel needle at its front end, is embedded.

The brass casing is reduced in diameter on the exterior at the front end.

The needle holder has an enlarged base and rests in the undercut recess and supports the pellet; the latter projects beyond the point of the needle and prevents it reaching the detonator.

The brass spiral spring fits round the smaller part of the brass casing and prevents rebound of the pellet.

Screw cap.—The screwed cap is threaded left-handed and closes the front end of the fuze; the rear portion is bored out to take the screw-plug, and through the centre of the front end a fire-hole is bored to allow the flash from the detonator to pass out.

The screw-plug contains the detonator, which is similar, except in dimensions and amount of composition, to the R.L. cap, described on p. 282. It screws into the rear of the screwed cap. The fuze is lacquered externally.

Action.—On the shock of discharge, the pellet sets back over the needle holder, thus allowing the steel needle to project beyond it. The alloy at the bottom of the pellet cushions against the bottom of the fuze, and a small portion of it dovetails into the undercut recess, round the base of the needle. This forms a weak connection between the pellet and fuze body, and assists the spring in checking rebound action. On graze or impact the pellet and needle set forward, the needle pierces the detonator, and the flash passes through to the bursting charge of the shell.

Mark III fuze.—Mark III fuze differs from Mark IV in the following particulars:—There is no undercut recess for the alloy of the pellet to dovetail into; the needle is shorter and is formed entirely of roughened brass wire, having no steel needle let into it; the spiral spring is weaker and is placed round the needle instead of round the pellet, and the front end of the brass casing of the pellet is consequently not decreased in diameter.

The Mark III fuze, when converted to Mark IV pattern, will be known as Mark III*.

Fuze, Percussion, Base, 9·45-inch Howitzer | L |.

This fuze is used with the common-pointed shell for the above howitzer; in external appearance it is similar to the Hotchkiss base fuze. (*See* Fig. 86.)

Description.—The body of the fuze is of gunmetal and is bored out and fitted with a loose needle plug with steel needle; fitting round the needle plug is a copper ring, with three clips, which prevent it moving forward till impact. A four-armed stirrup spring rests on the front of the needle plug, and supports a thimble; a brass cylinder fits round the thimble, so as to keep the needle central. The front of the fuze is closed by a screwed plug with anvil and detonator. This plug carries a screw cap at its front end, which holds a magazine of loose powder; a lead washer is placed under the flange of the head.

Action.—On shock of discharge the brass thimble sets back, straightening out the four clips of the stirrup spring; on graze or impact the needle plug and thimble displace the three lugs on the copper retaining ring in the base of the fuze, and fly forward and fire the detonator, which in turn fires the loose powder in the magazine of the fuze, and the bursting charge of the shell.

Fig. 86.

FUZE, PERCUSSION, BASE, 9·45-INCH HOWITZER, MARK I | L |.

Scale 1/1.

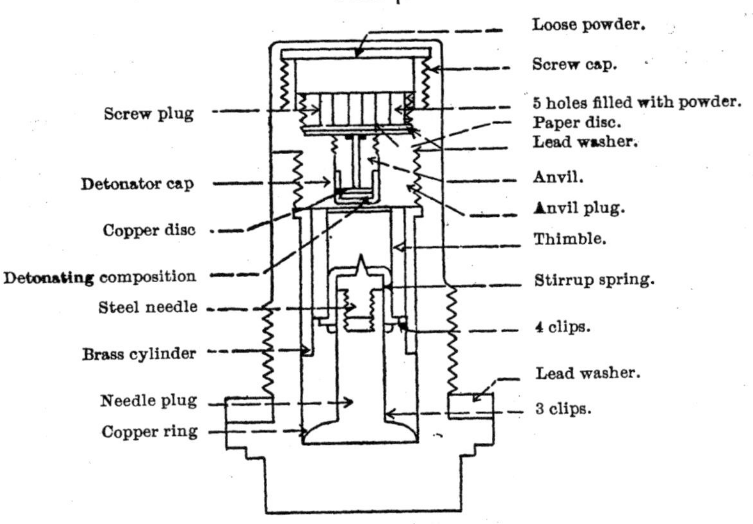

Drill Fuzes.

Drill fuzes are either burnt out fuze bodies, or solid gunmetal, of the same external dimensions as the fuze they represent; they are blackened, and stamped "DRILL."

TABLE No. 27.—*Percussion Fuzes used in Land Service.*

NOSE FUZES.

The Fuzes shown in Italics are Obsolete for future manufacture.

Fuzes.	Marks.	Paragraph in List of Changes.	Used with
PERCUSSION.			ORDNANCE.
Direct action, with cap, No. 1 ...	I*	5572	
		6740	
		9906	B.L., 60-pr. For common and lyddite shell
		10172	B.L., 4-inch, jointed. For common shell
		10173	B.L., 9·45-inch howitzer. For lyddite shell
		10322	B.L., 6-inch, 30-cwt. howitzer } For common and lyddite shell.
	I**	10297	B.L., 6-inch, 25-cwt. howitzer } For common and lyddite shell.
		10322	B.L., 5·4-inch howitzer } For common and lyddite shell.
			B.L., 5-inch howitzer } For common and lyddite shell.
	II	5216	B.L.C., 6-inch, on siege carriage. For lyddite shell
		9906	
		10087	Q.F., 4·7-inch, on travelling carriage. For common and lyddite shell
		10172	
		10322	Q.F., 4·5-inch howitzer. For lyddite shell
	III	11347	(All the above: For service only with lyddite shell.)
Direct action, delay, No. 10 ...	III	8871	B.L. 6-inch, 30-cwt. howitzer; B.L., 6-inch, 25-cwt. howitzer } For common shell, when specially ordered.

TABLE NO. 27.—*Percussion Fuzes used in Land Service*—continued.

NOSE FUZES.

The Fuzes shown in Italics are Obsolete for future manufacture.

Fuzes.	Marks.	Paragraph in List of Changes.	Used with	
PERCUSSION—*continued.*			ORDNANCE—*continued*—	
			B.L., 10-inch	For lyddite shell with picric powder exploders.
Direct-action, impact, No. 13 ...	I*	10321	B.L., 9·2-inch, IV to VIc	
	I**	11190	B.L., 9·2-inch, "B" VI and VIc	
	II	8630	B.L., 9·2-inch, IX	
		9620	B.L., 9·2-inch, X, Xv, X*	
		9673	B.L., 7·5-inch, II, II*	
		9721	B.L., 6-inch, VII, VIIv	
		10172	B.L., 30-pr.	
	II*	11190	B.L.C., 6-inch	
	III	10321	Q.F., 6-inch	
	III*	11190	Q.F., 4·7-inch	
	IV	11190	R.M.L., 10-inch, H.A.	
Direct-action, with cap, No. 17 ...	I	14561	B.L., 60-pr.	For service and practice with lyddite shell.
			B.L., 2·75-inch, converted	
			B.L., 2·75-inch	
			B.L., 9·45-inch howitzer	
			B.L., 6-inch, 30-cwt. howitzer	
	II	14897	B.L., 6-inch, 25-cwt. howitzer	
			B.L., 5·4-inch howitzer	
			B.L., 5-inch howitzer	
			B.L.C., 6-inch, on siege carriage	
			Q.F., 4·7-inch, on travelling carriage	
	III	15107	Q.F., 4·5-inch howitzer	

Direct-action, impact, No. 18 ...	I	15146	Q.F. 4·7-inch Q.F., 4-inch, III, III*	For lyddite shell with T.N.T. Exploders. For Coast Defences.
	II	16037	Q.F., 12-pr. 12-cwt.	
Direct-action, with Cap, No. 44 ...	I	—	Q.F., 3-inch—For H.E. shell.	
Direct-action, impact, No. 45 ...	I	—	B.L., 4-inch B.L., 6-inch B.L., 7·5-inch B.L., 9·2-inch Q.F., 12-pr. Q.F., 4·7-inch Q.F., 6-inch	For lyddite shell with T.N.T. Exploders.
Base Fuses.				
Base, large, No. 11	I	8099 8315	B.L., 12-inch B.L., 10-inch	For uncapped armour-piercing shell.
	I* II	12114 8652 9674 11873	B.L., 9·2-inch, IV to VIc B.L., 9·2-inch, IX B.L., 9·2-inch, X, Xv, X*	For uncapped armour-piercing and common pointed shell.
	II* III	12114 12114 12114	B.L., 6-inch, VII, VIIv B.L.C., 6-inch Q.F., 6-inch	For uncapped armour piercing shell.
	IV V	12114 14560	R.M.L., 9-inch, H.A.	For common pointed shell.
Base, large, bronze, No. 15... ...	I II III	12834 13126 14560	B.L., 9·2-inch, IV to VIc B.L., 9·2-inch, IX B.L., 9·2-inch, X, Xv, X* B.L., 7·5-inch, II, II* B.L., 6-inch, VII, VIIv	For capped armour-piercing shell, filled with gunpowder.
Base, large, bronze, No. 16... ...	I	15503 16255	B.L., 9·2-inch, X, Xv, X* shell. **B.L., 6-inch, Mark VII**	For armour-piercing capped steel, filled with lyddite.

TABLE NO. 27.—*Percussion Fuzes used in Land Service*—continued.

BASE FUZES.

The Fuzes shown in Italics are Obsolete for future manufacture.

Fuzes.	Marks.	Paragraph in List of Changes.	Used with
PERCUSSION—*continued*.			ORDNANCE—*continued*—
Base, medium, No. 12	I	8100	
		8315	
		9674	B.L., 4-inch, jointed } For common-pointed shell.
		11873	B.L., 30-pr. }
	II	12114	Q.F., 4·7-inch. For armour-piercing shell.
	III	12114	Q.F., 4-inch, III, III* } For common-pointed shell.
	IV	12249	Q.F., 12-pr., 12-cwt. }
	V	13126	Q.F., 2·95-inch. For double shell, except Mark I.
	VI	14463	
	VII	14560	
Base, Armstrong, No. 9	II	6193	
		8515	Q.F. 2·95-inch. For double shell, Mark I.
	III	7008	
		8515	
Base, Hotchkiss	III	7490	
		8229	Q.F., 6-pr. } For common and steel shell.
	IV	9451	Q.F., 3-pr. }
		9814	
Base, B.L., 9·45-inch howitzer ...	I	12355	B.L., 9·45-inch howitzer. For common-pointed shell.

TABLE No. 28.—*Percussion Fuzes used in Naval Service.*

NOSE FUZES.

Fuzes.	Marks.	Paragraph in List of Changes.	Used with	
PERCUSSION.			ORDNANCE—	
Direct-action, with plug, No. 3 ...	III	10322	B.L., 13·5-inch, I to IV	For nose-fuzed common shell.
			B.L., 10-inch, II to IV*	
			B.L., 9·2-inch	
			B.L., 6-inch, VII, VIII	
	IV	11396	Q.F., 6-inch	
Direct-action, impact, No. 13 ...			B.L., 15-inch	For lyddite shell with picric powder exploders.
	I*	10321	B.L., 13·5-inch, V	
	I**	11190	B.L., 12-inch	
	II	8630	B.L., 10-inch	
		9620	B.L., 9·2-inch	
		9673	B L., 7·5-inch	
		9721	B.L., 6-inch	
		10172	B.L., 4-inch, VII to VIII*	
			Q.F., 6-inch	
	II*	11190	Q.F., 4·7-inch	
	III	10321	Q.F., 4-inch, I to III*	
			Q.F., 4-inch, IV	
	III*	11190	Q.F., 14-pr.	
			Q.F., 12-pr. 18 cwt.	
			Q.F., 12 pr. 12 cwt.	
	IV	11190	Q.F., 12-pr. 8 cwt.	
	V			

TABLE No. 28.—*Percussion Fuzes used in Naval Service*—continued.

NOSE FUZES—*continued.*

Fuzes.	Marks.	Paragraph in List of Changes.	Used with
PERCUSSION—*continued.*			ORDNANCE—*continued.*
Direct-action, impact, No. 18 ...	I	15146	B.L., 4-inch, VII to VIII* Q.F., 4·7-inch Q.F., 4-inch Q.F., 14-pr. Q.F., 12-pr. 18 cwt. Q.F., 12-pr. 12 cwt. Q.F., 12-pr. 8 cwt. For lyddite shell with T.N.T. exploders.
	II	16037	
Direct-action, impact, No. 19 ...	I	15564	Q.F., 3-pr. For Mark I lyddite shell with T.N.T. exploders.
	II	16263	
	III	—	
Direct-action, impact, No. 19A ...	I	—	Q.F., 6-pr. Hotchkiss Q.F., 6-pr. Nordenfelt Q.F., 3-pr. For lyddite shell with T.N.T. exploders, except 3-pr. Mark I.
Direct-action, with cap, No. 44 ...	I	—	Q.F., 3-inch—For H.E. shell.
Direct-action, impact, No. 45 ...	I	—	B.L., 4-inch, VII to VIII* B.L., 6-inch B.L., 7·5-inch Q.F., 12-pr., 8, 12, and 18 cwt. Q.F., 14-pr. Q.F., 4-inch Q.F., 4·7-inch For lyddite shell with T.N.T. exploders.

BASE FUZES.

Fuze	Mark	No.	Nature of gun and use
Base, large, No. 11	I	8099	
		8315	B.L., 13·5-inch, I to IV } For uncapped armour-piercing and common pointed shell.
	I*	12114	B.L., 12-inch, VIII to X* } For uncapped armour-piercing and common pointed shell.
	II	8652	B.L., 10-inch, II to IV* } For uncapped armour-piercing and common pointed shell.
		9674	B.L., 10-inch, VI to VII*. For uncapped common pointed shell.
		11873	B.L., 9·2-inch. For uncapped armour-piercing and common pointed shell.
		12114	B.L., 7·5-inch. For uncapped common pointed shell.
	II*	12114	B.L., 6-inch, VII, VIII, XI, XI* } For uncapped armour-piercing and common pointed shell.
	III	12114	Q.F., 6-inch } For uncapped armour-piercing and common pointed shell.
	IV	12114	
	V	14560	
Base, large, bronze, No. 15... ...			B.L., 15-inch } For capped armour-piercing, filled with gunpowder, and common pointed shell; but can also be used with uncapped shell.
	I	12834	B.L., 13·5-inch } For capped armour-piercing, filled with gunpowder, and common pointed shell; but can also be used with uncapped shell.
			B.L., 12-inch } For capped armour-piercing, filled with gunpowder, and common pointed shell; but can also be used with uncapped shell.
			B.L., 10-inch } For capped armour-piercing, filled with gunpowder, and common pointed shell; but can also be used with uncapped shell.
	II	13126	B.L., 9·2-inch } For capped armour-piercing, filled with gunpowder, and common pointed shell; but can also be used with uncapped shell.
			B.L., 7·5-inch } For capped armour-piercing, filled with gunpowder, and common pointed shell; but can also be used with uncapped shell.
			B.L., 6-inch, VII, VIII, XI, XI* } For capped armour-piercing, filled with gunpowder, and common pointed shell; but can also be used with uncapped shell.
	III	14560	B.L., 6-inch, XII. For capped common pointed shell.
Base, large bronze, No. 16			B.L., 15-inch } For armour-piercing capped shell, filled with lyddite.
	I	15503	B.L., 13·5-inch, V } For armour-piercing capped shell, filled with lyddite.
			B.L., 12-inch, X to XII } For armour-piercing capped shell, filled with lyddite.

TABLE NO. 28.—*Percussion Fuzes used in Naval Service*—continued.

BASE FUZES—*continued.*

Fuzes.	Marks.	Paragraph in List of Changes.	Used with
PERCUSSION—*continued.*			ORDNANCE—*continued.*
Base, medium, No. 12		8100	
	I	8315	B.L., 4-inch, VII to VIII*
		9674	Q.F., 4·7-inch
		11873	Q.F., 4-inch
	II	12114	Q.F., 14-pr.
	III	12114	Q.F., 12-pr., 18 cwt.
	IV	12249	Q.F., 12-pr., 12 cwt.
	V	13126	Q.F., 12-pr., 8 cwt.
	VI	14463	Q.F., 12-pr., 4 cwt. — For common pointed shell, except Q.F. 4·7-inch, Marks I, II and III.
	VII	14560	
Base, Armstrong, No. 9	II	6193	
		8515	Q.F., 4·7-inch. For common pointed shell, Marks I, II and III.
	III	7008	
		8515	
Base, Hotchkiss	III	7490	
		8229	
	III*	14682	Q.F., 6-pr. — For steel shell, filled with gunpowder.
	IV	9451	Q.F., 3-pr. — For steel shell, filled with gunpowder.
		9814	

(E). TIME FUZES.

Time fuzes are used in the Land Service, generally for star shell only. (*See* page 112.)

In the Naval Service time fuzes are used with shrapnel shell from B.L. guns, 12-inch and above, when firing at short ranges. (*See* page 112.)

The following *time fuzes* will be met with :—

Land Service	No. 25. No. 24.
Naval Service	No. 30.

Time Fuzes Used in Land Service.

Fuze, Time, 15 Seconds, No. 25, Mark IV | L | .

(*Plate* LI.)

The Fuze, time, 15 seconds, No. 25, consists of the following parts :— Body, bottom plug, time ring, needle plug with steel needle, lighting pellet with R.L. cap and stirrup spring, safety pin, steel washer, cap and set-screw.

The body is made of aluminium ; the lower part is threaded to the G.S. gauge. Above the screw-threaded portion it is enlarged, forming a shoulder under which is placed a leather washer.

The top part of the body is reduced in diameter and screwed to receive the cap. The body is bored out from the bottom to take a charge of about 45 grains of R.F.G.^{2} powder, and is closed by a bottom plug with a central flash hole. This hole is covered on the inside of the plug by a muslin disc and a disc of brass, over which the metal of the plug is spun.

The body is bored out from the top, forming a lighting chamber in the stem ; in the bottom of this chamber is screwed a small needle plug of brass, having a steel needle on its upper face. Three flash-holes are bored through the stem, communicating from the lighting chamber into an annular groove cut round the exterior of the bottom of the stem. One of these holes is primed with a small pellet of gunpowder.

A setting mark on the body indicates the position of a small hole containing a perforated powder pellet which communicates with a fire channel leading to the magazine of the fuze.

The time ring is made of aluminium, and is graduated from 0 to 44.

On the underside of the ring a flange is formed, which fits into a groove cut in the face of the body round the bottom of the stem. The composition channel is driven with fuze composition, and then covered with a washer of vegetable paper. At the beginning of the fuze composition there is a gas escape hole closed by an aluminium disc, and another hole bored through to the inner circumference of the ring is intended to allow the flash from the holes in the stem to ignite the composition at the spot where it begins to burn. A cloth washer is shellaced to the top of the fuze body, on which rests the time ring.

The spaces between the cap, time ring and body, also set-screw recess of cap and gas escape hole disc in the ring, are waterproofed with a composition of beeswax, mineral jelly and French chalk.

Lighting pellet.—The lighting pellet is made of gunmetal and is supported in the top of the lighting chamber by means of a brass stirrup spring, the upper arms of which rest on two slots cut away on the top of the stem. The lighting pellet is bored out and contains a very small R.L. cap which is retained in position by means of a screw plug. *A steel spring washer* is placed on top of the time ring; the ring itself is then clamped by means of a cap of aluminium screwed on to the top of the stem and secured by a steel set-screw.

As the stirrup spring is of very weak construction the lighting pellet is secured in a safe position until the moment of loading by means of a phosphor-bronze safety pin which passes through the cap and the lighting pellet.

Action.—The fuze is set by means of the special key by turning round the ring until the graduation ordered is opposite the setting mark. At the moment of loading the safety pin is withdrawn, leaving the lighting pellet supported by means of the stirrup spring only.

On shock of discharge the pellet sets back, straightening out the arms of the stirrup spring, and carries its detonator on to the point of the needle. The flash from the exploding detonator passes through the holes in the stem and round the groove to the powder pellet at the commencement of the composition. The ring burns round until it comes to the pellet of powder in the body at the setting mark, which fires the magazine of the fuze and the shell.

This fuze was originally intended for use with the star shell for the B.L., 10-pr., but its use has now been extended to all B.L. and Q.F. guns and howitzers for use with star shell.

Mark III Fuze.

Mark III fuzes differ from the Mark IV in the bottom of the composition groove in the time ring being square instead of round, and in the groove having an asbestos lining.

Mark II Fuze.

The Mark II fuzes differ from Mark III in not being waterproofed.

Mark I Fuze.

Mark I fuzes differ from Mark II as follows :—

(1) The underside of the time ring is flat (instead of being provided with a lip and recess).
(2) The magazine channel is placed at a different angle.
(3) The aperture in the bottom plug is smaller.
(4) The external contour of the fuze is slightly different.

Fuze, Time, Sensitive Middle, No. 24, Mark I | L |.

(*See Plate* LII.)

This fuze can still be used with star and shrapnel shell with the B.L., 6-inch and 5·4-inch howitzers.

It will become obsolete when existing stock is expended.

Action.—The action of the fuze does not depend upon the shock of discharge, but immediately the shell begins to rotate, the retaining bolts, acted upon by centrifugal force, fly outwards, compressing the spiral springs and releasing the lighting pellet, which, acted upon by the same force, also flies outwards against the needle, firing the detonator, the flash passing through the body of the pellet to the axial magazine ; thence it passes out through the fire holes in the bottom of the stem and ignites the time ring. The ring burns round until it reaches the channel opposite the setting mark, when the magazine of the fuze is fired, so bursting the charge of the shell.

Time Fuzes Used in Naval Service.

Fuze, Time, No. 30, Mark II | N | .

(*Plate* LIII.)

This fuze is used with shrapnel shell having the 2-inch gauge, for B.L. 12-inch and above, when firing at short ranges.

The fuze is identical in construction with the Fuze, time, No. 25, Mark IV, but is made of metal instead of aluminium, is larger, and is threaded to the 2-inch gauge.

Mark I Fuze.

The Mark I fuze differs from the Mark II in the channel for the composition in the time ring being square instead of rounded at the bottom.

The channel is also lined with asbestos.

TABLE NO. 29.—*Time Fuzes used in Land Service.*

The Fuzes shown in Italics are Obsolete for future manufacture.

Fuzes.	Marks.	Paragraph in List of Changes.	Time of Burning at rest.	Used with.	
			Seconds.		
Sensitive, middle, No. 24...	I	5638 5982 7046 7231 8417	14·6 to 15·8	B.L., 6-inch, 30-cwt. howitzer B.L., 6-inch, 25-cwt. howitzer B.L., 5·4-inch howitzer	For star shell, but when stock of fuzes is exhausted, to be replaced by No. 25. For shrapnel shell, but when stock of fuzes is exhausted, to be replaced by No. 54.
15 seconds, No. 25	I II III IV	12592 13163 13352 14066 15760	15	B.L., 15-pr. B.L., 10-pr. B.L., 6-inch, 30-cwt. howitzer B.L., 6-inch, 25-cwt. howitzer B.L., 5·4-inch howitzer B.L., 5-inch howitzer B.L.C., 15-pr. Q.F., 18-pr. Q.F., 15-pr. Q.F., 13-pr. Q.F., 2·95-inch	For star shell.
Time Fuzes used in Naval Service.					
TIME.					
No. 30	I II	14872 16656	8·5 8·5	B.L., 13·5-inch, V B.L., 12-inch	For shrapnel shell having 2-inch fuze hole at short ranges.

(F). TIME AND PERCUSSION FUZES.

Time and percussion fuzes are here, for convenience, divided into three groups, namely :—

(1) The *small* type of T. and P. fuze with the G.S. gauge (1-inch).

Fuze, T. and P.	No. 60, two time rings,	burns at rest	20	sec.
,, ,,	No. 63, ,,	,, ,,	20·1	,,
,, ,,	No. 65 and 65A ,,	,, ,,	20·1	,,
,, ,,	No. 56, single time ring	,, ,,	13	,,

(2) The *large* type of T. and P. fuze with the G.S. gauge.

Fuze, T. and P.	No. 62, two time rings,	burns at rest	35	sec.
,, ,,	No. 64, ,,	,, ,,	30	,,
,, ,,	No. 66, ,,	,, ,,	34	,,
,, ,,	No. 54, single time ring,	,, ,,	16	,,

(3) T. and P. fuzes *having the* 2-*inch* fuze-hole gauge.

Fuze, T. and P.	No. 80, two time rings,	burns at rest	22	sec.
,, ,,	No. 81, ,,	,, ,,	30·25	,,
,, ,,	No. 82, ,,	,, ,,	40	,,
,, ,,	No. 83, ,,	,, ,,	30	,,
,, ,,	No. 84, ,,	,, ,,	30	,,

For nature of gun with which the above fuzes are used, *see* Table of Fuzes, page 334 to 338.

(1) SMALL T. AND P. FUZES OF G.S. GAUGE { No. 60. ,, 63 ,, 65 and 65A. ,, 56.

The above-mentioned T. and P. fuzes are similar in construction, differing only in minor details hereafter described.

The No. 56 fuze is fitted with only one time ring.

The No. 65A fuze is an entirely new fuze ; the No. 65 is a fuze which has been converted from old T. and P. No. 56, 60, or 63 fuzes.

The No. 65A fuze is here described in detail.

FUZE, T. AND P., No. 65A, MARK I | C | .

(*Plate* LIV.)

Parts.—The fuze consists of the following parts :—Body, percussion pellet with needle plug and retaining bolt, spiral spring, detonator plug, safety pellet, brass ball, bottom plug, two time rings, dome, washer and nut.

Description.—The bottom part of the body is screwed on the exterior to fit the G.S. fuze hole, and is bored out in the interior to take the percussion arrangement, and screwed to receive the bottom plug.

Above this the body is of larger diameter, and fits over the nose of the shell, a leather washer on the underside making the joint tight.

Above this again the body terminates in a stem, the top of which is threaded to receive the cap, and has two grooves cut in it to receive the feathers on the brass washer.

Two brass pins are secured into the stem, which engage with slots in the upper composition ring and prevent it turning.

In the enlarged diameter of the body are situated the safety pin of the percussion arrangement and a hole for the projection on the key, by which the fuze is screwed into the shell.

The percussion safety pin has a whipcord loop coloured black.

Time rings.—There are two time rings which contain the fuze composition. Each ring has a channel running nearly all round its under surface, and rings of vegetable paper are shellaced to the lower surface of each.

The lower ring is barrel-shaped and milled, having a setting pointer fixed at the commencement of the composition. The gas escape is external, *i.e.*, the gas escapes from the ring straight into the air. For this a hole is bored in the ring at the commencement of the composition, and is covered by a thin brass patch cemented over with Pettman's cement. The patch is blown out when the ring lights. This ring is movable, for setting the fuze. At the commencement of the composition in this ring there is a vertical hole, communicating with the upper ring.

The upper ring is pinned to the stem so that it cannot turn. It is cylindrical in shape and graduated from 0 to 44 divisions (half-divisions being shown by dots), and has an external gas escape, similar to the lower ring.

It is marked with an arrow, and when this is opposite the setting mark the fuze is set at safety, as the fire-hole in the body is covered by solid metal and not by fuze composition.

A cloth washer is secured by shellac to the top of the body, and another to the top of the bottom ring, a hole being pierced through each, so as to leave the powder pellet in the fuze body and in the composition ring exposed.

Lighting chamber.—On the upper side of the upper ring, at the beginning of the composition there is a small chamber containing the lighting arrangement, which consists of a gunmetal hammer having a steel needle, suspended by a thin copper shearing wire over ·35 grains of cap composition; this cap composition is surrounded by mealed powder, and covered by a thin brass disc which is kept in position by a small screw collar. This collar has a side groove cut away so as to form a channel for the flash to pass from the time detonator to the composition at the beginning of the ring. The top of the chamber is closed by a brass disc, the metal being spun over to retain it in position. A strong safety pin of phosphor-bronze wire passes through the ring from the outside, and underneath the hammer, which it supports. The letter "T" is stamped on the ring near the entrance of the safety pin.

Closing pellet.—When the safety pin is withdrawn, a small brass pellet with spiral spring behind it, closes the hole.

Dome.—The dome is made of brass; it fits over the upper ring and covers the lighting arrangement.

Washer.—Placed on top of the dome is a washer of sheet brass having two feathers which fit into grooves in the top part of the stem.

Nut.—The nut, which is hexagonal, screws on to the top of the stem and clamps the dome and time rings in position.

Percussion arrangement.—The percussion arrangement consists of a percussion pellet with steel needle, retaining bolt with spring, safety pellet, safety ball, detonator plug with R.L. cap, spiral creep spring and bottom plug.

Percussion pellet.—The percussion pellet is made of metal, cylindrical in shape, having a slot cut down one side for the safety pellet and ball to fall into; there is also a small groove cut down the body of the pellet into which projects a small pin in the fuze body to prevent the pellet from twisting round.

A hole is bored through the pellet, the top part of which is threaded to receive the needle plug with a hardened steel needle with six fire holes round it. Underneath the needle plug is a recess containing one grain of F.G. powder and a perforated pellet of powder, retained in position in the pellet by a brass washer and a muslin disc.

Creep spring.—A small recess is made on the underside of the body and a corresponding one on the top of the pellet, into which fits the spiral spring, intended to prevent the pellet moving forward in flight.

Retaining bolt.—The retaining bolt passes transversely through the percussion pellet, its small end projecting into a recess in the body, in which position it is kept by a small spiral spring. The other end of the bolt is heavier, and flies outward when the shell rotates, a slot being formed in the body into which this end can move.

Safety pellet.—The safety pellet is suspended in the body above the slotted out portion of the percussion pellet by means of a thin copper shearing wire, and by a safety pin passing through the body and the pellet. The letter "P" is stamped on the body near the entrance of this pin.

Closing pellet.—A small brass pellet, with a spiral spring behind it, closes the safety pin hole when the latter is withdrawn.

Safety pellet and ball.—The safety pellet retains in position a small brass ball lying on an inclined recess cut away on the top face of the pellet. The ball prevents any forward movement of the percussion pellet as long as it is kept in position by the safety pellet.

Detonator plug.—The detonator plug is a small cylinder of metal screwed on the exterior to fit into the body at the end of the horizontal powder channel, and immediately above the needle of the percussion pellet. It is recessed to receive the R.L. cap, and has a central fire hole.

Bottom plug.—The bottom plug is a short cylinder of metal, threaded externally to screw into the bottom of the fuze. It has a cavity filled by a perforated pellet of gunpowder covered on the top by a disc of paper and a disc of muslin, secured by a brass washer. The hole at the bottom is closed by a brass washer and shalloon disc, the base of the fuze and plug being waterproofed by a coating of Pettman's cement, and the bottom plug stabbed in three places to prevent it unscrewing.

Action of time portion.—The fuze is set by unclamping the nut by means of the "Key No. 5," and turning round the lower time ring until the graduation ordered is opposite the pointer on the body. The nut is then clamped securely. At the moment of loading the "T" pin is withdrawn and the closing pellet closes up the hole occupied by the pin and so prevents the gas on discharge getting into the fuze. On the shock of discharge the hammer in the lighting chamber sets back, shearing its suspending wire, and fires the detonating composition in the bottom of the lighting chamber and ignites the top ring, the brass disc being blown out to allow the gas to escape. The top ring burns round the reverse way to which the shell is rotating until it comes to the powder pellet at the beginning of the lower ring, which is then fired, blowing out the brass disc, and igniting the lower ring. The lower ring burns back the opposite way to the top until it comes to the pellet of powder in the body; this is fired, and ignites the powder in the horizontal channel, firing the percussion detonator and the bursting charge of the shell.

Action of percussion part.—At the moment of loading the "P" pin is withdrawn, the closing pellet closing the hole occupied by the pin. On shock of discharge the safety pellet sets back to the bottom of the slot in the percussion pellet, shearing its suspending wire, the brass ball following it on the first motion of rotation. The spiral spring prevents the percussion pellet rebounding, and the anti-boring pin prevents the pellet from turning.

Owing to the rotary motion of the shell the heavy end of the retaining bolt overpowers the spring and withdraws the smaller end from the recess, so that the percussion pellet is free to move forward, which it does on graze or impact, compressing the spiral spring; the needle striking the percussion detonator fires the fuze.

Fuze, T. and P. No. 65, Marks I and II.

Fuze, T. and P. No. 65, Marks I and II, are similar in design to the above, but are conversions from old No. 56, 60 and 63 fuzes.

The Mark II differs from the Mark I in the upper time ring, which is graduated in a slightly different manner.

Fuze, T. and P. No. 63, Mark I.

(*Plate* LV.)

The fuze is, in construction, generally similar to the Fuze, T. and P. No. 65, Mark I, but is a quick-setting fuze; the cap on the stem is circular and is not unclamped before nor clamped up after setting. It differs from the No. 65, Mark I, in the following particulars:—

The composition channels in the time rings are lined with asbestos paper.

The lower ring is milled in a different manner. The cap securing the dome is fixed in position by a steel set-screw.

This fuze is set by means of a special key (Key, No. 14, *see* page 343), shaped to fit the milling of the lower ring.

Fuze, T. and P. No. 60, Mark II | L | .

(*Plate* LVI.)

The Fuze, T. and P. No. 60, Mark II, differs from the No. 65 fuze as follows :—

(1) The lower, instead of the upper time ring, is graduated.
(2) The setting pointer is attached to the body of the fuze instead of to the lower ring.
(3) The composition channels are lined with asbestos paper.
(4) Calf-skin washers are used under the time rings instead of cloth washers.
(5) The time detonator in the upper time ring contains ·2 grains instead of ·35 grains of composition.

Fuze, T. and P. No. 60, Mark I | L | .

Fuze, T. and P. No. 60, *Mark I,* differs from Mark II in having a blackened notch instead of a pointer. In the first issues of Mark I the graduations showed the odd numbers.

Fuzes, T. and P. No. 60, *Marks I* and II*.*—A certain number of Marks I and II, T. and P. fuzes, No. 60, have been issued with ·35 instead of ·2 grains of detonating composition in the Time portion; such fuzes will be distinguished by a (*) added to their numeral.

Fuzes, T. and P. No. 60c, *Marks I and II,* are conversions from Nos. 56, 57 and 61 (single ring) fuzes. No more of No. 60c will be made.

They have the old numeral, lot number, and date of filling barred out and new substituted.

Fuze, T. and P. No. 56, Mark IV | C | .

(*Plate* LVII.)

T. and P. No. 56.—This fuze differs from the T. and P. No. 60, in having a *single,* barrel-shaped, time ring with lighting arrangement, graduated from 0 to 18 and subdivided to read in quarters. The gas from the burning composition escapes into the dome through a hole bored into the ring from the inner side, near the commencement, and through another at the top. A groove is cut in the top face of the body close to the stem and halfway round it, and a hole is bored through the body into this groove at an angle reverse to the spin of the rifling, for the gas to escape from the dome.

The percussion pellet is filled with F.G. powder instead of having a perforated pellet of gunpowder.

Shalloon is used in place of muslin for the discs. Time of burning at rest, 13 sec. :—

There are no earlier Marks of this fuze.

Fuze, T. and P. No. 56, *Mark IV.**—Certain lots of No. 56 fuzes have been re-fitted with ·35 grains instead of ·2 of detonating composition for the time portion, and are distinguished by having a (*) added to their numeral.

(2) LARGE T. AND P. FUZES OF G.S. GAUGE { No. 66. / „ 62. / „ 54. / „ 64. }

The Nos. 66, 62 and 54, T. and P. fuzes, are similar in construction ; the No. 54 is fitted with only one, instead of two, time rings.

The No. 64 fuze is entirely different in design from the above, and is described in detail.

FUZE, T. AND P. No. 66, MARK I.

(*Plate* LVIII.)

This fuze is similar in mechanism to the T. and P. No. 65A ; but the body above the screw-threaded portion is larger in diameter, the stem being hollow (for lightness) and thickened at the base to ensure the centering of the lower time ring. The upper ring is graduated from 0 to 60. The percussion part only differs from that already described in having two, instead of one spiral creep spring in front of the percussion pellet.

FUZE, T. AND P. No. 62, MARK II | L | .

(*Plate* LIX.)

This fuze is very similar to the No. 66, differing from it in the lower time ring, which is barrel-shaped and milled in a different manner.

The composition channels in the time rings are lined with asbestos paper.

The fuze composition in the rings is a mixture of black and brown mealed powder instead of F.G.

The fuze burns about a second longer than the No. 66.

FUZE, T. AND P. No. 62, MARK I.

A few T. and P. No. 62, Mark I fuzes have been issued ; they differ from the Mark II in the time rings, both of which are graduated, and both rings are capable of being moved in setting the fuze. The upper ring is graduated up to 30 ; the lower ring from 30 to 60.

The lower ring is fitted with a spring pawl, which, in the operation of setting the fuze, acts as follows :—For graduations up to 30, the pawl secures the lower ring to the body, so the upper ring is the only one which is moved ; for graduations *above* 30, the pawl no longer keeps the lower ring fixed, but secures it to the upper ring, so that the rings move round together. (*See* Fig. 87.)

FUZE, T. AND P. MIDDLE No. 54, MARK III.

(*Plate* LX.)

This is a single ring fuze of the same size as No. 62, but in construction and action it is similar to the T. and P., No. 56, Mark IV, except in marking and length of composition; the stem is hollow and the ring is graduated up to 30 divisions.

Fuze, T. and P. No. 54, *Mark III**.—Certain lots of No. 54 fuzes have been re-fitted with ·35 grains instead of ·2 of detonating composition for the time portion, and are distinguished by having a (*) added to their numeral.

No more No. 54 or 62 fuzes will be manufactured.

Fig. 87.

FUZE, T. AND P. NO. 62, MARK I.

Scale $\frac{1}{1}$.

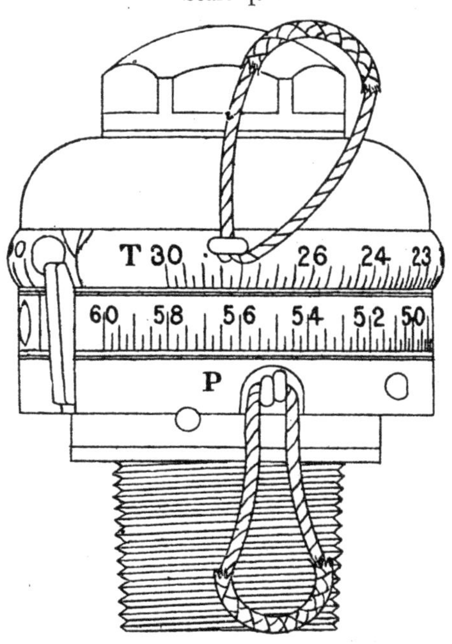

FUZE, T. AND P. NO. 64, MARK II | C | .

(*Plate* LXI.)

This fuze is used with shrapnel shell having the G.S. gauge, Naval Service—B.L. and Q.F., 6-inch to 9·2-inch.

Land Service—Q.F., 4·7-inch and 6-inch., B.L. and B.L.C., 6-inch to 10-inch.

Parts.—It consists of the following parts:—Body, bottom plug, percussion pellet with creep spring, detonator, and screw plug, stirrup spring, brass ferrule with safety pin, centrifugal bolt, two steel needle plugs, two time rings, lighting pellet with detonator and stirrup spring, safety pillar, safety pin and clamping nut.

Body.—The body is made of metal; externally it resembles that of the Fuze, T. and P. No. 66; the upper part of the stem is bored out to take the lighting arrangement of the time portion of the fuze; the lower part of the body is bored out to take the percussion arrangement, and fitted with a central steel needle plug.

Percussion pellet.—The percussion pellet is made of brass, cylindrical in shape, with a flanged base; this flange supports the phosphor-bronze creep spring; it has grooves cut round its circumference to allow the flash from the time rings to reach the powder in the bottom plug. The pellet is bored out to take a detonator, which is held in position by a detonator plug screwed into position beneath it; this plug has a central hole containing a perforated powder pellet.

Stirrup spring.—Fitting over the front end of the pellet is a small brass stirrup spring with two projecting arms, intended to support the ferrule.

Ferrule.—The ferrule is a short brass tube; it is supported by the arms of the stirrup spring, and projects beyond the front of the pellet, and so prevents any forward movement of the latter towards the needle plug.

The ferrule is further supported by a safety pin passing through it and the body of the fuze.

Centrifugal bolt.—In addition to the above safety arrangements, there is a centrifugal bolt placed into a radial recess in the fuze body; the inner end of this bolt projects in front of the pellet and so prevents it moving forward on to the needle, until the bolt has been spun out of the way by the rotary motion of the shell. The centrifugal bolt is held in the safe position by the upper part of the ferrule projecting into a slot cut in the bolt.

Creep spring.—A phosphor-bronze creep spring, resting on the flange of the pellet, prevents any rebound action and creeping forward of the pellet during flight.

Bottom plug.—The lower part of the body is closed by a gunmetal bottom plug; this plug supports the percussion pellet; its upper surface is grooved and filled with loose gunpowder communicating with a central perforated powder pellet. The plug has a flash hole closed by a muslin disc and a brass washer spun in; a disc of paper and a disc of muslin are also shellaced to the upper face of the plug to retain the powder in position.

Cloth washer.—Shellaced to the flat portion of the body is a cloth washer; a small hole in this washer leaves exposed a perforated powder pellet communicating by means of two other perforated powder pellets with the percussion chamber.

Lower time ring.—Resting on this washer is the lower time ring, which is movable; it is of gunmetal, milled on the exterior and fitted with a blackened cupro-nickel pointer. On the underside a channel is cut nearly all the way round, which is varnished, and driven with fuze composition (special black powder) and then covered with a washer of vegetable paper. At the beginning of the composition in the ring is a vertical hole containing a perforated powder pellet, to communicate with the composition in the upper ring. The gas escape hole is similar to that of the No. 66 fuze, but is fitted with a small pellet of gunpowder contained in an asbestos cup; the brass disc closing the hole is stabbed in four places.

Upper time ring.—Resting on a cloth washer shellaced to the top of the lower ring is the upper time ring; it is graduated from 0 to 60, sub-divided to read in quarters; it has a composition channel, and gas escape hole similar to the lower ring.

At the beginning of the composition a small hole is bored through to the inner circumference of the ring; this hole is in direct communication with another hole (fitted with a perforated powder pellet), bored through the stem into the lighting chamber.

The top ring is fixed to the stem by two metal pins fitting into slots cut down the stem and the inner circumference of the ring.

Lighting pellet.—The lighting pellet is made of delta metal bored out to take a detonator and screw plug; it is held in position in the top of the lighting chamber, directly over the needle plug, by means of a brass stirrup spring with two projecting arms, which fit into a recess bevelled off round the top of the stem.

Safety pillar.—In the event of the supporting arms of the stirrup spring being broken, the lighting pellet is prevented from coming in contact with the needle plug by means of a brass safety pillar.

This pillar is supported over a vertical hole in the fuze body by means of a safety pin, and projects up into the lighting chamber beyond the point of the needle.

Safety pin.—The safety pin is of phosphor bronze, and passes diagonally through the body of the fuze, the ends being splayed out into a groove to retain it in position; it is fitted with a becket of scarlet cord. A closing pellet is provided to close the hole on the removal of the pin.

Cap.—The time rings are clamped by a metal nut or cap, recessed to fit over the top of the lighting pellet. A hole in the side of the cap is intended to take the special tommy for loosening or tightening the cap before and after setting the fuze.

Action of time portion.—The fuze having been set at the required graduation, and the cap tightened by means of the setting tommy, at the last moment of loading the safety pin is withdrawn, leaving the safety pillar free to fall into its recess clear of the lighting pellet. On discharge, the lighting pellet sets back, straightening the arms of the stirrup spring, and carries its detonator on to the needle. The flash from the detonator fires the powder pellet in the flash hole in the stem, igniting the top time ring, and blows out the covering disc of the gas escape hole. The top ring burns round until it comes to the pellet of powder at the beginning of the composition in the lower ring; this pellet is ignited, blows out the gas escape hole disc and ignites the lower ring, which burns back in the opposite direction until it reaches the pellets of gunpowder in the body, thus firing the powder in the magazine of the fuze and the bursting charge of the shell.

Action of percussion portion.—At the moment of loading the "P" pin is withdrawn, the closing pellet closing the hole from which the pin has been removed. On shock of discharge the brass ferrule sets back, straightening the arms of the stirrup spring, unmasks the front of the pellet, and withdraws itself from the slot in the centrifugal bolt. As soon as the shell starts to rotate the centrifugal bolt is spun out into its recess by the rotary motion of the shell, clear of the front of the pellet; the spiral spring prevents the pellet creeping forward during flight.

On graze or impact the pellet is dashed violently forward on to the needle, compressing the creep spring. The flash from the exploding detonator fires the powder pellet in the percussion pellet, the powder in the magazine of the fuze, and the bursting charge of the shell.

Mark I fuze.

The Mark I fuze, T. and P., No. 64, differs from the Mark II in the following particulars :—

(*a*) The gas escape powder pellet in the lower time ring has no asbestos cup.

(*b*) The shape and direction of gas escape hole in the lower time ring differ ; in Mark I the hole is circular, and leads out at an angle against the spin of the shell ; in Mark II the hole is oval, and leads out almost at right angles to the axis of the fuze.

(3) T. and P. Fuzes of 2-inch Gauge { No. 80. ,, 81. ,, 82. ,, 83. ,, 84.

Fuze, T. and P., No. 80, Mark V | L | .

(*Plate* LXII.)

Use.—At present with B.L., 2·75-inch and Q.F., 13-pr., 18-pr., and 4-inch, Mark IV guns.

Parts.—The fuze consists of the following parts :—Body, bottom plug, percussion holder with steel needle, brass washer and cap, percussion pellet with detonator, detonator plug, stirrup spring, ferrule, creep spring, two time rings, lighting pellet with time detonator and screw-plug, stirrup spring, cap and set-screw and fuze cover.

Body.—The body is made of aluminium, the lower part screw-threaded on the exterior to the 2-inch gauge. Above the screwed portion the body is of larger diameter, forming a flange to which is screwed a brass ring ; the lower edge of this ring is tinned to facilitate the soldering on of the brass fuze cover.

The body is bored out to form three chambers of different diameters ; the lower chamber forms the magazine and is screw-threaded to take the bottom plug.

Graduations.—The brass ring on the body of the fuze is graduated from 0 to 22, reading in tenths. Near the " 0 " graduation a small vertical hole is bored in the top of the fuze body, into which is placed a perforated powder pellet which communicates with the magazine by means of a small channel filled with powder.

The hole and the channel are fitted with brass linings to prevent the powder coming in contact with the aluminium body.

A steel percussion holder is screwed into the centre of the body, and in its front end is a double-pointed needle ; the lower point of the needle projects into the percussion chamber, while the upper point is intended to fire the time portion of the fuze.

Percussion pellet.—The percussion pellet is made of brass, cylindrical in shape, reduced in diameter at the front end, and bored out to take a special detonator, which is held in position by a detonator plug screwed into position beneath it. This plug has a central hole for the flash from the detonator to pass to the magazine.

Detonator.—The detonator consists of a small copper cup with a central hole closed by a copper disc. It contains about $1\frac{1}{4}$ grains of detonating composition, and also a small pellet of gunpowder; these are retained in position by another covering disc of copper.

Stirrup spring.—Fitting over the front end of the percussion pellet is a small stirrup spring with two projecting arms which support the ferrule.

The ferrule is a short brass tube which projects beyond the pellet, and so prevents its forward movement on to the needle.

A steel spiral spring is placed around the lower point of the needle and in front of the pellet; it prevents any rebound action, and prevents "creeping."

The holder is closed by a brass washer, a disc of muslin, and a cap of brass with a central flash hole fitting tightly over the lower part of the holder.

Bottom plug.—The bottom plug is made of aluminium and has a small hole for the purpose of filling the magazine; this hole is closed by a screw plug.

The magazine of the fuze is filled with F.G. powder; the flash hole in the bottom plug is closed by a linen disc and aluminium washer spun in.

Lower time ring.—Shellaced to the top face of the body is a cloth washer having a small hole which leaves the powder pellet at the "0" graduation exposed. Resting on this washer is the lower time ring, which is made of brass and fits round the stem, and is free to turn. On the underside of the ring a channel is cut which is driven with fuze composition (special F.G. powder), and the bottom of the ring covered with a washer of vegetable paper. At the beginning of the composition there is a small hole bored through to the top face of the ring, containing a perforated powder pellet; at the same place there is also a gas escape hole bored through to the exterior. A perforated pellet of powder is placed into this hole, and the hole closed by a disc of aluminium retained in position by the metal of the ring being stabbed over the disc in four places. The beginning of the composition is indicated on the ring by a setting mark; there is also a metal stud attached to the ring to engage with the key or fuze setter in setting the fuze. The top face of the lower ring has shellaced to it a cloth washer, a small hole being left in the washer to leave the powder pellet in the ring exposed.

Top time ring.—The top ring is also made of brass; it is smaller in diameter than the lower; it has a composition channel and gas escape hole similar to the lower ring, but at the beginning of the composition a small hole is bored through to the inner circumference which is filled with mealed powder and closed by a patch of special paper. This hole is in direct communication with an oblique channel bored through the stem, leading into the lighting chamber in the top of the fuze.

The top ring is fixed to the stem by two aluminium pins fitted into vertical holes cut down the stem of the fuze and the inner circumference of the ring.

Lighting pellet.—The lighting pellet is made of brass, cylindrical in shape bored out to receive a detonator generally similar to that used in the percussion part. The detonator is secured in the pellet by a screw plug.

The lighting pellet is held in position in the top of the lighting chamber by a brass stirrup spring ; this spring has two projecting arms which fit into a recess bevelled off around the top of the stem, and so support the pellet with the time detonator over the upper point of the needle.

Cap.—The time rings are clamped by the cap of aluminium, which is recessed to fit over the top of the lighting pellet ; the cap, after being screwed home, so as to tension the rings, is secured by a set-screw.

Waterproofing.—The spaces between cap, time ring, body, &c., are waterproofed. (*See* page 270.)

Fuze cover.—The cover consists of a cap and a " tear off " strip.

The cover is made of brass, dome-shaped to fit over the fuze ; soldered to its lower edge is a brass strip, to the end of which is secured a ring of brass wire, intended to give a grip for the fingers when removing the cover.

The lower edge of this strip is soldered to the brass ring screwed to the flange of the fuze body. A projection is also provided on the cover to fit over the setting stud on the lower time ring, thus enabling the fuze with cover to be screwed into the shell by means of the Key No. 17.

Action of time portion.—The fuze cover having been removed, the fuze is set by turning round the lower time ring by means of the fuze key, or by the setter, until the setting mark on the ring is opposite the graduation ordered. On shock of discharge the lighting pellet sets back, straightening out the arms of the stirrup spring, carrying its detonator on to the upper point of the needle. The flash from the exploding detonator passes through the hole in the stem communicating with the top ring, and ignites the fuze composition, blowing out the aluminium disc covering the gas escape hole. The top ring burns round in the same direction as the spin of the shell until it comes to the exposed powder pellet at the beginning of the lower time ring, which is fired, igniting the lower ring, and blowing out the covering disc for the gas escape hole. The lower ring now burns back the reverse way to the upper ring until it arrives at the powder pellet in the body, which is fired, igniting the powder in the magazine of fuze, and the bursting charge of the shell.

Action of percussion portion.—On shock of discharge the brass ferrule sets back, straightening out the arms of the stirrup spring, and so unmasks the front end of the percussion pellet. The creep spring prevents rebound action on shock of discharge and creeping action during flight. On graze, or impact, the percussion pellet, with the ferrule, is dashed violently forward, compressing the creep spring, and carries its detonator on to the lower point of the needle. The flash from the exploded detonator passes down through the hole in the screw plug and fires the magazine of the fuze, the flash from which, passing through the hole in the bottom plug, explodes the bursting charge of the shell.

Fuze, T. and P., No. 80, Mark IV | L | .

The Mark IV fuze differs from the Mark V, described above, in the time rings which are made of aluminium instead of brass. The recess in the body of the fuze near the "0" graduation and the diagonal channel are not lined with brass; the cap closing the percussion holder is made of aluminium instead of brass; the fuze is lighter.

Fuze, T. and P., No. 80, Mark III | L | .

The Mark III fuze differs from the Mark IV in the body, which has no brass ring screwed to its flange as this, and early marks of the fuze take a different pattern of fuze cover, which is not soldered on. (*See* page 326.)

The bottom plug has no filling hole.

Fuzes issued prior to 3/1910 had a different lighting pellet for the time portion. The detonator was inserted into a recess in the bottom of the pellet and kept in position by the metal being spun over it instead of being inserted from the top and secured by a screw plug.

Fuze, T. and P., No. 80, Mark II.

This fuze is identical in construction to the Mark III, but the spaces between top cap, rings, body, &c., are not waterproofed.

Fuze, T. and P., No. 80, Mark I.

The Mark I fuze differs from the Mark II in having two slots cut in the top of the stem for the arms of the time stirrup spring instead of having the interior of the stem levelled all the way round at the top.

Fuze, T. and P., No. 80, Mark I* and II*.

Fuze, T. and P., No. 80, Mark I and II, when waterproofed as described on page 270, will be distinguished by the addition of a star (*) to the numeral as shown above.

Refilling of T. and P., No. 80 Fuzes.

All No. 80 T. and P. fuzes when refilled and refitted will have the letter "R" placed after the existing numeral.

The fuzes when refilled will also be waterproofed; a new Lot No. and date of repair will be stamped on the cap, the original Lot No. and date of manufacture being barred out.

Clip Safety Fuze, Time and Percussion, No. 80, Mark I.

This clip was used with Marks I to III, No. 80 fuzes ; it is obsolete for future manufacture.

The clip was intended to prevent the lower time ring from moving when the fuze was set at " safety."

The clip is made of steel, horse-shoe shaped, and fits round the graduated portion of the fuze body ; it has a slot cut in it to fit over the setting stud, and a tongue piece to fit into the fixing slot ; the ends of the clip have projections which fit under the edge of the body.

Old Pattern Fuze Covers.

Fuze covers.—The No. 80, Marks I to III fuzes when carried in shell are protected against damp by means of a brass cover. The fuze cover consists of a cap, a screwed ring with base ring and tin band, and a leather washer. The cap is conical in shape to fit over the fuze, and has a screw thread at its mouth to engage with the thread on the screwed ring.

The base ring is shaped to fit into the groove around the outside edge of the fuze socket of the shell, and is attached to the screwed ring by means of the soldered tin band.

Fuze, Time and Percussion, No. 81, Mark I.

(*Naval.*)

(*Plate* LXIII.)

In external appearance this fuze is somewhat similar to the T. and P., No. 80, Mark V.

It is an improvement on No. 80 ; the lighting pellet is not central, as in No. 80, for the following reasons :—Firstly, if a fuze such as No. 80 were to receive a very violent blow on the head, there is a risk of the lighting pellet being driven on to the needle ; secondly, the lighting pellet being placed at the side enables a recess to be made in the nut, which recess could be used with an automatic fuze setter.

Use.—(*See* Table No. 31, page 338.)

Parts.—The fuze consists of the following principal parts :—Body, bottom plug, percussion holder with needle, percussion plug, percussion pellet with creep spring, detonator, and screw plug, ferrule, stirrup spring and safety ball, two time rings, lighting pellet with detonator, stirrup spring and safety pin, cap and set screw.

Body.—The body is of aluminium, the lower part threaded to the 2-inch gauge ; above the threaded portion the body is enlarged and graduated from 0 to 22, subdivided to read in tenths. Above this enlarged portion the body terminates in a stem, the top of which is reduced in diameter and threaded for the clamping nut ; the stem has a flat formed on one side, leaving a step into which is screwed a needle plug with steel needle. The fuze body is bored out from the bottom and screw-threaded to take the percussion arrangement and bottom plug.

Percussion holder.—The percussion holder is of steel and is fitted with a steel needle; it screws into the fuze body. When screwed home a hole is bored radially through it and the stem, intended for the safety ball when released.

Percussion pellet.—The percussion pellet is similar to that of the No. 80 fuze, but has a larger flange at the base, on which rests a brass spiral creep spring.

Percussion detonator.—Percussion detonator is the latest type of R.L. cap described and illustrated on page 282; it is retained in position by a hollow screw plug.

Stirrup spring and ferrule.—Fitting over the front of the pellet is a brass stirrup spring with two projecting lugs to support the ferrule.

Safety ball.—Inside the ferrule, and in front of the percussion pellet, is a brass ball; this ball prevents the forward movement of the pellet until the ferrule has been set back by the shock of discharge, and until the ball itself, acted upon by centrifugal force, has been spun out into the radial groove in the stem.

Percussion plug.—Underneath the percussion pellet are placed a linen disc and an aluminium washer, the percussion arrangement being retained in the holder by a hollow screw plug fitting into the bottom of the screwed recess.

Bottom plug.—The lower part of the body is filled with powder, and is closed by an aluminium bottom plug having a central flash hole with a linen disc and a brass washer spun in.

Lower time ring.—Shellaced to the flat portion of the body is a cloth washer; a small hole in this washer leaves exposed a perforated powder pellet communicating by means of a diagonal channel, which is also provided with a pellet of powder, with the magazine.

Resting on this washer is the lower time ring, which is movable. It is made of gunmetal, with composition channel, gas-escape hole, and setting stud, similar to the No. 80 fuze.

Upper time ring.—The upper ring rests on a cloth washer shellaced to the lower ring; a flat solid portion is formed on the inner circumference which engages with the flat on the stem; this solid portion contains the lighting arrangement.

Lighting arrangement.—The lighting arrangement consists of a lighting pellet with detonator, screw plug, and stirrup spring.

The lighting pellet is made of brass, and is suspended, by means of the stirrup spring, immediately over the steel needle secured to the step on the stem; it contains a detonator secured in position by a screw plug.

Safety pin.—A safety pin of phosphor bronze, fitted with a becket of scarlet cord, passes through a hole in the ring and underneath the lighting pellet to support the latter in transit, &c.

Closing pellet.—To seal the hole on the removal of the safety pin a closing pellet with spiral spring is fitted to the ring as shown in the plate.

Cap.—The cap is of metal and screws on to the top of the stem; a steel set screw prevents it from unscrewing.

Action of time portion.—The safety pin is removed at the last moment of loading. On the shock of discharge the lighting pellet sets back, straightening out the arms of the stirrup spring, and carries its detonator on to needle. The flash from the exploding detonator ignites the top ring, which burns round the same way as the shell is rotating until it arrives at the perforated powder pellet at the beginning of the lower ring; the lower ring burns back in the opposite direction until it arrives at the perforated powder pellets in the body, which fire the magazine of the fuze and the bursting charge of the shell.

Action of percussion portion.—On shock of discharge the ferrule sets back, straightening out the arms of its stirrup spring, unmasking the front of the percussion pellets, and releasing the safety ball. The rotary motion of the shell causes the ball to spin out into the recess in the stem, leaving the pellet free to move forward; the creep spring prevents any rebound action, and also prevents the pellet from creeping forward during flight. On graze or impact the pellet is carried forward on to the needle, firing the detonator, the magazine of the fuze, and the bursting charge of the shell.

Fuze, Time and Percussion, No. 82, Mark III.

(*Land Service.*)

(*Plate* LXIV.)

This fuze was introduced for shrapnel shell with the Q.F. 4·5-inch howitzer.

As howitzers use a very small charge and also fire varying charges it is necessary to have a fuze that will act with as much certainty with the smallest charge as with the largest.

The fuze does not depend upon the shock of discharge to set the time portion in action, but on the rotation of the shell.

The use of No. 82 fuze has now been extended to Q.F., B.L., and B.L.C. 6-inch shrapnel having the 2-inch gauge.

Parts.—The fuze consists of the following principal parts:—Body, needle plug, percussion pellet with detonator, creep spring, three centrifugal bolts with spiral springs, brass cap, and bottom plug, lower time ring with setting mark, upper time ring with needle plug, lighting pellet with detonator, three retaining bolts with spiral springs and closing plugs, safety pellet with shearing wire, two safety pins, cap, and set screw.

Body.—The *body* is of aluminium and, externally, is similar to the Fuze, T. and P., No. 80, Mark III, but is graduated from 0 to 40, instead of from 0 to 22. It is bored out from the bottom to form a percussion chamber, into the top of which is screwed a needle plug with steel needle.

Percussion pellet.—The *percussion pellet* is made of metal, cylindrical in shape; the front end is reduced in diameter, forming a shoulder to support the creep spring. A hole bored vertically through the pellet contains at its upper end a detonator, below which is a perforated powder pellet retained in position by a disc of paper, and a brass washer spun in.

Percussion safety pin.—The percussion safety pin passes through the body of the fuze and the percussion chamber, and so prevents any forward movement of the pellet until the pin is withdrawn. A small metal closing pellet with spring, fitting into a vertical recess closes the safety pin hole when the latter is withdrawn.

Centrifugal bolts.—Passing through the side of the body are three *centrifugal bolts* ; each bolt is pressed inwards by means of a small spiral spring into a hole in the side of the percussion pellet, and so prevents the latter from moving forward until the bolts have been spun out by the rotary motion of the shell.

Slot and pin ; Creep spring.—The pellet is prevented from turning round during flight by means of a small pin fitting into a slot cut down the side of the pellet ; *creeping action* is checked by the brass spiral spring.

Brass cap.—The percussion chamber is closed by a flanged cap, having a central flash hole ; a linen disc shellaced to the inside of the cap prevents the powder in the magazine working through this hole into the percussion chamber

The cap is held in position by a fixing screw and Pettman's cement.

Bottom plug.—The bottom plug is made of metal and screws into the base ; it retains the percussion arrangement in position.

The upper surface of this plug is grooved out to form the magazine, which is filled with F.G. powder ; a central flash hole in the plug is closed by a linen disc and a brass washer spun in.

Shellaced to the flat portion of the body is a cloth washer ; a small hole in this washer leaves exposed a perforated powder pellet resting in a brass-lined recess.

A diagonal chamber also lined with brass and filled with F.G. powder, leads from the powder pellet to the magazine.

Lower time ring.—Resting on the cloth washer is the lower time ring, which fits round the stem and is free to turn.

On the underside of the ring a channel is cut which is driven with fuze composition (special F.G. powder), and the bottom of the ring covered with a washer of vegetable paper. At the beginning of the composition there is a small hole bored through to the top face of the ring, containing a perforated powder pellet ; at the same place there is also a gas escape hole bored through to the exterior. A perforated pellet of powder is placed into this hole, and the hole closed by a disc of brass retained in position by the metal of the ring being stabbed over the disc in four places. The beginning of the composition is indicated by a setting mark ; a small recess in the ring is intended for the projecting arm of the keys (Nos. 19 and 36) used in setting the fuze.

The top face of the lower ring has shellaced to it a cloth washer, a small hole being left in the washer to leave the powder pellet in the ring exposed.

Upper time ring.—The upper time ring is prevented from turning by means of two brass pins fitting into recesses in the ring and top of the stem ; it has a composition channel, and gas escape hole similar to the lower ring. This ring is fitted with a needle plug, underneath which is a perforated powder pellet communicating with the beginning of the fuze composition.

Lighting arrangement.—The lighting arrangement consists of a lighting pellet and three retaining bolts; all four (when released, and acted upon by centrifugal force) move outwards in radial holes bored through the ring and the stem. The radial holes in the stem are lined with brass.

Lighting pellet.—The lighting pellet contains, in its end facing the needle plug, a detonator similar to the time detonator of the Fuze, T. and P., No. 80; it is kept from moving outwards against the needle by two retaining bolts; the latter are locked in the safe position by a third retaining bolt; this bolt is secured by means of a phosphor bronze safety pin passing through it and the time ring, and by a safety pellet.

Fig. 88.

Scale $\frac{2}{1}$.

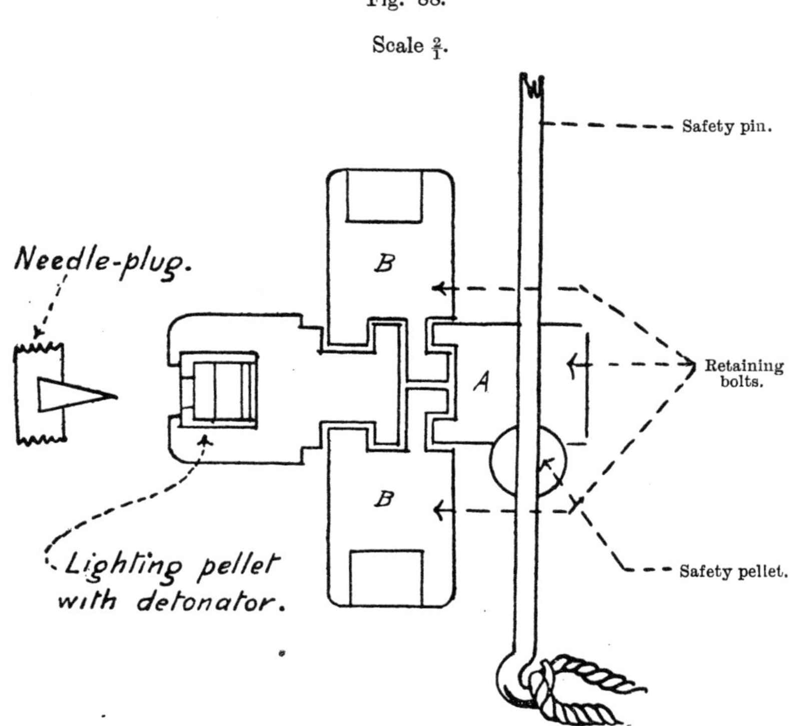

The radial holes in the upper time ring are closed by brass screw plugs; spiral springs are placed between them and the outer ends of the retaining bolts.

Safety pellet.—The safety pellet is of brass, cylindrical in shape, with a deep groove formed round its circumference; it is suspended by means of a copper shearing wire and the safety pin, in a vertical hole in the top of the stem. It acts as an additional safety arrangement; its lower end engages with a radial groove formed in the retaining bolt and so prevents the latter from moving until it has been set back by the shock of discharge.

Cap.—The cap is of aluminium ; it screws on to the top of the stem, and clamps the time rings ; it is prevented from unscrewing by means of a metal set screw.

Waterproofing.—The fuze is waterproofed as described on page 270.

Action of time portion.—The fuze having been set to the required graduation, the safety pin is withdrawn before loading.

On shock of discharge the shearing wire in the safety pellet is broken, and the pellet sets back to the bottom of its recess. In this position the pellet no longer prevents the bolt (A) from moving. (*See* Figs 88 and 89.)

When the shell rotates, the bolt (A) moves outwards, and unlocks the other two bolts (B, B) ; these now fly outwards, and unlock the lighting pellet ; this therefore flies outwards and carries its

Fig. 89.
Scale $\frac{2}{1}$.

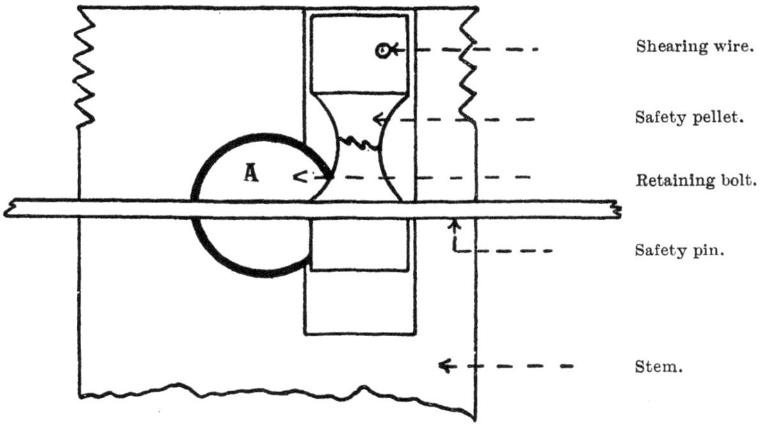

detonator on to the needle. The flash from the detonator fires the perforated powder pellet below the needle, and ignites the upper time ring.

The top ring burns round in the same direction as the spin of rotation until it comes to the exposed powder pellet at the beginning of the lower time ring, which is fired, igniting the lower ring and blowing out the covering disc for gas escape hole. The lower ring now burns back the reverse way to the upper ring until it arrives at the powder pellet in the body, which is fired, igniting the powder in the magazine and the bursting charge of the shell.

Action of the percussion portion.—At the moment of loading the " P " pin is withdrawn, the closing pellet closing the hole from which the pin has been removed.

On rotation, the centrifugal bolts of the percussion arrangement are spun out of their recesses in the percussion pellet ; the spiral spring prevents any rebound action, or tendency for the pellet to creep forward during flight ; the anti-boring pin prevents it turning round. On graze, or impact, the percussion pellet is carried forward, the needle firing the detonator and the perforated powder pellet, the flash from which ignites the magazine of the fuze and the bursting charge of the shell.

Fuze, T. and P., No. 82, Mark II.

The Mark II fuze differs from the Mark III in the following particulars :—

(1) The bottom plug and the flanged cap closing the percussion chamber are made of aluminium instead of brass ; this cap is not secured by a fixing screw.

(2) The diagonal channel and the hole for the powder pellet in the fuze body near the " 0 " graduation are not fitted with brass linings.

Fuze, T. and P., No. 82, Mark I.

In the Mark I fuze the percussion detonator is not central in the pellet, and the needle plug in the top of the percussion chamber is placed eccentrically to correspond.

Fuze, T. and P., No. 83, Mark I | L | .

(*Plate* LXV.)

Use.—This fuze is used with shrapnel shell for B.L., 60-pr.

It is similar in design to the Fuze, T. and P., No. 81, described on page 326, differing from it in being made of metal. The stirrup spring for the time pellet is of a stronger pattern ; the percussion arrangement of the fuze fits into the fuze body instead of into a percussion holder, and the hole in the top of the cap is cylindrical instead of conical as in No. 81.

Action.—For description of the action of the fuze, *see* Action of No. 81 fuze, page 328.

Fuze, T. and P., No. 84, Mark I | C | .

(*Plate* LXVI.)

This fuze has been approved for Naval Service with shrapnel shell, having the 2-inch gauge, B.L. and Q.F. 7·5-inch and under. In the Land Service it is used with the Q.F. 3-inch shrapnel.

Description.—The body of the fuze is made of aluminium, the percussion arrangement being identical with the T. and P., No. 82 fuze, already described on page 328.

Time rings.—The time rings are made of brass, the lower movable, the upper fixed to the stem by two metal pins fitting into slots cut down the stem and the inner circumference of the ring.

Lighting arrangements.—The lighting arrangement for the time portion of the fuze is contained in a recess bored out in the top of the percussion chamber to one side of the needle plug.

In the upper part of this recess is the lighting pellet. The lighting pellet is suspended in the recess by means of a phosphor bronze safety pin which passes through the upper time ring, the pellet, and the body of the fuze.

Underneath the lighting pellet is placed a steel spiral spring intended to support the pellet when the safety pin is withdrawn.

The time detonator is contained in a recess in the front end of a metal plug which screws into the bottom of the lighting chamber.

When screwed home, this plug compresses the steel spring, the needle of the lighting pellet being directly over the detonator.

A flash hole bored through the body from the lighting chamber allows the flash from the time detonator to fire the top time ring.

Top cap.—The rings are clamped by a cap similar to the T. and P. No. 81 fuze ; a set screw prevents the cap from unscrewing.

Action of time portion.—The fuze having been set to the required graduation, at the last moment of loading, the " T " pin is withdrawn, leaving the steel spiral spring supporting the pellet over the time detonator.

On shock of discharge the lighting pellet sets back, compresses the spring and carries its needle on to the time detonator.

The flash from the exploding detonator ignites the top ring, which burns round the same way as the shell is rotating until it arrives at the perforated powder pellet at the beginning of the lower ring ; the lower ring burns back in the opposite direction until it arrives at the perforated powder pellets in the body, which fire the magazine of the fuze and the bursting charge of the shell.

Action of percussion portion.—At the moment of loading the " P " pin is withdrawn, the closing pellet closing the hole occupied by the pin.

On rotation, the centrifugal bolts of the percussion arrangement are spun out of their recesses in the percussion pellet ; the spiral spring prevents any rebound action, or tendency for the pellet to creep forward during flight ; the anti-boring pin prevents it turning round. On graze or impact, the percussion pellet is carried forward, the needle firing the detonator and the perforated powder pellet, the flash from which ignites the magazine of the fuze and the bursting charge of the shell.

Drill Fuzes.

Drill fuzes are either burnt out fuze bodies, or solid metal, of the same external dimensions as the fuze they represent ; they are blackened, and stamped " Drill."

Time, and T. and P. drill fuzes with safety pins, are fitted with special pins ; the time rings and the safety mark are not blackened.

The T. and P. No. 80 and 81 drill fuzes are fitted with a steel setting pin, and a steel piece in the flange of the body in which a fixing notch is cut.

Table No. 30.—*Time and Percussion Fuzes used in the Land Service.*

The Fuzes shown in Italics are Obsolete for future manufacture.

Fuzes.	Marks.	Paragraph in List of Changes.	Time of Burning at rest.	Used with.	
Time and Percussion.			Seconds.	Ordnance—	
Middle, No. 54 ...	I* II III III*	8417, 10743 8417 8912, 9809, 9856, 11874 14562	16	B.L., 10-inch B.L., 9·2-inch, IV to VIc B.L., 9·2-inch, IX B.L., 9·2-inch, X, X^{v}, X* B.L., 6-inch, VII, VII^{v} B.L., 6-inch, 30-cwt. howitzer B.L., 6-inch, 25-cwt. howitzer B.L., 5·4-inch howitzer B.L., 5-inch howitzer B.L.C., 6-inch B.L.C., 6-inch, on siege carriage Q.F., 6-inch Q.F., 4·7-inch Q.F., 4·7-inch, on travelling carriage	For shrapnel shell, but when stock of fuzes is exhausted, to be replaced by No. 62, or No. 64 for fixed armaments.
No. 56	IV IV*	7716, 9088, 9194, 9809, 9856 13775	13	B.L., 4-inch jointed B.L., 30-pr. B.L., 15-pr. B.L., 12-pr. 6 cwt. B.L., 10-pr. B.L.C., 15-pr. Q.F., 12-pr. 12 cwt. Q.F., 2·95-inch	For shrapnel shell, but when stock of fuzes is exhausted, to be replaced by Nos. 60, 63 or 65.

No. 60	I I* II II*	11019 11137 11226 13989 11692 13989	20	B.L., 4-inch, jointed B.L., 30-pr. B.L., 15-pr. B.L., 12-pr. 6 cwt. B.L., 10-pr. B.L.C., 15-pr. Q.F., 15-pr. Q.F., 12-pr. 12 cwt. Q.F., 2·95-inch	For shrapnel shell.
No. 60C	I II	11691 11692	20·1		
No. 62	I II	11874 12921 13127	35	B.L., 10-inch B.L., 9·2-inch, IV to VIC B.L., 9·2-inch, IX B.L., 9·2-inch, X, X^{v}, X* B.L., 6-inch, VII, VII^{v} B.L., 60-pr. B.L., 6-inch, 30-cwt. howitzer B.L., 6-inch, 25-cwt. howitzer B.L., 5·4-inch howitzer B.L., 5-inch howitzer B.L.C., 6-inch B.L.C., 6-inch, on siege carriage Q.F., 6-inch Q.F., 4·7-inch Q.F., 4·7-inch, on travelling carriage	For shrapnel shell, but when stock of fuzes is exhausted, to be replaced by :— No. 64 for B.L., 10-inch, 9·2-inch, IX, X, X^{v}, X*, 6-inch, VII, VII^{v}, B.L.C. 6-inch, Q.F. 6-inch and 4·7-inch. No. 66 for B.L. 6-inch, 30-cwt., and 5-inch howitzers, and Q.F. 4·7-inch (travelling). No. 83 for B.L., 60-pr.
No. 63	I	12502 12969	20·1	B.L., 4-inch, jointed B.L., 30-pr. B.L., 15-pr. B.L., 12-pr. 6 cwt. B.L., 10-pr. B.L.C., 15-pr. Q.F., 15-pr. Q.F., 12-pr. 12 cwt. Q.F., 2·95-inch	For shrapnel shell.
No. 65 and 65A ...	I II	15452 15626 16160	20·1		

TABLE NO. 30.—*Time and Percussion Fuzes used in the Land Service*—continued.

Fuzes.	Marks.	Paragraph in List of Changes.	Time of Burning at rest.	Used with.	
TIME AND PERCUSSION.			Seconds.	ORDNANCE—	
No. 64	I II	15330 16117	30	B.L., 10-inch B.L., 9·2-inch, IX B.L., 9·2-inch, X, X^v, X* B.L., 6-inch, VII, VII^v B.L.C., 6-inch B.L.C., 6-inch, on siege carriage Q.F., 6-inch Q.F., 4·7-inch	For shrapnel shell.
No. 66	I	—	34	B.L., 6-inch, 30-cwt. howitzer B.L., 5-inch howitzer Q.F., 4·7-inch, on travelling carriage	For shrapnel shell.
No. 80	I I* II II* III IV V	12800 14670 13879 14670 14317 16008	22	Q.F., 4-inch, Mark IV Q.F., 18-pr. Q.F., 13-pr. B.L., 2·75-inch, converted B.L., 2·75-inch	For shrapnel shell.

Fuzes.	Marks.	Paragraph in List of Changes.	Time of Burning at rest.	Used with.
No. 82	I II III	15434 16080 16685	40	B.L., 6-inch, VII, VII* B.L.C., 6-inch B.L.C., 6-inch, on siege carriage Q.F., 6-inch } For shrapnel shell with 2-inch fuze-hole. Q.F., 4·5-inch howitzer. For common and shrapnel shell.
No. 83	I	15553	30	B.L., 60-pr. For shrapnel shell.
No. 84	I	—	30	Q.F., 3-inch. Shrapnel shell.

TABLE No. 31.—*Time and Percussion Fuzes used in the Naval Service.*

Fuzes.	Marks.	Paragraph in List of Changes.	Time of Burning at rest.	Used with.
TIME AND PERCUSSION.			Seconds.	ORDNANCE.
Middle, No. 54 ...	I* II III III*	8417 10743 8417 8912 9809 9856 11874 14562	16	B.L., 13·5-inch, I to IV B.L., 10-inch, II to IV* B.L., 9·2-inch B.L., 7·5-inch, except III to IV* B.L., 6-inch, VII, VIII, XI, XI* Q.F., 6-inch Q.F., 4·7-inch Q.F., 4-inch, I to III* } For shrapnel shell (G.S. fuze-hole).
No. 56	IV IV*	7716 9088 9194 9809 9856 13775	13	Q.F., 12-pr. 12 cwt. ,, 12-pr. 8 cwt. } For shrapnel shell (G.S. fuze-hole).

TABLE No. 31.—*Time and Percussion Fuzes used in the Naval Service*—continued.

Fuzes.	Marks.	Paragraph in List of Changes.	Time of Burning at rest.	Used with.
TIME AND PERCUSSION.			Seconds.	ORDNANCE.
No. 63	I	12502 12969	20·1	Q.F., 4-inch, I to III* Q.F., 12-pr. 18 cwt., on Field mountings Q.F., 12-pr. 12 cwt. Q.F., 12-pr. 8 cwt. For shrapnel shell (G.S. fuze-hole).
Nos. 65 and 65A ...	I II	15452 15626 16160	20·1	Q.F., 4-inch, I to III* Q.F., 12-pr. 8 cwt. For shrapnel shell (G.S. fuze-hole).
No. 64	I II	15330 16117	30	B.L., 9·2-inch B.L., 7·5-inch, except III to IV* B.L., 6-inch, VII, VIII, XI, XI* Q.F., 6-inch For shrapnel shell (G.S. fuze-hole) at long ranges.
No. 81	I	14913	30	B.L., 13·5-inch, V B.L., 12-inch B.L., 9·2-inch B.L., 7·5-inch, except III to IV* B.L., 6-inch B.L., 4-inch, VII to VIII* Q.F., 4-inch, IV, V For shrapnel shell (2-inch fuze-hole) at long ranges. Q.F., 12-pr. 18 cwt. Q.F., 12-pr. 4 cwt. For shrapnel shell (2-inch fuze-hole).
No. 84	I	—	30	Replaces No. 81 for future manufacture for B.L. 7·5-inch and below.

(G).—KEYS, &c., USED IN CONNECTION WITH FUZES.

Implements, Ammunition.

Key, No. 5 (Mark III).—G.S. Fuze-hole Fuzes | C | is a steel key with lanyard, and is used with fuzes of G.S. gauge, except Nos. 25 and 64.

Fig. 90.

KEY, NO. 5, MARK III, G.S. FUZE-HOLE FUZES | C |.

Scale ½.

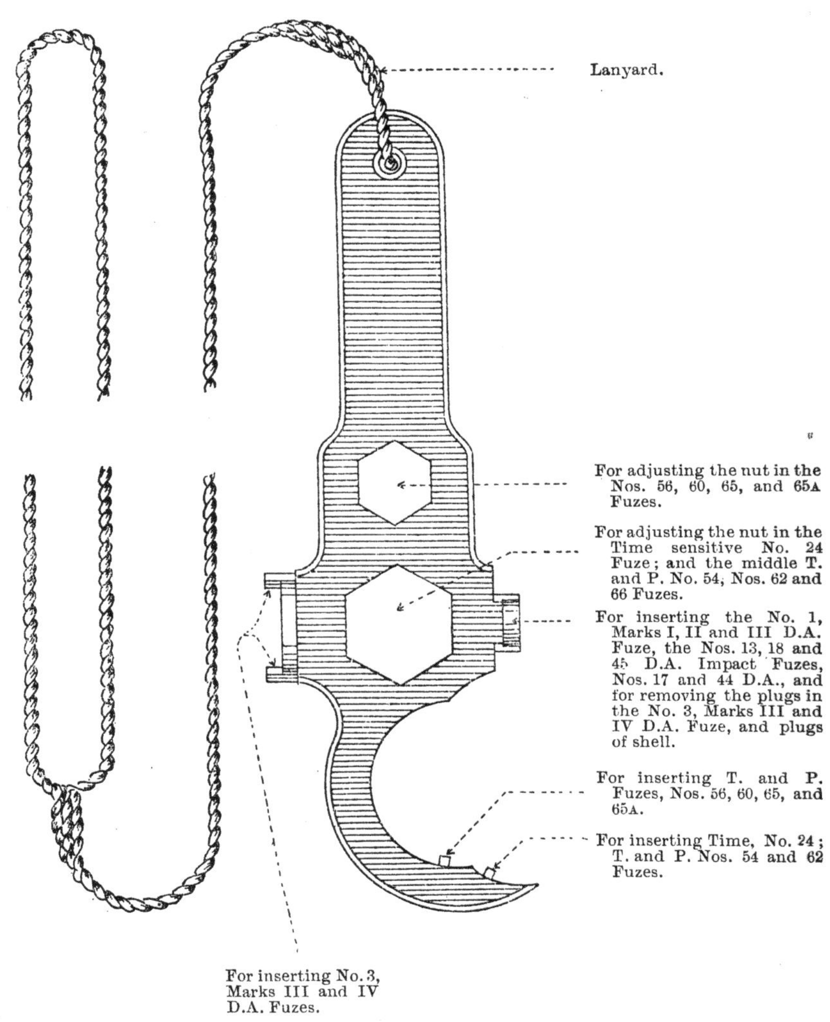

Key, No. 6 (Mark I), Extracting D.A. Fuzes | C | is made of steel to the form and dimensions shown in Fig. 91. It is fitted with a circular hinged clamp to fit round the head of the fuze in unscrewing.

Instructions for use :—

(1) Remove safety or securing pins.
(2) Remove the cap of the fuze.
(3) Apply the key so that :—
  (*a*) The pins on the body of the fuze fit in key in an axial direction.
  (*b*) The rim of the key fits into the recess in the head of the shell.
(4) Grip the key firmly, and unscrew the fuze. This should be done slowly and not with a jerk.
(5) The head of the fuze should be watched, and should distortion be noticed, the attempt to remove the fuze should be abandoned.
(6) Replace cap and safety or securing pins.

Fig. 91.

KEY, NO. 6, EXTRACTING D.A. FUZES.

Scale ½.

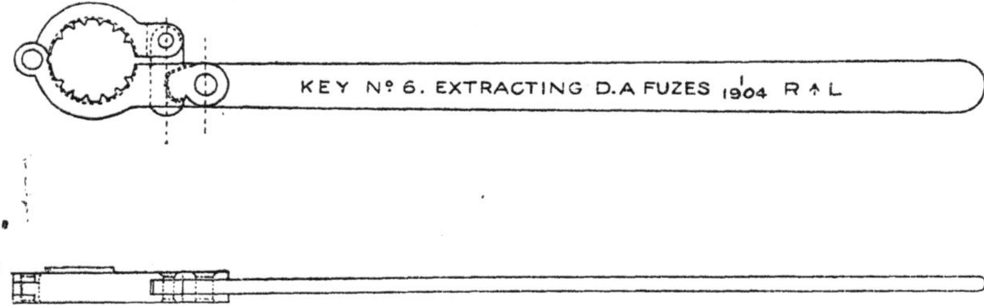

KEY, No. 7 (MARK I), ARMSTRONG FUZE | N | is a steel key in the form of a round bar, 0·6-inch in diameter and 15 inches long, having a pair of pins on one side, about the centre, for the Armstrong base fuze, and a single projection on the opposite side for the Armstrong base plug.

KEY, No. 8 :—

The under-mentioned Marks of the Key, No. 8, may all be met with :—

KEY, No. 8 (MARK IV).—BASE FUZES | C | is a steel key of the form shown in the figure. It is suitable for inserting or removing any Mark of No. 11, 12, 15 or 16 base fuzes or their corresponding plugs.

KEY, No. 8 (MARK III).—BASE FUZES | C | differs from the Mark IV in having the pair of projections for the latest issues of base fuzes and the projection for the base plugs nearer the longitudinal axis. It is suitable for use with any Mark of No. 11, 12 or 15 base fuzes or their corresponding plugs, but is not suitable for the No. 16 base fuze or the No. 6 plug, base, shell. The Mark III key can be converted to Mark IV pattern, and is then known as Mark III*.

KEY, No. 8 (MARK II).—BASE FUZES | C | differs from the Mark III in being without the pair of projections suitable for the key slots in the latest issues of the Nos. 11, 12 and 15 base fuzes. (§ 15105.) Mark II can be converted to Mark III pattern, when it becomes Mark II*. It can also be converted to Mark IV pattern, and then becomes Mark II**. (§§ 15108 and 16011.)

KEY, No. 8 (MARK I).—BASE FUZES | C | differs from Mark II in the shape of the bar, which is 1 inch wide and ½ inch thick at the centre. It is not so strong as the later Marks, and is not converted to conform with them.

Fig. 92.

KEY, NO. 8, MARK IV, BASE FUZES | C |.

Scale ½.

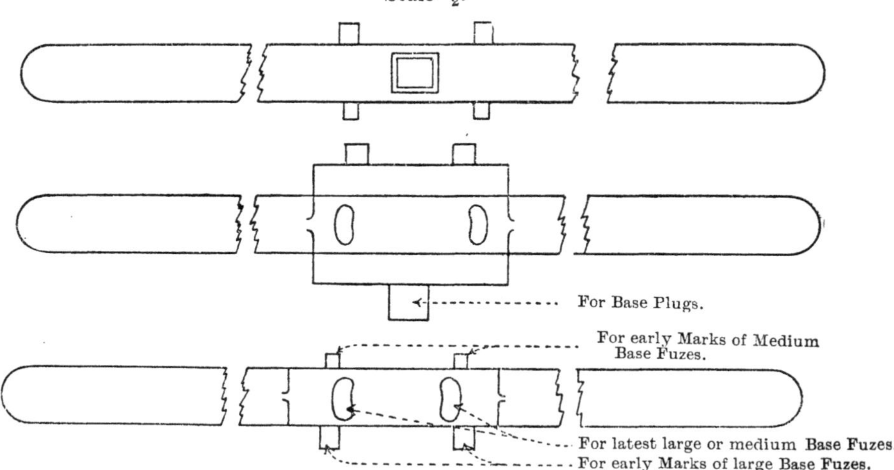

KEY, No. 10 (MARK I), No. 17 FUZE | L | is of steel with a screw-driver at one end and a projection on one side to fit the square recess in the cap of the No. 17 fuze. It is fitted with a lanyard of white cotton line. This key will be issued when existing stock of "*Keys, No.* 19, *Mark I*" is used up.

Fig. 93.

KEY, NO. 10, MARK I, NO. 17 FUZE | L |.

Scale ½.

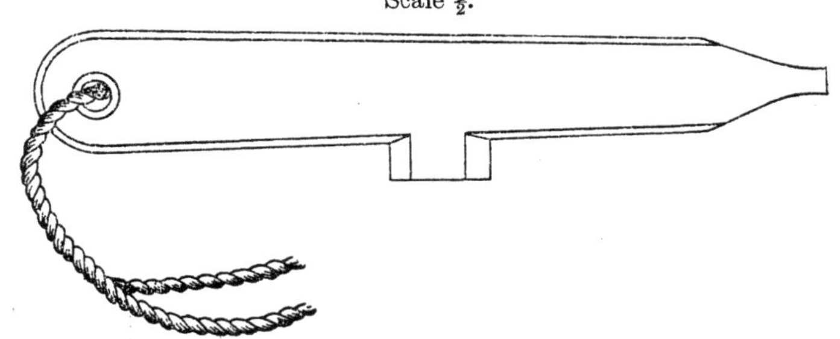

KEY, No. 11 (MARK I).—No. 19 FUZE | N | is of steel, T-shaped, with a square end to fit the recess in the safety cap of the Nos. 19

and 19A D.A. Impact fuzes, or the Nos. 4 and 4A, plug, fuze hole, special. It is fitted with a lanyard of white cotton line.

Fig. 94.

Scale $\frac{1}{1}$.

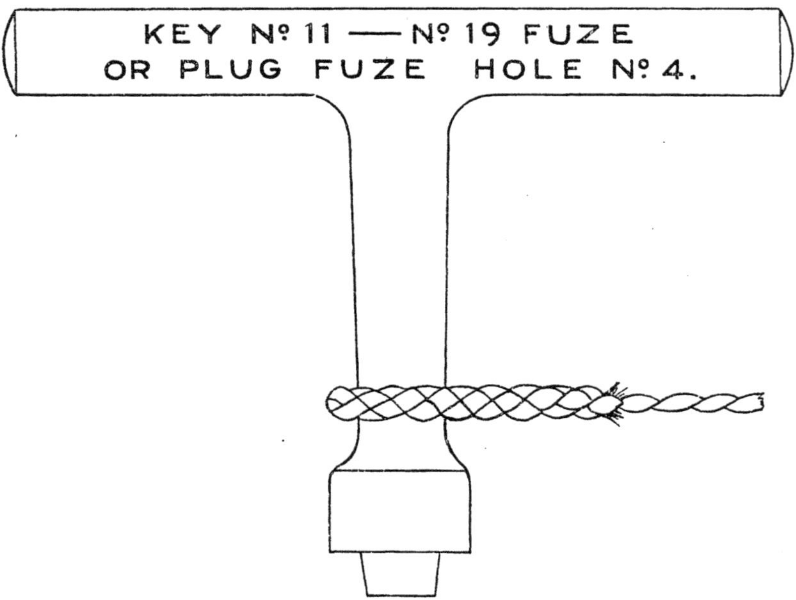

KEY, No. 13 (MARK I), No. 25 FUZE | L | .—The annular end of the key is for fixing and the other end for setting the " Fuze, time, No. 25."

Fig. 95.

KEY, NO. 13, MARK I, NO. 25 FUZE.

Scale $\frac{1}{2}$.

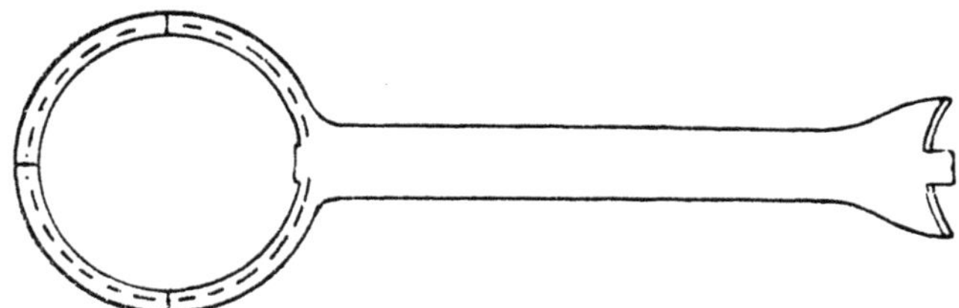

KEY, No. 14 (MARK I), SETTING, No. 63 FUZE | C | is a short steel fork; the arms of the fork are semi-circular, and teeth to grip the milled ring of the fuze are formed at the ends of the arms and centre portion of the key. It is fitted with a lanyard.

Fig. 96.

Full Size.

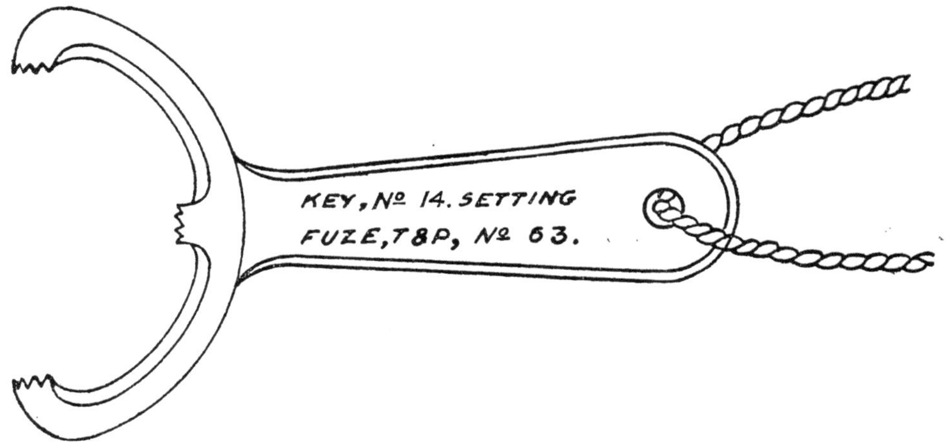

KEY, No. 16 (MARK I), ADAPTER | C | is made of steel, shaped at one end to fit the hole in the adapter described on page 259. It is used for screwing or unscrewing the latter into or out of the shell.

KEY, No. 16 (MARK II).—No. 64 FUZE AND ADAPTER | L | differs from Mark I in not having the horns on each side of the projection at the end of the key. It is used for fixing, and for adjusting the top cap of "Fuze, T. and P., No. 64," and for "Adapter, 2-inch fuze-hole."

Fig. 97.

Scale $\frac{2}{3}$.

Fig. 98.

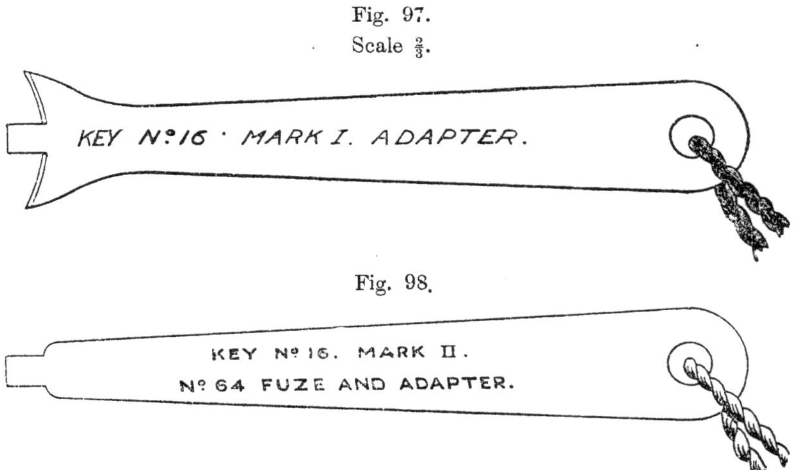

KEY, No. 17 (MARK II), FIXING NOS. 80 AND 83 FUZES | L |.—The Mark II key is of steel, one end being shaped to fit over the fuze; the lower edge of the ring portion is bevelled to suit all Marks of No. 80 fuze without cover, and is provided with a projection to fit the square notch in the flange of the fuze body.

The upper edge of the ring is provided with a slot to fit over the projection on the cover when screwing in Marks IV or V, No. 80, fuze with cover.

The Mark I key differs from the Mark II in the upper edge not being prepared for use with Marks IV or V fuze with cover.

All Mark I keys are to be brought to Mark II pattern and will then be known as Mark I*. (L. of C., § 16608.)

Fig. 99.

KEY, NO. 17 (MARK II), FIXING NOS. 80 AND 83 FUZES | L |.

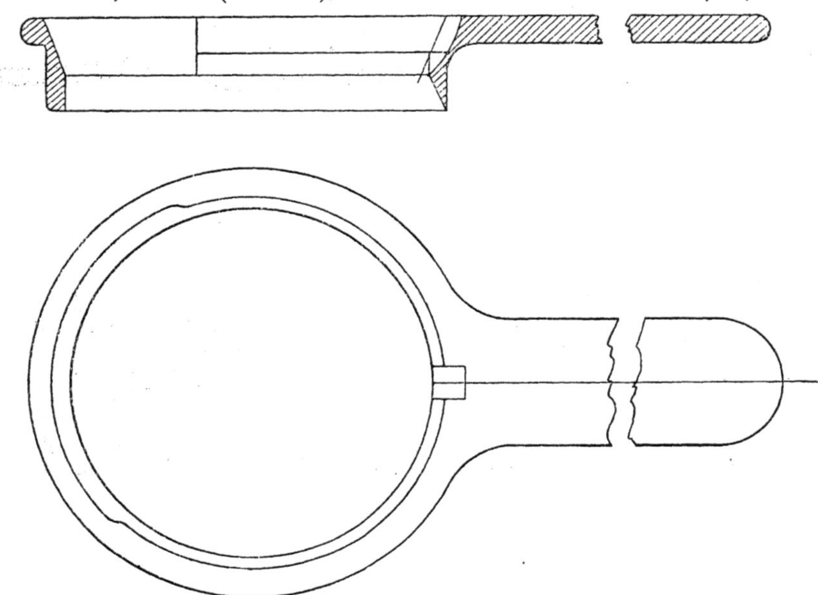

KEY, NO. 18 (MARK II).—SETTING NOS. 80 AND 83 FUZES | L | is similar in shape to the "Key, fixing" but is provided with a prong on the underside (as shown in Fig. 100) to engage with the stud on the lower time ring. The setting key is fitted with a lanyard.

Fig. 100.

KEY, NO. 18, MARK II.

Scale ½.

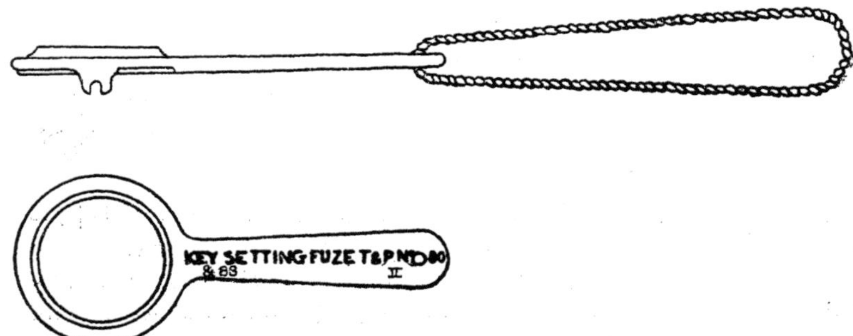

The Mark I key differs from the Mark II in the ring portion being of less depth, and consequently does not take such a good seating on the fuze.

Key, No. 19 (Mark I), Fixing Nos. 17 and 82 Fuzes | L | .—The key is of steel, the ends being shaped and provided with projections to suit the flange on the body of the No. 82 T. and P. fuze, one end being used for inserting and the other for removing the fuze from the shell. The ends are marked accordingly. A screwdriver is formed at one end for the fixing screw of the shell and a projection at the other end to fit the setting hole in the lower ring of the fuze. It has also a projection on the centre to fit the recess in the caps of D.A. fuzes. The key is fitted with a lanyard of white cotton line.

Fig. 101.

KEY, NO. 19, MARK I, FIXING NOS. 17 AND 82 FUZES | L | .

Scale ½.

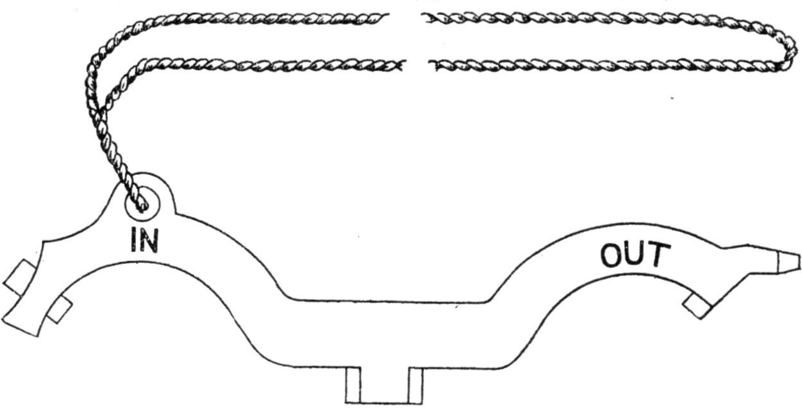

Key, No. 19 (Mark II), No. 82 Fuze | L | .—The Mark II key differs from the Mark I in not having the screwdriver end, or the projection to fit the recess in the caps of D.A. fuzes.

Fig. 102.

KEY, NO. 19, MARK II, NO. 82 FUZE | L | .

Scale ½.

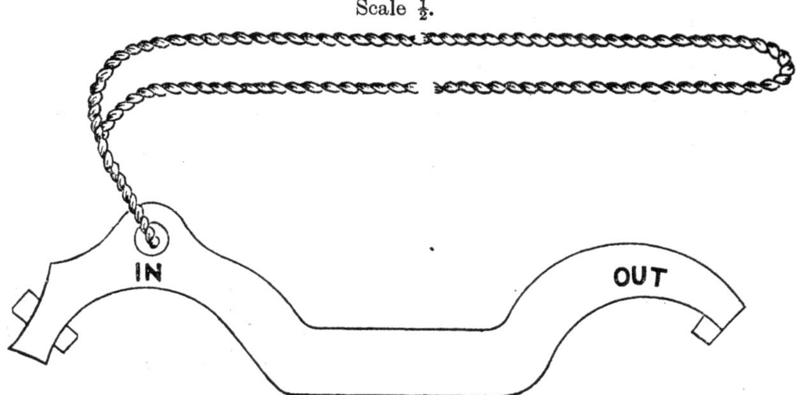

Key, No. 19 (Mark III), Fixing No. 82 Fuze | L | differs from Mark II in not having the projection for setting the No. 82 fuze.

This reduces the use of this key to "*Fixing* the No. 82 fuze" only. The No. 36 key is now used for "Setting the No. 82 fuze."

Key, No. 20 (Mark I), Time, T. and P. and D.A. Fuzes | C | is a steel key with lanyard.

It is used in the *Naval Service* for fixing and setting fuzes, time No. 30 and T. and P. No. 81; and for fixing Fuze, T. and P. No. 64, and for adjusting its top cap; also for D.A. and D.A. impact fuzes, except Nos. 3, 19, and 19A, and for all fuze-hole plugs except "Plug, fuze-hole special No. 4." It is also used for set-screws of shells taking fuzes of 2-inch gauge.

Fig. 103.

KEY, NO. 20, MARK I, TIME, T. AND P. AND D.A. FUZES | C |.

Scale ½.

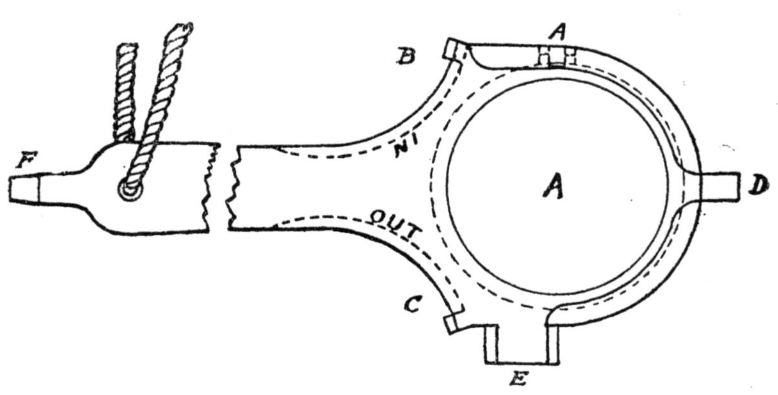

Part.	Use.
A	Setting No. 81 fuze.
B	Fixing No. 81 fuze.
C	Unscrewing No. 81 fuze.
D	Fixing and setting No. 30 and fixing and for top cap of No. 64 fuze.
E	Fixing and unscrewing plugs, fuze-hole and D.A. fuzes.
F	Set screws.

Key, No. 21 (Mark I), Hotchkiss Fuze | C | is made of metal, to the form and dimensions shown, and is for use in inserting or removing Hotchkiss base percussion fuzes, and "*Plugs, base, shell, No.* 3."

Fig. 104.

KEY, NO. 21 (MARK I), HOTCHKISS FUZE | C |.

Scale ½.

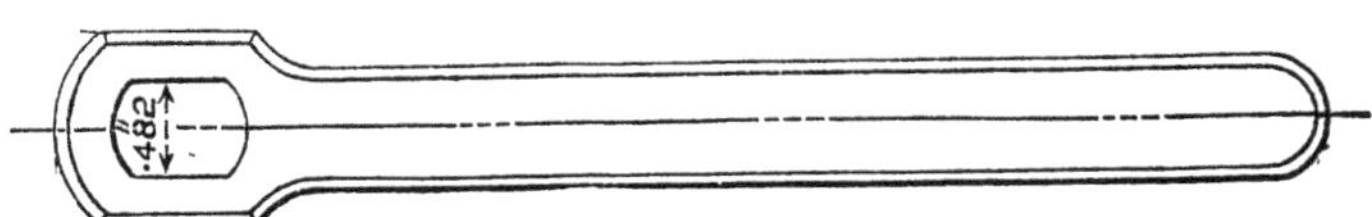

KEY, No. 22 (MARK I).—BASE FUZE, 9·45-INCH | L | is a steel bar about 20·5 inches long, having a slot formed in the centre to fit over the head of the special base fuze, and also a projecting stud, as shown in woodcut, to fit the recess in the base plug.

Fig. 105.

KEY, NO. 22 (MARK I), BASE FUZE 9·45-INCH | L |.

Scale ½.

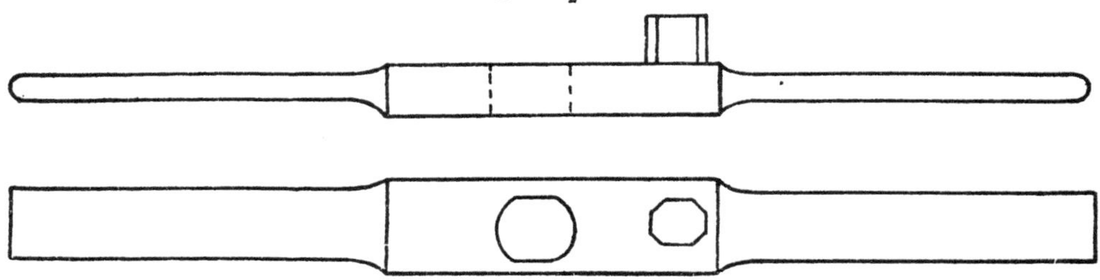

KEY, No. 23 (MARK I), PLUGS AND D.A. FUZES | C | is a steel key (T-shaped) for use with Plugs, fuze-hole G.S. and 2-inch; and fuzes D.A. and D.A. Impact, except Nos. 3 and 19.

KEY, No. 35 (MARK I), 1-PR., NOSE FUZE | C | is a steel key used for inserting or removing the "Nose, percussion, fuze," or the "Plug, fuze-hole, Q.F., 1-pr."

KEY, No. 36 (MARK I), SETTING No. 82 FUZE | L | is a steel key with lanyard. It is fork-shaped; having a pin in the centre of the fork to fit the setting hole in the lower time ring. The left arm of the fork has a portion raised to admit of seeing the setting mark.

Fig. 106.

Scale ½.

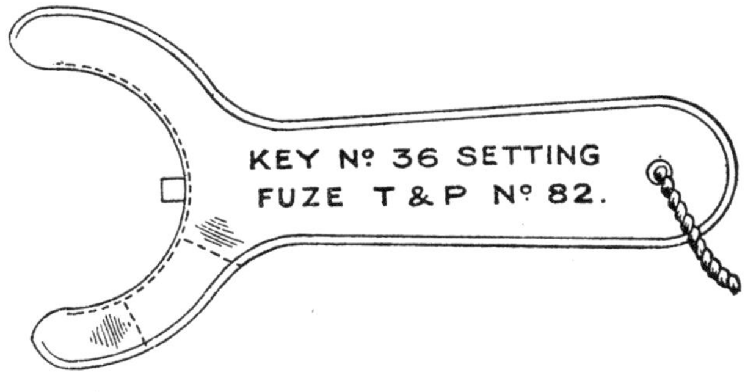

KEY, No. 37 (MARK I), 2·75-INCH | L | is a steel key with lanyard. (*See* Fig. 107, and table which explains its use.)

Fig. 107.

Scale ½.

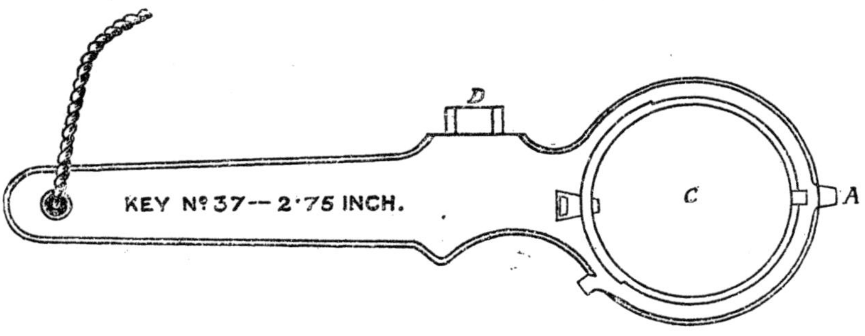

Part.	Uses.
A	Shrapnel set screws.
C	Fixing No. 80 fuze.
D	Fixing No. 17 fuze and plugs and clamping and releasing obturator box screw.

CHAPTER XIV.—TUBES.

(A) General Remarks on Tubes; (B) Vent-sealing Percussion "P" Tubes; (C) Vent-sealing, Electric, and V.S. Electric Wireless "P" and "S" Tubes, and Tube, Electric, Copper; (D) Friction Tubes:—Copper and Quill; (E) "T" Tubes:—Friction and Electric; (F) Impulse Torpedo Tubes:—Percussion and Electric; (G) Drill Tubes.

(A).—GENERAL REMARKS.

Use.—Tubes are employed to ignite the charge in B.L. and M.L. guns and howitzers; they are also used for firing Q.F. guns when the latter have their cartridges fitted with adapters. (*See* page 440.)

The action of the tube may affect the shooting considerably, since any variation in the rate at which the charge is ignited alters the pressures in the gun.

All tubes used with modern guns are "Vent-Sealing"; this effect is obtained as follows:—

(1) The tube fits a seating in the vent with great accuracy; on firing, the tube is expanded, and thus seals the escape of gas between itself and the vent.

(2) The interior construction of the tube prevents any gas escaping through the head.

(3) A lock or other contrivance holds the tube firmly and prevents it being blown out when the gun is fired.

With reference to (2), this is generally accomplished by one or more of the following methods:—

(*a*) A brass cone forced back into a cone seating.

(*b*) A shoulder in the tube against which the interior parts are forced by the pressure of gas.

(*c*) A small copper ball driven back into a cone seating.

(*d*) A brass or copper gas check.

Different Natures of V.S. Tubes.

There are at present three distinct natures of vent-sealing tubes employed with modern guns:—

"P" pattern, "S" pattern, and the "T" pattern tubes.

The "P" pattern tubes were first used with guns having "percussion locks," and consequently were called "P" tubes.

The "T" tubes are those used with guns and howitzers having "T" vents.

The "S" pattern tube is used with guns fitted with "strikerless locks."

Friction and percussion tubes are simpler and stronger in construction than electric tubes; there is no trouble with firing batteries, leads, &c.; but electric tubes are safer than percussion tubes, greater rapidity of fire is obtained, and the tubes can be tested.

"P" Tubes.

There are five natures of the "P" pattern vent-sealing tubes in the Service, namely :—

Tube, V.S., Percussion.
,, ,, *Percussion, Large. (Navy only.)*
,, ,, *Electric, Wireless, P.*
,, ,, *Electric, Wireless, Large. (Navy only.)*
,, ,, *Electric P.*

"S" Tubes.

There is only one nature of the "S" pattern tube, namely :—

Tube, V.S., Electric, "S," Large. (Navy only.)

"T" Tubes.

There are three natures of the "T" pattern tubes :—

Tube, Friction T.
,, *Friction T. (push).*
,, *V.S., Electric T., double wired.*

Tubes used for Firing Torpedoes.

In addition to the above, in the Naval Service there are two tubes employed in connection with the cartridge impulse torpedo :—

Tube, Electric, Wireless, Impulse Torpedo.
,, *Percussion, Impulse Torpedo.*

There is also an electric tube used in connection with the torpedo dropping gear, *i.e.*,

Tube, Electric, Wireless, Torpedo Dropping Gear, Mark I.

Marking, Lacquering, &c.

V.S. percussion tubes blackened and head notched.—All V.S. percussion tubes are now blackened, and have four notches cut in the rim of the head, so that they can readily be distinguished from wireless electric tubes; some old Mark IV tubes are without these distinguishing features.

Wireless electric tubes not lacquered externally.—The bodies of V.S. electric *wireless* tubes are not lacquered externally, so that the firing current may easily pass from the tube to metal of the gun; these tubes are therefore very liable to deterioration if exposed to damp; hence they should be kept in a dry store.

Large tubes.—The large pattern of V.S. tube was designed to prevent "hang fires," that is, a delay in the ignition of the cartridge such as might occur with a $\frac{3}{4}$ charge: the last $\frac{1}{4}$ charge, if pushed home, leaves a considerable space between itself and the mushroom head; these tubes are larger in diameter and have a more powerful charge of powder.

Marking on "P" and "S" tubes.—"P" and "S" pattern V.S. tubes are marked on the head with the contractor's initials and Mark of the tube.

The letter "P" is stamped on the "P" pattern electric tubes and the letter "S" on the "S" pattern.

Packing of V.S. Tubes.

All vent sealing tubes are packed in tin boxes painted black; the lids are secured by means of a tin strip *soldered* on. As the tubes themselves are not waterproof, they will probably deteriorate rapidly once the box has been opened.

On the exterior of the lid of each box is a label giving the number of tubes the box contains, the nature, Mark, and date of manufacture, and date of filling, &c.

These labels are printed in red for *percussion* and *friction* tubes, and have the additional information, "Not to be placed in the magazine on any pretence whatever."

Boxes containing *electric* tubes have their label printed in black.

The following show the labels that will be found on a box containing vent-sealing percussion tubes :—

OUTSIDE LABEL.

Label 101.

N

Group II. Division IA.

10 TUBES.

VENT-SEALING, PERCUSSION,

Mark VII.

Made by R.L. 27 / 8 / 1911

Re-packed by R.L. / /

INSIDE LABEL.

INSTRUCTIONS

For the use of

TUBES, VENT-SEALING, PERCUSSION.

If there is any appearance of fouling, the vent must be carefully cleaned with the rimer supplied for the purpose.

The vent-sealing action of the tube depends upon an accurate mechanical fit, and consequently a very slight amount of fouling in the vent will derange it.

Vent-sealing tubes are sometimes considered to be too high to gauge, when, in fact, the only defect is that the vent has not been properly cleaned.

(B).—VENT-SEALING PERCUSSION TUBES.

Tube, V.S. Percussion, Mark VII | C |.

(*Plate* LXVII.)

Use.—This tube is used with all B.L. guns having percussion locks except the 15-inch, 13·5-inch, Mark V, the 12-inch, Marks XI to XII, and the 4-inch, Marks VII and VIII guns; it may also be used with Q.F. guns as follows:—12-pr. of 8, 12, and 18 cwt., 14-pr., 4-inch, 4·7-inch, and 6-inch.

The *body* of the tube is of brass; the head is enlarged to prevent it being pushed too far into the vent, and is bevelled underneath to enable the tube to be readily extracted. The body tapers slightly towards the front end, and is made to fit the vent with great accuracy. The head is bored out and screw-threaded to take a "striker-holder" and a "cap-holder." The centre of the body is left solid to form an anvil, on which rests a copper cap containing cap composition; two fire holes are bored through the anvil to allow the flash from the cap to ignite the rifled pistol powder with which the front part of the tube is filled.

The *mouth* of the tube is closed by a cork plug having a disc of paper shellaced to each side; the metal of the tube is burred over the cork plug.

The *percussion cap* is held on the anvil by the cap-holder, which is a small screw plug having a central hole for the point of the striker.

The *striker-holder* is made of bronze, threaded to screw into the head of the tube, and bored out to take the striker; a shoulder is formed inside it, which fits a corresponding shoulder on the striker. At the rear end is a recess, into which the washer on the outer end of the striker is forced on firing; this allows for eccentricity of the firing pin of the breech mechanism.

The *striker* is also of bronze. It has a small cylindrical point which fits into the hole in the cap-holder. The outer end is reduced in diameter and is riveted to a brass washer. A copper shearing wire passes through the striker and the striker-holder.

The striker-holder is prevented from unscrewing by being stabbed in two places.

The tube is blackened, and has four notches cut on the head; these features are provided to distinguish percussion tubes from wireless electric tubes.

Action.—The striker is forced inwards by the firing pin of the percussion lock, shears its copper wire, and fires the cap. The flash passes through the two fire holes in the anvil to the powder, which ignites, blows out the cork plug, and fires the charge in the gun.

The explosion of the powder expands the tube against the walls of the vent and so prevents the rush of gas between the tube itself and the vent, while the copper cap and the shoulder of the striker prevent any escape of gas through the head of the tube.

Mark VI Percussion Tube.

Tube, V.S., Percussion, Mark VI | C | .

This Mark differs from the Mark VII as follows:—The anvil is *screwed* into the interior of the tube; it supports a cap (R.L. type), and is prevented from unscrewing by a copper washer, which is inserted in a cup-shaped form and is expanded into the threads underneath the anvil.

There is neither striker-holder nor cap-holder.

The shearing wire passes right through the body of the tube.

The striker has a needle-point, and supports a small copper gas-check.

Note.—In consequence of the liability of this mark of percussion tube to be fired prematurely by a sudden jar its manufacture was discontinued, and the manufacture of Mark IV was reverted to, until the introduction of the Mark VII tube.

Fig. 108.

TUBE, V.S., PERCUSSION, MARK VI.

Full Size.

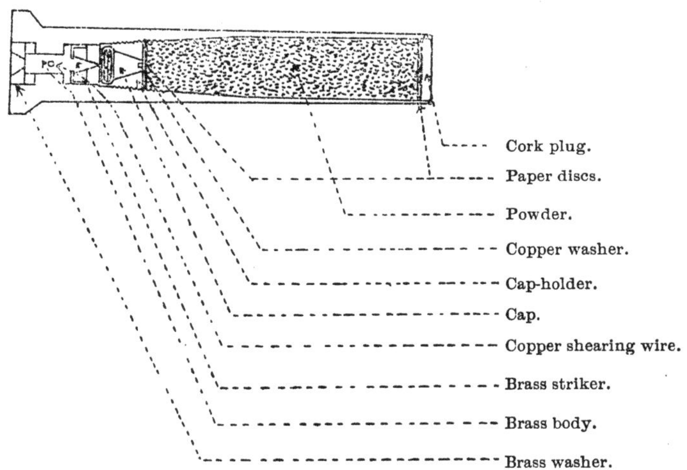

Mark V Percussion Tube.

Tube, V.S., Percussion, Mark V | C | .

A very limited number of Mark V tubes were made; they differ from Mark VI in the shape of the striker, and in having no copper gas-check round the needle-point.

When altered to agree with Mark VI these tubes were distinguished by having a star added to their numeral, being known as Mark V*.

Mark IV Percussion Tube.

Tube, V.S., Percussion, Mark IV | C | .

(*Plate* LXVIII.)

There are large numbers of this mark of tube in the Service.

The striker has no needle-point.

There is no copper cup to act as a gas-check.

There is no shearing wire.

The anvil carries a percussion cap and has three fire-holes.

There is no recess under the brass washer on the outer end of the striker such as is provided in the Mark VII to allow for eccentricity of the firing pin.

MARK III PERCUSSION TUBE.

TUBE, V.S., PERCUSSION, MARK III | C |.

The Mark III, V.S., percussion tube is similar to the Mark IV, except that the end is closed with a sulphur pellet in which is embedded a small *brass ball.*

This tube must not be used unless the range is clear.

Existing V.S., percussion tubes, Mark III, in the naval service, have been converted by the removal of the sulphur pellet and brass ball and reclosed by a cork plug and paper disc. Tubes so converted have a star added to their numeral, and boxes containing them are marked "For practice only; to be used up before tubes of a later date."

Certain Percussion Tubes not Blackened, &c.

In earlier issues of Marks IV and VI percussion tubes, the mouth was not burred over the cork plug; there were no notches on the head, and they were not blackened.

Packing.

Vent sealing percussion tubes are packed in flat tin boxes painted black; they contain 10 tubes, packed heads and tails, steadied by tin racks and brown paper.

LARGE PERCUSSION TUBE FOR NAVY.

TUBE, V.S., PERCUSSION, LARGE, MARK II | N |.

Use.—This tube is used with the B.L., 15-inch, 13·5-inch, Mark V, 12-inch, Mark XI to XII, and 4-inch, Marks VII and VIII guns.

It is similar in design to the Tube, V.S., Percussion, Mark VII, described on page 352, but is of larger diameter (0·525 instead of 0·4-inch); it is filled with small perforated pellets of gunpowder (53 grains) instead of rifled pistol powder; the mouth of the tube is recessed to form a seating for the cork plug.

The tube is capable of giving a much stronger flash than the small pattern.

Packed in flat tin boxes holding 10.

TUBE, V.S., PERCUSSION, LARGE, MARK I | N |.

The Mark I tube differs from the Mark II in being filled with rifled pistol powder instead of perforated gunpowder pellets; the mouth of the tube is not recessed to form a seating for the cork plug.

(C).—VENT-SEALING ELECTRIC P., AND V.S. WIRELESS ELECTRIC "P" AND "S" TUBES.

(i) Wireless "P" Tubes.

Tube, Vent-Sealing, Electric Wireless, "P," Mark VI | C |.

(*Plate* LXIX.)

Use.—This tube is used with all B.L. guns having wireless locks except the 13·5-inch, Mark V, the 12-inch, Mark XI to XII, and 4-inch, Marks VII and VIII guns; it may also be used with Q.F. guns as follows:—12-pr. of 8, 12, and 18 cwt., 14-pr., 4-inch, 4·7-inch, and 6-inch.

Body.—The body of the tube is made of solid drawn brass, the head enlarged to prevent it being pushed too far into the vent, and is bevelled underneath to enable it to be readily extracted; the body tapers slightly towards the front end, and is made to fit the vent with great accuracy.

Internal construction.—The interior is bored out from the front to take a charge of rifled pistol powder above which the walls are thickened forming a recess for a copper sealing plug.

A small central hole is bored through the upper part of the body leading into a hole in the head; this hole is screw-threaded and fitted with an ebonite plug, having an undercut recess.

Inserted through the mouth of the tube is a copper plug, through the centre of which passes a piece of insulated tinned copper wire. The copper plug is pressed on to the insulated wire, and a small recess formed at the bottom of the plug is sealed with special cement.

The end of the wire projecting through the cement is bared of its insulation and has soldered to it an iridio-platinum bridge giving a resistance of ·9 to 1·1 ohms.

The other end of the bridge is soldered to a small pole piece attached to the copper plug.

The lower end of the copper plug is shaped so as to act as a gas-check; the upper end has a ball formed on it, which fits into a seating in the body.

The rear end of the insulated wire, after passing through the central hole in the tube and through a small copper washer in the recess in the ebonite plug, is bared of its insulation and coiled down in the recess. A small plug of tin and antimony is then pressed into this recess to form the contact disc.

The contact disc is slightly below the head of the tube, and the inner edge of the ebonite plug is bevelled off all round.

The tube is lacquered internally, and contains a small charge of priming composition (guncotton dust and mealed powder) placed round the bridge, kept in position by a perforated glazeboard disc with paper disc attached.

The tube is then filled up with rifled pistol powder, the mouth being closed by a cork plug on each side of which is shellaced a paper disc.

The cork plug is secured in position by the mouth of the tube being burred over it.

Action.—On contact being made the current passes from the battery through the striker to the contact disc, down the insulated wire in the copper plug, through the iridio-platinum bridge to the small copper pole and the body of the tube, back to the battery. The bridge becomes incandescent, fires the priming composition and the F.G. powder.

The body of the tube expands against the vent and prevents the gas escaping between itself and the vent, and the copper sealing plug is forced inwards, the upper portion of the plug fitting into the small cone seating in the body, prevents the gas escaping through the head of the tube. The gas-check portion expands and helps to seal.

Tube, V.S., Electric, Wireless, "P," Mark VII.

The Mark VII tube differs from the Mark VI in the following particulars :—

(1) The copper sealing plug is not fitted with a small pole piece.

(2) The tube has the Nobel's vulcan bridge, one end of which is attached to the insulated wire leading from the contact disc, the other is attached to the end of the copper sealing plug. The resistance of this bridge is ·8 to 1·5 ohms.

(3) The bridge is surrounded with a special composition, instead of being primed with guncotton dust and mealed powder.

(4) No glazed board disc is used.

Tube, Vent-Sealing, Electric, Wireless, "P," Mark V | C | .

(*Plate* LXX.)

The body of this tube is similar, externally, to the Mark VI tube already described.

Internal construction.—In the centre of the upper portion of the body is a brass centre-piece. This centre-piece is reduced in diameter at each end, leaving a flange near the centre. On top of this flange is placed an insulating washer of mica, on which rests the flanged portion of a cup-shaped copper gas-check. In the interior of the gas-check, and round the lower part of the flange on the centre-piece, is pressed a small ring of asbestos. On this asbestos ring rests a small brass gas-check, held in position by a plug of ebonite screwing on to the lower portion of the centre-piece.

Formed on the lip of the copper gas-check is a small copper pole which is connected to the end of the centre-piece by an iridio-platinum bridge, giving a resistance of ·9 to 1·1 ohms.

Over the upper portion of the centre-piece, but insulated from it by an inner cylinder of ebonite, is a brass socket, on top of which are two insulating washers, one of mica and the other of ebonite; a brass nut screwing on to the end of the centre-piece secures the whole firmly in position.

A small brass washer is placed over the metal nut, and the whole of the internal arrangement is retained in position in the tube by means of a crown metal nut, which forms the contact piece; this nut fits into the ebonite insulator in the head of the tube, and screws on to the top of the centre-piece. The surface of the contact disc is just below the level of the head of the tube, and the edge of the ebonite insulator is bevelled off on to the contact disc.

The tube is primed with priming composition, of guncotton dust and mealed powder, kept in position by a perforated glazed board disc, with a paper disc attached. The remainder of the tube is filled with rifled pistol powder, and the end closed by a cork plug, on each side of which is shellaced a paper disc. The cork plug is secured in position by the mouth of the tube being burred over it.

Action.—On contact being made, the current passes from the battery, through the striker, to the crown metal contact disc, through the centre-piece, over the iridio-platinum bridge to the copper pole, the copper gas-check and body of the tube, back to the battery. The bridge becomes incandescent, fires the priming composition and the powder. The gas expands the copper gas-check, which, in conjunction with the small brass gas-check, the asbestos packing and the flange on the centre-piece, prevents any escape of gas through the head.

TUBE, VENT-SEALING, ELECTRIC, WIRELESS, "P," MARK IV | C |.

(*Plate* LXXI.)

The Mark IV tube in exterior dimensions is identical with the Mark V already described. The interior cavity terminates in a cone into which fits a conical brass plug at about ·25-inch from the head; this is insulated from the body by an ebonite cone. The front end of the cone is *cupped out* to form a gas-check, and a hole is bored in the centre, into which screws an ebonite plug. Into the centre of this plug fits a tinned copper pole which extends the full length of the tube. This pole is bent, and attached to the body of the tube at the mouth with pure tin. The pole and the edge of the brass cone are connected by an iridio-platinum bridge of ·9 to 1·1 ohms resistance attached with pure tin. At the rear end of the cone a small hole is drilled, a little out of centre, to receive the bared end of an insulated copper wire. A turn is taken in this short wire and it is passed through a hole in the head, also drilled a little out of centre and communicating with a recess, into which screws an ebonite plug. This plug is recessed in the centre and undercut, and the end of the wire passes into the recess. It is bared and coiled down upon a cup of pure tin ·025-inch thick at the bottom of the recess, the remainder of which is then filled in with molten tin. This is kept in by the undercut and forms the contact piece; its surface is slightly below that of the head of the tube. The tube is primed with the usual priming composition of guncotton dust and mealed powder, over which is placed a perforated glazed board disc with a paper disc attached, the remainder of the tube being filled with rifled pistol powder. The mouth of the tube is closed by a cork and paper disc shellaced in, and on the outside of the cork is a second paper disc to prevent the shellac sticking to the box in which the tubes are packed. The cork and paper discs are secured in position by the mouth of the tube being burred over.

Action.—On contact being made the current passes from the battery through the striker which is in contact with the tin disc, through the short wire, cone, the wire bridge, the long copper pole, the body of the tube and the metal of the gun, back to the battery again. The wire bridge becomes incandescent and fires the priming and the powder; the gas expands the cupped-out portion of the cone, which is driven back into its seating, and prevents any escape of gas through the head; the body expanding prevents any escape between it and the vent.

When used with full charges the cone was driven back so violently that the body of the tube became distorted and so became jammed in the vent.

Tube, V.S., Electric, Wireless, "P," Mark III.

The Mark III tube differs from the Mark IV as follows:—There is only a thin paper disc between the powder and the priming composition; and the bridge is made of platinum-silver having a resistance of 1·5 to 1·8 ohms.

Tube, V.S., Electric, Wireless, "P," Mark II.

The Mark II tube differs from Mark III in having a smaller contact disc, which in some of the earlier issues was made of solder instead of pure tin.

The Mark II tube is not to be used in adapters for Service practice in Q.F. guns.

The Mark I, wireless, "P" tube had two bridges giving a resistance of from ·6 to ·9 ohm, and in the majority of Mark I tubes the ebonite plug in the head was not screwed in.

Mark I tubes are used for instructional purposes only.

Packing of wireless "P" tubes.

Vent-sealing electric wireless "P" tubes are packed similarly to the percussion; additional tin fittings, to which pieces of glazed board are attached, steady the tubes, and keep the contact discs clean. Calf-skin was first used in place of glazed board, and the first boxes issued were lined with cork.

Tube, V.S., Electric, Wireless, Large, Mark IV | N |.

The Mark IV tube, V.S., electric, wireless, large, is similar in design to the Mark VI wireless, electric "P," tube (*see* Plate LXIX), but is of larger diameter (0·525 instead of 0·4-inch).

It is filled with perforated pellets of gunpowder.

Tube, V.S., Electric, Wireless, Large, Mark III | N |.

The Mark III tube is filled with perforated pellets of gunpowder, and is similar in internal construction to the Tube, V.S., electric, wireless, "P," Mark IV (*see* Plate LXXI), but a small ebonite washer is placed in the top of the cone seating in the body of the tube.

TUBE, V.S., ELECTRIC, WIRELESS, LARGE, MARK II | N | .

The Mark II tube differs from the Mark III in being filled with rifled pistol powder instead of perforated powder pellets.

TUBE, V.S., ELECTRIC, WIRELESS, LARGE, MARK I | N | .

The Mark I has a larger contact disc, and the recess in the ebonite plug in the head of the tube is not so deeply undercut, and so the discs are liable to blow out on firing.

(ii) WIRELESS "S" TUBES.

TUBE, V.S., ELECTRIC, "S," LARGE, MARK I | N | .

(*Plate* LXXII.)

The above-mentioned tube has been approved for Naval Service with guns having "strikerless locks."

The internal arrangement is similar to the "Tube, V.S., electric, wireless, large, Mark IV," but differs from it in the head, which has a contact piece instead of a contact disc.

This contact piece is made of brass and projects above the head of the tube; it screws into the ebonite insulating cup in the head.

Under the flange of the contact piece are placed two washers of vulcanized paper which insulate it from the body.

The wire passing through the copper sealing plug is led through a hole in the centre of the contact piece and is soldered to it.

(iii) WIRED "P" TUBE.

TUBE, V.S., ELECTRIC, "P," MARK VII | L | .

(See *Plate* LXXIII.)

This tube resembles the wireless electric tube in exterior form and dimensions.

The interior of the body near the head is conical, and a small hole is drilled through the head, through which pass two tinned copper wires twisted together and insulated by varnished silk. Outside the tube the wires are parted and led through a V-shaped groove across the head; they are then twisted together again and wrapped with oiled silk for a distance of 5 inches and terminate in spirals 22 inches from the tube; the spirals are covered with sarcenet.

In the interior near the head are two cones, the larger one fitting in the conical recess in the body, the small one fitting into a conical recess in the large cone; the cones are insulated from each other and from the body by ebonite. The front end of one of the wires is attached to the rear end of the large cone, and the front end of the other wire passes through the large cone, is insulated from it, and is attached to the rear end of the small cone. One pole of copper is attached to the front end of the large cone, and the other pole to the front end of the small cone. These poles are connected by a platinum-silver wire bridge, resistance 1·5 to 1·8 ohms, which is embedded in a priming composition of guncotton dust and mealed powder, contained in an ebonite cylinder closed at the end by a perforated glazed board disc with a paper disc attached.

The remainder of the tube is filled with rifled pistol powder, and the end is closed by a cork plug having a paper disc shellaced on each side. The mouth of the tube is burred over to hold the plug in position. The tube is lacquered inside and out.

Action.—The wires from the battery are connected to the wires of the tube, and when the circuit is completed the bridge becomes incandescent and fires the tube. The small cone is jammed into the larger cone, and the latter into the coned portion of the body, thus preventing any escape of gas through the head.

Tube, Vent-Sealing Electric, P, Mark VI | L |.

Mark VI in construction is identical with Mark VII, but the wires are led through a groove across the head, instead of the V-shaped groove, and it is without the extra 5 inches of oiled silk wrapping.

Fig. 109.

TUBE, ELECTRIC, NO. 10, MARK IV.

LOW TENSION, WITH INSULATED WIRES.

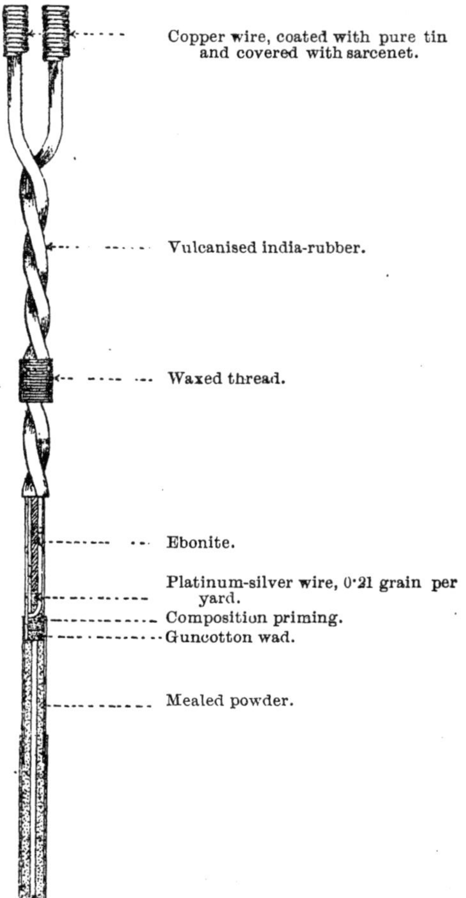

Packing of Wired Tubes.

Packing. — Electric P tubes are packed five in a box, which is almost square. The tubes are separated by tin partitions, and movement is prevented by two end pieces and one top piece of cork. The wires of the tubes are coiled up tightly close to the head. A piece of tape is placed under one of the tubes to facilitate extraction from the box.

Tube, Electric, No. 10, Mark IV | L |.

This tube was originally introduced for firing radial-vented R.M.L. guns. It is now used for carrying out the destruction of unserviceable cordite.

The construction of the tube is shown in Fig. 109.

Tube, Electric, No. 10a, Mark I | C |.

The No. 10a tube is similar in construction to the Tube, electric, No. 10, but is smaller.

It was introduced for firing time guns; it is obsolete for future manufacture.

Tube, Electric Copper, No. 10a, Mark I | C |.

The general arrangement of this pattern of tube is shown in the figure; it is for use in both Land and Naval Services for firing time guns.

This tube will replace the No. 10a *quill* tube.

Fig. 110.

TUBE, ELECTRIC COPPER, NO. 10A, MARK I | C |.

FOR TIME GUNS.

Full Size.

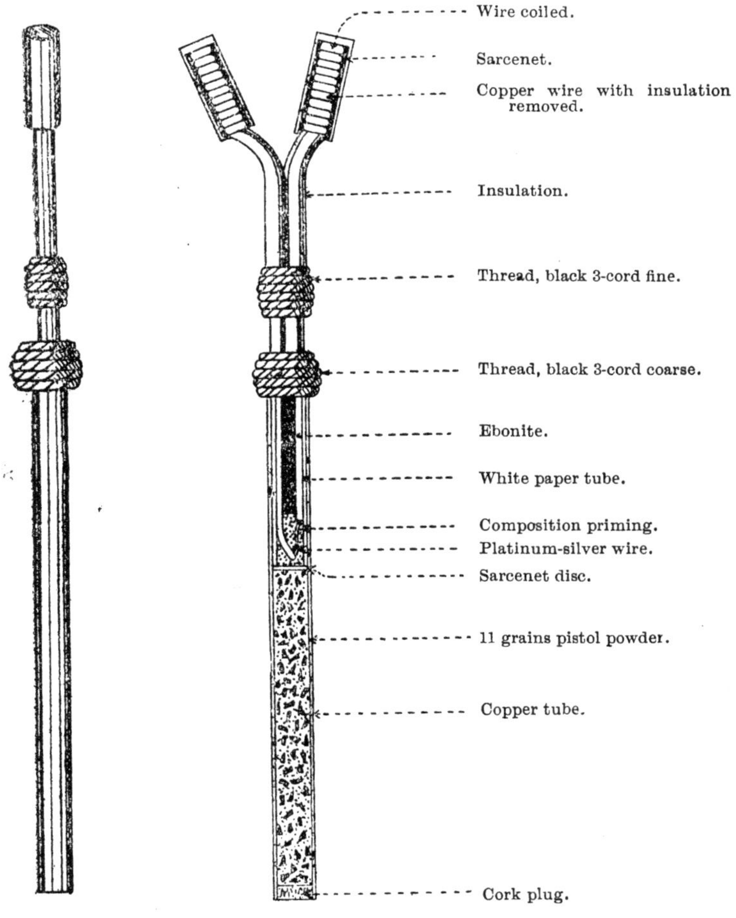

Wire coiled.
Sarcenet.
Copper wire with insulation removed.
Insulation.
Thread, black 3-cord fine.
Thread, black 3-cord coarse.
Ebonite.
White paper tube.
Composition priming.
Platinum-silver wire.
Sarcenet disc.
11 grains pistol powder.
Copper tube.
Cork plug.

(D.)—FRICTION TUBES (COPPER AND QUILL).

The *Copper* friction tubes in the *Land* Service are :—

Tube, Friction, Copper, Solid drawn, without ball, Mark III.
,, ,, ,, ,, ,, with ball, Mark II.
,, ,, ,, ,, ,, without ball, Mark I.
,, ,, ,, ,, ,, Special, Mark I.
,, ,, L.S., Short, Mark II.
,, ,, " T " for blank, Tube, Copper, Mark I.

There is also a Tube, Friction, Machine Rocket, Signal, Mark I used in both Land and Naval Service.

Copper Tubes.

Tube, Friction, Copper, Solid Drawn, without Ball, Mark III | L |.

The body of this tube is made from solid drawn copper. The head is solid; it is about 2 inches long and ·2 inch in diameter, and is lacquered inside and out. About a quarter of an inch below the head a small hole is bored through one side; a bulge is made on the inside opposite to it, forming a seat for the crown of the solid-drawn copper nib piece, which is inserted and soldered; a small hole is bored through the underside of the nib piece inside the body to enable the flash from the detonating composition to ignite the powder.

The nib piece contains a copper friction bar roughened on both sides and smeared with a detonating composition composed of chlorate of potash, sulphur, and sulphide of antimony. This composition is damped with shellac varnish, while it is being smeared on. The nib piece is pressed down on to the sides of the friction bar. The projecting part of the friction bar has a vertical eye, into which the hook of the lanyard fits; the junction of the nib piece and friction bar is sealed by shellac varnish. The body of the tube contains 9 grains of pistol powder, and the bottom is closed by a cork and paper disc secured with shellac.

Use.—This tube is used for firing radial vented R.M.L. guns, R.B.L. guns, and puffs.

Fig. 111.

TUBE, FRICTION, COPPER, SOLID-DRAWN, WITHOUT BALL, MARK III | L | .

Full Size.

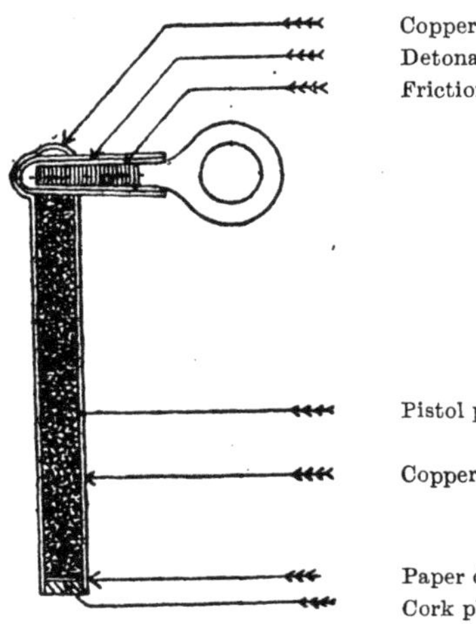

Tube, Friction, Copper, Solid-Drawn, with Ball, Mark II | L | .

This tube is similar to the Mark III fricton tube already described, but differs from it in having a small *brass ball* inserted into the mouth of the tube on top of the cork plug.

This tube can be used with the same guns as the Mark III tube, with the exception of R.B.L. guns other than the 40-pr. side closing.

Tube, Friction, Copper, Solid-Drawn, without Ball, Mark I | L | .

The Mark I tube differs from the Mark III in the nib piece, which is of sheet copper secured by a binding of fine wire and solder.

Tube, Friction, Copper, Solid-Drawn, Special, Mark I | L | .

This tube is obsolete for future manufacture, but may still be met with; it was used for firing *blank* charges in radially "T" vented guns; it has been superseded by the *Tube, friction "T" for blank, tube, copper, Mark I,* described below. The tube is similar to the Tube, friction, copper, solid-drawn, Mark III, but is shorter and contains about ·6 grain of powder.

Fig. 112.

TUBE, FRICTION, L.S., SHORT, MARK II | L |.

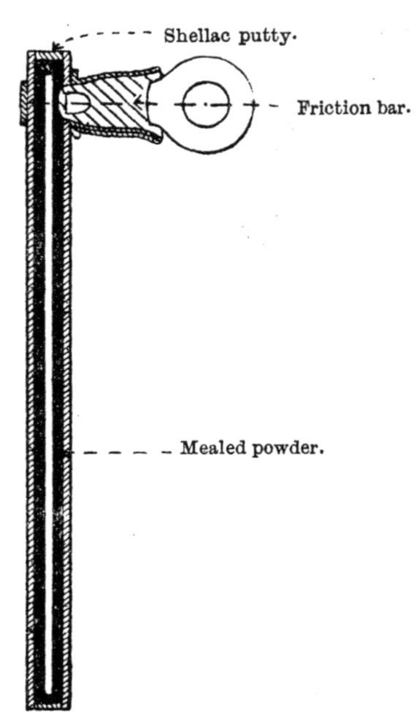

Tube, Friction, L.S., Short, Mark II.

This tube is still retained in the L.S. for firing war rockets as the solid-drawn friction tube is *too short* for this purpose.

The tube is made of sheet copper about 3 inches long, driven with mealed powder and pierced with a central hole. The top is closed with shellac putty and varnished paper, and the bottom by a disc of varnished paper. The nib piece is similar to that of the Mark I solid-drawn tube; the friction bar is roughened on both sides, turned up for ·1 inch at one end, and smeared with detonating composition. The exterior of the tube is varnished black.

Fig. 113.

TUBE, FRICTION, "T" FOR BLANK, TUBE, COPPER, MARK I | L |.

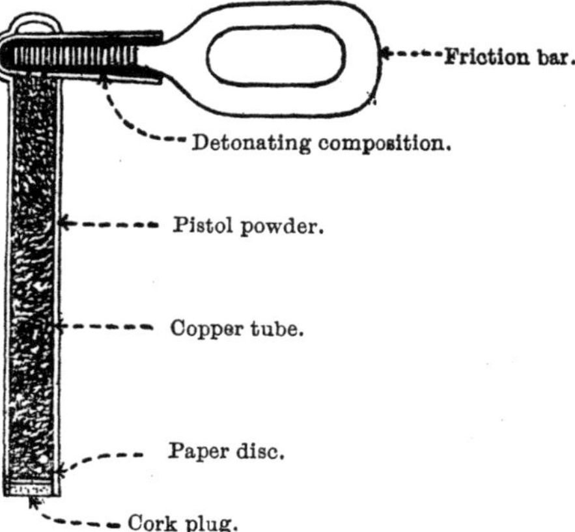

Tube, Friction, "T" for Blank, Tube, Copper, Mark I.

The above copper tube is used with the special adapter described on page 368, for firing blank from all B.L. guns and howitzers having "T" vents.

The tube differs from the solid-drawn Mark III tube in having a longer eye for the hook of the firing lanyard. (*See* Fig. 113.)

Tube, Friction, Machine, Rocket, Signal, Mark I | C | .

The above tube is generally similar to the "Tube, friction, copper, solid-drawn, Mark III," shown in Fig. 111 but is shorter, and the portion of the tube above the friction bar is slightly longer.

This tube will supersede the "Tube, friction, *quill*, short," when the stock of the latter is used up.

Issue and Packing of Copper Friction Tubes.

Issue of copper friction tubes.—All copper friction tubes, except the *solid drawn special*, are issued in hermetically sealed cylinders; 20 tubes, friction "T" for blank, tube, copper, Mark I in a cylinder; all other natures 25 in a cylinder.

There is a rack soldered to the inside of the cylinder made by a strip of tin corrugated so as to form a loop for each tube. Felt packing pieces are used to prevent the tubes shaking about in transport.

Quill Friction Tubes.

There are two sizes of quill friction tubes in use *in the Naval Service*, namely:—

"Long" and "Short."

Use.—The "long" tube is used for firing war rockets, the "short" tube is used for firing signal rockets when fired from a rocket tube. When stock is used up it will be superseded by the "Tube, friction, machine, rocket, signal, Mark I."

Tube, Friction, Quill, Short, Mark VI | N | .

Description.—The short quill friction tube is about $2\frac{3}{4}$ inches long, and consists of a goose quill rammed with mealed powder, pierced with a small channel. In the head is a little detonating composition, through which passes a roughened copper bar, called the friction bar, fitted with an eye for the hook of the tube lanyard.

A binding of fine copper wire serves to straighten and support the top when the tube is in the vent, and, to support it when the pull of the lanyard comes on it, a leather loop is attached to the head. The loop slips over the friction tube pin screwed into the gun near the vent. The composition contains, in addition to the ingredients used in the composition for copper tubes, a little mealed powder, and also ground glass to render it more sensitive.

Fig. 114.

TUBE, FRICTION, QUILL, SHORT, MARK VI | N | .

Scale Full Size.

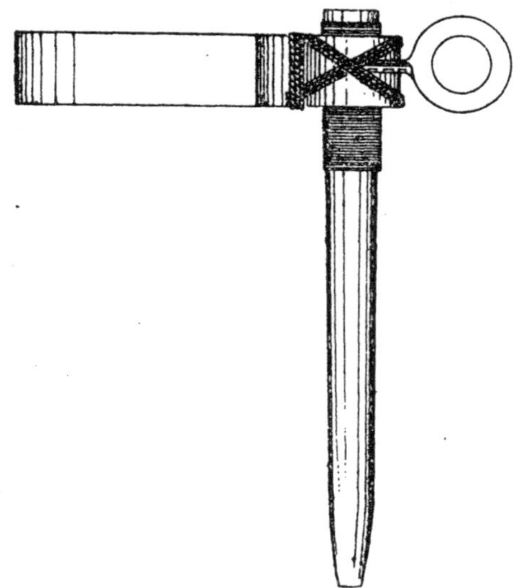

The Mark IV long quill friction tube differs from the short one in length only, being 4 inches long. The increased length is given by cementing two quills together.

Issue.—Twenty-five in tin cylinders.

(E.)—" T " TUBES.

Use.—There are three patterns of " T " tubes, namely :—

(1) *Tube, friction " T,"* for B.L. guns and howitzers having T-vents.

(2) *Tube, friction " T " (push)* for the B.L.C. 15-pr.

(3) *Tube, vent-sealing, electric " T " double wired,* for use with 6-pr. Q.F. sub-calibre on top of 9·2-inch Mark X guns on Mark V, barbette mountings.

Tube, Friction, " T," Mark IV | L | .

(See *Plate* LXXIV.)

Description.—The body of the tube is made of solid-drawn brass, slightly coned to fit the vent, and bored out to receive 8 grains of rifled pistol powder ; above this it is screwed to receive a screw plug having three fire-holes on top of which rests a small copper ball ; the upper part of the interior of the body above the ball terminates in a cone ; a small hole is bored through the upper part from the end of the cone to admit the flash from the detonating composition ; the exterior of the body is screwed to fit into the head.

The head is made of metal, rectangular in section, and screws on to the body and is secured by a brass pin; a hole is drilled into the head to receive the friction wire; the outer end of the hole is countersunk, as shown in the plate. The friction wire is made of half-round copper wire twisted and roughened. A large loop is formed at one end by two turns of the wire, which are fixed together by dipping the loop in solder. The wire does not extend right through the head, and is retained in place by a shearing wire of tin and antimony, passing through the head and friction wire, near the loop. A hole is bored into the side of the head at right angles to it, so as to come over the fire-hole of the body, and is screw-threaded at the end. Detonating composition is pressed into this hole, surrounding the roughened part of the friction wire, and the hole is closed by a screwed brass plug.

The bottom of the body is closed by a cork plug having a paper disc shellaced to it on the outside. The mouth of the tube is burred over to hold the plug in position.

The tube is lacquered inside and out.

Action.—The tube being held in the " T " vent, when the friction wire is withdrawn, the flash from the detonating composition passes down the small hole in the top of the body, over the copper ball, and through the fire-holes in screw plug, and ignites the powder; the gas forces the ball up into the cone seating and so prevents any escape of gas through the head; it also expands the body, and so prevents any escape of gas between it and the vent; the flash ignites the charge.

Earlier Marks of Tube, Friction, " T."

Tube, friction, " T " Mark III.—Mark III differs from Mark IV in the loop of the friction wire, which is smaller, and in having the hole in the side of the head closed by a gut-skin disc, a cork plug shellaced, and shellac cement outside it. Some of the earlier issues of Mark IV also had the hole closed in this way.

Tube, friction, " T " Mark II.—The Mark II tube has the hole for the friction wire bored completely through the head; there is no shearing wire.

Tube, friction, " T " Mark I.—Mark I differs from Mark II in having the friction wire made of ordinary twisted wire, and in the detonating composition being pressed into a hole in the head in prolongation with the body instead of at the side. The hole for the friction wire was not closed by a cork plug.

Remarks.—Marks I and II " T " friction tubes when converted to agree as far as possible with Mark III will have a (*) added to their numeral.

Marks I*, II*, III, and IV tubes, will, when fitted with new bodies and friction wires, be known as Marks I**, II**, III*, and IV* respectively.

The existing stock of " T " tubes will be used up as follows :—

Mark III, III*, IV and IV*, with Service ammunition.
Earlier Marks with blank ammunition.

Marking and Packing "T" Friction Tubes.

Certain "T" tubes which have been fired and refilled are stamped on the head thus (R) ; if emptied and refilled "R."

Packing.—"T" friction tubes are packed ten in a square tin box, painted black, and having both top and bottom removable which are secured by tin bands soldered over the joint. Inside the box at each end there is a partition, with a corrugated strip for holding five tubes. Movement of the tubes is prevented by a cork packing piece and a felt wad on top. There is a tape band under one of the tubes to facilitate removal. This makes a neat package, and if only one or two tubes are required, five out of ten remain hermetically sealed.

Tube, Friction, "T," for Blank, Adapter, Mark I.

The above adapter is used in conjunction with the "Tube, copper, Mark I" (*see* page 364) when firing blank ammunition with B.L. guns and howitzers having "T" vents.

The adapter is made in two parts, viz. :—

Head, and screwed stem.

Fig. 115.

TUBE, FRICTION, "T," FOR BLANK, ADAPTER, MARK I.

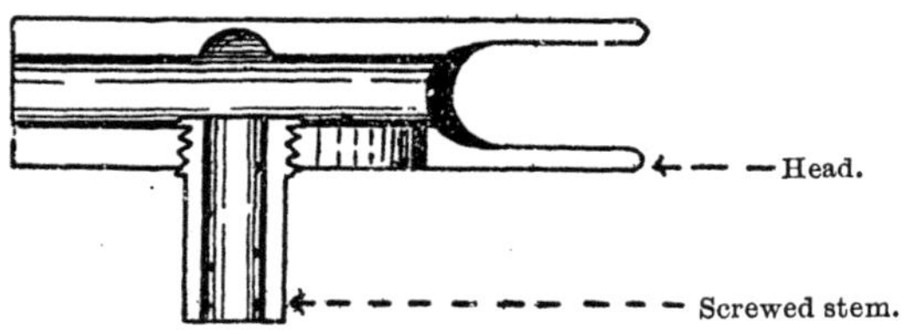

The head is made of brass, and is generally similar in shape to the head of the "T" friction tube; it has a long slot in its under surface, through which the tube can be inserted into its position. *The screwed stem* is about half an inch in length and screws into the head as shown in the woodcut.

The copper friction tube is inserted into the adapter as follows :— Unscrew the stem; insert the tube into the adapter, and secure it in that position by slipping the stem over the body of the tube and screwing it into the adapter.

Tube, Friction "T" (Push), Mark I | L |.

(*Plate* LXXV.)

Use.—This "T" tube was introduced with the 15-pr. B.L.C. guns; it enables the layer to fire the gun in a similar manner to a Q.F. gun, and abolishes the necessity of having a lanyard to hook to the tube after the breech has been closed.

Parts.—The main parts of the tube are :—Body, screw-plug, copper ball, head, screw collar with detonator, push bar with shearing wire and screw collar, and cork plug with paper discs.

Body.—The *body* is similar in all respects to that of the Tube, friction, " T," Mark IV.

Head.—*The head* is of brass, square in section, except one end, which is round to ensure the correct insertion of the tube in the vent, and also to facilitate extraction after firing. The square portion of the head is bored out to take the detonator and push bar, each of which is held in position by a screw collar ; the inner end of the recess is roughened and a gas escape hole (for the gas from the *detonator*) ·08 inch in diameter is bored at the further end as shown in the plate ; this escape hole is closed by means of a plug of shellac putty, coated with Pettman's cement.

Push bar.—The *push bar* is of bronze, the internal striker portion being conical and roughened, and terminates in a point ; the outer end is reduced in diameter to form a shoulder against which a screw collar bears and retains the push bar in position ; a shearing wire of lead and antimony passes through the collar and the push bar. A disc of paper coated with Pettman's cement is shellaced to the outer end of the bar to prevent the ingress of damp.

Cap.—*The cap* is similar to the R.L. cap, except for the form into which the composition is pressed ; it is held in position by a screw collar.

Action.—The friction bar, on being pushed in, shears the suspending wire and travels forward ; the roughened conical portion of the bar pierces the detonator and fires it. Ignition is doubly ensured by the push bar, after passing through the composition, carrying portions of it and grinding it against the roughened interior of the head. The flash passes down through the plug in the body, firing the powder in the tube and the charge. The copper ball is driven upwards by the explosion, and seals the escape of gas through the head.

The small escape hole in the head is intended to allow the escape of gas from the exploding cap, and thus prevents the head expanding and jamming in the vent.

Remarks.—Tubes manufactured prior to 1909 have no escape hole and will be used up for blank firing.

Packing.—Ten in a tin box similar to that used for the Tube, friction, " T."

Tube, V.S., Electric, " T," Double-Wired, Mark I.

This tube resembles the " T " friction tube in exterior form and dimensions. (*See* Fig. 116.)

It is furnished with two tinned copper wire terminals, insulated with silk—one coloured green and one red. The red terminal is called the " ground " wire and is merely attached to the head of the tube. The green terminal passes through a hole in the head and is secured with tin to a copper pole. This copper pole has a ball formed on it, immediately below which it passes through an ebonite plug. The ball and upper part of the pole are insulated from the body by silk and an ebonite cone. A strip of brass with two projections—one long and one short—is bent round the lower part of the ebonite plug.

Fig. 116.

TUBE, V.S. ELECTRIC "T," DOUBLE-WIRED, MARK I | L | .

Full size.

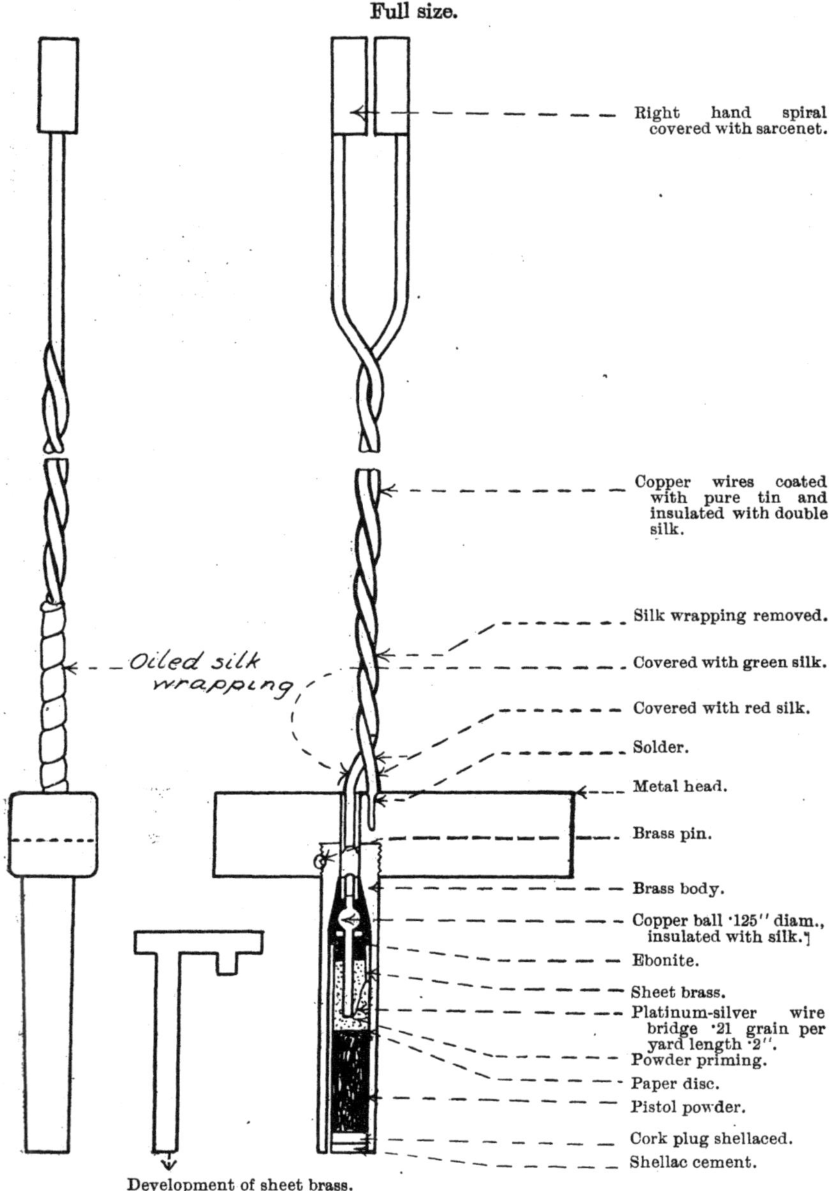

Development of sheet brass.

The short projection forms the second pole, while the long one is attached to the end of the body by tin. The two poles are joined by a platinum-silver bridge, ·2 inch long, resistance 1·0 to 1·3 ohms.

The bridge is surrounded by a priming of 1 grain of mealed powder, separated by a disc of paper from about 5 grains of rifled pistol powder. The tube is closed by a cork plug and shellac cement, to which is attached a paper disc to prevent sticking to the box. If this tube is used with an "earth return," it is necessary to connect the green terminal to the battery. The terminals are twisted together above the head, wrapped with oiled silk for about 1 inch, and terminate in spirals covered with sarcenet the same colour as their insulation.

Use.—Q.F., 6-pr., sub-calibre on top of 9·2-inch, Mark X guns on Mark V barbette mountings.

Action.—The current passes through the terminals, head and body, the poles and bridge, raises the latter to incandescence and so fires the priming and powder. The gas forces the ebonite plug and copper ball into the coned seating, thus sealing escape of gas.

Packing.—" T " electric tubes are packed 10 in a tin box similar to but larger than that used for the " T " friction tubes.

(F).—TUBES, IMPULSE, TORPEDO, AND TUBES, ELECTRIC, TORPEDO, DROPPING GEAR.

Impulse, Torpedo, Tubes.

There are two patterns of impulse torpedo tubes in the Naval Service, viz. :—

(1) *Tube, Electric, Wireless, Impulse, Torpedo.*
(2) *Tube, Percussion, Impulse, Torpedo.*

Fig. 117.

TUBE, ELECTRIC, WIRELESS, IMPULSE, TORPEDO, MARK III | N | .

Scale $\frac{1}{1}$.

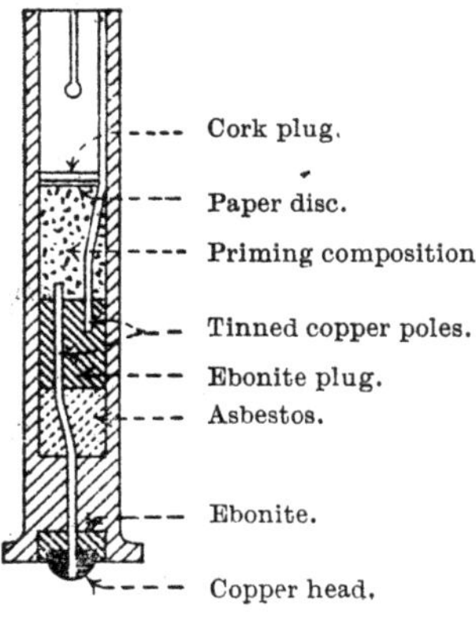

Tube, Electric, Wireless, Impulse, Torpedo, Mark III | N | .

This tube is made of solid-drawn brass, with an enlarged head; it is $1\frac{7}{8}$ inches long and 0·3 inch in diameter. The front end of the tube has two slots cut down it parallel to the axis, and is slightly enlarged to make it act as a spring so as to fit tightly into the central tube of the *Cartridge, impulse, torpedo,* with which the tube is used.

The *body* is bored out to take the electrical arrangements, and a recess formed in the head for a copper contact piece which is fitted into an ebonite plug; a copper pole passes into the interior through the contact piece, ebonite plug, and head of the tube, and also through a plug of asbestos and one of ebonite fitted into the interior. The lower end of this pole is connected by means of a platinum-silver bridge 1·5 to 1·8 ohms resistance to a second pole, one end of which fits into the ebonite plug; the other end is soldered to the mouth of the tube with pure tin.

Priming composition surrounds the bridge and poles, and is kept in position by a cork plug and paper disc.

The *front portion* of the tube is empty.

Tube, Percussion, Impulse, Torpedo, Marks I and II | N |.

The percussion impulse torpedo tubes are of the same external dimensions as the electric wireless tube above described; they are used for above-water tubes of single revolving light type.

Fig. 118.

TUBE, PERCUSSION, IMPULSE, TORPEDO, MARK II.

Scale $\frac{1}{1}$.

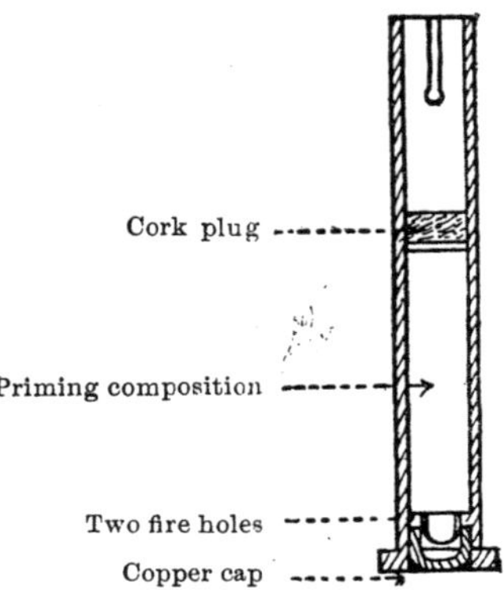

The Mark II tube consists of a body, copper cap, cork plug, paper discs and priming composition.

The body is made of brass, recessed at the head to receive a percussion cap, and the bottom of this recess is formed into an anvil perforated with two fire holes. The tube is partly filled with priming composition, retained in position by a cork plug and paper disc. Two slots are cut in the mouth of the tube.

A certain number of tubes have been issued which differ from the Mark II tubes in having a solid head, a brass cap held in position by a screwed plug, which forms an anvil and also a sealing chamber; the latter, which contains a soft copper ball, is closed by a perforated screwed plug. These tubes are known as Mark I.

The existing stock of Mark I tubes will be used up in the torpedo schools, and no more will be made. The tubes are packed 10 in a tin box.

Tube, Electric, Wireless, Torpedo Dropping Gear, Mark I | N |.

(Torpedo dropping gear, also for fore-bridge firing.)—The above tube is almost identical in construction with the V.S. electric wireless "P," Mark IV, differing from it in being filled with a 15-inch strand of No. 1 guncotton yarn instead of having a charge of powder. In order to distinguish it *from the wireless "P" tube the letters "T.D.G."* are stamped on the head.

EXPLOSIVE STORES USED IN CONNECTION WITH TUBES.

Primer, vent, cordite, Mark I.—The Primer, vent, cordite, is for use with V.S. tubes without ball in B.L. guns, 8-inch and upwards, with *powder charges only.* It consists of a stick of cordite, size 20, 4¾ inches long; it is placed in the vent after the breech is closed prior to inserting the tube.

The primers are packed 10 in a tin box lined with felt, and closed with a tape band shellaced on.

Primers, vent piece, were used in R.B.L. guns for extending the flash from the tube; they were declared obsolete, but have been re-introduced for use with the R.B.L., 40-pr. (other than the side closing) for firing salutes. Each primer consists of a tube of paper about 2½ inches long, coloured black and filled with a compressed powder pellet, perforated with a central hole; strands of red worsted are attached, which keep the primer in the hole in the vent piece.

They are packed 25 in tin cylinder, No. 7P.

(G).—DRILL TUBES.

For each pattern of Service tube there has been a corresponding drill tube issued.

The V.S. *electric* drill tubes which were generally issued to Naval Service only, have been declared obsolete for future manufacture, and will be used up in gunnery schools and drill batteries; a description of V.S. electric drill tubes will be found in earlier editions of the Treatise on Ammunition.

Fig. 119.

TUBE, V.S. PERCUSSION, DRILL, MARK II.

Scale 1/1.

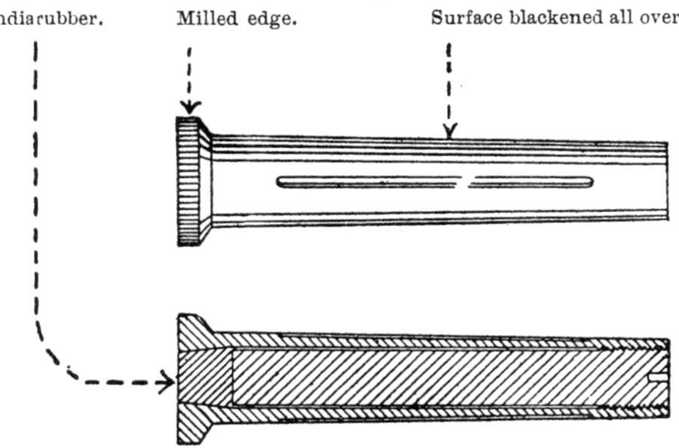

Tube, Vent-Sealing Percussion, Drill, Mark II.

The V.S. percussion drill tube is made of gunmetal to the same external dimensions as the Service V.S. percussion tube, but differs from it in external appearance in having four longitudinal grooves on the body, and the rim of the head is milled.

The head of the tube is fitted with a coned indiarubber plug kept in position by a long screwed metal plug.

Mark I percussion drill tube.—The Tube, V.S. percussion, drill, Mark I, was much shorter; it was without the milled head and fluted body, and was not blackened.

Fig. 120.

TUBE, V.S. PERCUSSION, DRILL, MARK I.

Scale ½.

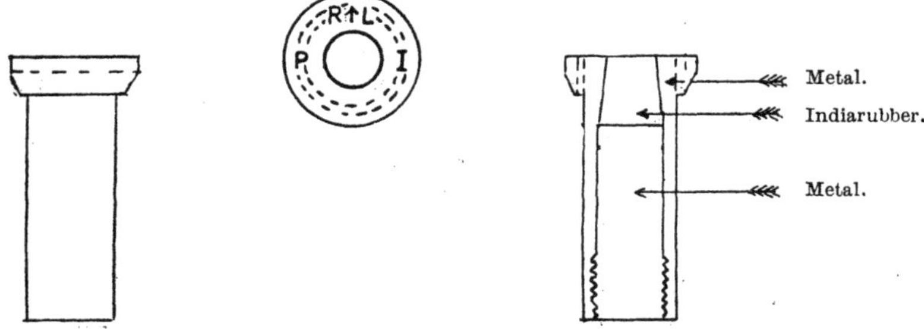

Tube, Friction, "T," Drill, Mark I.

This tube is for drill purposes with all guns and howitzers fitted with "T" vents except the B.L.C., 15-pr., which takes a drill push tube.

The tube is made of hardened steel, and resembles in form the Service tube; the head is slotted out and a lip formed on the lower part, and a curved spring is fitted into the slot in the head which offers the same resistance to the pull of the lanyard as the bar in a friction tube.

Fig. 121.

TUBE, FRICTION, "T," DRILL, MARK I.

Full Size.

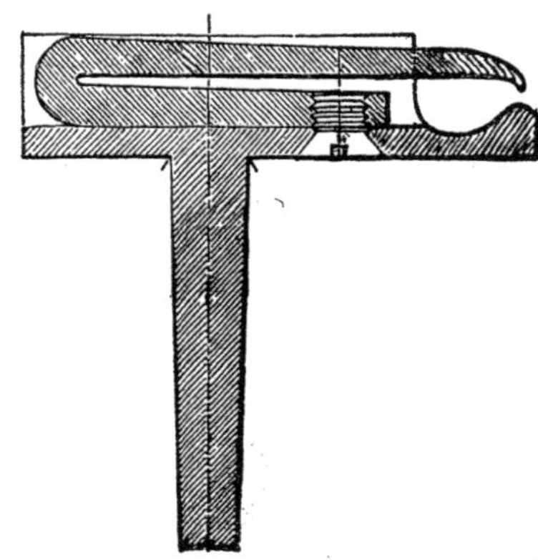

TUBE, FRICTION, " T," DRILL, CONVERTED, MARK II.

The Mark II converted drill tube differs from the Mark I drill tube above mentioned in the following particulars :—

The curved spring is of a different shape and is attached in a different manner. The head of the tube is shorter and ends at the base of the spring instead of running the whole length as in the previous pattern.

Fig. 122.

TUBE, FRICTION, " T," DRILL, CONVERTED, MARK II.

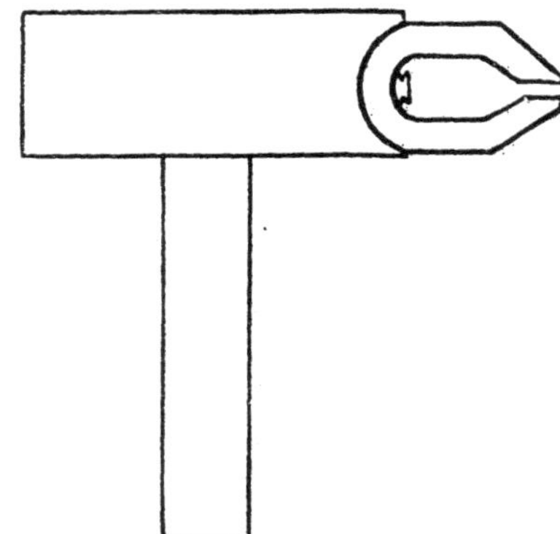

The Mark I converted drill " T " tube differs from the above in having a weaker form of clip spring, which fits inside the slotted-out portion of the " T " head.

TUBE, FRICTION, " T " (PUSH), DRILL, MARK I.

The above tube is for drill purposes with the B.L.C., 15-pr. guns. The head of the tube is fitted with a push bar and spiral spring, the body with a wood plug as shown in the figure.

Fig. 123.

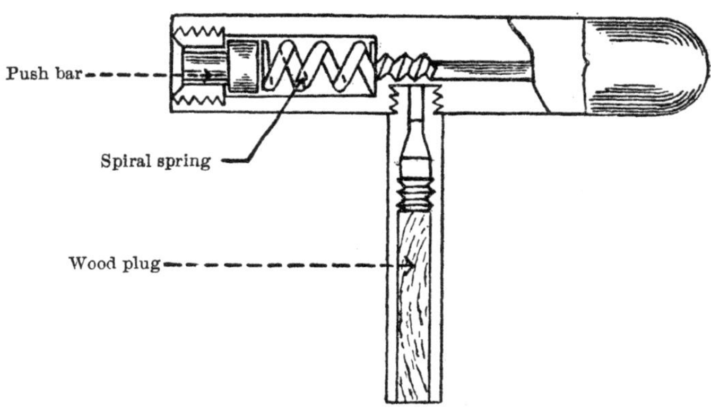

CHAPTER XV.—PRIMERS.

(A) General Remarks ; (B) Electric Primers ; (C) Percussion Primers ; (D) Keys used with Primers.

(A).—GENERAL REMARKS.

Use.—Primers are used for igniting the charge in Q.F. and Q.F.C. guns, and Q.F. howitzers ; the B.L., 9·45-inch howitzer is also fired by means of a percussion primer.

Primers differ from tubes in being screw-threaded externally to screw into the cartridge case or end of vent instead of accurately fitting into a vent seating.

The electric primers that will be met with in the Service are :—

Cartridge, Q.F. or Q.F.C., Primer, Electric, Large | C | .

Cartridge, Aiming Rifle, 1-inch Electric, Primer, Mark I | C | .

Percussion primers used with Q.F. cartridges are placed in a numbered series as follows :—

Primer, Percussion, Q.F. Cartridges,	No. 1, for	Q.F., 12-pr. of 4 cwts. Q.F., 13- and 18-pr. Q.F., 3-inch. Q.F., 4-inch, Mks. IV and V. Q.F., 4·5-inch howitzer.
,, ,, ,,	No. 2, for	Q.F., 3- and 6-pr.
,, ,, ,,	No. 3, for	Q.F., 15-pr.
,, ,, ,,	No. 4, for	Q.F., 2·95-inch.

There is also the *Primer, percussion, B.L., 9·35-inch howitzer.*

There is a *special percussion primer* for firing blank from 3- and 6-pr. and 2·95-inch Q.F.

All percussion primers are fitted with a percussion cap ; these caps contain cap composition, which, with the exception of the Primer, percussion, No. 3, contains no fulminate of mercury.

The composition of percussion caps for Primers, Q.F., Nos. 1, 2 and 4, is as follows :—

	Parts.
Potash, chlorate of	12
Antimony, sulphide of	18
Mealed powder	1
Sulphur	1
Ground glass	1

(B).—ELECTRIC PRIMERS.

Cartridge, Q.F. or Q.F.C., Primer, Electric, Large, Mark V | C |.

(See *Plate* LXXVI.)

Use.—This primer is used with the Q.F. 12-pr. of 8, 12, and 18 cwt., Q.F. 14-pr., 4-inch, 4·7-inch, and 6-inch guns, until the existing stock is used up, when an adapter with V.S. percussion, or V.S. electric "P" tube will be used.

In the Naval Service the use of this primer is now restricted to gunnery ships and for firing salutes.

Parts.—The primer consists of the following parts :—Body, cone, contact disc, ebonite insulator, two crown metal poles, iridio-platinum wire bridge, ebonite washer, screw-collar, brass cylinder, priming composition and glazeboard closing disc.

The body is of manganese bronze or aluminium bronze, screwed below the shoulder to fit the hole in the base of the Q.F. cartridges; the end of the body is reduced in diameter, and threaded to receive a brass cylinder; the face of the body between the screwed portions is cupped out to a depth of ·25 inch to form a gas-check. The head is recessed in the centre and screw-threaded to receive the insulator and contact disc; the inside is bored out and the part near the head coned, and a small hole is drilled through to the recess for the contact disc.

Two slots are formed in the head to take a key by which the primer is screwed into the case.

The ebonite insulator is screwed into the recess in the body and is hollowed out and undercut. The contact disc is of pure tin; the top is smooth, and slightly below the surface of the body.

The letter "T," denoting pure tin, was stamped upon the head and also marked upon all packages containing such primers, but this is now discontinued.

The cone is made of brass, cupped out in front to form a gas-check and insulated from the body by oiled silk. To the base of the cone a piece of insulated copper wire is soldered; a turn is taken in this wire and it is passed through the hole to the recess for the contact disc. At the bottom of this recess a thin disc of tin is laid. The wire, bared of its insulation, is coiled down upon this and the recess is filled up with molten tin.

The poles are of crown metal. One screws into the front of the cone and is kept central by an ebonite washer which in turn is kept in place by a brass screwed collar.

The other pole is fixed to the face of the body; and joining the two poles is a bridge of iridio-platinum, ·25 inch long and having a resistance of from ·75 to ·95 ohm.

This bridge is attached to the poles by pure tin. It is stronger than those fitted to earlier primers.

The cylinder is made of brass; one end is screw-threaded inside to fit the body, on to which it screws, the joint being made tight by Pettman's cement. It is prevented from unscrewing by a small set-screw.

The mouth is recessed to receive the glazed-board disc.

A tuft of guncotton is wrapped completely round the bridge, and the interior of body and cylinder is filled with priming composition (2 parts guncotton dust, 3 parts mealed powder), and the top closed by a glazed-board disc cemented in and covered with Pettman's cement.

The primer is lacquered inside and out, except the exterior of the body below the shoulder. Primers have the initial of contractor or trade mark stamped on the head; those of Ordnance Factory manufacture have the number of thousand and year of manufacture, and those obtained from contractors, the month and year of manufacture stamped on them.

These primers are packed 10 in a tin cylinder, 20 cylinders in a packing case.

Action.—On the circuit being completed, the current flows through the wire to the brass cone and so through the bridge, which becomes incandescent and fires the guncotton and composition priming, which in turn fires the charge. The return path of the current is by the metal of primer, cartridge, gun, &c., to battery. The cone is driven by the force of the explosion into its seating, and the cupped-out portion expands, thus effectually sealing any escape of gas through the head. The cupped-out portion of the body expands and prevents any escape of gas over the exterior.

The principal improvements embodied in Mark V are :—

(*a*) Contact piece of tin instead of white metal which was found to corrode and become covered with a deposit of high resistance.
(*b*) The ebonite cup for the contact disc is screwed in.
(*c*) The cone plug is cupped out in front, to give better gas sealing.
(*d*) The central pole is steadied by an ebonite plug and a brass washer.
(*e*) The body pole is bent inward to clear the cylinder.
(*f*) Strengthened bridge.
(*g*) Waterproofing threads with Pettman's cement.
(*h*) Closing the end of primer with glazed-board disc and Pettman's cement.

Earlier Marks of Primer, Electric, Large.

Primers may be divided into "unconverted," "converted," and those with "strengthened bridges."

Unconverted primers are Marks I, I*, II, III, and IV. They have the double bridge of platinum-silver, resistance ·6 to ·9 ohm, body filled with mealed powder and the top closed by a brass washer and cork plug shellaced in.

Converted primers are Marks I**, II*, II**, III*, III** and IV*. They have a single bridge of platinum-silver, resistance 1·5 to 1·8 ohms, surrounded by a tuft of guncotton, body filled wth priming composition and top closed by a glazed-board disc secured with Pettman's cement.

Strengthened bridge primers are Marks I*$_*$*, II*$_*$*, III*$_*$*, IV**, and V. They have an iridio-platinum wire bridge, resistance ·75 to ·95 ohm.

PRIMER FOR 1-INCH AIMING RIFLE.

Cartridge, aiming rifle, 1-inch electric primer, Mark I.—The electric primer used with the 1-inch aiming rifle electric cartridge is described and illustrated with that ammunition. (*See* page 527.)

(C).—PERCUSSION PRIMERS.

PRIMER, PERCUSSION, Q.F. CARTRIDGES, No. 1, MARK II | C |.

This primer is used with the Q.F. 12-pr. of 4 cwts., Q.F. 13- and 18-pr., Q.F. 3-inch, Q.F. 4-inch, Marks IV and V; and Q.F. 4·5-inch howitzer.

(*Plate* LXXVII.)

Parts.—It consists of a body, percussion cap, anvil plug, copper ball, screw plug, and brass closing disc.

The body is made of metal with a flanged head; it is screw-threaded externally to screw into the base of the cartridge case; two slots are cut in the head for the key.

The interior is bored out, cupped, and screwed to take the percussion arrangement. The cap is of copper, and is placed into a recess in the body, over which screws an anvil plug having a cone-seating into which is placed a soft copper ball. Three fire holes are bored through the anvil plug to allow the flash from the cap to pass into the cone-seating.

The copper ball is retained in position by a perforated screw plug. A fillet of Pettman's cement is placed between the cap and the body to prevent the ingress of damp.

The front of the primer is filled with R.F.G.[2] gunpowder covered by a brass disc; this disc has six radial slits, and has a paper disc secured to its underside by Pettman's cement. The brass closing disc is secured in position by the metal of the primer being burred over it, and coated with Pettman's cement.

The primer is marked on the head with its serial number (No. 1), contractor's initials, or recognized trade mark, its Mark, and date of filling.

PRIMER, PERCUSSION, Q.F. CARTRIDGE, No. 1, MARK I.

This primer differed from the above in having a cap chamber with a percussion cap pressed into a recess in the base; no anvil plug

Fig. 124.

PRIMER, PERCUSSION, Q.F. CARTRIDGES, NO. I, MARK I*.

Full Size.

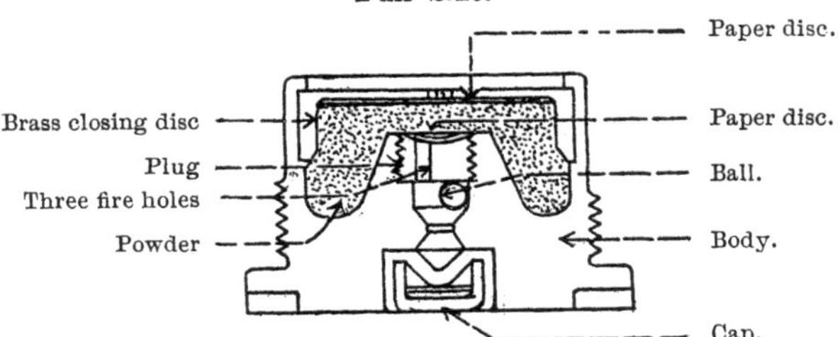

was used; the copper ball fitted into a cone seating in the body, and the front end of the primer was closed by a disc of cordite. Mark I primers have been declared obsolete; the existing stock has been returned to Woolwich and converted to Mark I*.

Mark I primer.*—The Mark I primer, when fitted with a brass closing disc similar to Mark II, is known as Mark I*.

Refilling of Primers.

Fired Mark II primers will, if considered suitable, be refilled once. They will be distinguished by having the original date of filling barred out, and the letter " R " and the date of refilling substituted; fresh key slots will also be prepared if necessary.

Packing.—Primers, Percussion, No. 1, are packed in two ways, *i.e.* :—

4 primers in " Box, percussion primers."
10 primers in Cylinder No. 103.

The " Box, percussion primers " is made of tin, square in shape, and the primers are packed between sheet cork packing pieces, top and bottom, and are kept from moving laterally by a thick piece of cork, through which four holes are cut for their insertion.

Cylinder No. 103 is made of tin, fitted with a diaphragm perforated with 10 holes to take the primers; the cylinder is also fitted with two discs of cork which are placed above and below the primers.

The cylinder is painted black; the lid is secured by means of a tin strip soldered on.

PRIMERS, PERCUSSION, No. 2 | C |.

The Marks IV and II primers are used with Q.F. 3-pr. and 6-pr. ammunition in the latest Marks of cartridge cases, namely, cartridge cases specially made to take this pattern of primer.

The Marks III and I primers are only used with 3-pr. and 6-pr. cartridge cases originally fitted with a percussion cap when altered to take a primer.

The Marks I and II are obsolete for future manufacture and are being converted to Marks III and IV.

PRIMER, PERCUSSION, Q.F. CARTRIDGES, No. 2, MARK IV | C |.

(*Plate* LXXVIII.)

The Mark IV primer consists of the following parts:—Body, percussion cap, anvil plug, copper ball, perforated brass disc and magazine cylinder.

Body.—The body is threaded externally to screw into the base of the cartridge case; two slots are cut in the head to take the Key No. 30. A hole is bored through the centre of the body, the upper part threaded to take the anvil plug, a shoulder being formed below the threaded portion to support a percussion cap.

Cap and anvil plug.—The cap is made of brass and contains 1·2 grains of composition pressed in and varnished, and covered with a disc of tin foil. The cap is held in position by the anvil plug, which is screwed into the body of the primer after the insertion of the cap. The anvil plug is bored out to form a cone-seating, in which is placed a soft copper ball to seal the escape of gas on firing and relieve the pressure on the cap; 3 holes bored through the bottom of the plug allow the flash from the cap to pass into the cone-seating.

Magazine cylinder.—The magazine cylinder is made of brass, the front end taperedand left solid as shown in Plate LXXVIII. The cylinder screws into the front of the primer body; it is perforated with 8 holes and contains a perforated pellet of R.F.G.^{2} gunpowder contained in a tube of varnished paper; the bottom is closed by a brass disc having 3 fire holes. The cylinder is prevented from unscrewing from the body by 3 punch dabs.

Packing.—10 in cylinder No. 114.

Primer, Percussion, Q.F. Cartridges, No. 2, Mark II | C |.

(*See* Fig. 125.)

The Mark II percussion primer (of which only a few thousand were issued) differs from the Mark IV in having a magazine *plug* filled with R.F.G.^{2} powder, instead of a magazine *cylinder* and perforated brass disc.

Packing.—10 in cylinder No. 106.

Fig. 125.

PRIMER, PERCUSSION, Q.F. CARTRIDGES, NO. 2, MARK II.

Scale $\frac{3}{1}$.

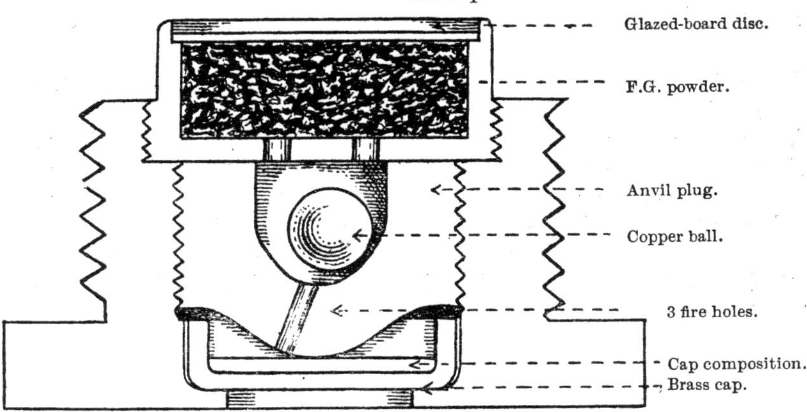

Primers, Percussion, Q.F. Cartridges, No. 2, Mark III | C |.

(*Plate* LXXVIII.)

Owing to the thinner base in the early Marks of 3-pr. and 6-pr. Q.F. cartridge cases, the primers used with these cases, when they are converted, are smaller than those described above; the *heads* of the primers are screw-threaded instead of the *bodies*, and a gas-check is formed on the front end. The Mark III has a magazine cylinder containing a perforated pellet of R.F.G.^{2} powder; the anvil plug has only 2 fire holes and has no copper ball.

Packing.—20 in cylinder No. 113.

Primer, Percussion, Q.F. Cartridges, No. 2, Mark I | C | .

(*See* Fig. 126.)

The Mark I primer differs from the Mark III in having a magazine plug instead of a cylinder. The magazine plug has a central fire hole and a copper ball.

Only a few thousand of the Mark I primers have been issued.

Packing.—20 in cylinder No. 105.

Fig. 126.

PRIMER, PERCUSSION, Q.F. CARTRIDGES, NO. 2, MARK I.

Scale $\frac{3}{1}$.

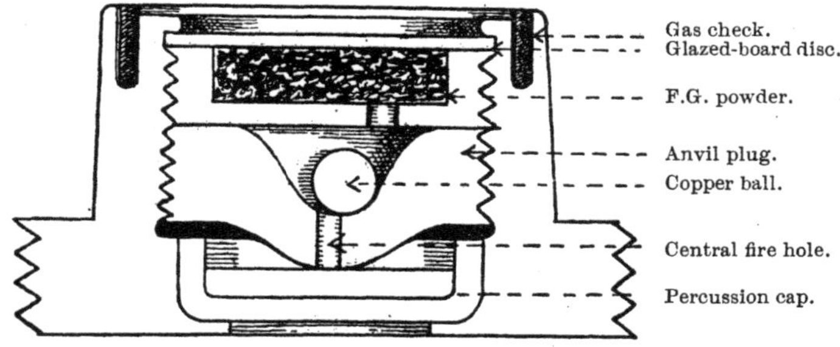

Primers, Percussion, Q.F. Cartridges, No. 3, Mark I | L | .

The Primer, percussion, No. 3, is used with the Q.F. 15-pr.

It consists of a body, percussion cap, anvil plug, charge of R.F.G.[2] powder and a glazed-board closing disc.

The primer externally is similar to the Primer, percussion, No. 2, Mark I, differing in the following particulars:—The anvil plug has 3 fire holes, there is no copper ball, and the body above the plug contains 1·3 grains of F.G. powder, covered by a glazed-board disc coated with Pettman's cement.

Packing.—Primers, percussion, No. 3, are packed 20 in cylinder No. 108.

Fig. 127.

PRIMER, PERCUSSION, Q.F. CARTRIDGES, NO. 3, MARK I.

Scale $\frac{4}{1}$.

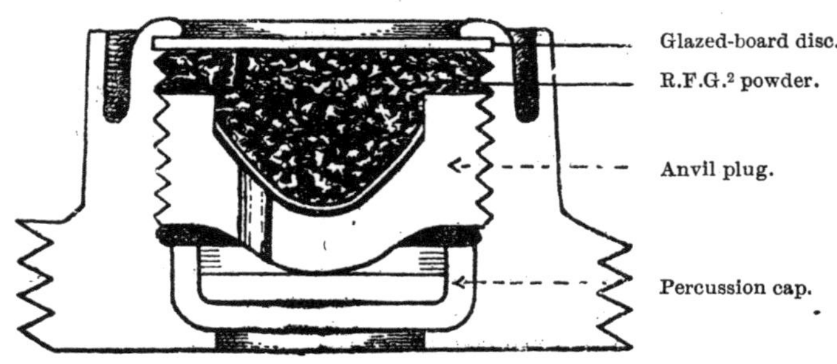

Primer, Percussion, No. 4, Mark I | L | .

Use.—The Primer, percussion, No. 4, is used with the Q.F. 2·95-inch.

The body of the primer is made of metal, screwed externally to fit the cartridge case; a central hole is bored through it which is enlarged at the base to take a cap chamber with percussion cap.

Fig. 128.

PRIMER, PERCUSSION, NO. 4, MARK I.

Scale $\frac{2}{1}$.

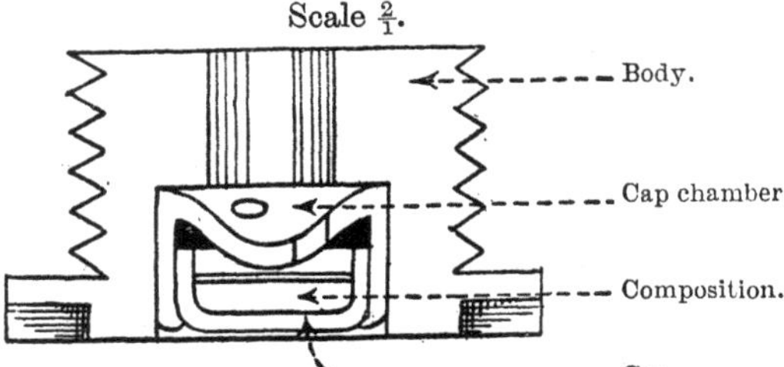

Cap chamber.—The cap chamber is made of brass with a raised anvil pierced with three fire holes; it contains a copper cap filled with 1·2 grains of composition, pressed in, varnished, and covered with a tinfoil disc.

The cap chamber with its percussion cap is pressed into the recess in the base of the primer.

Packing.—25 in cylinder No. 1.

Fig. 129.

PRIMER, PERCUSSION, B.L. 9·45-INCH HOWITZER, MARK I.

Scale $\frac{3}{1}$.

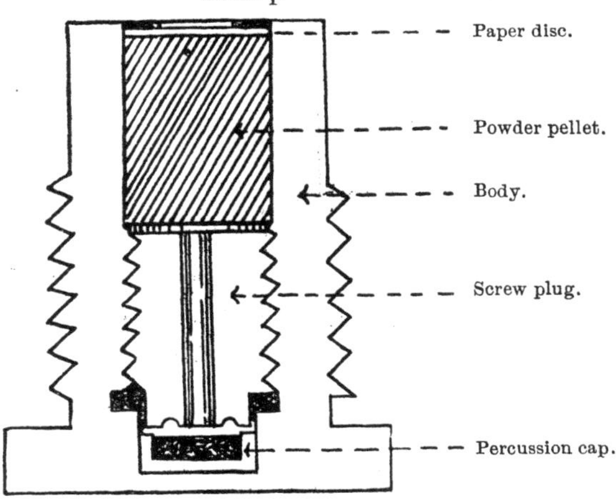

Primer, Percussion, B.L. 9·45-inch Howitzer, Mark I | L | .

The body of the primer is made of metal, screwed externally to fit the stem of the obturating cup of the howitzer. It is bored out from the front, and fitted with a small brass percussion cap, filled with 0·2 grains of cap composition, over which is screwed a metal anvil; the anvil has a central fire hole. The front end of the primer is filled with a pressed powder pellet, and closed by a varnished paper disc.

The metal in the head of the primer over the cap is very thin, so as to allow the striker to fire the cap on the face of the anvil.

Three recesses are cut in the head for the Key No. 31.

Packing.—The above percussion primers are issued 10 in a cardboard box.

Cartridge, Q.F., Blank, 2·95-inch and 6- or 3-pr. Primer, Percussion, Mark III.

The above primer is made of brass with a cap chamber and anvil formed in its rear end; three fire holes communicate the flash from the copper percussion cap to about $5\frac{1}{2}$ grains of R.F.G.[2] powder with which the body of the primer is filled. The front end is closed with a glazed-board disc, spun over and coated with shellac. A small brass pin projects from the head of the primer, by which it is secured to the case. There are also two slots in the head to take the "screw-driver, primer," for inserting or removing it.

The primers are packed in hermetically sealed cylinders containing 20.

Fig. 130.

Cartridges, Q.F. Blank, 2·95-inch and 6- or 3-pr. Primer, Percussion, Mark III.

Scale $\frac{2}{1}$.

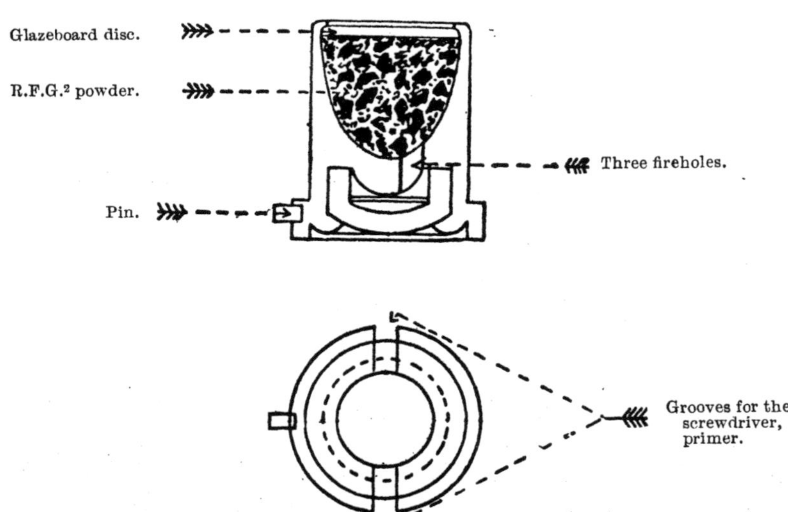

(D).—KEYS, &c., USED IN CONNECTION WITH PRIMERS.

Implements, Ammunition.

Key, No. 24 (Mark I), Adapter or Primer | C | is a small "T" shaped key used for screwing home electric primers or adapters. It may also be used for removing unfired primers.

Fig. 131.

KEY, NO. 24 (MARK I), ADAPTER OR PRIMER | C |.

Scale ½.

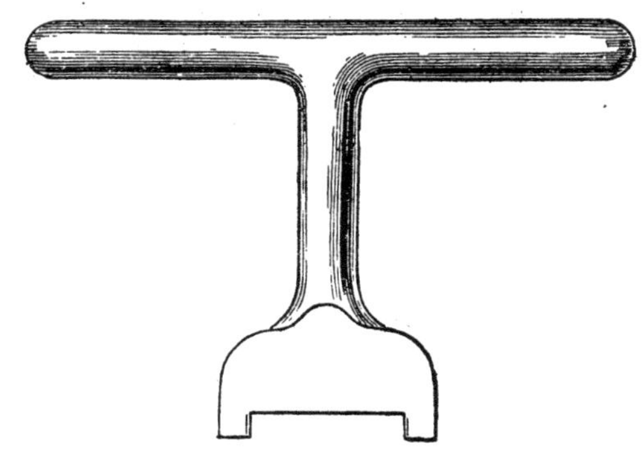

Key, No. 25, Adapter or Primer | C |.—There are three marks of this key (Marks II, III and IV), which may all be met with in the Service. They are used for removing fired "Primers, electric, large," or adapters.

Fig. 132.

KEY, NO. 25, ADAPTER OR PRIMER | C |.

Scale ½.

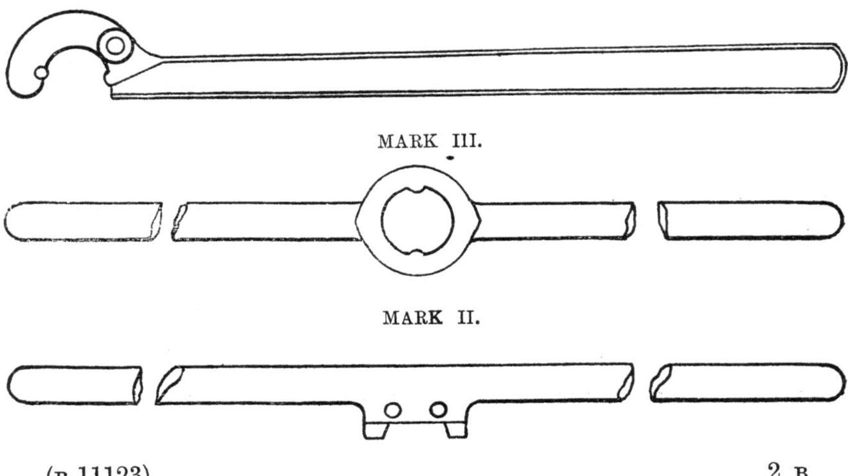

Key, No. 26 (Mark I), Primer, 4·5-inch | L | .

Key, No. 26 (Mark II), Primer, 4·5-inch and 4-inch | C | .

Key, No. 27 (Mark I), Primer, 13- and 18-pr., and 12-pr. 4 cwt. | C |

Key, No. 28 (Mark I), Primer, 15-pr. | L | .

Key, No. 30 (Mark I), Primer, 3-pr. and 6-pr. | C | .

Key, No. 34 (Mark I), Primer, 3-inch | C | .

The above keys are made of steel, cranked to fit over the head of the cartridge case, and have two steel pins screwed into the cranked portion which engage in the slots in the head of the primer.

The Nos. 26, 27 and 34 keys are fitted with lanyards of white line about 43 inches in length, as shown in Fig. 133.

The Nos. 28 and 30 keys are not fitted with a lanyard. The No. 30 key is a flat bar and has one pair of pins on one side for the "Primers, percussion, No. 2, Marks I or III," and another pair of pins on the other side for the "Primers, percussion, No. 2, Marks II or IV."

Fig. 133.

KEY, NO. 27 (MARK I), PRIMER, 13- AND 18-PR. AND 12-PR. 4 CWT. | C | .

Scale ½.

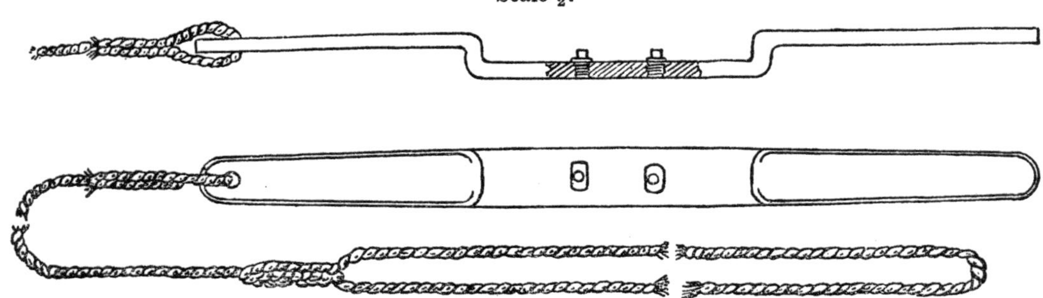

Key, No. 29 (Mark I), Primer, 2·95-inch | L | is a steel key, bent over at one end, where two projections are formed to fit the recesses in the head of the "Primer, percussion, No. 4."

Fig. 134.

KEY, NO. 29 (MARK I), PRIMER, 2·95-INCH | L | .

Scale ½.

Key, No. 31 (Mark I), Primer, 9·45-inch | L | is a small steel "T" shaped key having three studs as shown in Fig. 135. It is used for inserting or removing the "Primer, percussion," or the "Gauge, testing blow of striker, B.L., 9·45-inch howitzer."

Fig. 135.

KEY NO. 31 (MARK I), PRIMER, 9·45-INCH | L |.

Full size.

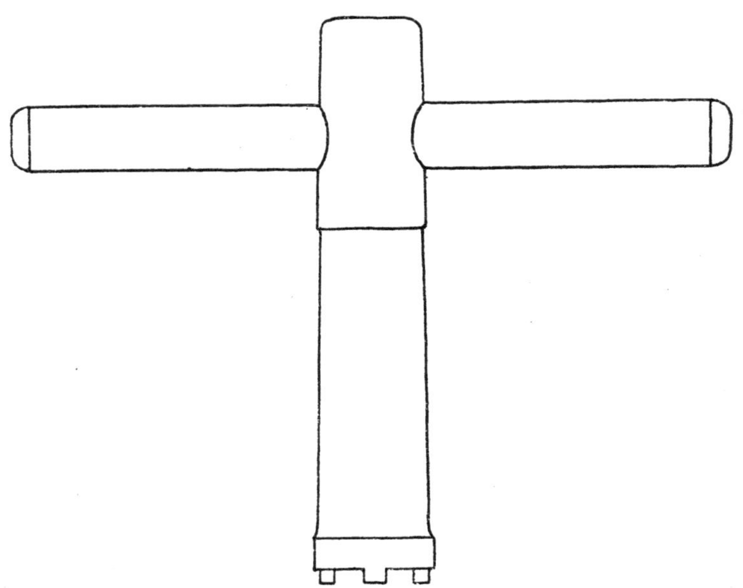

SCREWDRIVER, PRIMER (MARK I) | C | is a steel blade fitted with a wood handle.

The blade is shaped at the end to fit the slots in the head of the "Primer, percussion, 2·95-inch, 6- or 3-pr., for blank cartridges."

CHAPTER XVI.—IGNITERS, TORPEDO.

Torpedo igniters fit inside the air vessels of torpedoes, and on being fired, heat and expand the gases for driving the engines, so greatly increasing the range obtainable.

There are two types of igniters—R.G.F. and Weymouth. These are similar and differ in minor details only. The Weymouth type is obsolete, and no more will be made. For details, *see* Tables 32 and 33.

Igniter, Torpedo, R.G.F., Mark IV Z | N |.

(*Plate* LXXIX.)

The body of the igniter is made of steel, bored out and threaded at the rear end to receive the screwed enlarged end of the central tube. The central tube of steel has a cap chamber formed at its rear end, and into this is pressed a thin copper cap with its brass anvil. The front end of the central tube is closed by a disc of vegetable paper, shellaced on. The body of the igniter has a coarse thread cut on the inside, for about 0·6 inch from the front end, to enable the priming composition and sealing varnishes to obtain a hold. The steel body and the exterior of the central tube are lacquered. The space between the body and tube is filled with ground cordite, with 10 per cent. magnesium added, pressed in at a total pressure of 1,000 lbs. A priming of 15 grains S.F.G.[2] is pressed in on top, four holes having been stabbed in the pressed cordite for the priming to grip. The mouth of the igniter is closed by layers of shellac, Pettman's cement and fuze beeswax composition. The cap is waterproofed with Pettman's cement.

Action.—On the cap being struck the flash passes up the central tube, ignites the priming composition and starts the cordite *burning* from the front end, the hot gases from which heat and so expand the gases driving the torpedo engines.

Igniters are packed ten in a tin box, five boxes in a wood case with sliding lid.

TABLE 32.

Particulars of Igniters used in R.G.F. and R.N.T.F. Heater Torpedoes.

NOTE.—In all these igniters the body and tube are of steel.

—	Length.	Diameters.		Charged with.	Priming.	Cover for priming.	Case roughed at end (inside).	Absolute pressure used in filling.	
		External.	Internal.					Charge.	Priming.
								lbs.	lbs.
Mark IV O... ...	2·75″	·58″	·4″	Ground M.D. cordite with 10 per cent. magnesium.	Powder, S.F.G.[2], 10 grains.	Pettman's cement, shellac and wax filling.	No.	650	5
Mark IV P... ...	2·75″	·58″	·4″	Cordite, Size ·$\frac{1}{05}$, with 10 per cent. magnesium.	Sulphurless powder, S.F.G.[2], 10 grains.	Pettman's cement, shellac and wax filling.	No.	650	650
Mark IV S	2·75″	·58″	·4″	Ground M.D. cordite with 10 per cent. magnesium.	Powder, R.F.G.[2], 10 grains.	Pettman's cement, shellac and wax filling.	No.	650	650

NOTE.—The Mark IV Z is now the recognised Service igniter, and all the others are being altered to this pattern.

TABLE 32—*continued.*

Particulars of Igniters used in R.G.F. and R.N.T.F. Heater Torpedoes.

NOTE.—In all these igniters the body and tube are of steel.

—	Length.	Diameters.		Charged with.	Priming.	Cover for priming.	Case roughed at end (inside).	Absolute pressure used in filling.	
		External.	Internal.					Charge.	Priming.
Mark IV Z	2·75″	·58″	·5″	Ground M.D. cordite with 10 per cent. magnesium.	Powder S.F.G.², 15 grains.	Shellac and fuze water-proofing composition.	Yes.	lbs. 1,000	lbs. 1,000
Mark IV Z³	2·75″	·58″	·5″	Ground M.D. cordite with 10 per cent. magnesium.	Powder, S.F.G.², 15 grains.	Pettman's cement, shellac and wax filling.	Yes.	1,000	8

NOTE.—The Mark IV Z is now the recognised Service igniter, and all the others are being altered to this pattern.

TABLE 33.

Igniters used in Fiume and Weymouth Torpedoes.

—	Length.	Diameters.		Material.		Charged with.	Priming.	Cover for priming.	Case roughed at end (inside).	Absolute pressure used in filling.	
		Outside (Taper).	Inside.	Body.	Tube.					Charge.	Priming.
Mark IV... ...	3·503″	·524″ at top. ·5″ at bottom.	·441″	Steel.	Steel.	Cordite, size $\frac{1}{\cdot 05}$, mixed with 10 per cent. magnesium.	Sulphurless powder, S.F.G.², 15 grains.	Pettman's cement, shellac and wax filling paraffin mixture.	Yes.	lbs. 1,000	lbs. 1,000
Mark V	3·503″	·524″ at top. ·5″ at bottom.	·441″	Steel.	Steel.	Cordite, size $\frac{1}{\cdot 05}$, mixed with 10 per cent. magnesium.	Sulphurless powder, S.F.G.², 10 grains.	Pettman's cement, shellac and wax filling paraffin mixture.	No.	650	650
Mark VII ...	3·5″	H ·525″ } top. L ·519″ } H ·505″ } bottom L ·502″ }	H ·420″ L ·417″	Brass.	Paper.	Composition, brown powder (in 8 charges).	Black powder (stemmed).	Paper disc, covered with shellac and coated with Pettman's cement.	No.	2,000	1,550

CHAPTER XVII.—AMMUNITION FOR Q.F. GUNS.

Section A, General Remarks; Section B, Fixed Ammunition; Section C, Q.F. Separate Ammunition; Section D, Blank Cartridges for Q.F. Guns; Section E, Marking on Q.F. Cartridges; Section F, Drill Cartridges for Q.F. Guns; Section G, Packing and Storage of Q.F. Ammunition.

SECTION A.—GENERAL REMARKS.

A Q.F. gun, as regards its ammunition, differs from a B.L. gun in having its charge and the means of ignition contained in a brass case.

Q.F. guns are divided into two classes, viz. :—

(1) Those which fire "*Fixed Ammunition.*"
(2) Those which fire "*Separate Ammunition.*"

With "*Fixed Ammunition,*" the projectile is attached to the brass cartridge case, the projectile and cartridge being loaded in one operation.

With "*Separate Ammunition*" the projectile is not attached to the cartridge case, but is packed and loaded separately.

The following Q.F. guns fire "*Fixed Ammunition*" :—

<table>
<tr><td>Q.F. 1 Pr. .. ..</td><td>Fired by percussion cap.</td></tr>
<tr><td>Q.F. 3 Pr. .. ..</td><td rowspan="3">Fired by cap, or Primer Percussion No. 2.</td></tr>
<tr><td>,, 3 Pr. Vickers .. .. ..</td></tr>
<tr><td>,, 6 Pr. .. ..</td></tr>
<tr><td>Q.F. 6 Pr. Swiftsure and Triumph..</td><td>Fired by percussion cap.</td></tr>
<tr><td>Q.F. 13 Pr. .. ..</td><td rowspan="4">Fired by Primer Percussion No. 1.</td></tr>
<tr><td>,, 18 Pr. .. ..</td></tr>
<tr><td>,, 3-inch .. ..</td></tr>
<tr><td>,, 4-inch, Mark IV | L | ..</td></tr>
<tr><td>Q.F. 2·95-inch .. .. ..</td><td>Fired by cap, or Primer Percussion No. 4.</td></tr>
</table>

The following Q.F. guns fire "*Separate Ammunition*" :—

<table>
<tr><td>Q.F. 12 Pr. of 8, 12 and 18 cwt. ..</td><td rowspan="5">Fired by an Electric Primer, or V.S. Tube and Adapter.</td></tr>
<tr><td>,, 14 Pr. .. ..</td></tr>
<tr><td>,, 4-inch Marks I to III .. ..</td></tr>
<tr><td>,, 4·7-inch .. .. ..</td></tr>
<tr><td>,, 6-inch .. ..</td></tr>
<tr><td>Q.F. 12 Pr. of 4 cwts. .. ..</td><td rowspan="3">Fired by Primer Percussion No. 1.</td></tr>
<tr><td>,, 4-inch Marks IV and V | N |</td></tr>
<tr><td>,, 4·5-inch Howitzer .. ..</td></tr>
<tr><td>Q.F. 15 Pr. .. ..</td><td>Fired by Cap, or Primer Percussion No. 3.</td></tr>
</table>

Advantages and Disadvantages of Q.F. System.

Advantages.—Certain advantages are obtained by adopting a Q.F. system. A more rapid rate of fire can be attained in many cases, especially when fixed ammunition is used; the cartridge case acts as an obturator; it saves riming out the vent, sponging, and placing the electric tube in the vent separately; there is no fear of double loading, or of the cordite accidentally firing from a heated chamber, or from "back flash," or from smouldering fragments from previous rounds.

Disadvantages.—There are the following disadvantages to be considered:—There is a great increase in weight and magazine space, which are important considerations on board ship; the case is liable to split and allow the gas to escape to the rear and damage the gun; in the case of a miss-fire the breech must be opened after waiting a certain time, whereas with a B.L. gun a fresh tube may be inserted without opening the breech.

The system of enclosing the cartridge in a brass case was introduced into the British Service because it enabled greater rapidity of fire to be attained than was possible at that time from B.L. guns; this was due to the fact that the operations of inserting a tube and sponging out were dispensed with. Improvements in breech mechanisms, however, now enable a tube to be inserted during the loading of the gun, and to be fired immediately the breech is closed; and it was found, at any rate in the case of a 6-inch, that in spite of the sponging out necessary with the silk cloth cartridge the brass case is no longer a necessity for a rapid rate of fire. In consequence of this the 6-inch Q.F. gun was superseded by the 6-inch B.L., Mark VII. The saving in weight and magazine space in this case is considerable on board ship, but would not be so great with smaller Q.F. guns.

Shape of the case, &c.—All cartridge cases for Q.F. guns are made of brass, generally solid drawn, but some may be met with termed "built up," *i.e.*, the body and base made separately and screwed together.

All cases are coned to the front to facilitate loading and extraction. The metal of the case becomes gradually thinner towards the mouth.

A rim is formed at the rear end to prevent the case being forced too far into the chamber of the gun, and to facilitate extraction.

All Q.F. cartridge cases, with the exception of those used with Horse, Field and Siege equipment, are lacquered inside and out with transparent lacquer to prevent them deteriorating.

Cartridge cases for Horse, Field and Siege equipment are coated externally with a dull black lacquer, so that they will not glitter in the sunlight and thus show the position of the gun.

To enable this black lacquer to adhere to the case, the case is first given a rough surface by being sand-blasted.

Life of a Q.F. cartridge case.—Q.F. cases can be refilled after firing but, before this is done, as the case expands with the pressure on firing, they must be cleaned and re-formed, or *rectified.*

This operation of rectifying tends to weaken the case, as a considerable amount of metal is turned off the base and lower part of the body each time. To prevent accidents the number of rounds a Q.F. case is allowed to fire is limited to *six* when fired with cordite and *eleven* when fired with gunpowder.

To enable this limit to be adhered to a record is kept on the base of each case, showing the number of times filled. This record is termed the "*life of the case.*"

(For marking on the base see page 480.)

Cleaning Q.F. and 1-inch aiming rifle cartridges after firing.—Fired Q.F. and 1-inch aiming rifle cartridges must always be cleaned immediately after firing. If the cartridges have been fired with cordite or ballistite they must be soaked for 15 hours in clean fresh water containing ½ oz. of soda to the gallon, and afterwards well scrubbed until clean; they must then be rinsed in clean water and wiped perfectly dry. A longer soaking or the use of warm water will facilitate the cleaning and may be necessary in very cold weather. Soda may be used with cartridges which have been fired with powder, but its use is not essential in this case. After cleaning, cartridges must be mopped inside and outside with mineral jelly, or (in Naval Service) mineral grease, and repacked in the boxes in which they were supplied and returned to the Army or Naval Ordnance Department as soon as possible, the clips being replaced on the cartridges which have them. Fired cartridges are not on any account to be repacked in boxes containing unfired cartridges.

SECTION B.—FIXED AMMUNITION.

Cartridge, Q.F., 1-Pr. Common Shell, Mark III | C |.

Cartridge, Q.F., 1-pr. Common Shell, Mark III consists of a brass case with cap, igniter, charge, and fuzed projectile. The cartridge case is made of solid drawn brass with a projecting rim at the base, and is lacquered with transparent lacquer inside only.

The cap is made of copper, filled with 1·2 grains of special cap composition (containing no fulminate of mercury), pressed in, varnished, and covered with a tinfoil disc. The cap is contained in a brass cap chamber, having a raised anvil and three fire-holes, and is secured in position by the metal of the cap chamber being spun over it.

The cap chamber with cap is pressed into a hole in the centre of the base of the case.

The igniter, which is shellaced to the bottom of the case, is made of cambric and contains 5 drams of R.F.G.² powder.

The charge consists of 520 grains of cordite M.D., size 2¼, placed loose in the case.

The shell is made of cast iron, and is fitted for a nose fuze. It has a cannelure near the base by which it is secured in the case, and above it a copper driving band is pressed into a groove.

There is also a narrow copper steadying band pressed into a groove near the shoulder.

The shell is lacquered inside and contains a bursting charge of about 270 grains of F.G. powder. The weight of the shell is 1 lb. It is pressed into the mouth of the case and secured by the metal of the case being indented into the cannelure.

The fuze used with this ammunition is described and illustrated on page 280.

The Mark II cartridge is similar, but has a charge of 1 oz. 3¼ drams of cordite, size 3¾.

The Mark I cartridge differed from the above in having an igniter of nitrated canvas; a felt wad was used on top of the cordite charge.

CARTRIDGE, Q.F. 1-PR. DAY TRACER PROJECTILE, MARK II | C |.

The Cartridge, Q.F. 1-pr. with Day Tracer projectile is identical with Cartridge, Q.F. 1-pr. common shell, Mark III, described above, differing only in the projectile, which consists of common shell filled with Day Tracer liquid and plugged with the body of an old 1-pr. nose fuze, or a Plug, Fuze-hole, Q.F. 1-pr.

A hole in the base of the shell is bored as shown in the plate, for the "Plug, Day Tracer, Vent No. 2, Mark I."

The Mark I cartridge has a charge of 1 oz. 3¼ drams of cordite, size 3¾.

Packing of 1-pr. Q.F. Ammunition.

The cartridges are packed in a box, which holds 50 rounds, for Land Service. For Naval Service 66 rounds in "Box, Cartridge, Q.F., 12-pr. of 8 cwt."

Q.F. 6 AND 3-PR. AMMUNITION.

The following projectiles are used for making up ammunition for 6- and 3-pr. Q.F. guns :—

A *Steel* shell, filled and fuzed.

A Steel Practice shell, filled with salt and plugged.

A Steel Practice shot.

The 3-pr. also fires a common lyddite shell.

Q.F. 6- and 3-pr. Cartridge Cases.

The earlier marks of cartridge cases for Q.F. 6- and 3-pr. were prepared to take a percussion cap. A certain number of these cases, fitted with caps, are still being issued to service, but a large number of them have been altered by having a hole bored through the base and threaded to take a percussion primer (No. 2, Marks I or III). All new cartridge cases will be manufactured with a thicker base and will be fitted with a larger pattern of percussion primer than the converted cases (*i.e.*, No. 2, Marks II or IV).

LATEST MARKS OF SERVICE CARTRIDGES FOR Q.F. 6-PR. (STEEL SHELL).

CARTRIDGE, Q.F. 6-PR. CORDITE M.D., STEEL SHELL, MARK XIV | N |.

The Cartridge, Q.F. 6-pr. cordite M.D. steel shell, Mark XIV, consists of a brass case, percussion cap, charge of M.D. cordite, Mark VII igniter and fuzed shell.

The case.

The case, Mark III, is made of solid drawn brass; at the base a projecting rim is formed to prevent the cartridge case being pushed too far into the chamber, and also to admit of easy extraction; a hole is bored through the centre of the base, enlarged at the lower end to take a cap chamber with cap. The case is lacquered inside and out with transparent lacquer.

Percussion cap.

The cap is contained in a cap chamber; the latter is made of brass with a raised anvil pierced with three fire-holes. The cap is made of copper, and contains 1·2 grains of cap composition, pressed in and varnished and covered with a tinfoil disc; it is secured in the cap chamber by the metal of the latter being burred over it. The first cap used in making up 6- and 3-pr. ammunition (Mark I) was found to be rather thick, and not sensitive enough to the blow from the striker. It was superseded by a thinner cap (Mark II), and all cartridge cases fitted with these new caps are distinguished by being stamped on the base with the numeral *one* in a circle, thus— (I).

The charge.

The charge, Mark IV, consists of a bundle of M.D. cordite, size 4¼, 8 ozs. 11½ drs., cut about 9·5 inches long, tied together with silk sewing. Two fins, each containing 1 dram of cordite cut 2·5 inches long, tied loosely in the middle, pass through the charge (at right angles to each other) to keep it central in the case.

The igniter.

The igniter (Mark VII) consists of 4 drams of R.F.G.[2] powder contained between two discs of shalloon; it is secured to the base of the charge by silk sewing as shown in the drawing.

Steel shell.

The shell, Mark V, is made of forged steel, pointed, the head being struck with a radius of nearly three calibres. Near the base a cannelure is formed round the shell for the purpose of securing it to the case, and below this the diameter is slightly reduced to facilitate insertion into the mouth of the case. The shell is rotated by a broad Vavasseur band; it is oil hardened and the interior is velvrilled and filled with 4 ozs. of Q.F. shell F.G. powder; the hole in the base is threaded with a left-hand screw thread to take the Hotchkiss base percussion fuze (see page 298). The body of the shell below the driving band is varnished and the shell is pressed into the mouth of the case until the edge of the case bears against the driving band; it is then secured by three indents, which force the brass case into the cannelure round the base of the shell.

Protecting clip. (*See Fig.* 136.)

The percussion caps and primers in 6- and 3-pr. cartridges are protected by a "*Clip, cartridge, Q.F.*" Those of the present pattern

(Mark III) are of brass, with three arms and a central dome. The arms at the end are bent and clip on to the projecting base of the cartridge; the dome protects the cap from an accidental blow. Marks I and II clips are obsolete.

Fig. 136.

CLIP, CARTRIDGE, Q.F., 6-PR., MARK III | C |.

Scale $\frac{1}{1}$.

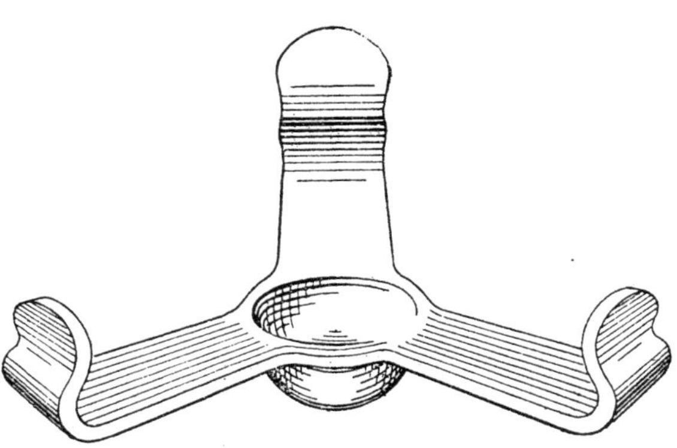

Cartridge, Q.F. 6-Pr., Cordite M.D., Steel Shell, Mark XIII | C |.

The *Cartridge, Q.F. 6-pr., Cordite M.D., Steel Shell, Mark XIII* consists of a Mark IV case fitted with a No. 2, Mark IV percussion primer, charge of M.D. cordite, and a fuzed shell.

The Mark IV case differs from the Mark III in having a thicker boss in the base of the case through the centre of which a hole is bored and screw threaded to receive a *Primer, percussion, No.* 2, *Mark IV*. (See Plate LXXXI.)

The charge.

The charge (Mark III) consists of a bundle of M.D. cordite, 8 ozs. 11½ drs., size 4¼, cut to the required length and tied with silk sewing in three places. The sticks of cordite in the centre of the charge are cut shorter than the outer layers so as to fit over the projecting boss in the base of the case. No igniter is used, and the charge is made up without the cordite fins, as the magazine cylinder on the end of the primer acts as an igniter to the charge, and, fitting into the base of the charge, retains it in position in the case.

Cartridge, Q.F. 6-Pr., Cordite M.D., Steel Shell, Mark XII | C |.

The *Cartridge, Q.F. 6-pr., Cordite M.D., Steel Shell, Mark XII*, consists of a Mark III* case fitted with a No. 2, Mark III percussion primer, a Mark II charge of M.D. cordite and a fuzed shell.

The cartridge cases used in making up this Mark of 6-pr. ammunition were originally intended to take percussion caps; they have been converted to take the small type of Percussion primer, No. 2, Mark III. (See Plate LXXVIII.)

The charge differs from that described for the Mark XIII cartridge in the sticks of cordite being all cut to the same length.

CARTRIDGE, Q.F. 6-PR., CORDITE, STEEL SHELL, MARK XI | L | .

CARTRIDGE, Q.F. 6-PR., CORDITE M.D., STEEL SHELL, MARK X | N | .

CARTRIDGE, Q.F. 6-PR., CORDITE, STEEL SHELL, MARK IX | C | .

In making up the above Marks of Q.F. 6-pr. cartridges the following may be used :—

(*a*) Mark III case with Mark II cap.
(*b*) Mark III* case with No. 2, Mark I primer.
(*c*) Mark IV case with No. 2, Mark II primer.

MARK XI CARTRIDGE.

The *Cartridge, Q.F. 6-pr., Cordite, Steel Shell, Mark XI | L |*, consists of a brass case with a cap or percussion primer, igniter, a charge of Mark I cordite, a paper cylinder and a fuzed shell.

The charge.

The charge consists of 7¾ ozs. of Mark I cordite, size 5. The sticks of cordite, about 12 inches in length, are folded double and tied together by shalloon braid in two places.

The igniter.

The igniter (Mark VI) consists of two outer and an inner disc of shalloon stitched together around the edge, containing 4 drams of S.F.G.[2] powder. Four ·35-inch shalloon braids are sewn radially to the igniter for securing it to the charge, two of the braids being passed through the charge at right angles to each other and tied to the braids on the opposite side.

Paper cylinder, 5·15-*inch, Mark I.*

The Mark I cylinder is made of brown paper, pierced with holes; to the rear end of the cylinder are attached two perforated millboard discs. A strip of glazed board passes transversely through the paper cylinder above the millboard discs, and is secured by a copper tack at each end. A glazed-board disc is then riveted to the base of the cylinder by three copper tacks.

NOTE.—The earlier patterns of paper cylinders used in making up the above cartridges had the millboard discs secured with glue and copper wire. They were known as "*Cartridge, Q.F. 6-pr., Cordite, Cylinder, Paper, Mark IV,*" instead of being designated according to the length of the cylinder.

Mark X Cartridge.

(*Plate* LXXXII.)

The *Cartridge, Q.F.* 6-*pr., Cordite M.D., Steel Shell, Mark X* | *N* |, consists of a case with cap or primer, igniter, a charge of M.D. cordite and a fuzed shell.

The charge.

The charge, Mark I, consists of a bundle of M.D. cordite, 8 ozs. 11½ drams, size 4¼, tied with shalloon braid in three places.

The igniter.

The igniter, Mark V, consisting of 4 drams of R.F.G.[2] powder contained between two discs of shalloon, to which two shalloon braids are sewn, is secured to the base of the charge by one of the braids being passed over the middle tie, through the centre of the charge, and tied on the opposite side to the other braid. No paper cylinder is used.

Mark IX Cartridge.

The *Cartridge, Q.F.* 6-*pr., Cordite, Steel Shell, Mark IX* | *C* |, is made up in the same way as the Mark XI cartridge (see Plate LXXXII), differing from it only in the igniter, which is an earlier pattern (Mark IV instead of Mark VI).

Mark IV Igniter.

The Mark IV igniter contains 1½ drams of waterproofed guncotton yarn, and has only one disc of shalloon on the underside, next the cap or primer, instead of two, as in the Mark VI.

Note.—A few Mark IV igniters have been issued filled with S.F.G.[2] powder in lieu of the 1¼ drams of guncotton yarn. Cartridges so fitted can be distinguished by "S.F.G.[2]" stencilled on the base and (S.P.) stencilled on the package.

In future filling, Mark IV igniters will be filled with 4 drams of R.F.G.[2] powder.

Mark VIII Cartridge.

The *Cartridge, Q.F.* 6-*pr., Cordite, Steel Shell, Mark VIII* | *C* |, differs from the Mark IX as follows :—

> None of the cartridges are fitted with a percussion primer; they all have a cap chamber with cap pressed into a recess in the base. The cordite charge is tied with silk sewing in three places, instead of having two ties of shalloon braid; the igniter is secured to the bend of the cordite charge by silk sewing, and a (Mark III) shorter paper cylinder is used.

For particulars of earlier marks, see Table 34.

Q.F. 6-Pr. Practice Cartridges.

The *practice* cartridges for the Q.F. 6-pr. are made up in a similar manner to the various *Service* cartridges already described, differing from them in the projectiles, which are either steel shell filled with salt and plugged, or a practice shot. The plugs used with the practice shell are "*Plugs, Base Shell, No.* 3," or an old Hotchkiss fuze body.

These plugs and fuze bodies are stamped on the base with the letter "P," to prevent fuzed shell on recovery being taken for plugged shell.

Practice shell or shot have a yellow band around the centre and the shell have stencilled on them in white paint the word "Salt."

The following Q.F. 6-pr. *Practice Cartridges*, with the exception of the projectiles used, correspond to the following Marks of *Service cartridges* :—

PRACTICE.		SERVICE.
CARTRIDGES, Q.F. 6-PR., CORDITE M.D., PRACTICE.		CARTRIDGES, Q.F. 6-PR., CORDITE M.D., STEEL SHELL.
Mark XV (with practice shot)	... ...	= Mark XIV.
" IX (with plugged shell)	... ...	
" XIV (with practice shot)	... ...	= Mark XIII.
" XII (with plugged shell)	... ...	
" XIII (with practice shot)	... ...	= Mark XII.
" XI (with plugged shell)	... ...	
" VIII (with plugged shell)	... ...	= Mark X.
CARTRIDGES, Q.F. 6-PR., CORDITE, PRACTICE.		CARTRIDGES, Q.F. 6-PR., CORDITE, STEEL SHELL.
Mark X (with plugged shell)	... ...	= Mark XI.
" VII (with plugged shell)	... ...	= Mark IX.

Q.F. 6-PR. SUB-CALIBRE CARTRIDGES.

In order to ensure that only ammunition having annealed shell or shot is used in sub-calibre guns, the nomenclature shown below has been approved for 6-pr. ammunition so fitted.

"*Cartridge, Q.F. Sub-calibre 6-pr. Cordite.*"

The following Q.F. 6-pr. sub-calibre cartridges, with the exception of the projectiles used, correspond with the following Marks of Service cartridges :—

SUB-CALIBRE.	SERVICE.
Cartridge, Q.F. Sub-calibre.	*Cartridge, Q.F. 6-Pr.*
6-Pr., Cordite, Mark I (with plugged shell) ...	Cordite, Steel Shell, Mark IX.
6-Pr., Cordite, M.D., Mark II (with plugged shell)	Cordite, M.D., Steel Shell, Mark XIV,
6-Pr., Cordite, M.D., Mark III (with plugged shell)	Cordite, M.D., Steel Shell, Mark XII.
6 Pr., Cordite, M.D., Mark IV (with plugged shell)	Cordite, M.D., Steel Shell, Mark XIII.

Packing of 6-pr. Ammunition.

The above 6-pr. Q.F. Service and practice cartridges are packed 11 in a box, which is painted stone colour ; the lids of boxes containing practice cartridges are painted yellow.

For particulars of these boxes, *see* page 486.

Special Q.F. 6-pr. Cartridges (Naval Service).

(For H.M.S. Swiftsure and Triumph.)

Cartridge, Q.F. 6-pr., Steel Shell, H.M.S. Swiftsure, Mark II | N | .

Cartridge, Q.F. 6-pr., Steel Shell, H.M.S. Triumph, Mark II | N | .

The above 6-pr. cartridges have a special brass case, and a charge of 1 lb. $2\frac{5}{16}$ oz. of cordite M.D., size 8, made up in a similar way to the charge for the Q.F. 6-pr., Mark X Service cartridge (Plate LXXXII), but a perforated paper cylinder, with a felt wad and glazed board disc attached, is used to fill up the space between the charge and the base of the shell. The cartridge issued to H.M.S. Swiftsure is fitted with a special shell, slightly shorter than the service steel shell, Mark V (§12251). This shell has a special type of driving band.

The cartridge issued to H.M.S. Triumph is fitted with the ordinary 6-pr. steel shell, Mark V.

The Mark I cartridge had a different pattern of paper cylinder on top of the cordite charge.

Practice Cartridges, Q.F. 6-pr.

H.M.S. Swiftsure and Triumph.

The charge consists of 14 oz. cordite M.D., size 8, made up in the same way as the service charge, but a longer paper cylinder is used.

The cartridges for H.M.S. Swiftsure have a solid shot with a truncated head, and a special type of driving band.

The cartridges issued to H.M.S. Triumph are fitted with any Mark of 6-pr. steel shell, filled with salt, and plugged.

Packing.

Q.F. 6-pr. cartridges for H.M.S. Swiftsure and Triumph are packed in " Box, ammunition, 6-pr., special | N | " holding 9 rounds. (*See* page 486.)

Q.F. 3-pr. Ammunition.

Cartridge, Q.F. 3-pr., Cordite M.D., Steel Shell, Mark XI | C | .

The *Cartridge, Q.F. 3-pr., Cordite M.D., Steel Shell, Mark XI | C |*, consists of a brass case, percussion primer, M.D. cordite charge, and a fuzed shell.

The Case and Primer.

In making up this Mark of cartridge, the following cases may be used :—

(*a*) A Mark II* case fitted with a No. 2, Mark III primer.
(*b*) A Mark III case fitted with a No. 2, Mark IV primer.

The Mark II 3-pr. cartridge cases were intended to take a percussion cap ; when converted to take the primer mentioned above, they are distinguished by having a star (*) added to their numeral.

The charge.

The charge (Mark II) consists of a cylindrical bundle of 7 ozs. 4 drams of Cordite M.D., Size 4½, secured by seven ties of silk sewing, with fins, to keep the charge central in the case. The fins are formed by two small bundles of short lengths of cordite, which pass through the cylindrical bundle near each end so as to project at right angles to each other. No igniter is used; the magazine cylinder on the percussion primer projects into the base of the cordite charge and acts as an igniter.

The shell.

The steel shell is similar in design to the 6-pr. steel shell; it has a bursting charge of 2 ozs. of Q.F. shell F.G. powder.

CARTRIDGES, Q.F. 3-PR., CORDITE, STEEL SHELL, MARK X | L |.

CARTRIDGES, Q.F. 3-PR., CORDITE M.D., STEEL SHELL, MARK IX | N |.

CARTRIDGES, Q.F. 3-PR., CORDITE, STEEL SHELL, MARK VIII | C |.

In making up the above Marks of Q.F. 3-pr. cartridges the following may be used :—

(*a*) A Mark II case with a Mark II percussion cap.
(*b*) A Mark II* case with a No. 2, Mark I primer.
(*c*) A Mark III case with a No. 2, Mark II primer.

The above cartridges are similar to, but a Mark behind the Q.F. 6-pr. cartridges, thus :—

The "*Cartridge, Q.F. 3-pr., Cordite, Steel Shell, Mark X*" is similar to the "*Cartridge, Q.F. 6-pr., Cordite, Steel Shell, Mark XI,*" and the "*Cartridge, Q.F. 3-pr., Cordite M.D., Steel Shell, Mark IX*" is made up in the same way as the "*Cartridge, Q.F. 6-pr., Cordite M.D., Steel Shell, Mark X,*" and so on.

The Marks VIII and X 3-pr. cartridges have a charge of 6⅝ ozs. Cordite, Mark I, Size 5; the Mark IX has a charge of 7¼ ozs. Cordite M.D., Size 4¼.

For particulars of earlier Marks, see Table 34.

CARTRIDGE, Q.F. 3-PR., LYDDITE SHELL.

The Cartridge, Q.F. 3-pr., Lyddite Shell, Mark III, *N.T.* | *N* |, is made up in the same way as the Mark XI 3-pr. cartridge already described.

The Marks III or V Lyddite shell, with No. 19A D.A. Impact fuze, and External Night Tracer are used.

The Cartridge, Q.F. 3-pr. Lyddite Shell, Mark II | *N* |, differs from the Mark III cartridge in being fitted with either a Mark II or IV shell, which are lighter and not fitted with an External Night Tracer.

The Cartridge, Q.F. 3-pr. Lyddite Shell, Mark I | *N* | is fitted with a Mark I shell, the fuze-hole bush of which has a parallel screw thread, and takes the No. 19 D.A. Impact fuze.

3-pr. Lyddite Shell.

There are five Marks of 3-pr. Lyddite shell; Mark I has a parallel screw-thread in the fuze-hole bush, and takes the No. 19 D.A. Impact fuze or No. 4 plug, fuze-hole, special.

It takes a bursting charge of $4\frac{4}{16}$ oz. of lyddite, and an exploder of T.N.T. in batiste bags in the nose as shown in Plate XXXVIII, page 209.

The Marks II to V shell have a tapered thread in the fuze-hole bush and take the No. 19A D.A. Impact fuze or No. 4A plug, fuze-hole, special.

The bursting charge is $4\frac{4}{16}$ oz. of lyddite; the exploder of 3 drams of T.N.T. in a batiste or vulcanised cashmere bag is placed in a small asbestos cylinder fitting into a cavity in the lyddite.

The Marks III and V shell are fitted with External Night Tracers.

(For particulars of shell, *see* Table 23, page 212.)

Q.F. 3-pr. Practice Cartridges.

The practice cartridges for the Q.F. 3-pr. are made up in a similar manner to the practice cartridges for the Q.F. 6-pr.; they are fitted with a plugged shell or a practice shot.

The Marks XIII and XIV practice cartridges have projectiles fitted with External Night Tracers.

The following Q.F. 3-pr. practice cartridges, with the exception of the projectiles, correspond to the following Marks of Q.F. 3-pr. Service cartridges:—

PRACTICE.	SERVICE.
Cartridges, Q.F. 3-Pr., Cordite M.D., Practice.	Cartridges, Q.F. 3-Pr., Cordite M.D., Steel Shell.
Mark XIV, N.T. (with practice shot and tracer)	Mark XI.
" XIII, N.T. (with plugged shell and tracer)	
" XII (with practice shot)	
" XI (with plugged shell)	
" VIII (with plugged shell)	Mark IX.
Cartridges, Q.F. 3-Pr., Cordite, Practice.	Cartridges, Q.F. 3-Pr., Cordite, Steel Shell.
Mark XV (with practice shot)	Mark X.
" X (with plugged shell)	
" VII (with plugged shell)	Mark VIII.

Q.F. 3-pr. Sub-calibre Cartridges.

The following Q.F. 3-pr. sub-calibre cartridges, with the exception of the projectiles used, correspond with the following Marks of Service cartridges:—

Sub-Calibre.	Service.
Cartridges, Q.F. Sub-calibre.	*Cartridges, Q.F. 3-Pr.*
3-Pr. Cordite, Mark I (plugged shell) ...	Cordite, Steel Shell, Mk. VIII.
3-Pr. Cordite, M.D., Mark II (plugged shell)	Cordite, M.D., Steel Shell, Mk. IX.
3-Pr. Cordite, M.D., Mark III (plugged shell)	Cordite, M.D., Steel Shell, Mk. XI.

Packing of Q.F. 3-pr. Ammunition.

Q.F. 3-pr. cartridges are packed 16 in a box, which is painted lead colour for Practice and Steel Shell Cartridges, and yellow for Lyddite Cartridges. Boxes containing Practice Ammunition have the lids only painted yellow. (See page 487.)

Q.F., 3-pr. Vickers' Cartridges.

Service.

Cartridge, Q.F., 3-pr., Vickers', Steel Shell, Mark IV | N |.

The above cartridge consists of a case, percussion primer, charge of M.D. cordite, paper cylinder and a fuzed shell.

The Case and Primer.

The brass cartridge case is a Mark II. It differs from the Mark I in having a thicker boss, and the base being bored and screw-threaded to take a Primer, Percussion, No. 2, Mark IV.

Charge and Paper Cylinder.

The charge (Mark III) consists of a cylindrical bundle of 13 ozs. 6 drams Cordite M.D., size 8, tied with silk sewing in three places. The sticks of cordite in the centre of the charge are shorter than the outer layer of sticks, forming a recess at one end to suit the boss in the case. An igniter is not used; the magazine cylinder of the percussion primer projecting into the base of the charge keeps it central.

In making up this cartridge, a short paper cylinder, 3·75 inches long, is used to fill up the space between the top of the charge and the base of the shell.

The Shell.

The shell is the same as that used with the 3-pr. Hotchkiss, and takes the Mark IV Hotchkiss fuze.

Cartridge, Q.F., 3-pr., Vickers', Steel Shell, Mark III | N |.

The Mark III cartridge differs from the Mark IV described above, in the charge, case and primer.

The Mark II charge is used; it differs from that in the Mark IV cartridge in the sticks of cordite being all cut to the same length.

The case is a Mark I, converted to Mark I*, by being altered to take a primer instead of a cap. The primer used is No. 2, Mark III.

Cartridge, Q.F., 3-pr., Vickers', Steel Shell, Mark II | N |.

The Mark II cartridge differs from the Mark III in the charge (Mark I) being fitted with an igniter of 4 drams of powder. The primer used is No. 2, Mark II.

The Mark I cartridge had a percussion cap instead of a percussion primer, and the paper cylinder on top of the charge was ·4 inch longer.

Cartridge, Q.F., 3-pr., Vickers', Lyddite Shell, Mark III N.T. | N |.

The Cartridge, Q.F., 3-pr., Vickers', Lyddite Shell, Mark III N.T. | *N* | consists of a Mark II case fitted with a No. 2, Mark IV percussion primer, a Mark III charge of M.D. cordite, paper cylinder and a fuzed Mark III or V lyddite shell, No. 19A D.A. Impact fuze and an External Night Tracer.

The cartridge is made up in the same way as the Mark IV 3-pr. Vickers' cartridge with steel shell. The lyddite shell is the same as that used with the Mark III 3-pr. Hotchkiss cartridge and takes the same fuze.

CARTRIDGE, Q.F. 3-PR., VICKERS', LYDDITE SHELL, MARK II | N |.

The Mark II cartridge differs from the Mark III in the shell (Mark II), which is lighter and is not fitted with an External Night Tracer.

The Mark I cartridge is fitted with a Mark I shell, which differs from the Mark II shell in the fuze-hole bush, which has a parallel screw thread instead of being tapered; it takes the No. 19 D.A. Impact fuze.

Q.F. 3-PR., VICKERS', PRACTICE CARTRIDGES.

There are two practice charges for use with the Q.F. 3-pr., Vickers', viz., a "Full" and a "Reduced."

The "Reduced charge" is for use with 3-pr. guns mounted on top of turrets or barbette mountings.

To distinguish readily between the full and reduced practice charges, the word "Full" or "Reduced" is stencilled in red on the base of the respective cartridges.

CARTRIDGES, Q.F. 3-PR., VICKERS', PRACTICE, FULL CHARGE.

The practice cartridges for the Q.F. 3-pr., Vickers', are made up in the same way as the Service cartridges already described, but are fitted with a plugged shell or a practice shot.

The following *full charge* practice cartridges, with the exception of the projectiles, correspond to the following Marks of Q.F. 3-pr., Vickers', Service cartridges :—

CARTRIDGES, Q.F. 3-PR., VICKERS'.

PRACTICE (FULL CHARGE).	STEEL SHELL.
Mark VII (with practice shot)	Mark IV.
" V (with plugged shell)	
" VI (with practice shot)	Mark III.
" IV (with plugged shell)	
" III (with plugged shell)	Mark II.
" II (with practice shot)	Mark I.
" I (with plugged shell)	

CARTRIDGE, Q.F. 3-PR., VICKERS', PRACTICE, REDUCED CHARGE, MARK VI | N |.

The following are used in making up this cartridge :—

A Mark I empty case with a Mark II percussion cap.
A Mark IV charge with igniter.
A Mark I 3-pr. practice shot.

The charge is similar to the Mark II charge, with the following differences :—

(i) All ties are of silk sewing.
(ii) A *cambric* bag with shalloon igniter is used to enclose the short sticks of cordite at the base of the charge.
(iii) A practice shot (§ 15384 L. of C.) is used.

Cartridge, Q.F. 3-pr., Vickers', Practice, Reduced Charge, Mark V | N | .

The Mark V cartridge differs from the Mark VI described above in being fitted with a plugged shell instead of a practice shot.

Cartridge, Q.F. 3-pr., Vickers', Practice, Reduced Charge, Mark IV | N | .

The charge, Mark III, for the above cartridge consists of 6 ozs. 14 drams of Cordite M.D., size $4\frac{1}{4}$, made up into a cylindrical bundle, secured by seven ties of silk sewing. The charge is made up with fins to keep it central in the cartridge case; these are formed by two small bundles of cordite, which pass at right angles to each other through the central bundle, near each end. No igniter is used. The following may be used in making up this Mark of cartridge :—

Mark I* case with No. 2, Mark III primer.
Mark II case with No. 2, Mark IV primer.

The cartridge is fitted with a practice shot.

Cartridge, Q.F. 3-pr., Vickers', Practice, Reduced Charge, Mark III | N | .

The Mark III cartridge differs from the Mark IV in being fitted with a plugged shell instead of a practice shot.

Cartridge, Q.F. 3-pr., Vickers', Practice, Reduced Charge, Mark II | N | .

The following may be used in making up the Mark II cartridge :—

(*a*) Mark I case with Mark II cap.
(*b*) Mark I* case with No. 2 Mark, I primer.
(*c*) Mark II case with No. 2 Mark, II primer.

The charge is made up in a different manner to the Marks III and IV reduced practice cartridges described above. It consists of a bundle of M.D. cordite about $14\frac{1}{2}$ inches in length tied with silk sewing near the lower end; around the lower end of the charge are placed several layers of cordite about $2\frac{3}{4}$ inches in length, also tied with silk sewing. The upper part of the charge is tied in three places with shalloon braid. The lower portion of the charge is enclosed in a shalloon bag carrying an igniter of 4 drams of R.F.G.[2] powder.

The Mark I cartridge differs from the Mark II in the charge, the whole of which was enclosed in shalloon.

Packing of 3-pr., Vickers' Cartridges.

Q.F. 3-pr., Vickers' cartridges are packed 16 in "Box, Ammunition, Q.F. 3-pr., Vickers'" (*see* page 486).

Q.F. 13 and 18-pr. Ammunition.

Cartridge, Q.F. 13-pr., Shrapnel Shell, Mark II | L |.
(*Plate* LXXXIII.)

Cartridge, Q.F. 13-pr., Shrapnel Shell, Mark II | L | Fuzed, consists of a brass case, percussion primer, charge of M.D. cordite, shrapnel shell, and No. 80, T. and P. fuze, with fuze cover.

The case.—The case is made of solid drawn brass, the head stamped into shape, and formed with a projecting rim to prevent it being pushed too far into the chamber and to admit of extraction. A hole through the centre of the base is screw-threaded to take the primer, the case is sand blasted inside and out, and coated externally with a dull black lacquer so as not to glitter in the sun and so reveal the position of the gun.

The primer.—The percussion primer used is No. 1. For details of Marks I*, and II primers, *see* page 379.

The charge.—The charge consists of a bundle of cordite M.D., size 8, 1lb. $4\frac{1}{16}$ oz. tied in three places with shalloon braid. The sticks in the centre of the charge are slightly shorter than the outer layer, thus forming a recess at one end to fit over the projecting part of the primer.

The shrapnel shell.—The shrapnel shell (Mark III) is made of F.S. Its length is about 2·6 calibres; the walls near the base are thickened, forming a shoulder on which rests a steel disc; below the disc is placed a tin cup for the bursting charge. A hole is bored through the centre of this disc, and screwed into it is the lower end of a brass central tube, which passes through the disc and projects into the mouth of the tin cup. The shell contains about 236 mixed metal bullets (7 parts lead, 1 part of antimony) 41 to the lb., the spaces between the bullets being filled up with resin. The front end of the shell is closed by a flanged gunmetal bush, or "fuze socket," screwed in. The fuze socket is screw-threaded in the interior to the 2-inch gauge to take the T. and P., No. 80 fuze, a hole being bored through the bottom of the socket for the top of the central tube. To prevent the resin working up into the fuze socket the top of the central tube is soldered to the latter.

The shell is rotated by means of a copper band pressed into an undercut groove with waved ribs near the base. The driving band is of the special narrow type (No. 14, Plate XVI.); it is slightly enlarged near the lower edge to form a bearing for the mouth of the case, below which a groove is formed for the front of the case to be coned into all the way round.

The bursting charge is $1\frac{1}{4}$ oz. of F.G. powder, which fills the tin cup, and the central tube is filled with six perforated powder pellets;

the bottom ends of the two lower powder pellets are covered with discs of muslin, and a disc of shalloon shellaced in the bottom of the fuze-hole socket prevents the pellets and F.G. powder working up into the fuze hole. The shell is painted lead colour.

The fuze.—The fuze used with the shrapnel shell is the T. and P., No. 80. For particulars of this fuze and fuze cover, *see* page 322.

Safety clip.—The clip is made of brass, cross-shaped so as to form four arms, the ends of which are turned in to form clips to engage with the rim of the cartridge case. One arm is painted red, and is slightly longer than the others, the clip portion being differently shaped, so as to spring over the rim of the cartridge. The other three arms are sand blasted and black lacquered. It has a canvas loop for withdrawing the cartridges from the baskets in the ammunition boxes of the limbers and wagons.

The clip protects the cap of the percussion primer.

Marking.—The following information will be found stamped and stencilled on the base of the cartridge.

Stamped on the base of the case :—

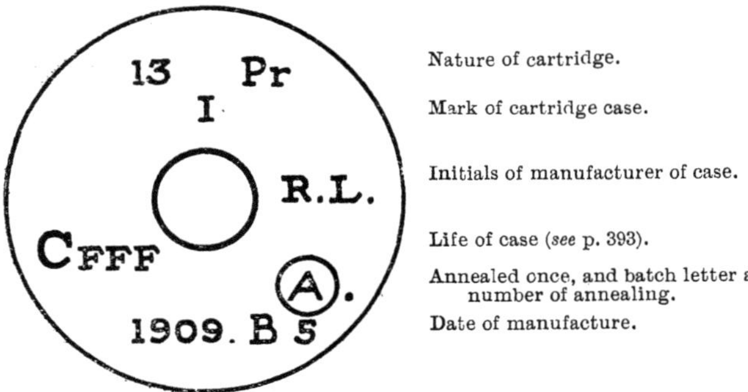

Stencilled in red on the base of the case :—

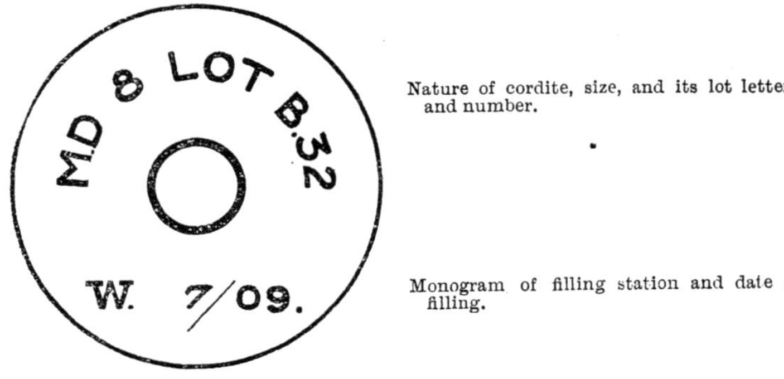

Stamped on the body of the shell :—

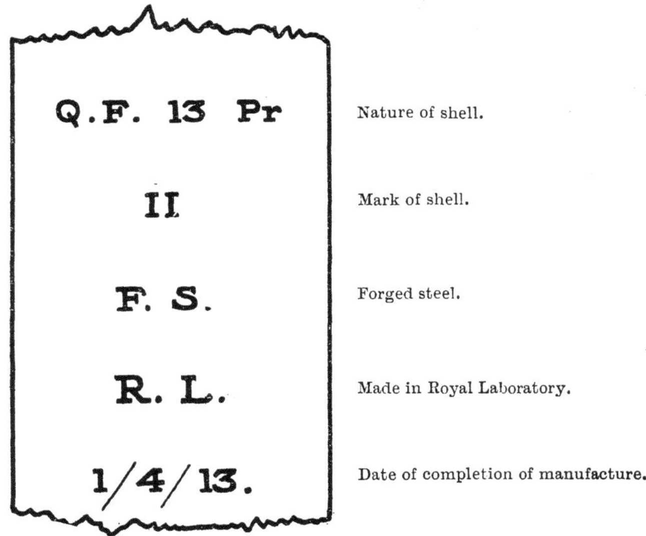

Stencilled in red on the body of the shell :—

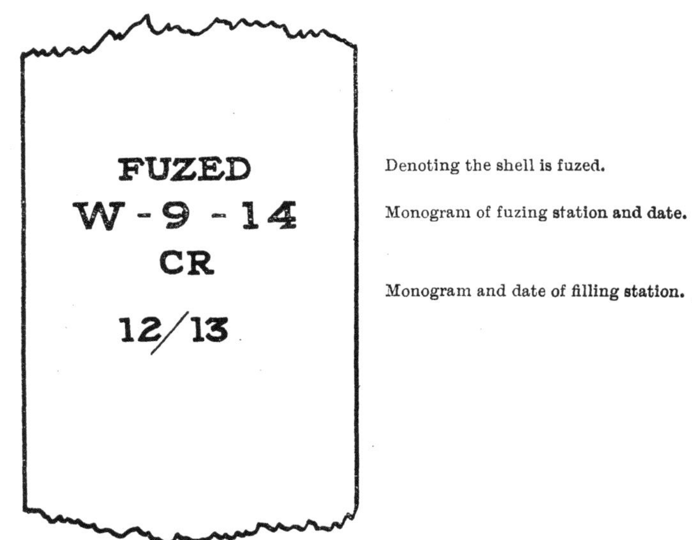

The total weight of the round for the 13-pr. Q.F. is 16 lb. 7 oz., the weight of the filled and fuzed shell being $12\frac{1}{2}$ lb.

Mark I 13-pr. Cartridge.

Cartridge, Q.F. 13-pr., shrapnel shell, Mark I | L | *fuzed,* differs from the Mark II in having a charge consisting of a bundle of cordite M.D., $2\frac{4}{16}$ oz., size $2\frac{1}{4}$, surrounded by 1 lb. $1\frac{10}{16}$ oz., size 8, cordite M.D.

Cartridge, Q.F. 13-pr. Shrapnel, Mark III | S.I. | .

The Mark III cartridge for the Q.F. 13-pr. differs from the Mark II in the charge, which consists of 1 lb. 1 oz. 13 dr. of Cordite, Mark I, size $7\frac{3}{4}$, instead of a charge of Cordite M.D.

This cartridge is Special for India.

Earlier Marks of Q.F. 13-pr. Shrapnel Shell.

Mark II Shell.

The Mark II shell differs from the Mark III in having a different type of driving band (No. 13), and a cannelure is formed round the shell near the base into which the cartridge case is indented by four indents $1\frac{1}{4}$ inches in length.

Mark I Shell.

The Mark I shrapnel shell for 13-pr. Q.F. differs from the Mark II already described, in the head being struck with a radius of $1\frac{1}{2}$, instead of two diameters; the walls at the base are weaker, and the tin cup and steel disc are slightly different in shape.

The Mark I shell is painted black, the Marks II and III lead colour, in order to make them easily distinguishable from each other, as the Marks II and III shell will range further than the Mark I.

Earlier Marks of Fuze Covers.

The fuze cover used with shell fitted with Marks I to III, No. 80 T. and P. fuzes is made of brass, in two parts with a leather washer; the top portion consists of a dome-shaped cap which fits over the fuze; the lower portion consists of a screwed ring and a base ring held together by a tin band soldered on.

The fuze and fuze cover are fitted to the shell as follows:—

The dermatine washer is placed on the fuze socket; on it is placed the lower portion of the fuze cover, the base ring resting on the dermatine washer, with its rim fitting in the groove in the fuze socket; the fuze itself is then screwed into the shell. The fuze body bears against the base ring, compressing the dermatine washer so as to make a damp-proof joint. The fuze is then kept from unscrewing by a set screw, which passes through the fuze socket and enters a recess drilled in the threaded part of the fuze body; the threads of the cap are then coated with Pettman's cement and the cap screwed tightly on to the screwed ring.

Immediately before loading the tin band is torn off; the cap and the screwed ring thus become detached, and the shell is ready for loading.

Packing.

Q.F. 13-pr. cartridges are packed in Box, Ammunition, Q.F. 13-pr., Marks I to III, containing four complete rounds.

(For description of box, *see* page 488.)

Cartridge, Q.F. 18-pr., Shrapnel Shell, Mark I.

The cartridge, Q.F. 18-pr., shrapnel shell, Mark I, is made up in the same way as that already described for the Q.F. 13-pr., Mark II, differing only in weight and dimensions.

The charge.—The charge consists of 1 lb. $6\frac{15}{16}$ oz. of cordite, M.D., size 8.

The shrapnel shell.—The Marks II and III shrapnel shells contain about 375 mixed metal bullets (41 to the lb.), and the central tube contains 8 instead of 6 perforated powder pellets.

The Marks II and III shells are painted *lead colour*, and when filled weigh $18\frac{1}{2}$ lb. The total weight of the complete round is 22 lb. $13\frac{15}{16}$ oz.

The Mark III shell has no cannelure round the base; the mouth of the case is coned into the groove in the rear portion of the driving band.

The Mark II shell differs from the above in having a cannelure round the base into which the cartridge case is indented. It has a different type of driving band.

The Mark I shell is similar to the Mark I, Q.F. 13-pr. shrapnel shell, differing only in dimensions (*see* Table 34).

Cartridge, Q.F. 18-pr. Shrapnel, Mark II | S.I. | .

The above cartridge is Special for use in India; it differs from the Mark I cartridge in having a charge of 1 lb. 4 oz. Cordite, Mark I, size $7\frac{3}{4}$, instead of a charge of Cordite M.D.

Packing.

Cartridges for the Q.F. 18-pr. are packed in a similar way to the Q.F. 13-pr. cartridges, in "Box, Ammunition, Q.F. 18-pr., Mark I," containing 4 rounds.

Cartridge, Q.F. 13-pr., Star Shell.

The Cartridge, Q.F. 13-pr., Star Shell, Mark I | L | consists of a brass case and primer, charge, paper cylinder and filled shell.

The same cartridge case and primer are used as with the shrapnel shell. The charge consists of 6 oz. 10 dr. of M.D. cordite, size $4\frac{1}{4}$, tied in two places with shalloon braid. The central sticks of cordite are shorter at the lower end to fit over the primer, and the charge is kept in position by a perforated paper cylinder to the underside of which is attached a glazed board disc.

The shell, Mark II, is made of steel recessed in the base to receive a bursting charge of $3\frac{1}{4}$ dr. of R.F.G.[2] powder contained in a shalloon bag primed with quick match. A metal central tube perforated with 12 holes is screwed into a wrought iron disc resting over the bursting charge.

The interior of the shell is velvrilled and lined with brown paper; it contains 10 stars in two tiers of 5; a perforated iron disc covered with a felt washer separates the tiers. The iron disc is supported by wood wedges placed between the stars, and is prevented from turning by means of two projections or feathers fitting into two featherways cut down the inside of the shell.

The head is lined with wood, and is attached to the body by 4 brass screws and 4 steel twisting pins. A felt wad is placed between the wood block and the top tier of stars. A gunmetal fuze hole bush is fitted to the head, threaded internally to the G.S. gauge to take the No. 25 time fuze. The shell is painted black, and is secured to the case in a similar manner to the shrapnel shell. Stencilled round the head is a red ring denoting "filled," and on the body a white disc with a red star denoting a star shell.

The Mark I Star shell differs from the Mark II in having a cannelure near the base, and is fitted with a different type of driving band.

The weight of the shell is 7 lb. 5 oz.

Total weight of cartridge, 11 lb. 0$\frac{6}{16}$ oz.

Cartridge, Q.F. 18-pr., Star Shell.

Cartridge, Q.F. 18-pr., Star Shell, Mark I | *L* | .—This cartridge differs from the Cartridge, Q.F. 13-pr., star shell, in dimensions and weight (*see* Plate LXXXIV.) The Mark I shell has a cannelure near the base, and is secured by indenting ; the Mark II shell has no cannelure, and is secured by coning the mouth of the case into a groove in the driving band. The charge is 8 oz. of cordite M.D., size 4$\frac{1}{4}$. The shell weighs 10 lb. 3$\frac{1}{4}$ oz. Total weight of cartridge is 14 lb. 2$\frac{1}{4}$ oz.

Q.F. 3-inch Ammunition.

The following projectiles are used in making up ammunition for the Q.F. 3-inch gun, viz. :—

Shrapnel, with Day and Night Tracer | C | .
,, ,, Day Tracer only | N | .
High explosive, with Day and Night Tracer | C | .
,, ,, ,, Day Tracer only | N | .

The above-mentioned cartridges are made up in the same manner.

The H.E. shell when fuzed takes the Fuze, percussion, D.A., No. 44 ; when plugged, it takes the Plug, fuze-hole, Special, No. 1.

The shrapnel shell takes the Fuze, T. and P., No. 84 ; when plugged, the Plug, fuze-hole, 2-inch, No. 2.

Cartridge, Q.F. 3-inch, Shrapnel Shell, with Day and Night Tracer, Mark I, N.T. | C | .

(*Plate* LXXXV.)

The cartridge consists of a brass case, Primer Percussion No. 1, cordite M.D. charge, shrapnel shell with day and night tracers, and safety clip.

The case and primer.—The cartridge case is similar to the Q.F. 13-pr. case, but longer, and the front end of the case is reduced in diameter forming a neck to grip the base of the shell.

The same pattern of percussion primer is used. The safety clip is similar to that used with the Q.F. 3-pr. ammunition, but is coated with black lacquer.

The charge.—The charge consists of 2 lb. $7\frac{6}{16}$ oz. of cordite M.D., size 11.

It is made up of an inner bundle of cordite secured by silk sewing, round which is placed the remainder of the charge, tied in two places with shalloon braid.

The lower ends of the outside sticks of cordite are divided into a number of bundles, and tied with silk sewing, so as to form a firm enlarged base to the charge.

A small coned paper cylinder is inserted into the centre of the upper end of the charge, into which the external tracer on the base of the shell fits.

Shrapnel shell.—The shell is made of forged steel in two parts, screwed together. The lower portion is fitted with a copper driving band (Type No. 14) pressed into an undercut groove near the base.

It is bored out to form a cavity for the tracer liquid (turpentine and aniline dye) ; a small tapered hole to one side of the base, leading into the cavity, is fitted with a Plug, Day Tracer, Vent, No. 1.

The External Night Tracer (described on page 148) screws into a boss formed on the base of the shell. A steel set screw prevents it from unscrewing.

The bottom of the upper portion of the shell is reduced in diameter and threaded to screw into the lower ; a copper ring fitting into a groove in the latter, on which the upper portion bears, is intended to make a liquid-tight joint.

The two portions of the shell are prevented from unscrewing by a set screw.

The bursting charge is contained in a tin cup, above which rests a steel disc.

The central tube is made of brass and screws into the centre of the steel disc.

The shell has no paper lining ; it contains about 79 mixed metal bullets (41 to the lb.).

The space between the bullets is filled in with resin.

To the mouth of the shell is screwed a metal fuze-hole bush, which is bored out and threaded to the 2-inch gauge ; a hole is bored through the centre of the bush for the top of the central tube.

To prevent the resin working up into the fuze socket the top of the central tube is soldered to the bush.

The bursting charge is $1\frac{1}{4}$ oz. F.G. powder ; the central tube is empty.

The shell is secured to the cartridge case by the mouth of the latter being coned into a groove in the driving band as shown in the plate.

The shell weighs $12\frac{1}{2}$ lb.

Total weight of complete cartridge, 20 lb. $11\frac{14}{16}$ oz.

Cartridge, Q.F. 3-inch, Shrapnel Shell, with Day Tracer, Mark II | N | .

The Mark II cartridge differs from the Mark I in the shrapnel shell, which has no night tracer, and contains 83 mixed metal bullets (41 to the lb.).

Cartridge, Q.F. 3-inch, High Explosive Shell with Tracer, Mark I, N.T.

The above-mentioned cartridge is made up in the same manner as the cartridge fitted with a shrapnel shell, and is, therefore, not described in detail.

Shell, Q.F. High Explosive, with Tracer, 3-inch, Mark II, N.T. | C |.

The shell is made of forged steel 3 calibres in length; the head struck with a radius of 2 calibres. In the centre of the base is screwed a *Tracer* socket with fixing screw for the Night Tracer.

The lower part of the body which forms the Day Tracer is filled with turpentine and aniline dye.

A small conical hole is bored through the base of the shell for the *Plug, Day Tracer, Vent.*

The front of the shell is threaded internally to take a "*gunmetal container,*" a small groove being formed for a copper washer intended to seal the joint between the container and the body.

The front of the container is closed by a metal fuze-hole bush secured by a steel locking screw.

The fuze-hole bush is bored out and threaded to take the No. 44 D.A. fuze.

The shell is fitted with the same type of driving band as the shrapnel shell, and is attached to the cartridge case in the same way.

The Mark I shell differs from the Mark II in dimensions only.

Q.F., 4-inch, Ammunition.

Shrapnel and lyddite shell have been approved for the *fixed* ammunition for the Q.F., 4-inch guns, Mark IV, for Land Service.

The ammunition in the Naval Service for the Mark IV guns is *separate.*

Cartridge, Q.F., 4-inch, Fixed, Shrapnel, and Day Tracer, Mark I, N.T. | L |.

The cartridge consists of a brass case, Primer, Percussion, No. 1, Cordite M.D. charge, shrapnel shell with Day and Night Tracers, and safety clip.

The *case* is similar to that for the 3-inch, but larger.

The *charge* consists of 4 lb. 1½ oz. Cordite M.D., Size 11, and is made up in a manner similar to that for the 3-inch. The charge does not fill the whole case, the space between it and the base of the shell being filled in by a "Cylinder, paper, 3·8-inch, Mark I," with a felt ring stitched to its upper end. The External Night Tracer on the base of the shell fits into the paper cylinder.

Shrapnel Shell.—The shell is similar to that for the 3-inch, differing only in dimensions. A special driving band is employed which is wider than that for the 3-inch. The shell below the driving band is provided with a cannelure, into which the cartridge case is indented as well as being coned on to the lower part of the band.

The shell weighs 31 lb., and the complete cartridge 45 lb. 1½ oz.

Cartridge, Q.F. 4-inch, Fixed, Lyddite Shell, Mark I, N.T. | L |.

The cartridge is made up in the same manner as that fitted with a shrapnel shell, but the lyddite shell has a Night Tracer only.

Q.F. 2·95-inch Ammunition.

The following projectiles are fired from the Q.F. 2·95-inch gun :—

Shrapnel with a charge of 5 oz. 4 drs. cordite, size 5.
Double shell with a charge of 5 oz. cordite, size 5.
Case shot with a charge of $5\frac{3}{4}$ oz. cordite, size 5.
Star shell with a charge of $2\frac{3}{4}$ oz. cordite, size $3\frac{3}{4}$.

Cartridge, Q.F. 2·95-inch, Cordite, Shrapnel, Mark IV | L |.

The Mark IV cartridge consists of a case with percussion primer, charge, shrapnel shell and fuze.

The case, Mark III, is of solid drawn brass, slightly tapered towards the mouth, and has a hole in the base, screwed and recessed to take the primer.

The charge consists of $5\frac{1}{4}$ oz. of cordite, size 5, in a circular bundle, enclosed in a shalloon bag, with an igniter of 4 dr. of S.F.G.[2] powder at one end in contact with the primer. A paper cylinder is placed between the charge and the shell.

The primer.—The No. 4 Percussion Primer is used ; for description of primer, *see* page 383.

Shrapnel shell.—The shell has a forged steel body with a recess in the base to contain the tin cup for the bursting charge.

The head is made of steel and is screwed to the body; it is struck with a radius of $1\frac{1}{2}$ diameters and is fitted with a metal G.S. fuze-hole bush.

The lower end of the fuze-hole bush is reduced in diameter, and screwed to receive a shrapnel primer. A metal central tube in two parts screwed together conveys the flash of the fuze to the bursting charge; the upper end of the tube is larger in diameter and fits round the bottom of the fuze socket, the lower end screws into a steel disc placed over the tin cup.

A wood block is fitted to the interior of the head.

The shell contains 203 mixed metal bullets (41 to the lb.).

It is secured in the cartridge case, by the latter being indented in four places into a groove just below the driving band.

Mark III Cartridge.

The Mark III cartridge differs from the Mark IV in having a case (Mark I* and II) fitted to take a cap instead of a percussion primer.

Mark II Cartridge.

The Mark II cartridge is similar to the Mark IV, but has a gun-cotton yarn instead of an S.F.G.[2] powder igniter.

Fig. 137.
CARTRIDGE, Q.F. 2·95-INCH, CORDITE, SHRAPNEL, MARK IV | L |.
Scale $\frac{3}{8}$.

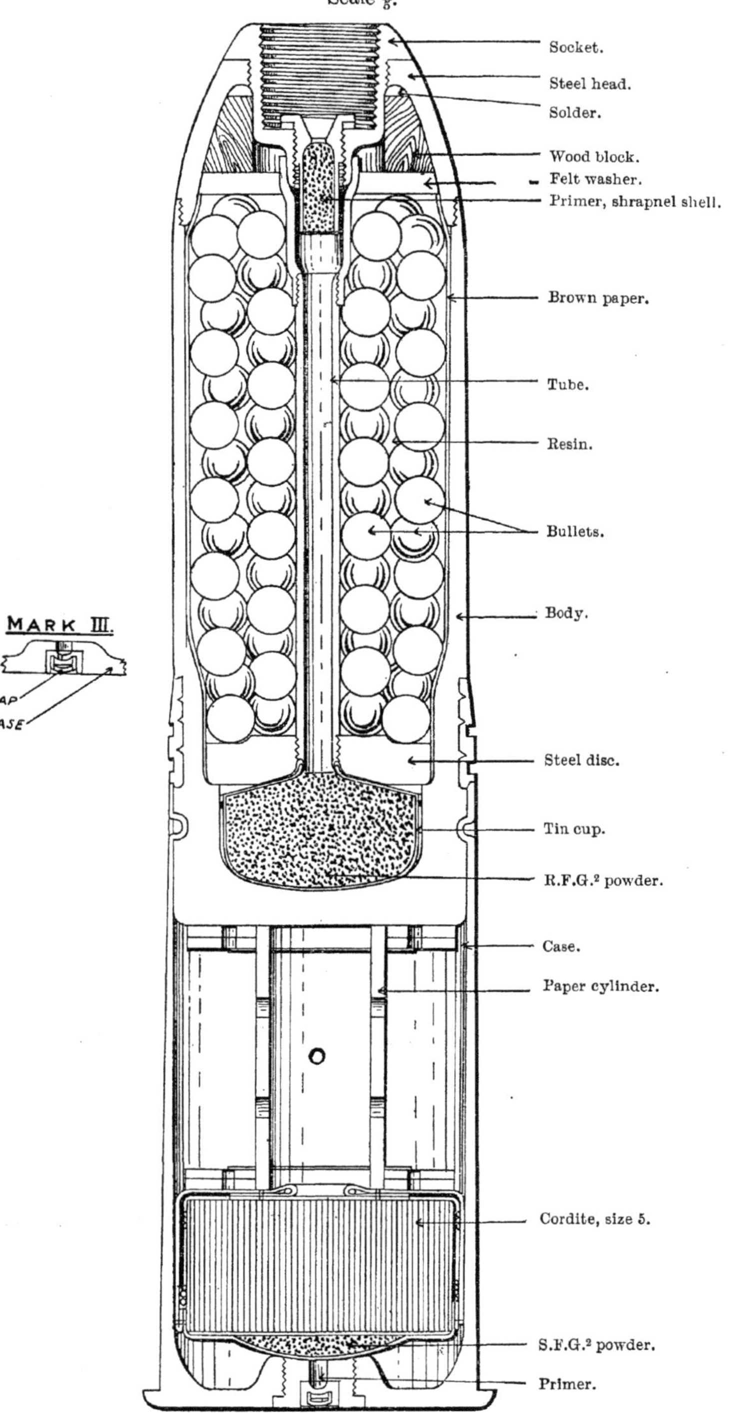

Mark I Cartridge.

The Mark I cartridge is similar to the Mark III, but has a guncotton yarn igniter.

Cartridge, Q.F. 2·95-inch, Cordite, Double Shell.

The Mark IV cartridge, Q.F. 2·95-inch cordite, double shell, is made up similarly to the Mark IV shrapnel cartridge, except that it contains a charge of 5 oz. cordite, and a *double shell.* (*See* Fig. 138.)

The shell is made of iron with a flat nose, and is threaded at the base to receive a No. 12 medium, base, percussion fuze; it contains a bursting charge of 14 oz. of "P" mixture.

Mark III Cartridge.

The Mark III cartridge differs from the Mark IV in having a cartridge case fitted to take a cap instead of a primer.

Mark II Cartridge.

The Mark II cartridge is similar to Mark IV, but has a guncotton yarn igniter.

Mark I Cartridge.

The Mark I cartridge has a case fitted to take a cap; the cordite charge has a guncotton yarn igniter.

Cartridge, Q.F. 2·95-inch, Cordite, Star Shell.

The Mark III cartridge, Q.F. 2·95-inch, star shell, is made up in the same way as the Mark IV shrapnel cartridge, except that it contains a charge of $2\frac{3}{4}$ oz. cordite, size $3\frac{3}{4}$.

The star shell for the 2·95-inch is described on page 196.

Mark IV Cartridge.

The Mark IV cartridge is made up in the same way as the above, but the cartridge case is fitted with a cap instead of a primer.

Mark II Cartridge.

The Mark II cartridge differs from the Mark IV in having a guncotton yarn igniter.

Mark I Cartridge.

The Mark I cartridge is similar to the Mark III, but has a guncotton yarn instead of R.F.G.[2] powder igniter.

Fig. 138.

CARTRIDGE, Q.F. 2·9-INCH, CORDITE, DOUBLE SHELL, MARK IV | L | .

Scale $\frac{3}{5}$.

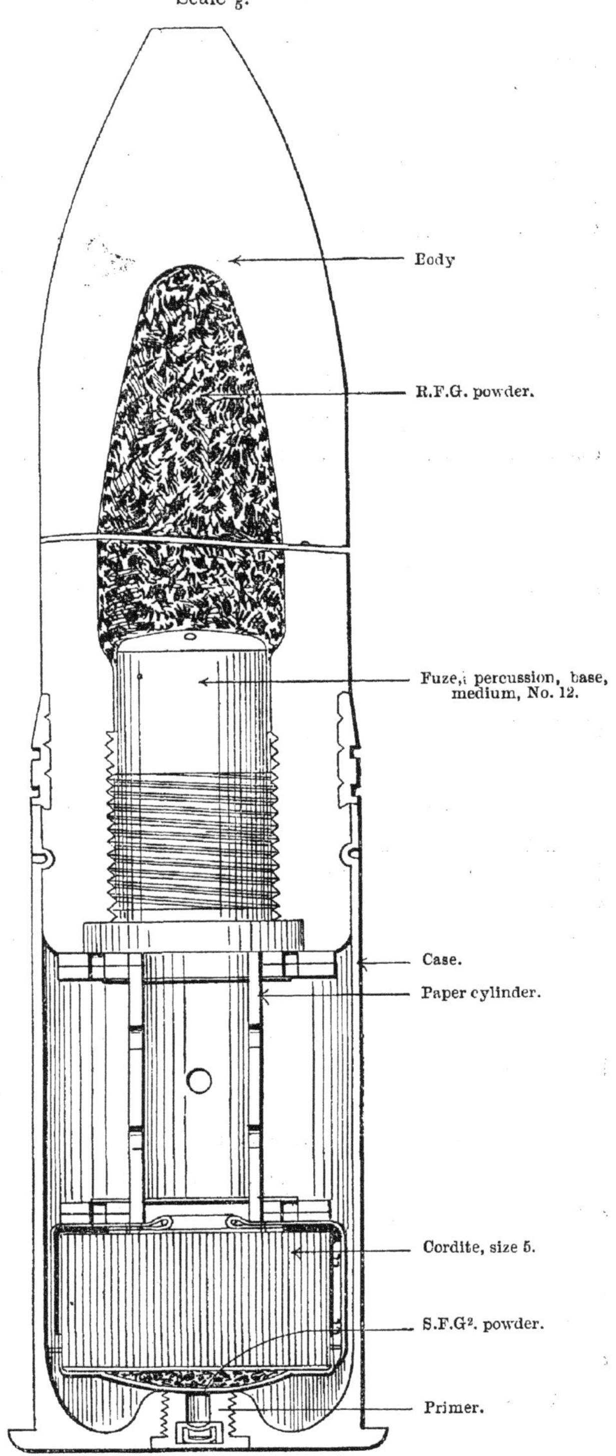

Cartridge, Q.F. 2·95-inch, Cordite, Case Shot.

The latest Mark of cartridge is Mark V; it is made up in the same way as the Mark IV cartridge with shrapnel shell; the charge is $5\frac{3}{4}$ oz. cordite, size 5.

The case shot contains 330 mixed metal bullets (41 per lb.), and weighs 15 lb.

Mark IV Cartridge.

The Mark IV cartridge differs from the Mark V in having a case fitted to take a cap instead of a primer.

Packing of Q.F. 2·95-inch Ammunition.

The ammunition for the 2·95-inch, Q.F., is carried in boxes holding *six* rounds for "*Pack Transport*," and *two* or *three* rounds for "*Man Transport.*"

For description of packages *see* page 487.

Storage of Fixed Ammunition.

The ammunition described in Chapter IX, Section (B) (Fixed Ammunition) is stored in Group III, Division IIA, in a magazine or an explosive store, but always in a separate compartment and never along with any other explosives.

TABLE NO. 34.—*Q.F. 1-Pr. Ammunition.*

Para. in List of Changes.	Mark of Cartridge.	Service.	Case, Empty, Mark.	Cap 1 Pr. Q.F. Mark.	Charge, Nature and Mark.	Igniter, Nature and Mark.	Projectile, Nature and Mark.	Bursting Charge.	Fuze, Nature and Mark.	Remarks.
					Common Shell.					
10963	I	L	I	I	1 oz. 90 grains Cordite, Mark I, Size 3¾	Nitrated Canvas	Q.F. 1-pr. common, Mark I	About 270 grains F.G. powder	Nose Percussion, 1-pr. Q.F., Mark I	
16812	II	C	I	I	1 oz. 3¼ drams Cordite, Mark I, Size 3¾	5 grains R.F.G.^{2}, Mark I	Q.F. 1-pr. common, Mark I	Do.	Do.	
	III	C	I	I	520 grains Cordite, Mark I, Size 2¼	5 grains R.F.G.^{2}, Mark I	Q.F. 1-pr. common, Mark I	Do.	Do.	
					Day Tracer Projectile.					
16812	I	C	I	I	1 oz. 3¼ drams Cordite, Mark I, Size 3¾	5 grains R.F.G.^{2}, Mark I	Q.F. 1-pr. day tracer projectile, Mark I	Filled day tracer liquid	Plug, Fuze-hole Q.F. 1-pr.	
	II	C	I	I	520 grains Cordite M.D., Size 4¼	5 grains R.F.G.^{2}, Mark I	Q.F. 1-pr. day tracer projectile, Mark I	Filled day tracer liquid	Plug, Fuze-hole Q.F. 1-pr.	

TABLE NO. 34—*continued.*—3-*Pr. Ammunition* (*Service*).

Para. in List of Changes.	Mark of Cartridge.	Service.	Case, Empty, 3-Pr., Mark.	Cap, 2·95″, 3 or 6-Pr., Mark.	Primer Percn., No. 2, Mark.	Charge, Mark.	Igniter, Mark.	Cylinder Paper, Mark.	Projectile, Nature and Mark.	Bursting Charge.	Fuze, Nature and Mark.	Remarks.
9299	IV	C	II	I	...	$6\frac{3}{8}$ ozs. Cordite, Mark I, Size 5, Mark II	4 drams R.F.G.^{2}, Mark III	III	Steel shell, II	About 2 ozs. Q.F. Shell, F.G.	Base, Hotchkiss, Mark III	...
9450	V	C	II	I	...	Do.	Do.	III	Do.	Do.	Base, Hotchkiss, Mark IV	Numeral advanced owing to introduction of Mark IV fuze.
9957	VI	C	II	I or II	...	$6\frac{3}{8}$ oz. Cordite, Mark I, Size 5, Mark II	4 drams R.F.G.^{2}, Mark III, or $1\frac{1}{2}$ drams G.C. yarn, Mark IV	III	II or III	Do.	Do.	Numeral advanced owing to introduction of Mark III shell, which differs from Mark II in having groove for driving band undercut.
12295	VII	C	II	II	...	Do.	$1\frac{1}{2}$ drams G.C. yarn, Mark IV	III	II to V	Do.	Do.	Mark IV shell introduced having a different shaped base and a slightly greater capacity. Mark V shell introduced, having cannelure for indents $\frac{1}{4}$ inch further forward.

TABLE No. 34—*continued*—3-*Pr. Ammunition* (*Service*)—*continued.*

Para. in List of Changes.	Mark of Cartridge.	Service.	Case, Empty, 3 Pr., Mark.	Cap, 2·95″ 3 or 6-Pr., Mark.	Primer Percn., No. 2, Mark.	Charge, Mark.	Igniter, Mark.	Cylinder Paper, Mark.	Projectile, Nature and Mark.	Bursting Charge.	Fuze, Nature and Mark.	Remarks.
13422 15700	VIII	C	II II* III	II	... I II	6⅜ ozs., Cordite, Mark I, Size 5, Mark II	1½ drams G.C. yarn, 4 drams S.F.G.² or R.F.G.², Mark IV	IV (changes numeral)	Steel shell, II to VI	About 2 ozs. Q.F. Shell, F.G.	Base, Hotchkiss, IV	§ 11524. Mark II cap introduced. § 12031. All 3 and 6-pr. ammunition made up between 1.6.99 and 29.11.01 to be recapped and reprimed with gun-cotton. § 12595. Mark VI shell introduced, which is slightly longer and has slightly greater capacity.
15724	IX	N	II II* III	II	... I II	7 ozs. 4 drams Cordite, M.D., Size 4¼, Mark I	4 drams S.F.G.² or R.F.G.², Mark V	Nil	Do.	Do.	Do.	...
15701	X	L	II II* III	II	... I II	6⅜ ozs. Cordite, Mark I, Size 5, Mark II	4 drams S.F.G.², Mark VI	Mark I, N.P.	Do.	Do.	Do.	...
15703	XI	C	II* III	... ...	III IV	7 ozs. 4 drams Cordite, M.D., Size 4¼, Mark II	Nil	Nil	Do.	Do.	Do.	The Mark II charge has fins near each end to keep the charge in position.

3-*Pr. Lyddite Ammunition (Service).*

15725	I	N	III	...	IV	7 ozs. 4 drams Cordite, M.D., Size 4¼, Mark II	Nil	Nil	Shell Q.F., Common Lyddite, Mark I, § 15566	About 4 ozs. 14 drams.	D.A. Impact, No. 19, Marks I or II	The Mark II charge is the same as that used with the Mark XI cartridge with steel shell.
	II	N	III	...	IV	Do.	Nil	Nil	Shell Q.F., Common Lyddite, Mark II	Do.	D.A. Impact, No. 19A, Mark I	
	III	N	III	...	IV	Do.	Nil	Nil	Shell Q.F., Common Lyddite, Mark III	Do.	Do.	

3-*Pr. Ammunition (Practice).*

13422 15700	VII	C	II II* III	II	... I II	6⅜ ozs. Cordite, Mark I, Size 5, Mark II	1½ drams G.C. yarn, 4 drams S.F.G.[2] or R. F. G.[2], Mark IV	IV	Steel shell, II-VI	Filled salt	Plugged	Corresponds to Mark VIII service.
15724	VIII	N	II II* III	II	... I II	7 ozs. 4 drams Cordite, M.D., Size 4¼, Mark I	4 drams S.F.G.[2] or R. F. G.[2], Mark V	Nil	Do.	Do.	Do.	Corresponds to Mark IX service.
	IX											
15701	X	L	II II* III	II	... I II	6⅜ ozs. Cordite, Mark I, Size 5, Mark II	4 drams S. F. G.[2], Mark VI	Mark I, N.P.	Do.	Do.	Do.	No issues have been made of Mark IX. Mark X corresponds to Mark X service.

TABLE NO. 34—*continued*—3-*Pr. Ammunition* (*Practice*)—*continued.*

Para. in List of Changes.	Mark of Cartridge.	Service.	Case, Empty, 3-Pr., Mark.	Cap, 2·95″ 3 or 6-Pr., Mark.	Primer Percn., No. 2, Mark.	Charge, Mark.	Igniter, Mark.	Cylinder Paper, Mark.	Projectile, Nature and Mark.	Bursting Charge.	Fuze, Nature and Mark.	Remarks.
15703	XI	C	II* III	... ...	III IV	7 ozs. 4 drams Cordite, M.D., Size 4¼, Mark II	Nil	Nil	Steel shell, II–VI	Filled salt	Plugged	Corresponds to Mark XI service.
15703	XII	C	II* III	... ...	III IV	Do.	Nil	Nil	Practice shot	...	...	Do.
16632	XIII N.T.	C	II* III	... ...	III IV	Do.	Nil	Nil	Converted steel shell and Night Tracer	Filled sand or salt	Plugged	A paper cylinder, 1 inch long, serrated at lower half, is inserted into top of charge to receive the External Night Tracer.
16632	XIV N.T.	...	II* III	... ...	III IV	Do.	Nil	Nil	Shot Practice 3-pr., Mark II and Night Tracer	...	...	Do.
16631	XV	L	II II* III	II	... I II	6⅜ ozs. Cordite, Mark I. Size 5, Mark II	4 drams S. F. G.², Mark VI	Mark I, N.P.	Shot Practice, 3-pr., Mark I	...	...	Differs from Mark X Practice only in having a Practice Shot.
						3-*pr. Ammunition* (*Sub-Calibre*).						
16722	I	N	II II* III	II	... I II	6⅜ ozs. Cordite, Mark I, Size 5, Mark II	1½ drams G.C. yarn, 4 drams S.F.G.² or R. F. G.², Mark IV	IV	Shot Practice, 3-pr., Mark I	...	...	Corresponds to Mark VIII Service, or Mark VII Practice,

16722	II	N	II II* III	II	... I II	7 ozs. 4 drams Cordite, M.D., Size 4¼, Mark I	4 drams S.F.G.[2] or R.F.G.[2], Mark V	Nil	Do.	...	...	Corresponds to Mark IX Service, or Mark VIII Practice.
16722	III	N	II* III	... ...	III IV	7 ozs. 4 drams Cordite, M.D., Size 4¼, Mark II	Nil.	Nil	Do.	...	...	Corresponds to Mark XI Service, or Mark XI Practice.

3-Pr. Vickers' Ammunition (Service).

Para. in List of Changes.	Mark of Cartridge.	Service.	Case, Empty, 3-Pr., Vickers', Mark.	Cap, 2·95″ 3 or 6-Pr., Mark.	Primer Percn., No. 2, Mark.	Charge, Mark.	Igniter, Mark.	Cylinder Paper, Mark.	Projectile, Nature and Mark.	Bursting Charge.	Fuze, Nature and Mark.	Remarks.
13440 15727	I	N	I	II	...	$13\frac{6}{16}$ ozs. Cordite, M.D., Size 8, Mark I	Special, 4 drams S.F.G.[2] or R.F.G.[2]	I, Vickers', 3·5 inches long	Steel shell III to VI	About 2 ozs. Q.F. Shell, F.G.	Base, Hotchkiss, Mark IV	...
15727	II	N	II	...	II	Do.	Do.	I, Vickers', 3·1 inches long	Do.	Do.	Do.	...
15726	III	N	I*	...	III	$13\frac{6}{16}$ ozs. Cordite, M.D., Size 8, Mark II	Nil	I, Vickers', 3·75 inches long	Do.	Do.	Do.	...
15726	IV	N	II	...	IV	$13\frac{6}{16}$ ozs. Cordite, M.D., Size 8, Mark III	Nil	Do.	Do.	Do.	Do.	...

TABLE No. 34—*continued*—3-*Pr. Vickers' Lyddite Ammunition (Service).*

Para. in List of Changes.	Mark of Cartridge.	Service.	Case, Empty, 3-Pr., Vickers', Mark.	Cap, 2·95", 3 or 6-Pr., Mark.	Primer Percn., No. 2, Mark.	Charge, Mark.	Igniter, Mark.	Cylinder Paper, Mark.	Projectile, Nature and Mark.	Bursting Charge.	Fuze, Nature and Mark.	Remarks.
15725	I	N	II	...	IV	$13\frac{6}{16}$ ozs. Cordite, M.D., Size 8, Mark III	Nil	I, Vickers', 3·75 inches long	Shell, Q.F., Common Lyddite, Mark I, § 15566	About 4 ozs. 14 drams	D.A. Impact No. 19, Marks I or II	...
	II	N										
	III N.T.	N										
						3-Pr. Vickers' Ammunition (Full Practice).						
13785	I	N	I	II	...	$13\frac{6}{16}$ ozs. Cordite, M.D., Size 8, Mark I	Special, 4 drams S.F.G.[2] or R.F.G.[2]	I, Vickers', 3·5 inches long	Steel shell, III to VI	Filled salt	Plugged	For guns not mounted on top of barbette or turret mountings. Corresponds to Mark I, Service.
13785	II	N	I	II	...	Do.	Do.	Do.	Practice shot	...	...	(as above)
15727	III	N	II	...	II	Do.	Do.	I, Vickers', 3·1 inches long	Steel shell, III to VI	Filled salt	Plugged	Corresponds to Mark II, Service.
15726	IV	N	I*	...	III	$13\frac{6}{16}$ ozs. Cordite, M.D., Size 8, Mark II	Nil	I, Vickers', 3·75 inches long	Do.	Do.	Do.	Corresponds to Mark III, Service.

15726	V	N	II	...	IV	$13\frac{6}{16}$ ozs. Cordite, M.D., Size 8, Mark III	Nil	Do.	Do.	Do.	Do.	Corresponds to Mark IV, Service.
15726	VI	N	I*	...	III	$13\frac{6}{16}$ ozs. Cordite, M.D., Size 8, Mark II	Nil	Do.	Practice shot	...	...	Corresponds to Mark III, Service.
15726	VII	N	II	...	IV	$13\frac{6}{16}$ ozs. Cordite, M.D., Size 8, Mark III	Nil	Do.	Do.	...	...	Corresponds to Mark IV Service.
	VIII N.T.	N										
	IX N.T.	N										
	X N.T.	N										
	IX N.T.	N										
3-Pr. Vickers' Ammunition (Reduced Practice).												
13519	I	N	I	II	...	$6\frac{14}{16}$ ozs. Cordite, M.D., Size $4\frac{1}{4}$, Mark I	Special. 4 drams S.F.G.[2] or R.F.G.[2]	Nil	Any Service Q.F. 3-Pr. steel shell	Filled salt	Plugged	For guns on top of turret or barbette mountings.
13519 15728	II	N	I I* II	II	... I II	$6\frac{14}{16}$ ozs. Cordite, M.D., Size $4\frac{1}{4}$, Mark II	Do.	Nil	Do.	Do.	Do.	...

TABLE NO. 34—*continued*—3-*Pr. Vickers' Ammunition* (*Reduced Practice*)—*continued.*

Para. in List of Changes.	Mark of Cartridge.	Service.	Case, Empty, 3-Pr., Vickers', Mark.	Cap, 2·95″ 3 or 6-Pr., Mark.	Primer Percn., No. 2, Mark.	Charge, Mark.	Igniter, Mark.	Cylinder Paper, Mark.	Projectile, Nature and Mark.	Bursting Charge.	Fuze, Nature and Mark.	Remarks.
15729	III	N	I* II	... ...	III IV	$6\frac{14}{16}$ ozs. Cordite, M.D., Size $4\frac{1}{4}$, Mark III	Nil	Nil	Any Service Q.F. 3-pr. steel shell	Filled salt	Plugged	This charge is made up on the same design as the Q.F. 3-pr., Mark II charge.
15729	IV	N	I* II	... ...	III IV	Do.	Nil	Nil	Practice shot	...	...	...
15975	V	N	I	II	...	$6\frac{14}{16}$ ozs. Cordite, M.D., Size $4\frac{1}{4}$, Mark IV	Special. 4 drams R.F.G.[2]	Nil	Any Service Q.F. 3-pr. steel shell	Filled salt	Plugged	...
15975	VI	N	I	II	...	Do.	Do.	Nil	Practice shot	...	...	...
	VII N.T.	N	I* II	... ...	III IV	$6\frac{14}{16}$ ozs. Cordite M.D., Size $4\frac{1}{4}$, Mark III	Nil	Nil	Converted steel shell and Night Tracer	Filled sand or salt	Plugged	A paper cylinder, 1 inch long, serrated at lower half, is inserted into top of charge to receive the External Night Tracer.
	II T.	N	I* II	... ...	III IV	Do.	Nil	Nil	Shot practice, 3-pr., Mark II and Night Tracer	...	...	Do.

6-*Pr. Ammunition* (*Service*).

Para. in List of Changes.	Mark of Cartridge.	Service.	Case, Empty, 6-Pr., Mark.	Cap 2·95″ 3 or 6-Pr., Mark.	Primer Percn., No. 2, Mark.	Charge, Mark.	Igniter, Mark.	Cylinder Paper, Mark.	Projectile, Nature and Mark.	Bursting Charge.	Fuze, Nature and Mark.	Remarks.
9299	V	C	III	I	...	$7\frac{3}{4}$ ozs. Cordite, Mark I, Size 5, Mark II	4 drams R.F.G.[2], Mark III	III	Steel shell, III	About 4 ozs. Q.F. shell, F.G.	Base, Hotchkiss, III	
9450	VI	C	III	I	...	Do.	Do.	III	Do.	Do.	Base, Hotchkiss, IV	Numeral advanced owing to introduction of Mark IV fuze.
9957	VII	C	III	I or II	...	Do.	4 drams R.F.G.[2], Mark III, or $1\frac{1}{2}$ drams G.C. yarn, Mark IV	III	III or IV	Do.	Do.	Numeral advanced owing to introduction of Mark IV shell, which differs from Mark III in having groove for driving band undercut. § 11524. Mark II cap introduced.
12295	VIII	C	III	II	...	Do.	$1\frac{1}{2}$ drams G.C. yarn, Mark IV	III	III, IV or V	Do.	Do.	§ 12031. All 3 and 6-pr. cartridges, made up between 1. 6. 99 and 29. 11. 01 to be recapped and reprimed with G.C. yarn.

TABLE No. 34—*continued*—6-*pr. Ammunition* (*Service*)—*continued.*

Para. in List of Changes.	Mark of Cartridge.	Service.	Case, Empty, 6-Pr., Mark.	Cap, 2·95″ 3 or 6-Pr., Mark.	Primer Percn., No. 2, Mark.	Charge, Igniter.	Igniter, Mark.	Cylinder Paper, Mark.	Projectile, Nature and Mark.	Bursting Charge.	Fuze, Nature and Mark.	Remarks.
13422 15700	IX	C	III III* IV	II	... I II	$7\frac{3}{4}$ ozs. Cordite, Mark I, Size 5, Mark II	$1\frac{1}{2}$ drams G.C. yarn, 4 drams S.F.G.[2] or R.F.G.[2], Mark IV	IV (changes numeral)	Steel shell, III, IV or V	About 4 ozs. Q.F. shell, F.G.	Base, Hotchkiss, IV	§ 12251. Mark V shell introduced which has a slightly differently shaped base and slightly greater capacity. Mark IV paper cylinder introduced.
15724	X	N	III III* IV	II	... I II	8 ozs. $11\frac{1}{2}$ drams Cordite, M.D., Size $4\frac{1}{4}$, Mark I	4 drams S.F.G.[2] or R.F.G.[2], Mark V	Nil	Do.	Do.	Do.	
15701	XI	L	III III* IV	II	... I II	$7\frac{3}{4}$ ozs. Cordite, Mark I, Size 5, Mark II	4 drams S.F.G.[2], Mark VI	Mark I, N.P.	Do.	Do.	Do.	
15702	XII	C	III*	...	III	8 ozs. $11\frac{1}{2}$ drams Cordite, M.D., Size $4\frac{1}{4}$, Mark II	Nil	Nil	Do.	Do.	Do.	
15702	XIII	C	IV	...	IV	8 ozs. $11\frac{1}{2}$ drams Cordite, M.D., Size $4\frac{1}{4}$, Mark III	Nil	Nil	Do.	Do.	Do.	

15904	XIV	N	III	II	...	8 ozs. 11½ drams Cordite, M.D., Size 4¼, Mark IV	4 drams R.F.G.^{2}, Mark VII	Nil	Do.	Do.	Do.	

6-*Pr. Ammunition* (*Practice*).

13422 15700	VII	C	III III* IV	II	... I II	7¾ ozs. Cordite, Mark I, Size 5, Mark II	1½ drams G.C. yarn, 4 drams S.F.G.^{2} or R.F.G.^{2}, Mark IV	IV	Steel shell, III, IV or V	Filled salt	Plugged	Corresponds to Mark IX service.
15724	VIII	N	III III* IV	II	... I II	8 ozs. 11½ drams Cordite, M.D., Size 4¼, Mark I	4 drams S.F.G.^{2} or R.F.G.^{2}, Mark V	Nil	Do.	Do.	Do.	Corresponds to Mark X service.
15904	IX	N	III	II	...	8 ozs. 11½ drams Cordite, M.D., Size 4¼, Mark IV	4 drams R.F.G.^{2}, Mark VII	Nil	Do.	Do.	Do.	Corresponds to Mark XIV service.
15701	X	L	III III* IV	II	... I II	7¾ ozs. Cordite, Mark I, Size 5, Mark II	4 drams S.F.G.^{2}, Mark VI	Mark I, N.P.	Do.	Do.	Do.	Corresponds to Mark XI service.
15702	XI	C	III*	...	III	8 ozs. 11½ drams Cordite, M.D., Size 4¼, Mark II	Nil	Nil	Do.	Do.	Do.	Corresponds to Mark XII service.
15702	XII	C	IV	...	IV	8 ozs. 11½ drams Cordite, M.D., Size 4¼, Mark III	Nil	Nil	Do.	Do.	Do.	Corresponds to Mark XIII service.

TABLE No. 34—*continued*—6-*Pr. Ammunition* (*Practice*)—*continued*.

Para. in List of Changes.	Mark of Cartridge.	Service.	Case, Empty, 6-Pr., Mark.	Cap, 2·95″, 3 or 6-Pr., Mark.	Primer Percn., No. 2, Mark.	Charge, Mark.	Igniter, Mark.	Cylinder Paper, Mark.	Projectile, Nature and Mark.	Bursting Charge.	Fuze, Nature and Mark.	Remarks.
15702	XIII	C	III*	...	III	8 ozs. 11½ drams Cordite, M.D., Size 4¼, Mark II	Nil	Nil	Practice shot	...	...	Corresponds to Mark XII service.
15702	XIV	C	IV	...	IV	8 ozs. 11½ drams Cordite, M.D., Size 4¼, Mark III	Nil	Nil	Do.	...	...	Corresponds to Mark XIII service.
15904	XV	N	III	II	...	8 ozs. 11½ drams Cordite, M.D., Size 4¼, Mark IV	4 drams R. F. G.², Mark VII	Nil	Do.	...	...	Corresponds to Mark XIV service.

6-*Pr. Ammunition* (*Sub-calibre*).

Para. in List of Changes.	Mark of Cartridge.	Service.	Case, Empty, 6-Pr., Mark.	Cap, 2·95″, 3 or 6-Pr., Mark.	Primer Percn., No. 2, Mark.	Charge, Mark.	Igniter, Mark.	Cylinder Paper, Mark.	Projectile, Nature and Mark.	Bursting Charge.	Fuze, Nature and Mark.	Remarks.
16722	I	N	III III* IV	II	... I II	7¾ ozs. Cordite, Mark I, Size 5, Mark II	1½ drams, G.C. yarn, or 4 drams S.F.G.² or R. F. G.², Mark IV	IV	Annealed steel shell III, IV or V	Filled sand or salt	Plugged	Corresponds to Mark IX Service or Mark VIII Practice.

Para. in List of Changes.	Mark of Cartridge.	Service.	Case, Empty, 6-Pr., Special, Mark.	Cap, 2·95″, 3 or 6-Pr., Mark.	Primer Percn., No. 2, Mark.	Charge, Mark.	Igniter, Mark.	Cylinder Paper, Mark.	Projectile, Nature and Mark.	Bursting Charge.	Fuze, Nature and Mark.	Remarks.
16722	II	N	III	II	...	8 ozs. $11\frac{1}{2}$ drs. Cordite, M.D., Size $4\frac{1}{4}$, Mark IV	4 drams R. F. G.[2], Mark VII	Nil	Do.	Do.	Do.	Corresponds to Mark XIV Service, or Mark IX Practice.
16722	III	N	III*	...	III	8 ozs. $11\frac{1}{2}$ drs. Cordite, M.D., Size $4\frac{1}{4}$, Mark II	Nil	Nil	Do.	Do.	Do.	Corresponds to Mark XII Service, or Mark XI Practice.
16722	IV	N	IV	...	IV	8 ozs. $11\frac{1}{2}$ drs. Cordite, M.D., Size $4\frac{1}{4}$, Mark III	Nil	Nil	Do.	Do.	Do.	Corresponds to Mark XIII Service, or Mark XII Practice.
6-Pr. Ammunition (Special Service), H.M.S. "Swiftsure."												
12986	I	N	I	II	...	1 lb. $2\frac{5}{16}$ ozs. Cordite, M.D., Size 8, Mark I	Special, 4 drams F.G.	Special, with felt wad at base, Mark I	Steel shell, Mark I, Special	About 4 ozs. Q.F. Shell, F.G.	Base, Hotchkiss, Mark IV	The shell is special for H.M.S. "Swiftsure," and is fitted with a special plain copper driving band.
13861	II	N	I	II	...	do.	Special, 4 drams S.F.G.[2] or R.F.G.[2]	Special, with felt wad and glazed-board disc at base, Mark II	do.	do.	do.	The Mark II paper cylinder is ·25 inch longer than the Mark I. The glazed-board disc prevents the felt wad coming in contact with the cordite.

TABLE No. 34—*continued.*

Para. in List of Changes.	Mark of Cartridge.	Services.	Case, Empty, 6-Pr., Special, Mark.	Cap, 2·95", 3 or 6-Pr., Mark.	Primer Percn., No. 2, Mark.	Charge, Mark.	Igniter, Mark.	Cylinder Paper, Mark.	Projectile, Nature and Mark.	Bursting Charge.	Fuze, Nature and Mark.	Remarks.
						6-Pr. Ammunition (Special Practice), H.M.S. "Swiftsure."						
12986	I	N	I	II	...	14 ozs. Cordite, M.D., Size 8, Mark I (reduced)	Special, 4 drams F.G.	Special, with felt wad at base, Mark I	Practice shot with truncated head	...	...	The Practice shot is special for H. M. S. "Swiftsure," and is fitted with a special plain copper driving band.
13860	II	N	I	II	...	1 lb. $2\frac{5}{16}$ ozs. Cordite, M.D., Size 8, Mark I	Special, 4 drams S.F.G.[2] or R.F.G.[2]	Special, with felt wad and glazed-board disc at base, Mark II	do.	...	...	Same paper cylinder as used with Mark II Service.
						6-Pr. Ammunition (Special Service), H.M.S. "Triumph."						
12986	I	N	I	II	...	1 lb. $2\frac{5}{16}$ ozs. Cordite, M.D., Size 8, Mark I	Special, 4 drams F.G.	Special, with felt wad at base, Mark I	Steel shell, Mark V, § 12,251	About 4 ozs. Q.F. Shell, F.G.	Base, Hotchkiss, Mark IV	...
13861	II	N	I	II	...	do.	Special, 4 drams S.F.G.[2] or R.F.G.[2]	Special, with felt wad and glazed-board disc at base, Mark II	do.	do.	do.	...

6-*Pr.* *Ammunition (Special Practice), H.M.S. "Triumph."*

12986	I	N	I	II	...	14 ozs. Cordite, M.D., Size 8, Mark I (reduced)	Special, 4 drams F.G.	Special, with felt wad at base, Mark I	Any Service pattern 6-pr. steel shell	Filled salt	Plugged	...
13860	II	N	I	II	...	1 lb. $2\frac{5}{16}$ ozs. Cordite, M.D., Size 8, Mark I	Special, 4 drams, S.F.G.[2] or R.F.G.[2]	Special, with felt wad and glazed-board disc at base, Mark II	do.	do.	do.	...

Cartridges, Q.F. 13-*Pr.*

Para. in List of Changes.	Mark of Filled Cartridge.	Service.	Case, Empty, 13-Pr., Mark.	Primer Percn., No. 1, Mark.	Charge.	Shell.	Fuze.	Total Weight of Round.	Remarks.
					Shrapnel Shell.				
12775 13345	I	L	I	I* or II	1 lb. 3 ozs. 14 drs. Cordite, M.D., $2\frac{4}{16}$ ozs., Size $2\frac{1}{4}$., and 1 lb. $1\frac{10}{16}$ ozs., Size 8	Shrapnel, 13-Pr., Marks I to III	T. and P., No. 80	16 lbs. $5\frac{14}{16}$ ozs.	
13347 13497	II	L	I	I* or II	1 lb. 1 oz. 14 drams Cordite, M.D., Size 8	Shrapnel, 13-Pr., Marks I to III	do.		
15234	III	S.I.	I	I* or II	1 lb. 1 oz. 13 drams Cordite, Mark I, Size $7\frac{3}{4}$	do.	do.		
					Star Shell.				
13811 14095	I	L	I	I* or II	6 ozs. 10 drams Cordite, M.D., Size $4\frac{1}{4}$	Star, 13-pr., Marks I to II	Time, No. 25		

TABLE NO. 34—*continued*—*Cartridges, Q.F. 18-Pr.*

Para. in List of Changes.	Mark of Filled Cartridge.	Service.	Case, Empty, 18-Pr., Mark.	Primer, Percussion, No. 1, Mark.	Charge.	Shell.	Fuze.	Total Weight of Round.	Remarks.
					Shrapnel Shell.				
12774 13497	I	L	I	I* or II	1 lb. 6 ozs. 15 drs. Cordite, M.D. Size 8	Shrapnel, 18-pr., Marks I to III	T. and P. No. 80	22 lbs. $13\frac{15}{16}$ ozs.	
15234	II	SI	I	I* or II	1 lb. 4 ozs. Cordite, Mark I, Size $7\frac{3}{4}$	Do.	Do.		
13811 14095	I	L	I	I* or II	8 ozs. Cordite, M.D., Size $4\frac{1}{4}$	Star, 18-pr., Marks I or II	Time, No. 25		

Cartridges, Q.F. 3-inch.

Para. in List of Changes.	Mark of Filled Cartridge.	Service.	Case, Empty, 3-inch, Mark.	Primer, Percussion, No. 1, Mark.	Charge.	Shell.	Fuze.	Total Weight of Round.	Remarks.
					Shrapnel and Day Tracer.				
	IN.T.	C	I to II	II	2 lbs. 7 ozs. 6 drs. Cordite, M.D., Size 11	IN.T.	No. 84	20 lbs. $11\frac{14}{16}$ ozs.	With day and night tracers.
	II	N	I to II	II	Do.	II	No. 84	Do.	Day tracer only.

High Explosive Shell.

In.T.	N	I to II	II	2 lbs. 7 ozs. 6 drs. Cordite, M.D., Size 11	I	No. 44	20 lbs. $11\frac{14}{16}$ ozs.	No day tracer.
In.T.	C	I to II	II	Do.	I to III	No. 44	Do.	With day tracer.
II	N	I to II	II	Do.		No. 44	Do.	Do.
				Cartridges, Q.F. 4-inch, Shrapnel and Day Tracer.				
I	L	I	II	4 lbs. 1 oz. 8 drs. Cordite, M.D., Size 11	In.T.	...	45 lbs. $1\frac{8}{16}$ ozs.	With day and night tracers.
				Lyddite Shell.				
I	L	I	II	4 lbs. 1 oz. 8 drs. Cordite, M.D., Size 11	In.T.	No. 44	45 lbs. $1\frac{8}{16}$ ozs.	Night tracer only.

Cartridges. Q.F. 2·95-inch Shell.

Para. in List of Changes.	Mark of Filled Cartridge.	Service.	Case, Empty, 2·95-inch, Mark.	Cap, 2·95-inch, 3 or 6-Pr., Mark.	Primer, Percussion, No. 4, Mark.	Charge.	Shell.	Fuze.	Total Weight of Round.	Remarks.
						Shrapnel Shell.				
11979	I	L	I* or II	II	...	$5\frac{1}{4}$ ozs. Cordite, Mark I, Size 5, with 1 dram G.C. yarn Igniter	I	Nos. 56, 60, 63, or 65	...	Mark I paper cylinder over charge.
11979 11792	II	L	I* or II	II	...	Do.	II	Do.	...	Do.

TABLE No. 34—*continued*—*Cartridges, Q.F.* 2·95-*inch Shell*—*continued.*

Para. in List of Changes.	Mark of Filled Cartridge.	Service.	Case, Empty, 2·95-inch, Mark.	Cap, 2·95-inch, 3 or 6-pr., Mark.	Primer, Percussion, No. 4, Mark.	Charge.	Shell, Mark.	Fuze.	Total Weight of Round.	Remarks.
14824	III	L	I* or II	II	...	$5\frac{1}{4}$ ozs. Cordite, Mark I, Size 5, with 4 drams S.F.G.[2] Igniter	II	Nos. 56, 60, 63 or 65	...	Paper cylinder over charge = 2·8 inches long.
14824	IV	L	III	...	I	Do.	II	Do.	...	Paper cylinder over charge = 2·65 inches long.
						Double Shell.				
11979	I	L	I* or II	II	...	5 ozs. Cordite, Mark I, Size 5, with 1 dram G.C. yarn Igniter	II	Base, Percussion, No. 12	...	Mark I paper cylinder over charge.
11979	II	L	I* or II	II	...	Do.	III	Do.		Do.
14824	III	L	I* or II	II	...	5 ozs. Cordite, Mark I, Size 5, with 4 drams S.F.G.[2] Igniter	III	Do.	...	Paper cylinder over charge = 2·8 inches long.
14824	IV	L	III	...	I	Do.	III	Do.	...	Paper cylinder over charge = 2·65 inches long.

						Star Shell.				
13384 13773	I	L	III	...	I	$2\frac{3}{4}$ ozs. Cordite, Mark I, Size $3\frac{3}{4}$ with 1 dr. G.C. yarn Igniter	II	Time No. 25	...	Mark I paper cylinder over charge.
14824	II	L	I* or II	II	...	Do.	II	Do.	...	Paper cylinder over charge=3·3 inches long.
14824	III	L	III	...	I	$2\frac{3}{4}$ ozs. Cordite, Mark I, Size $3\frac{3}{4}$ with 4 drs. S.F.G.[2] Igniter	II	Do.	...	Paper cylinder over charge=3·15 inches long.
14824	IV	L	I* or II	II	...	Do.	II	Do.	...	Paper cylinder over charge=3·3 inches long.
						Case Shot.				
11979	I	L	I* or II	II	...	$5\frac{3}{4}$ ozs. Cordite, Mark I, Size 5 with 1 dr. G.C. Yarn Igniter	Shot, Case, II	...	...	Mark I paper cylinder over charge.
14824	II	L	III	...	I	$5\frac{3}{4}$ ozs. Cordite, Mark I, Size 5 with 4 drs. S.F.G.[2] Igniter	Do.	...	...	Paper cylinder over charge=2·65 inches long.
14824	III	L	I* or II	II	...	Do.	Do.	...	...	Paper cylinder over charge=2·8 inches long.
16007	IV	L	I* or II	II	...	Do.	Shot, Case, III	...	...	Paper cylinder over charge=2·8 inches long.
16007	V	L	III	...	I	Do.	III	...	...	Paper cylinder over charge=2·65 inches long.

SECTION C.—Q.F., SEPARATE AMMUNITION.

In the Q.F. ammunition described in this section, the projectiles are not attached to the brass case, and are loaded separately. As such projectiles have already been described in Chapter XI, they are merely referred to here, and only the cartridges described.

With the exception of the Q.F., 12-pr. of 4 cwts., 15-pr., 4-inch Mark IV and V guns and the 4·5-inch Howitzer, the latest marks of Q.F. cartridges are made up on the same model, differing only from each other in minor details.

They are fired electrically, either by an electric primer, or by using an adapter (*see* page 445) and a V.S. *electric* wireless P tube; but in case of a breakdown of the electrical arrangements an adapter and *percussion* tube must be used.

Exceptions.—The latest marks of Q.F., 15-pr. cartridges are fitted with "*Primer, Percussion, No.* 3"; earlier marks with a percussion cap.

The Q.F. 12-pr., of 4 cwts., 4-inch, Marks IV and V, and the 4·5-inch Howitzer are fitted with "*Primer, Percussion, No.* 1."

Ammunition for Q.F., 12-pr. Guns.

There are four natures of 12-pr., Q.F. guns:—

12-pr. of 4 cwts. (Naval).
12-pr. of 8 cwts. (Naval).
12-pr. of 12 cwts. (Common).
12-pr. of 18 cwts. (Naval).

Q.F., 12-pr. of 4 Cwts.

(*Plate* LXXXVI.)

Cartridge, Q.F., 12-pr. of 4 Cwt. filled, 1 lb. 0 oz. 12 drs., Cordite, M.D., Size 8, Mark I | N |.

The above cartridge consists of a case, Primer, Percussion, No. 1, charge of M.D. cordite, felt wad, lid, and safety clip.

The case.—The case is made of solid drawn brass with a projecting rim; it has a central hole bored through the base, screw-threaded for the percussion primer. Three tongues are formed at the mouth to secure the lid. The case is coated inside and out with transparent lacquer.

The primer.—For description of the primer *see* page 379.

The charge.—The charge consists of a cylindrical bundle of Cordite M.D., Size 8, about 7·6 inches in length, tied with shalloon braid in two places; a recess is formed at one end of the charge to allow for the projecting end of the primer.

The wad and glazed board disc.—The wad (Mark I) consists of a disc of felt on top of which are placed two rings of felt; a disc of glazed board is then placed underneath the felt disc, and the whole stitched together, thus making the wad thicker round the edge than it is in the centre.

The lid.—The lid (Mark I) is concave, and consists of two discs of white metal soldered together, the space between being filled with a strawboard disc having a central perforation; the strawboard disc tends to strengthen the lid. To enable the lid to break up easily it is weakened by radial and concentric grooves, and it has a small projecting flange by which it is supported on the mouth of the case; three notches are cut on the rim of the lid to take the tongues on the mouth of the case.

Safety clip.—The safety clip is similar to that used with the Q.F., 6-pr. ammunition (*see* page 396).

Making up the cartridge.—In making up the cartridge the wad is placed on top of the charge with the glazed board disc downwards; the edge of the white metal lid is coated with Pettman's cement The lid is then pressed home into the mouth of the case and secured by the three tongues being bent over it.

Q.F., 12-pr. of 12 Cwt.

The following are the latest Marks of cartridges used with the Q.F., 12-pr. of 12 cwt. gun:—

Cartridge, Q.F., 12-pr. of 12 cwt., 2 lb. Cordite M.D., Size 11, Mark III | C |.

Cartridge, Q.F., 12-pr. of 12 cwt., 1 lb. 15 oz. Cordite, Size 15, Mark IV | C |.

Cartridge, Q.F., 12-pr. of 12 cwt., 13⅞ oz. Cordite M.D., Size 4¼, Mark II | N |.

The above cartridges are made up in the same manner; the last mentioned is a reduced charge for Naval Service.

Cartridge, Q.F., 12-pr. 12 cwt., filled, 2 lbs., Cordite, M.D., Size 11, with Adapter, Mark III.

(*Plate* LXXXVII.)

The cartridge consists of a case, charge, metal igniter, felt wad with glazed board disc and paper cylinder, white metal lid, and adapter.

The case.—The case is made of solid drawn brass with a projecting rim, and has a central hole through the base, screw-threaded for the adapter. Three tongues are formed at the mouth to secure the lid. The case is coated inside and out with transparent lacquer.

The charge.—The charge consists of an inner bundle of cordite secured by shalloon braid, fitted in the centre at the lower end, with a metal igniter. The remainder of the cordite charge is placed round the bundle thus formed, and tied in three places with a clove hitch of shalloon braid secured with a thumb knot. The lower ends of the outside sticks of cordite are divided into six bundles, and tied with silk sewing, so as to form a firm enlarged base to the charge.

Igniter, metal, Mark I.—The igniter consists of a flanged metal thimble, a sheet brass container and a charge of R.F.G.^{2} powder.

The thimble is screw-threaded internally to take the tapered portion of the Mark VI adapter; the front end of the thimble is closed by a paper disc, secured with Pettman's cement.

The container is hexagonal in shape, the sides pierced with flash holes, and lined with paper.

The lower portion of the container is secured by means of three brass rivets to the flange of the thimble.

It contains a charge of ½oz. of R.F.G.[2] powder, retained in position by the tapered fringe on the upper part of the container being bent inwards, and then coated with Pettman's cement.

Wad and glazed board disc.—The wad (Mark IV) is made of felt; it has a disc of glazed board stitched to the underside. The wad and glazed-board disc have a central perforation 1·5 inches in diameter, into which is secured a paper cylinder.

The projecting end of the paper cylinder fits into the upper end of the cordite charge.

It is intended to allow the gases from the fired charge to pass unobstructed through the wad to the Night Tracer.

The lid.—The lid (Mark VII) consists of a corrugated disc of white metal, perforated in the centre and weakened by a number of radial slits. The upper portion is flanged so as to form a lip to rest on the mouth of the cartridge case; three notches are cut in the rim for the tongues on the case.

To the upper face of the disc is soldered a corrugated ring of white metal which tends to strengthen the lid.

The hole in the centre of the lid is closed by a disc of batiste and a disc of paper secured to the underside of the lid by shellac.

Adapter.—For description of the Mark VI adapter, *see* page 445.

Making up the cartridge.—The charge is placed into the cartridge case, igniter end first, care being taken to keep it central, the adapter is then screwed into the base of the case and into the metal igniter. Over the charge is placed the felt wad, the paper cylinder attached to the wad fitting into a recess formed by the cordite sticks being bent outwards.

The edge of the white metal lid before insertion is coated with Pettman's cement to make a damp-proof joint; the lid is pressed home in the mouth of the case and retained in position by the three tongues being bent over it.

For marking on cartridge, *see* page 480.

Earlier Marks of Q.F., 12-pr., of 12 cwt. Cartridges.

Cartridge, Q.F., 12-*pr.,* 12 *cwt., Filled* 2 *lb. Cordite, M.D., Size* 11.

With Primer, Mark II.
With Adapter, Mark II.

Cartridge, Q.F., 12-*pr.,* 12 *cwt., Filled* 1 *lb.* 15 *oz. Cordite, Size* 15.

With Primer, Mark III.
With Adapter, Mark III.

Cartridge, Q.F., 12-*pr.,* 12 *cwt., Filled* 13¾ *oz., Cordite, Size* 4½.

With Primer, Mark I.
With Adapter, Mark I.

The above-mentioned cartridges are all made up in the same manner; they differ from the latest type, already described, in having a shalloon igniter stitched into the end of a cordite cylinder; an earlier pattern of adapter, metal lid and felt wad are used. (The reduced charge has a paper cylinder.)

The Cartridge, Q.F., 12-pr., 2 lb. M.D. charge, Mark II, is described below.

Cartridge, Q.F., 12-pr., 12 cwt., 2 lb., M.D., Size 11, with Adapter, Mark II.

(*Plate* LXXXVII.)

The cartridge consists of a brass case, charge, igniter, cordite cylinder, felt wad with glaze board disc, and adapter.

The case.—The case is made of solid drawn brass with a projecting rim, and has a central hole through the base, screw-threaded for the adapter or electric primer. Three tongues are formed at the mouth to secure the lid. The case is coated inside and out with transparent lacquer.

The charge.—The charge consists of an inner bundle of cordite secured by silk or shalloon braid, fitted at the centre of the rear end with a cordite cylinder and igniter. The remainder of the cordite charge is placed round the bundle thus formed, and tied in three places with shalloon braid. The lower ends of the outside sticks of cordite are divided into six bundles, and tied with silk sewing, so as to form a firm enlarged base to the charge.

The igniter.—The Mark II igniter consists of 10 drams of R.F.G.[2] powder contained in a cylindrical shalloon bag choked with silk sewing. It is stitched to the inside of a cordite cylinder 4 inches in length, 1 inch internal diameter, and ·05 inch thick. The function of the cordite cylinder is to form a recess for the projecting end of the adapter or primer; it is included in the weight of the charge.

The wad and glazed board disc.—The wad (Mark III) consists of a disc of felt on top of which are placed two rings of felt; a disc of glazed board is then placed underneath the felt disc, and the whole stitched together, thus making the wad thicker round the edge than in the centre.

The lid.—The latest Mark of lid is Mark VI; it is concave, and consists of two discs of white metal soldered together, the interior being filled up with a strawboard disc having a number of perforations. The strawboard disc tends to strengthen the lid; the perforations are intended to allow the gas pressure (when the cordite charge is fired) free access through the holes in the protecting plate of the latest Marks of base fuzes, so as to ensure that the copper pressure plate and spindle are crushed in. To enable the metal lid to break up easily, it is weakened by radial and concentric grooves, and it has a small projecting flange by which it is supported on the mouth of the case; three notches are cut in the rim of the lid for the tongues on the mouth of the case.

Making up the cartridge.—The charge is placed into the cartridge case, igniter-end first, and then covered with the felt wad with the glazed-board disc next to the cordite. The edge of the white metal lid before insertion is coated with Pettman's cement to make a damp-proof joint; the lid is pressed home into the mouth of the case and retained in position by the three tongues being bent over on to it. An electric primer, or an adapter (*see* Plate LXXVI.) is screwed into the base, the front end projecting into the cordite cylinder.

For information on the cartridge, *see* page 480.

The *Cartridge, Q.F.*, 12-*pr.*, 12 *cwt.*, 1 *lb.* 15 *oz.*, *Cordite, size* 15, *Mark III*, differs from the above in the igniter which contains 1¼ oz. instead of 10 drs. of R.F.G.[2] powder, and the cordite cylinder is longer.

The outer layer of cordite sticks forming the charge are divided into five instead of six bundles.

Cartridge, Q.F., 12-*pr. of* 12 *cwt.*, 2 *lb. Cordite M.D.*, *Size* 11.

WITH PRIMER, MARK I.

WITH ADAPTER, MARK I.

Cartridge, Q.F., 12-*pr. of* 12 *cwt.*, 1 *lb.* 15 *oz.*, *Cordite, Size* 15.

WITH PRIMER, MARK II.

WITH ADAPTER, MARK II.

The above cartridges for the 12-pr. of 12 cwt. differ from those already described in the method of making up the charge, which is divided into two portions. The larger portion consists of sticks nearly the full length of the case, which are secured together by two pieces of shalloon braid. Round the bottom of these sticks a short shalloon bag, with a pocket for the igniter, is placed, and the short sticks, which form the second portion of the charge, are packed in this bag, thus enlarging the diameter of the charge at the bottom. The bag is provided with a draw string of shalloon braid which is pulled in and tied tightly above the short sticks. Two pieces of shalloon braid, attached to the bag, are passed up round the securing braid on the long sticks and fastened in the centre of the cordite by a reef knot. Silk braid was used with earlier issues.

1¼ oz. IGNITERS.

The Mark IV igniter consists of 1¼ oz. of R.F.G.[2] powder contained in a cylindrical shalloon bag choked with silk sewing. It is stitched to the inside of a cordite cylinder 4 inches in length, 1 inch internal diameter, and ·05 inch thick. The function of the cordite cylinder is to form a recess for the projecting end of the adapter or primer; it is included in the weight of the charge.

Mark III igniter.—The Mark III igniter differs from the Mark IV above described in the shalloon bag which was closed with a disc of shalloon top sewn all round, instead of being choked with silk sewing.

Mark II igniter.—The Mark II igniter consists of a paper and calico dome choked at one end, and perforated. Secured to the inside of the dome is a shalloon bag containing 8½ drs. R.F.G.[2] powder.

Mark I igniter.—The Mark I igniter was a paper cylinder pointed at one end, filled with 1¼ oz. of R.F.G.[2] powder and closed with a perforated millboard wad and shalloon disc. It screwed on the front end of the electric primer.

10-dram Igniters.

Mark II igniter.—The Mark II igniter is similar to the Mark IV 1¼ oz. igniter, but a smaller cordite cylinder is used.

Mark I igniter.—The Mark I 10-dram igniter differs from the Mark II in the shalloon bag which was closed with a disc of shalloon top sewn all round, instead of being choked with silk sewing.

Earlier Marks of Lids.

The latest Mark of lid for the Q.F., 12-pr., 12 and 18 cwt., is Mark VII, intended for use with shell fitted with Night Tracers; it is described on page 442.

The Mark VI lid.—The Mark VI lid is concave; it consists of two discs of white metal, soldered together, the space between being fitted with a strawboard disc having a number of perforations. The perforations are intended to allow the gas pressure (when the cordite charge is fired) free access through the holes in the protecting plate of the Mark VII medium base fuze.

Mark V lid.—The Mark V lid differs from the Mark VI in the strawboard disc which has only one central perforation.

This Mark of lid was declared obsolete, owing to the introduction of the new pattern base fuze with the pressure plate to the side, instead of in the centre of, the base.

Mark IV lid.—The Mark IV lid was flat instead of being concave; an earlier Mark of felt wad is used with this lid.

Mark III lid.—The Mark III lid was similar to the Mark IV, but the strawboard disc inside the lid was not perforated. It prevented the pressure plate in medium base fuzes from acting properly as it sealed the holes in the steel plate and thus caused blinds.

All Mark III lids have had a hole bored in the centre of the underside of the lid and strawboard wad, and are known as Mark III* lids.

Mark II lid.—The Mark II lid contained a lubricant of beeswax and tallow instead of a disc of strawboard.

Mark I lid.—The Mark I lid had a stronger bottom disc of white metal than Mark II.

Cartridge, Q.F. or Q.F.C. Adapters.

(For use in Q.F. Cartridges, to enable tubes to be used instead of electric primers.)

Q.F. and Q.F.C. cartridges, 12-pr., 14-pr., 4-inch, 4·7-inch and 6-inch are now issued with adapters in place of electric primers; but the primers will still be met with both in Land and Naval Service.

Mark VI adapter.—The latest Mark of adapter is Mark VI; it is intended for use with cordite charges fitted with metal igniters.

It is made of aluminium bronze or manganese bronze similar to the "Primer, electric, large" in external appearance, but has no gas-check portion; the front end is screw-threaded to fit the metal igniter.

It is bored out to take a V.S. tube; the front end is closed by a disc of foolscap paper shellaced on and coated with Pettman's cement.

The adapter is fitted with a small spring plunger to prevent the tube from falling out during loading.

Fig. 139.

MARK VI ADAPTER (WITH METAL IGNITER).

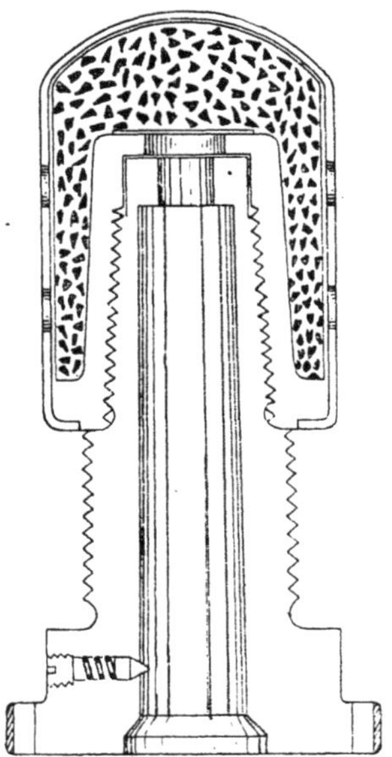

Marks IV and V adapter.—The Marks IV and V adapters are used with Q.F. cartridges when the charge is fitted with a cordite cylinder igniter.

They differ from the Mark VI in the front end being left plain instead of being screw-threaded.

The Mark IV has no spring plunger to hold the tube from falling out during loading.

Fig. 140.

MARK V ADAPTER.

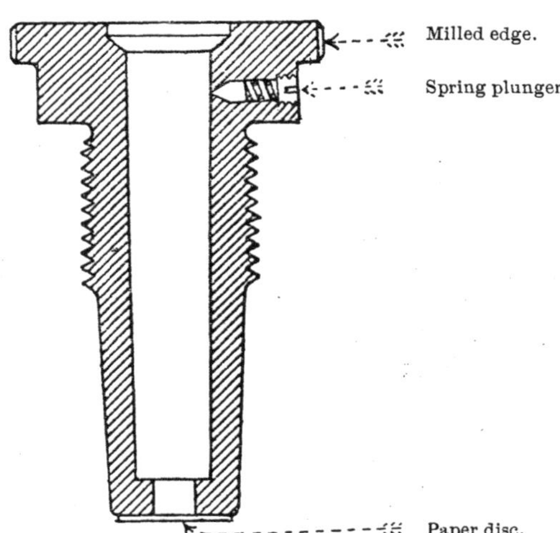

Marks I and II adapters.—The Mark I and II adapters may still be met with in the Service; they are made of hardened steel and must not be stored in cartridge cases.

There is no Mark III adapter in the Service.

Marks I and II adapters (converted).—The converted adapters are the bodies of old "*Primers, Electric, Large*" (Marks II to V) bored out to take a V.S. tube.

A brown paper ring is shellaced into the recess at the small end, which is then closed by a disc of foolscap paper shellaced on, and coated with Pettman's cement.

The *Mark II* is fitted with a spring plunger to hold the tube in position during loading.

The converted adapters are used only in cartridges where the charge is fitted with a cordite cylinder igniter.

They can only be used once.

Q.F. 12-pr. of 8 cwt.

Cartridge, Q.F., 12-pr. of 8 cwt., Filled 13¾ ozs. Cordite, Size 10. With Adapter, Mark III.

The Mark III cartridge for the Q.F., 12-pr. of 8 cwt. is made up in the same way as the latest Mark of 12-pr. of 12 cwt. cartridge already described (*see* Plate LXXXVII.)

The cartridge case is shorter and is closed with the 12-pr. of 8 cwt. lid, Mark VI.

This lid is similar to the Mark VII lid used with the 12-pr. of 12 cwt. cartridge, but is larger in diameter.

Mark II Cartridge.

The Mark II differs from the above in the charge, which has a cordite cylinder igniter, and an earlier pattern of lid and adapter (*see* Plate LXXXVII.)

Mark I Cartridge.

The Mark I cartridge differs from the Mark II in the charge, which is made up in two portions; the lower part is enclosed in a shalloon bag with a pocket for the cordite cylinder igniter.

Q.F., 12-pr. of 18 cwt.

Full Charge.

Cartridge, Q.F., 12-pr. of 18 cwt., Filled 2 lbs. 12 ozs. 2 drs. Cordite, M.D., Size 11. With Adapter, Mark III | N |.

The cartridge consists of a case with Mark VI adapter, cordite M.D. charge, with metal igniter, paper cylinder with glazeboard disc and felt wad, and a white metal lid.

The charge.—The charge is made up in the same way as the 12-pr. of 12 cwt. cartridge illustrated on Plate LXXXVII.

Paper cylinder.—The paper cylinder is perforated, and is about 5 inches long; near the lower end of the cylinder, secured by copper wire, is a millboard ring with a ring of glazeboard on the underside.

The projecting end of the paper cylinder fits into the top of the cordite charge. A millboard and a felt ring are attached to the upper end of the paper cylinder.

The white metal lid (Mark VII) employed in making up the 12-pr. of 12 cwt. cartridge is used.

Cartridge, Q.F. 12-pr., 18 cwt., filled, 2 lb. 12$\frac{2}{16}$ oz. Cordite, M.D., Size 11.

With Adapter, Mark II. With Primer, Mark II.

The above cartridge is made up in the same way as the Mark II 2 lb. M.D. charge for the 12 cwt. gun; the same cordite cylinder and 10 dram igniter are used. The case is, however, much longer and a paper cylinder is placed on top of the charge. The Mark VI lid is used.

The Mark I cartridge is obsolete.

Reduced Charge for 12-pr. of 18 cwt.

Cartridge, Q.F. 12-pr., 18 cwt., filled, 1 lb. 11 oz. 14 drams Cordite M.D., Size 8, with Adapter, Mark III.

The above cartridge is made up in the same way as the Mark III, *full charge;* as the charge is shorter, a longer paper cylinder is used.

Mark II Cartridge.

The Mark II reduced cartridge for the 12-pr. of 18 cwt. differs from the Mark III in having a cordite cylinder igniter, instead of a metal igniter; an earlier pattern of adapter, paper cylinder and lid, is used.

Q.F. 14-pr.

Cartridge, Q.F. 14-pr., filled, 2 lb. 12 oz. Cordite, M.D., Size 11, with Adapter, Mark III | N |.

The Mark III cartridge for the Q.F. 14-pr. consists of a case, Mark IV adapter, charge with metal igniter, paper cylinder, wad and lid.

The case.—The case is similar to the 12-pr. of 12 cwt., but is longer and is necked at the top.

The charge.—The charge with metal igniter is made up in the same way as the 12-pr. of 12 cwt. charge.

The paper cylinder.—The cylinder is made of brown paper, perforated and fitted at each end with double perforated millboard rings. A disc of glazed-board is attached by copper tacks to the end of the cylinder, which is placed next the cordite charge.

Wad.—The wad (Mark III) consists of a disc of felt on top of which are placed two rings of felt; a disc of glazed-board is then placed underneath the felt disc, and the whole stitched together, thus making the wad thicker round the edge than in the centre.

The lid.—The Mark V white metal lid is used; it is concave, and is strengthened by a perforated strawboard disc; it is similar to the Mark VI lid for the Q.F. 12-pr., 12 and 18 cwt.

Marks I and II Cartridges.

The Marks I and II cartridges are made up in the same way as the Marks I and II (full charge) cartridges for the 12-pr. of 18 cwt.

(For particulars, *see* Table 35.)

Q.F. 15-PR.

The following charges are used with the Q.F. 15-pr. gun, *i.e.* :—

1 lb. $2\frac{10}{16}$ oz. Cordite, M.D.T., Size 20–10. (For shrapnel shell and case shot.)

4 oz. Cordite, Size 5. (Reduced charge for star shell.)

CARTRIDGE, Q.F. 15-PR., FILLED, 1 LB. $2\frac{10}{16}$ OZ. CORDITE, M.D.T., SIZE 20–10, MARK III | L |.

(*Plate* LXXXIX.)

The cartridge consists of case, with percussion cap or primer, cordite M.D.T. charge with igniter and glazed-board lid.

The case.—There are three marks of case for the Q.F. 15-pr., and either may be used in making up this cartridge.

The Mark I case takes a special percussion cap, which is pressed directly into a recess in the base of the case.

The Mark II case takes a cap in a cap chamber which is similar in form to the 2·95-inch, 6, or 3-pr. cap and cap chamber illustrated on Plate LXXXI. It differs in the cap being made thinner.

The Mark III case takes the Primer, Percussion Q.F. Cartridges, No. 3. *See* page 382.

Marks I and II cases can be prepared to take the percussion primer, and then become Marks I* or II* respectively.

The charge is made up of a core of size 20–10 M.D.T. cordite, hooped in two places with No. 1 sewing silk, and having a Mark III igniter secured at one end by the ties being passed over the top hoop, through the core and tied together. Around this core longer sticks of similar cordite are bundled and secured by shalloon braid.

The igniter consists of 4 drams of R.F.G.² gunpowder, contained between two discs of undyed shalloon, to which are attached four ties made up of two strands of No. 1 sewing silk, doubled and knotted.

The glazed-board lid is cup-shaped. The outer edge is coated with Pettman's cement and the lid is inserted as shown in the plate.

Three tongues on the case are then turned over to secure the lid, these joints being also coated with Pettman's cement.

A label giving particulars of the charge is shellaced to the top of the lid. *See* page 480.

Mark II Cartridge.

Fig. 141.

CARTRIDGE, Q.F., 15-PR., FILLED, 1 LB. $2\frac{10}{16}$ OZ. CORDITE M.D.T., SIZE 20–10, MARK II.

Scale $\frac{1}{1}$.

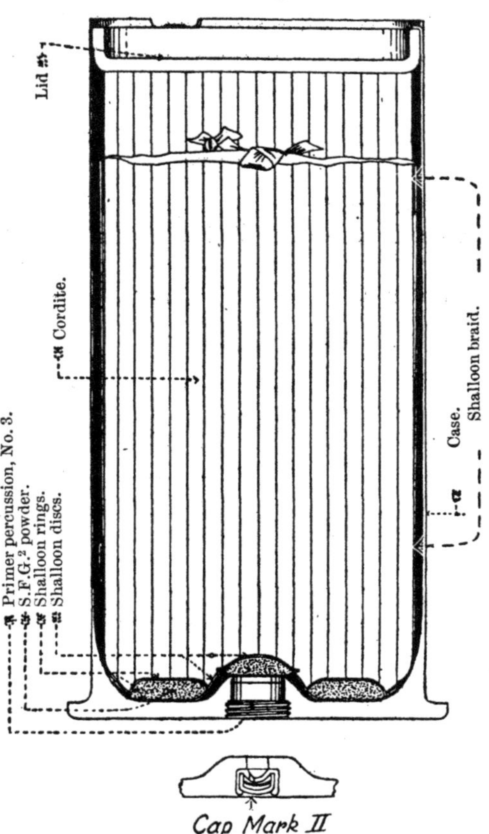

Cap Mark II

The 1 *lb.* $2\frac{10}{16}$ *oz. cartridge, Mark II* differs from the Mark III only in the method of making up the charge, and in the igniter. The charge consists of a bundle of cordite M.D.T. tied with shalloon braid at each end ; to the base is attached the igniter. The igniter is made in two parts so as to cover the bottom of the charge, one part being in the form of an annular ring of shalloon to fit round the bottom of the case, and the other in the form of a disc to rest on the front end of the primer. The disc portion is sewn to the ring, and both are filled with S.F.G.[2] powder. Four pieces of shalloon braid attached to the igniter pass up under the lower and over the upper securing braid on the sticks, and are fastened in the centre by reef knots.

The Mark I cartridge was ordered to be converted locally to Mark II pattern, and to be then known as Mark I* (§ 15208).

STAR SHELL CHARGE FOR 15-PR., Q.F.

CARTRIDGE, Q.F. 15-PR., FILLED, 4 OZS. CORDITE, SIZE 5, MARK II. (FOR STAR SHELL.)

The charge consists of 4 oz. of cordite, size 5, cut about ·9 inch in length, made up into a circular bundle and enclosed in a shalloon bag, the mouth of which is drawn in by a drawstring and closed by a disc of shalloon stitched to the bag. It has an igniter at one end filled with 4 drams of S.F.G.[2] powder contained in a pocket formed at the bottom of the shalloon bag.

The charge is held in position in the case by a perforated paper cylinder with two perforated discs of paper secured to each end, and by a glazed-board lid.

The same cases may be used, as are used in making up the full charge.

The Mark I cartridge differed from the Mark II in the igniter being filled with 2 drams of guncotton yarn.

Q.F. 4-INCH, MARKS I TO III* AND Q.F.C. 4-INCH.

There are at present four cartridges for the above-mentioned guns, viz. :—

FULL CHARGES.

CARTRIDGE, Q.F. 4-INCH, MARKS I TO III* GUNS, FILLED, 3 LBS. 9 OZS. CORDITE, SIZE 15, WITH ADAPTER, MARK IV | C | .

CARTRIDGE, Q.F. 4-INCH, MARKS I TO III* GUNS, FILLED, 3 LBS. 8 OZS. 11 DRS. CORDITE, M.D.T., SIZE 20–10, WITH ADAPTER, MARK II | C | .

REDUCED CHARGES.

CARTRIDGE, Q.F. 4-INCH, MARKS III TO III* GUNS, FILLED, 1½ LBS. CORDITE, SIZE 5, WITH ADAPTER, MARK IV | L | .

CARTRIDGE, Q.F., MARKS I TO III* OR Q.F.C. 4-INCH GUNS, FILLED, 1 LB. 10 OZS. 14 DRS. CORDITE, M.D., SIZE 4¼, WITH ADAPTER, MARK III | C | .

The above-mentioned cartridges are all made up in the same way as the Mark III 2-lb. Cordite, M.D. cartridge for the Q.F. 12-pr. of 12 cwts., described on page 441.

They differ only in weight, dimensions, and nature and size of cordite used. The reduced charges are cut shorter than the full charges. This necessitates the use of a paper cylinder.

For earlier Marks of Q.F. or Q.F.C. 4-inch cartridges, *see* Table 35.

Q.F. 4-INCH, MARK IV.

(*Plate* XC.)

CARTRIDGE, Q.F. 4-INCH, MARK IV GUN, FILLED, 5 LBS. 1 OZ. 12 DRS. CORDITE, M.D., SIZE 16, MARK II | N | .

The above-mentioned cartridge consists of a Case, Primer, Percussion, No. 1, Charge of M.D. Cordite, Felt Wads, Lid and Safety Clip.

The case, Mark II, is made of solid drawn brass. It has a projecting rim at the base to prevent it being pushed too far into the gun and to facilitate extraction. A hole is bored through the centre of the base, screw-threaded to take the Primer, Percussion, No. 1. Three tongues are formed at the mouth to secure the lid. The case is coated inside and out with transparent lacquer.

The primer.—For description of the primer, *see* page 379.

The charge, Mark II, consists of a central bundle of cordite, M.D., tied together with silk sewing. Around this central bundle is arranged another layer of cordite sticks of greater length than the central bundle so as to leave a small central recess at the lower end of the charge to fit over the primer and the boss on the inside of the case.

The remainder of the cordite charge is placed around the core thus formed, and tied with shalloon braid in three places.

The lower ends of the outside layer of cordite sticks are divided into six bundles and tied with silk sewing, so as to form a firm enlarged base to the charge.

Wad and glazed-board disc.—The wad (Mark I) is made of felt, with a disc of glazed-board stitched to the underside.

The wad and glazed-board disc have a central perforation 1·4 inches in diameter, into which is secured a paper cylinder. The projecting end of the cylinder fits into the upper end of the cordite charge.

It is intended to allow the gases from the fired charge to pass unobstructed through the wads to the Night Tracer.

Two or more rings of felt are placed on top of the wad to fill up the space in the mouth of the case.

The lid.—The lid (Mark VII) consists of a corrugated disc of white metal, perforated in the centre and weakened by a number of radial slits. The upper portion is flanged, so as to form a lip to rest on the mouth of the cartridge case; three notches are cut in the rim for the tongues on the case.

To the upper face of the disc is soldered a corrugated ring of white metal, which tends to strengthen the lid.

The safety clip is similar, except in size, to that for the 6-pr. illustrated on page 397.

Mark I Cartridge.

The Mark I cartridge differs from the Mark II in the case and charge which are both Mark I pattern.

The Mark I case has a thinner boss inside the base. It was found that the gases from the propellant charge escaped to the rear between the case and the primer.

Mark I cases are being brought approximately to Mark II pattern by riveting in a bush and preparing it for the primer.

Such cases are known as Mark I* and will be used in making up Mark II cartridges.

The Mark I charge has a smaller recess at the base than the Mark II.

Q.F. 4-INCH, MARK V.

(*Plate* XCI.)

CARTRIDGE, Q.F. 4-INCH, MARK V GUN, FILLED, 7 LBS. 11 OZS. CORDITE, M.D., SIZE 16, MARK I | N |.

The above cartridge differs from that for the 4-inch, Mark IV gun chiefly in dimensions, and in the weight of the charge.

The wad consists of a ring of felt, and a disc of glazed-board stitched together.

Q.F. 4·5-INCH HOWITZER.

There are two cartridges for the Q.F. 4·5-inch Howitzer, viz. :—

CARTRIDGE, Q.F. 4·5-INCH HOWITZER, FILLED, 15 OZS. 14 DRS. CORDITE, M.D., SIZES 2¼ AND 4¼, MARK I | L |.

CARTRIDGE, Q.F. 4·5-INCH HOWITZER, FILLED, 14 OZS. CORDITE, SIZES 5 AND 3¾, MARK I | S I |.

The 15 *ozs.* 14 *drs. charge* is here described, and is illustrated on Plate XCII.

CARTRIDGE, Q.F. 4·5-INCH HOWITZER, FILLED, 15 OZS. 14 DRS. CORDITE, M.D., SIZES 2¼ AND 4¼, MARK I | L |.

The cartridge consists of a case, percussion primer, charge and leatherboard lid.

The case.—The case is made of solid drawn brass of the form and dimensions shown on Plate XCII. A hole is bored through the centre of the base, screw-threaded to take the *Primer, Percussion, No.* 1; the case is sand blasted inside and out, and coated externally with a dull black lacquer, similar to the Q.F. 13-pr. and 18-pr. cases.

Charge.—The charge is made up in five portions, each portion being enclosed in a cambric bag. Each portion has printed on it the calibre of the howitzer, the number of the portion and its weight, the numeral, the lot number of the cordite, and manufacturer's initials, or trade mark.

The weights of the five portions are :—

1st portion (core)	{ 4 ozs.	13 drs.	Size 2¼
	1 ,,	1 ,,	,, 4¼
2nd ,, ...	1 ,,	5½ ,,	,, 4¼
3rd ,, ...	1 ,,	12 ,,	,, 4¼
4th ,, ...	2 ,,	9½ ,,	,, 4¼
5th ,, ...	4 ,,	5 ,,	,, 4¼

The 1st portion (core) is made up as follows :—1 ozs. 2 drs. of cordite, M.D., size 2¼, is cut 1·75 inches long and tied in two places with machine thread; round this bundle is tied a layer of size 2¼, cut 2·25 inches long, thus forming a recess at one end to fit over the primer when the charge is inserted into the case. Around the base of this bundle is placed a ring of 1 oz. 1 dr., size 4¼, and the whole is then placed in a cambric bag.

Rings.—The remaining portions are first formed into rings, tied across in three places with machine thread, and then enclosed in cambric bags; the rings fit over the stem of the core.

Leatherboard cup or lid.—This is inverted when placed in position, and closes the mouth of the cartridge case. A tape becket is fitted to the outside of the cup, by means of which the latter can be withdrawn.

Packing.—The cartridges are packed, one in a tin box, the box being lined with leatherboard.

There are two Marks of the box. Mark I has the lid secured by a tin band soldered on, and Mark II has the lid secured by a bayonet joint.

Cartridge, Q.F. 4·5-inch Howitzer, filled, 14 ozs. Cordite, Size 5 and 3¾, Mark I | S I |.

The 14-*oz. charge* is made up in exactly the same way as the 15-oz. 14-dr. charge. It is used in India only.

The weights of the five portions are :—

1st portion (core) ..	..	5 ozs. 5 drs.	Size 3¾
2nd ,,	.. ..	1 ,, 1 ,,	,, 5
3rd ,,	.. ..	1 ,, 7 ,,	,, 5
4th ,,	.. ..	2 ,, 5 ,,	,, 5
5th ,,	.. ..	3 ,, 14 ,,	,, 5

Q.F. 4·7-inch, Marks I to IV* Guns.

There are two cordite cartridges for the above-mentioned guns, viz. :—

Full Charge.

Cartridge, Q.F. 4·7-inch, Marks I to IV* Guns, filled, 5 lbs. 7 ozs. Cordite, Size 20, with Adapter, Mark VII | C |.

Reduced Charge.

Cartridge, Q.F. 4·7-inch, Marks I to IV* Guns, filled, 2 lbs. 5¾ ozs. Cordite, Size 7½, with Adapter, Mark IV | C |.

The above-mentioned cartridges are made up in the same way as the "Cartridge, Q.F. 12-pr., 12 cwts., filled, 2 lbs. Cordite, M.D., size 11, Mark III | C |" described on page 441.

They differ only in weight, dimensions, and nature and size of cordite used. The latest Mark of lid for the 4·7-inch is Mark IX.

The empty case for the 4·7-inch, Marks I to IV* guns is Mark II. Mark I is obsolete.

The reduced charge is used in Naval Service, and with the 4·7-inch on travelling carriage (L.S.) for practice purposes with all projectiles. The cartridge is made up in the same way as the Mark VI full charge, but a paper cylinder is used to fill up the space in the front of the case. (*See* Plate XCIII.)

Earlier Marks of 4·7-inch Cartridges.

Full Charge.

The Mark VI 5-lb. 7-oz. cordite cartridge for the Q.F. 4·7-inch, Marks I to IV* guns is similar in design to the Q.F. 12-pr. cartridge described on page 443.

The outer layer of cordite sticks are divided into six bundles and tied with shalloon braid instead of silk sewing.

Fig. 142.

CARTRIDGE, Q.F. 4·7-INCH, MARKS I TO IV* GUNS, FILLED, 5 LBS. 7 OZS. CORDITE, SIZE 20, MARK VI.

Scale $\frac{1}{3}$.

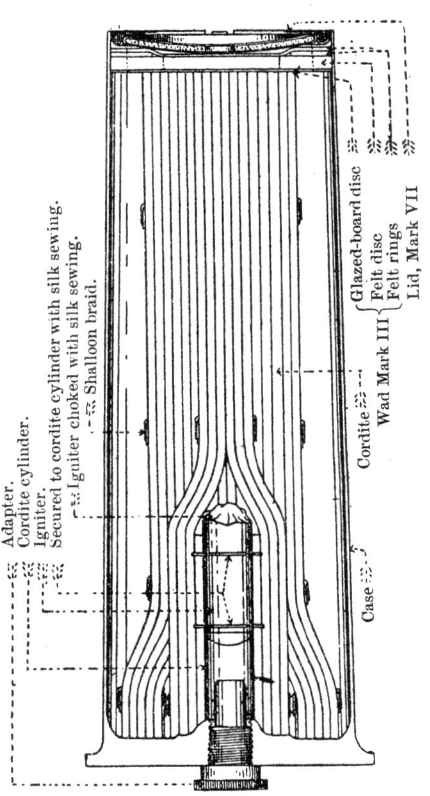

The Mark VIII lid is used in making up this Mark of cartridge; it is concave, and contains a perforated strawboard disc; the Mark VII lid has a strawboard disc with only one central perforation; the Mark VI lid was flat; in the Mark V lid the strawboard disc had no central perforation; the Mark IV contained a lubricant of bees-wax and tallow.

The *Mark V cartridge* differs from the Mark VI in the charge, which is made up in a different manner, a shalloon bag with pocket for the cordite cylinder and igniter being used.

Mark IV cartridge differs from Mark V in the following points: the lengths to which the cordite is cut; the shalloon bag is longer, and above the charge is a paper cylinder, with a disc of silk cloth on its underside, and above it a felt wad.

A certain number of Mark IV cartridges have been altered to Mark IV* by omitting the paper cylinder, and filling up the space by a glazed-board disc and felt wads.

Mark III 4·7-inch cartridge differs from Mark IV in having the Mark II igniter and no shalloon bag.

Owing to the fact that portions of this igniter sometimes remained in the chamber, unconsumed, and caused the shell to jam, when loading the next round, Mark III cartridges have been altered to Mark III*.

Mark III* cartridges are made to conform as far as possible to Mark V, but the cordite being shorter, the space above the felt and glazed-board wad is filled in by one or more felt wads.

Earlier Marks of Reduced Charge.

The earlier Marks of the 2 lb. $5\frac{3}{4}$ oz. cordite cartridge are fitted with cordite cylinder igniters and an older pattern of adapter.

The Mark III cartridge has a concave lid, the Mark II a flat lid.

The Mark I has the cordite charge made up in a different way; a shalloon bag with pocket for the cordite cylinder igniter is used.

Q.F. 4·7-inch, Mark V Gun (L.S.).

Cartridge, Q.F. 4·7-inch, Mark V Gun, filled, 8 lb. 10 oz. Cordite, M.D., Size 16. With Adapter, Mark I. With Primer, Mark I.

This cartridge is made up in the same way as the Q.F. 12-pr., 2 lb. Cordite, M.D., Mark I cartridge described on page 444.

The cartridge case, Mark II, differs from that for Marks I to IV* guns in being longer; it takes the same lid.

The charge is divided into two portions. The larger portion consists of sticks nearly the full length of the case, which are secured together by two pieces of shalloon braid. Round the bottom of these sticks a short shalloon bag, with a pocket for the igniter, is placed and the short sticks, which form the second portion of the charge, are packed in this bag, thus enlarging the diameter of the charge at the bottom. The bag is provided with a draw-string of shalloon braid which is pulled in and tied tightly above the short sticks. Two pieces of shalloon braid, attached to the bag, are passed up round the securing braid on the long sticks and fastened in the centre of the cordite by a reef knot.

The Mark I empty case differed from the Mark II in being wider at the mouth, and took a special white-metal lid.

Q.F. 6-inch Guns.

There are two cordite cartridges for the Q.F. 6-inch guns, the latest Marks of which are :—

Full Charge.

Cartridge, Q.F. 6-inch Gun, Short, filled, 13 lbs. 4 ozs. Cordite, Size 30. With Adapter, Mark X | C | .

Reduced Charge.

Cartridge, Q.F. 6-inch Gun, Short, filled, 5 lbs. 8 ozs. Cordite, Size 10. With Adapter, Mark III | N | . With Primer, Mark III | N | .

Full Charge.

The 13 lbs. 4 ozs. charge is made up in the same way as the 12-pr. of 12 cwts., 2 lbs. M.D. cordite. Mark III, cartridge, described on page 441, differing from it only in weight, dimensions, nature and size of cordite used.

Q.F. 6-inch cases.—There are two different cases for the 6-inch—the " long " and the " short." The former was introduced for powder charges, and is still so used.

It has also been filled with cordite charges, but in future these will be made up in short cases. In consequence of this the word " long " or " short " is found in the name of each filled cartridge.

Long cases which are not required for powder charges are cut down to the same length as the short cases, and take the same white-metal lid.

The latest Mark of the empty " short " case is Mark IV ; it has 6 tongues at the mouth. The Mark III case has only 3 tongues ; when fitted with 3 additional tongues it will be known as Mark III*.

The latest pattern of " long " case is Mark II ; when cut down to make a short case it is known as Mark II* if provided with 3 tongues, and as a Mark II** if provided with 6 tongues.

Mark I cases were not cut down.

The latest Mark of lid for the 6-inch short cartridge is Mark V. For earlier Marks of 6-inch cartridges, *see* Table 35.

Reduced Charge.

The 5 lbs. 8 ozs. cartridge is illustrated in Fig. 143.

Fig. 143.

CARTRIDGE, Q.F. OR Q.F.C. 6-INCH GUN, SHORT, FILLED, 5 LB. 8 OZ. CORDITE, SIZE 10, MARK III.

Scale $\frac{1}{3}$.

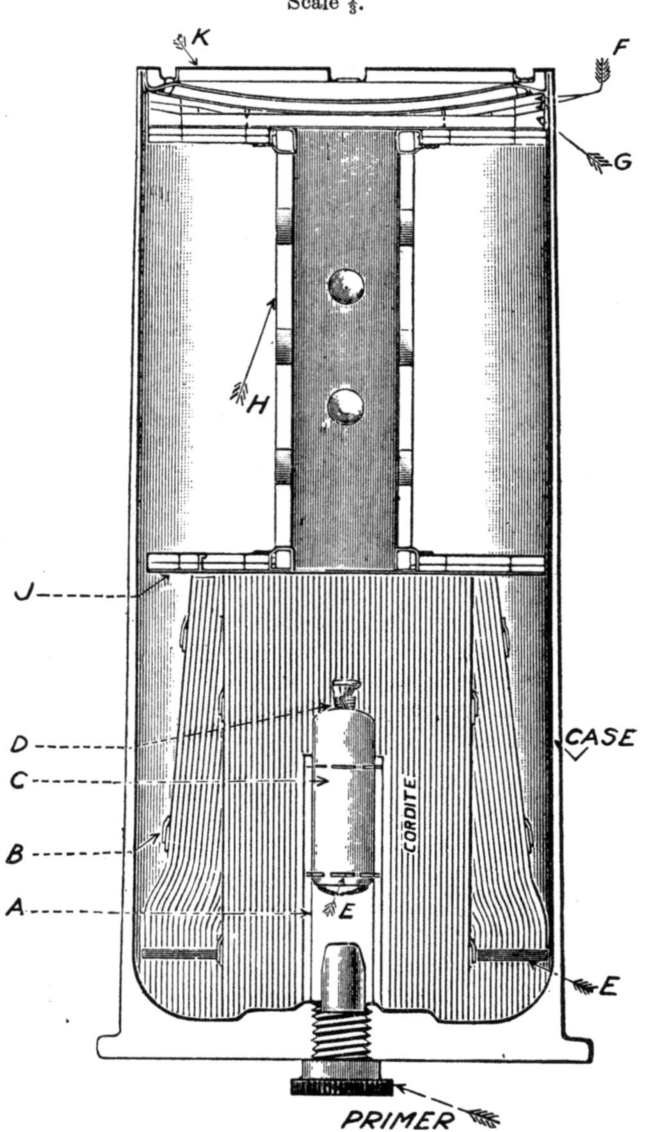

A.—Cordite cylinder. B.—Shalloon braid. C.—Igniter.
D.—Silk, sewing, No. 2. E.—Silk, sewing, No. 1. F.—Felt rings.
G.—Felt disc. H.—Paper cylinder. J.—Glazed-board disc.
K.—Lid, Mark III.

It consists of a Mark IV, III*, or II** case with "Primer, Electric, Large" or "Adapter."

The charge is made up by tying a portion of the cordite with shalloon braid, a Mark IV igniter being fitted in the centre at the base. The remainder of the charge is divided into six portions, which are tied around the central bundle with shalloon braid.

A paper cylinder, a Mark III felt and glazed-board wad and a Mark III lid are placed on top of the charge, the 6 tongues of the case being turned over to secure the lid.

Q.F. Cartridges for Paper Shot.

(*Plate* XCIV.)

The under-mentioned cartridges are used for firing paper shot from guns in the Land Service, which are so situated that they cannot fire their *Service*, or even a *Practice* projectile, in time of peace.

Cartridge, Q.F. 12-pr., 12 cwts., filled, 3¾ lbs., S.P. With Adapter, Mark I | L | . With Primer, Mark I | L | .

The 3¾ lbs. S.P. powder cartridge consists of a Service pattern case, with an adapter or primer.

The charge is contained in a silk cloth bag which is hooped with silk braid and is choked with silk sewing ; in the base of the bag is secured a Mark II 8½ drams R.F.G.[2] igniter in a perforated calico and paper dome.

The mouth of the case is closed by a felt wad and a white-metal lid.

Cartridge, Q.F. 4·7-inch Marks I to IV* Guns, filled, 12½ lbs., S.P. With Adapter, Mark I | L | . With Primer, Mark I | L | .

The 12½ lbs. S.P. powder cartridge consists of a Service pattern case, the interior being coated with brown lacquer.

The charge is placed in loose, and the mouth of the case is closed by a felt and glazed-board wad and a brass lid.

Cartridge, Q.F. 6-inch Gun, Long, filled, 29¾ lbs., E.X.E. With Adapter, Mark I | L | . With Primer, Mark I | L | .

The 29¾ lbs. E.X.E. powder cartridge is made up in a Mark I or II long case, the interior being coated with brown lacquer.

The charge is made up in two portions. The first portion consists of 13 lbs. of powder placed in the case by hand, in 9 layers, the bottom and second layers having a priming of prism[1] black in the centre. The second portion, consisting of 16¾ lbs. of powder, is built up in 13 layers and enclosed in a shalloon bag choked with silk sewing ; a felt wad is placed over the second portion, and the mouth of the case closed by a brass lid.

TABLE NO. 35.—*Q.F. Separate Ammunition.*

Para. in List of Changes	Nature of Gun.	Ser-vice.	Mark of Filled Cart-ridges.	Charge.	Igniter.	Bag.	Number Packed.	Package.	Remarks.
16652 16653	12-pr., 4 cwt.	N	I	1 lb. 0 oz. 12 drams cordite M.D., Size 8	Nil	Nil	8	Box, Cartridge, Q.F. 12-pr., 4 cwt.	
7738 8610	12-pr., 8 cwt.	N	I	13¾ ozs. cordite, Mark I, Size 10	1¼ ozs. Mark III	Shalloon	10	Box, Cartridge, Q.F. 12-pr., 8 cwt.	
14599	Do.	N	II	Do.	1¼ ozs. Mark IV	Nil	10	Do.	Charge:— 1 central bundle. 6 outer bundles.
	Do.	N	III	Do.	Metal	Nil	10	Do.	Metal igniter introduced.
7904 8420	12-pr., 12 cwt.	C	I	1 lb. 15 ozs. cordite, Mark I, Size 15	8½ drams Mark II	Nil	10	Box, Cartridge, Q.F. 12-pr., 12 cwt.	
8876	Do.	C	II	Do.	1¼ ozs. Mark III	Shalloon	10	Do.	Cordite cylinder introduced.
14599	Do.	C	III	Do.	1¼ ozs. Mark IV	Nil	10	Do.	Charge:— 1 central bundle. 5 outer bundles.
	Do.	L	IV	Do.	Metal	Nil	10	Do.	Metal igniter introduced.
7904 12030	Do.	C	I	2 lbs. cordite M.D., Size 11	10 drams Mark I	Shalloon	10	Do.	M.D. cordite introduced.
14599	Do.	C	II	Do.	10 drams Mark II	Nil	10	Do.	Charge:— 1 central bundle. 6 outer bundles.
	Do.	C	III	Do.	Metal	Nil	10	Do.	Metal igniter introduced.

14611	Do.	N	I Reduced	13 ozs. 14 drams cordite M.D., Size $4\frac{1}{4}$	$1\frac{1}{4}$ ozs. Mark IV	Nil	10	Do.	Charge :— 1 central bundle. 6 outer bundles. For Gunnery Schools.
16637	Do.	N	I Reduced	13 ozs. cordite Mark I, Size $7\frac{1}{2}$	Do.	Nil	10	Do.	Do.
12888 12923	12-pr., 18 cwt.	N	I	2 lbs. 12 ozs. 2 drams cordite M.D., Size II	Special, 10 drams. No cordite cylinder	Shalloon	8	Box, Cartridge, Q.F. 12-pr., 18 cwt.	Charge :— 1 central bundle. 8 outer bundles.
14599	Do.	N	II	Do.	10 drams Mark II	Nil	8	Do.	Charge :— 1 central bundle. 6 outer bundles. Paper cylinder.
	Do.	N	III	Do.	Metal	Nil	8	Do.	Metal igniter introduced.
13102	Do.	N	I Reduced	1 lb. 11 ozs. 14 drams cordite M.D., Size 8	Special, 10 drams. No cordite cylinder	Shalloon	8	Do.	Charge :— 1 central bundle. 6 outer bundles. Paper cylinder.
14599	Do.	N	II Reduced	Do.	10 drams Mark II	Nil	8	Do.	Do.
12757 13783	14-pr. ...	N	I	2 lbs. 12 ozs. cordite M.D., Size 11	Special, 10 drams. No cordite cylinder	Shalloon	10	Box, Cartridge, Q.F. 14-pr.	Charge :— 1 central bundle. 12 outer bundles. Special felt and paper cylinder.
14599	Do. ...	N	II	Do.	8 drams Mark I	Nil	10	Do.	Charge :— 1 central bundle. 6 outer bundles.
16821	Do. ...	N	III	Do.	Metal	Nil	10	Do.	Metal igniter introduced.

TABLE No. 35.—*Q.F. Separate Ammunition*—continued.

Para. in List of Changes	Nature of Gun.	Service.	Mark of Filled Cartridges.	Charge.	Igniter.	Bag.	Number Packed.	Package.	Remarks.
13166 15208	15-pr. ...	L	I*	1 lb. 2 ozs. 10 drams cordite M.D.T., Size 20–10	4 drams Mark II	Nil	22	Box, Cartridge, Q.F. 15-pr.	
16318 15208	Do. ...	L	II	Do.	Do.	Nil			
16319	Do. ...	L	III	Do.	4 drams Mark III	Nil			
13346	Do. ...	L	I	4 ozs. cordite Mark I, Size 5	2 drams G.C. yarn	Shalloon	40	Case, Powder, M.L., whole	
	Do. ...	L	II	Do.	4 drams S.F.G.[2]	Do.			
8193 8477	4-inch Marks I to III* guns	N	I	3 lbs. 9 ozs. cordite Mark I, Size 15	$1\frac{1}{4}$ ozs. Mark II	Nil	8	Box, Cartridge, Q.F. 4-inch, Marks I to III* guns	To be converted to Mark II (§10323).
9532	Do.	N	II	Do.	$1\frac{1}{4}$ ozs. Mark III	Shalloon	8	Do.	Mark III igniter.
14599	Do.	C	III	Do.	$1\frac{1}{4}$ ozs. Mark IV	Nil	8	Do.	Charge:— 1 central bundle. 5 outer bundles.
	Do.	C	IV	Do.	Metal	Nil	8	Do.	Metal igniter introduced.
15142 16636	Do.	C	I	3 lbs. 8 ozs. 11 drams cordite M.D.T., Size 20–10	$1\frac{1}{4}$ ozs. Mark IV	Nil	8	Do.	Charge:— 1 central bundle. 6 outer bundles.

	Do.	C	II	Do.	Metal	Nil	8	Do.	Metal igniter introduced.
8193 14601 16638 14599 16651	4-inch Marks III and III*	L	II Reduced	1 lb. 8 ozs. cordite Mark I, Size 5	$1\frac{1}{4}$ ozs. Mark IV	Nil	8	Do.	Flat lid.
	Do.	L	III Reduced	Do.	Do.	Nil	8	Do.	Concave lid.
	Do.	L	IV Reduced	Do.	Metal	Nil	8	Do	Paper cylinder, wad and lid suitable for shell with night tracer.
16638 15329 14608	4-inch Marks I to III* (Q.F. or Q.F.C.)	C	1 Reduced	1 lb. 10 ozs. 14 drams cordite M.D., Size $4\frac{1}{4}$	$1\frac{1}{4}$ ozs. Mark IV	Nil	8	Do.	Flat lid.
	Do.	C	II Reduced	Do.	Do.	Nil	8	Do.	Concave lid.
	Do.	C	III Reduced	Do.	Metal	Nil	8	Do.	Paper cylinder, wad and lid suitable for shell with night tracer.
16638 16651	4-inch Mark IV	N	I	5 lbs. 1 oz. 12 drams cordite M.D., Size 16	Nil	Nil	6	Box, Cartridge, Q.F. 4-inch Mark IV gun	
	Do.	N	II	Do.	Nil	Nil	6	Do.	Differs from Mark I in size of recess at bottom of charge.
	Do.	N	I	5 lbs. 1 oz. 12 drams cordite M.D., Size 16	Nil	Nil	6	Do.	
16694	4-inch Mark V	C	I	7 lbs. 11 ozs. cordite M.D., Size 16	Nil	Nil	6	Box, Cartridge, Q.F. 4-inch Mark V gun	
15434 15486	4·5-inch Howitzer	L	I	15 ozs. 14 drams cordite M.D., Sizes $2\frac{1}{4}$ and $4\frac{1}{4}$	Nil	Cambric	1	Box, Cartridge, Q.F., 4·5-inch Howitzer	
	Do.	S I	I	14 ozs. cordite Mark I, Sizes $3\frac{3}{4}$ and 5	Nil	Do.	1	Do.	

TABLE NO. 35—*Q.F. Separate Ammunition*—continued.

Para. in List of Changes.	Nature of Gun.	Ser-vice.	Mark of Filled Cart-ridges.	Charge.	Igniter.	Bag.	Number Packed.	Package.	Remarks.
9511	4·7-inch, Marks I to IV*	C	IV	5 lbs. 7 ozs. cordite Mark I, Size 20	1¼ ozs. Mark III	Shalloon	6	Box, Cartridge, Q.F., 4·7-inch, Marks I to IV* guns; also Boxes, Cartridge, Q.F., 4·7-inch, "Naval outfit" and "Naval transport"	
9511	Do.	C	V	Do.	Do.	Do.	6		
14599	Do.	C	VI	Do.	1¼ ozs. Mark IV	Nil	6		Charge:— 1 central bundle. 6 outer bundles.
	Do.	C	VII	Do.	Metal	Nil	6		Metal igniter introduced.
13307	Do.	C	I Reduced	2 lbs. 5¾ ozs. cordite Mark I, Size 7½	1¼ ozs. Mark III	Shalloon	6		
14601	Do.	C	II Reduced	Do.	1¼ ozs. Mark IV	Nil	6		Flat lid.
14599	Do.	C	III Reduced	Do.	Do.	Nil	6		Concave lid.
	Do.	C	IV Reduced	Do.	Metal	Nil	6		Paper cylinder, wad and lid suitable for shell with night tracer.
11698	4·7-inch, Mark V	L	I	8 lbs. 10 ozs. cordite M.D., Size 16	1¼ ozs. Mark III	Shalloon	4	Box, Cartridge, Q.F. 4·7-inch Mark V gun	

8875	6-inch ...	C	VI	13 lbs. 4 ozs. cordite Mark I, Size 30	Do.	Do.	4		
9677	Do. ...	C	VII	Do.	Do.	Do.	4	Box, Cartridge, Q.F. 6-inch; also Boxes Cartridge, Q.F., "Naval outfit" and "Naval transport"	
14065	Do. ...	C	VIII	Do.	$1\frac{1}{4}$ ozs. Mark IV	Nil	4		
	Do. ...	C	IX	Do.	Do.	Nil	4		
	Do. ...	C	X	Do.	Metal	Nil	4		Metal igniter introduced.
14065	Do. ...	N	III Reduced	5 lbs. 8 ozs. cordite Mark I, Size 10	$1\frac{1}{4}$ ozs. Mark IV	Nil	4		

TABLE NO. 36.—*Powder Cartridges for use with Q.F. Guns when firing Paper Shot in the Land Service.*

Para. in List of Changes	Nature of Gun.	Mark of Cart-ridge.	Weight and Nature of Powder.	Length. / Diameter. (Inches.)	No. in Package.	Package.	Remarks.
10357 8420	12-pr. 12 cwt.	I	$3\frac{3}{4}$ lbs. S.P.	Made up in Service Brass Cases closed by a felt wad and a white metal lid	Charges. 30 14	Case powder, metal-lined :— Whole Half	
10960 10961 9624	4·7-inch, Marks I to IV*	I	$12\frac{1}{2}$ lbs. S.P.	Made up in Service Brass Cases closed by a wad of felt and glazed-board and a lid	The powder is placed in the brass cases loose		
11697 9624	6-inch Q.F.	I	$29\frac{3}{4}$ lbs. E.X.E.	Made up in Service Long Brass Cases, 13 lbs. loose, 160 lbs. in a shalloon bag, closed by a felt wad and lid			

SECTION D.—BLANK CARTRIDGES FOR Q.F. GUNS.

Q.F. 1-pr. Blank.

Cartridge, Q.F., Blank, 1-pr., filled, Mark II | *C* | consists of the Service case fitted with a percussion cap, and containing an igniter of 12 grains of nitrated canvas on top of which a charge of size 20 sliced cordite is placed loose.

The charge is covered by a thick paper wad, supported by four indents in the walls of the case. On top of this wad, a paper disc is placed and the mouth of the case is turned in all round.

The Mark I cartridge had a charge of F.G. powder, *see* Table 37.

Q.F. 3 and 6-pr. Blank.

Cartridge, Q.F., Blank, 3-pr., filled, Mark IV | C | .

Cartridge, Q.F., Blank, 3-pr., Vickers', filled, Mark I | N | .

Cartridge, Q.F., Blank, 6-pr., filled, Mark IV | N | .

Cartridge, Q.F., Blank, 6-pr., filled, H.M.S. " Swiftsure " and " Triumph," Mark II | N | .

(*Plate* XCV.)

The above-mentioned blank cartridges are made up in the same way; they consist of a case, primer, charge, felt jacket and wads.

The cases are special for blank, and are shorter than the Service ones.

They are prepared at the base to take the primer mentioned below. Two slots are cut near the mouth to receive the braid used to secure the charge.

The cases are lacquered inside and out with transparent lacquer.

The numeral of the case corresponds with that of the filled cartridge.

The primer used in making up the above cartridges is the ***Primer,*** *Percussion, Blank,* 2·95-inch, 6, or 3-pr., described on page 384.

The charge consists of 11 ozs. of Blank L.G. for the 3-pr., and 3-pr. Vickers; 15 ozs. for the 6-pr., Mark IV, and 1½ lbs. for the " Swiftsure " and " Triumph." It is contained in a shalloon bag (except for the " Swiftsure " and " Triumph," which is of silk cloth with a shalloon base); the bag is hooped and choked with silk sewing.

The felt jacket encloses the upper half of the charge and is sewn to it. A silk braid loop is attached to the top of the felt jacket.

A felt wad and *millboard wad* are placed above the charge. These wads have slots in the centre through which the silk braid loop of the felt jacket is passed. A piece of silk or shalloon braid is passed through the slots in the case and through the loop on the felt jacket, and tied on top, so securing the charge in the case.

Reduced Blank Cartridge, Q.F., 6-pr.

(*Mark II Cartridge.*)

Fig. 144.

CARTRIDGE, Q.F., BLANK, 6-PR., REDUCED, MARK II.

Scale $\frac{1}{3}$.

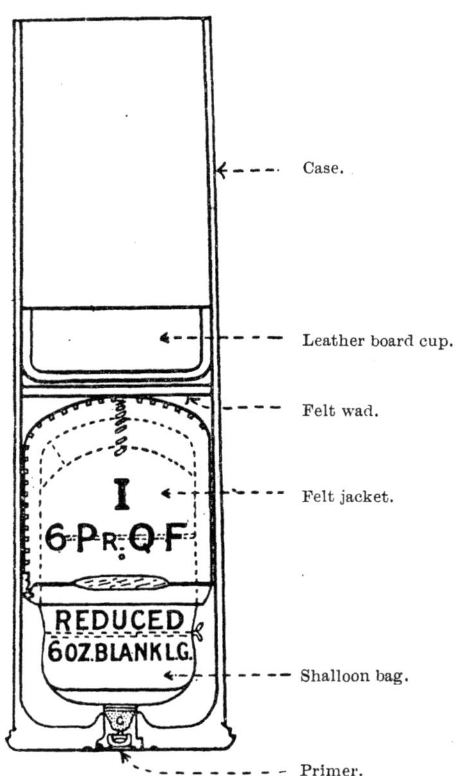

Cartridge, Q.F., Blank, 6 pr., Reduced, Mark II | *C* | consists of a brass case, primer, charge of blank L.G. powder, felt jacket, felt wad, and a leather board cup.

The *case* and *primer* are the same as used with the 15 ozs., Mark IV cartridge, described above.

The charge consists of 6 ozs. of blank L.G. in a shalloon bag, hooped and choked with silk sewing.

The felt jacket is placed over the top of the charge, and is secured near the base by a silk braid draw string.

A silk braid loop is secured to the top of the felt jacket.

A felt wad is placed above the charge when in the case.

A leather board cup is driven in from the mouth of the case and secures the charge in position.

The Mark I Cartridge.

The Mark I cartridge differed from the Mark II in the method of securing the charge.

A split paper cylinder and millboard wad were placed above the charge, and the silk braid loop of the felt jacket passed through them. The whole were then secured by a piece of braid passed through the slots in the case and tied on top.

Q.F., 2·95-INCH BLANK.

CARTRIDGE, Q.F., BLANK, 2·95-INCH, FILLED, MARK I | L | .

Fig. 145.

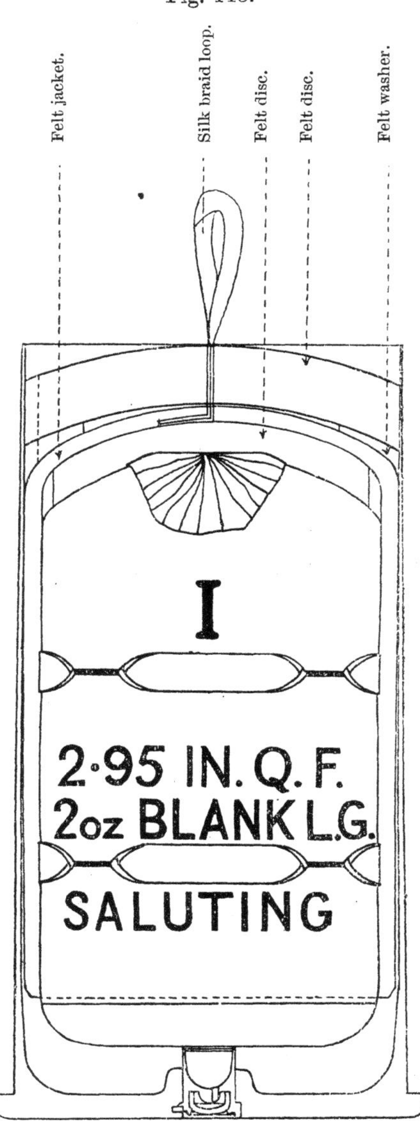

The Cartridge, Q.F., Blank, 2·95-inch, filled, Mark I | L | consists of a case, primer, charge, felt jacket and wad.

The case is specially prepared to take the "Primer, Percussion, Blank, 2·95-inch, 6 or 3-pr." described on page 384.

The charge consists of 12 ozs. of Blank L.G. in a silk cloth bag, having a shalloon base. It is choked with silk sewing and hooped with silk braid.

The felt jacket is placed over the top of the charge and is secured near the base with a draw-string of silk sewing; a silk braid loop is attached to the top of the felt jacket.

The wad consists of a disc of felt, having a ring of felt stitched to the underside.

A slot is cut in the wad through which the loop of the felt jacket is passed.

The wad is a tight fit in the mouth of the case and is the only means of securing the charge in position.

Q.F. 15-pr. Blank.

Cartridge, Q.F., Blank, 15-pr., filled, Mark I | L |.

The Cartridge, Q.F., Blank, 15-pr., filled, Mark I | L | consists of a case, charge, felt jacket and wad.

The case is a fired Service case, Mark I or II, or a case sentenced for blank.

The charge, felt jacket and wad are similar to those described for the 2·95-inch, but a charge of 1 lb. of Blank L.G. is used.

Q.F., 12-pr. of 4 cwt., 13 and 18-pr. Blank.

Cartridge, Q.F., Blank, 12-pr., 4 cwt., filled, Mark I | N |.

Cartridge, Q.F., Blank, 13-pr., filled, Mark II | L |.

Cartridge, Q.F., Blank, 18-pr., filled, Mark II | L |.

(*Plate* XCVI.)

The above-mentioned blank cartridges consist of a case, primer, charge, felt jacket and leather board cup. The 18-pr. has also a split paper ring.

The case is a fired Service case, or one sentenced for blank.

The primer is the same as that used with Service ammunition; it is described on page 379.

The charge consists of 1 lb. of Blank L.G. in a silk cloth bag, having a shalloon base. It is hooped with silk braid and choked with silk sewing. The same bag is used for the 13 and 18-pr. cartridges, but in the charge for the 12-pr. of 4 cwt. cartridge, the hoops pass under strips of silk braid sewn to the bag instead of passing through holes in the bag itself.

The felt jacket is placed over the top of the charge and is secured near the base with a draw-string of silk sewing. A silk braid loop is attached to the top of the jacket.

The leather board cup.—The same cup is used in each case, but owing to the 18-pr. case being larger in diameter, a split paper ring is inserted on top of the charge, into which the cup fits.

Q.F., Blank, 12-pr., 8, 12, and 18 cwt.

Q.F. 12-pr., 8 cwt., Blank.

Cartridge, Q.F., Blank, 12-pr., 8 cwt., filled. With Primer, Mark I | N | . With Adapter, Mark I | N | .

(*Plate* XCVII.)

The Cartridge, Q.F., Blank, 12-pr., 8 cwt., filled, Mark I | N | consists of case, electric primer or adapter, charge with igniter, felt jacket and wad.

The case is a Service case, or one sentenced for blank, fitted with a primer, electric, large, or an adapter.

The charge with igniter consists of 1½ lbs. of Blank L.G. in a silk cloth bag, hooped with silk braid and choked with silk sewing. In the base of the bag a pocket is formed into which is sewn the igniter.

The igniter, Mark II, consists of a perforated calico and paper dome, into which is secured a shalloon bag containing 8½ drams of R.F.G.^{2} powder.

The felt jacket and *wad* are similar to those used with the 2·95-inch blank cartridge and described on page 470.

Q.F. 12-pr., 12 cwt., Blank.

Cartridge, Q.F., Blank, 12-pr., 12 cwt., filled. With Primer, Mark II | C | . With Adapter, Mark II | C | .

(*Plate* XCVIII.)

The Cartridge, Q.F., 12-pr., 12 cwt., filled, Mark II | C | consists of a case, primer or adapter, charge with igniter, felt wad and leather board cup.

The case is a Service case or one sentenced for blank.

The electric primer or adapter and the blank charge with igniter are the same as used with the 12-pr., 8 cwt. blank cartridge.

A felt wad is placed on top of the charge, and a leather board cup pressed into the mouth of the case retains the charge in position.

Q.F. 12-pr., 18 cwt., Blank.

Cartridge, Q.F., Blank, 12-pr., 18 cwt., filled. With Primer, Mark I | N | . With Adapter, Mark I | N | .

The Cartridge, Q.F., Blank, 12-pr., 18 cwt., filled, Mark I | N | is made up in the same way as the 12-pr. 12 cwt. blank cartridge, but differs from it in the following particulars :—

(i) The felt wad is thicker and larger in diameter.

(ii) A split paper ring is placed on top of the charge, into which the leather board cup is forced. (This is due to the same leather board cup being used for both 12-pr. of 12 cwt. and 12-pr. of 18 cwt. blank cartridges.)

Q.F. and Q.F.C., 4-inch, Blank.

(4-inch Marks I to III* Guns.)

Cartridge, Q.F. or Q.F.C., Blank, 4-inch, Marks I to III*, filled. With Primer, Mark IV | C | . With Adapter, Mark IV | C | .

The blank cartridge for the above guns is made up in the same way as the Mark II blank cartridge for the 12-pr. of 12 cwt.

The charge is 3 lbs. of Blank L.G. enclosed in a silk cloth bag, with calico and paper dome igniter.

The hoops around the charge pass under strips of braid sewn to the bag instead of passing through holes in the bag itself.

The Mark III blank cartridge for the Q.F. or Q.F.C., 4-inch, differs from the Mark IV in having an asbestos cylinder placed over the charge instead of a felt wad and leather board cup ; the mouth of the case is closed by a Service lid ; if a dished lid is used a Mark III wad is required.

Plate XCIX illustrates this method of filling.

(The 4-inch, Marks IV and V Gun Cartridges.)

Cartridge, Q.F., Blank, 4-inch, Mark IV Gun, filled, Mark I | N | .

Cartridge, Q.F., Blank, 4-inch, Mark V Gun, filled, Mark I | N | .

The blank cartridge for the Q.F., 4-inch, Marks IV and V guns consists of a case, primer, charge, felt jacket, split ring, and leather board cup.

The case is a Service case or one sentenced for blank, fitted with a " Primer, Percussion, No. I, Mark II. "

The charge consists of 3 lbs. of Blank L.G. in a silk cloth bag, having a shalloon base. It is choked with silk sewing and hooped with silk braid, the hoops passing under strips of silk braid sewn to the bag instead of through holes in the bag itself.

The felt jacket is placed over the top of the charge, and is secured to the charge near the base by a draw-string of silk sewing.

A silk braid loop is secured to the top of the jacket.

The split ring is made of leather board in the cartridge for the Mark IV gun and of paper in the cartridge for the Mark V gun.

The rings have the lower edge coned, and are split vertically ; they are placed on top of the charge. The leather board cup is pressed into the split ring by means of a wooden drift.

Plate XCVI illustrates the method of making up the *blank* cartridges for the Q.F., 4-inch, Marks IV and V guns.

Q.F., 4·5-inch Howitzer, Blank.

Cartridge, Q.F., Blank, 4·5-inch Howitzer, filled, Mark II | L | .

Th Cartridge, Q.F., Blank, 4·5-inch Howitzer, filled, Mark II | L | consists of a case, primer, charge, felt jacket and leather board lid.

The case is a Service case or one sentenced for blank, fitted with a "Primer, Percussion, No. 1."

The charge, consists of $1\frac{1}{4}$ lbs. of R.F.G.[2] contained in a silk cloth bag having a shalloon base. The bag is not choked, but is closed by a disc of silk cloth, sewn to the mouth of the bag.

The felt jacket is passed over the top of the charge and lightly stitched to it in four places.

The leather board lid is the same as is used with the Service cartridge.

Blank for Q.F., 4·7-inch and 6-inch.

The latest Marks of blank cartridges for the 4·7-inch and 6-inch Q.F. guns are made up in the same way, consisting of a Service case or a case sentenced for blank, and electric primer or adapter, charge of blank L.G. powder, asbestos cylinder, wad, and concave lid.

Q.F., 4·7-inch, Blank (Plate XCIX.)

Cartridge, Q.F., Blank, 4·7-inch, filled. With Adapter, Mark V. With Primer, Mark V.

The charge of 3 lbs. blank L.G. is enclosed in a silk cloth bag with a dome-shaped igniter into which is secured a shalloon bag containing $8\frac{1}{2}$ drams R.F.G.[2] or new blank F.G. powder. The bag is hooped with silk braid in the ordinary way, but the hoops pass under strips of braid sewn to the bag so as to form loops, instead of through the silk cloth.

This prevents the escape of powder dust from the bag into the interior of the brass case.

The space above the charge is filled up by an asbestos cylinder, consisting of two perforated sheets of asbestos, each cut down the centre for half its length and fitted one across the other at right angles, with a disc of asbestos secured at each end.

The case is closed at the mouth with the latest pattern of lid and wad as used with the Service cartridge.

The Mark IV blank cartridge had a longer asbestos cylinder and a flat lid. With a flat lid, a wad is not used.

Q.F., 6-inch, Blank.

Cartridge, Q.F., Blank, 6-inch Gun, Short, Filled. With Adapter, Mark V | C | . With Primer, Mark V | C | .

The Mark V blank cartridge is made up in exactly the same way as the 4·7-inch blank; the charge is 7 lbs. of blank L.G.

The Mark IV has a longer asbestos cylinder, and a flat instead of a concave lid. With a flat lid, a wad is not used.

TABLE No. 37.—*Blank Cartridges for Q.F. and Q.F.C. Guns and Howitzers.*

Para. in List of Changes.	Nature of Gun.	Service.	Mark of Cartridge.	Charge.	Igniter.	Cylinder.	Means of closing mouth of case.	Charges.		Made up Cartridges.		Remarks.
								No. packed.	Package.	No. packed.	Package.	
11877	1-pr. Q.F.	L	I	2 ozs. 30 grs. F.G. powder	—	Card-board	Coned over a jute wad	—	—	—	—	Service pattern case and cap.
13386	1-pr. Q.F.	C	II	265 grs. cordite, size 20 sliced	12 grs. nitrated canvas	—	Case turned in all round over a paper wad	—	—	—	—	Do.
8299 9235 15182	3-pr. Q.F.	C	IV	11 ozs. Blank L.G.	—	—	Felt and mill-board wads tied with silk braid	50 20	Case, powder, M.L., Half Case, powder, M.L., Quarter	20	Box, cartridge, Q.F., 3-pr., Blank § 5954	Special blank case and Primer, Percussion, Blank, 2·95-inch, 6 or 3-pr.
13423 13438 15182	3-pr., Q.F. Vickers	N	I	Do.	—	—	Do.	43	Case, powder, M.L. Half	20	Box, cartridge, Q F., Blank	Do.
8299 8440 9235 14316 15182	6-pr. Q.F.	N	IV	15 ozs. Blank L.G.	—	—	Do.	37	Do.	20	Box, cartridge, Q.F., 6-pr. Blank § 5944	Do.

10116 12985 14316 15182	6-pr. Q.F. Reduced	C	I	6 ozs. Blank L.G.	—	Paper split	Felt wad, paper cylinder and a millboard wad tied with silk braid	144 60 28	Case, powder, M.L. :— Whole Half Quarter	20	Box, cartridge, Q.F., Blank, 6-pr.	Do
15183	Do.	C	II	Do.	—	—	Felt wad and leather - board cup	144 60 28	Case, powder, M.L. :— Whole Half Quarter	20	Do.	Do.
12985 15182 16334	6-pr. Q.F., H.M. Ships "Swift-sure" and "Triumph"	N	I	$1\frac{1}{2}$ lbs. Blank F.G.	—	—	Felt and mill-board wads tied with silk braid	60 30 9	Case, powder, M.L. :— Whole Half Quarter	—	—	With cap, 2·95-inch, 6 or 3-pr.
12985 16334	Do.	N	II	$1\frac{1}{2}$ lbs. Blank L.G. Mark I	—	—	Do.	60 30 9	Case, powder, M.L. :— Whole Half Quarter	—	—	Special blank case and Primer, Percussion, Blank, 2·95-inch, 6 or 3-pr.
10899 10962 15182	2·95 - inch Q.F.	L	I	12 ozs. Blank L.G.	—	—	Felt wad ...	90 40 18	Case, powder, M.L. :— Whole Half Quarter	—	—	Do.

TABLE NO. 37.—*Blank Cartridges for Q.F. and Q.F.C. Guns and Howitzers*—continued.

Para. in List of Changes.	Nature of Gun.	Service.	Mark of Cart-ridge.	Charge.	Igniter.	Cylinder.	Means of closing mouth of case.	Charges.		Made up Cartridges.		Remarks.
								No. packed.	Package.	No. packed.	Package.	
16633	12-pr. Q.F. 4 cwt.	N	I	1 lb. Blank L.G. Mark I	—	—	Leather - board cup	—	—	—	—	Service pattern case with Primer, Percussion No. 1.
8915 9235 12250 12371 14277 15182	12-pr. Q.F. 8 cwt.	N	I	1½ lbs. Blank L.G.	8½ drams R.F.G.[2]	—	Felt wad	50 20	Case, powder, M.L.:— Whole Half	10	Box, cartridge, Q.F. 12-pr. 8 cwt.	Service pattern case, with Primer, Electric, Large, or Adapter.
12250 12371 14277	12-pr. Q.F. 12 cwt.	C	II	Do.	Do.	—	Felt wad and leather - board cup	65 30 11	Case, powder, M.L.:— Whole Half Quarter	10	Box, cartridge, Q.F. 12-pr. 12 cwt.	Do.
13167 14277	12-pr. Q.F. 18 cwt.	N	I	Do.	Do.	—	Felt wad, split paper ring and leather - board cup	65 30 11	Case, powder M.L.:— Whole Half Quarter	8	Box, cartridge, Q.F. 12-pr. 18 cwt.	Do.

14931	13-pr. Q.F.	L	I	1 lb. Blank L.G.	—	—	Leather - board cup	75 40 18	Case, powder, M.L. :— Whole Half Quarter	20	Box, cartridge case, Q.F. 13-pr. § 14859	Service pattern case and Primer, Percussion No. 1.
15068	Do.	L	II	Do.	—	—	Do.	75 40 18	Case, powder, M.L. :— Whole Half Quarter	20	Do.	Differs from Mark I in dimensions of charge only.
14931	18-pr. Q.F.	L	I	Do.	—	—	Leather - board cup and split paper ring	60 30 11	Case, powder, M.L. :— Whole Half Quarter	20	Box, cartridge case, Q.F. 18-pr. § 14859	Service pattern case and Primer, Percussion No. 1.
15068	Do.	L	II	Do.	—	—	Do.	60 30 11	Case, powder, M.L. :— Whole Half Quarter	20	Do.	Differs from Mark I in dimensions of charge only.
10726 12029 14029 15068	15-pr. Q.F.	L	I	Do.	—	—	Felt wad ...	75 40 18	Case, powder, M.L. :— Whole Half Quarter	—	—	Service pattern, Marks I or II cases, with cap.
13540	4-inch Q.F. or Q.F.C., Marks I to III* guns	C	III	3 lbs. Blank L.G.	8½ drams R.F.G.[2]	—	Asbestos lid ...	30 12 5	Case, powder, M.L. :— Whole Half Quarter	8	Box, cartridge, Q.F. 4-inch, Marks I to III* guns	Service pattern, case, with Primer, Electric, Large, or Adapter.

TABLE NO. 37.—*Blank Cartridges for Q.F. and Q.F.C. Guns and Howitzers*—continued.

Para. in List of Changes.	Nature of Gun.	Service.	Mark of Cartridge.	Charge.	Igniter.	Cylinder.	Means of closing mouth of case.	Charges.		Made up Cartridges.		Remarks.
								No. packed.	Package.	No. packed.	Package.	
14459	4-inch Q.F. or Q.F.C., Marks I to III* guns	C	IV	3 lbs. Blank L.G.	$8\frac{1}{2}$ drams R.F.G.[2]	—	Felt wad and leather - board cup	30 12 5	Case, powder, M.L. :— Whole Half Quarter	8	Box, cartridge, Q.F. 4-inch, Marks I to III* guns	Service pattern case, with Primer, Electric, Large, or Adapter.
16639	4-inch, Mark IV gun	N	I	Do.	—	—	Split leather board ring and leather - board cup	—	—	—	—	Service pattern case, with Primer, Percussion No. 1.
	4-inch, Mark V gun	N	I	Do.	—	—	Split paper ring and leather-board cup	—	—	—	—	Do.
15434	4·5 - inch, Q.F. Howitzer	L	I	$1\frac{1}{2}$ lbs. R.F.G.[2]	—	—	Leather - board cup	90 42 17	Case, powder, M.L. :— Whole Half Quarter	10	Box, cartridge, case, Q.F., 4·5 - inch Howitzer	Do.
16503	Do.	L	II	$1\frac{1}{4}$ lbs. R.F.G.[2]	—	—	Do.	—	—	10	—	Has a felt jacket over charge.

13540	4·7 - inch Q.F. Marks I to IV*	C	IV	3 lbs. Blank L.G.	8½ drams R.F.G.[2]	Asbestos	Lid	30 12 6	Case, powder, M.L. :— Whole Half Quarter	6	Box, cartridge, Q.F. 4·7-inch, Marks I to IV* guns; also	Service pattern case, and Primer, Electric, Large, or Adapter.
15886	Do.	C	V	Do.	Do.	Do.	Dished lid and wad	30 12 6	Whole Half Quarter	6	Boxes, cartridge, Q.F. 4·7 - inch, Naval Outfit and Naval Transport	Do.
13540	6-inch Q.F.	C	IV	7 lbs. Blank L.G.	Do.	Do.	Lid	15 5	Case, powder, M.L. :— Whole Half	4	Box, cartridge, Q.F. 6-inch; also	Do.
14065	Do.	C	V	Do.	Do.	Do.	Dished lid and wad	15 5	Whole Half	4	Boxes, cartridge, Q.F., Naval Outfit and Naval Transport	Do.

SECTION E.—MARKING ON Q.F. CARTRIDGES.

(i) *Marking Common to both Fixed and Separate Ammunition.*

Stamped on the base :—

(*a*) The mark of the empty case.

(*b*) The year of manufacture of the case.

(*c*) The trade mark or initials of the maker of the case.

(*d*) The letter " C " if cordite or ballistite is used, " P " if powder; followed by an F or R for each time the case has been *filled* with a full or reduced charge; F or R barred out thus, F̶, means that the case has been emptied without being fired.

(*e*) The symbol (A) followed by a punch-mark, indicates that the case has been annealed after firing; an additional punch-mark is placed on for each subsequent annealing.

(*f*) All cartridge cases are now annealed in batches, the cases from each manufacturer and each nature of case being kept in distinct batches. The year of annealing, denoted by a letter, and the batch, denoted by a serial number, is then stamped on the base.

Marking stamped on the base of Q.F. case :—

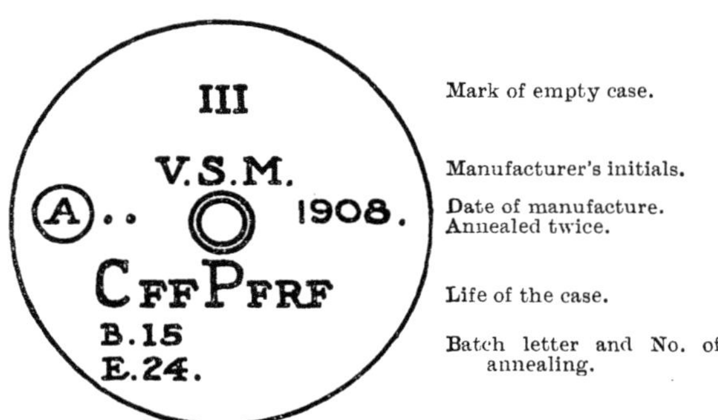

Additional markings stamped on the base.—3-pr. Vickers' cases have the word " Vickers." Q.F., 4·7-inch, Mark V have " For Mark V gun only ", all 6- and 3-pr. cases having a Mark II cap, (1).

The Q.F., 12-pr. of 4 cwt., 13-pr., 15-pr., 18-pr., 3-inch, 2·95-inch, 4-inch, Marks IV and V, and the 4·5-inch Howitzer cartridges have the nature of the gun stamped on them.

(ii) *Marking only found on separate ammunition.*—Separate ammunition has the following information printed on a circular label attached to the lid of the case.

The information is printed in *red* for cordite cartridges, *black* for powder filled cartridges.

CORDITE.

POWDER.

The letters " F " or " F.F." on the circular label on cordite cartridges denote " *First use* " or " *Fire first* " cordite.

(iii) *Marking only found on fixed ammunition.*—Fixed ammunition has the following information *stencilled* on the base of the case :—

(*a*) M.D. or M.C. when M.D. or M.C. cordite is used.

(*b*) Lot letter and No. of cordite, and A.C., if the charge is adjusted.

(*c*) Monogram of filling station.

(*d*) Date of filling.

Additional marking stencilled on the base of 3- and 6-pr. cases :—

(*a*) The mark of the complete filled cartridge.

(*b*) The nature of explosive used in the igniter, *e.g.*, G.C. or S.F.G.².

(*c*) 3-pr. Q.F. Vickers' practice cartridges have the words " Full," or " Reduced " (as the case may be).

(*d*) 3-pr. Q.F Vickers' full charges (Service and Practice) have the word (Spl.) denoting Special Igniter.

Marking stencilled on 3-pr. and 6-pr. shell.—The word " Fuze " and the Mark of fuze used is stencilled in red on the body of the shell. " A " denotes that the shell has been annealed. " N " denotes Naval Service. For special marking on 3-pr. lyddite shell, *see* page 202.

13-pr. and 18-pr. ammunition.—For marking on 13-pr. and 18-pr. ammunition, *see* page 408.

The Colonial Markings are shown on page 105.

SECTION F.—DRILL CARTRIDGES FOR Q.F. GUNS (FIXED AMMUNITION).

Cartridge, Dummy, Q.F., 1-pr., Mark I | C | .

The cartridge is made of brass of the same size as the Service case, but the rim is milled. It is prepared at the base to take the "*Primer, Percussion, Dummy, Q.F. Cartridges, No.* 2, *Mark I.* (*See* Fig. 146.)

Two holes, ·375 inch in diameter, are bored through the case, so that it can be readily distinguished. It is fitted with a hollow teak block, which contains a lead core, by means of which the weight is adjusted. The teak block is held in the case by three brass screws which are long enough to enter the lead core and so secure it in position.

The portion of the teak block projecting from the mouth of the case is shaped like a common shell.

Fig. 146.

PRIMER, PERCUSSION, DUMMY, Q.F. CARTRIDGES, NO. 2, MARK I | C | , 6-, 3-, AND 1-PR.

Scale = $\frac{2}{1}$.

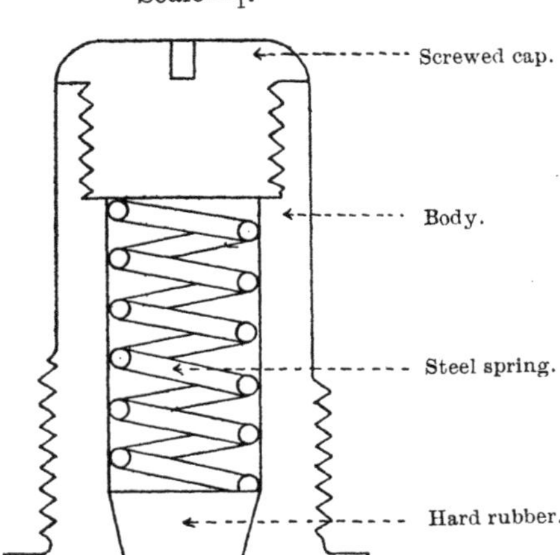

Primer, Percussion, Dummy, Q.F. Cartridges, No. 2, Mark I | C | , 6-, 3- and 1-pr.

The primer consists of a metal body about 1·1 inch long having a flange at the base, above which the primer is screw-threaded for ·35 of an inch, the remainder of the body being plain.

The body is bored out internally, the recess terminating in a tapered portion at the base. Inserted from the front of the primer is a tapered piece of hard rubber, which is kept back in the base of the recess by a steel spring.

The spring is kept compressed by a metal screwed cap, which screws into the front end of the primer.

CARTRIDGE, Q.F., 6-PR., DUMMY DRILL, MARK III | C | .

CARTRIDGE, Q.F., 3-PR., DUMMY DRILL, MARK III | C | .

These cartridges consist of a Service pattern case which is milled around the rim.

They are prepared at the base to receive the "*Primer, Percussion, Dummy, Q.F. Cartridges, No. 2, Mark I.*" (*See* Fig. 146.)

Four holes, 1 inch in diameter, are bored through the case.

The case is fitted with a hollow teak block, which is secured by six screws. The portion of the teak block projecting from the mouth of the case is shaped like a common shell.

CARTRIDGE, DUMMY, Q.F., 18-PR., MARK II | L | .

CARTRIDGE, DUMMY, Q.F., 13-PR., MARK II | L | .

To Practice Fuze Setting.

These cartridges are empty Service cases having four holes bored through the body and three holes bored through the base.

Each case is fitted with a beech wood block having a central perforation.

A steel bolt shaped at the base like the "Primer, Percussion, No. 1, Mark II" passes through the beech block and screws into the base of the shell.

The shell consists of a Service pattern body, filled with dust shot and lead ash.

It is fitted with a 2-inch fuze-hole bush and a "Plug, fuze-hole, 2-inch, No. 1, Mark II."

The shell is inserted into the mouth of the case, the wood block acting as a support.

The shell is secured by indenting or coning, according to Mark of shell body used; also by the steel bolt, which is screwed up tight and secured by stabbing at the base.

CARTRIDGE, DRILL, Q.F., 3-INCH (MARK I) | C | .

The cartridge consists of a metal base 2 inches long, shaped like the base of the cartridge case, and milled around the rim.

Into the base is fitted, and secured by brass pins, a hollow teak block, shaped externally to represent both the body of the case and shell.

A metal top is secured by brass pins to the teak block to form the head of the shell. The metal top is prepared to receive fuzes of the 2-inch gauge or the "Plug, fuze-hole, 2-inch, No. 2, Mark II."

A steel bolt passes through the hollow teak block and further holds the metal base and top together.

CARTRIDGE, DRILL, Q.F., 2·95-INCH, MARK II | L | .

The cartridge consists of a Service pattern case, having three holes ·5 inch diameter bored through the base.

Into the base is fitted a "Primer, Percussion, Dummy, Q.F., Cartridge, No. 4, Mark I," which differs from the No. 2 primer (Fig. 146) only in being slightly larger in diameter from the top of the screwed portion to the base.

The case is fitted with a wood block of the same dimensions as a shrapnel shell. The wood block is secured to the case by three screws.

A G.S. fuze-hole bush is fitted to the top of the wood block, being secured by a 2½-inch iron screw.

DRILL CARTRIDGES, SEPARATE AMMUNITION.

(*Plate* C.)

CARTRIDGE, Q.F., 12-PR. OF 12 AND 8 CWT., MARK IV | C | .

CARTRIDGE, Q.F., 12-PR. OF 18 CWT., MARK I | N | .

CARTRIDGE, Q.F., 14-PR., MARK I | N | .

CARTRIDGE, Q.F. OR Q.F.C., 4-INCH, MARKS I TO III* GUNS, MARK IV | L | .

CARTRIDGE, Q.F., 4·7-INCH, MARK VI | C | . MARK II | L | .

CARTRIDGE, Q.F., 6-INCH, MARK IV | C | .

The above Drill cartridges are made of wood with metal ends which are secured to the teak by brass screws, and a central tube which is screwed into the primer end by being passed down from the opposite end of the cartridge and secured by a nut; this tube is to enable the electrical testing of primers and tubes to be carried out. The cartridge is brought up to weight by a lead cylinder under the metal top. The base is prepared for the adapter or primer in the usual way, and to facilitate handling the cartridge the base flange is milled.

The earlier Marks of the above Drill cartridges and the latest Mark of Q.F. or Q.F.C., 4-inch, for Naval Service (*i.e.*, Mark V) are fitted with a central rod instead of a central tube.

CARTRIDGE, Q.F., 4-INCH, MARK IV GUN, MARK I.

CARTRIDGE, Q.F., 4-INCH, MARK V GUN, MARK I.

The above Drill cartridges are made of teak with metal ends secured by brass rivets.

A metal rod passes through the centre of the teak block, and is secured and riveted to the metal ends; in a recess in the centre of the base is fitted a small plug of hard rubber for the firing pin to impinge against.

The cartridge is brought up to the same weight as a Service cartridge with lead.

CARTRIDGE, DRILL, Q.F., 4·5-INCH HOWITZER, MARK I | L | .

The Drill cartridge consists of a Service cartridge case filled with a wood block which is secured by three screws through the base of the case. Four holes are bored through the walls, and three smaller holes through the base of the case to facilitate identification. The word "DRILL" is also stamped on the base.

The base of the cartridge is fitted with a dummy primer containing a hard rubber coned plug in the head held in position by a screwed plug.

CARTRIDGE, DUMMY, Q.F., 4·5-INCH HOWITZER, MARK I | L | .

The Dummy cartridge is for instructional purposes and consists of a charge made up to represent the Service charge, enclosed in a Service cartridge case having holes similar to the Drill cartridge, the case being closed with a leatherboard lid.

The portions of the charge are leather covered and marked "ONE" to "FIVE," as the case may be, on the upper surface.

It is fitted with a Dummy primer.

CARTRIDGE, DRILL, Q.F., 15-PR., MARK I.

The Drill cartridge consists of a Service case lined with teak. It is closed at the mouth with a lid soldered to the case.

The base of the cartridge is fitted with a dummy primer containing a spiral spring and rubber pad for the firing pin to impinge against. Two holes are bored through the cartridge, at right angles to each other, to facilitate identification.

SECTION G.—PACKING OF Q.F. AMMUNITION.

(i) Service fixed ammunition	Boxes for
(ii) Service separate ammunition	
(iii) Blank ammunition for Q.F. guns	

(i) PACKING OF FIXED AMMUNITION. BOXES FOR Q.F., 1-PR. POM-POM.

Box, Ammunition, Q.F., 1-pr., 50 rounds, Mark I | C | is of deal, with elm ends, having cleats and copper wire handles. The lid is fastened by a hasp and turnbuckle. The box is provided with a lining of tinned steel, having three diaphragms and a lid. The diaphragms are perforated in 50 places ; the lid of the lining fits into a luting groove.

Boxes, Cartridge, Q.F., 12-pr., 8 cwt. or 1-pr., Marks I to II* | N |* .—These early Marks of 12-pr. 8 cwt. boxes have been approved for use in packing Q.F., 1-pr. ammunition for Naval Service. They hold 50 rounds in two belts.

For description of these boxes, *see* page 490.

Boxes for Q.F., 6- and 3-pr. Ammunition.

The latest boxes for 6- and 3-pr. ammunition are :—

Boxes, Ammunition, Q.F. :—		
6-pr. Naval, Mark VII ...	N	Wood; tinned copper lined; to hold 11.
6-pr., Mark VI	C	Wood; zinc lined; to hold 11.
6-pr. Special, Mark IV ...	N	Wood; tinned copper lined; to hold 9; H.M.S. "Swiftsure" and "Triumph."
3-pr. Naval, No. 2, Mark I...	N	Wood; tinned copper lined; for lyddite shell ammunition; to hold 16.
3-pr. Naval, No. 1, Mark IX...	N	Wood; tinned copper lined; for steel shell or practice ammunition; to hold 16.
3-pr., Mark VIII	C	Wood; zinc lined; for steel shell or practice ammunition; to hold 16.
3-pr. Vickers, No. 2, Mark I...	N	Wood; tinned copper lined; for lyddite shell ammunition; to hold 16.
3-pr. Vickers, No. 1, Mark II	N	Wood; tinned copper lined; for steel shell or practice ammunition; to hold 16.

The boxes are made of teak or mahogany, with a lid working on gunmetal hinges and fastened by a hasp and turnbuckle of the same material, the latter being secured by a short lanyard of white line. Two strong cleats secured by copper rivets extend across the front and back and to these are attached handles of copper wire, 24 inches long, and covered with leather at the upper portion. These cleats also strengthen the box.

The latest Marks of Naval boxes are fitted with a tinned copper lining.

The earlier Marks of Naval boxes and all Land Service boxes are fitted with a zinc lining. Naval boxes having zinc linings are to be fitted with tinned copper ones when the stock of zinc linings is used up. When tinned copper linings are fitted to boxes which originally had zinc ones, a star is added to the numeral.

The linings have a channel formed round the top to receive luting and so make a watertight joint. The linings are closed by a lid, having a flanged edge to rest in the luting channel. Two diaphragms are soldered to the inside of the lining.

These diaphragms are perforated to support the number of cartridges which the box will hold. Underneath the lower diaphragm is a wood bottom, consisting of a separate section for each round, each of which has a recess lined with metal to take the point of a projectile. These sections are kept in position by a thin zinc plate soldered to the lining, and are capable of slight lateral movement so as to accommodate themselves to the points of the projectiles.

Earlier Marks of 6- and 3-pr. ammunition boxes differ from the above only in small details, and they are brought up to the latest pattern when passing through Ordnance Factories for repair, one or more stars being added to the numeral.

The wooden boxes are painted inside and out as under. The linings are neither painted nor varnished :—

6-*pr. Boxes* are painted stone colour.

3-*pr. Boxes,* containing steel shell, filled gunpowder, or practice ammunition are painted lead colour.

3-*pr. Boxes,* containing lyddite shell ammunition, are painted yellow.

6- *or* 3-*pr. Boxes,* containing practice ammunition, have the lids painted yellow.

Fig. 147.

BOX, AMMUNITION, 6-PR., Q.F.

Q.F., 2·95-inch Ammunition Boxes.

Q.F., 2·95-*inch : Pack transport.*—No. 1 box takes 6 shrapnel or 6 case shot ; No. 2 box takes 4 double shell ; No. 3 box takes 6 star shell.

Man transport.—" A " box takes 3 shrapnel, " B " box 2 double shell, " C " box 3 case shot, " E " box 4 star shell.

Q.F., 13- and 18-pr. Ammunition Boxes.

The latest Marks of 13- and 18-pr. ammunition boxes are :—

Box, Ammunition, Q.F., 13-pr., Mark III | L | .

Box, Ammunition, Q.F., 18-pr., Mark III | L | .

The boxes are made of deal with elm ends, and are provided with a cleat and galvanised iron wire handles at each end.

For India and East and West Africa, the boxes are to be made of teak throughout.

Steel strap bolts, with hexagonal nuts for securing the lid, are attached near each end of the box. Fittings are provided to carry four complete fuzed rounds with fuze covers in each box.

The boxes are not fitted with tinned iron linings, and are unpainted.

Fig. 148.

BOX, AMMUNITION, Q.F., 18-PR.

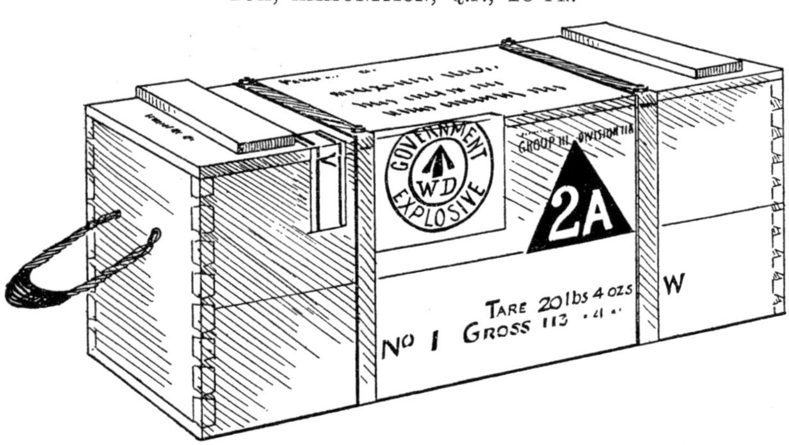

The Mark II boxes were slightly longer and were provided with fittings to take the four plugged rounds, and four fuzes in tin cylinders or four fuzed rounds with fuze covers.

The Mark I boxes differed from either Mark II or III in being lined with tinned iron.

3-inch Ammunition Boxes.

Boxes, Ammunition, Q.F., 3-inch, Land, Mark I | L | .

The box is made of teak throughout.

The sides are provided with top, bottom and vertical battens.

Gunmetal brackets, with studs and wing nuts for securing the lid, are let into the top and vertical battens.

The lid is provided with brass strengthening strips and wood battens, and is fitted with two brass T-shaped handles for removing the lid.

The box is fitted with two copper wire lifting handles.

The box and the lid are lined with zinc. The body lining projects slightly above the body of the box, and has a rounded edge which fits into a luting groove formed in the lining of the lid.

Each box is provided with internal fittings to carry four complete rounds.

Paint.—The boxes are painted stone colour for shrapnel shell ammunition, and yellow for high explosive shell ammunition.

Box, Ammunition, Q.F., 3-inch, Naval, Mark I | N | .

The Naval 3-inch box differs from the Land Service one in having a sliding locking plate for securing the lid instead of wing nuts. It is also fitted with a copper wire handle at one end in addition to the handles on the side, for the purpose of hoisting the boxes from the ammunition room to the gun floor.

(ii) Packing of Q.F., Separate Ammunition.

Cartridges for the Q.F., 12-pr., 14-pr., 15-pr., 4-inch, 4·7-inch and 6-inch guns are packed in "*Boxes, Cartridge.*"

There are four sizes of boxes for the 12-pr. cartridges :—

- The 12-pr. of 4 cwt. box, which holds 8 cartridges, packed vertically.
- The 12-pr. of 8 cwt. box, which holds 10 cartridges, packed horizontally.
- The 12-pr. of 12 cwt. box, which holds 10 cartridges, packed horizontally.
- The 12-pr. of 18 cwt. box, which holds 8 cartridges, packed horizontally.

The 14-pr. box holds 10 cartridges, packed horizontally.

The 15-pr. box holds 22 cartridges, packed vertically. (15-pr. cartridges are also issued 40 in a case, powder, metal lined, whole, packed horizontally.)

There are three sizes of boxes for the 4-inch cartridges :—

- The 4-inch, Marks I to III* and Q.F.C. gun box, which holds 8 cartridges, packed horizontally.
- The 4-inch Mark IV gun box, which holds 6 cartridges, packed vertically.
- The 4-inch Mark V gun box, which holds 6 cartridges, packed vertically.

There are two sizes of boxes for the 4·7-inch cartridges :—

- The 4·7-inch Marks I to IV* gun box, which holds 6 cartridges, packed horizontally.
- The 4·7-inch Mark V gun box, which holds 4 cartridges, packed horizontally.

The 6-inch box, which holds 4 cartridges, packed vertically.

For Land Service the boxes, with the exception of the 15-pr. box described on page 490, are painted *Service colour*, and the lids of all the later Marks are secured with winged nuts; for Naval Service they are painted *stone colour*, and the lids of all the later Marks are secured by a locking plate engaging with bolts on the box.

Q.F., 12-pr. Boxes.

The latest Marks of Q.F., 12-pr. boxes are :—

Box, Cartridge, Q.F., 12-pr., 4 cwt., Mark II | N | .

Box, Cartridge, Q.F., 12-pr., 8 cwt., Mark IV | N | .

Box, Cartridge, Q.F., 12-pr., 12 cwt., Mark III | N | .

Box, Cartridge, Q.F., 12-pr., 12 cwt., Mark III | L | .

Box, Cartridge, Q.F., 12-pr., 18 cwt., Mark II | N | .

These boxes are generally similar in construction. They differ in dimensions and in the number of packing pieces and lifting bands.

The Box, Cartridge, Q.F., 12-pr. of 12 cwt., Mark III | L | is described in detail, and is illustrated in Fig. 149.

The box is made of teak or mahogany, with strengthening battens along the sides, top and bottom. Two wire handles are secured to the upper battens. Four gunmetal plates are attached to the upper battens, and hinged to these plates are studs carrying wing-nuts. The body has a zinc lining, the top edges of which project slightly above the box.

The lid is lined with zinc, and in it a luting groove is formed, into which the top of the lining fits. It is strengthened by battens. Four slots are cut in the edge, and strengthened by metal plates for the wing-nuts to screw down on.

A T-shaped lifting handle is secured to each end of the lid by split pins. Formerly rope handles were attached to the ends of the lid for the purpose of removing the lid from the box. Copper wire lifting handles are provided for lifting the boxes. Formerly rope handles were used for Land Service boxes.

Packing pieces are provided, fitted with brass spring cups to fit over the mouth of each cartridge.

Hemp lifting bands are provided where the cartridges are packed horizontally to facilitate their removal from the boxes.

(The latest Naval boxes have tinned copper linings. Earlier Naval boxes have zinc linings, but will be fitted with tinned copper linings when the stock of zinc ones is used up, a star being added to the numeral of the box when this change takes place.)

The Naval boxes have the lids secured by a sliding locking plate, having four arms which engage with a slot in each of four studs fixed to the body of the box.

The locking plate is moved by an eccentric turned by means of a special key.

The Q.F., 14-pr. Box.

The latest Mark of 14-pr. Box is the *Box, Cartridge, Q.F., 14-pr., Mark II | N |* , which is similar to the Q.F., 12-pr. Naval boxes.

Q.F., 15-pr. Box.

Box, Cartridge, Q.F., 15-pr., Mark I | L | is made of deal with cleats and copper wire handles, the top and bottom secured by tinned iron screws. It is lined with tinned iron and is fitted with two diaphragms, perforated to take the 22 cartridges which the box contains. A thick piece of felt is placed above and below the cartridges; the lining is closed with a lid of tinned iron soldered on.

Fig. 149.
BOX, CARTRIDGE, Q.F., 12-PR. OF 12 CWT., MARK III.
(LAND SERVICE BOX.)

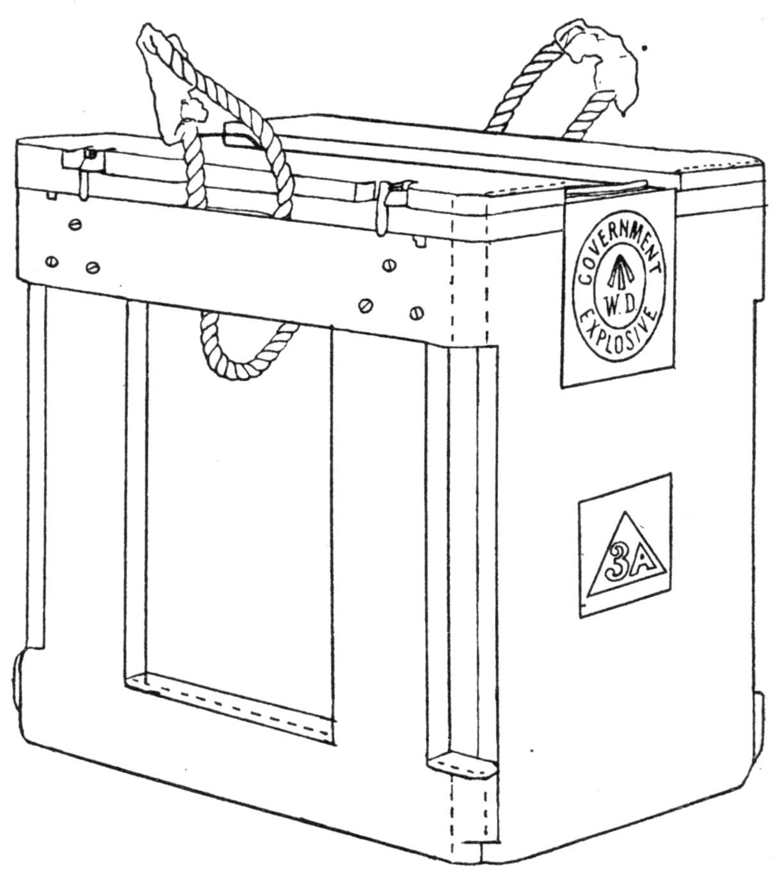

Q.F. OR Q.F.C., 4-INCH BOXES.

The latest Marks of Q.F. or Q.F.C., 4-inch boxes are :—

Box, Cartridge, Q.F., 4-inch, Land, Mark I | L |.

Box, Cartridge, Q.F., Marks I to III* or Q.F.C., 4-inch Guns, Naval, Mark IV | N |.

Box, Cartridge, Q.F., 4-inch, Mark IV Gun, Mark I | N |.

Box, Cartridge, Q.F., 4-inch, Mark V Gun, Mark I | N |.

The two first-named boxes are generally similar to those for the Q.F., 12-pr. guns, described on page 490, differing only in dimensions and detail of packing pieces and lifting bands.

The two last-named are stronger boxes, the lids being strengthened top and bottom by brass strips.

Since the cartridges are packed vertically, lifting bands are not required.

Q.F., 4·7-inch Boxes.

The latest Marks of 4·7-inch Q.F. boxes are :—

Box, Cartridge, Q.F., 4·7-inch, Marks I to IV* Guns, Mark II | L | .

Box, Cartridge, Q.F., 4·7-inch, Naval Outfit, Mark III | N | .

Box, Cartridge, Q.F., 4·7-inch, Mark V Gun, Mark I | L | .

The 4·7-inch boxes differ from the 12-pr. boxes described on page 490 only in dimensions and detail of packing pieces and lifting bands.

Q.F., 6-inch Boxes.

The latest Marks of 6-inch Q.F. boxes are :—

Box, Cartridge, Q.F., 6-inch, Mark III | L | .

Box, Cartridge, Q.F., 6-inch, Naval Cordite, Mark IV | N | .

The 6-inch boxes differ from the 12-pr. boxes described on page 490 only in dimensions and detail of packing pieces.

The cartridges being packed vertically, lifting bands are not required.

The packing pieces for 6-inch Q.F. cartridges are of two kinds and are known as :—

No. 1 for flat lids. No. 2 for concave lids.

Boxes, Ammunition.

The ammunition for 4·5-inch Howitzers and 4·7-inch guns on travelling carriages is carried in "Boxes, Ammunition."

Q.F., 4·5-inch Howitzer Boxes.

Box, Ammunition, Q.F., 4·5-inch Howitzer, Shrapnel, Mark II | L | .

Box, Ammunition, Q.F., 4·5-inch Howitzer, Lyddite, Mark I | L | .

The ammunition boxes for the 4·5-inch Howitzer are made of deal with elm ends. Wood cleats and galvanised iron wire handles are provided at each end of the boxes. Steel strap bolts, with hexagonal nuts for securing the lid, are secured near each end. The boxes are not painted, and are fitted internally to hold the following :—

The box for shrapnel shell holds 2 shrapnel, plugged or fuzed, 2 cartridges each in a tin box (*see* page 454), and, if the shell are carried plugged, 2 T. and P. fuzes in tin cylinders.

The box for lyddite shell holds 2 lyddite shell, plugged, 2 cartridges, each in a tin box (*see* page 454), and 2 fuzes, each in a tin cylinder.

Q.F., 4·7-inch Boxes.

Box, Ammunition, Q.F., 4·7-inch, Converted, Shrapnel, Mark I | L |.

Box, Ammunition, Q.F., 4·7-inch, Converted, Lyddite, Mark I | L |.

The "*Boxes, ammunition, Q.F.* 4·7-*inch, converted*," "*shrapnel,*" *and* "*lyddite,*" are Mark II "Boxes, cartridge" fitted to hold 3 cartridges and 3 fuzed shell of any Mark, but all the shell in one box are to be of the same Mark. These converted boxes have been designed to enable a number of fuzed shell to be carried in the ammunition wagons of the Q.F., 4·7-inch Territorial Batteries.

(iii) Packing of Q.F. Blank Cartridges.

In some instances blank cartridges are issued filled, while in others the components are issued and are made up by the unit requiring them.

The brass cases for Q.F. blank cartridges, whether filled or empty, are issued in boxes; certain guns have special boxes for blank cases, as shown below; the others use the ordinary Service ammunition boxes.

Paint.—Boxes containing blank cartridge or blank charges are painted red.

Q.F., 6-pr.—Box for Blank Cases.

Box, Cartridge Cases, Q.F., 6-*pr., Mark II, blank* is made of deal, with elm ends, having cleats and rope handles; the lid is hinged and fastened by a hasp and turnbuckle in front. The box is fitted with two mahogany fittings, perforated with 20 holes for the cases, which stand base upwards. The underside of the lid is recessed so that safety clips may remain on the bases of the cases.

Printed instructions for filling, and lithograph showing method of filling, are fixed on the inside of the lids of these boxes.

Q.F., 3-pr.—Box for Blank Cases.

Box, Cartridge Cases, Q.F., 3-*pr., Mark I, blank,* is similar to the 6-pr. box, but is slightly smaller.

Q.F., 3-pr. Vickers.—Box for Blank Cases.

"*Box, Cartridge, Q.F., Blank,* 3-*pr., Vickers, Cartridge cases, converted,*" to hold 20.

This box is converted from the "*Box, Cartridge cases, Q.F.,* 6-*pr., Mark II, Blank,*" described above. The conversion consists of removing the existing diaphragms and substituting others suitable for the Vickers' cases.

Q.F., 13 and 18-pr.—Box for Blank Cases.

The cases are issued in :—

Box, Cartridge case, Q.F., 13-pr.	Wood, to hold 20 cases for blank with clips.
,, ,, ,, Q.F., 18-pr.	

The boxes are fitted with hinged lids secured by hasps and turnbuckles, and with wire rope lifting handles. The interior of each is fitted with partitions to form a separate compartment for the cases.

Instructions for making up the cartridges are affixed to the lid of each box.

Q.F., 4·5-inch Howitzer.—Box for Blank Cases.

The *Box, Cartridge case, Q.F., 4·5-inch Howitzer, Mark I | L |* is converted from the Mark I box for Shrapnel Ammunition (§ 15434) by the removal of the internal fittings and the substitution of a wood frame to take 10 cases on edge in two rows.

Instructions for making up the cartridges are affixed to the lid of the box.

Storage of Q.F. Ammunition.

(1) Q.F. ammunition filled with cordite and fitted with a percussion cap or a percussion primer, viz. :—

All fixed ammunition,
Q.F., 12-pr. of 4 cwt.,
Q.F., 15-pr.,
Q.F., 4-inch, Marks IV and V guns,
Q.F., 4·5-inch Howitzer,

are classified for storage in Group III, Division IIA, Blank Ammunition powder for the above guns, in Group III, Division IIB.

(2) All other Q.F. ammunition fitted with an adapter or electric primer when filled cordite in Group III, Division IIIA. When filled with *powder*, in Group III, Division IIIB.

CHAPTER XVIII.

Cartridges, Impulse, Torpedo.

Cartridges, impulse, torpedo, with the exception of those for H.M.S. " Triumph " and " Swiftsure," are only used for firing torpedoes from *above water tubes*, and are all similar in construction, differing only in the weight of propellant and minor details, *see* Table 38.

The latest cartridge for 21-inch, Mark II torpedoes is described in detail.

Cartridge, Impulse, Torpedo, 10 oz., Mark Ia | N |.

(*Plate* CI.)

The cartridge consists of a brass case (a 3-pr. Q.F. case cut down); a ¾-inch hole is bored in the base and bushed. Into this bush is screwed a brass tube which reaches to within ½ inch of the mouth of the case. The walls of this tube are perforated near the front end, the end itself being closed by a brass disc spun in. The other end of this tube is prepared to take a Tube, impulse, torpedo. In the base of the case is a filling-in piece of wood which is prevented from moving forward by a split brass ring snapped into a recess cut in the tube. The charge consists of alternate layers of pebble powder and R.L.G.^{2}, the latter contained in circular muslin bags. The mouth of the case is closed by a millboard disc, covered with shellac, held in place by the mouth of the case being turned over.

Cartridges, impulse, torpedo, are issued one in a hermetically sealed tin cylinder with tear-off band. Ten cylinders in a wood box.

Charge, Impulse, Torpedo, 2 lb. Cordite, Mark II | N |.

The above charge is special for H.M.S. " Swiftsure " and " Triumph," for 18-inch submerged tubes.

The charge consists of short lengths of cordite, sizes 15 and 20, contained in a muslin bag choked with silk sewing.

A separate igniter is used.

The igniter consists of 1 oz. cordite, size 10, cut to 1½-inch lengths, and ½ oz. of S.F.G.^{2}, enclosed in a flat Japanese silk bag.

The igniter and charge are inserted into a special brass case (similar to the Q.F., 12-pr., 12 cwt. case, but longer), which is fitted with a bronze adapter.

The charge is fired by an electric wireless " P " tube.

The charges and igniters are issued separately, wrapped in brown paper, in metal-lined cases.

The Mark I charge differed from the above by having the igniter *attached* to it.

TABLE 38.—*Cartridges, Impulse, Torpedo.*

Mark.	Charge.	No. packed.	Package.	Remarks.
IA	10-oz. pebble and R.L.G.[2]	10 in cylinders No. 112.	Box, cartridge, Impulse, Torpedo, 21-inch.	For 21-inch, Mark II Torpedo.
I	10-oz. pebble and R.L.G.[2]	10 in cylinders No. 112.	Box, cartridge, Impulse, Torpedo, 21-inch.	For 21-inch, Mark II Torpedo. No split brass collar on central tube.
I	9-oz. pebble and R.L.G.[2]	10 in cylinders No. 112.	Box, cartridge, Impulse, Torpedo, 21-inch.	For 21-inch, Mark I Torpedo.
I	8½-oz. pebble and R.L.G.[2]	10 in cylinders No. 96, Mark II.	Box, cartridge, Impulse, Torpedo, 18-inch, Marks I^{x} and II.	For 18-inch, Marks VII and VII^{x}, and 18-inch, Weymouth, Mark I Torpedoes.
I II I I	7½-oz. pebble 7½-oz. pebble and R.L.G.[2] 6-oz. pebble and R.L.G.[2] 5½-oz. pebble and R.L.G.[2]	10 in cylinders No. 96, Mark I.	Box, cartridge, Impulse, Torpedo, 18-inch, Mark I^{x} and II.	For 18-inch, Marks I^{x} to VI^{xx} H Torpedoes.
II	4½-oz. pebble ...	10 in cylinders No. 95.	Box, cartridge, Impulse, Torpedo, 14-inch, Marks I^{x} and II.	For 14-inch Torpedoes.

CHAPTER XIX.—IMPLEMENTS, STORES, &c., USED IN CARRYING OUT LABORATORY WORK.

In this chapter the implements, &c., used in carrying out laboratory work are described in alphabetical order.

This list also includes those stores which come under the heading of "Tools, re-forming cartridges, Q.F. and Q.F.C." and "Fuze keys" used by Inspecting Ordnance Officers.

Adze, cooper's, magazine.—This is an ordinary metal adze; it is used for heading, unheading, and hooping powder barrels.

Bar, lifting, shell, Mark I, is for use with shell holders in laboratories. It consists of a steel bar about 2 feet long with a central hole for the hook of the raising tackle, and three holes at each end, to any of which can be attached a shackle holding chains about 3 feet long.

Fig. 150.

BAR, LIFTING, SHELL, MARK I.

Scale $\frac{1}{8}$.

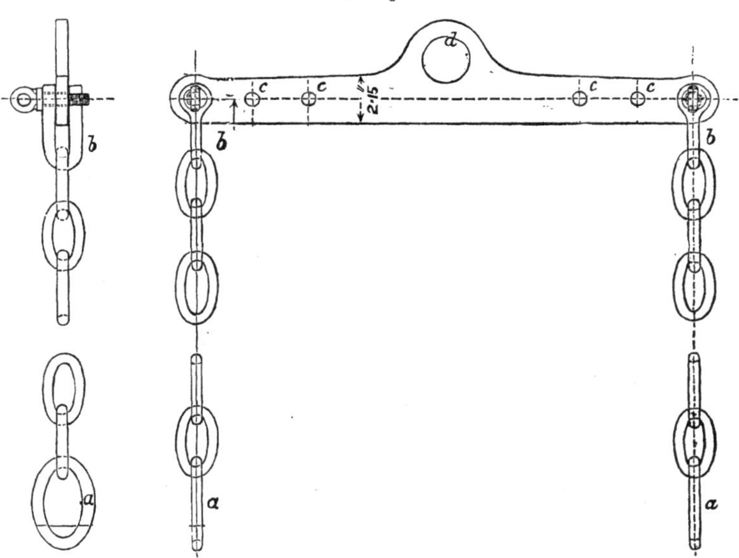

Method of use.

Place the holder over the nose of the projectile, and fix it at the centre of gravity, by means of the screws and spanner supplied with the holder. Pass the bottom loops (*a*) of the chain on the bar over the trunnions of the holder, adjusting the chains so as to keep them parallel by attaching the shackles (*b*) in the required holes (*c*). The raising tackle will then be hooked in the bar at (*d*), and the projectile lifted and placed in position.

Blocks, Nos. 1 *and* 2, *Mark I* | *C* | are cubical blocks of wood measuring 1 foot each way, bored out to take the heads of the projectiles. No. 1 is for 12-inch calibre and above, No. 2 is for 8-inch to 10-inch calibre. They are intended for use in laboratories and shell filling rooms, a hole being made in the floor to receive the block, so that it will lie flush with the floor.

Fig. 151.

BLOCK, NO. 1, MARK I | C | .

Section on KK.

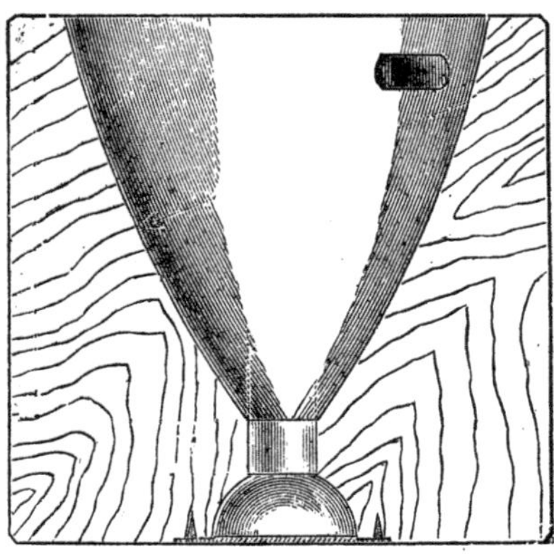

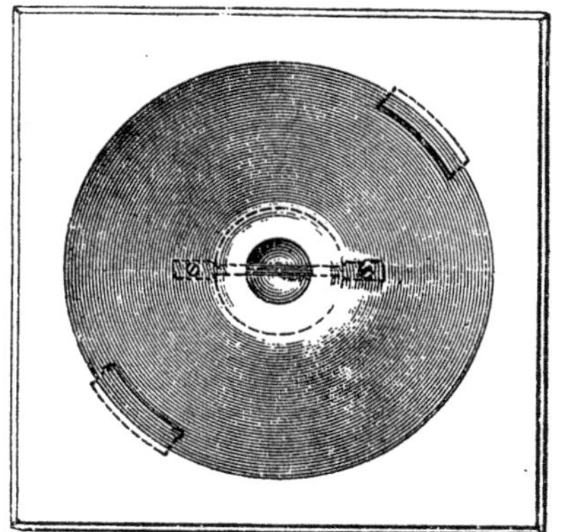

Plan.

When not required for use, the block will be inserted in the floor, with the hole for shell downwards. Two recesses are provided in the sides of the cavity, and a bar across the hole in the bottom (*see* drawing) to facilitate lifting the block out of the floor.

Blocks for projectiles under 8-inch will be made locally as required, no fixed size being necessary, as they will stand on the floor or a bench at a most convenient height.

Block, No. 3, *Mark I* | *L* | is made of hard wood, wedged shaped and hollowed out to receive the projectile. The base of the block is grooved as shown in sketch to grip the floor when the projectile is being up-ended.

Fig. 152.

BLOCK, NO. 3, MARK I | L | .

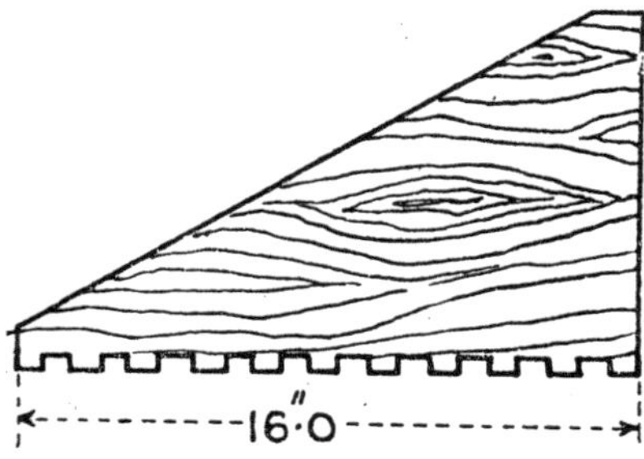

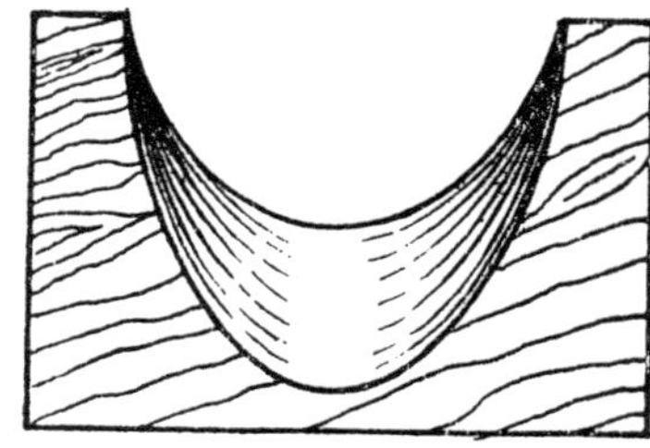

They are used for fixing base fuzes in shell 9·2-inch calibre and above, except when using No. 14 projectile barrow.

Block, No. 4, Mark I | *L* | is similar to the No. 3 block, but is intended for use with 9·2-inch projectiles only, in conjunction with the No. 14 projectile barrow for guns on Mark V mountings.

Bearer, cartridge cylinder, Mark III.—This is an ash stave 4 feet long, 1 inch thick, 2½ inches broad in the centre, tapering off to each end. It has three grooves cut in the side near the centre to take the handles of the cartridge cylinders. The bearer is also used as a lever in unscrewing and screwing home the lids of the cylinders.

Brush, fuze-hole, Mark I | *C* | .—This is a small cylindrical shaped brush fitted with a copper wire handle about 3·75 inches long. It is issued for cleaning the fuze-holes of 3 and 6-pr. Q.F. shell when breaking up ammunition.

Fig. 153.

DISC, CLEANING FUZE-HOLE, MEDIUM, 2·1-INCH DIAMETER.

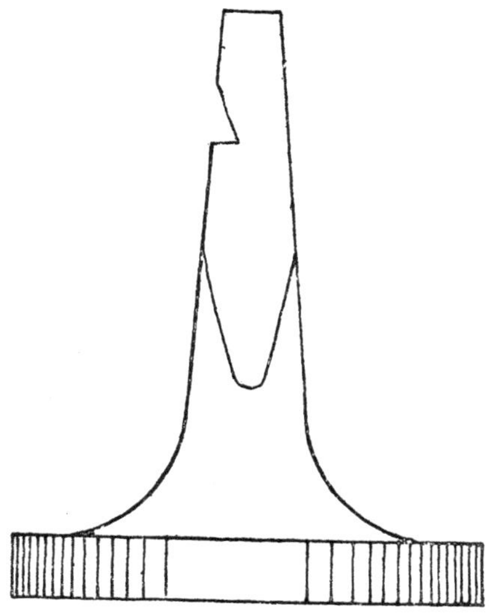

Brush, primer hole, Mark I | *C* | is similar to the above, but larger. It is used for cleaning the primer holes of Q.F. cartridges previous to filling.

Brace, magazine, Mark II, with bit, is for use in removing screws of powder and cordite cases, &c., for which the "Driver, screw, magazine" is not suitable. The *bit* is made of phosphor bronze and is fixed to the brace by a set screw.

Chisel, preparing cannelures, Mark I, is made of steel, about 6 inches long and ·4 inch wide at the point, which is slightly curved.

It is intended for undercutting the cannelures of the early type of broad Vavasseur driving-bands to take augmenting strips.

Chisel, metal, Mark I.—This is a cross-cut chisel about 7½ inches long and ¼ inch wide, and is intended for cutting out the lead rings and base discs from shell.

Discs, cleaning, fuze holes, medium, Mark I, for shells taking medium base fuze, and *Discs, cleaning, fuze holes, large, Mark I,* for shell with large base fuzes. These are made of phosphor bronze formed with a stem which is shaped at the end to fit the "brace magazine," and are for use in cleaning the recess for the flange of the base fuze in the bases of shells.

Drift, G.S., long, Mark I is a round piece of boxwood about 4 inches long; the lower portion for a length of 1·6 inch is coned to fit the G.S. fuze hole. It is used for inserting G.S. wads in shrapnel shell, 4 inch, Marks IV to VI, and 4·7-inch, Mark IV.

Drifts, Q.F., 12-pr., 4-inch, 4·7-inch, and 6-inch are rings of gunmetal, the lower edge being reduced and bevelled inside. They are intended for use with the white metal lids when inserting the latter into the mouth of the cartridge case.

Drifts, Q.F., blank, 6-pr., 12-pr., 13-pr., 18-pr., and 4-inch are made of wood cylindrical in shape, and are used in driving the leather-board cups into position in the cartridge case in making up blank ammunition for the above guns.

Drifts, 6- and 3-pr., Q.F., are bars of wood about 2 feet long, used for turning 6- and 3-pr. cases in the rectifying press.

Driver, grummet, Mark I, is made of hard wood, and is used in the L.S. for removing the early pattern of rope grummets from projectiles with gas-check driving-bands.

Driver, cooper's, socket, metal, with wood handle, is used in conjunction with the "Adze, cooper's, magazine" in removing the hoops, &c., from powder barrels.

Funnels, shell, copper, are issued in two sizes (large and small). They are intended for use in filling shell which have their bursting charges in bags.

Funnels, cartridge, are made of copper, similar to, but larger than "Funnels, shell"; they were introduced for filling cartridges with granulated or cubical powder.

Gauges, cartridge, ring, consist of rings of gunmetal, with straight handles; they are marked on the handles with the internal diameter of the gauge. The ring gauges test the diameter and are all "high" gauges, that is to say, the greatest diameter of the cartridge must pass through them.

Gauges, shell, ring, are plain wrought-iron rings with handles formed on them. They are issued for gauging the diameter of projectiles, and are of three kinds, namely—"Body high," "Band high and low" and "Gas-check high."

Note.—Great care is required to use these ring gauges; unless held quite square they will not pass over the projectiles.

Gauges, chamber, low, are issued for gauging the Q.F., 3-pr., 6-pr., 13-pr., and 18-pr., and 3-inch cartridges. They are made of cast iron of the same internal dimensions as the low limit for the chamber of their respective guns. The cartridge must enter the gauge freely to ensure loading. Similar gauges are issued for Q.F., 12-pr., 4-inch and above, for use in gauging the cartridge cases after rectifying.

Gauge, screw, primer hole, No. 1, Mark I | C | is a steel screw gauge used with Q.F., 12-pr., 14-pr., 4-inch, Marks I to III, 4·7-inch and 6-inch cartridge cases. When the primer hole admits of the gauge being screwed through it, the cartridge requires bushing.

Gauge, screw, primer hole, No. 2, *Mark I* | *L* | is similar to the above, but is for use with 12 pr. of 4 cwt., 13- and 18-pr., 3-inch, 4-inch, Mark IV and V, and 4·5-inch Howitzer, Q.F., cartridges.

Gauge, depth, fuze hole, base percussion fuzes.

No. 1, for shell with No. 11, Mark V, and No. 15, Marks II and III fuzes.

No. 2 for shell with No. 11, Marks I to IV, and No. 15, Mark I fuzes.

No. 3, for shell with No. 12, Marks V and VII fuzes.

No. 4, for shell with No. 12, Marks I to IV fuzes.

The above gauges are made of hard wood to the form shown on accompanying drawing; they have their name and number stamped

Fig. 154.
GAUGE, DEPTH, FUZE HOLE.
Scale ½.

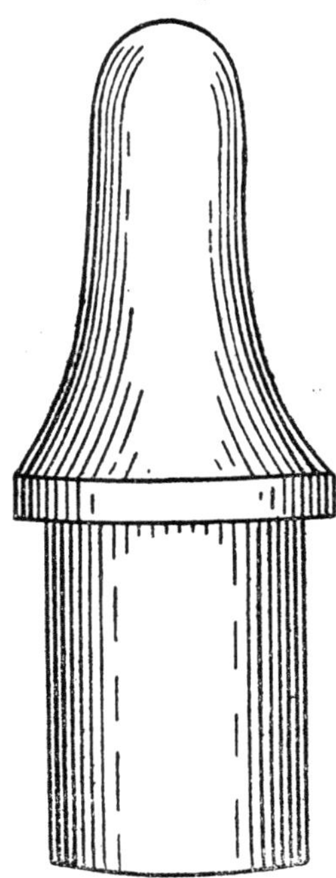

on the handle as a means of identification. When using these gauges care is to be taken that the flange of the gauge entirely enters the recess in the base of the shell, otherwise the shell is not fit for fuzing.

Gauges, depth of cavity, filled Lyddite Shell.

No. 1 for shell without asbestos lining } And taking picric powder
No. 2 for shell with asbestos lining .. } long exploders.
No. 3 for shell taking long exploders of T.N.T.

The above gauges are made of beech wood, tipped at the lower ends with aluminium; they are marked with the *high* and *low* limits of the depth of the cavity in lyddite shell, necessary to take the different lengths of cylindrical exploders.

Hammers, metal, 1 *lb.* 12 *oz., Mark I,* is an ordinary metal hammer fitted with a wood handle. It is used with the "chisel, metal."

Holder, shell, for laboratories, is intended for use in holding and slinging shell in laboratories.

Fig. 155.

HOLDER, SHELL, FOR LABORATORIES

Scale $\frac{1}{8}$.

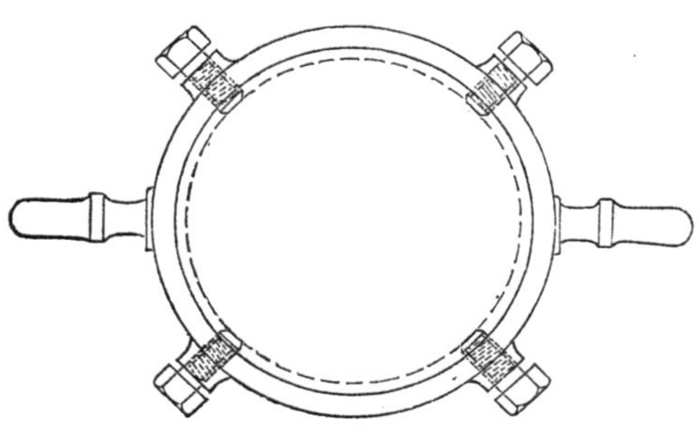

Holder, shell, B.L. | *C* | *with spanner attached* consists of handles attached to a jointed ring, which is tightened by means of a nut working on a screw bolt. For use with the nut a spanner is attached by a chain to one of the handles.

Fig. 156.

HOLDER, SHELL, Q.F., 12-PR., MARK I.

Scale $\frac{1}{4}$.

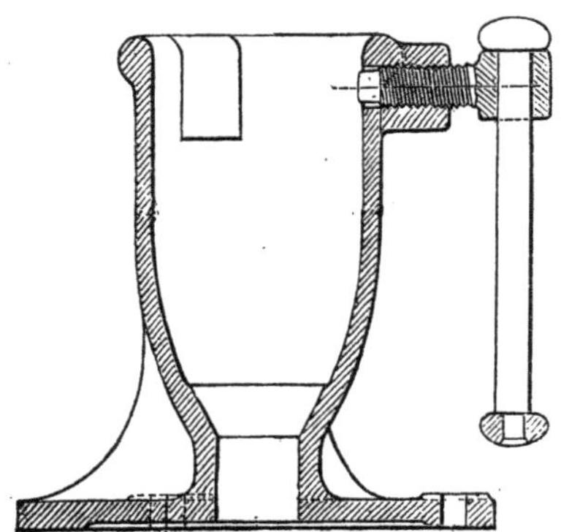

Holder, shell, Q.F., 12-pr., Mark I; B.L., 5-inch, or Q.F., 4·7-inch, Mark II; and B.L., Q.F., or Q.F.C., 6-inch, Mark II, are made of cast iron, to the form shown in Fig. 156, those for 12-pr. 5-inch and 4·7-inch having flat bases with three screw holes for securing them to a bench or table, a clamping screw (with lever handle attached) being provided for retaining the shell in position.

Holder, shell, B.L., Q.F., or Q.F.C., 6-inch, Mark II, is similar to the above but has a wood block fixed in the base for the reception of the point of the shell, and a semi-circular band, with fly nut, hinged to the top for retaining the shell in position.

Fig. 157.

HOLDER, SHELL, B.L., Q.F., OR Q.F.C., 6-INCH, MARK II.

Scale $\frac{1}{6}$.

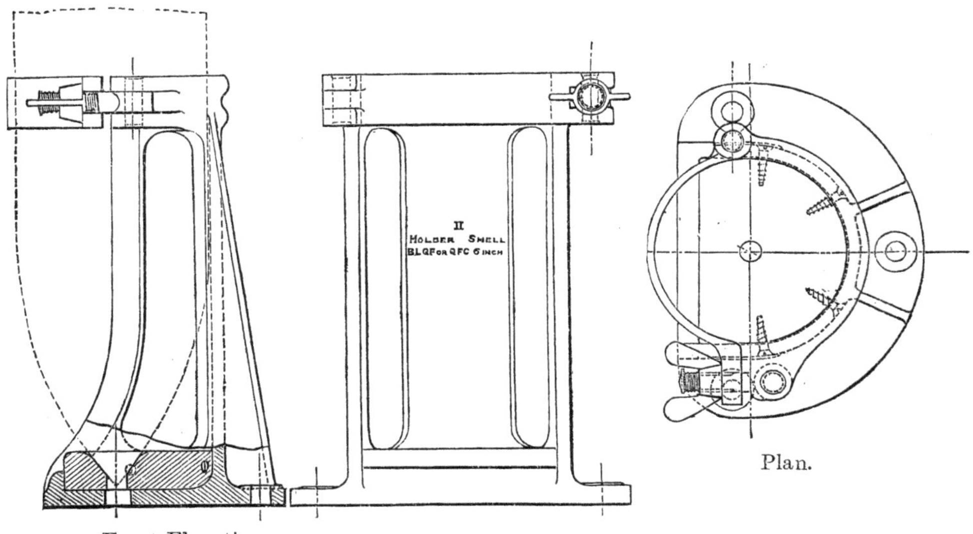

Holder, cartridge, Q.F., Mark I | *C* | consists of a steel band which encircles the cartridge case. The steel band terminates at each end in long handles, which, being forced together, clip the band tight round the case. A link working on these handles keeps them together when forced away from the cartridge. This tool is used to hold the case when the primer or adapter is being actuated by the "Key, removing primer."

Holder, cordite charges, 6- and 3-pr. Q.F., Mark I | *N* | consists of a gunmetal pillar and stand which may be screwed to a table. It has a projecting arm fitted with a circular slot, the outer part being hinged and fitted with a spring catch for closing it. It is for use in tying up the cordite charges for 6- and 3-pr. cartridges.

Hook, G.S. wad is made of copper, and has a wooden handle; it is used to remove the G.S. wad from the fuze hole of common shell filled without bag, and shrapnel shell, 4-inch, Marks V and VI, and

4·7-inch, Mark IV. The hook is forced through the hole in the wad and the point brought up under the wad to one side; the wad can then be pulled out.

Fig. 158.

HOOK, G.S. WAD.

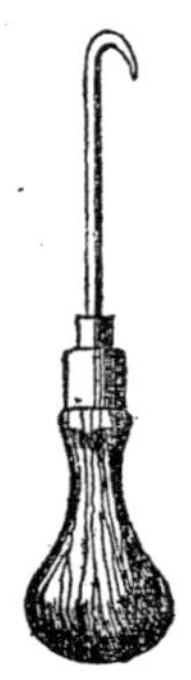

The hook, G.S. wad is also used for extracting exploders from lyddite shell.

Fuze Keys used by Inspecting Ordnance Officers.

Keys, No. 1 (*Mark I*), *D.A. Fuzes* | *C* | for bottom plugs of D.A. fuzes.

No. 2 (*Mark I*), *Nos.* 11 *and* 15 *Fuzes* | *C* | for magazine plugs of Nos. 11 and 15 fuzes.

No. 3 (*Mark I*), *No.* 12 *Fuzes* | *C* | for magazine plugs of No. 12 fuzes.

No. 4 (*Mark I*), *T. and P. Fuzes* | *C* | for removing bottom plugs of Fuzes, T. and P., Nos. 54, 56, 60 and 62 to 66.

The above-mentioned keys are for the use of Inspecting Ordnance Officers.

They are steel keys with wooden handles, and are fitted with a clamping arrangement and nut, so that the pins may be readjusted or replaced. A steel spanner is used in conjunction with these keys.

Key, No. 9 (*Mark III*), *Base Fuzes* | *C* | is for the use of Inspecting Ordnance Officers, in removing the screwed caps of large and medium base fuzes. It is made of steel. A clamping arrangement at the centre holds two sizes of pins for the above-mentioned fuzes.

Mark II is a flat bar of steel, with several pins on each side, arranged so as to fit the plugs of several different fuzes. It is being superseded by Keys, Nos. 1 to 4 and 9, Mark III.

Key, No. 12 (*Mark I*), *Sensitive Fuzes* | *C* | is a steel key with wooden handle for the use of Inspecting Ordnance Officers in removing or replacing the side plugs in sensitive time fuzes.

Key, No. 15 (*Mark I*), *Cap, No.* 63 *Fuze* | *C* | is a flat steel bar about 9 inches long, and fitted with two pins to enter the recesses in the top cap of the No. 63 fuze.

It is for the use of Inspecting Ordnance Officers in removing or fixing the cap.

Key, No. 32 (*Mark I*), *Fixing, Nos.* 80 *and* 83 *Fuzes* | *C* | , is a steel key, similar to the No. 17, Mark II key (*see* page 344), except that it is double handed. It is also heavier and stronger than the No. 17 key. It is for the use of Inspecting Ordnance Officers in fixing or removing the fuzes.

Fig. 159.

MACHINE, EXTRACTING SHELL, Q.F. 6- AND 3-PR.

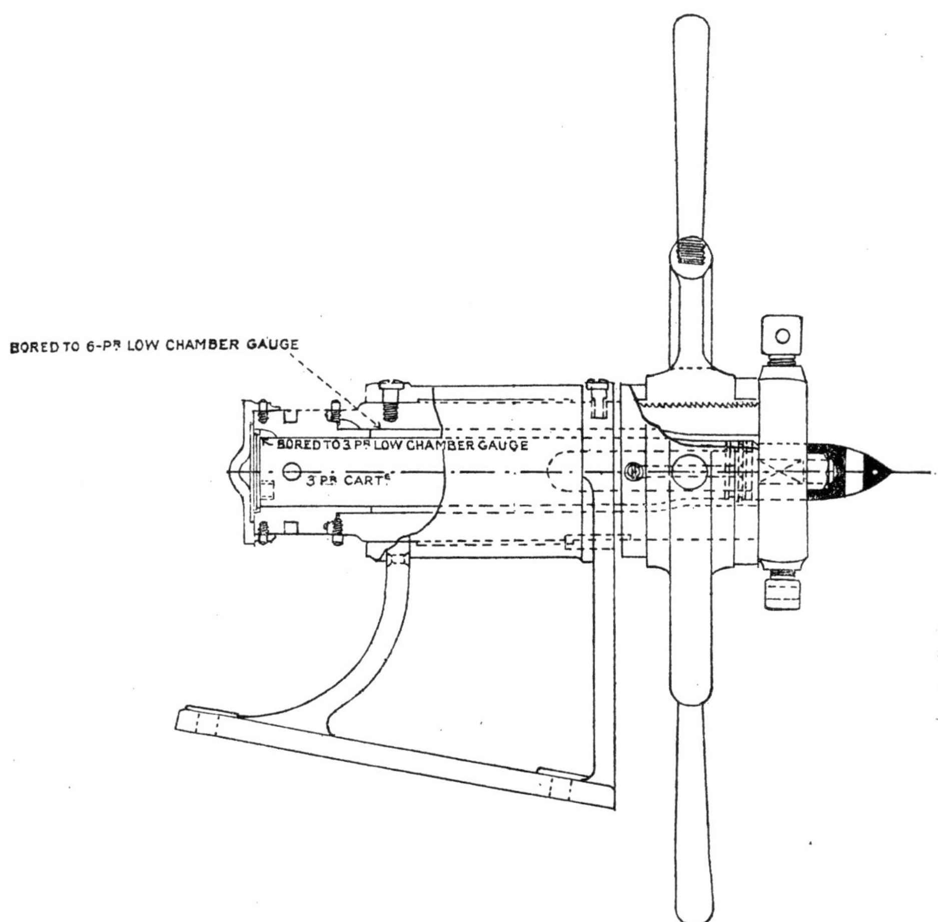

Lever, grummet, is a steel rod about 19 inches in length and 0·6-inch diameter; one end is tapered to a point, and slightly bent; the other end is flattened like a chisel and also slightly bent. It is intended for removing the rope grummets from projectiles and is also used for extracting eyebolts.

Mandrels are used in connecting with the re-forming of Q.F. cartridge cases. The cartridge case is placed over the mandrel while bulges, &c., are removed by means of a wooden mallet.

Mallets, Light, Mark I | *C* | .

Mallets, Heavy, Mark II | *C* | .

The above mallets are made of wood, with plain ash handles, and are for use in re-forming cartridges with the above-mentioned mandrels.

The light mallet is used with all natures Q.F., 3-pr. to 3-inch, the heavy mallet is used with all natures 4-inch to 6-inch.

Machine, extracting shell, 6-pr., and 3-pr., Mark I | *C* | , consists of a steel bush bored out to take a 6-pr. cartridge ; the bush fits into a cast-iron stand, to which it is secured by set screws ; a feather fitting into a featherway prevents it turning. The front end of the bush is threaded and carries a handwheel. Two guide bolts fit into slots at the front end of the bush, and are free to slide in and out. At the front end they are square and support a clamp, which is secured by nuts. Each guide bolt is provided with two projections, between which the handwheel fits ; a collar fits on the rear projections and is fitted with a set screw ; the front projections are also fitted with a collar which bears against the clamp. When the handwheel is turned it moves along the bush, the guide bolts sliding in the slots of the bush, thus forcing out the clamp and extracting the shell from the case. The clamp is provided with two clamping blocks to grip the shell ; these blocks are worked by screws. At the rear of the bush are two studs for securing a metal protecting cap. A bush bored to receive the 3-pr. cartridge can be screwed on the rear of the 6-pr. bush.

Machine, extracting Shell, 2·95-inch, Q.F., Mark I | *L* | .

Machine, extracting Shell, Q.F., 13 or 18-pr., Mark I | *L* | .

Machine, extracting Shell, Q.F., 3-inch, Mark I | *C* | .

The above machines are generally similar in design to the 6 and 3-pr. extracting machine, but differ in dimensions and in not being fitted with a metal cap.

Machine, indenting, 6-pr., Q.F., consists of a cylindrical steel body, bolted to a bed plate, which can be bolted to a bench. The body is chambered to take the cartridge, and fitted with three spring indenting pins, actuated by a short lever and cam. The shell can be forced into the cartridge case, before the latter is indented, by a screw bolt and handwheel. The machine is arranged so that the cartridge can be placed in it with its clip on, and the clip is always to be on the cartridge when it is placed in the machine.

Machine, indenting, 3-pr., Q.F., is similar to the above, but differs in dimensions.

Machine, coning or indenting, 18-pr., Mark I | *L* | .—The coning or indenting machine for the Q.F., 18-pr., is shown in Fig. 160.

The four indenting tools C are simultaneously actuated by the revolving of the cam wheel B, the cams in which are shaped to permit of the indenting tools springing back at each quarter of a revolution of the cam wheel. The bodies of the machines are chambered to take the respective cartridges, the shell being forced into the cartridge case, before the latter is indented, by the screw bolt and hand-wheel A.

Fig. 160.

MACHINE, CONING OR INDENTING, Q.F., 18-PR., MARK I.

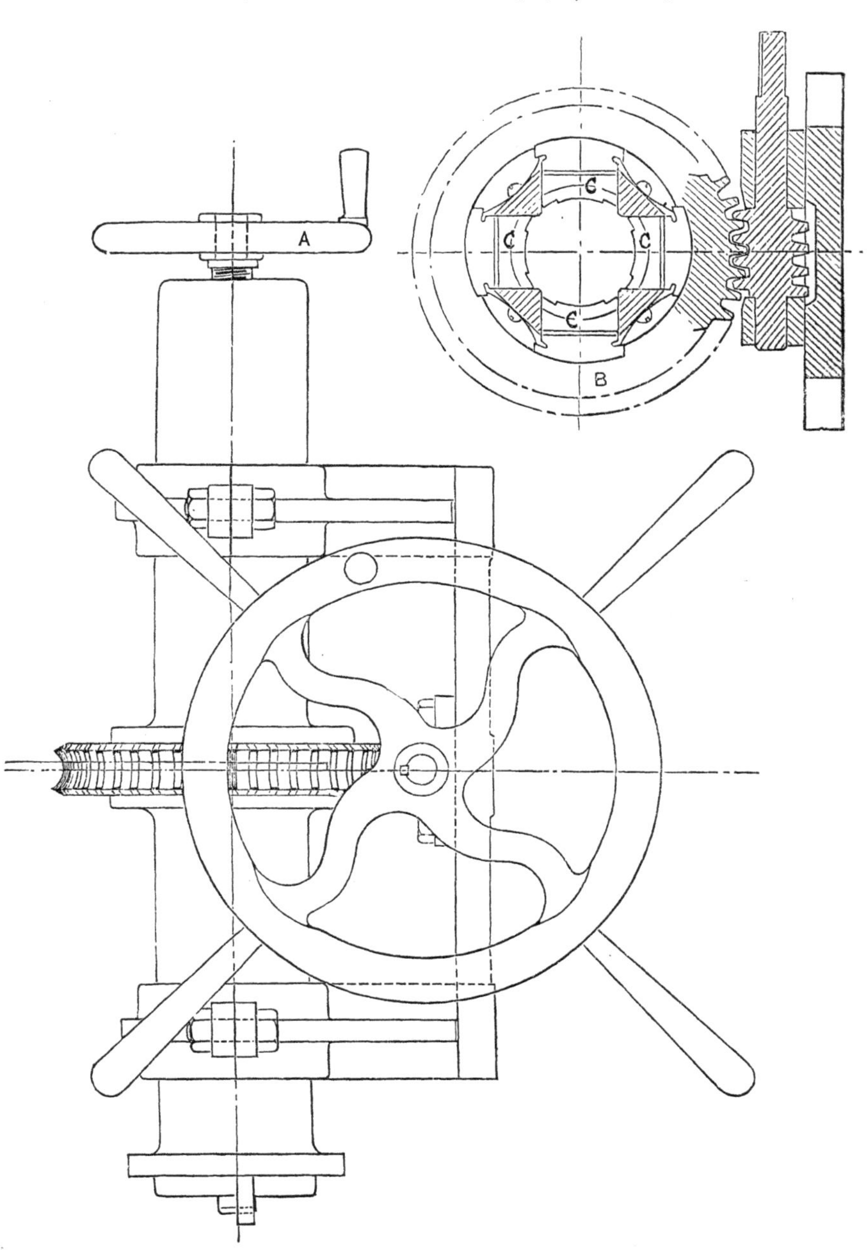

The coning or indenting machines for the Q.F., 13-pr. and 3-inch, differ from the above in dimensions.

Needle, magazine, phosphor bronze, 9-inch, is used for inserting silk braid hoops of cartridges.

Needle, magazine, phosphor bronze, 4-inch, is used for choking cartridges containing grain powder.

Needle, magazine, phosphor bronze, 1¾-inch.—For use in fine sewing on cartridge bags, &c. There is also a 1¾-inch needle made of nickel silver, but this has been superseded by the phosphor bronze needle.

Needle, magazine, curved, phosphor bronze, is used for threading silk or shalloon braid under other hoops or bindings, when tying together the sticks of cordite in making up charges for Q.F., 12-pr., 14-pr., 4-inch, 4·7-inch and 6-inch cartridges.

Pincers, shrapnel, primers, Mark II, are made of stout brass wire, and resemble a pair of sugar tongs. At the end of each branch of the fork the wire is flattened on the inner side so as to grip the brass primer and so enable it to be withdrawn after it has been unscrewed by the screwdriver, thus preventing the necessity of turning the shell over, and causing the powder to run out.

Planks, stacking, projectiles, Mark I, are made of elm 7 feet by 12 inches by 3 inches, bevelled at one end and strengthened at both ends by iron bands.

Pricker, cartridge, is made of bronze and is fitted with a wooden handle; it is used in making up cartridges with the 4-inch and 1¾-inch magazine needles when several thicknesses of cartridge bag have to be pierced.

Plugs, rectifying, Q.F. cartridges are made of steel and are for use in rectifying the mouth of the cases if they are deformed. The plug is driven in with a wooden mallet; a lip on the upper edge prevents it being inserted too far.

Rods, filling shell, "large," Mark I and "small," Mark I.—The large rod is 5 feet long and ·4 inches in diameter, the small 3 feet long and ·3 inches in diameter. They are made of brass, fitted with a wooden handle, a brass knob being formed on the other end of the rod for pressing down the powder.

Rod, 17-inch (driving caps from Service cases, 6 and 3-pr.) is made of steel 17 inches long, and reduced in diameter at the point. It is used for driving out the fired caps in 6 and 3-pr. Q.F. cartridge cases.

Rod, 4-inch (inserting caps in cases, 6 and 3-pr.).—The rod is of steel, ·625 inch in diameter, and has a slight projection at one end to fit over the cap in driving it in.

Scrapers, shell, are copper rods having both ends flattened out; one end is turned up at right angles, the other has a slight bend in the opposite direction.

They are used in removing powder from filled shell, or for searching empty shell.

There are four sizes, viz. :—

42-inch long for R.M.L. 10-in. and B.L. 10-in. to 9·2-in.
32 ,, ,, ,, R.M.L. 9-in., B.L. 7·5-in. to 5-in., and Q.F. 4·7-in.
20 ,, ,, ,, B.L. 4-in. and 12-pr.
17 ,, ,, ,, B.L. 12-pr.

The 17-inch rod will become obsolete when used up.

Screwdriver, shrapnel, large, Mark IV.—This is an ordinary screwdriver ; the blade is of phosphor bronze. It is used for inserting or removing the primers from shrapnel shell that are fitted with them.

Screwdrivers, magazine, are made of phosphor bronze, fitted with wood handles. They are issued in three sizes, namely :—

12-inch long with $\frac{3}{8}$-inch end.
6-inch ,, ,, $\frac{5}{16}$,, ,,
$1\frac{7}{8}$-inch ,, ,, $\frac{9}{16}$,, ,,

Sets, Q.F., are used in turning the tongues at the mouth of Q.F. cases when securing the lid. They are made of steel with a small notch for turning over the tongue.

Fig. 161.

SET, Q.F.

Tongs, Mark III (for extracting exploders from lyddite shell).—The tongs are made of manganese bronze, hinged, and fitted with handles similar to a pair of scissors. They are issued to Inspecting Ordnance Officers for extracting exploders when the waterproof cylinder is not fitted with a silk loop.

Fig. 162.

TONGS, MARK III.

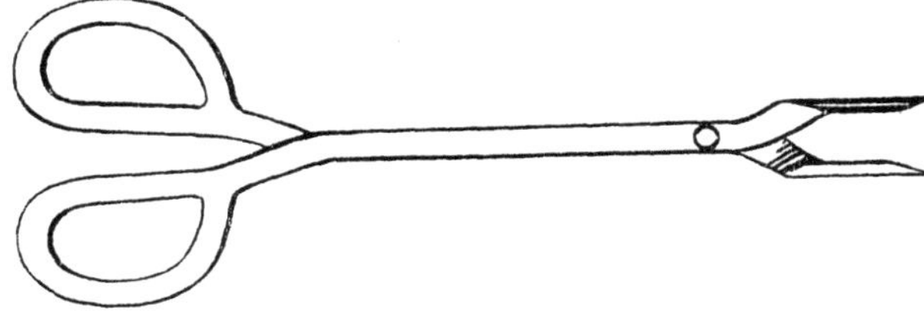

The *Mark II* tongs differ from the above in having longer and thinner jaws, and in the hinge being placed nearer the handles.

The *Mark I* tongs were made of $\frac{1}{4}$ brass rod and are now obsolete.

Tool, opening, cartridges, Q.F. or Q.F.C., Mark I | *C* | is a pair of curved pincers made of steel (Fig. 163), and is intended for use with Q.F. cartridges for prising up the tongues over the metal lid when it is necessary to remove the latter.

Fig. 163.

TOOL, OPENING CARTRIDGE, Q.F. OR Q.F.C.

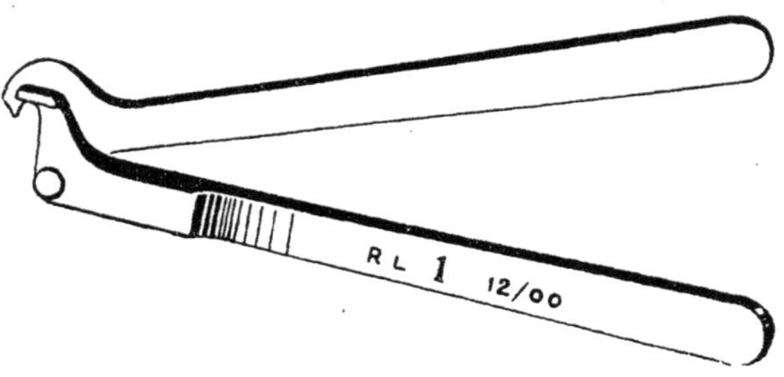

CHAPTER XX.—DETONATORS AND ELECTRIC FUZES.

Detonators are used for detonating charges of guncotton ; they all contain *Fulminate of Mercury* and are of two sorts :—

Non-electric.—This sort is fired by a length of safety fuze, or instantaneous fuze. (*See* page 512.)

NOTE.—For TORPEDO DETONATORS (*see* page 521), a special detonator is used in the HAND GRENADE. (*See* page 523.) Special detonators are also used for SOUND ROCKETS. (*See* page 521.)

Electric.—(For firing currents, *see* Table 40.)

Electric fuzes contain *gunpowder, but no fulminate,* and are of two sorts :—

(*a*) Those intended to fire the powder charges of land mines.

(*b*) "Disconnectors," used in connection with submarine mines ; these contain very little powder and are not to be relied on to fire powder charges.

Detonators and electric fuzes are distinguished by numbers, as follows :—

TABLE 39.

Service.	Detonators.			Electric Fuzes.	
	Non-Electric.	Electric.	Drill.	Disconnector.	Fuzes for firing Gunpowder.
Naval	No. 15	No. 9	—	No. 19	—
Land	No. 8	No. 13	No. 20	—	No. 14
Submarine	—	No. 12	No. 18	No. 16	—

The colours with which detonators and electric fuzes are painted have a special significance :—

Red—denotes the presence of *Fulminate of Mercury ;* in electric detonators, the portion containing the fulminate is coloured red ; but the non-electric detonators are coloured red all over, though only the thinner portion contains fulminate.

Yellow—denotes Naval Service.

White—denotes Land Service.

Blue—denotes Submarine Mining Service.

Label.—On every detonator and electric fuze is pasted a small label showing its number and Mark.

Non-Electric Detonators.

No. 8 | L | and No. 15 | N | .

Detonator No. 8, Mark VII | L | .

The No. 8, Mark VII Non-electric Detonator consists of a solid drawn copper tube about 2·18 inches long and ·23 inch in diameter.

It contains 30·8 grains of Fulminate of Mercury composition (Fulminate of Mercury, 80 parts; Chlorate of Potash, 20 parts)

Fig. 164.

DETONATOR NO. 8, MARK VII | L | .

Scale $\frac{2}{1}$.

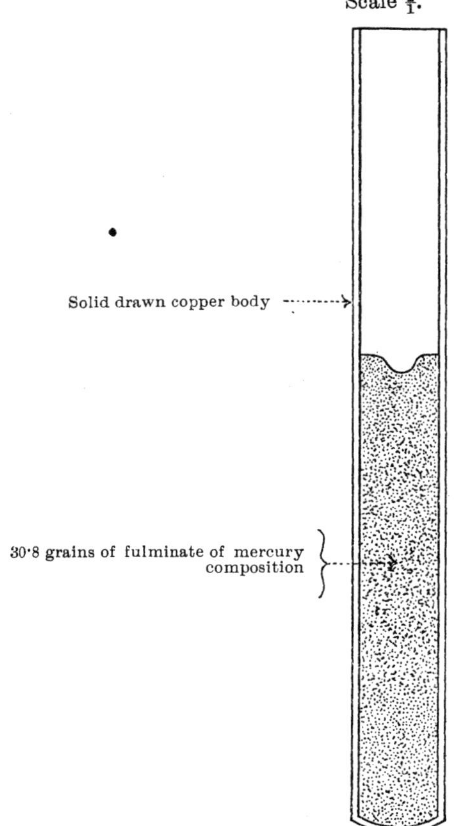

pressed in to the lower part of the tube; the upper portion of the tube is left empty for the reception of a piece of safety or instantaneous fuze (*see* page 36).

The exterior of the body is painted red and a small label bearing the number and numeral of the detonator is affixed to it.

The detonators are packed 25 in a tin cylinder (No. 8 D, Mark III).

For issue to cavalry the detonators are fitted with 2 feet of safety fuze, and, when so fitted, they will be packed in a tin cylinder (No. 49, Mark V).

Instructions for fitting Detonator No. 8, *Mark VII., with two feet of safety fuze.*—The outside covering of the safety fuze is to be removed for a length of about 1 inch to enable the fuze to be fitted into the detonator. The detonator is then secured to the safety fuze by means of shellaced tape wound round the detonator and the fuze. About 5 inches of $\frac{11}{20}$-inch tape are used, and the tape extends about $\frac{3}{4}$ inch up the fuze and $\frac{3}{4}$ inch on to the detonator. The tape and the whole of the detonator are then coated with vermilion paint.

The other end of the safety fuze is covered with shellac tape secured by two turns of thread.

Fig. 165.

DETONATOR NO. 8, MARK VI | L | .

Scale $\frac{2}{1}$.

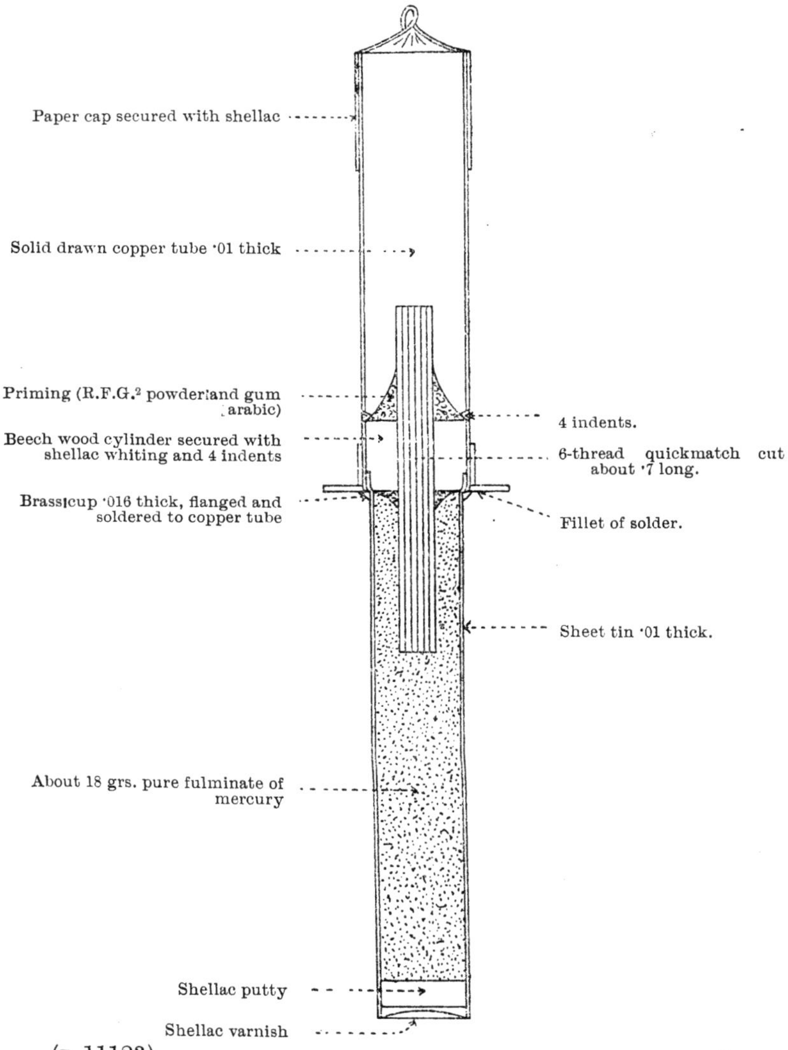

Paper cap secured with shellac

Solid drawn copper tube ·01 thick

Priming (R.F.G.[2] powder and gum arabic)

4 indents.

Beech wood cylinder secured with shellac whiting and 4 indents

6-thread quickmatch cut about ·7 long.

Brass cup ·016 thick, flanged and soldered to copper tube

Fillet of solder.

Sheet tin ·01 thick.

About 18 grs. pure fulminate of mercury

Shellac putty

Shellac varnish

Detonator, No. 8, Mark VI | L |.

The Mark VI detonator consists of a tin tube containing a charge of about 18 grains of fulminate of mercury. To the top of the tin tube is soldered a copper socket, for the reception of the end of the safety or instantaneous fuze.

A beechwood cylinder, secured by four indents, is placed on top of the fulminate of mercury.

A piece of 6-thread quickmatch passes through the hole in the beechwood cylinder, and carries the flash from the fuze to the fulminate; a priming of gunpowder and gum is placed round the quickmatch at each end of the cylinder to prevent the fulminate working through into the socket.

A ring of brass with two projecting lugs is soldered round the centre of the detonator to prevent it being pushed too far into a guncotton primer.

The top of the copper socket is closed by a paper cap (except when issued with 2 feet of safety fuze) and the bottom of the tin tube is closed with shellac putty coated with shellac varnish.

The detonator is painted red.

Detonator, No. 8, Mark V.

The Mark V detonator differs from the Mark VI in the body, which is made of solid drawn copper, instead of a tin tube and copper socket soldered together.

It contains about 20 grains of fulminate of mercury.

No. 8, Mark IV.

Mark IV differs from the Mark V in the tube being made of brass instead of copper, and the quickmatch is pressed down on top of the wooden plug.

Packing.

For R.E. units.—Twenty-five No. 8, Mark VII detonators in "Cylinder, No. 8 D," Mark III.

Earlier Marks (IV, V and VI) are packed in "Cylinder, No. 8 D," Mark II, with a "*Rectifier, Guncotton Primers, Mark V.*"

The cylinders are lined throughout with asbestos paper, and are closed by means of a lid, secured by a tin band soldered on.

For other Services.—Six No. 8, Mark VII detonators having 2 feet of safety fuze attached, in a "Cylinder, No. 49, Mark VI."

This cylinder is painted red. The lid is secured by a bayonet joint and tied to the body by whipcord.

Earlier Marks of No. 8 detonator, when fitted with 2 feet of safety fuze, are packed 6 in a "Cylinder, No. 49, Mark V," with a rectifier.

Four of these cylinders are packed in a "Box, No. 8, detonators," which is a deal box with elm ends having elm cleats and copper wire lifting handles. It has wooden fittings to prevent damage to the cylinders in transit and is painted Service colour.

Detonator No. 15 (*Naval Service*).

This consists of a body, tube, neck, wooden cylinder for quick-match, charge and paper cap. (*See* Fig. 166.)

The body is made of brass, tinned inside and out. The tube is made of tin, the top being flanged and soldered in the bottom of the body. The neck is made of tin, flanged at the bottom, having a flanged collar soldered over it for securing it to the body, and is intended to receive the end of the safety fuze. It has 4 indentations, so that when the fuze is inserted, and the neck compressed round it, the points of the indentations will grip the fuze.

The neck is covered with a paper cap, which is torn off before inserting the safety fuze.

Fig. 166.

DETONATOR NO. 15, MARK III.

Scale $\frac{1}{1}$.

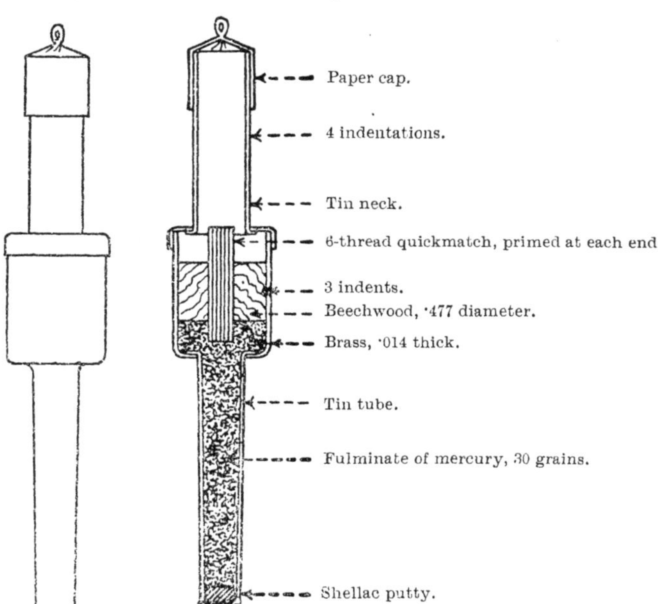

The cylinder is made of beechwood, and a piece of 6-thread quick-match is passed through it, the ends projecting about ·2 inch on either side. The ends of the quickmatch are coated with a priming of mealed powder, gum, and water to prevent the fulminate of mercury working through.

The cylinder is coated with shellac putty, and secured in the centre of the body by 3 indents.

The tube and lower part of the body are filled with a charge of about 30 grains of fulminate of mercury. The bottom of the tube is closed with shellac putty.

Colour.—Red all over.

Use.—In the Naval Service with safety fuze.

Packing.—No. 15 detonators are packed for issue as follows :—

25 in "Cylinder, No. 15, Mark II."

Note.—All detonators Nos. 8 and 15, which have attained the age of five years, will be destroyed or used up for instructional purposes.

Electric Detonators.

Bridges and firing currents.—There are three kinds of bridges used in detonators and fuzes, known as Naval, Field and Siege, and Submarine, the bridge for all N.S. fuzes, &c., being made of platinum and silver, which is very fuzible, and consequently fuzed with certainty when fired. For land and submarine services an alloy of platinum with iridium is employed. The latter metal increases the resistance and thus allows the bridge to be made rather thicker and stronger.

Details of the various bridges are given in Table 40.

Firing currents.—Detonators with Naval bridges should fire without appreciable delay with a current of 1 ampère, but should not fire when a current of ·32 ampère is applied for 4 seconds ; those with L.S. bridges with a current of ·8 ampère, but should not fire with a current of ·32 ampère for 4 seconds ; while the S.M. detonators are tested, all for over-sensitiveness with a current of ·8 ampère, which must not fire the detonator within 4 seconds, and 2 per cent. are tested for under-sensitiveness with a current of 1 ampère which must fire the detonator within 4 seconds.

The current for firing these detonators and fuzes is derived from a battery of voltaic cells, or the Exploder, Dynamo, Electric Quantity.

Detonator, Electric, No. 13, *Mark III* (*Land Service*).

(*Plate CII.*)

The Detonator, Electric, No. 13, *Mark III* | *L* | consists of the following parts :—

Head, 2 insulated wire terminals, 2 copper pole pieces, iridio-platinum wire bridge, brass socket, tin tube and charge.

The head is of ebonite with a hemispherical top ; two holes are bored through it lengthwise for the reception of the wires and pole pieces. The conducting wires are composed of 3 strands of tinned copper ; they are covered with pure indiarubber, and then with an outer cover of vulcanized indiarubber. To the ends of the wires, which fit into the head of the detonator, are soldered two short tinned copper poles, which are flattened to prevent them turning round after they are fixed and so breaking the bridge.

The pole pieces are dipped into hot guttapercha cement, and forced into the holes prepared for them in the head ; they project ·1 inch and are ·25 inch apart at the ends, and are connected by a bridge of iridio-platinum wire soldered with pure tin. The bridge is surrounded by a tuft of guncotton yarn.

The conducting wires are brought together above the head and frapped with waxed thread; the ends are bared and are of unequal length to minimise the risk of a short circuit when connected up.

Attached to the lower part of the head is a flanged brass socket; this supports a tin tube. The tube and lower part of the socket are filled with a charge of about 43 grains of fulminate of mercury, which surrounds the bridge and the ends of the pole pieces. The bottom of the tube is closed by shellac putty.

Colour.—The head and socket are painted white and the tube red.

Mark II was filled with 37 grains of fulminate of mercury; the wires were twisted above the head and were not frapped with thread.

Use.—For field and siege operations.

Packing.—25 in " Cylinder, No. 13 D, Mark I."

Detonator, Electric, No. 9. (*Naval Service.*)

(*Plate CIII.*)

Detonator, Electric, No. 9, *Mark IV* | *N* | .—The head is of ebonite, with a hemispherical top. The lower part of the head receives a brass socket attached to a tin tube. The conducting wires are composed of 3 strands of tinned copper, which are easier to manipulate than a thick single wire. To the ends of these, which fit into the head of the detonator, are soldered 2 short tinned copper poles, flattened to prevent them from turning round after they are fixed and so breaking the bridge.

The 2 solid poles, thus treated, are dipped into hot guttapercha cement, and forced down into the holes prepared for their reception in the ebonite head. When cool the conducting wires are twisted together above the head, and whipped near the head with waxed thread. The two poles above-mentioned project about ·1-inch beyond the bottom of the head, and are ·25-inch apart at the ends. They are connected by a bridge of fine platinum-silver wire (·21 grains to the yard), the ends of the bridge being carefully soldered with pure tin on to the flat ends of each pole. Round the bridge is priming composition.

This composition is separated from the fulminate by a thin brass cup having a small hole in the base covered with paper attached by shellac.

The tube contains about 32 grains fulminate of mercury.

Colour.—The head and socket are painted yellow, the tube red.

Packing.—25 in " Cylinder No. 9, Mark I."

Detonator, Electric, No. 12.

(For Submarine mining for India and Colonies.)

Detonator, Electric, No. 12, *Mark VI* | *L* | resembles No. 9, Mark IV in construction; the wires, however, are straight, being whipped above the head with waxed thread; it contains 24 grains of fulminate of mercury. The bridge is of iridio-platinum wire.

The wires are insulated by pure indiarubber covered by vulcanized indiarubber, and are of unequal length.

Colour.—The head is painted white, socket blue, and tube red.

Mark V differs from Mark VI in having the wires of equal length.

Note.—All electric detonators which have attained the age of 10 years will be destroyed or used up for instructional purposes.

Packing Electric Detonators.

All electric detonators are packed 25 in a tin cylinder around a *rectifier.* (*See* page 524.)

The latest cylinders have a central tube for the rectifier, so that it is easily got at without disturbing the detonators. These cylinders are closed by a tin band soldered on, and have their bodies and lids lined with sheet asbestos, which extends down the bodies as far as the tin diaphragm, which supports the detonators. The lining is to prevent heat from the soldering iron reaching the contents of the cylinder.

To obviate any risk from rough usage in transit, cotton wool is placed inside the cylinders, and each detonator fits into a separate hole in the tin diaphragm.

The cylinders are numbered to correspond with their contents.

Electric Detonators for Drill.

[Note.—These contain *no* fulminate.]

Detonator, Electric, No. 20 (*Mark II*) | *L* | is the representative for drill purposes of No. 13, which it resembles in dimensions. The tube is empty and is not closed at the bottom. A small disc of beech, having a ·2-inch hole in the centre with a disc of paper on the top and a disc of calico shellaced on the bottom, fits in the bottom of the brass socket. About 1 grain of priming composition is placed in a recess in the body between the poles, which are connected by the "Field and Siege" bridge surrounded by a tuft of guncotton yarn.

Painted white all over, to distinguish it from the Service detonator.

Mark I differs from Mark II in having the wires twisted instead of straight.

Detonator, Electric, No. 18, *Drill* | *L* | . (*Submarine Mining.*)

Detonator, Electric, No. 18, *Drill, Mark VI* | L | is the representative for drill purposes of No. 12, which it resembles in dimensions. The tube is empty and is not closed at the bottom. A small disc of beech, having a ·2-inch hole in the centre, fits in the bottom of the brass socket. An ebonite washer and two white fine paper discs are placed over, and an oiled silk disc under this wood disc; $2\frac{1}{2}$ grains of priming composition are placed in a recess in the body between the poles, and surround the bridge, which is identical with that of No. 12. The body fits into the brass socket, and is secured by a brass

Fig. 167.

DETONATOR, ELECTRIC, NO. 20, MARK II, DRILL, LAND SERVICE.

Scale $\frac{2}{3}$.

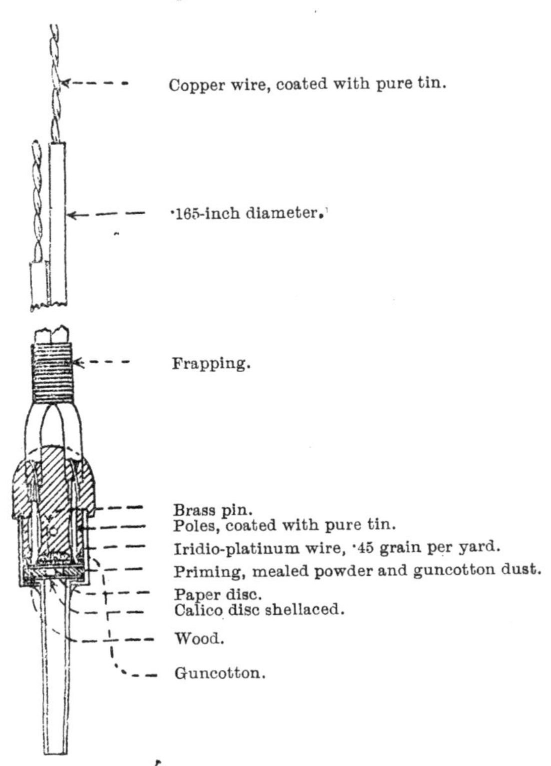

pin, passing through it and the socket, which can thus be removed when necessary for rebridging and repriming. These drill detonators are to be packed and treated as explosives.

The head is painted white, the socket blue, and the tube white; the object of the latter being to distinguish it from the Service detonator, which has a red tube.

Mark V differs from Mark VI in the length of the insulated wires.

Mark IV differs from Mark V in having the insulated wires twisted above the head.

TABLE No. 40.—*Bridges for Electric Detonators and Fuzes.*

Bridge.	Material.	Length, Inches.	Diameter, Inches.	Resistance in Ohms.		Firing Current in Ampères.	Fuzing Current in Ampères.	Remarks.
				Cold.	At Fuzing Point.			
Naval	Platinum-Silver	·25	0·0014	1·5 to 1·8	2·9	0·33	0·48	Used in No. 9 Detonator and No. 19 Fuze.
Field and Siege ...	Iridio-Platinum	·25	0·0014	0·9 to 1·1	2·6	0·48	0·8	Used in Nos. 13 and 20 Detonators and No.14 Fuze.
Submarine	Iridio-Platinum	·25	0·003	0·3 to 0·35	0·74	0·85 to 0·9	1·65	Used in Nos. 12 and 18 Detonators and No.16 Fuze.

Miscellaneous Detonators.

Torpedo Detonator.

Detonator, torpedo, small flange, Mark III | *C* | .—77 grains for all pistols (S.F. and A.W.) in all torpedoes, 18-inch and 14-inch, also for Brennan.

The above detonator is a solid drawn copper tube, about 1·4 inches in length and ·3 inch in diameter; the end of the detonator is round and solid, and the mouth is formed into a flange about ½ inch in diameter. The detonator is filled with 77 grains of fulminate of mercury pressed in and covered by a brass disc, retained in position by the metal at the mouth being burred over it. On the flange is the contractor's initials, month and year of manufacture, and Mark.

Packing.—Packed 5 in a tin cylinder, and 2 cylinders in a "Box, detonator, torpedo, or Sound Rocket."

Fig. 168.

DETONATOR, TORPEDO, SMALL FLANGE, MARK III | C | .

Scale $\frac{2}{1}$.

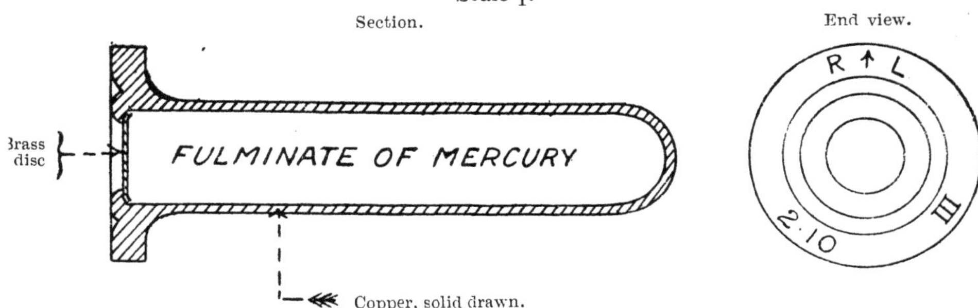

Detonator, Sound Rocket.

Detonator, Sound Rocket, No. 1, *Mark III* | *C* | consists of a tin tube which is slightly coned. At the larger end a tin flange is soldered to the tube. Inserted into the large end of the tube is a beechwood cylinder which is secured by shellac. Through the centre of this cylinder a hole is bored, into which is placed a piece of 6-thread quickmatch; around the outer end of the quickmatch is a priming of mealed powder, which in making up the detonator, is moistened with gum and water. The tin tube contains 12 grains of fulminate of mercury, and is closed at the small end by a glazed-board disc and solder.

Colour.—Red all over.

Use.—For detonating the guncotton primer in the Rocket, sound, ½ lb., Mark II.

Packing.—5 in "Cylinder No. 63"; 2 cylinders in "Box, Detonator, Torpedo, or Sound Rocket."

Fig. 169.

DETONATOR, SOUND ROCKET, NO. 1, MARK III | C |.

Full size.

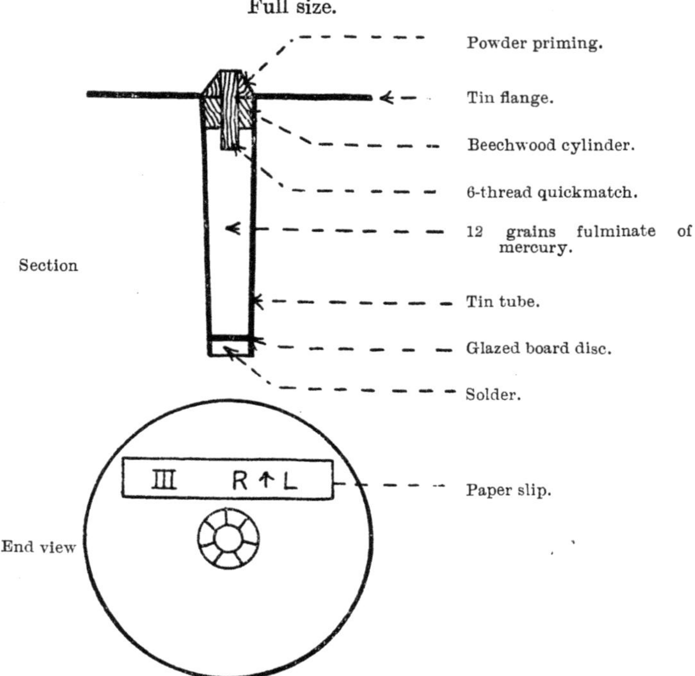

Detonator, Sound Rocket, No. 2, *Mark I | C |* consists of a solid drawn copper tube open at one end only. The tube contains 15·5 grains of fulminate composition (fulminate of mercury, chlorate of potash, and guncotton) strongly pressed in. A white paper disc is then inserted, over which is pressed 2·4 grains of mealed powder.

Fig. 170.

DETONATOR, SOUND ROCKET, NO. 2, MARK I | C |.

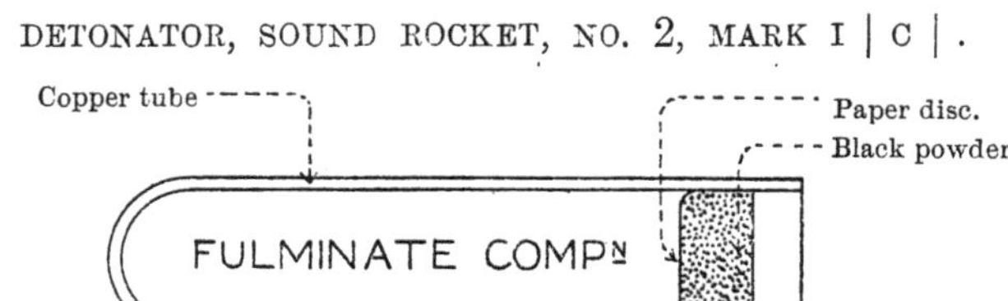

This detonator is not painted.

Use.—For detonating the tonite charge in Rockets, sound, ½ lb., Mark III, also in Rockets, light and sound, 1 lb., Mark I.

Packing.—5 in "Cylinder, No. 71"; 2 cylinders in "Box, Detonators, Torpedo, or Sound Rocket."

Detonator for Hand Grenade.

Detonator, Grenade, hand, Mark I | L | consists of a solid drawn copper tube, on the top of which is formed a flange having two upturned projections and two semi-circular portions cut away. A brass spring plate is soldered in a recess on top of the flange, and when compressed is flush with the top. In the lower end of the tube is placed about 30 grains of fulminate of mercury, over which is a brass cup. The cup is secured by four indents in the body of the tube.

Use.—For detonating the lyddite charge of the hand grenade.

Packing.—10 in "Box, Detonators, Grenade, hand," Mark I | L | .

The above-mentioned box is made of tin, painted red externally, lined with asbestos and fitted with a cork block bored to receive 10 detonators.

Fig. 171.

DETONATOR, GRENADE, HAND, MARK I | L | .

Scale $\frac{3}{1}$.

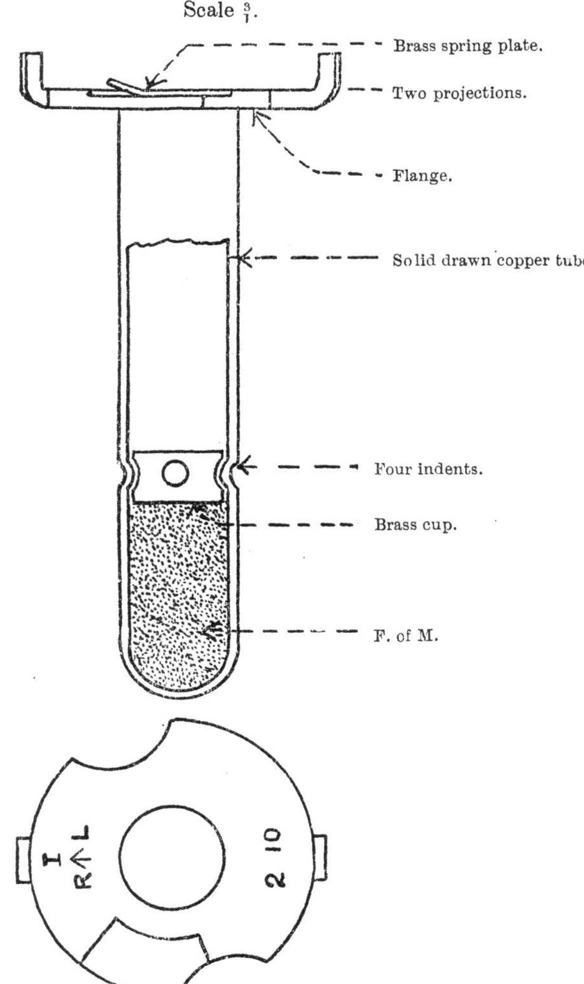

A sheet of cork fitted with a tape lifting loop is placed over the detonators.

The box is closed by a soldered tin band.

Five boxes are packed in a "*Case, Wood Packing.*"

General Notes on Detonators.

It is especially important to remember that detonators contain a large proportion of fulminate of mercury, and that it is most dangerous to treat them roughly, or to entrust them to unskilled persons. They should be handled with care, and only by persons well conversant with their properties. In testing electric detonators for continuity they should always be either removed to some distance, or strongly confined, in case of one being accidentally fired.

It must be remembered that no amount of immersion in water will destroy fulminate of mercury; when wet it is harmless, but when dried again it is as active and dangerous as ever.

On no account is any detonator to be taken to pieces for examination or any other purpose.

Any detonator that may have missed fire, or that may be found distorted or injured in any way that would appear to render it unfit for use, should at once be destroyed.

Inserting detonators.—Before inserting a detonator into a guncotton primer, force the rectifier (*see* below) into the hole in the primer, up to the full extent to which the detonator is to enter, and then withdraw the rectifier by twisting; the detonator should then be inserted *gently*. On *no account* should it be forced into the guncotton; screwing or twisting it should be particularly avoided.

Rectifier.

Rectifier, guncotton primers, Mark V | C | consists of a piece of lignum vitæ wood; the handle is round and has a notched rim, the lower part being of the same form and dimensions as the body and tube of the detonator. This portion is inserted in the perforation made in the primer to receive the detonator and "rectifies" the hole. It is to be used *in all cases* before inserting the detonator into guncotton.

Mark IV differs in the shape of the handle, which is flat.

Previous patterns differed in dimensions and are obsolete.

Rimer.—The *rimer, rectifying guncotton primers, Mark I | L |* is made of hard wood, and is for use in enlarging the perforations of the "Charges, priming, 2¼ lb. and 4½ lb." Care must be taken to use the rimer carefully and slowly, and to see that it is not applied too vigorously for too long a period, as otherwise there may be a danger from heating. The rectifier will still be required to press the rough edges into shape.

Case, Transport, Detonators.

Case, transport, detonators, Mark III, is similar in construction to the "Case, transport, explosives, Mark II," *see* Fig. 3, page 43, but differs in dimensions, which are 26·48 inches × 9·48 inches × 12·262 inches. Weight 63 lbs. It is painted red.

Dimensions :—Marks I* and II, 26·6 inches × 20·6 inches × 15·5 inches.

Mark I case has galvanized wire handles.

Use.—It is used for the conveyance of detonators and almost all sorts of explosive stores.

Electric Fuzes.

There are at present three electric fuzes in the Service :—

No. 14—Land Service—for land mines.
,, 16—Submarine mining—a disconnector. (For India and the Colonies.)
,, 19—Naval—a disconnector.

Fuze, Electric, No. 14. (*Land Service.*)
(*Plate CIV.*)

Fuze, electric, No. 14, *Mark III.*—The body is of ebonite, cylindrical with a hemispherical top ; the poles, wires, and bridge with guncotton tuft, are similar to those of No. 13 detonator, Mark III.

The part of the body reduced in diameter is fitted with a lengthening piece secured by Pettman's cement. The interior is charged with pellet powder which surrounds the poles and bridge, then closed by a glazed-board disc coated with Pettman's cement ; a brass cup is fitted over the lower portion of the body.

A label is fixed on the body, bearing the number and mark.

Colour.—It is painted white all over.

Use.—This fuze is used in field and siege operations for exploding charges of gunpowder, and for instructional purposes.

Packing.—25 in a tin cylinder similar to that for detonators. (Cylinder No. 14.)

Eight cylinders in a "Case, Packing, Electric Fuzes, Nos. 14 and 16."

Fuze, Electric, No. 16, *Mark II.*

This fuze has an ebonite body, and the general arrangements are shown in Fig. 172.

The bridge is of iridio-platinum wire and has the same resistance as those of other submarine electric stores. The centre of the lower part of the body is recessed between the wires to form a cavity of ·2 inch diameter. This hole contains a small quill driven with mealed powder and pierced in the ordinary way. Above the quill is a small quantity of fine grain powder. Immediately below the quill is the bridge, the object of this arrangement being to ensure the breaking of the bridge and consequent interruption of the current when the fuze is fired.

The heads fit into a short brass socket whose base is closed with a disc of paper, secured with shellac. The socket is lined with paper to prevent contact with the poles.

Use.—To disconnect one of a series of electro-contact mines when fired, without interfering with the current to the other mines yet unfired.

The details do not come within the scope of this work.

Colour.—The head is painted white and the body blue.

Packing.—25 in "Cylinder No. 16."

Eight cylinders in a "Case, Packing, Electric Fuzes, Nos. 14 and 16."

Fig. 172.

FUZE, ELECTRIC, NO. 16, MARK II, DISCONNECTING (SUBMARINE).

Scale ½.

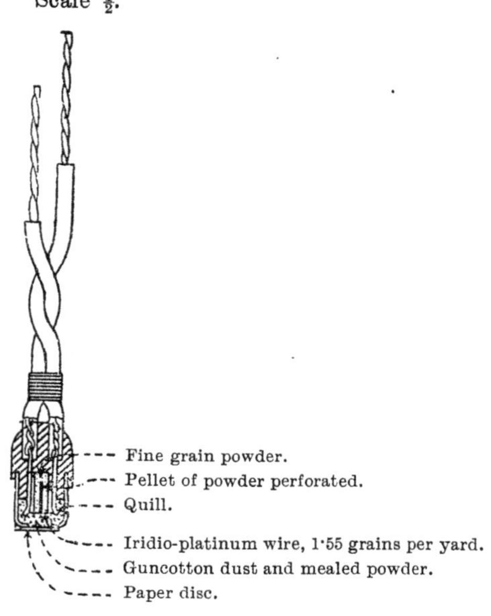

Fuze, Electric, No. 19.

(*Plate CV.*)

Fuze, Electric, No. 19, *Mark III* | *N* | consists of a mushroom-headed body, insulated wire terminals, two copper pole pieces, bridge, brass socket, and priming composition.

The body is made of ebonite, the lower portion being screw-threaded externally; two holes are bored through the body to receive the wires and pole pieces; the latter have double flanges to prevent them turning.

The bridge is of platinum-silver and is soldered to the pole pieces with pure tin; it is surrounded by 2½ grains of priming composition, retained by a brass socket, which screws on to the body.

In the bottom of the socket is a small hole, covered by a piece of oiled silk, which is itself covered by a paper disc.

Use.—This fuze is used in the Naval Service to disconnect a mine when the latter is fired, without interfering with the current to the other mines.

Colour.—The head is painted black and the body yellow.

Packing.—15 in a tin cylinder—No. 19.

CHAPTER XXI.—AIMING RIFLE, MACHINE GUN, AND SMALL ARM AMMUNITION.

Aiming Rifle Ammunition.

(i) Electric Ammunition. (With Gunpowder Charge.)

There are two makes of 1-inch Aiming Rifle Electric Ammunition with gunpowder charge—Morris (M.), and King's Norton (K.N.).

The Morris Pattern (Plate CVI.)

Cartridge, Aiming Rifle, 1-inch, Electric, Mark V, M | C | , consists of a solid drawn brass case, with a hole in the base tapped to receive the primer.

The interior of the case, except that part which envelops the bullet, is coated with a hard brown varnish. The exterior is not lacquered.

The *primer* consists of a brass tube with an enlarged head; it is threaded near the head, so as to screw into the case; the head fits into a recess, a fibre washer making a tight joint. The tube is bored out, the metal being thinned at the front end.

Fitting in the tube is a brass contact pin, which is insulated with ebonite plugs, the front plug being coned to suit the coned seating in the primer. An iridio-platinum wire bridge (resistance 1 to 1·5 ohms) is soldered, with pure tin, to the point of the contact pin and front edge of the body, the bridge being surrounded with guncotton dust or cotton powder, and the primer is closed with a card wad shellaced in. Two slots are cut in the head for the key removing and inserting primer.

The charge, which consists of 400 grains of R.F.G.[2] or other suitable gunpowder, is covered by a grease-proof card wad, a felt wad lubricated with beeswax, and a white card wad on top, next the bullet.

The bullet is made from an alloy of 12 parts lead, 1 part tin. It has 3 cannelures round it filled with beeswax, and the base is hollowed out. The bullet is partly covered with a patch of fine white paper, which is lubricated with beeswax at the base after being crimped over.

The bullet weighs 9 ozs. 408 grains $\pm$ 70 grains. It is firmly pressed into the case, which is then reduced at the mouth by coning to hold the bullet tightly.

The cartridge is stamped on the base with the numeral and contractor's initials.

Key, inserting and removing, primer cartridge, aiming rifle, 1-inch, electric, Mark I | C | is used with the cartridge.

The Mark IV cartridge differs from the above in the primer, which is pressed instead of being screwed in. Most of these cartridges were without the paper patch, and the earlier issues had the bullet secured by indenting the case into the rear cannelure, and the cases were weaker.

The King's Norton Pattern. (See *Plate CVI.*)

Cartridge, Aiming Rifle, 1-inch, Electric, Mark IV KN | *C* | differs from the Mark V M in the primer only, which is of different dimensions and internal arrangements.

The primer consists of a brass tube with enlarged head, bored out to receive a copper contact piece, which is insulated from the body by ebonite. The contact piece is cupped out in front, and into this fits a brass centre piece insulated from the body by ebonite. An iridio-platinum wire bridge (resistance 1 to 1·5 ohms) is soldered, with pure tin, to the centre piece and into a slot in the front edge of the body, the bridge being surrounded with guncotton dust or cotton powder, and the primer is closed with a card wad shellaced in.

In order to facilitate identification of the pattern of primer in the Mark IV cartridges, the letters "M" or "KN" are stamped on the cartridge, and printed on the wrapper, after the numeral, to indicate that the primers are, respectively, of the Morris, or King's Norton Company's pattern.

Early issues similar to the Mark IV Morris cartridge were made for Naval Service and designated Mark III. The Mark II cartridge differs in the primer.

The Mark I cartridge only differed from the Mark II in the bullet being made of brass, flat headed, with a lead core, and weighing 11 ozs. 130 grains.

(ii) Electric Ammunition. (With Cordite Charge.) (*Plate CVII.*)

The "*Cartridge, Aiming Rifle, 1-inch, Electric, Cordite, Mark I* | *C* | , consists of a case, electric primer, charge and bullet.

The case is of the same pattern as that used for gunpowder filled cartridges, but is *lacquered* internally, except that part which envelopes the bullet and the threads of the primer hole.

The primer consists of a brass body having an enlarged head; the body is screw-threaded near the head to screw into the case; the head fits into a recess, a fibre washer making a tight joint.

The body is bored out, the metal being thinned at the front end.

Fitted into the body is a brass contact pin, insulated from the body by two ebonite plugs, the front plug being coned to suit the coned seating in the primer.

An iridio-platinum wire bridge (resistance 1 to 1·5 ohms) is soldered with pure tin to the front of the contact pin and front edge of the body. Two slots are cut in the head for the "*Key, inserting or removing primer.*"

A paper tube is secured outside the plain part of the body by shellac varnish.

The body of the primer is filled with dry guncotton or cotton powder dust and the mouth is closed by a card disc pressed in.

The paper tube is charged with about 14 grs. of pistol powder and closed by a card disc, the end of the tube being turned over and secured with shellac.

The charge consists of about 160 grains of cordite, size 3, cut about 2·4 inches long and tied near the front end by a single tie of silk sewing.

The bullet is made of an alloy composed of 98 parts lead and 2 parts antimony; two cannelures are formed around it, and are filled with pure beeswax; the base is reduced in diameter, to receive a copper cup.

The cup is made of solid drawn copper and is firmly pressed on and indented. The bullet weighs 10 ozs. ± 70 grs.

Marking.—The cartridges are stamped on the base with the numeral, contractor's initials and the letter C.

If the cases used were originally filled with gunpowder, the old marking is barred out and the above substituted.

Key, inserting and removing, primer, cartridge, aiming rifle, 1-*inch, electric, Mark I* | *C* | is made of steel to the form shown in Fig. 173. It is used with the cartridge, aiming rifle, 1-inch, electric.

Fig. 173.
Full Size.

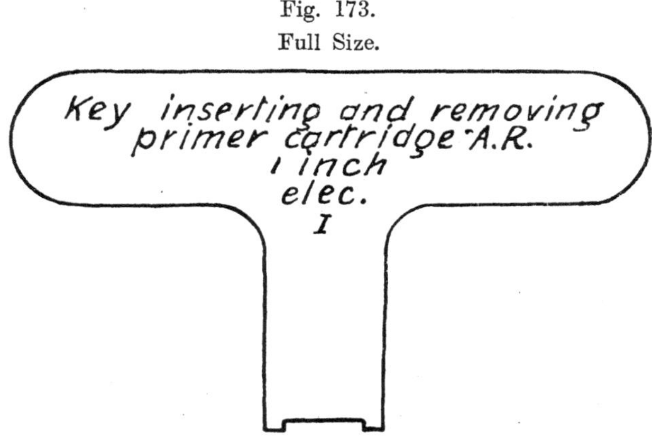

(iii) Percussion Ammunition.

(*Plate* CVIII.)

Cartridge, aiming rifle, 1-*inch, percussion, Mark I* | *L* | consists of a brass case, cap, charge, wads and bullet.

The case is of solid drawn brass, with a cap chamber formed in the base, in which is an anvil made by a projection of the material, round the head of which there are three fire-holes.

The interior of the case, except that part which envelops the bullet, is varnished with hard brown varnish.

The cap is a double one, the outer of brass, the inner of copper, and contains ·3 grain cap composition pressed in and varnished, and may be covered with a tinfoil disc.

The charge, which consists of 400 grains of R.F.G.[2] or other suitable gunpowder, is covered by a grease-proof card wad, a felt wad lubricated with beeswax, and a white card wad on top, next the bullet.

The bullet, of 12 parts lead and 1 part tin, weight 9 ozs. 408 grains, is pointed, and has three cannelures, which are filled with beeswax. It is fitted with a paper patch, the same as the bullet for the electric cartridge. It is secured in the mouth of the case by being firmly pressed in and by coning the mouth of the case. Earlier issues were without the paper patch, and the case was indented into the rear cannelure.

(iv) Blank Electric.

Cartridge, Aiming Rifle, 1-*inch, Electric, Blank, Mark I* | *L* | .

The cartridge consists of the Service charge and the Morris pattern case and primer, the charge being covered by two asbestos discs, which are coated with Pettman's cement on the top and edges.

The mouth of the case is turned in.

Fired cases of Mark V, M, or Mark IV, M, pattern may be used for making up these cartridges, the original numeral and contractor's initials (if necessary) being barred out and the new numeral and initials substituted.

Blank Percussion.

Cartridge, Aiming Rifle, 1-*inch, Percussion, Blank, Mark I* | *L* | .

The " Cartridge, Machine Gun, Nordenfelt, 1-inch, Blank, Mark I," declared obsolete, has been re-introduced for blank firing with 1-inch aiming rifles. The name has therefore been altered to that shown above.

Fired 1-inch percussion cases may be used for making up these cartridges, the original contractor's initials being barred out (if necessary) and those of the firm refilling substituted.

Cartridge for Instruction.

The Cartridge for Instruction, Aiming Rifle, 1-*inch,* has a wood block below the bullet, instead of the charge. Two holes are bored through the case to distinguish it from the Service cartridge.

Packing and Issue of 1-*inch Ammunition.*

Packing.—Both electric and percussion cartridges are packed in bundles of 12, wrapped in brown paper.

Issue.—Eight bundles in Box, Ammunition, S.A., Mark XI, or 9 bundles in Box, cartridge, aiming tube, C.F.

Blank electric or percussion cartridges are packed in bundles of 12, wrapped in brown paper and issued :—

Nine bundles in Box, Ammunition, S.A., Mark XI.

Distinguishing marks on boxes containing aiming rifle ammunition. (*See* Plate CXIV.)

Refilling, 1-*inch Aiming Rifle Cartridges.*

Cartridges, Aiming Rifle, 1-*inch, Percussion,* in the L.S. may be refilled ; they are cleaned and returned to Woolwich.

The cases of fired cartridges, Mark V, M, electric, may be used for refilling for both Naval and Land Services. Cases should be cleaned after firing, in the manner laid down in paragraph 204, "Regulations for Magazines, &c., 1913," and returned to store, where they will be examined, and any found to be unserviceable set aside.

Mark V, M, cases which are fit for refilling, and also Mark IV, M, which are fit for refilling for blank, as mentioned in § 12410, should be sent to Woolwich at the first opportunity.

Electric aiming rifle cartridges not to be refilled for L.S. will now be understood as applying to cartridges of earlier Marks than Mark V, M, excepting that the cases of Mark IV, M, fired cartridges will be used for refilling as blank cartridges.

Marking cases when refilling.—A centre punch mark will be stamped on the base of the case each time it is refilled.

Small Arm Ammunition.

·303-inch Ammunition.

Cartridge, S.A. Ball, ·303-inch (Mark VII) | *C* | .

Solid case, all suitably sighted and fitted ·303-inch small arms and machine guns.

(*Plate* CIX.)

The cartridge consists of a case, percussion cap, charge, glazed-board disc, and bullet.

The case is of solid drawn brass, with a cap chamber formed in the base, in which an anvil is made by a projection of the material, and two fire-holes are drilled. The case is not lacquered.

The cap is of copper, and contains ·6 grain of cap composition pressed in and varnished, and it may be covered with a tinfoil disc.

The charge consists of about 35·5 to 37·5 grains of cordite M.D.T., size 5–2. The number of tubes varies from about 36 to 44.

A glazed-board disc is placed on top of the charge.

The bullet weighs 174 grains. It is more pointed than earlier Marks, the head being struck with a radius of nearly 8 calibres. The envelope consists of an alloy of about 80 per cent. copper and 20 per cent. of nickel; the core is in two parts, the front portion consisting of an alloy of 90 per cent. aluminium and 10 per cent. zinc, or pure aluminium, the rear portion of 98 per cent. of lead and 2 per cent. of antimony. A cannelure is formed around the bullet near the base and this is filled with beeswax. The bullet is secured in the case by the necking of the latter, which is also indented in three places into the cannelure.

Cartridge, S.A. Ball, ·303-inch (Mark VI), | *C* | .

Solid case, all suitably sighted and fitted ·303-inch small arms and machine guns.

(*Plate* CX.)

The cartridge consists of a case, percussion cap, charge, glazed-board disc, and bullet.

The case and *percussion cap* are the same as those used with the Mark VII cartridge.

The charge consists of about 31 grains of cordite, size $3\frac{3}{4}$, in 60 strands.

A glazed-board disc is placed on top of the cordite.

The bullet consists of a core (98 parts lead, 2 parts antimony), enclosed in a cupro-nickel envelope, and weighs about 215 grains. The envelope is solid drawn from an alloy of about 80 per cent. copper, 20 per cent. nickel, and the core is secured inside it by turning over the end of the envelope; a cannelure runs round the bullet near the base. The bottom part of the bullet, except the base, but including the cannelure, is coated with beeswax. It is secured in the case by the latter being necked and indented in three places into the cannelure.

Cartridge, S.A., Ball, ·303-*inch, Mark II,* differs from Mark VI in the bullet which has the envelope thicker at the nose, and ·5 per cent. of iron was permitted in its alloy. The later issues have the rim made to suit the charger.

Marks III, IV and V cartridges have hollow-nosed bullets and have been ordered to be used up for practice.

·303 Blank.

Cartridge, S.A., Blank, ·303-*inch, without Bullet,* (*Mark V*) | *C* | .

Solid Case.

(*Plate* CXI.)

The Mark V ·303-inch blank cartridge consists of a Service pattern case and cap.

The case contains a charge of 10 grains of sliced cordite, size 20, on top of which is placed a strawboard wad. The mouth of the case is then necked and crimped.

The Mark VI ·303-inch blank cartridge originally had a mock bullet, but all mock bullets were ordered to be removed.

Marking on Case.

The following markings are now stamped on the base of the cartridge case:—

1. Initials of manufacturer.
2. Mark of cartridge.
3. Last two figures of year of manufacture.
4. Either one or two broad arrows.

With reference to the above, the following details may be noted:—

Since 1907, the last two figures of the year of manufacture have been stamped on the base of the case, the year dating from 1st April to March 31st.

The letter C, denoting cordite, is now omitted.

Contract supplies since 1908 have two broad arrows stamped on the base of the case, in addition to the initials of the contractor. Cartridges of Royal Laboratory manufacture will be, as hitherto, stamped R ↑ L.

In ·303-inch blank cartridge cases the numeral on the base may not always be applicable to pattern of cartridge. For instance, any Mark of case may be used for a Mark V blank cartridge.

If ball cartridge cases are emptied or rejected as unsuitable for filling as ball, the numeral remains if the cases are used for blank.

Charger for ·303-inch, S.A.A.

Charger, ·303-inch cartridges, Mark II, is made of steel, and holds five rounds.

It has a spring stop formed at each end to prevent the cartridges falling out, and is strengthened by having three ridges on the base.

The numeral II is shown on the side of the charger.

The Mark I charger has neither the spring stops nor the strengthening ridges on the base.

Fig. 174.
CHARGER, MARK II.
Scale $\frac{1}{1}$.

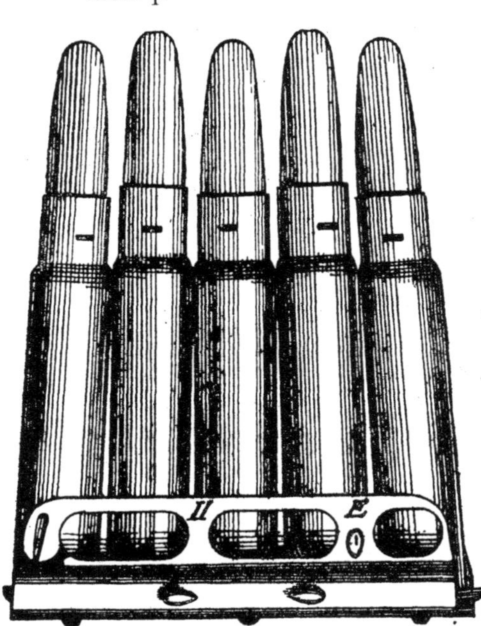

NOTE.—A " charger " is not loaded into a rifle: a " clip " on the other hand, *is* loaded with the cartridges, but clips are not used in the British Service.

Cases, Charger, ·303-inch Cartridges | C | .

The " *Cases, charger* " are made of leather board. After forming, the parts are secured together by wire stitches.

Bands of " buckram " are passed around the cases at right angles to each other, and are secured by glue. They are used to pack 20 rounds of ·303 inch cartridges in 4 chargers.

Bandolier, Cotton, 50 Rounds, Mark II | *C* | .

Bandoliers, cotton, 50 rounds, Mark II, are used for packing ·303-inch Mark VII ammunition in chargers. They are made of khaki-coloured jean, and consist of a body divided into five pockets and provided with a carrying strap. The pockets, each of which holds 10 rounds in chargers, are closed by means of fasteners.

Marks I and I* bandoliers differ from the Mark II in being made of drill and, in the case of the Mark I, in having larger pockets divided by single rows of stiching. No more will be made and, as soon as existing stocks are used up, they will be regarded as obsolete.

Packing ·303-*inch Service Ball Cartridges.*

Mark VII cartridges are usually packed 5 rounds in a charger and 10 chargers in a bandolier.

Mark VI cartridges are usually packed either :—

(i) 5 rounds in a charger and 20 rounds in a " Case, charger " or

(ii) 10 rounds in a brown-paper wrapper.

Both Marks VI and VII cartridges have, however, been packed in all three of the above-mentioned ways.

Marks II to V cartridges were all packed 10 rounds in a brown-paper wrapper.

The ammunition in packages as above is further packed :—

For Land Service.

Packages	Packed in
1,000 rounds, in chargers, packed either in " Cases, charger " or bandoliers. 1,400 rounds in " brown-paper wrapper."	In Box, Ammunition, Small Arms, ·303-inch, 1,000 rounds.
850 rounds in chargers packed in " bandoliers." 840 rounds in chargers packed in " Cases, charger." 1,100 rounds in " brown-paper wrappers."	In Box, Ammunition, Small Arms, Marks XI to XV.
600 rounds in chargers packed in " Cases, charger." 840 rounds in " brown-paper wrappers." 800 rounds in " brown-paper wrappers," for West African Field Force.	In Box, Ammunition, Small Arms, 600 rounds.

For Naval Service.

850 rounds in chargers, packed in "bandoliers." 840 rounds in chargers, packed in "Cases, charger." 1,100 rounds in "brown-paper wrappers."	In Box, Ammunition, Small Arms, Mark XI, Naval.
350 rounds in chargers, packed in "bandoliers." 360 rounds in chargers, packed in "Cases, charger." 500 rounds in "brown-paper wrappers."	In Box, Ammunition, Small Arms, ·303-inch, Half, Naval.

For distinguishing marks on packages, *see* Plate CXIV.
For weights, *see* Table 41.

Packing ·303-inch Blank Cartridges.

The cartridges are packed in bundles of 10 in purple paper wrappers.

The bundles are packed in Boxes, Ammunition, Small Arms; Barrels, Cartridge, and Cases, Powder, Metal-lined.

For numbers, &c., *see* Table 42, page 551.

·303-inch Dummy Drill Cartridges.

Cartridge, S.A., dummy drill, ·303-inch rifles or carbines, Mark III | *C* | consists of a Service case without cap; the bullet is of boxwood secured to the case by coning and three indents. Four holes are drilled through the case. The bullet is coloured red.

These cartridges are issued as required, loose in a packing case.

The distinguishing mark on the wrapper and box is a black rectangle with two uncoloured bars, and the letter D, in black, across the centre.

Cartridge, S.A., dummy drill, ·303-inch rifles or carbines, Mark IV | *C* | differs from the Mark III described above only in the shape of the boxwood bullet, which is of the same shape as the Mark VII Service bullet. No more Mark IV dummy drill cartridges will be made. The existing stock will be used up, after which the pattern will become obsolete.

·303-inch Dummy Cartridges for Inspectors.

(Also for Armourers and Serjeant-Instructors.)

Cartridge, Small Arm, Dummy, ·303-inch, for Inspectors, Mark III | *C* | consists of a tinned Service case without cap or charge, and a Mark VI Service bullet.

The weight is brought up to that of the Service cartridge by putting coal dust into the case.

Cartridge, Small Arm, Dummy, ·303-inch, for Inspectors, Mark IV | *C* |, differs from the Mark III cartridge in being fitted with a Mark VII Service bullet.

The cartridges are issued in bundles of 10.

Short Range Practice Cartridge.

Cartridge, S.A. Ball, ·303-inch, Short Range, Practice, Mark IV | *N* | . *Solid case, for use at certain Coastguard ranges.*—The cartridge consists of a Service pattern case and percussion cap.

The case is blackened for a length of 1·4-inch from the mouth and contains a charge of about 18 grains of cordite, M.D.T., size 4–2. A glazed-board disc is placed on top of the charge. The bullet consists of a cupro-nickel envelope, having a lead core ; it is 1·076 inch long, and weighs 188 grains.

It is secured in the case by 3 indents. The total length of the cartridge is from 2·9 to 2·975 inches.

The Mark III cartridge differs from Mark IV in the core of the bullet, being made of lead and antimony, and in the bullet being slightly longer.

No more Mark III cartridges will be made.

Packing and issue.—Short-range cartridges are packed heads and tails, in bundles of 10, in yellow paper wrappers. The labels on the boxes are of yellow paper also, and the distinguishing mark on both is a rectangle containing two diagonal lines and the letter C near each end. The letters and distinguishing marks are printed in black ; 500 cartridges are packed in " Box, A.S.A., ·303-inch, half, Naval."

Webley Pistol Ammunition.

(*Plate* CXII.)

Cartridge, S.A., Ball, Pistol, Webley, Mark II | *C* | , consists of a case, cap, charge, glazed-board disc and bullet.

The case is of solid drawn brass, with a cap chamber formed in the base, in which is an anvil made by a projection of the material, pierced with two fireholes. It has the manufacturer's initials or recognised trade mark, the numeral, and one or two broad arrows stamped on the base.

The cap is of copper, and contains ·4 grains of cap composition pressed in and varnished and may be covered with a tin-foil disc.

The charge consists of about 6½ grains of cordite, size 1, length ·05 inch.

The glazed-board disc is placed next the charge.

The bullet is made of 12 parts lead and 1 part antimony, and weighs 265 grains. It has a cavity formed in the base and 3 cannelures round the body ; these cannelures are filled with beeswax, and the bullet is secured in the case by choking the latter into the front cannelure all round.

In the early issues, the bullets were of lead and tin.

Cartridge, S.A., Ball, Pistol, Webley, Cordite, Mark III | *C* | , differs from Mark II in the bullet having a cavity in the head and weighing only 218½ grains. The existing stock, in the L.S., at home stations, will be used up for practice.

Cartridge, S.A., Ball, Pistol, Webley, Marks IV and V | *C* | , differs from Mark II in the bullet, which has a flat head and weighs 220 grains. A slightly heavier charge is used.

Mark V differs from Mark IV in the bullet being made of lead and antimony, instead of lead and tin.

Packing, Webley Pistol, Ball Cartridges.

Land Service.

Twelve rounds in a brown-paper wrapper, *Marks III and IV*, are then packed 300 rounds, and *Mark II* 276 rounds, in a "*Box, Ammunition, Small Arms, Pistol, Mark III* | *L* | ."

Naval Service.

Six rounds in a brown-paper wrapper, *Marks III and IV*, are then packed 936 rounds, and *Mark II* 828 rounds, in a "*Box, Ammunition, Small Arms,* ·303 *inch, half, Naval, Mark I* | *N* | ."

(For distinguishing mark, *see* Plate CXIV.)

Cartridge, S.A., Blank, Pistol, Webley, Mark II | *L* | , consists of a case, cap, charge and two wads.

The case is of solid-drawn brass, with cap chamber and anvil formed in the base and two fire-holes.

The cap is of copper and contains ·25 grain of cap composition pressed in and varnished, and may be covered with a tin-foil disc. It is secured in the case by four indents.

The charge consists of 8 grains R.F.G.² powder covered with two *felt wads*, over which the mouth of the case is crimped.

These cartridges are packed, in bundles of 12, in purple paper wrappers, 420 in Box, Ammunition, S.A., Pistol, Mark III.

Fig. 175.

CARTRIDGE, S.A., BLANK, PISTOL, WEBLEY, MARK II | L | .

Full Size.

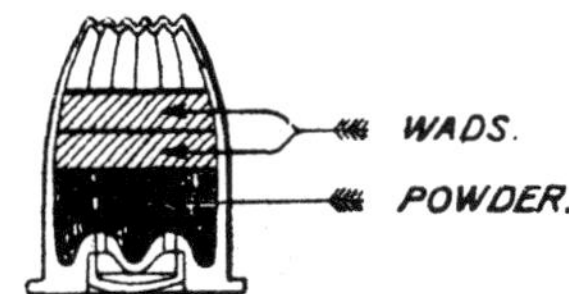

Cartridge, S.A., Ball, Pistol, Self-loading, Webley and Scott, 0·455-*inch, Mark I* | *N* | .

(*Plate* CXIII.)

The cartridge consists of a case, cap, cordite charge and bullet.

The case is made of solid-drawn brass, a cap chamber and anvil being formed in the base. Three fire-holes communicate between the cap chamber and the charge.

The percussion cap is of copper and contains 0·4 grain of cap composition, pressed in and varnished, which may be covered with tinfoil if desired.

The charge consists of about 7 grains of cordite, size 1, length 0·05-inch.

The bullet consists of a solid-drawn copper envelope, which is nickel-plated. A cannelure is formed around the bullet into which the mouth of the case is coned. The envelope contains a core consisting of an alloy of 98 parts lead and 2 parts antimony, secured by turning over the base of the envelope. The bullet weighs about 224 grains.

Packing.—7 rounds in a brown-paper wrapper and 840 rounds in a "*Box, A.S.A.*, 0·303-*inch, half, Naval.*"

(For distinguishing mark on packages, *see* Plate CXIV.)

Aiming Tube Ammunition.

Cartridge, Aiming Tube, C.F., Mark I | *C* | .

The case is of solid drawn brass with the cap chamber and anvil formed in the base. Two fire-holes in the latter communicate from the brass cap to the powder charge.

Fig. 176.

CARTRIDGE, AIMING TUBE, C.F., MARK I | C | .

Scale $\frac{2}{1}$.

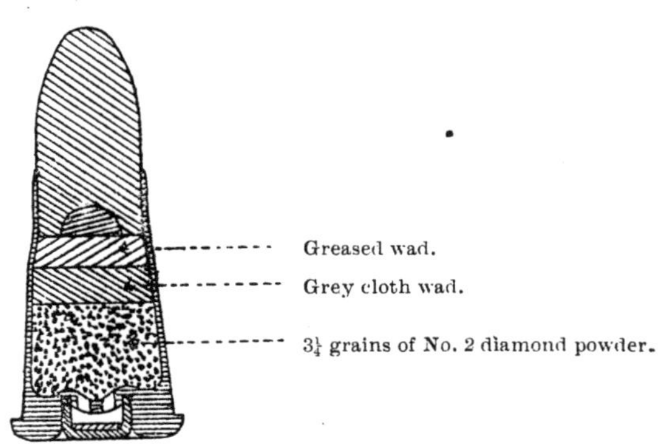

The charge is $3\frac{1}{4}$ grains Curtis and Harvey's Diamond No. 2.

The bullet is of 12 parts lead and 1 part tin, and weighs 37 ± 3 grains, and there are 2 wads fastened to its base, that next the bullet being greased and the other of grey cloth.

Mark II differs from Mark I only in the arrangement of the wads and in having the base of the bullet slightly recessed.

Issue.—100 rounds are packed in a cardboard box about the size of a bundle of 10 rounds of the Service ammunition, tied with string.

The distinguishing mark on the wrapper or box is a black circle with a black dot in the centre.

Supplies of cartridges, aiming tube, will be made in tin-lined boxes, each containing 10,000 rounds, but when required for the Royal Navy they will be repacked in quarter metal-lined cases locally as required, containing 9,100 rounds.

Rim Fire Cartridge.

Cartridge, ·22-inch, R.F., No. 1, Mark I | C |.

The case is made of solid drawn copper zinc alloy, the fold in the rim being charged with cap composition as shown in the figure.

The bullet is of lead with 1 to 1·5 per cent. of tin, and weighs about 40 grains, and has 3 cannelures round it to retain the lubricant.

The charge is 4·7 grains of black powder.

This is the commercial 0·22 calibre long rim fire cartridge.

Fig. 177.

CARTRIDGE, AIMING TUBE, R.F., NO. I, MARK I | C |.

Scale $\frac{2}{1}$.

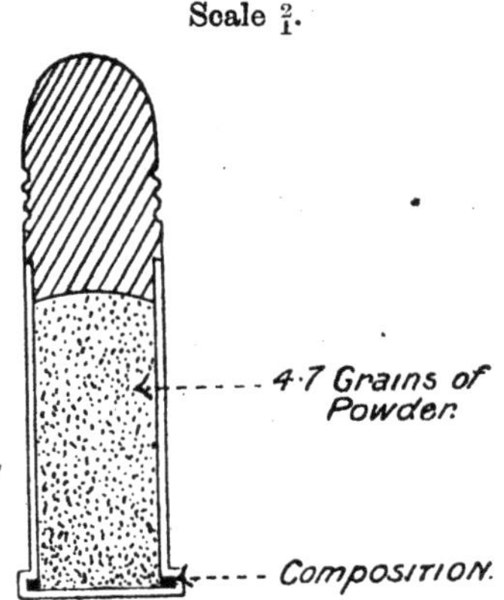

Packing for issue :—

Land Service.

100 rounds in a cardboard box.

10 cardboard boxes in a tin box hermetically sealed.

10 tin boxes in a Box, Cartridge, ·22-inch, R.F., No. 1, 10,000 rounds. (*See* page 546.)

Naval Service.

100 of the cardboard boxes are packed in a Case, Powder, M.L., Quarter, *i.e.*, 10,000 rounds.

Cartridge, ·22-inch, R.F., No. 2, Mark I | N |.

This cartridge has been introduced for use in Webley pistol aiming tubes in Naval Service, and differs from the No. 1 cartridge in being shorter and in having a 30-grain bullet. It corresponds to the commercial short ·22-inch cartridge. The cartridges are packed 100 in a cardboard box, 100 boxes in a Case, Powder, M.L., Quarter, 1,000 rounds in all, or 52 boxes in a "Box, A.S.A., Half, Naval," *i.e.*, 5,200 rounds.

Boxes for aiming tube ammunition.—See page 546.

Cartridge, machine gun, ball, ·45-inch, powder, Mark IV | C | Gardner and Nordenfelt. (*See* Fig. 178.)

The case is made from solid drawn brass, with a cap chamber with raised anvil formed in the metal of the base; two fireholes pass from the bottom of the chamber to the interior. The inside of the case where the powder charge rests is coated with hard brown varnish, which is removed from that part which envelops the bullet. This was in consequence of the liability of the empty case to stick in the gun after firing, which was attributed to the surplus varnish at the end of the case.

The cap is of copper, and contains ·3 grain of cap composition.

The charge is 85 grains of R.F.G.[2] powder, over which is placed a glazed-board disc, then a beeswax wad cupped out in front, and then two glazed-board discs.

The bullet is of 12 parts lead to 1 part tin; a cannelure is formed near the base; the bullet is covered for about two-thirds of its length with fine white paper; the paper is crimped over at the base and lubricated. The case is choked into the cannelure.

Fig. 178.

CARTRIDGE, MACHINE GUN, BALL, ·45-INCH, MARK IV | C |, GARDNER AND NORDENFELT.

Full Size.

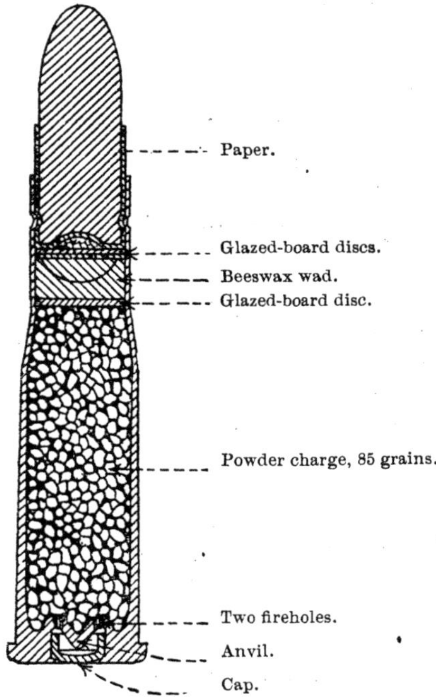

Packing and issue.—These cartridges are packed by tens, heads and tails, in brown-paper wrappers, and issued in S.A.A. boxes containing 680 rounds each. The distinguishing mark is a solid black triangle.

Cartridge, Machine Gun, Ball, ·45-inch, Cordite, Mark I | N | .—The cartridge consists of a case, cap, charge, wad and bullet. The case is similar to the powder case described above, but is not varnished inside, and the letter C is stamped on the base.

The cap is of copper, and contains ·7 grain of cap composition pressed in and varnished, and may be covered with a tinfoil disc.

Fig. 179.

CARTRIDGE, S.A., BALL, M.-H. RIFLE, SOLID CASE, CORDITE, MARK I.

Full Size.

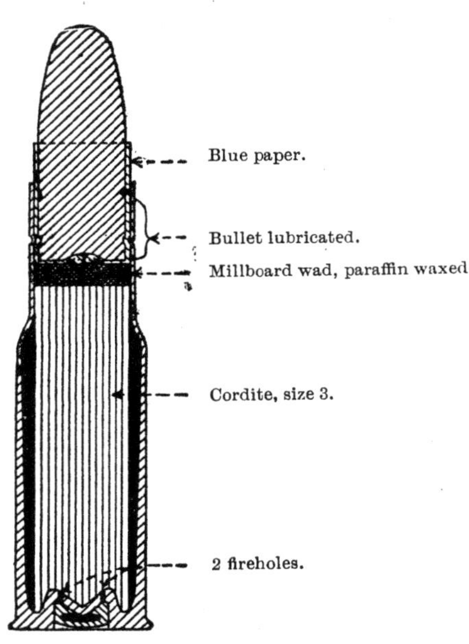

The charge is about 38 grains of size 3 cordite in 100 strands, covered by a millboard wad paraffin waxed.

The bullet is similar to the one for the powder cartridge, but the paper patch is orange coloured; the mouth of the case is choked into the cannelure, the edge fitting against a small shoulder which is formed on the bullet.

Packing and issue.—The cartridges are packed in bundles of 10; 680 rounds in a Mark XI, S.A.A. box. The distinguishing mark is a red triangle with the letter C in white in the centre.

Cartridge, S.A., Ball, M.-H. Rifle, Solid Case, Mark II, is similar to the *Cartridge, M.G., Ball, ·45-inch, Mark IV,* but the case is more bottle shaped. The distinguishing mark is a solid rectangle in red on the box and in black on the brown-paper wrappers.

Cartridge, S.A., Ball, M.-H. Rifle, Solid Case, Cordite, Mark I.—The case is solid drawn, bottle shaped, with the usual rim for extraction; an anvil with two fire holes is formed in the base. The cap is of copper, and contains ·7 grains of cap composition pressed in and varnished; a tinfoil disc may be used.

The mark of the case "I," contractor's mark, and the letter C are stamped on the base of the cartridge.

The charge consists of 35·8 grains of cordite, size 3, on top of which is placed a waxed millboard wad.

The bullet weighs 480 grains, and is provided with two cannelures, the neck of the case being choked into the rear cannelure. The bullet is covered for about two-thirds of its length with paper which is coloured *blue*, to facilitate identification.

Certain issues have been made in which the bullets were provided with *orange* coloured paper, but no more will be issued so covered. The wrappers of these cartridges were stamped "For rifle only."

Issue 10 in a bundle, 580 in a Mark XIV box. Distinguishing mark, solid red rectangle with C in white.

Cartridge, S.A., Ball, M.-H. Carbine, Solid Case, Cordite, Mark I, differs from that for the rifle as follows:—The charge consists of 34 grains of cordite, size 3, on top of which is placed a waxed millboard wad. The bullet weighs 410 grains, and is provided with one cannelure, into which the neck of the case is choked. The bullet is covered for about two-thirds of its length with paper, which is coloured *green* to facilitate identification.

Issue 10 in a bundle, 600 in Mark XIV box. Distinguishing mark, red rectangle with C in red.

Cartridge, S.A., blank, M.-H., or Snider, Rifle or Carbine, Mark IV.—The case is made of brown paper; it has a base cup and the base of the Service cartridge. It contains about 68 grains of blank F.G. powder.

Marking on S.A.A. Boxes.

(*A*) *Marking generally.*

In order that the particular kind of ammunition packed in S.A.A. boxes may be readily distinguished, all such boxes issued from Woolwich now have labels with distinguishing marks (*see* Plate CXIV); these distinguishing labels are placed on each side and on each end of the box.

The labels are, as a rule:—

White, for ball cartridges.
Blue ,, blank ,,
Purple ,, dummy ,,

The distinguishing mark is also printed on the descriptive labels, which are fixed, one on the closing plate of the lining, and one on the top of the box.

The manufacturer's initials will always be found in the lower line of the descriptive labels, in the left-hand corner.

(*B*) *Marking on Boxes containing* ·303-*inch Ammunition.*

There is now only one date marked on a box containing ·303-inch ammunition, and that is the date of manufacture; this date, together with the manufacturer's initials, is printed on the descriptive labels, and stencilled on each end of the box.

One other detail is required to completely identify the ammunition, and that is the *number* of the box, which will be found stencilled on the top, and also marked on the closing plate of the lining.

The gross weight is stencilled on one end.

In addition to the distinguishing and descriptive labels, the following are placed on every box:—

Government explosive label ..	On junction of lid and box.
Small label, with a number in red	In the circular recess over the wire attached to the pin.
A label with directions for opening	Over the string of boxes having the T-shaped brass handle for pulling out the pin.
Classification label	Where convenient.

Since 1st January, 1908, the particulars of the cordite (*i.e.*, Lot number) have not been given. Before that date this information used to be stencilled on the box and printed on the label on the lining.

S.A. Ammunition Barrels, Boxes, &c.

Barrels, Cartridge | *C* |, are of three sizes—half, quarter and eighth; they have no copper hoops, and are used for conveyance and storage of blank S.A. ammunition.

Box, Ammunition, S.A., 1,000 *rounds,* ·303-*inch in chargers, Mark I* | *L* |, *see* Fig. 180.

Originally the above box, when made of deal, was known as No. 1; when made of teak or mahogany, as No. 2. No. 2 was intended for foreign climates, but No. 1 can now be made suitable for this purpose by treatment with oil, so No. 2 is obsolete for manufacture, and the names No. 1 and No. 2 have been abolished.

This box, which is used for packing 1,000 rounds of ·303-inch cartridges in chargers, whether in cases, charger, or in bandoliers, and 1,400 rounds in brown-paper wrappers, is of deal with elm ends, the sides and ends being glued and dovetailed together, the bottom secured by tinned iron screws. The top is in three parts, the centre part forming a sliding lid. The sliding lid is secured by a split brass pin, which passes through a hole in the lid and into a similar hole in the side of the box. This pin is secured to the lid by a piece of twine, and is fitted with a T-shaped handle which fits into a recess in the lid, a calico label with instruction for opening being pasted over it; a second recess is for a sealing label. A tarred rope handle passes through holes in each end of the box. These handles are parcelled with strips of leather to prevent wear on the hooks of pack saddles.

The box has a removable tin lining coated with black varnish; the lining is closed by a tin lid soldered on.

Dimensions, 17″×8·312″×10·85″.

This box will eventually supersede the Boxes, A.S.A., Marks XI to XV for Land Service.

Fig. 180.

BOX, AMMUNITION, S.A., 1,000 ROUNDS, ·303-INCH, IN CHARGERS, MARK I | L |.

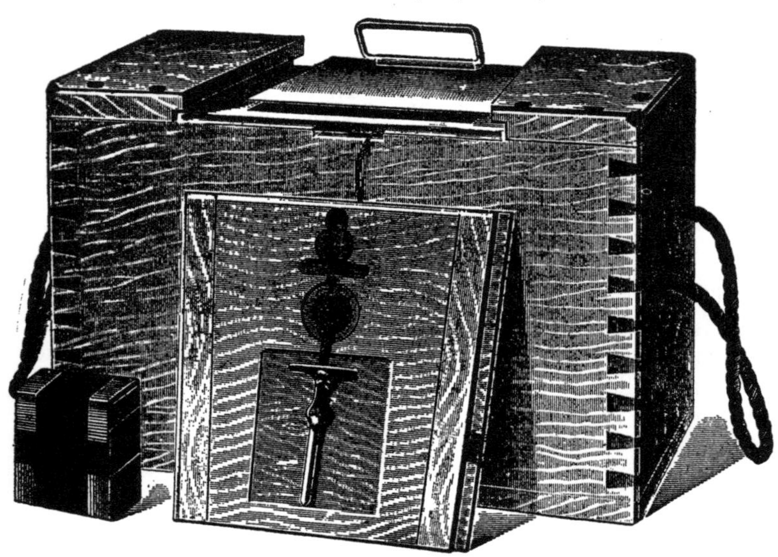

Boxes, Ammunition, Small Arms, G.S. (Land, Mark XI). (Naval, Mark XI.)

The Boxes, A.S.A., G.S., Mark XI, are made of teak or mahogany. The Land and Naval boxes differ from each other in the Land box having rope handles, parcelled with leather, and the Navy having copper-wire handles covered with leather for about 4 inches at the joint, and in having the letter N branded on each end.

Box, Ammunition, Small Arms, Home and Special, Mark XII.

The Mark XII box is similar to the Mark XI in appearance, but is made of deal with elm ends.

Box, Ammunition, Small Arms, Mark XIII.

The Mark XIII box was introduced for packing 1,000 rounds of Martini-Henry cartridges.

It is obsolete for future manufacture and the stock is being used up for packing blank ·303-inch cartridges. It is larger than Marks XI, XII, XIV or XV.

Box, Ammunition, Small Arms, Home and Special, Mark XIV.

The Mark XIV box differs from the Mark XII chiefly in having no cleats for the rope handles, which pass through the ends of the box.

Box, Ammunition, Small Arms, Home and Special, Mark XV.

The Mark XV box differs from Mark XIV in having a cheaper lid. Only a few were made.

Dimensions of Boxes, Ammunition, Small Arms, for Storage.

Mark	XI	21·812	×	8·312	×	6·937	inches.
,,	XII	20·625	×	8·325	×	6·962	,,
,,	XIII	20·75	×	10·25	×	10	,,
,,	XIV	22·187	×	8·5	×	7·02	,,
,,	XV	22·212	×	8·437	×	7·02	,,

Box, Ammunition, S.A., 600 *rounds*, ·303-*inch, in Chargers.*—The box is for use in packing either 600 rounds of ·303-inch cartridges in chargers (§ 11753), or 840 rounds in ordinary paper packets, for Colonial Service.

When issued to the West African Field Force the number of rounds in paper packets is 800.

There are two Marks of this box: the Mark I is made of mahogany and the Mark II of soft wood, which, when required for use abroad, will be treated with oil, mineral, preserving wood. These boxes have a tin lining, and are provided with a sliding lid, which is secured by a half-round brass split pin, having a T-shaped handle attached to it.

The tin lining is provided with a closing plate fitted with a handle.

Each end of the box is provided with a rope handle for lifting purposes.

Dimensions, 12·375″ × 9·812″ × 8·625″.

A certain number of 780-*round boxes* have been issued for Colonial Service.

This box differs from the 600-round box in dimensions, in having a rope handle at one end only, and in being provided with three tin boxes, instead of a single tin lining. The tin boxes will each contain 260 rounds of ·303-inch cartridges in ordinary paper packets, or 200 rounds in chargers.

There is also a 750-*round box*, which is similar to the 780-*round box*, but has only one tin lining.

No more 780 or 750-round boxes will be provided. They will be superseded by the 600-round box.

Camel and bullock boxes are Special for India; but the first is also used for the issue of S.A.A. to Colonial Governments.

Handles of Small Arm Boxes.—In future, manufacture and repair of S.A.A. Boxes in the Land Service, the hole in the boxes, for the handles, will be chamfered, and the handles will be parcelled with strips of leather, to prevent wear on the hooks of pack saddles.

Box, A.S.A., ·303-*inch, half, Naval, Mark I* | *N* |, is similar in material and construction to the Mark XI box, but has only one cleat and one handle. It is used in the N.S. for packing ·303-inch and pistol ammunition, of which it will contain :—

500 rounds, S.A. Ball, ·303-inch, or ·303-inch Short Range Practice in brown-paper wrappers.
350 rounds, S.A. Ball, ·303-inch in chargers, packed in bandoliers.
360 rounds, S.A. Ball, ·303-inch in chargers, packed in cases, charger.
936 rounds, Webley Pistol Ammunition, Marks III, IV or V.
828 rounds, Webley Pistol Ammunition, Mark II.
840 rounds, Webley and Scott, Mark I.
Stowage dimensions, 10·875″ × 8·375″ × 7·0″ ± ·05″.

Box, Ammunition, S.A., Pistol, Mark III | L | is similar in construction to the Box, A.S.A., G.S., Mark XI, but is much smaller.

It has a cleat and rope handle at one end only.

The sides and bottom are made of deal, the ends of elm, and the top, including the sliding lid, of teak.

Stowage dimensions, 8·625″ × 6·5″ × 4·625″.

Boxes for Aiming Tube Cartridges.

Box, Cartridge, Aiming Tube, "C.F.," Mark I | L |, is of deal with elm ends ; the lid is secured by brass screws with leather washers. It has a tin lining and a closing plate with handle. Cleats and rope handles are attached to each end of the box. Dimensions, 18⅞″×7¼″ ×11″.

This box is used for packing 10,000 cartridges, aiming tube ; or 108 cartridges, aiming rifle, 1 inch.

Box, Cartridge, ·22-inch, "R.F." No. 1, Mark I | L |, is of deal with elm ends ; the sides and ends are dovetailed, and the top and bottom are secured by brass screws. The ends are fitted with elm cleats and rope handles. The box contains 10 tin boxes, each having a lid secured by a tin band soldered on. Each tin contains 1,000 rounds in 10 cardboard boxes. Wooden partitions are placed in the box to prevent movement of the tins during transit. Dimensions, 18″×13·8″ × 7·2″.

Tool, Extracting Bullets, Small Arm, Mark I.

Tool, Extracting Bullets, Small Arm, Mark I, is for use of Inspecting Ordnance Officers in opening S.A. and ·45-inch M.G. cartridges.

It consists of a steel body ; gunmetal socket, fitted with milled handwheel steel flanged bush, secured in the body by a small screw ; steel milled-headed clamping screw, and two steel sleeves ·8 inch and 3 inches long respectively.

The body has a shoulder on the outside, a screw thread being cut behind the latter to take the socket. The body is bored out (the diameter being greater in rear) to take the steel flanged bush. In this form the tool is adapted for use in extracting bullets from M.H. rolled and solid case cartridges.

Fig. 181.

TOOL, EXTRACTING BULLETS, SMALL ARM, MARK I.

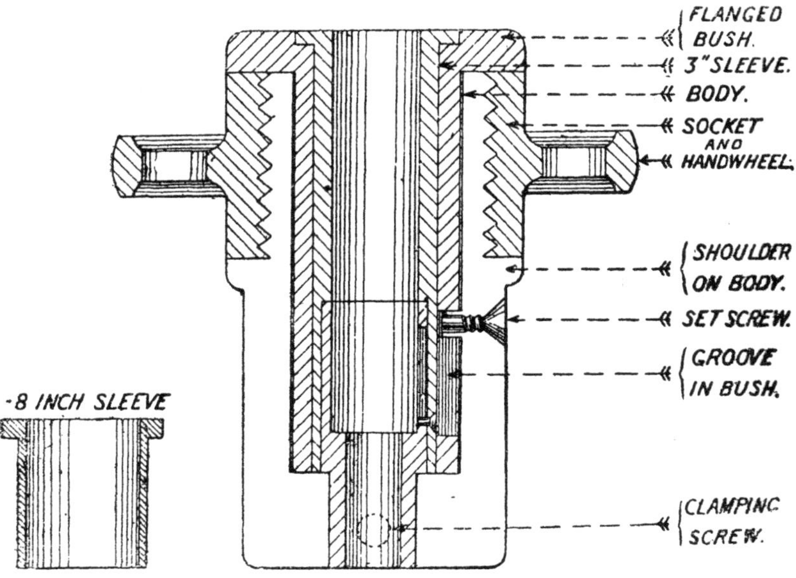

Instructions for Use.

1. *M.H. rolled and solid case cartridges.*—Insert the cartridge in the bush of the tool, and grip the bullet with the clamping screw; turn the handwheel, and withdraw the body of the cartridge from the bullet.

2. *G.G. cartridges.*—Insert the ·8-inch steel sleeve in the bush, and proceed as in 1.

3. ·303-*inch cartridges.*—Insert the 3-inch steel sleeve in the bush and proceed as in 1.

Care must be taken in inserting the 3-inch sleeve that the lines on it and on the head of the body of the tool coincide, so as to ensure that the clamping screw will pass through the hole in the sleeve and thus be free to grip the bullet.

TABLE NO. 41.—*Packages of Cartridges, Aiming Rifle, S.A., and M.G. (Ball).*

Nature.	Charge.	Bullet.	Bundle.		Box.			Remarks.
			No.	Weight.	No.	Mark.	Approximate Gross Weight.	
	grs.	ozs. grs.		lbs. ozs.			lbs. ozs.	
1-inch Aiming Rifle, Electric (Gunpowder)	400 R.F.G.[2]	9 408	12	10 10¾	96	Box, A.S.A., XI	97 10	
1-inch Aiming Rifle, Percussion (Gunpowder)					108	Box, Cartridge tube, C.F.	107 6	
1-inch Aiming Rifle, Electric (Cordite)	160 Size 3	10 0	12	10 0	96	Box, A.S.A., XI	93 8	
1-inch Aiming Rifle, Percussion (Cordite)					108	Box, Cartridge aiming tube, C.F.	102 10	
·303-inch, Mark VII	35·5 to 37·5 M.D.T. Size 5·2	0 174	50	3 3	1,000	Box, A.S.A., 1,000 rounds	74 8	In Bandoliers.
					850	Boxes, A.S.A., XI, XII, XIV or XV	66 8	
					350	Box, A.S.A., half, Naval	28 12	
					600	Box, A.S.A., 600 rounds	48 4	In Bandoliers.
			20	1 6	1,000	Box, A.S.A., 1,000 rounds	80 8	In "Cases, Charger."
					840	Boxes, A.S.A., XI, XII, XIV or XV	69 0	
					360	Box, A.S.A., half, Naval	30 8	
					600	Box, A.S.A., 600 rounds	51 8	In "Cases, Charger."
			10	0 9	1,400	Box, A.S.A., 1,000 rounds	—	In "brown - paper wrappers."

							1,100	Boxes, A.S.A., XI, XII, XIV or XV	75	0	
							500	Box, A.S.A., half, Naval	37	4	
							840	Box, A.S.A., 600 rounds	61	0	
·303-inch, Marks II to VI	31·5 Size 3¾	0	215	50	3	6	1,000	Box, A.S.A., 1,000 rounds	75	8	In "Bandoliers." (Mark VI only.)
							850	Boxes, A.S.A., XI, XII, XIV or XV	67	0	
							350	Box, A.S.A., half, Naval	28	14	
							600	Box, A.S.A., 600 rounds	48	8	
				20	1	6¼	1,000	Box, A.S.A., 1,000 rounds	81	0	In "Cases, Charger."
							840	Boxes, A.S.A., XI, XII, XIV or XV	70	12	
							360	Box, A.S.A., half, Naval	30	10	
							600	Box, A.S.A., 600 rounds	51	12	
				10	0	10	1,400	Box, A.S.A., 1,000 rounds	98	8	In "brown - paper wrappers."
							1,100	Boxes, A.S.A., XI, XII, XIV or XV	79	6	
							500	Box, A.S.A., half, Naval	37	8	
							840	Box, A.S.A., 600 rounds	61	8	
·303-inch, Short Range Practice, Marks III and IV	18 M.D.T. Size 4-2	0	188	10	0	9	500	Box, A.S.A., half, Naval	34	8	
Webley Pistol, Mark II	6½ Size 1	0	265	6	0	4¾	828	Box, A.S.A., half, Naval	43	0	
				12	0	9½	276	Box, A.S.A., Pistol III | L |	16	8	
Webley Pistol, Marks III or IV ...	6½ Size 1	0	220	6	0	4¼	936	Box, A.S.A., half, Naval	47	8	
				12	0	8¼	300	Box, A.S.A., Pistol III | L |	16	12	
Webley and Scott, Self-loading ...	7 Size 1	0	224	7	0	5¼	840	Box, A.S.A., half, Naval	47	6	

TABLE No. 41.—*Packages of Cartridges, Aiming Rifle, S.A., and M.G. (Ball)*—continued.

Nature.	Charge.	Bullet.	Bundle.		Box.			Remarks.
			No.	Weight.	No.	Mark.	Approximate Gross Weight.	
	grs.	ozs. grs.		lbs. ozs.			lbs. ozs.	
Aiming tube, Central Fire, Mark II ...	$3\frac{1}{4}$ C. H. Diamond No. 2	0 37	100	1 $0\frac{1}{5}$	10,000	Box, Cartridge, aiming tube, C.F.	113 0	
Aiming tube, Rim Fire, No. 1, Mark I	4·7 R.F.G.[2]	0 40	100	0 $13\frac{1}{2}$	10,000	Box, Cartridge, ·22-inch, R.F., No. 1	100 8	
					10,000	Case, Powder, M.L., Quarter	103 0	
Aiming tube, Rim Fire, No. 2, Mark I	—	0 30	100	—	10,000	Do.	—	
					5,200	Box, A.S.A., half, Naval	—	
M.G., ball, 4·5-inch, Mark IV	85 R.F.G.[2]	480	10	1 $3\frac{5}{16}$	680	Box, A.S.A., XI, XII, XIV or XV	94 8	
M.G., ball, 4·5-inch, Mark I (cordite)...	38 Size 3	480	10	1 $2\frac{1}{2}$	680	Box, A.S.A., XI	90 8	
M.H., Rifle, solid case, Cordite, Mark I	35·8 Size 3	480	10	1 $2\frac{3}{4}$	580	Box, A.S.A., XIV ...	81 10	
M.H., Carbine, solid case, Cordite, Mark I	34 Size 3	410	10	—	600	Do. ...	79 8	

TABLE No. 42.—*Packages of Cartridges, Aiming Rifle, S.A. and M.G. (blank).*

Nature.	Charge.	Bundle.		Box.			Remarks.
		No.	Weight.	No.	Mark.	Gross Weight.	
	grs.		lbs. ozs.			lbs. ozs.	
Aiming Rifle, 1-inch	400	—	—	108	XI	40 0	
M.G., G.G., chamber powder, Mark I, solid case...	65	10	0 $8\frac{3}{16}$	1,670	Half barrel	95 12	
				2,000	Half-metal lined case ...	134 8	
·303-inch cordite, Mark V, solid case	10 Size $\frac{20}{\text{S.C.}}$	10	0 $4\frac{1}{4}$	1,400	XI, XII, XIV	48 0	
				2,500	XIII	81 0	
				3,200	Half barrel	103 8	
				1,900	Quarter barrel	59 6	
				7,400	Whole M.L. case ...	257 0	
				3,400	Half „ ...	124 4	
				1,450	Quarter „ ...	58 4	
M.H., or Snider Rifle or Carbine, Mark IV ...	68 Blank F.G.	10	0 4	960	XI, XII, XIV	36 4	
				1,500	XIII	52 4	
				2,000	Half barrel	62 2	
				1,300	Quarter barrel	40 4	
				2,400	Half M.L. case	90 0	
				1,020	Quarter „	43 8	
Webley Pistol	10	6	0 $2\frac{1}{4}$	420	Box, A.S.A., Pistol, III | L |	8 8	

CHAPTER XXII.—ROCKETS AND LIGHTS.

Rockets are employed for signalling, for display, as weapons of war, and in conjunction with the life-saving apparatus.

There are also in the Service sundry port-fires and lights, the manufacture of which is analogous to that of rockets, and which will accordingly be described in this chapter.

A rocket *propels itself* through the air. In all cases a rocket consists of a cylinder closed at the head and having a vent at the rear end; in this cylinder is a quick-burning composition, up the centre of which is a conical hole; the ignition of this composition causes a pressure of gas in the rocket; the gas escapes out of the vent, presses against the air, and so drives the rocket in whatever direction the head may be pointing.

It is necessary to provide some means of keeping a rocket travelling in the direction in which it is started; for, if the rocket were a simple cylinder, it would tend to turn over and over. As the composition burns away the centre of gravity is always altering. The rocket is kept straight in one of two ways, either by—

(1) *Rotation* (as in the war rocket) imparted by an application of the force of the gas escaping from the rocket; or

(2) *The attachment of a long stick,* or a short stick with a tail of rope. The rocket in this case is kept fairly straight by the rush of air past the stick or rope. In any breeze this has the effect of turning the point of the rocket to the wind, and it will therefore be found to travel to windward.

War Rockets.

Paint.—All war rockets are painted red.

War rockets are issued in wooden cases fitted to hold three 24-pr. or four 9-pr. rockets respectively.

Age of rockets.—New war rockets will not be issued from Woolwich until one year after the date of their manufacture.

Any war rockets more than ten years old, reckoning from date of manufacture, will be destroyed locally.

Range.—The average range of the 24-pr. rocket is about 1,800 yards; the 9-pr. 1,200 yards.

Rocket, War, 24-pr.

Rocket, War, 24-pr., Mark VII | *C* | , consists of a body, head, base piece, tail piece, and safety cap.

The body is made of steel tubing, cut to the required length, and tested internally by hydraulic pressure to 1 ton on the square inch. The interior is then roughened by scoring it spirally and scoured till it is quite clean and bright.

The head is of cast iron lined with wood, and is fastened to the body by screws.

The base piece of wrought iron or mild steel fits into the end of the case and is secured by screws. It is bored out and screwed to receive the tail piece and also to receive the safety cap as shown in the woodcut.

Fig. 182.

ROCKET, WAR, 24-PR., MARK VII | C | .

Scale $\frac{1}{8}$.

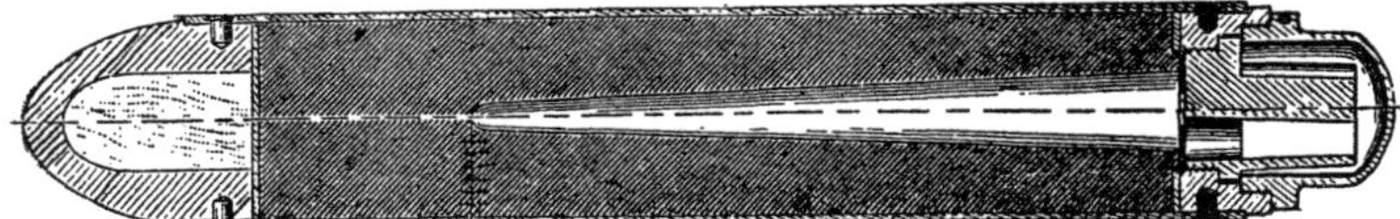

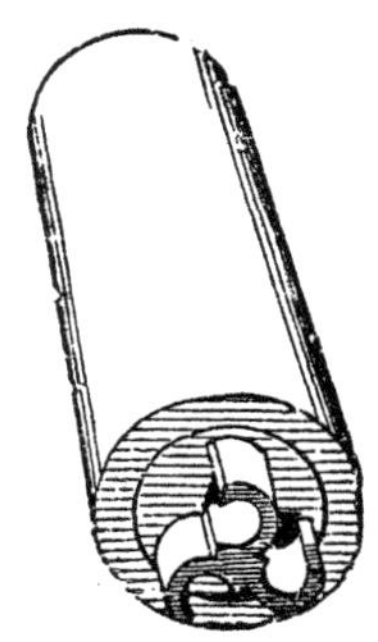

The tail piece is of cast iron, and contains three conical vents, the larger part of the cone being towards the interior of the rocket. The vents are cut away on one side ; hence the gas issuing from the vents meets with resistance on the side where they are prolonged, and, there being no counterbalancing resistance where they are cut away, rotation is given to the rocket.

The safety cap is of mild steel and is screwed on to the base piece. Its object is to cause the rocket to burst, instead of becoming a dangerous missile, should it be accidentally ignited. Between the face of the cap and the base piece an asbestos washer makes a water-tight joint.

A special key, known as the "Key, rocket, war," is issued for unscrewing these caps.

The composition consists of ground saltpetre, sulphur and alder charcoal.

This is first pressed into pellets for convenience of handling, and the hole for the tail piece having been filled up with a false base, the rocket is placed base downwards in the press and the pellets put in from the top and subjected to pressure.

An asbestos disc is placed over the top of the composition, the edges of the disc being turned up round a disc of millboard which is placed over the asbestos. The upper surface of the millboard and the interior of the top part of the case are coated with thinned luting, and the head is pressed in on top of the millboard and fastened on. The false base piece is then removed and a conical hole 11·875 inches long drilled in the composition. Round the edge of the base piece inside there is a lead ring, so as to seal the joint and prevent the escape of gas in that direction, and between the base piece and the composition is a millboard washer.

When screwed in, the tail piece is retained in position by a keep screw.

Before screwing the safety cap into the base piece the threads are lubricated with thinned luting.

The numeral and date are stamped on the base of the rocket.

Each rocket has also a letter of the alphabet and a number stamped on both head and case. The numbers run up to 1,000, and then the letter is changed.

Rocket, War, 9-pr.

Rocket, War, 9-pr., Mark VII | *C* | , is similar in construction to the 24-pr., Mark VII, but the proportions of the ingredients of the composition are slightly different.

Fig. 183.
Scale $\frac{1}{12}$.

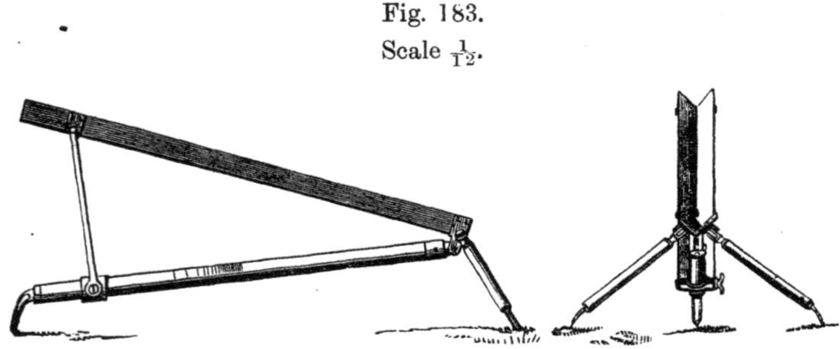

Method of firing the above Rockets.

The safety cap is removed and the rocket placed at the rear end of the rocket machine; it is fired by a friction tube which is placed in a slot on the machine so that the flash passes through one of the vents and so ignites the composition.

Tube, friction, quill, long, is used in the N.S., and *Tube, friction, copper, short,* in the L.S. (The "*Tube, friction, copper, solid drawn*" is too short for this purpose.)

Rocket Machines (for War Rockets).

Machines, rocket, war, 24-pr. and 9-pr. | *L* | , differ from each other in dimensions and weight.

The following general description applies to both:—

Each size consists of a sheet-iron V trough, supported at rear by three legs made of wrought-iron tubing, two short ones opening right and left, and one long one to the front beneath the trough, each terminating in a prong. On the front one runs a gunmetal ring connected by two bars with a V near the front of the trough, the bars pivot on V and ring; the elevation is given by slipping the ring up and down the front leg, and clamping it with the arrow on the rear edge of the ring at the required line of graduation up to 15° of elevation for 9-pr., and 25° for 24-pr. machine, with references to the plane on which the machine stands.

At the back end of the trough is an iron stop preventing the rocket sliding back; it is slotted to form a crutch for the copper friction tube.

There is also a Machine, Rocket, War, 24-pr., for Naval Service.

Life-Saving Rockets.

Rocket, Life-Saving, Boxer. (*See* Plate CXV.)

Rocket, Life-Saving, Boxer, Mark VI | *L* | , consists of body with pins and screws, two clips, front disc, head, and closing plug.

The body is of solid-drawn steel, in two parts, connected by a wrought-iron or mild-steel connecting piece, which is secured to the front portions by pins and brazing, and to the rear portion by screws.

A wrought-iron or mild-steel *base piece*, threaded internally to receive a closing plug, is secured in the end of the rear portion by pins and brazing and burring over the edge of the case. The two parts of the body and the screws are blackened in boiled linseed oil and the interior of the body receives three coats of white paint.

The clips for securing the stick are of sheet iron and connected to the body by screws.

The front disc is of wrought iron or mild steel, the head of wood, and the closing plug of gunmetal, with a square recess to fit the G.S. key.

The composition is pressed into pellets for convenience of handling and inserted into the front and rear parts of the body. A millboard disc is placed on top of the composition in the front part, the top of the disc and inner surface of case being coated with thinned luting. A lead disc, coated with thinned luting, is placed above the millboard disc, and over this the front disc, which is secured to the body by screws. A conical cavity is formed up the centre of the composition. The rear part of the body has the composition pellet at the top made quicker burning than the remainder. This is covered by a millboard washer, with its upper surface and the interior of the case coated with thinned luting. Two lead washers, coated with thinned luting, are placed on top. A conical cavity is formed in the rear part and the bottom of the cavity is covered by a paper disc, giving instructions to "break through before firing," shellaced on. The threads of the closing plug are slightly greased with thinned luting. The plug is to ensure the rocket bursting, instead of being projected, if the composition be accidentally ignited.

The front clip is secured by screws, and the wood head by tacks. The rear clip has its rear edge turned over the end of the body and the clip secured by screws.

These rockets range from about 300 to 470 yards, giving a mean range of 375 yards, and a mean deviation of 37·5 yards down wind.

Marking.—The numeral and contractor's initials, or trade mark, are stamped on the base piece, and the numeral and date of manufacture stencilled on the body.

Paint.—Exterior of rocket, except the plug, painted red.

The reason for having the two rockets, one fixed in prolongation of the other, is to give great length of burning and continue the propulsion through a much longer period without any excessive strain upon the line.

Issue.—Six rockets in a packing case.

Method of firing " Rocket, Life-Saving, Boxer."

The rocket is attached to a stick 9½ feet long, to which the rope is attached. (*See* Plate CXV.) It is fired from a tripod stand (the " rocket machine ") somewhat similar to those used with the war rockets.

To fire.—Remove the closing plug by means of the key, and break the paper disc covering the cavity in the composition. Insert the " Fuze, Rocket, Boxer," the body of which is conical, to fit the vent of the rocket. Light the fuze by the flame of a portfire, and stand clear.

Caution.—The life-saving rockets must, like other stores of a similar nature, be treated with care. If accidentally ignited when pointed in the wrong direction, or when lying about on the ground, they may become dangerous missiles.

Stick for Life-Saving Rocket.

Rocket, stick, 9 *feet* 6 *inches, Mark IV* | *L* | , is of deal, square, with the corners rounded off. The upper part is recessed to fit close to the rocket, and has two iron plates—one close to the end, the other where the stick fits into the bottom clip. The latter has a flange to rest against the base clip of the rocket. Below this the rocket stick is plated with tin to prevent its being burned by the flame of the rocket. On the back of the stick is fitted in a slot a flat spring with a catch to hold the stick in position when inserted.

The bottom end of the stick is bound with an iron ring, and the line is passed through a hollow in each end of the stick. After coming out through the top it passes through two indiarubber and one brass washer, and is secured by an overhand knot.

The indiarubber washers are intended to reduce the effect of the sudden jerk when the rocket is fired. A second knot is usually made in the rope near the hinder end of the stick in case the upper part of the line should be burnt through by the flame of the rocket.

Issue.—Nine sticks in a bundle.

Rocket machine and stores used with Life-Saving Rocket.

Machine, rocket, life-saving, Boxer, Mark V | *L* | , forms part of the life-saving apparatus ; it is intended for firing the life-saving rocket.

Other combustible stores.—The following combustible stores are also issued in connection with the life-saving rocket, viz. :—

Fuze, Rocket, Boxer, Marks III and IV.

Light, Long, G.S., Mark I.—When for use with the life-saving rocket, it is issued with metal handle.

Portfire, Life-Saving.—A metal handle is issued with this portfire.

Fuze for Life-Saving Rocket.

Fuze, Rocket, Boxer, No. 20, *Mark III.*—The exterior is conical to fit into the vent of the life-saving rocket. The body is of paper, 2·75 inches long, and is driven with composition, which burns 10 seconds. The exterior of the fuze is covered with kamptulicon. Over the priming there is a waterproof paper cap tied on with twine, which need not be removed before firing.

Fuze, Rocket, Boxer, No. 20, *Mark IV.*—The Mark IV fuze differs from the Mark III in being slightly longer, and in having the contour of the body altered. The leading portion of the body only is tapered, the taper being sharper than in the Mark III fuze, to allow of the fuze going further into the rocket, whilst the rear portion is of greater diameter and has parallel sides to more completely fill the mouth of the rocket and prevent premature firing due to "flash over" of the portfire.

It has also a perforated grain powder pellet instead of mealed powder for the priming.

The time of burning is the same as the Mark III fuze, viz., 10 to 11 seconds.

The fuzes, &c., are issued in a "Box, Life-Saving Fuze," made of tin, containing 12 fuzes, 12 indiarubber washers and 6 brass washers.

Portfires, Primers and Lights, for use with Life-Saving Rocket.

See page 572. The lights are issued in a deal box, closed with hinged lid secured by hasp and staple, called *Box, life-saving lights, Mark II*, which contains 12 lights, 2 handles, and 15 G.S. primers in a tin cylinder.

A similar *Box, Life-Saving, Portfire, Mark I*, contains 24 portfires, 2 handles, and 30 G.S. primers in a tin cylinder.

SIGNAL ROCKETS.

(1) Rockets, Signal.
(2) Rockets, Light.
(3) Rockets, Sound.
(4) Rockets, Light and Sound.
(5) Rockets, Flash and Sound.

The difference between the above is as follows :—

A "*Rocket, Signal*" contains a number of stars.

A "*Rocket, Light*" contains a single star.

A "*Rocket, Sound*" contains no stars, but a primer of guncotton or other detonating explosive.

A "*Rocket, Light and Sound,*" is a combination of the last two above mentioned.

A "*Rocket, Flash and Sound,*" is a combination of a "Rocket, Signal," and a "Rocket, Sound."

ROCKETS, SIGNAL.

Rocket, Signal, 1 *lb. Service, Mark III* | *L* | .

Rocket, Signal, 1 *lb., Service, Mark III,* has a case made of brown paper, rolled into a cylinder. The composition is driven by hand, and has a conical hollow at the rear.

A paper case is attached to the head, terminating in a cone; this serves to contain the stars and some mealed powder which serves to open the case and scatter the 28 stars. The star chamber is separated from the rocket composition by some clay driven in at the top of the composition, having a central hole forming a communication; the rocket is choked near the base, and has a priming made up of L.G. powder and isinglass.

The vent is closed by a wooden screw-plug, intended to reduce the area over which the destructive effect of the accidental ignition of a store of rockets would extend, as rockets so fitted will burst, instead of being projected in the usual way.

Stick.—Stick, 5 feet, with notch, Mark I | C | , is used with "Rocket, Signal, Service"; "Rocket, Signal," "Blue," "Red," or "Green"; also with the combined "Light and Sound Rocket."

The sticks are 5 feet long, tapered to the end. For Naval Service, the stick is only about 1 foot 6 inches long, and has a rope tail 5 feet long. This is more convenient than the long stick for use in confined spaces, such as boats, &c.

Issue of Rocket, Signal, 1 *lb., Service, Mark III.*—One in a tin cylinder, No. 75; 15 cylinders in "Box, Rocket, Signal, 1 lb." (*See* page 561.)

Rocket, Signal, ½ *lb., Service.*—*Rocket, Signal,* ½ *lb., Service,* resembles the 1-lb. rocket, except in size and in having the paper case larger in diameter than the body of the rocket and containing only 20 stars.

Issue of Rocket, Signal, ½ *lb., Service.*—One in a tin cylinder, No. 76; 14 cylinders in box, rocket, signal, ½ lb., or 60 cylinders in a whole metal lined case.

Coloured Rockets.

Coloured rockets are made for the purpose of display; they are issued in two sizes—1 lb. containing 49 and ½ lb. containing 30 stars (blue, green or red).

There is also a 1-lb. rocket containing red and white stars.

Rocket, Signal, 1 *lb. Red* | *C* | .
,, ,, 1 *lb. Blue* | *C* | .
,, ,, 1 *lb. Green* | *C* | .

The above resemble the Rocket, Signal, 1 lb. Service, but the heads are more rounded, and are painted the same colour as the stars they contain.

The heads of the rockets are in all cases opened by quickmatch packed in with the stars.

Issue of Rocket, Signal, 1 *lb., Red, Blue, or Green.*—The 1 lb. coloured rocket is issued in a tin cylinder, No. 98, 42 cylinders in a whole metal-lined case for Naval Service, 50 cylinders in an ordinary packing case for Land Service.

Rocket, Signal, $\frac{1}{2}$ *lb. Red* | *C* | .
,, ,, $\frac{1}{2}$ *lb. Blue* | *C* | .
,, ,, $\frac{1}{2}$ *lb. Green* | *C* | .

The $\frac{1}{2}$-lb. size coloured rockets are similar to the above, but smaller, and contain 30 stars.

Issue of Rocket, Signal, $\frac{1}{2}$ *lb., Red, Blue, or Green.*—The $\frac{1}{2}$-lb. rocket is issued one in a tin cylinder, No. 99, 60 cylinders in a whole metal-lined case for Naval Service; for Land Service, 50 cylinders in an ordinary packing case.

Rocket, Signal, 1 *lb., Red and White, Mark II.*

The *Rocket, Signal,* 1 *lb., Red and White, Mark II,* differs somewhat from the above. It is the same in external form as the 1-lb.

Fig. 184.

ROCKET, SIGNAL, 1-LB., RED AND WHITE, MARK II.

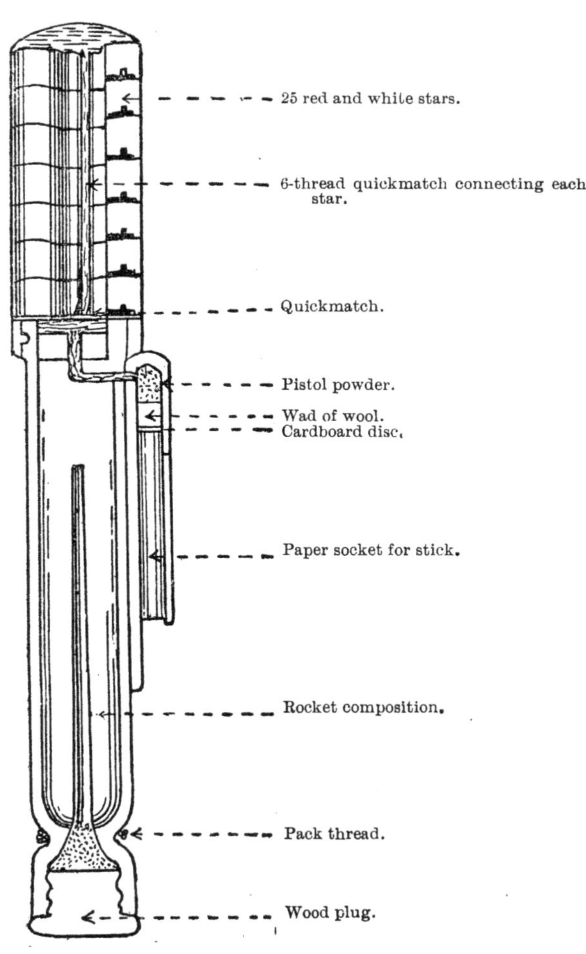

red signal rocket, Mark II, and has the head painted in longitudinal stripes of red and drab. It contains 25 red and white stars. In order to avoid the danger involved in firing the other signal rockets over the heads of crowds, from the metal socket remaining on the rocket stick and causing it to fall swiftly point first like an arrow, the socket in this case is made of paper instead of copper, and contains at the top end a puff of powder which communicates by a fire-hole with the top of the composition, and is exploded at the same time that the rocket opens, and thus separates the stick from the case. Mark I of this rocket, which was not mentioned in "L. of C.," had this arrangement for ejecting the stick, but no wooden plug in the base. The red and white rocket is not made up in the $\frac{1}{2}$-lb. size.

Method of Firing "*Rockets, Signal.*"

To prepare a "Rocket signal" for firing, it should be attached to the stick and the screw-plug removed from the vent (a label showing the method of fixing the stick will be found on the body); the rocket may be fired with the stick merely stuck into sand; it is usually convenient to support the rocket in the required position (almost vertical) by passing the stick through a couple of iron loops on a post in such a manner that no resistance is offered to the stick going up with the rocket.

The "*Machine, Rocket, Signal*" may also be used (*see* below) when firing rockets from boats.

To fire the rocket, ignite the priming in the vent by means of a portfire; with the "Machine, Rocket, Signal," a "Tube, Friction Quill, Short," or a "Tube, Friction, Machine, Rocket, Signal" is used instead of a portfire.

Caution.—Care should be taken to stand clear when firing the rocket, as the back rush of flame may injure the operator. The rocket comes down point first with considerable force. In order to cause it to come down at a safe distance, incline the rocket very slightly away from the vertical, remembering that the rocket tends to travel to *windward.*

Machine, Rocket, Signal, Mark II.

The *Machine, Rocket, Signal, Mark II,* consists of an oval tube of sheet iron to take the rocket with the portion of the stick at its side, a round tube of sheet iron being fixed on to it to take the remainder of the stick in its interior.

The two tubes are joined together by a middle piece of gunmetal, to which both are riveted.

The larger part of the finished tube is about 1 foot 8 inches and the smaller 4 feet 6 inches long.

The metal at the mouth of the finished tube is wire edged; at the opposite end is a ground spike.

A vent is made in the close portion of the base of the oval tube opposite to the vent of the rocket to take a friction tube for firing, which is prevented from falling out when the tube machine is pointed up into the air, by a hinged piece of gunmetal which shuts in behind the head.

Use.—For firing signal rockets from boats, and under circumstances when the back rush of flame might do injury.

Packing of "Rockets, Signal."

Boxes, Rocket, Signal, 1 *lb., Mark I* | *C* | and *Boxes, Rocket, Signal,* ½ *lb., Mark I* | *C* | .—These are ordinary deal boxes having elm ends fitted with elm cleats and galvanised iron wire handles.

The sides, ends, and bottoms are secured by iron nails and the top is secured by iron screws. Internally they are fitted with two diaphragms, the large box being prepared to take 15 "Rockets, Signal, 1 lb., Service," and the small box 14 "Rockets, Signal, ½ lb., Service." The rockets are packed in cylinders which fit in perforations in the diaphragms.

Cases, Packing, Rocket, Signal, 1 *lb., Red and White.*—This is generally similar to the "Boxes, Rocket, Signal," 1 lb. and ½ lb. It differs in dimensions; in having rope instead of wire handles, and in having two sets of movable fittings to support 16 cylinders containing the rockets, horizontally.

TABLE NO. 43.—*Rockets, Signal.*

Description.	Mark.	Colour.	Contents.	Packing.
Rocket, Signal, ½ lb., Service ...	III	Drab	20 stars ...	1 in a tin cylinder, No. 76. 60 cylinders in a whole metal lined case. 14 in a box, Rocket, Signal, ½ lb. §§4909, 14655.
Rocket, Signal, 1 lb., Service ...	III	Drab	28 stars ...	1 in a tin cylinder, No. 75. 15 in a box, Rocket, Signal, 1 lb. §§4909, 14655.
Rocket, Signal, ½ lb., Red	III	Drab, with red point ...	30 stars ...	1 in a tin cylinder, No. 99. 60 cylinders in a whole metal lined case. 50 cylinders in an ordinary packing case. §13164.
Rocket, Signal, ½ lb., Blue	II	Drab, with blue point...	30 stars ...	Do.
Rocket, Signal, ½ lb., Green	II	Drab, with green point	30 stars ...	Do.
Rocket, Signal, 1 lb., Red	III	Drab, with red point ...	49 stars ...	1 in a tin cylinder, No. 98. 42 cylinders in a whole metal lined case. 50 cylinders in an ordinary packing case. §13164.
Rocket, Signal, 1 lb., Blue	II	Drab, with blue point...	49 stars ...	Do.
Rocket, Signal, 1 lb., Green	II	Drab, with green point	49 stars ...	Do.
Rocket, Signal, 1 lb., Red and White	II	Drab. Head painted red and white	25 red 24 white } stars	1 in a tin cylinder, No. 98. 42 cylinders in a whole metal lined case. 16 cylinders in a case, packing, Rocket, Signal, 1 lb., Red and White.

NOTE.—The cylinders are all closed by a lid secured by a tape band shellaced on.

ROCKETS, LIGHT, AND ROCKETS, SOUND. (*See Table* 44.)

Rocket, Light, ½ *lb., Mark II,* and *Rocket, Sound,* ½ *lb., Mark II,* are issued to the Board of Trade; the Rocket, Sound, ½ lb., Mark II, is also issued to R.G.A. for examination vessels, &c.

These two rockets are similar and the following remarks apply to both. The body of the rocket proper, paper case, and the fitting of the stick are similar to those of the ordinary ½-lb. signal rocket. They have, however, a different arrangement for firing. About 4 inches of safety fuze is laid up alongside the copper socket for the stick. One end of this fuze passes into the vent round the lower edge of the case; the other end, protected by a paper cap, may be ignited by a vesuvian or other convenient means.

The base of the rocket is closed by the usual wooden plug screwed in.

Fig. 185.

ROCKET, LIGHT, ½ LB., MARK II.

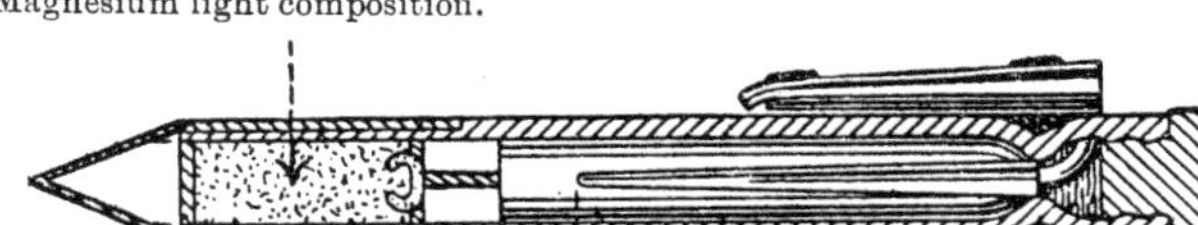

ROCKET, SOUND, ½ LB., MARK II.

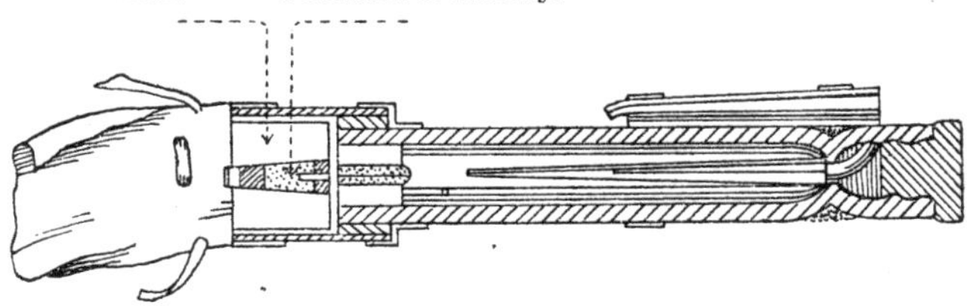

Rocket, Light.—The light rocket has the head filled with a single star of magnesium light composition contained in a paper case, and matched and primed. It is ignited and blown off when the rocket has reached its maximum height. The star burns about 15 seconds.

Rocket, Sound.—The sound rocket has, instead of a star, a 2-oz. primer of dry guncotton, coated with paraffin, and a detonator. These are carried separate from the rocket and from each other until required for use. The head is a cylindrical paper case, rather larger in diameter than the body of the rocket. It has a piece of calico at the top fitted with a tape. When the guncotton is inserted into the head the calico and tape fasten it in, the calico being tied up by the tape like a bag. The detonator is a small tin tube, containing fulminate of mercury, and is ignited by quick-match passing into the top of the rocket composition. (*For detonator, see* page 522.)

Stick.—The stick used with these rockets is similar to that used with the ordinary ½ lb. rocket, but is 12 inches shorter, for more convenient firing in a restricted space, such as the gallery of a lighthouse, &c.

The above rockets are intended for use in giving signals from lighthouses, lightships, &c.

Issue of Rockets, Light, and Rockets, Sound.

The Rocket, Light, ½ lb., Mark II, is issued 1 in a tin cylinder No. 76; 60 cylinders in a "Whole metal lined case."

The Rocket, Sound, ½ lb., Mark II, is issued 1 in a tin cylinder No. 110 for stations abroad, and 50 (loose) in a "Whole metal lined case" for Home Stations.

ROCKET, SOUND, ½ LB., MARK III.

(*Plate* CXVII.)

This rocket differs from the Mark II sound rocket, in having a tonite charge instead of guncotton, also the detonator is not flanged, as in the Mark II; the head is closed by a wooden plug, held by a bayonet joint, instead of the calico and tape of the Mark II. There is no plug in the base, as in the Mark II; the base has a paper covering.

Fig. 186.

ROCKET, SOUND, ½ LB., MARK III, CHARGE AND DETONATOR.

Scale 1/1.

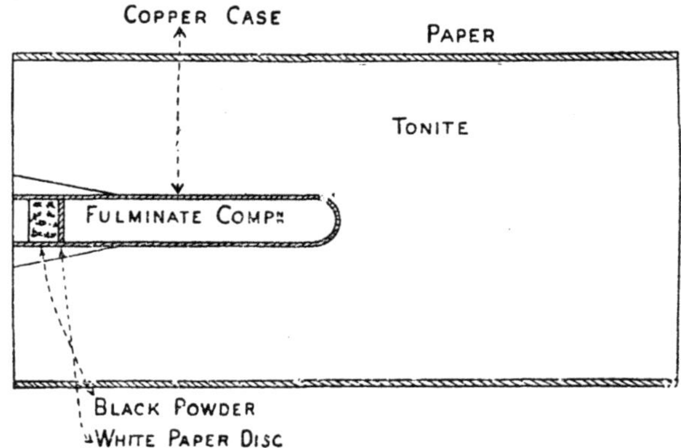

The tonite charge.—The tonite charge for the Mark III sound rocket consists of 1,729 grains of tonite (equal parts of guncotton and barium nitrate) recessed to receive a detonator, as shown in Fig. 186.

The *Detonator, No.* 2, is made of copper, and is filled with fulminate composition (fulminate of mercury, chlorate of potash, and guncotton) strongly pressed in. It is primed with mealed powder, a disc of paper being placed between the mealed powder and composition. (*See* page 522.)

Issue of Rocket, Sound, ½ lb., Mark III | C |.—The rockets, tonite charges, and detonators are issued and stored separately :—

1 *Rocket* to a cylinder, No. 111, for stations abroad, or 40 loose rockets in a " Whole metal-lined case."

The *tonite charges* are issued 50 in a " Quarter metal-lined case."

The *Detonators, Sound Rocket, No.* 2, are issued 5 in a tin cylinder, No. 71.

Rockets, Light and Sound.

Rocket, Light and Sound, 1 *lb., Mark I | L |* , is similar to the ½-lb. Sound, Mark III; the body is painted drab, and the head red.

The star consists of a brown-paper cylinder filled with a composition to give a red light, and has a piece of quick-match at one end. The star is inserted in the head of the rocket, primed end downwards, after having two pieces of quick-match passed round it. On the top of the star is placed a felt wad, with a central hole for the flash of quick-match round the star to pass to the detonator.

The tonite charge is similar to that of the " Rocket, Sound, ½ lb., Mark III," but smaller.

The detonator is the same as that for the " Rocket, Sound, ½ lb., Mark III."

The tonite charge and detonator are issued separately, and the space in the rocket is occupied by a brown-paper cylinder until the rocket is required for use.

The 5-ft. stick is used. The height of burst should be at least 600 feet.

Care should be taken in inserting the detonator and replacing the charge, and in subsequent handling.

Issue of Rockets, Light and Sound.—The Rocket, light and sound, is issued to stations abroad 1 in a tin cylinder, No. 109, for home stations 50 rockets (loose) in a " Whole metal lined case."

Method of firing the Rocket, Sound, ½ lb., Mark III.—A rocket machine is supplied for firing the sound rocket (*see* Plate CXVII.), consisting of a piece of wood having a handle about two-thirds the way up, placed at right angles to the stick; there are two metal eyes placed on the stick, as shown, for the rocket and stick to rest in.

The rocket should first be placed on the stick, being careful to see that the tongue bites into one of the notches on the rocket stick, then place the stick through the two metal eyes so that the base of the rocket rests on the upper eye; take the tonite charge and place the detonator, with the open end up, into the recess in the bottom of the tonite, and place the charge into the tonite chamber with the detonator down; next place the wood plug in the head and secure it with the bayonet joint, tear off the paper covering the safety fuze and ignite it with a portfire or any available means of ignition, keeping the rocket machine in an upright position.

On the safety fuze being ignited, it will burn away quietly for a few seconds until the flame reaches the priming and rocket composition inside the rocket, when the latter will ascend into the air. The burning composition will eventually reach the strands of quick-match, which will in turn ignite the detonator and so detonate the tonite charge.

The "Rocket, light"; "Rocket, Sound, Mark II"; and the "*Rocket, light and sound,*" are fired in the same way as the above, but when the rocket has a plug in the base, this must be removed before firing.

Rocket, Flash and Sound, 1 lb., Mark I | N |.

This rocket will supersede the "Rocket, Signal, 1 lb., Service," as the existing stock of the latter is used up.

Fig. 187.

ROCKET, FLASH AND SOUND, 1 LB., MARK I | N |.

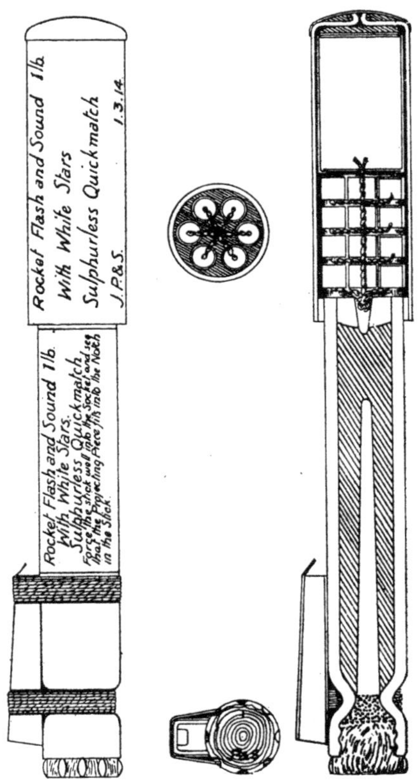

It consists of a cylindrical brown-paper case containing a lifting charge of rocket composition.

The lower end of the rocket is choked with pack thread, so as to form a vent, and is closed by means of a wooden screw plug.

The rocket composition which is driven into the case by hand, has a conical hole, bored from the base.

The vent of the rocket is primed with mealed powder, damped with a solution of isinglass and methylated spirits and sprinkled with L.G. powder while damp.

A copper socket is glued to the rear end of the rocket case, and further secured with pack thread; a tongue piece projects from the top of the copper socket for securing the rocket stick in position.

A clay plug is driven into the case on top of the rocket composition. This plug has a central hole filled with sulphurless powder and sulphurless quick-match.

A brown-paper cylinder, choked at the top with twine, is secured with glue to the front end of the rocket case.

It contains 24 stars, consisting of paper cylinders driven with star composition. Each star is primed at the lower end with sulphurless quick-match.

Placed on top of the stars is a "*flash explosive*" charge contained in a tin cylinder, primed with sulphurless quick-match.

Quick-match is also arranged in the cylinder, so that it communicates with the quick-match of each star and the explosive charge.

The rocket is painted drab colour.

A label giving instructions for fixing the stick and firing the rocket is pasted on the body. The date of manufacture and manufacturer's initials are stamped on the wood plug.

The rocket stick, 5 ft., with notch, is used with this rocket.

Issue.—One in a tin cylinder, No. 117.

Method of Firing Rocket, Flash and Sound.

The rocket is fired in the same way as the " Rocket, Signal, Service, 1 lb., Mark III (*see* page 560).

When ignited, the composition in the case burns, and lifts the rocket and in turn ignites the quick-match leading to the stars and the " flash explosive " charge. The latter explodes with a bright flash, scattering the stars, which burn with a bright white light.

TABLE No. 44.—*Rockets: Light; Sound; Light and Sound; Flash and Sound*

Description.	Mark.	Colour.	Contents.	Detonators.	Packing.
Rocket, light, ½ lb. ...	II	Drab	Single star of magnesium, burns 15 secs.	Nil	1 in a tin cylinder, No. 76. 60 cylinders in a whole metal lined case.
Rocket, sound, ½ lb. ...	II	Drab	2 ozs. guncotton primer	*Detonator, Sound Rocket, No. 1, Mark III | C |	1 in a tin cylinder, No. 110, for Stations abroad, §14751; or 50 (loose) in a whole metal lined case.
Rocket, sound, ½ lb. ...	III	Black	Tonite primer, 1,729 grains	†Detonator, Sound Rocket, No. 2, Mark I | C |	1 in a tin cylinder, No. 111, for Stations abroad, §14751; or 40 (loose) in a whole metal lined case.
Rocket, light and sound, 1 lb.	I	Tail stone colour, and head red	Red star and tonite charge	Do.	1 in a tin cylinder, No. 109, for Stations abroad, §14751; or 50 (loose) in a whole metal lined case.
Rocket, flash and sound, 1 lb.	I	Drab	24 stars and flash explosive charge	Nil	1 in a tin cylinder, No. 117.

**Detonator, Sound Rocket, No.* 1, *Mark III* | *C* | consists of a tin tube slightly tapered, and has a tin flange soldered round its top end; the bottom end is closed with solder. The tube contains 12 grains of fulminate of mercury, a glazed-board disc being underneath the charge, and a wood plug threaded with quick-match on top. (*See* page 521.)

†*Detonator, Sound Rocket, No.* 2, *Mark I* | *C* | consists of a solid drawn copper tube; it has no flange, and is filled with fulminate composition and primed with meal powder. (*See* page 521.)

The rocket cylinders are all closed by a lid secured by a tape band shellaced on.

Lights and Portfires.

Lights.

Very Signal Cartridges | *C* | .—These cartridges are similar in appearance to the blank cartridges of 1-inch aiming-rifle ammunition. Each cartridge contains a single star, which is fired from the cartridge in the same way as a projectile. The cartridges are fired in a Very pistol, and the star should ignite and rise to a height of 300 feet without breaking up, and should burn brightly for about 9 seconds.

Cartridges, Signal, Very, Marks II, III and IV, are issued for signalling purposes and contain a single green, red, or white star. The cartridge consists of a brass case, rolled for Mark II, solid drawn for Marks III and IV, lined with brown paper which projects beyond the mouth. The case is provided with a percussion cap and the charge consists of gunpowder. Above the charge is the star, the mouth of the case being secured by felt and cardboard wads.

The Mark IV differs from the Mark III in the composition of the star.

The portion of the lining which projects beyond the case is painted the same colour as that given by the star in the cartridge.

The rim of the base of the cartridge for a green star is smooth; for a red star is milled all round; and for a white star is milled halfway round.

Light, Coastguard.

Light, Coastguard, Mark II | *N* | , burns about five minutes. The spike at the end is to enable the light to be stuck in the ground.

The composition is contained in a paper case fitted with Brock's patent igniting arrangement. A wooden plug with composition at

Fig. 188.

LIGHT, COASTGUARD, MARK II | N | .

Scale ¼.

one end is contained in a paper cylinder attached to one side of the light. The top of the light is primed with composition and covered and protected by a paper disc. To ignite the light tear off the disc, pull out the plug, and draw its primed end smartly across the exposed surface of the light, holding the latter so that it points away from the body.

Issue.—One in a tin cylinder.

Mark I differed in the igniting arrangement.

Light, Long, G.S.

Light, Long, G.S., Mark III, consists of a rolled paper case into which is pressed a column of light-giving composition consisting of :—

Saltpetre, ground	.. .. ..	$17\frac{1}{2}$ parts.
Sulphur ..	.. ..	$4\frac{6}{16}$,,
Red orpiment	.. ..	$1\frac{1}{4}$,,

The top of the composition is covered with a calico disc smeared with igniting composition covered and protected by a paper disc and cap under which is glued a piece of tape.

Fig. 189.

LIGHT, LONG, G.S., MARK III.

Scale $\frac{1}{4}$.

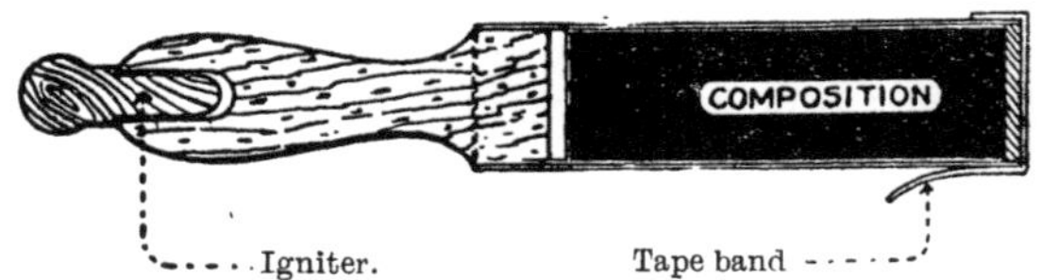

At the bottom end of the light-giving composition is a plug of ground clay. A beech-wood handle, having a recess at the lower end for a plug, is then inserted into the bottom of the case till it bears against the ground clay plug and is secured to the case by shellac.

In the recess at the lower end of the handle fits a beech-wood plug, which is covered at its top end with igniting composition.

Paint and Marking.

The whole of the light is painted *drab.*

The numeral and date of manufacture are stencilled in white on the body ; the contractor's initials or recognised trade mark, numeral and date of manufacture are stamped on the handle. Two labels, one showing the nature of light and time of burning, and the other giving directions for lighting, are pasted on the light.

To ignite the light, tear off the paper disc and cap, pull out the plug and draw its primed end across the exposed surface of the igniting composition, holding the light so that it points away from the body. *On no account is the igniting composition to be struck with the primed end of the plug.*

The long light burns about 5 minutes.

Packing.

One in a tin cylinder, No. 34 ; 24 cylinders in Box, fireworks, small. (*See* page 573.)

Light, Short, G.S.

The short light, Mark II, is similar in construction to the Mark III long light, except that it is much shorter and only burns from $1\frac{1}{2}$ to 2 minutes.

Supplied one in a tin cylinder.

Lights, Long, Blue, Green and Red, Mark III.

The above coloured lights are similar to the Light, Long, G.S., already described; they are painted externally according to their colour.

The red and blue lights burn 2 minutes, the green a minute and a half.

Packing.—Coloured lights are issued 1 in a tin cylinder, No. 39; the "tear-off" band closing the cylinder is painted the same colour as the light the cylinder contains.

Light, Illuminating Wrecks.

Light, Illuminating Wrecks, Mark IV.—The present pattern of this light is 2·65 inches diameter and about 30 inches long. It is made up of 10 rings of sheet iron, each about 3·25 inches long. One end of each ring is increased in diameter so as to form a cap to fit over the other end of the next ring, to which it is attached by solder. At the point where the enlargement begins there is an internal diaphragm in each ring having a central hole 2·1 inches diameter, the function of which is to prevent the composition burning up the side. Each ring is filled with composition. (*See* page 46.)

A small hole is made in the adjoining faces of the composition to ensure the continuous burning of the whole, and the faces are themselves roughened with the same object.

One end of the light is fitted with a hemispherical piece of wood, through which passes a loop of iron wire, by which to suspend the light from the stand when burning, as shown in Fig. 190; the other end is primed with mealed powder and covered with a disc of millboard and a cap of kit plaster.

LIGHT, ILLUMINATING WRECKS, MARK V | L |.

The Mark V light differs from the Mark IV in the case consisting of a drawn tube of aluminium, which burns away at the same speed as the composition and so frees the light better than the iron segments of the previous pattern.

Stand, Light, Illuminating Wreck, Mark I.

Stand, Light, Illuminating Wreck, Mark I, is a simple tripod, consisting of three wooden legs, about 6 feet in length, connected at the top by a piece of iron wire, having a small hook attached to it, on which the light is suspended; there are three iron rods which are hooked to and connect two of the legs, forming an incline for the light to rest on, so as to hang in a sloping direction, not vertically downwards.

The light, if hung as described, clears itself of dross when burning, and is kept further clear by the case separating each joint, as the heat of the burning composition successively melts the soldering of the rings.

Fig. 190.

STAND, LIGHT, ILLUMINATING WRECK, MARK I.

Scale $\frac{1}{12}$.

PORTFIRES.

Portfire, Common, consists of a cylinder about 16 inches long, and rather more than ½ inch diameter. It is made of stout brown paper pasted, rolled, and, when dry, turned in at one end to form a bottom. The case or cylinder is driven with portfire composition.

The top has a small hole bored in the composition, and is primed with mealed powder to make it light easily. They burn from 12 to 15 minutes, and are generally lighted by a slow match.

They may be lit also by any means handy, as a vesuvian, a burning stick, &c. In the field, if no other means are handy, put a friction tube on the ground, under a brick or stone, leaving the ends out. With one hand pull the lanyard to ignite the tube, keeping the stone firmly pressed down under the foot, and with the other hold the end of the portfire to the end of the tube, so that the flash of the latter may ignite it.

Painted flesh colour.

Packing.—In bundles of 12, packed in deal boxes; the exposed ends are secured by a paper cap tied on with twine.

Portfire, Life-saving.—Differs from a common portfire in being 8 inches long, and in being made so as to ignite by means of a detonating primer, the end being closed by a paper cap, and strengthened by a tin band, perforated to take the detonating primer, which enters into a small space beneath the paper cap. The composition is primed in the usual method with mealed powder, perforated in the centre. This portfire is used with a metal handle similar to that for the G.S. long light.

Primers, Portfire, Life-saving.—The primer is made on a similar plan to the head of the copper friction tube; the pin is roughed and coated with the friction tube composition, and the blow driving it through the wedge-shaped copper case explodes it. The case is open at one end and protected by varnished paper.

Fig. 191.

PRIMER, PORTFIRE, LIFE-SAVING.

Scale $\frac{1}{1}$.

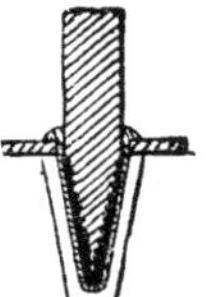

BOXES, FIREWORKS, LARGE, MARK II | N |.

and

BOXES, FIREWORKS, SMALL, MARK II | C |.

These boxes are generally similar to each other, differing only in dimensions and internal fittings. They are made of deal, having elm cleats and galvanized iron wire handles at each end. The sides and ends are dovetailed and the bottom is secured by brass screws. The lid is hinged to the box by two metal hinges and is secured by two metal bolts which are prepared on top for the "Case, powder, metal lined" key which is used with these boxes.

Internally they are fitted to take the following stores:—

Stores.	Large Box.	Small Box.
Lights, Long	24	8
Lights, Short	24	16
Portfires, Common	14	10
Rockets, Flash, and Sound, 1 lb.	24	12

NOTE.—In the Box, Fireworks, Small, Mark II, for Land Service, "Rocket, Signal, 1 lb. Service" is issued instead of the Rocket, Flash and Sound.

Signal, Fog.

Signal, Fog, Mark I | L | consists of two concentric compartments of tin each containing gunpowder. The inner one is fitted with three, and the outer with four percussion caps as shown in Fig. 192.

A clip of lead is soldered to the top by which the signal is attached to a rail.

Issue :—In a tin cylinder containing 12.

The cylinder No. 66 is of tin with a lid closed by a bayonet joint, and it is furnished with tin D's to take a strap for transport purposes.

Fig. 192.

SIGNAL, FOG, MARK I | L | .

Scale ½.

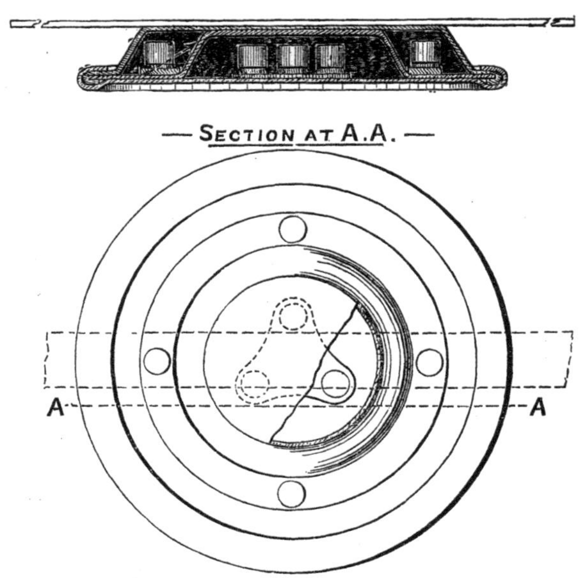

Plan, with outer cover removed.

TABLE NO. 45.—*Tin Cylinders with Contents.*

No.	To contain.	Mark.	Service.
1	5 Fuzes, percussion, D.A., with cap No. 1, or 25 primers, percussion, No. 4	II	C
1F	1 Fuze, percussion, D.A., with cap No. 1	I	L
2	5 Fuzes, percussion, B.L., plain No. 3	II	C
3	5 Fuzes, percussion, D.A., with plug No. 3, or ½ lb. Mark III luting and mineral jelly mixed	II	C
4	5 Fuzes, graze, No. 4	II	C
5	5 Fuzes, percussion, Pettman, G.S., or beeswax for torpedoes, ½ lb.	I	C
6	8 long or 125 7-dram exploders for I.O.O.	I	C

TABLE NO. 45.—*Tin Cylinders with Contents*—continued.

No.	To contain.	Mark.	Service.
7	5 Fuzes, percussion, N.L., No. 7 ; also 25 primers, vent	I	C
8D	25 Detonators, No. 8, Marks IV and V	II	L
8D	25 Detonators, No. 8, Mark VII	III	L
8F	1 Fuze, percussion, small, No. 8, Mark IV	I	L
9	25 Detonators, electric, No. 9	I	N
10F	1 Fuze, percussion, D.A. delay, No. 10	I	L
10T	25 Tubes, electric, No. 10	I	C
11F	1 Fuze, base, percussion, large, No. 11, Marks I to IV or No. 15, Mark I	I & II	C
11F	1 Fuze, base, percussion, large, No. 11, Mark V, No. 15, Marks II, or III, and 16, Mark I	II	C
11T	25 Tubes, electric, No. 11	I	N
12D	25 Detonators, electric, No. 12	I	L
12F	1 Fuze, percussion, base, medium, No. 12, Marks I to IV	I & II	C
12F	1 Fuze, percussion, base, medium, No. 12, Marks V to VII	II	C
13D	25 Detonators, electric, No. 13	I	L
13F	1 Fuze, percussion, D.A. impact, No. 13, or 1 fuze, percussion, D.A., with cap No. 17	I	C
14	25 Fuzes, electric, No. 14	I	L
15	25 Fuzes, electric, No. 15	II	N
16	25 Fuzes, electric, No. 16	III	L
18	25 Detonators, electric, No. 18	II	L
18F	1 Fuze, percussion, D.A. impact, No. 18	I	C
19	15 Fuzes, electric, No. 19	II	N
19F	1 Fuze, percussion, D.A. impact, No. 19	I	N
20	25 Detonators, electric, No. 20	II	L
21	5,000 Discs, paper, tubes, electric, No. 11	I	C
22	1 Fuze, time, E., No. 22	II	C
23	25 Portfires, life-saving, or 4 lights long, G.S. ...	II	L
24	25 Tubes, friction, copper, L.S., long	II	L
25	25 Tubes, friction, copper, L.S., short	II	L
25F	1 Fuze, time, No. 25	I	L
27	25 Tubes, friction, copper, solid drawn, or tubes, friction, machine, rocket, signal, or 20 tubes, copper for tube, friction "T" for blank ...	I & II	C
28	25 Tubes, friction, quill, time gun	I	L
29	25 Tubes, friction, quill, short	I	N
30	10 Washers, leather, plug insulating	I	C
30F	1 Fuze, time, No. 30	I	N
31	5 Primers, light, G.S., or 30 primers, portfire, life-saving, or 2 ozs. vermilion paint prepared for use	I	L
32	8 Fathoms safety fuze, or 25 tubes, friction, quill, long	II	C
33	1 2-oz. coil, Indiarubber, tape	II	C
34	1 Light, long, G.S., except Mark I	I	C
35	1 Light, coastguard, Mark II	I	N
36	1 lb. luting, common, Mark III	I	C
37	8 ozs., 6 thread quickmatch	I	C
38	1 Light, short, G.S.	I	N
39	1 Light, red, blue, or green, Mark II	I	C
40	5 Fuzes, time, 30 seconds, M.L., No. 40	I	L

TABLE No. 45.—*Tin Cylinders with Contents*—continued.

No.	To contain.	Mark.	Service.
41	5 Fuzes, time, 15 seconds	I	L
42	5 Fuzes, time, 15 seconds (with special priming) ...	III	L
43	5 Fuzes, time, 15 seconds, with detonator, No. 43 ...	II	L
44	100 Cartridges, pistol, safety fuze, or 4 cylinders, No. 65	I	C
44F	1 Fuze, percussion, D.A., with cap, No. 44... ...	I	C
45	6 Washers, leather, paraffined, for mouthpiece, countermine | N | 500 lbs., Mark II	I	N
45F	1 Fuze, percussion, D.A. impact, No. 45	I	C
46	5 Detonators, torpedo, 38 grains	I	N
47	150 Caps, percussion, 3 and 6-pr. Q.F.	I	C
48	8 Detonators, No. 8, Mark III, with 2 ft. of safety fuze	III	L
49	6 Detonators, No. 8, Mark IV to VI, with 2 ft. of safety fuze	V	L
49	6 Detonators, No. 8, Mark VII with 2 ft. of safety fuze	VI	L
50	9 Guncotton, primers, rocket, 2 ozs....	II	C
51	5 Guncotton, primers "F" (obsolete)	I	L
52	5 Guncotton, primers "H" (obsolete)	I	N
53	1 lb. luting, common, Mark I	I	N
54	1 Fuze, time and percussion, No. 54	I	C
56	1 Fuze, time and percussion, No. 56	II	C
57	1 " " " " No. 57	II	L
60F	1 Fuze, time and percussion, No. 60	I	L
61F			
62	8 Washers, Indiarubber, fabric, 3-cell battery, Le-Clanche	I	L
62F	1 Fuze, time and percussion, No. 62 or 66... ...	I	L
63	5 Detonators, rocket, sound, No. 1	I	C
63F	1 Fuze, time and percussion, No. 63	I	C
63F	1 Fuze, time and percussion, No. 56, 60, 63 or 65 ...	II	C
64F	1 Fuze, time and percussion, No. 64	I	N
65	25 Cartridges, pistol, safety fuze	I	N
66	12 Fog signals	II	L
67	10 Igniters, Q.F., or Q.F.C. Cordite, 1¼ ozs. ...	II	C
68	10 Igniters, Q.F., or Q.F.C. cordite, 8½ drams ...	II	C
68F	1 Fuze, time and percussion, No. 68	I	L
69	10 Primers, electric, large	III	C
70	20 Primers, 6 and 3-pr. (blank)	I	C
71	5 Detonators, sound rocket, No. 2, or 5 detonators, torpedo, small flange	I	C
72	10 Primers, shrapnel shell	I	C
73	1 Guncotton, primer, 16 ozs., Berlin torpedo ...	I	N
75	1 Rocket, signal, 1 lb., service	I	L
76	1 Rocket, signal, ½ lb., service	I	C
77	25 Tubes, electric, No. 10A	I	L
79	10 Primers, field, G.C., 1 oz., dry	I	C
80F	1 Fuze, time and percussion, No. 80	III to V	L
81	Tubing, Indiarubber, ⅜-inch	I & II	L
81F	1 Fuze, time and percussion, No. 81, 83 or 84 ...	I & II	N
82	Tubing, Indiarubber, ½-inch	I & II	L
82F	1 Fuze, time and percussion, No. 82	I & II	L

TABLE No. 45.—*Tin Cylinders with Contents*—continued.

No.	To contain.	Mark.	Service.
83	Wire covered, C 20, 20 yards	I	L
84	24 Discs, charge, priming, warhead, 1 lb. 1 oz. ...	I	N
85	24 Discs, torpedo primer, D	I	N
86	12 Washers, leather, paraffined for plugs	I	N
87	12 Washers, charge, priming warhead, 1 lb. 1 oz. ...	I	N
88	5,000 Discs, paper, for tubes, V.S., electric P (drill)...	I	C
89	5,000 Discs, fine white paper, for tubes, V.S., electric P (drill)	I	C
90	10 Washers, leather, paraffined, for mouthpiece, countermine, naval, 500 lbs., Mark III	I	N
91	Washers, dermatine, cases, powder, cylindrical ...	I	N
93	6 Washers, dermatine, manhole, countermine, naval, 500 lbs.	I	N
95	1 Cartridge, impulse torpedo, 4½ ozs. or 4 ozs. ...	I	N
96	1 Cartridge, impulse torpedo, 7½, 6, or 5½ ozs. ...	I	N
97	1 lb. mineral jelly	I	N
98	1 Rocket, signal, 1 lb., blue, green, or red	I	N
99	1 Rocket, signal, ½ lb., blue, green, or red	I	N
101	12 Discs, Indiarubber, G.C. dry, charge, priming, mine, 2¼ lbs.	I	N
102	6 Washers, dermatine, mines, spherical	I	N
103	10 primers, percussion, Q.F., No. 1	I	C
104	5 Washers, dermatine, for pistol, mines, spherical...	I	N
105	20 Primers, percussion, Q.F., No. 2, Mark I ...	I	C
106	10 Primers, percussion, Q.F., No. 2, Mark II ...	I	C
107	5 Washers, dermatine, mooring rope, spherical, mines	I	N
108	20 Primers, percussion, Q.F., No. 3	I	L
109	1 Rocket, light and sound, or 1 rocket, flash and sound	I	L
110	1 Rocket, sound, ½ lb., Mark II	II	C
111	1 Rocket, sound, ½ lb., Mark III	III	C
112	1 Cartridge, impulse, torpedo, 9 ozs. or 10 ozs. ...	I	N
113	20 Primers, percussion, No. 2, Mark III	I	C
114	10 Primers, percussion, No. 2, Mark IV	I	C
115	10 Tracers, shell, night, internal	I	C
116	10 Tracers, shell, night, external	I	C
117	1 Rocket, flash and sound	I	N

INDEX.

A.

B.

C.

D.

F.

F.

G.

H.

I.

PAGE.

L.

M.

N.

P.

Q.

T.

www.ingramcontent.com/pod-product-compliance
Ingram Content Group UK Ltd.
Pitfield, Milton Keynes, MK11 3LW, UK
UKHW051130290726
14058UKWH00003BB/553

9 781843 425601